Undergraduate Lecture Notes in Physics

Series Editors

Neil Ashby, University of Colorado, Boulder, CO, USA

William Brantley, Department of Physics, Furman University, Greenville, SC, USA

Michael Fowler, Department of Physics, University of Virginia, Charlottesville, VA, USA

Morten Hjorth-Jensen, Department of Physics, University of Oslo, Oslo, Norway

Michael Inglis, Department of Physical Sciences, SUNY Suffolk County Community College, Selden, NY, USA

Barry Luokkala📵, Department of Physics, Carnegie Mellon University, Pittsburgh, PA, USA

Undergraduate Lecture Notes in Physics (ULNP) publishes authoritative texts covering topics throughout pure and applied physics. Each title in the series is suitable as a basis for undergraduate instruction, typically containing practice problems, worked examples, chapter summaries, and suggestions for further reading.

ULNP titles must provide at least one of the following:

- An exceptionally clear and concise treatment of a standard undergraduate subject.
- A solid undergraduate-level introduction to a graduate, advanced, or non-standard subject.
- A novel perspective or an unusual approach to teaching a subject.

ULNP especially encourages new, original, and idiosyncratic approaches to physics teaching at the undergraduate level.

The purpose of ULNP is to provide intriguing, absorbing books that will continue to be the reader's preferred reference throughout their academic career.

More information about this series at https://link.springer.com/bookseries/8917

Lev Kantorovich

Mathematics for Natural Scientists

Fundamentals and Basics

Second Edition

 Springer

Lev Kantorovich
Physics
King's College London
London, UK

ISSN 2192-4791   ISSN 2192-4805   (electronic)
Undergraduate Lecture Notes in Physics
ISBN 978-3-030-91221-5   ISBN 978-3-030-91222-2   (eBook)
https://doi.org/10.1007/978-3-030-91222-2

This Springer imprint is published by the registered company Springer Nature Switzerland AG
The registered company address is: Gewerbestrasse 11, 6330 Cham, Switzerland

Introduction to the Second Edition

In the second edition the length of the book grew almost by half. Changes happened in all chapters including improved explanations, corrected typos and occasional errors; a substantial number of new problems were added across the whole book. However, Chaps. 1, 3, 6–8 experienced most changes and contain most of the additional material.

In Chap. 1, elementary mathematics has been significantly expanded by adding more problems across the whole chapter; new sections appeared and the existing ones were significantly expanded and rewritten. Among more noticeable additions, we mention detailed discussions on algebra; algebraic equations and systems of algebraic equations; functional inequalities; areas of simple plane figures; polynomials and their manipulations; roots of polynomials, their factorisation; Vieta's formulae; golden ratio and golden triangle, Fibonacci numbers; more on ellipses and hyperbolas; a fascinating arithmetic (turned algebraic) problem developed by my father; Cardano's formula to justify introduction of complex numbers; the square root of a complex number; *summae potenstatum* and Bernoulli numbers; prime numbers; classical probability theory; various types of averages of number sequences and their comparison; intersections of lines, planes and spheres.

In Chap. 2, a noticeable addition is concerned with a detailed discussion on a rational function of polynomials and their partial fraction decomposition into simpler fractions.

In Chap. 3, we discuss in more detail complex numbers including Moivre formula, roots of complex numbers, a more careful introduction of Euler's formulae. Then we discuss solving cubic and quartic algebraic equations, higher order derivatives of inverse functions and introduce the differentiation operator.

The first three chapters should be accessible to school students and I hope would serve as a useful source of wonderful exciting mathematics for school teachers.

Main improvements in Chap. 4, apart from new problems and various changes throughout, include a more careful consideration of integrals depending on a parameter, their convergence; convolution and correlation functions; probability distributions.

There are only a few small changes in Chap. 5.

In Chap. 6, we shall note several physics applications added, specifically kinetic theory of dilute gases (Maxwell distribution, gas equation, diffusion, heat transport and viscosity). Line integrals are introduced with a bit more care. We have also added a comparison of surface and line integrals to stress their interconnections. Many problems from physics have been also added.

Chapter 7 has undergone significant changes. We have added a discussion on multiple series; uniform convergence of improper integrals; considered lattice sums encountered in solid-state physics and a rigorous derivation of the Ewald formula for the Coulomb potential inside a crystal; Bernoulli numbers in relation to summation of powers of integers; generation function for Fibonacci numbers; complex exponential; Taylor expansion of a function of many variables; diffusion as a physics application.

Finally, Chap. 8 has undergone substantial changes as well. Many more problems have been added. Non-linear first- and second-order differential equations are discussed, and existence and uniqueness of their solutions; Picard's method; we have considered in much greater detail the method of undetermined coefficients. A section on systems of two linear differential equations has been also added. Among added applications are orthogonal trajectories, curve of pursuit and catenary curve, celestial mechanics including a proof of the inverse Kepler's law and a finite amplitude pendulum.

There are now nearly 800 problems that is by far more than 300 new ones compared to the first Edition.

To avoid possible errors and typos Mathematica was used where possible to check algebraic calculations and answers to problems.

To prepare this Edition, I have used a number of brilliant (mostly Russian) texts which I have found extremely useful, and I'd like to take this opportunity to acknowledge them here:

1. И. С. Соминский, "Элементарная алгебра. Дополнительный курс." Москва, Наука, 1967 (in Russian). [I. S. Sominsky, High algebra. Additional material., Moscow, Nauka, 1967.]
2. С. И. Туманов, "Поиски решения задачи", Москва, Просвещение, 1969 (in Russian). [S. I. Tumanov, In search of solutions to problems, Moscow, Prosvezhenie, 1969.]
3. Н. Я. Виленкин, Р. С. Гутер, С. И. Шварцбурд, Б. В. Овчинский, В. Г. Ашкинузе. "Алгебра", Москва, Просвещение, 1972 (in Russian). [N. Y. Vilenkin, R. S. Guter, S. I. Sharcburd, B. V. Ovchinskii, and V. G. Ashkinuze, Algebra, Moscow, Prosvezhenie, 1972.]
4. Т. А. Агекян, "Теория вероятностей для астрономов и физиков", Москва, Наука, 1974 (in Russian). [T. A. Agekyan, Probability theory for astronomers and physicists., Moscow, Nauka, 1974.]
5. R. Kent Nagle, E. B. Saff, and A. D. Snider, Fundamentals of differential equations, Addison-Wesley, Boston, 8th ed., 2012.
6. Э. Б. Винберг, "Начала алгебры", Москва, МЦНМО, МК НМЫ, "УРСС", 1998 (in Russian). [E. B. Vinberg, Beginnings of algebra, Moscow, МЦНМО, МК НМЫ, "УРСС", 1998.]

Introduction to the First Edition

The idea to write a mathematics textbook for physics undergraduate students presented itself only after more than 10 years of teaching a mathematics course for the second year students at King's College London. I discovered that students starting their studies at university level are lacking elementary knowledge of the fundamentals of mathematics; yet in their first year, this problem is normally left untreated. Instead, the assumption is made that a baseline understanding of numbers, functions, numerical sequences, differentiation and integration were adequately covered in the school curriculum and thus they remain poorly represented at university. Consequently, the material presented to first years is almost exclusively new and leaves the students wondering where it had all come from and struggling to follow the logic behind it. Students, for instance, have no understanding of "a proof". They are simply not familiar with this concept as everything they have learned so far was presented as "given", which is a product of the spoon-feeding teaching methods implemented in so many schools. In fact, the whole approach to teaching mathematics (and physics) at school is largely based on memorising facts. How can one learn, appreciate and apply mathematics in unknown situations if the foundations of what we now call mathematics, formed from lines of thought by some of mankind's best minds over the centuries, remain completely untouched? How can we nurture independent thinkers, thoughtful engineers, physicists, bankers and so on if they can only apply what is known with little ability to create new things and develop new ideas? We are raising a generation of dependents. This can only be avoided if students are taught how to think, and part of this learning must come from understanding the foundations of the most beautiful of all sciences, mathematics, which is based on a magnificent edifice of logic and rigour.

One may hope, naturally, that students can always find a book and read. Unfortunately, very little exists in the offering today for physics students. Indeed, there are some good books currently being used as textbooks for physics and engineering students [1–5], which give a broad account of various mathematical facts and notions with a lot of examples (including those specifically from physics) and problems for students to practice on. However, these books in their presentation lack rigour; most of the important statements (or theorems) are not even mentioned while others are discussed rather briefly and far too "intuitively". Most worryingly, all these books assume previous knowledge by the student (e.g. of limits, derivatives and integrals) and start immediately from the advanced material leaving essentially an insurmountable vacuum in students' knowledge of fundamental principles.

I myself studied physics and mathematics between 1974 and 1979 at the Latvian University in Riga (the capital of Latvia, formerly USSR) where a very different approach to teaching students was adopted. We were taught by active research mathematicians (rather than physicists) and at a very different level to the one which is being widely used today at western universities. The whole first year of studies was almost completely devoted to mathematics and general physics; the latter was an introductory course with little mathematics, but covering most of physics from mechanics to optics. Concerning mathematics, we started from the very beginning: from logic, manifolds and numbers, then moving on to functions, limits, derivatives and so on, proving every essential concept as we went along. There were several excellent textbooks available to us at the time, e.g. [6, 7], which I believe are still among the best available today, especially for a student yearning for something thought-provoking. Unfortunately, majority of these books are inaccessible for most western students as they are only available in Russian, or are out of print.

In this project, an attempt has been made to build mathematics from numbers up to functions of a single and then of many variables, including linear algebra and theory of functions of complex variables, step by step. The idea is that the order of introduction of material would go gradually from basic concepts to main theorems to examples and then problems. Practically everything is accompanied by proofs with sufficient rigour. More intuitive proofs or no proofs at all are given only in a small number of exceptional cases, which are felt to be less important for the general understanding of the concept. At the same time, the material must be written in such a way that it is not intimidating and is easy to read by an average physics student and thus a decision was made not to overload the text with notations mathematicians would find commonplace in their literature. For instance, symbols such as $\in$, $\ni$, $\subset$, $\supset$, $\cup$, etc. of mathematical logic are not used although they would have been enabled in many places (e.g. in formulating theorems and during proofs) to shorten the wording considerably. It was felt that this would make the presentation very formal and would repel some physics students who search for something straightforward and palpable behind the formal notation. As a result, the material is presented by a language that is more accessible and hence should be easier to understand but without compromising the clarity and rigour of the subject.

The plan was also to add many examples and problems for each topic. Though adding examples is not that difficult, adding many problems turned out to be a formidable task requiring considerable time. In the end, a compromise was reached to have problems just sufficient for a student reading the book to understand the material and check him/herself. In addition, some of the problems were designed to illustrate the theoretical material, appearing at the corresponding locations throughout the text. This approach permitted having a concise text of reasonable size and a correspondingly sensible weight for the hardcopy book. Nevertheless, there are up to a hundred problems in each chapter, and almost all of them are accompanied by answers. I do not believe that not having hundreds of problems for each section (and therefore several hundreds in a chapter) is a significant deficiency of this book as it serves its own specific purpose of filling the particular gap in the whole concept of teaching mathematics today for physics students. There are many other books available now to students, which contain a large number of problems, e.g. [1–5], and hence students should be able to use them much more easily as an accompaniment to the explanations given here.

As work progressed on this project, it became clear that there was too much material to form a single book and so it was split into two. This one you are holding now is the first book. It presents the basics of mathematics, such as numbers, operations, algebra, some geometry (all in Chap. 1), and moves on to functions and limits (Chap. 2), differentiation (Chaps. 3 and 5), integration (Chaps. 4 and 6), numerical sequences (Chap. 7) and ordinary differential equations (Chap. 8). As we build up our mathematical tools, examples from physics start illustrating the maths being developed whenever possible. I really believe that this approach would enable students to appreciate mathematics more and more as the only language which physics speaks. This is a win-win situation as, on the other hand, a more in-depth knowledge of physics may follow by looking more closely at the mathematics side of a known physics problem. Ultimately, this approach may help the students to better apply mathematics in their future research or work.

In the second book, more advanced material will be presented such as linear algebra, Fourier series, integral transforms (Fourier and Laplace), functions of complex variable, special functions including general theory of orthogonal polynomials, general theory of curvilinear coordinates including their applications in differential calculus, partial differential equations of mathematical physics and finally calculus of variation. Many more examples from physics will be given there as well. Both volumes taken together should comprise a comprehensive collection of mathematical wisdom hopefully presented in a clear, gradual and convincing manner, with real illustrations from physics. I sincerely hope that these volumes will be found sufficient to nurture future physicists, engineers and applied mathematicians, all of which are desperately lacking in our society, particularly those with a solid background in mathematics.

I am also convinced that the book should serve well as a reach reference material for lecturers. In spite of a relatively small number of pages in this volume, there really is a lot of in-depth material for the lecturers to choose from.

In preparing the book, I have consulted extensively a number of excellent existing books [6–9], most of them written originally in Russian. Some of these are the same books I used to learn from when I was a student myself, and this is probably the main reason for using them. To make the reading easier, I do not cite the specific source I used in the text, so I'd like to acknowledge the above-mentioned books in general here. A diligent student may want to get hold of those books for further reading to continue his/her education. In addition, I would also like to acknowledge the invaluable help I have been having along the way from Wikipedia, and therefore my gratitude goes to all those scholars across the globe who contributed to this fantastic online resource.

I would also like to thank my teachers and lecturers from the Latvian University who taught me mathematics there and whose respect, patience and love for this subject planted the same feeling in my heart. These are, first of all, Era Lepina who taught me the foundations of analysis during my first year at University (I still keep the notes from her excellent lectures and frequently consulted them during the work on this book!), Michael Belov and Teodors Cirulis (theory of functions of complex variables and some other advanced topics such as asymptotic series). I would also like to mention some of my physics lecturers such as Boris Zapol (quantum mechanics), Vladimir Kuzovkov (solid-state physics) and Vladimir Irvin (group theory) who made me appreciate mathematics even more as the universal language of physics.

Finally, I would like to apologise to the reader for possible misprints and errors in the book which are inevitable for the text of this size in spite of all the effort to avoid these. Please send your comments, corrections and any criticism related to this book either to the publisher or directly to myself (lev.kantorovitch@kcl.ac.uk). Happy reading!

Literature

1. D. McQuarrie, Mathematical methods for scientists and engineers, Univ. Sci. Books, 2003 (ISBN 1-891389-29-7).
2. K. F. Riley, M. P. Hobson, and S. J. Bence, Mathematical Methods for Physics and Engineering, Cambridge Univ. Press, 2006 (ISBN 0521679710).
3. M. Boas, Mathematical methods in the physical sciences, Wiley, 2nd Edition, 1983 (ISBN 0-471-04409-1).
4. K. Stroud, Engineering mathematics, Palgrave, 5th Edition, 2001 (ISBN 0-333-919394).
5. G. B. Arfken, H. J. Weber and F. E. Harris, Mathematical Methods for Physicists. A Comprehensive Guide, Academic Press, 7th Edition, 2013 (ISBN: 978-0-12-384654-9).
6. В. И. Смирнов, "Курс высшей математики", т. 1-5, Москва, Наука, 1974 (in Russian). Apparently, there is a rather old English translation: V. I. Smirnov, A Course of Higher Mathematics: Adiwes International Series in Mathematics, Vols. 1 (ISBN 1483123944), 2 (ISBN 1483120171), 3-Part-1 (ISBN B0007ILX1K) and 3-Part-2 (ISBN B00GWQPPMO), Pergamon, 1964.
7. Г. М. Фихтенголц, "Курс дифференциального и интегрального исчисления", Москва, Физматлит, 2001 (in Russian). [G. M. Fihtengolc, A Course of Differentiation and Integration, Vols. 1–3., Fizmatgiz, 2001] I was not able to find an English translation of this book.
8. В. С. Шипачев, "Высшая математика", Москва, Высшая Школа, 1998 (in Russian). [V. S. Shipachev, Higher Mathematics, Vishaja Shkola, 1998.] I was not able to find an English translation of this book.
9. D. V. Widder, Advanced calculus, Dover, 2nd Edition, 1989 (ISBN-13 978-0-486-66103-2).

Famous Scientists Mentioned in the Book

Throughout the book various people, both mathematician and physicists, who are remembered for their outstanding contribution in developing science, will be mentioned. For reader's convenience, their names (together with some information borrowed from their Wikipedia pages) are listed here in an approximate order they first appear in the text:

René Descartes (Latinized: **Renatus Cartesius**) (1596–1650) was a French philosopher, mathematician and writer.

Étienne Bézout (1730–1783) was a French mathematician.

François Viète (**Franciscus Vieta**) (1540–1603) was a French mathematician.

Pythagoras of Samos (c. 570 BC–c. 495 BC) was an Ionian Greek philosopher, mathematician and founder of the religious movement called Pythagoreanism.

Hero (Heron) of Alexandria (10 AD–c. 70 AD) was a Roman Egyptian mathematician.

Leonardo of Pisa, also known as **Fibonacci** (1170–1240/50), was an Italian mathematician.

Jacques Philippe Marie Binet (1786–1856) was a French mathematician, physicist and astronomer.

Tullio Levi-Civita (1873–1941) was an Italian mathematician.

Niccolo Fontana Tartaglia (1499/1500–1557) was an Italian mathematician.

Sir **William Rowan Hamilton** (1805–1865) was an Irish physicist, astronomer and mathematician.

Leopold Kronecker (1823–1891) was a German mathematician who worked in the areas of number theory and algebra.

Rafael Bombelli (1526–1572) was an Italian mathematician.

Lodovico Ferrari (1522–1565) was an Italian mathematician.

Gerolamo (or **Girolamo** or **Geronimo**) **Cardano** (1501–1576) was an Italian mathematician, physician, astrologer and gambler.

Paolo Ruffini (1765–1822) was an Italian mathematician.

Niels Henrik Abel (1802–1829) was a Norwegian mathematician. He is famously known for proving that the solution of the quintic algebraic equation cannot be represented by radicals.

Blaise Pascal (1623–1662) who was a French mathematician, physicist, inventor, writer and Christian philosopher.

Abu Bakr ibn Muhammad ibn al Husayn al-Karaji (or **al-Karkhi**) was a Persian mathematician and engineer.

Jacob Bernoulli (also known as James or Jacques) (1655–1705) was one of the many prominent mathematicians in the Bernoulli family.

Pierre de Fermat (1607–1665) was a French mathematician.

Leonhard Euler (1707–1783) was a famous Swiss mathematician and physicist. He made important discoveries in various fields of mathematics and introduced much of the modern mathematical terminology and notation. He is also renowned for his work in mechanics, fluid dynamics, optics, astronomy and music theory.

John Venn (1834–1923), who was an English mathematician, logician and philosopher.

Andrey Nikolaevich Kolmogorov (1903–1987) was a Soviet mathematician, one of the founders of modern probability theory.

Baron **Augustin-Louis Cauchy** (1789–1857) was a famous French mathematician who contributed in many areas of mathematics.

Viktor Yakovych Bunyakovsky (1804–1889) was a Ukrainian mathematician.

Karl Hermann Amandus Schwarz (1843–1921) was a famous German mathematician.

Guido Grandi (1671–1742) was an Italian mathematician.

Oliver Heaviside (1850–1925) was a self-taught English electrical engineer, mathematician and physicist.

Heinrich Eduard Heine (1821–1881) was a German mathematician.

Bernhard Placidus Johann Nepomuk Bolzano (1781–1848) was a Czech mathematician, philosopher and theologian.

Sir **Isaac Newton** PRS MP (1642–1726/7) was an outstanding English scientist who contributed in many areas of physics (especially mechanics) and mathematics (calculus).

Johannes Diderik van der Waals (1837–1923) was a Dutch theoretical physicist and thermodynamicist.

Abraham de Moivre (1667–1754) was a French mathematician.

Gottfried Wilhelm von Leibniz (1646–1716) was a German mathematician and philosopher.

Brook Taylor (1685–1731) was an English mathematician who is best known for Taylor's theorem and Taylor's series.

Michel Rolle (1652–1719) was a French mathematician.

Colin Maclaurin (1698–1746) was a Scottish mathematician.

Joseph-Louis Lagrange (1736–1813) was an Italian Enlightenment Era mathematician and astronomer. He made significant contributions to the fields of analysis, number theory, and both classical and celestial mechanics.

Guillaume François Antoine, Marquis de l'Hôpital (1661–1704) was a French mathematician.

Lev Davidovich Landau (1908–1968) was a prominent Soviet physicist who made fundamental contributions to many areas of theoretical physics, 1962 Nobel laureate. He is a co-author of a famous ten-volume course of theoretical physics.

Vitaly Lazarevich Ginzburg (1916–2009) was a Soviet and Russian theoretical physicist, astrophysicist, 2003 Nobel laureate.

Georg Friedrich Bernhard Riemann (1826–1866) was a German mathematician.

Thomas Joannes Stieltjes (1856–1894) was a Dutch mathematician.

Henri Léon Lebesgue (1875–1941) was a French mathematician.

Jean-Gaston Darboux (1842–1917) was a French mathematician.

Jean Baptiste Joseph Fourier (1768–1830) was a French mathematician and physicist, best known for Fourier series and integral.

Robert Brown (1773–1858) was a Scottish botanist and paleobotanist.

Paul Langevin (1872–1946) was a prominent French physicist.

Paul Adrien Maurice Dirac (1902–1984) was an English theoretical physicist who is regarded as one of the most significant physicists of the twentieth century.

Johann Carl Friedrich Gauss (1777–1855) was a German mathematician and physicist.

Enrico Fermi (1901–1954) was an Italian (later naturalised American) physicist.

Ludwig Eduard Boltzmann (1844–1906) was an Austrian physicist and philosopher famously known for his fundamental work in statistical mechanics and kinetics.

Lorenzo Romano Amedeo Carlo Avogadro di Quaregna e di Cerreto (1776–1856) was an Italian scientist.

Jean-Baptiste Biot (1774–1862) was a French physicist, astronomer and mathematician.

Félix Savart (1791–1841) was a French physicist.

Pierre-Simon, marquis de Laplace (1749–1827) who was a French mathematician and astronomer.

Hermann Ludwig Ferdinand von Helmholtz (1821–1894) was a German physician and physicist.

Josiah Willard Gibbs (1839–1903) was an American physicist, chemist and mathematician.

James Clerk Maxwell (1831–1879) was a Scottish mathematical physicist best known for his unification of electricity and magnetism into a single theory of electromagnetism.

Karl Theodor Wilhelm Weierstrass (1815–1897) was an outstanding German mathematician contributed immensely into modern analysis.

Carl Gustav Jacob Jacobi (1804–1851) who was a German mathematician.

Andrey (Andrei) Andreyevich Markov (1856–1922) was a Russian mathematician.

Albert Einstein (1879–1955) was a German-born theoretical physicist famously known for his relativity and gravity theories.

Marian Smoluchowski (1872–1917) was an Austro-Hungarian Empire scientist of a Polish origin.

Sydney Chapman (1888–1970) was a British mathematician and geophysicist.

Norbert Wiener (1894–1964) was an American mathematician and philosopher.

Adolf Eugen Fick (1829–1901) was a German-born physician and physiologist.

George Green (1793–1841) was a British mathematical physicist.

August Ferdinand Möbius (1790–1868) was a German mathematician and theoretical astronomer.

Johann Benedict Listing (1808–1882) was a German mathematician.

George Gabriel Stokes (1819–1903) was an Irish and British mathematician, physicist, politician and theologian.

Mikhail Vasilyevich Ostrogradsky (1801–1862) was a Russian–Ukrainian mathematician, mechanician and physicist.

Archimedes of Syracuse (c. 287 BC–c. 212 BC) was a Greek mathematician, physicist, engineer, inventor and astronomer.

Erwin Rudolf Josef Alexander Schrödinger (1887–1961) was an Austrian physicist, one of the founders of quantum theory, he is famously known for Schrödinger (wave) equation of quantum mechanics. He also contributed in other fields for physics such as statistical mechanics and thermodynamics, physics of dielectrics, colour theory, electrodynamics, general relativity and cosmology.

Siméon Denis Poisson (1781–1840) was a French mathematician, geometer and physicist.

Hendrik Antoon Lorentz (1853–1928) was a Dutch physicist who shared the 1902 Nobel Prize in Physics with Pieter Zeeman for the discovery and theoretical explanation of the Zeeman effect. He also derived the transformation equations underpinning Albert Einstein's special theory of relativity.

André-Marie Ampère (1775–1836) was a French physicist and mathematician.

Michael Faraday (1791–1867) was a famous English physicist.

Oleg Dmitrovich Jefimenko (1922–2009) was a Ukrainian and American physicist.

Jean-Baptiste le Rond d'Alembert (1717–1783) was a French mathematician, physicist and philosopher.

Lorenzo Mascheroni (1750–1800) was an Italian mathematician.

Douglas Rayner Hartree (1897–1958) was an English mathematician and physicist.

Max Karl Ernst Ludwig Planck (1858–1947) was a German theoretical physicist.

Satyendra Nath Bose (1894–1974) was an Indian Bengali physicist specialising in mathematical physics.

Adrien-Marie Legendre (1752–1833) was a French mathematician.

Charles Hermite (1822–1901) was a French mathematician.

Edmond Nicolas Laguerre (1834–1886) was a French mathematician.

Pafnuty Lvovich Chebyshev (1821–1894) was a Russian mathematician.

Augustin-Jean Fresnel (1788–1827) was a French engineer known for his significant contribution to the theory of wave optics.

Léon Nicolas Brillouin (1889–1969) was a French physicist.

Charles-Augustin de Coulomb (1736–1806) was a French officer, engineer and physicist.

Paul Peter Ewald (1888–1985) was a German crystallographer and physicist, a pioneer of X-ray diffraction methods.

Charles Émile Picard (1856–1941) was a French mathematician.

Józef Maria Hoene-Wroński (1776–1853) was a Polish Messianist philosopher who worked in many fields of knowledge, including mathematics and physics.

Friedrich Wilhelm Bessel (1784–1846) was a German mathematician and astronomer.

Pierre Bouguer (1698–1758) was a French mathematician, geophysicist, geodesist and astronomer.

George Boole (1815–1864) was a largely self-taught English mathematician, philosopher and logician.

Christiaan Huygens (1629–1695), also spelled Huyghens, was a Dutch mathematician, physicist, astronomer and inventor.

Johann Bernoulli, also known as Jean or John (1667–1748), was a Swiss mathematician and was one of the many prominent mathematicians in the Bernoulli family.

Ferdinand Georg Frobenius (1849–1917) was a German mathematician.

Gustav Robert Kirchhoff (1824–1887) was a German physicist who contributed to the fundamental understanding of electrical circuits, spectroscopy and the emission of black-body radiation by heated objects.

Johannes Kepler (1571–1630) was a German astronomer, mathematician, astrologer, natural philosopher and writer on music.

Konstantin Eduardovich Tsiolkovsky (1857–1935) was a Russian and Soviet rocket scientist and pioneer of the astronautic theory.

Svante August Arrhenius (1859–1927) was a Swedish physicist, one of the founders of physical chemistry.

Contents

Part I
Fundamentals

Basic Knowledge

1

This is an introductory chapter which sets up some basic definitions we shall need throughout the book. We shall also introduce some elementary concepts (some of which will be refined and generalised later on in forthcoming chapters) and derive simple formulae which the reader may well be familiar with; however, it is convenient to have everything we will need under "one roof". Hence, the content of this chapter will enable the reader to work on the forthcoming chapters without the need of consulting other texts or web pages, and should make the book more or less self-contained.

1.1 Logic of Mathematics

In mathematics (and in other natural sciences, especially in physics), we would like to establish relations between various properties A, B, C, etc. In particular, we might like to know if property A is equivalent to B, or, less strictly, whether B follows from A (i.e. B is true if A is, i.e. A⇒B). This kind of argument is what we will be referring to as a "proof". It must be based on known facts. For instance, if we would like to prove that A⇒B, then we use known properties of A, and via a logical argument, the property B should follow.

What needs to be done in order to prove that some property A is *equivalent* to B? In this case, it is not sufficient to show that B follows from A, i.e. A⇒B; it is also necessary to show that, conversely, A follows from B, i.e. B⇒A. If it was possible to provide a logical path in both directions, then one can say that B is true *"if and only if"* A is true. It is also said that there is *one-to-one correspondence* between A and B, or that for B to be true it is a *necessary and sufficient* for A to be true. For instance, A could be the statement that "a number consists of three digits", while B is that the number lies inclusively between 100 and 999. Both these statements are equivalent.

© The Author(s), under exclusive license to Springer Nature Switzerland AG 2022
L. Kantorovich, *Mathematics for Natural Scientists*, Undergraduate Lecture Notes
in Physics, https://doi.org/10.1007/978-3-030-91222-2_1

Indeed, from A (three digits) obviously B follows (between and including 100 and 999), and, conversely, from B immediately follows A.

If, however, it is only possible to provide a proof in one direction, e.g. A⇒B, but not in the other (e.g. by providing a contradicting example), then it is said that for B to happen it is *sufficient* that A is true, i.e. A implies B. If for B to be true A is necessarily required (but there are also some additional conditions required for B to be true as well), then it is said that A is a *necessary* condition for B. If A is necessarily required for B to be true, then B⇒A. Therefore, if both statements are proven, i.e. that A is both necessary and sufficient for B to be true, then obviously A and B are equivalent.

For instance, we know that any even number is divisible by 2; conversely, any number divisible by 2 can be written as $2n$ with n being an integer, and so it is even. Hence, these two conditions are indeed equivalent. However, not every even number is divisible by 4, i.e. it is a necessary condition for the number to be even in order to be divisible by 4, but not sufficient. Similarly, all numbers ending with 2 are divisible by 2, this is a sufficient condition; however, there are also other even numbers divisible by 2. Therefore, it is not possible to show that any number divisible by 2 ends with the digit 2, i.e. having the digit 2 at the end of the number is sufficient for the number to be even, but not necessary.

In some cases, several conditions are equivalent. In such cases, the logic of a proof goes in a cycle. If we would like to prove the equivalence of three conditions A, B and C, then we prove that: A⇒B, then B⇒C and, finally, C⇒A.

How is the proof of say A⇒B done in practice? Normally, one constructs using algebra or geometry, or maybe some other argument, a logical path to B based on accepting A, and this process may contain several steps. If the reverse is to be proven, the process is repeated, but this time assuming that B is true, and, on the acceptance of that, a logical path is built towards A. Alternatively, an example may be proposed (just one such example would be enough) that corresponds to B, but contradicts A, in which case the reverse passage B⇒A is not possible, and hence only the sufficient condition is proven. Other methods of proof exist in mathematical logic as well but will not be used in this book.

If one makes a statement, it can be proven wrong by suggesting a single example which contradicts it.

In some cases, a statement can be proven right by assuming an opposite and then proving that the assumption was wrong—this is called *proving by contradiction*.[1]

There is one rather famous logical method of a proof called *mathematical induction* (or simply *induction*), and since we are going to use it quite often, let us discuss it here. Suppose, there is a property which depends on the natural number $n = 1, 2, 3, \ldots$. We would like to establish a formula for that property which would be valid for any $n = 1, 2, 3, \ldots$ and up to infinity. We perform detailed calculations for some small values of n, say $n = 1, 2, 3, 4$, and see a pattern or a rule, and devise a formula for a general n. But would it be valid for any n? Obviously no one

---

[1] The corresponding Latin expression *"reductio ad absurdum"* is also frequently used.

can repeat the calculation for all possible integer values of n as there are an infinite number of them. Nevertheless, there is a beautiful logical construction which allows one to prove (or disprove) our formula. The method consists of three steps and runs as follows:

- check that the formula is valid for the smallest values of n of interest, e.g. for $n = 1$;
- *assume* that the formula is valid for some integer value n;
- prove then that it is also valid for $n + 1$ by performing the necessary calculation assuming the formula for the previous integer value (i.e. for n); if the result for $n + 1$ looks exactly the same as for n but with n replaced by $n + 1$, the formula is correct; if this is not the case, the formula is wrong.

If the last step is successfully proven, then that means the formula is valid for *any* value of n. Indeed, let us assume our formula is trivially verified for $n = 1$. Then, because of the last two steps of the induction, it must also be valid for the next value of $n = 2$. Once it is valid for $n = 2$, by virtue of the same argument, it must also be valid for the next value of $n = 3$, and so on till infinity.

As an example, consider a trivial case of even natural numbers. Let us prove that any even natural number can be written using the formula $N_k = 2k$, where k is an integer. Indeed, for $k = 1$ we get the number $N_1 = 2 \cdot 1 = 2$, which is obviously even. Now we *assume* that the number $N_k = 2k$ is even for any natural number k. Let us prove that the number associated with the next value of k is also even. We add 2 to the even number N_k to get the next even number $N_k + 2 = 2k + 2 = 2(k + 1)$, which is exactly of the required N_{k+1} form. **Q.E.D.**[2]

Problem 1.1 Demonstrate that any odd number can be written as $N_k = 2k - 1$, where $k = 1, 2, 3, \ldots$.

There will be more examples, less trivial than these ones, later on in this and other chapters.

1.2 Real Numbers

In applications, we often talk about various quantities having a certain value. For this, we use *numbers*.

---

[2] Which means (from Latin "*quod erat demonstrandum*") " which was to be demonstrated".

1.2.1 Integers

The simplest numbers arise from countable quantities which may only take one of the positive integer numbers $1, 2, 3, \ldots$ called natural numbers. These numbers form a sequence (or a set) containing an infinite number of numbers (members). By *infinite* here we mean that the set never ends: after each, however big, number there is always standing the next one bigger by one. We shall denote this set with the symbol $\mathbb{N}at$, it is formed only by positive integer numbers. We also define the simplest *operations* on the numbers such as addition, subtraction, multiplication and division. Addition and multiplication of any two numbers from $\mathbb{N}at$ is a number from the same set; subtraction and division of two numbers may be either the number of the same set or not.

Using the subtraction operation, negative integer numbers and the zero number are created forming a wider set $\mathbb{Z} = \{0, \pm1, \pm2, \pm3, \ldots\}$ of positive and negative integers (including zero). Specifically, the number zero, 0, has a property that when added to (or subtracted from) any other number belonging to $\mathbb{Z}$, the number does not change.[3] The integer numbers have a property that summation, subtraction or multiplication of any two members of the set $\mathbb{Z}$ is a member of the same set, i.e. it is an integer number. An integer power operation n^k (where k is from $\mathbb{N}at$) is then defined for any $n \neq 0$ as a multiplication of n with itself k times: $n^1 = n, n^2 = n \cdot n$, $n^3 = n \cdot n \cdot n$, and so on. We also define that zero power gives always one: $n^0 = 1$.

Consider a division of two integer numbers N and M. If N is divisible by M, then we can write $N = Md$, where $d = N/M$ is also an integer. If, however, N is not divisible by M, then one can find r, that is, between 0 and M, such that $N - r$ is divisible by M. In other words, one can write $N - r = Md$ or $N = Md + r$. Here $0 < r < M$ is an integer called residue, while d is called quotient. For instance, when $N = 10$ is divided by $M = 3$, we can write $10 - 1 = 3 \times 3$ or $10 = 3 \times 3 + 1$. Here $d = 3$ and $r = 1$. Hence, an integer number N is divisible by another integer N if there is no residue, i.e. $r = 0$.

Any integer number can be represented as a sum containing the base ten:

$$\pm a_n a_{n-1} \cdots a_1 a_0 = \pm \left(a_n 10^n + a_{n-1} 10^{n-1} + \cdots + a_1 10^1 + a_0 \right), \quad (1.1)$$

where on the left we wrote an integer consisting of $n + 1$ digits (with the corresponding plus or minus sign), and on the right the decimal representation of the whole number. For instance, $375 = 300 + 70 + 5 = 3 \cdot 10^2 + 7 \cdot 10^1 + 5$ and $-375 = -\left(3 \cdot 10^2 + 7 \cdot 10^1 + 5 \right)$. The representation above allows proving familiar divisibility rules. For instance, it follows from the above that any number is divisible by 2 or 5 if the last digit a_0 is divisible or is 0 since any base 10 is divisible by 2 or 5.

---

[3] Often 0 is considered belonging to natural numbers $\mathbb{N}at$, but this is not so important for us here.

Problem 1.2 Prove that a number is divisible by 3 or 9 if the sum of all its digits is divisible by 3 or 9, respectively. [Hint: *note that the number* $10^k - 1$ *is divisible by 3 or 9 for any integer k*.]

Problem 1.3 Prove the following divisibility rule for 7: multiply the first, the second, the third, etc. digits by 1, 3, 2, 6, 4 and 5, respectively (cut or continue the same sequence as necessary depending on the number of digits in the given number), then sum up all the numbers thus obtained. If the sum is divisible by 7, so is the whole number.

Problem 1.4 Using the method of mathematical induction, prove that the number $5^n - 1$, where n is an integer, is divisible by 4.

Problem 1.5 Similarly, prove that $3^{2n} + 7$, where n is an integer, is divisible by 8.

1.2.2   Rational Numbers

Rational numbers form a set $\mathbb{Q}$. These are formed by dividing all possible numbers m and n from $\mathbb{Z}$, i.e. they are represented via the ratio m/n with $n \neq 0$. Integer numbers are a particular case of the rational numbers formed by taking the denominator $n = 1$. Any sum, difference, product (including power) or division of two rational numbers is a rational number, i.e. it belongs to the same set over all four operations. For instance, in the case of the summation:

$$z_1 + z_2 = \frac{m_1}{n_1} + \frac{m_2}{n_2} = \frac{m_1 n_2 + m_2 n_1}{n_1 n_2} = \frac{m_3}{n_3} = z_3 \,,$$

where the result, z_3, is obviously a rational number as well since $m_3 = m_1 n_2 + m_2 n_1$ and $n_3 = n_1 n_2$ are each integers.

Problem 1.6 Prove that a product or division of two rational numbers, or an integer power of a rational number, is also a rational number.

> **Problem 1.7** Prove that there exist an infinite number of rational numbers between any two rational numbers. [Hint: *construct a set of n equidistant numbers lying along a linear interpolation between the two numbers and show that they are all rational.*]

Any rational number $z = m/n$ can be represented as an integer i plus a rational number r, which is between -1 and 1. For instance, consider $z > 0$ with $0 < m < n$. In this case, $0 < m/n < 1$ and hence $i = 0$ and $r = z$. If $m > n$, then it can always be written as $m = i \cdot n + m_1$, where m_1 is the remainder lying between 0 and n excluding n (i.e. $0 \le m_1 < n$). Then

$$z = \frac{i \cdot n + m_1}{n} = i + \frac{m_1}{n} = i + r .$$

Similarly for negative rational numbers. For instance, $13/6 = (2 \cdot 6 + 1)/6 = 2 + 1/6$; here $i = 2$ and $r = 1/6$, while $-13/6 = -2 - 1/6$.

In the *decimal representation*, positive rational numbers are represented as a number with the dot: the integer part is written on the left of the dot, while the rest of it (which is a number between 0 and 1, called fractional part of the number) on the right. This is done in the following way: consider a number $m/n = i + m_1/n$ with $0 \le m_1 < n$. Let us *assume* that there exists the smallest positive integer k such that $n \cdot k = 1\underbrace{0 \cdots 0}_{l} = 10^l$, i.e. it is represented as 1 followed by l zeroes. For instance, if $n = 2$, then the smallest possible k is 5 as $2 \cdot 5 = 10 = 10^1$; if $n = 8$, then the smallest possible $k = 125$ yielding $8 \cdot 125 = 1000 = 10^3$. Then,

$$\frac{m_1}{n} = \frac{m_1 \cdot k}{n \cdot k} = \frac{m_1 \cdot k}{10^l} .$$

Since $m_1 < n$, then $m_1 \cdot k < 10^l$ and hence the number m_1/n is represented by *no more* than l digits. These are these digits that are written on the right of the dot in the decimal number. For instance, $3/8 = (3 \cdot 125)/(8 \cdot 125) = 375/1000 = 0.375$, $12/5 = 2 + 2/5 = 2 + 4/10 = 2.4$. For negative numbers, the same procedure is applied first to its positive counterpart, and then the minus sign is attached: $-3/8 = -0.375$, $-12/5 = -2.4$.

It is easy to see that if a positive rational number has an integer part i and the rest $r = m_1/n$ consists of digits $d_1 d_2 \cdots d_l$, i.e. $r = 0.d_1 d_2 \cdots d_l$, then one can always write:

$$z = \frac{m}{n} = i + \frac{m_1}{n} = i + \frac{d_1}{10^1} + \frac{d_2}{10^2} + \frac{d_3}{10^3} + \cdots + \frac{d_l}{10^l} . \tag{1.2}$$

For instance,

$$\frac{3}{8} = 0.375 = 0 + \frac{3}{10^1} + \frac{7}{10^2} + \frac{5}{10^3} \quad \text{and} \quad \frac{12}{5} = 2.4 = 2 + \frac{4}{10^1} .$$

If a rational number has a finite number of digits after the dot (e.g. $0.415 = 415/1000$), then its representation (1.2) via a sum of inverse powers of 10 will contain a finite number of such terms. However, in some cases, this is not the case: some rational numbers may have an *indefinitely repeating* sequence of digits (called *periods*), e.g. $1/3 = 0.33333\ldots = 0.(3)$ (which means it is of the period of 3), $1/6 = 0.16666\ldots = 0.1(6)$, and hence an infinite number of terms with inverse powers of 10 are needed:

$$\frac{1}{3} = 0.(3) = 0 + \frac{3}{10^1} + \frac{3}{10^2} + \frac{3}{10^3} + \frac{3}{10^4} + \cdots$$

or

$$\frac{1}{6} = 0.1(6) = 0 + \frac{1}{10^1} + \frac{6}{10^2} + \frac{6}{10^3} + \frac{6}{10^4} + \cdots .$$

These are examples of infinite numerical series which we shall be studying in other chapters of this course. The numbers m/n with periods in the decimal representation arise if it is impossible to find an integer k such that $n \cdot k$ is represented as 10^l with some positive integer l.

1.2.3 Irrational Numbers

One can also introduce numbers which cannot be represented as a ratio of two integers; in other words, there are also numbers z, for which it is impossible to find m and n from $\mathbb{Z}$ such that $z = m/n$. These are called *irrational* numbers. For instance, the rational number $1/4$ can be also written as a square of $1/2$, i.e. $(1/2)^2 = 1/4$, or, inversely, $1/2 = \sqrt{1/4}$ is expressed via the square root operation $\sqrt{\ldots}$. However, not every square root of a rational number is also a rational number, e.g. $\sqrt{1/3}$ is not; this is because the integer number 3 cannot be represented as a square of another integer number. In the decimal representation, irrational numbers are constructed from a sequence of an infinite number of digits which is never repeated; $\sqrt{2} = 1.414213562373095\ldots$ is an example of an irrational number.

Any irrational number z can be bracketed by two rational numbers r_1 and r_2, i.e. $r_1 < z < r_2$, and this can be done with arbitrary precision, with the two rational numbers being arbitrarily close to the irrational number z. Indeed, suppose $z > 0$ is represented in the decimal representation as $z = n.d_1 d_2 d_3 d_4 d_5 \cdots d_{i-1} d_i d_{i+1} \cdots$, where n is its integer part and $0.d_1 d_2 d_3 d_4 d_5 \cdots d_{i-1} d_i d_{i+1} \cdots$ is its fractional part. Then, we can always write, choosing a particular decimal at the position $i = 1, 2, 3, \ldots$ in the fractional part:

$$n.d_1 d_2 d_3 d_4 d_5 \cdots d_{i-1} d_i < z < n.d_1 d_2 d_3 d_4 d_5 \cdots d_{i-1} (d_i + 1) \ ,$$

where $r_1 = n.d_1 d_2 d_3 d_4 d_5 \cdots d_{i-1} d_i$ was constructed by terminating the decimal representation of z by the ith digit and hence is a rational number (as it has a finite decimal representation), while $r_2 = n.d_1 d_2 d_3 d_4 d_5 \cdots d_{i-1} (d_i + 1)$ is obtained similarly by terminating on the same digit, and then increasing it by one. For instance, for $\sqrt{2}$ we can write the following bracketing approximations each being closer to the actual value of $\sqrt{2}$ because of choosing a larger position i of the digit in question:

$$1.414 < \sqrt{2} < 1.415 \quad (i = 3) \; ,$$

$$1.4142 < \sqrt{2} < 1.4143 \quad (i = 4) \; ,$$

$$1.41421 < \sqrt{2} < 1.41422 \quad (i = 5) \; .$$

If the chosen digit is 9, then r_2 can be chosen using the next nearest digit that is less than 9, for instance:

$$1.41421356237309 < \sqrt{2} < 1.414213562373096 \quad (i = 14) \; .$$

Similarly, one can bracket a negative irrational number $z < 0$.

1.2.4 Real Numbers

All positive and negative rational and irrational numbers form a complete set of *real* numbers $\mathbb{R}$. Addition, subtraction, multiplication (and power) and division of any two real numbers (avoiding division by zero) always result in a real number. Operations of summation and multiplication are *commutative*, i.e. they do not depend on the order of terms ($x + y = y + x$ and $xy = yx$), and *associative* (($x + y) + z = x + (y + z)$ and $(xy)z = x(yz)$); they are also *distributive*: $x(y + z) = xy + xz$. The subtraction operation $x - y$ is understood as a summation of x and $-y$; division of x and y is understood as a multiplication of x and $1/y$. Here x, y and z are any real numbers.

Real numbers can be represented as points on the number axis as shown in Fig. 1.1a. The distance between a point on the axis and zero represents an absolute value $|x|$ of the number x; its sign is plus if the point is on the right of 0, while the sign is minus if it is on the left. Between any two real numbers x_1 and x_2, no matter how close they are to each other, there is always an infinite number of real numbers x between them, $x_1 < x < x_2$. Numbers run in both directions from zero indefinitely, i.e. for any number $x > 0$ there is always another number $x' > x$ on the right of it; similarly, for any number $x < 0$ on the left of the zero there can always be found a number $x' < x$ on the left of it. It is convenient to introduce symbols on the right and the left edges of the set of real numbers, called plus and minus infinities, ∞ and $-\infty$, which designate the largest possible positive and negative numbers, respectively. These are not actual numbers (since for any real number there is always a real number either larger or smaller), but special symbols indicating unbounded limits of the sequence of real numbers at both ends. Although the symbols $\pm\infty$ are shown at the right and left edges of the number axis, they are actually located indefinitely far away from zero, and any number x satisfies $-\infty < x < \infty$. Division of a positive real number by 0 is defined to be $+\infty$, while division of a negative number by 0 is defined to be $-\infty$.

1.2.5   Intervals

We shall frequently be dealing with continuous subsets of numbers enclosed between two real numbers a and b. These subsets are called *intervals* and could be either: (i) unbounded, when $a < x < b$ (both ends are open, i.e. x could be as close as desired to either a or b, but cannot be equal to them); (ii) semi-bounded, if either $a \le x < b$ or $a < x \le b$ (one end is open and one closed, i.e. x could also be equal to either a or b) or (iii) bounded, if $a \le x \le b$ (both ends are closed, so all numbers including the boundaries a and b are accepted). Intervals symmetric with respect to zero can be written using the absolute value symbol which is defined as follows:

$$|x| = \begin{cases} x\,, & \text{if } x \ge 0 \\ -x\,, & \text{if } x < 0 \end{cases} \tag{1.3}$$

Then, $|x| < a$ corresponds to the unbounded interval between $-a$ and a, and $|x| \le a$ to the same interval with the addition of the upper and lower bounds $-a$ and a (assuming here that $a > 0$). A set of real numbers which are on the left of $-a$ or on the right of a can shortly be written as $|x| > a$. All real numbers can be represented as an interval between $\pm\infty$, i.e. $-\infty < x < \infty$. The interval $|x - x_0| < a$ corresponds to $-a < x - x_0 < a$ or $-a + x_0 < x < a + x_0$, while $|x - x_0| > a$ to $x < -a + x_0$ and $x > a + x_0$.

   Frequently other notations for the intervals are also used: $[a, b]$ is equivalent to $a \le x \le b$, $[a, b)$ is equivalent to $a \le x < b$, $(a, b]$ is equivalent to $a < x \le b$ and, finally, (a, b) is equivalent to $a < x < b$. In the mathematics literature, the fact that an object "belongs" to a particular set is expressed with a special symbol $\in$ (the membership relation), e.g. the fact that x is somewhere between a and b, $a < x < b$, is expressed compactly as $x \in (a, b)$; however, we shall not use these notations here in the book.

1.3   Basic Tools: An Introduction

1.3.1   Cartesian Coordinates in 2D and 3D Spaces

The number axis in Fig. 1.1a corresponds to a one-dimensional (often denoted as 1D) space $\mathbb{R}$. Any operation between two numbers from $\mathbb{R}$ such as summation, subtraction, multiplication (and, hence, power) and division results in another number from the same set $\mathbb{R}$.

   By taking an ordered pair (x, y) of two real numbers x and y, a two-dimensional (or 2D) space is formed, see Fig. 1.1b. In this case, one draws two perpendicular number axes, one for x and another for y. A point on thus constructed plane, P, is said to have Cartesian[4] coordinates x and y, if two projections (perpendiculars) drawn from the point onto the two axes will cross them at the real numbers x and

---

[4] Due to Renè Descartes.

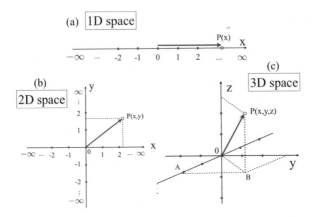

Fig. 1.1 Graphical representation of **a** one-, **b** two- and **c** three-dimensional spaces of real numbers

y, respectively. This point is denoted $P(x, y)$. The 2D space formed by pairs of real numbers is denoted $\mathbb{R}^2$.

Similarly, a set (x, y, z) of three real numbers, x, y and z, corresponds to a point in a three-dimensional space $\mathbb{R}^3$ (or 3D). It is built by three mutually orthogonal number axes for the x, y and z, see Fig. 1.1c, and a point P is said to have Cartesian coordinates x, y and z if the corresponding projections onto the three axes cross them at points x, y and z; the point is denoted $P(x, y, z)$. Obviously, the 2D Cartesian system is a particular case of the 3D space corresponding to the $z = 0$ plane in it.

1.3.2   Algebra

A more detailed description of functions will be given in the next chapter. However, for the purposes of this chapter, some elementary knowledge is necessary. Although the reader should be familiar with what is to follow, we shall concisely summarise it here.

We shall start by reminding the reader the main ideas of *algebra*. Actually, we have already benefited from it above when using letters to designate numbers; here we shall consider it more systematically. In mathematics it is often useful (and convenient) to use letters instead of numbers as many operations on numbers such as multiplication, addition, division, etc. result in a single answer in a form of another number and no general rules are seen through this arithmetic calculation. Replacing numbers by letters allows one to perform calculations in a general form obtaining the answer (we call it *formula*) in the form of a combination of letters. Replacing the letters by numbers in the answer (i.e. in the resulting formula) gives the required numerical value. However, the convenience of this "algebraic way", as opposed to manipulating directly with numbers, is in the fact that the calculation from the original point (initial input) to the final result only needs to be done once and then the resulting expression (formula) will be valid for any numerical values of the input numbers, and hence can be used many times. Moreover, using algebraic manipulations it is often discovered that many terms may cancel out in the final result yielding a simple final expression. Therefore, using algebra the same final formula can be applied for different initial values and this may save time and effort.

However, we need to work out rules for algebraic manipulations. These follow from the rules which are valid for numbers. Since multiplication of two numbers does not depend on their order, e.g. $5 \cdot 3 = 3 \cdot 5$, we establish a general rule that multiplication of two letters, say a and b, is *commutative*: $a \cdot b = b \cdot a$. The dot here indicates the multiplication operation explicitly, it will be omitted from now on as only a single letter is used for a single number, so two letters one after another imply two numbers multiplied with each other. Similarly summation is also commutative, e.g. $3 + 5 = 5 + 3$, and so is summation of letters in algebra: $a + b = b + a$. A combination of addition and multiplication, e.g. $3 \cdot (5 + 6)$, may also be performed as $3 \cdot 5 + 3 \cdot 6$; we say that this binary operation is *distributive*. Here brackets indicate the order in which the operations are to be performed: expressions within the brackets must be performed first. Therefore, the same can be then stated for letters: using distributivity, one can write $a(b + c)$ as $ab + ac$, which is called "open the brackets". Next, attaching a minus sign to a number a makes the number to change sign, $-a$. Adding a and $-b$ is equivalent to subtracting a and b, i.e. $a + (-b) = a - b$. The division $c = a/b$ of a and b is defined such that $cb = a$. In fact, the ratio a/b is equivalent to a multiplication of a and $1/b$, and

$$\frac{a}{b} : \frac{c}{d} = \frac{a}{b} \frac{d}{c} = \frac{ad}{bc} .$$

Next, we mention two special numbers: 0 and 1. Their properties are summarised below:

$$a + 0 = a ; \quad a + (-a) = 0 ; \quad a \cdot 0 = 0 ; \quad a \cdot 1 = a \quad \text{and} \quad a\frac{1}{a} = \frac{a}{a} = 1 .$$

Division by 0 results in infinity: $a/0 = \infty$ if $a > 0$ and $a/0 = -\infty$ if $a < 0$. Expressions $0/0$, $0 \cdot \infty$ and ∞/∞ are not defined and in each particular case must be specifically analysed. However, $a/\infty = 0$ for any $a \neq 0$.

We can also use powers to indicate multiplication of a number with itself, e.g. $a^0 = 1$, $a^2 = aa$, etc. Both summation and multiplication are also *associative*, i.e. they do not depend on the order in which they are performed: $(a + b) + c = a + (b + c)$ and $(ab) c = a (bc)$. Finally, the following rules related to integer powers are true:

$$a^n a^m = a^{n+m} \quad \text{and} \quad (a^n)^m = \underbrace{a^n \cdots a^n}_{m} = a^{nm} .$$

Negative powers are defined via $\frac{1}{a^n} = a^{-n}$, where $a \neq 0$ and n is a positive integer. Obviously,

$$\frac{a^n}{b^m} = a^n b^{-m} \quad (b \neq 0) .$$

In particular,

$$\left(\frac{a}{b}\right)^n = \frac{a^n}{b^n} = a^n b^{-n} \quad \text{and} \quad \frac{a^n}{a^m} = a^{n-m} = \frac{1}{a^{m-n}} \ .$$

Also,

$$\left(a^n\right)^{-m} = \frac{1}{(a^n)^m} = \frac{1}{a^{nm}} = a^{-nm} \quad \text{and} \quad \left(a^{-n}\right)^{-m} = \frac{1}{(a^{-n})^m} = \frac{1}{\left(\frac{1}{a^n}\right)^m} = \frac{1}{\frac{1}{a^{nm}}} = a^{nm} \ .$$

We see that the natural powers taken one after another are multiplied algebraically.

The operation that is inverse to the square is the square root: if $x^2 = a > 0$, then $x = \sqrt{a}$. In fact, it is easy to see that the negative root value $x = -\sqrt{a}$ also satisfies the equation $x^2 = a$. Hence, there are two roots of this equation, which can be conveniently written as $x = \pm\sqrt{a}$.

It is easy to generalise the notion of the root and introduce a rational power. Indeed, let us define the square root as the $\frac{1}{2}$ power: $\sqrt{a} = a^{1/2}$. Taking square of both sides of this identity, we obtain:

$$\left(\sqrt{a}\right)^2 = \left(a^{1/2}\right)^2 \ .$$

In the left-hand side we will have a, so we should have the same in the right-hand side as well. We saw above that natural powers multiply if taken one after another. Extending this rule to rational powers as well, we shall get in the right-hand side $a^{2/2} = a^1 = a$, as required. Similarly, one can define the power $1/n$ for any integer n as the operation inverse to the nth power: if $x^n = a$, then $a = x^{1/n} = \sqrt[n]{x}$. Finally, the rational power n/m (with both n and m being positive integers) is defined as

$$a^{n/m} = \left(a^n\right)^{1/m} = \sqrt[m]{a^n} = \left(a^{1/m}\right)^n = \left(\sqrt[m]{a}\right)^n \ ,$$

while

$$a^{-n/m} = \frac{1}{a^{n/m}} = \frac{1}{\sqrt[m]{a^n}} = \frac{1}{\left(\sqrt[m]{a}\right)^n} \ .$$

For any rational power, the following identities hold:

$$a^{p/q} a^{m/n} = a^{p/q+m/n} \quad \text{and} \quad \left(a^{p/q}\right)^{m/n} = a^{pm/qn} \ .$$

We shall discuss powers in more detail in Sect. 2.3.3.

As an example, let us first consider an expression $(a + b)^2$. By definition of the power, this is $(a + b)(a + b)$. Then we use distributivity, treating one bracket as another number $c = a + b$:

$$(a + b)^2 = (a + b)c = ac + bc = a(a + b) + b(a + b) \ .$$

Next, we use the distributivity again:

$$(a+b)^2 = a(a+b) + b(a+b) = a^2 + ab + ba + b^2 .$$

Because a and b commute, $ab = ba$, we can sum up ab and ba together as $ab + ba = ab + ab = 2ab$, which finally allows us to arrive at the familiar result:

$$(a+b)^2 = a^2 + 2ab + b^2 .$$

Similarly,

$$\frac{a}{b} + \frac{c}{d} = \frac{ad}{bd} + \frac{cb}{db} = \frac{ad}{bd} + \frac{cb}{bd} = \frac{1}{bd}(ad) + \frac{1}{bd}(cb) = \frac{1}{bd}(ad + cb) = \frac{ad + cb}{bd} .$$

Here we multiplied and divided the first fraction by d and the second by b (this is permissible as, e.g. $d/d = 1$), so that in both fractions the denominator becomes the same and equal to bd. Next, we combined the two fractions into one by taking the common denominator out.

Problem 1.8 Prove the following algebraic identities:

$$a^2 - b^2 = (a-b)(a+b) ; \quad (a-b)^2 = a^2 - 2ab + b^2; \tag{1.4}$$

$$(a \pm b)^3 = a^3 \pm 3a^2 b + 3ab^2 \pm b^3; \tag{1.5}$$

$$a^3 \pm b^3 = (a \pm b)\left(a^2 \mp ab + b^2\right). \tag{1.6}$$

In the last two cases either upper or lower signs of $\pm$ or $\mp$ should be consistently used on both sides.

Problem 1.9 Prove more algebraic identities by expanding expressions in the right-hand sides and simplifying:

$$a^4 - b^4 = (a-b)(a+b)\left(a^2 + b^2\right); \tag{1.7}$$

$$a^6 - b^6 = (a-b)(a+b)\left(a^2 - ab + b^2\right)\left(a^2 + ab + b^2\right); \tag{1.8}$$

$$a^6 + b^6 = \left(a^2 + b^2\right)\left(a^4 - a^2 b^2 + b^4\right)$$

$$= \left(a^2 + b^2\right)\left(a^2 - \sqrt{3}ab + b^2\right)\left(a^2 + \sqrt{3}ab + b^2\right) . \tag{1.9}$$

Problem 1.10 Simplify the following expressions:

$$\frac{(a+b)^2 + (a-b)^2}{(a+b)^2 - (a-b)^2} \; ; \quad \frac{(a+b)^3 + (a-b)^3}{(a+b)^3 - (a-b)^3} \; ;$$

$$\left(1 + \frac{x^2 + y^2 - z^2}{2xy}\right) \frac{\frac{1}{z} - \frac{1}{x+y}}{\frac{1}{z} + \frac{1}{x+y}} \; ; \quad \left(\frac{b}{a}\left(\frac{a^n b^m}{c^g d^l}\right)^3\right)^{-5} .$$

Next, let us use the method of mathematical induction to prove the following formula:

$$(a_1 + \cdots + a_n)^2 = a_1^2 + \cdots + a_n^2 + 2\,(a_1 a_2 + a_1 a_3 + \cdots + a_2 a_3 + a_2 a_4 + \cdots + a_{n-1} a_n) \; . \tag{1.10}$$

Here in the right-hand side we have first the sum of squares of all numbers $a_1, \ldots, a_n$, and in the last term—twice the sum of all their possible pairs. Obviously, this formula is valid for $n = 2$ as only one pair exists in this case: $(a_1 + a_2)^2 = a_1{}^2 + a_2^2 + 2a_1 a_2$. It is easy to check the case of three numbers as well:

$$(a_1 + a_2 + a_3)^2 = [(a_1 + a_2) + a_3]^2 = (a_1 + a_2)^2 + a_3^2 + 2\,(a_1 + a_2)\,a_3$$

$$= \left(a_1^2 + a_2^2 + 2a_1 a_2\right) + a_3^2 + 2a_1 a_3 + 2a_2 a_3 = a_1^2 + a_2^2 + a_3^2 + 2\,(a_1 a_2 + a_1 a_3 + a_2 a_3) \; .$$

In this case only three pairs exist.

Now we shall prove the general case. Let us assume that formula (1.10) is valid for any integer $n \geq 2$, and let us consider the case of $n + 1$:

$$(a_1 + \cdots + a_n + a_{n+1})^2 = (a_1 + \cdots + a_n)^2 + a_{n+1}^2 + 2\,(a_1 + \cdots + a_n)\,a_{n+1}$$

$$= \left(a_1^2 + \cdots + a_n^2\right) + 2\,(a_1 a_2 + \cdots + a_{n-1} a_n) + a_{n+1}^2 + 2\,(a_1 + \cdots + a_n)\,a_{n+1} \; ,$$

where we have made use of the formula for n terms that we assumed to be true. In the second term in the right-hand side, we have the sum of all pairs that one can build using the first n terms $a_1, a_2, \ldots, a_n$, while in the last term all the pairs with a_{n+1} are present. Rearranging the above expression, we arrive at the required result for the case of $n + 1$ terms, which has an identical structure as in Eq. (1.10), as required.

The above proven formula can also be written in a much shorter way using the summation symbol $\sum$ (called sigma) that is very useful and we shall be employing it a lot:

$$\left(\sum_{i=1}^{n} a_i\right)^2 = \sum_{i=1}^{n} a_i{}^2 + 2\sum_{i<j} a_i a_j \; . \tag{1.11}$$

In the left-hand side and in the first term on the right-hand side, the sign $\sum_{i=0}^{n}$ encodes a single summation of an expression next to it on the right with respect to the index i; it is called the summation index and it runs here from 1 (shown underneath the sum symbol) to the value of n (shown on top of it), so that, for instance,

$$\sum_{i=1}^{n} a_i = a_1 + a_2 + \cdots + a_n .$$

The summation symbol $\sum_{i<j}$ in the second term on the right-hand side of Eq. (1.11) is actually a double sum $\sum_{i=1}^{n-1} \sum_{j=i+1}^{n}$ in which the first summation is performed with respect to i from 1 to $n-1$, while the second summation (run by index j)—with respect to all j that are larger than i. This way all distinct pairs of $a_i a_j$ can be listed.

In practice, it is sometimes useful to be able to *factorise* a particular expression, i.e. write it down as a product of simpler expressions. Formulae (1.7)–(1.9) give examples of such factorisation. These can be obtained by certain algebraic manipulations using simple formulae, e.g. such as in Eq. (1.4). Indeed,

$$a^4 - b^4 = \left(a^2\right)^2 - \left(b^2\right)^2 = \left(a^2 - b^2\right)\left(a^2 + b^2\right) = (a - b)(a + b)\left(a^2 + b^2\right) .$$

Factorisation of $a^4 - a^2 b^2 + b^4$ used in Eq. (1.8) can be performed using a special trick:

$$a^4 - a^2 b^2 + b^4 = \left(a^4 + b^4 + 2a^2 b^2\right) - 3a^2 b^2$$

$$= \left(a^2 + b^2\right)^2 - \left(\sqrt{3}ab\right)^2 = \left(a^2 + b^2 - \sqrt{3}ab\right)\left(a^2 + b^2 + \sqrt{3}ab\right) ,$$

i.e. the right-hand side of Eq. (1.9) should now have more sense.

Problem 1.11 Prove factorisations written below using appropriate algebraic manipulations of the left-hand sides:

$$a^4 + b^4 = \left(a^2 + b^2 + \sqrt{2}ab\right)\left(a^2 + b^2 - \sqrt{2}ab\right) ;$$

$$a^8 - b^8 = (a - b)(a + b)(a^2 + b^2)\left(a^2 + b^2 + \sqrt{2}ab\right)\left(a^2 + b^2 - \sqrt{2}ab\right) .$$

Problem 1.12 Prove the following formulae:

$$x^n - a^n = (x - a)\left(x^{n-1} + ax^{n-2} + a^2 x^{n-3} + \cdots + a^k x^{n-k-1} + \cdots + a^{n-1}\right)$$

$$= (x - a) \sum_{k=0}^{n-1} a^k x^{n-k-1} \; ; \tag{1.12}$$

$$x^{2n+1} + a^{2n+1} = (x + a) \sum_{k=0}^{2n} (-1)^k a^k x^{2n-k} \; . \tag{1.13}$$

Here an expression $(-1)^k$ is a device meant to generate an alternating sign for the terms in the sum since $(-1)^0 = 1, (-1)^1 = -1, (-1)^2 = 1$ and so on. Correspondingly, prove:

$$x^8 - a^8 = (x - a) \left(x^7 + a x^6 + a^2 x^5 + a^3 x^4 + a^4 x^3 + a^5 x^2 + a^6 x + a^7 \right) \; ;$$

$$x^9 + a^9 = (x + a) \left(x^8 - a x^7 + a^2 x^6 - a^3 x^5 + a^4 x^4 - a^5 x^3 + a^6 x^2 - a^7 x + a^8 \right) \; .$$

Problems below are a good exercise on using the method of induction.

Problem 1.13 Prove that $(n = 1, 2, 3, \ldots)$:

$$n \left(2n^2 + 1 \right) \quad \text{is divisible by 3} \; ;$$

$$n(n + 1)(2n + 1) \quad \text{is divisible by 6} \; ;$$

$$7^n + 3n - 1 \quad \text{is divisible by 9} \; ;$$

$$n^3 + 3n^2 + 2n \quad \text{is divisible by 3} \; .$$

The last problem can also be solved by factorising the expression and then deducing that a product of three consecutive integer numbers is always divisible by 3.

Let us also look at a simple trick that allows simplifying certain expressions with radicals. Consider the ratio

$$\frac{A}{B + \sqrt{C}} \; .$$

We can multiply and divide it by $B - \sqrt{C}$ and then use the left identity in Eq. (1.4):

$$\frac{A}{B + \sqrt{C}} = \frac{A}{B + \sqrt{C}} \frac{B - \sqrt{C}}{B - \sqrt{C}} = \frac{A \left(B - \sqrt{C} \right)}{B^2 - C} \; .$$

This simple trick enables one to get rid of the radical in the denominator. For instance,

$$\frac{1}{1+\sqrt{5}} = \frac{1-\sqrt{5}}{-4} = \frac{\sqrt{5}-1}{4}.$$

1.3.3 Inequalities

We must also introduce *inequalities* as these are frequently used in mathematics and will be met very often in this book: these are relations between numbers (or corresponding letters) stating their relation to each other. If a number a is strictly *larger* than b, it is written as $a > b$; by writing $a \geq b$ we mean that a could also be equal to b, i.e. a is *not smaller* than b. Similarly, $a < b$ and $a \leq b$ state that either a is strictly *less* than b or it is *not larger* than it. Obviously, $a > b$ and $b < a$ are equivalent, as are $a \geq b$ and $b \leq a$.

One can also perform some algebraic manipulations on inequalities. Firstly, one may add to both sides of an inequality the same number, i.e. if $a < b$, then $a + c < b + c$. One can take a number from one side of the inequality to the other by changing the sign of the number, i.e. $a < b$ is equivalent to $a - b < 0$ (here b was taken from the right to the left side) or $0 < b - a$ (conversely, here a was taken from the left to the right side). The latter inequality can also be written as $b - a > 0$. For instance, let us prove that from $a < b$ follows $a - b < 0$: adding to both sides of $a < b$ the number $-b$ yields $a - b < b - b$ or $a - b < 0$, as required.

It is possible to multiply both sides of the given inequality by some number $c \neq 0$: if $c > 0$, the sign of the inequality does not change; however, if $c < 0$, it changes to the opposite one. For instance, if $a < b$, then $5a < 5b$ as the number 5 is positive. However, $-5a > -5b$. Indeed, the latter inequality can be easily manipulated into the first one:

$$-5a > -5b \quad \Longrightarrow \quad 0 > 5a - 5b \quad \Longrightarrow \quad 5b > 5a \ ,$$

which we know is correct.

Problem 1.14 Prove that if $a < b$ and $0 < c < d$ (i.e. both c and d are positive numbers), then $ac < bd$, i.e. one can "multiply both inequalities" side by side.

Problem 1.15 Show that if $0 < a < b$, then $\sqrt{a} < \sqrt{b}$ and $a^2 < b^2$.

Two inequalities of the same type (i.e. both are either "less than" or " larger than" types) can be "summed up", i.e. from $a < b$ and $c < d$ follows $a + c < b + d$. Indeed, if we add to both sides of the inequality $a < b$ the *same* number c, the inequality will not change: $a + c < b + c$. However, since $c < d$, then we can similarly state that $b + c < b + d$, which proves the above made statement:

$$a + c < b + c \quad \text{and} \quad b + c < b + d \quad \Longrightarrow \quad a + c < b + d \, .$$

Similarly, from $a \geq b$ and $c \geq d$ follows $a + c \geq b + d$.

Two inequalities of the opposite types can be subtracted, e.g. if $a < b$ and $c > d$, then $a - c < b - d$. Indeed, $c > d$ is equivalent to $d < c$, so that summing up this one with $a < b$, we get $a + d < c + b$ or $a - c < b - d$, as required.

Problem 1.16 Prove the following identity ($a^2 \geq b \geq 0$ and $a \geq 0$):

$$\sqrt{a + \sqrt{a^2 - b}} \pm \sqrt{a - \sqrt{a^2 - b}} = \sqrt{2 \left(a \pm \sqrt{b} \right)} \, . \qquad (1.14)$$

Note that here there are two identities: one with the plus sign on both sides and another with the minus. [Hint: *first, verify that the left-hand side is always positive, even in the case of the minus sign. Then square both sides and simplify.*]

Problem 1.17 Prove that

$$\sqrt{a \pm \sqrt{a^2 - b^2}} = \sqrt{\frac{a + b}{2}} \pm \sqrt{\frac{a - b}{2}} \, .$$

You may either prove this independently or using Eq. (1.14).

If we have an inequality between two positive numbers a and b, the same sign of the inequality would also be valid for their squares. For instance, if $a < b$, then $a^2 < b^2$. Indeed, multiplying both sides of the original inequality by $a > 0$, we get $a^2 < ab$. However, as $a < b$, then $ab < b^2$ and hence $a^2 < ab < b^2$, as required.

Problem 1.18 Prove that for any two positive numbers $a, b > 0$,

$$\frac{a + b}{2} \geq \sqrt{ab} \, . \qquad (1.15)$$

At which values of a and b does the equality sign hold? [Hint: *make the substitutions $a = x^2$ and $b = y^2$ and take the root into the left-hand side.*]

Problem 1.19 Prove that

$$2\left(a_1^2 + a_2^2\right) \ge (a_1 + a_2)^2 .$$

Problem 1.20 Show that the inequality $r + R < 2L$, where $R > r$, is equivalent to $|L - r| > |L - R|$. [Hint: *consider instead $(L - r)^2 > (L - R)^2$.*]

1.3.4 Functions

Next we define a real *function*. A real function $y = f(x)$ establishes a correspondence between two real numbers x and y, both from $\mathbb{R}$. There must be a single value of y for each value of the x. Note, however, that the same value of y may exist for different values of x. The absolute value (or modulus) function introduced above by Eq. (1.3) is one example of a function: each value of the x is related to its absolute value $|x|$. The other simplest class of functions are *polynomials*:

$$f(x) = a_n x^n + a_{n-1} x^{n-1} + \cdots + a_1 x + a_0 = \sum_{i=0}^{n} a_i x^i. \tag{1.16}$$

This is a general way of writing a polynomial of degree n, where a_0, a_1, etc. are the $(n + 1)$ real coefficients; also, it is assumed that at least $a_n \ne 0$. Again, we have used the symbol of summation $\sum$, and it is a matter of simple exercise to check that the full expression just after the $f(x)$ is exactly reproduced by allowing i to run between its two limiting values (note that $x^0 = 1$ and $x^1 = x$) under the summation symbol.

If we divide one polynomial $P_n(x)$ of degree n on another one $Q_m(x)$ of degree m, we arrive at the class of *rational functions*.

The simplest polynomials are the first- and the second-order ones. Their graphs are shown in Fig. 1.2. When plotted, a linear polynomial $P_1(x) = a_0 + a_1 x$ appears as a straight line with $a_0 = P_1(0)$ being the point where its graph crosses the y-axis, while a_1 shows its slope: if $a_1 > 0$, then the curve goes up from left to right (green curves 1 and 2 in Fig. 1.2a), while when $a_1 < 0$ the line goes down (blue curves 3 and

Fig. 1.2 Graphs of linear **a** and square **b** polynomials

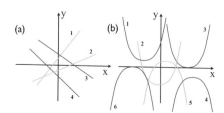

4). The straight line crosses the x-axis at the point $x = -a_0/a_1$, when $P_1(x) = 0$. It happens always as long as $a_1 \neq 0$. If $a_1 = 0$, then the line is horizontal and parallel to the x-axis, hence never will cross it.

Graphs of various second-order polynomials (called *parabolas*), $P_2(x) = a_0 + a_1x + a_2x^2$, are symmetrical curves with either a minimum (if $a_2 > 0$) or maximum ($a_2 < 0$), see Fig. 1.2b. Depending on the values of the constants a_0, a_1 and a_2 the parabola may either cross the x-axis in 1 (blue curves 3 and 6) or 2 (green curves 2 and 5) points or do not cross at all (red curves 1 and 4). This property can be established by completing the square:

$$
\begin{aligned}
P_x(x) = a_2x^2 + a_1x + a_0 &= a_2\left[x^2 + 2\left(\frac{a_1}{2a_2}\right)x\right] + a_0 \\
&= a_2\left[\left(x^2 + 2\left(\frac{a_1}{2a_2}\right)x + \left(\frac{a_1}{2a_2}\right)^2\right) - \left(\frac{a_1}{2a_2}\right)^2\right] + a_0 \\
&= a_2\left[\left(x + \frac{a_1}{2a_2}\right)^2 - \left(\frac{a_1}{2a_2}\right)^2\right] + a_0 \\
&= a_2\left(x + \frac{a_1}{2a_2}\right)^2 + \left(-\frac{a_1^2}{4a_2} + a_0\right) .
\end{aligned}
\tag{1.17}
$$

If the free term $a_0 - a_1^2/4a_2 = 0$, then the parabola just touches the x-axis and hence has only a single *root* at $x = -a_1/2a_2$. If $a_0 - a_1^2/4a_2 > 0$ and $a_2 > 0$, then the entire graph of this function is above the x-axis and there will be no roots (0 crossings of the x-axis); if however $a_2 < 0$, then there will be 2 roots. Similarly, in the case of $a_0 - a_1^2/4a_2 < 0$ there will be 2 roots if $a_2 > 0$ and no roots if $a_2 < 0$. The roots can be established by setting to zero the complete square expression obtained above[5]:

$$
a_2x^2 + a_1x + a_0 = 0
$$

$$
\Rightarrow \quad a_2\left(x + \frac{a_1}{2a_2}\right)^2 + \left(-\frac{a_1^2}{4a_2} + a_0\right) = 0
$$

$$
\Rightarrow \quad \left(x + \frac{a_1}{2a_2}\right)^2 = \frac{a_1^2 - 4a_0a_2}{4a_2^2}
$$

$$
\Rightarrow \quad x + \frac{a_1}{2a_2} = \pm\sqrt{\frac{a_1^2 - 4a_0a_2}{4a_2^2}}
$$

$$
\Rightarrow \quad x = -\frac{a_1}{2a_2} \pm \sqrt{\frac{a_1^2 - 4a_0a_2}{4a_2^2}} ,
$$

---

[5] Here we wrote both roots at the same time using the combined symbol $\pm$. One has to understand this as two separate equations: one root is obtained with the upper sign (which is "+"), and another one with the lower sign "−".

which results in the final formula:

$$x = \frac{-a_1 \pm \sqrt{a_1^2 - 4a_0 a_2}}{2a_2}. \tag{1.18}$$

No real roots (i.e. within the manifold of real numbers $\mathbb{R}$) are obtained if the *discriminant* $\mathcal{D} = a_1^2 - 4a_0 a_2 < 0$, one root is obtained if $\mathcal{D} = 0$, and, finally, two roots exist if $\mathcal{D} > 0$. These are the same results as previously worked out by inspecting the graphs of the parabolas.

Problem 1.21 Show by direct calculation that the square polynomial can also be written as

$$P_2(x) = a_2 (x - x_1)(x - x_2), \tag{1.19}$$

where x_1 and x_2 are its two roots (1.18).

Problem 1.22 Does the square polynomial $x^2 + x + 1$ have real roots? [Answer: *no.*]

Problem 1.23 Solve $x^2 + 7x + 1 = 0$. [Answer: $\left(-7 \pm 3\sqrt{5}\right)/2$.]

Problem 1.24 Solve $2x^2 + x - 6 = 0$. [Answer: -2; $3/2$.]

Problem 1.25 Solve $3x^2 + 6x - 13 = 0$. [Answer: $-1//3 \left(3 \pm 4\sqrt{3}\right)$.]

Problem 1.26 Solve $2x^2 - x - 6 = 0$. [Answer: 2; $-3/2$.]

Problem 1.27 Solve $x^2 + x - 6 = 0$. [Answer: 2; -3.]

Problem 1.28 Solve $x^2 - x - 6 = 0$. [Answer: -2; 3.]

1.3.5   Simple Algebraic Equations

In applications one frequently need to solve an equation $f(x) = 0$ with respect to the unknown x. Here $f(x)$ is a function that could be very complicated, and x is called its root (if exists). In this section, we shall consider examples of equations with relatively simple functions and how in those cases one can find the unknown x.

Consider first a linear equation $P_1(x) = a_1 x + a_0 = 0$, where $a_1 \neq 0$ and a_0 are numbers. We need to find that x that satisfies this equation. The solution (see Sect. 1.3.4) is based on a set of elementary manipulations which do not change the equality. First, we take a_0 into the right-hand side: $a_1 x = -a_0$. Then we divide both sides on the same number a_1 (note that $a_1 \neq 0$) that yields $x = -a_0/a_1$. So, the solution is found, and we find only a single value of x that satisfies that equation, the unique solution.

The next case is the square equation $P_2(x) = a_2 x^2 + a_1 x + a_0 = 0$ with the square polynomial $P_2(x)$ in the left-hand side; there could be up to two solutions, see Sect. 1.3.4. According to Eq. (1.18), two real roots of this equation are obtained if $\mathcal{D} = a_1^2 - 4a_0 a_2 > 0$. If $\mathcal{D} = 0$, a single root $x = -a_1/2a_2$ is to be found, and if $\mathcal{D} < 0$, no real roots exist anymore as one has to take the square root of a negative number which is impossible within the realm of real numbers $\mathbb{R}$. We shall see later on in Sect. 1.11 that real numbers can be extended and complex numbers defined within which the square root of a negative number is well defined. In that sense, even if $\mathcal{D} < 0$, we get two complex roots.

Considering higher order polynomials $P_n(x)$ with $n \geq 3$ in the equation $P_n(x) = 0$ is less trivial. Explicit solutions via radicals exist in the general case only for cubic and quartic equations. These solutions were discovered in the sixteenth century by Italian mathematicians. We shall consider those in Sects. 1.11.1, 3.4.5 and 3.4.6. Here, in the rest of this subsection we shall study several types of algebraic equations with respect to one variable. One also frequently faces a problem of finding several unknown variables x, y, etc. that are defined by more than one algebraic equation. These problems are called systems of equations. We shall consider examples of simple systems of equations here as well.

It is reasonable to start by stressing certain general principles which could be used while solving equations. Consider a general equation $f_1(x) = f_2(x)$, where $f_1(x)$ and $f_2(x)$ are some functions of x. First of all, we check at which x those functions are defined. For instance, if $f_1(x) = 1/(x - 2)$, then the value $x = 2$ must be excluded when solving the original equation.

Once we defined the acceptable values of x, we can approach solving the equation. This could be done by a set of transformations that ideally preserve the equation. In other words, they are supposed to transform the equation into an equivalent form without losing existing or acquiring additional solutions. One method is to add a function $g(x)$ to both sides (and making sure that $g(x)$ is defined at least for the same x as the original equation) and another to multiply both sides of the equation by some $h(x)$. In the latter case, one has to ensure that this operation does not either bring new roots or remove existing ones; the former may happen if $h(x) = 0$ for some values of x within the acceptable set of x values, and the latter—if $1/h(x) = 0$ for the value(s) of x that is(are) solution(s) of the original equation.

If the equation has the form $f_1(x) = f_2(x) + f_3(x)$, then adding $-f_2(x)$ to both sides results in an equivalent equation $f_1(x) - f_2(x) = f_3(x)$. In effect, in this operation we took the term $f_2(x)$ from the right-hand side and moved it into the left-hand side, and by doing so we attached an opposite sign to it.

Problem 1.29 Solve $\sqrt{5}x + \sqrt{6} = 3x$. [Answer: $\sqrt{6}/(3 - \sqrt{5})$.]

Problem 1.30 Solve equation

$$\frac{ax + b}{c} + \frac{\alpha x + \beta}{\gamma} = d$$

with respect to x (all other symbols are non-zero constants). [Answer: $x = (cd\gamma - b\gamma - \beta c)/(a\gamma + \alpha c)$.]

If the equation has the form $f_1(x)/f_2(x) = f_3(x)/f_4(x)$, then multiplying both sides of the equation by $f_4(x)$, it transforms into $f_1(x)f_4(x)/f_2(x) = f_3(x)$, i.e. in this case the denominator in the right-hand side was moved to the numerator in the left-hand side; no change of sign happened in this process. Similarly, we could have multiplied both sides of the original equation by $1/f_3(x)$, in which case $f_3(x)$ from its numerator position in the right-hand side would move into the denominator position in the left-hand side, $f_1(x)/[f_2(x)f_3(x)] = 1/f_4(x)$.

For instance, consider equation $x + 6 = 8$. All values of the x in $\mathbb{R}$ are acceptable $(-\infty < x < \infty)$. If we multiply both sides of it by $x^2 + 4$, which is not equal to zero for any x from $\mathbb{R}$, then this operation is permissible as not bringing new roots:

$$(x + 6)(x^2 + 4) = 8(x^2 + 4) .$$

The obtained equation is equivalent to the original one, i.e. it has the same roots (in fact, a single root $x = 8 - 6 = 2$). Of course, by performing this transformation we have got something unnecessary complicated, and only have done this here to state the point. Similarly, given the above equation instead, we can multiply both sides of it by $1/(x^2 + 4)$, which results in a simpler (but equivalent) equation $x + 6 = 8$, yielding the solution $x = 2$.

On the other hand, equation $x^3 = -27$ cannot be multiplied on both sides by $1/(x + 3)$ as this operation would exclude the value of $x = -3$ from the permissible set of x values leading to an equation that has no solutions (as the solution of the original equation is precisely $x = -3$).

Another example: consider equation $x + 6 = 7$. Let us multiply both sides of it by $x - 2$. We will get:

$$(x + 6)(x - 2) = 7(x - 2) , \tag{1.20}$$

which has two solutions: $x = 1$ and $x = 2$. Since the original equation has a single root $x = 1$, the multiplication operation resulted in acquiring an additional solution not present in the original equation.

On the other hand, suppose we are given Eq. (1.20). Because of its form, it is clear that at least $x = 2$ is its solutions. However, if we divide both sides by $x - 2$ (or, equivalently, multiply both sides by $1/(x - 2)$), then we shall loose this solution as the resulting equation $x + 6 = 7$ only has the $x = 1$ solution.

The above discussion is meant to caution the reader in performing seemingly equivalent operations on the algebraic equations; if not careful, either new solutions can be acquired or some (or all) solutions could be removed.

Next, consider equation

$$\frac{1}{x + 1} + \frac{1}{x - 2} = 0 . \tag{1.21}$$

This equation is defined for all values of x apart from $x = -1$ and $x = 2$. Hence, the permissible values of x are: $-\infty < x < -1$, $-1 < x < 2$ and $2 < x < \infty$. Multiplying both sides of the equation by $(x + 1)(x - 2)$ (which is not equal to zero for any of the permissible values of x), we obtain:

$$\frac{(x + 1)(x - 2)}{x + 1} + \frac{(x + 1)(x - 2)}{x - 2} = 0$$
$$\Rightarrow \quad (x - 2) + (x - 1) = 0$$
$$\Rightarrow \quad 2x - 1 = 0 ,$$

whose solution is provided by the root $x = 1/2$, and it is a value that belongs to the permissible set and so must be accepted as the solution.

Problem 1.31 Solve $6/x - 1/x^2 = 3$. [Answer: $1 \pm \sqrt{2/3}$.]

Problem 1.32 Show that equation

$$\frac{1}{x + 1} = \frac{x}{x^2 + 2x + 1}$$

does not have solutions.

Problem 1.33 Find all solutions of

$$\frac{1}{x + 1} = \frac{1}{x^2 - 2x + 1} .$$

[Answer: 0; 3.]

Problem 1.34 Solve

$$\frac{x+1}{x-9} = \frac{x-9}{x+1}.$$

[Answer: 4.]

Problem 1.35 Solve

$$\frac{2x+3}{x-3} + \frac{4x+4}{x+1} = 1.$$

[Answer: 6/5.]

Problem 1.36 Solve

$$\frac{2x+3}{x-3} + \frac{x+4}{x+1} = 1.$$

[Answer: $-2 \pm \sqrt{7}$.]

Problem 1.37 Solve in real numbers the equation

$$\frac{x^2+1}{x^2-1} - \frac{x^2-1}{x^2+1} = 1.$$

[Answer: $\pm\sqrt{2+\sqrt{5}}$.]

Problem 1.38 Solve in real numbers the equation

$$\frac{x^2+2}{x^2-2} - \frac{x^2-2}{x^2+2} = 4.$$

[Answer: $\pm\sqrt{1+\sqrt{5}}$.]

When solving equations, it is usually convenient to transform the equation into the form $f(x) = 0$ and then factorise the left-hand side, i.e. represent the function $f(x)$ as a product of possibly simpler functions yielding the equation: $f_1(x)f_2(x)\cdots f_n(x) = 0$. This form of the equation may be advantageous. Indeed, instead of a single equation $f(x) = 0$ one has to solve a set of separate equations $f_1(x) = 0$, $f_2(x) = 0$, etc. If functions $f_1(x)$, $f_2(x)$, etc. are simpler than the original function $f(x)$, then solving each equation $f_i(x) = 0$ (with $i = 1, \ldots, n$) must be easier. Obviously, this way no solutions are lost or extra solutions acquired. Indeed, if a is a root of the original equation, $f(a) = 0$, then, since $f(x) = f_1(x)\cdots f_n(x)$, there will at least be one of the functions $f_1(x)$, $f_2(x)$, etc. that is zero at $x = a$.

Alternatively, if a is a root of one of the functions of $f_i(x)$, then obviously $f(a) = 0$ as well.

Example 1.1 ▶ Solve equation

$$(x + 2)(x - 1) - 2(x - 1) = 0 .$$

Solution. Here $x - 1$ is a common multiplier to both terms in the left-hand side, so that the expression can be factorised:

$$(x + 2)(x - 1) - 2(x - 1) = (x - 1)\,[(x + 2) - 2] = (x - 1)x = 0 ,$$

giving two solutions: $x = 0$ and $x = 1$. ◀

Problem 1.39 Solve $x^2 - 1 + 3(x + 1) = 0$. [Answer: $-1, -2$.]

Problem 1.40 Solve $2(x + 1) + 3(x^3 + 1) + 6(x^2 - 1) = 0$. [Hint: *use one of the identities* (1.6). Answer: $-1; \left(-3 \pm \sqrt{21}\right)/6$.]

Problem 1.41 Solve $2(x^4 - 16) - 9(x^2 - 4) = 0$. [Hint: *use identity* (1.7). Answer: $\pm 2; \pm 1/\sqrt{2}$.]

In some cases, equations could be solved by adopting a new variable. Let us consider a few examples to illustrate this point.

Example 1.2 ▶ Find solutions of the equation

$$2\left(x + \frac{1}{x}\right) + 4\left(x^2 + \frac{1}{x^2}\right) = 4$$

in real numbers.

Solution. If we try to multiply both sides of the equation by x^2, we would have obtained an algebraic equation of the fourth order, which is very difficult to solve. However, if we introduce a variable $t = x + 1/x$, then the second bracket in the left-hand side can be written as $t^2 - 2$, as can easily be verified. Hence, in terms of t our equation reads

$$2t + 4(t^2 - 1) = 4 \quad \Longrightarrow \quad 2t^2 + t - 6 = 0 .$$

This is a quadratic equation, whose solutions are $t = -2$ and $t = 3/2$. In order to find x, we need to solve the equation

$$x + \frac{1}{x} = t \implies x^2 - tx + 1 = 0$$

for both values of t. It is again a quadratic equation. Using $t = -2$, we obtain one root $x = -1$. Using $t = 3/2$, no real solutions are obtained as formula (1.18) contains a root from a negative number. ◀

Problem 1.42 Solve in real numbers the equation

$$x^2 + 4x - 5 + \left(x^2 + 4x + 1\right)^2 = 1 .$$

[Answer: $-2 \pm \frac{1}{2}\sqrt{2\left(5 + \sqrt{29}\right)}$.]

Problem 1.43 Solve equation $x^4 + 4x^2 + 1 = 0$. [Answer: *no real solutions.*]

Problem 1.44 Solve equation $x^4 + 2x^2 - 8 = 0$. [Answer: $\pm\sqrt{2}$.]

Problem 1.45 Solve equation

$$\left(\frac{x+1}{x-1}\right)^2 + 4\frac{x+1}{x-1} + 4 = 0 .$$

[Answer: 1/3.]

Problem 1.46 Solve equation

$$\frac{2x+1}{3x-1} - 2\frac{3x-1}{2x+1} = 1 .$$

[Answer: 0, 3/4.]

Next we shall consider another class of equations which contain the module function $|f(x)|$, Eq. (1.3). Since this expression is either $f(x)$ or $-f(x)$ depending on the sign of $f(x)$, i.e. whether it is positive or negative, one has to carefully analyse all possible cases.

Example 1.3 ▶ Solve equation $|x + 1| + |x - 1| = 1$.

Solution. We have to consider the cases of $x + 1$ and $x - 1$ being either positive or negative. It is convenient then to split the x-axis into three intervals: $-\infty < x \le -1$, $-1 < x < 1$ and $1 \le x < \infty$, and consider each of them separately. Consider the first interval, $-\infty < x \le -1$. In this case, $x + 1 \le 0$ and $x - 1 \le -2 < 0$, so that we can replace each absolute value with the minus expression in it, and the equation is rewritten as

$$- (x + 1) - (x - 1) = 1 \quad \Longrightarrow \quad 2x = -1 \quad \Longrightarrow \quad x = -\frac{1}{2}.$$

However, the obtained value is outside the chosen interval and hence has to be rejected. Next we shall consider the second interval, $-1 < x < 1$. In this case $x + 1 > 0$ and $x - 1 < 0$, so that we obtain

$$+ (x + 1) - (x - 1) = 1 \quad \Longrightarrow \quad 2 = 1 \quad \Longrightarrow \quad \text{not possible.}$$

Finally, in the case of the last interval, $x \ge 1$, we get

$$+ (x + 1) + (x - 1) = 1 \quad \Longrightarrow \quad 2x = 1 \quad \Longrightarrow \quad x = \frac{1}{2},$$

that is again outside the interval. So, the equation does not have solutions.

Note also that the fact that both values $\pm 1/2$ cannot be accepted follows immediately from substituting them directly into the equation. For instance, if we take $x = -1/2$, we obtain:

$$\text{LHS} = \left| -\frac{1}{2} + 1 \right| + \left| -\frac{1}{2} - 1 \right| = \frac{1}{2} + \frac{3}{2} = 2,$$

which is not 1 as required. ◀

Problem 1.47 Solve $|x + 1| - |x - 1| = 1$. [Answer: $1/2$.]

Problem 1.48 Find real solutions of $|x^2 + 4x + 4| = 3|x - 1|$. [Answer: $\left(-7 \pm 3\sqrt{5} \right)/2$.]

Problem 1.49 Find real solutions of $|x^2 + 2x + 4| = 3|x - 1|$. [Answer: $\left(-5 \pm \sqrt{21} \right)/2$.]

Problem 1.50 Find real solutions of $\left|x^2 - 9\right| = 6x$. [Answer: $\pm 3 + 3\sqrt{2}$.]

Finally, we shall consider equations containing radicals, e.g. square or cubic roots. A general method of solving such equations is to eliminate radicals by taking both sides of the equation into an appropriate power as many times as necessary. These steps may require some rearrangement to be made prior to doing this in order to perform the full elimination of radicals in as little number of steps as possible. Another important point is that the operation of taking both sides of the equation into an even power may acquire new solutions. Indeed, an equation $a = b$, when squared, becomes $a^2 = b^2$. However, the same result is obtained by squaring the equation $a = -b$, hence solving $a^2 = b^2$ may give solutions that are not solutions of the original equation. Hence, care is needed in performing such operations. A general rule is to substitute the solutions into the original equation to check if they satisfy it.

Example 1.4 ▶ Solve in real numbers the equation

$$\sqrt{x+1} - \sqrt{x-1} = 3 \ .$$

Solution. Here the expression under the square roots must be non-negative, hence we have two conditions: $x \geq -1$ and $x \geq 1$, which are both satisfied if we require that $x \geq 1$. Let us square both sides of the equation:

$$(x+1) - 2\sqrt{(x-1)(x+1)} + (x-1) = 9 \quad \Longrightarrow \quad 2x - 2\sqrt{x^2 - 1} = 9 \ .$$

Then we take the square root term to the right-hand side and the 9 to the left-hand side; this step would then enable one to have just one more square to do as otherwise this operation would not eliminate the root that would still remain in the left-hand side. We also have to assume that $2x - 9 \geq 0$ or $x \geq 9/2$ as the square root in the right-hand side is non-negative. Therefore,

$$2x - 9 = 2\sqrt{x^2 - 1} \quad \Longrightarrow \quad (2x - 9)^2 = 4\left(x^2 - 1\right)$$

$$\Longrightarrow \quad 4x^2 - 36x + 81 = 4x^2 - 4 \quad \Longrightarrow \quad 36x = 85 \quad \Longrightarrow \quad x = \frac{85}{36} \ .$$

The obtained value is larger than one and hence seems to be fully acceptable. However, after substituting into the equation this value we find that it does not satisfy it:

$$\text{LHS} = \sqrt{\frac{85}{36} + 1} - \sqrt{\frac{85}{36} - 1} = \sqrt{\frac{121}{36}} - \sqrt{\frac{49}{36}} = \frac{11}{6} - \frac{7}{6} = \frac{4}{6} = \frac{2}{3}$$

and not 3 as required. Hence, the steps we have taken resulted in acquiring a spurious solution. Hence, this equation does not have solutions within real numbers. This is because the solution does not satisfy the condition of x being not less than $9/2$ that was required to do the second square. ◄

Example 1.5 ► Solve in real numbers the equation

$$\sqrt{x+1} + \sqrt{x-1} = 3 .$$

Solution. This equation differs from the previous one only by the plus sign between the two roots. Using the same method as above, we arrive at exactly the same value of $x = 85/36$. However, this time, it is actually the solution as after substitution into the equation we get

$$\text{LHS} = \sqrt{\frac{85}{36} + 1} + \sqrt{\frac{85}{36} - 1} = \frac{11}{6} + \frac{7}{6} = 3 ,$$

as required. ◄

Problem 1.51 Solve $\sqrt{2x+5} = 8$. [Answer: $59/2$.]

Problem 1.52 Solve $\sqrt{5x} + \sqrt{6} = x$. [Answer: $\sqrt{6} + 5/2 \pm 1/2\sqrt{5\left(4\sqrt{6}+5\right)}$.]

Problem 1.53 Solve equation $\sqrt[3]{x^2 + 1} = 2$. [Answer: $\pm\sqrt{7}$.]

Problem 1.54 Solve in real numbers the equation $\sqrt{x+1} + \sqrt{x-1} = \sqrt{x+3}$. [Answer: $-1 + 4/\sqrt{3}$.]

1.3.6   Systems of Algebraic Equations

If more than a single variable are to be determined, there should be at least as many equations given as the number of the variables, i.e. we have to solve a system of equations.

Suppose there are n unknown quantities $x_1, x_2, \ldots, x_n$, and we are given m equations to find them:

$$\begin{cases} f_1(x_1, x_2, \ldots, x_n) = 0 \\ f_2(x_1, x_2, \ldots, x_n) = 0 \\ \quad \cdots \\ f_m(x_1, x_2, \ldots, x_n) = 0 \end{cases} . \tag{1.22}$$

Here $f_i (x_1, x_2, \ldots, x_n)$ (where $i = 1, \ldots, m$) are functions of n variables, i.e. for each particular set of values of the variables each such function has a particular real value. If there exist a set of values $\{x_1, x_2, \ldots, x_n\}$ that satisfy all equations at the same time, then we say that this set is a solution of Eq. (1.22). A given system of equations may have several (or even an infinite number of) solutions. If, however, such a set does not exist, then the equations are inconsistent, they do not have solutions.

One can perform certain operations on the equations that do not change their solutions (if exist); these are equivalent operations. First of all, as in the case of single algebraic equations, one can multiply (divide) both sides of each of the equations $f_i = 0$ by a non-zero expression, or add (subtract) an expression. Secondly, one can sum (subtract) any two equations in the system and replace one of the two with the result. Combining both types of operations is also a permissible operation which leaves the set of the equations to remain equivalent (with the same solutions). For instance, we can multiply the first equation by α, the second equation by β, subtract from each other, and replace the first equation with the result, $\alpha f_1 - \beta f_2 = 0$. The obtained modified set of equations is equivalent to the original set. Indeed, the second equation, $f_2 = 0$, together with the modified first immediately results in $f_1 = 0$ as before. Normally, to find n variables one requires n equations. If the number of equations $m < n$, then one, many or even an infinite number of solutions are possible, or none. Indeed, a single equation $x^2 + y^2 + z^2 = 0$ with respect to three variables x, y and z (i.e. $m = 1$ and $n = 3$) in real numbers has a single solution $x = y = z = 0$, while the equation $x^2 + y^2 + z^2 = -1$ has none. At the same time, a set of two equations ($m = 2$),

$$\begin{cases} x + y + z = 0 \\ x + y = 0 \end{cases},$$

with respect to three ($n = 3$) variables x, y and z has an infinite number of solutions given by $x = t$, $y = -t$ and $z = 0$, where t is any real number between $-\infty$ and ∞. If $m \geq n$, then the solutions are only possible if the equations are consistent. It is possible that some of the equations in the set are redundant and hence could be removed. An equation is redundant, if it can be obtained by a set of permissible operations from other equations in the set. In this case, it contains the same "information" as the others and hence can be omitted.

One of the simplest methods of solving a system of equations is the method of substitution. Let us illustrate this method on a system of two equations with respect to two variables x and y:

$$\begin{cases} f_1(x, y) = 0 \\ f_2(x, y) = 0 \end{cases}.$$

Suppose we are able to solve the first equation with respect to y in terms of x, i.e. we find $y = f(x)$. Substituting this into the second equation, we obtain a single equation $f_2 (x, f(x)) = 0$ with respect to x, which is to be solved. Hence, the substitution here allowed to remove one unknown variable and reduce the system of two equations to a simpler problem of solving just a single equation. If the single equation has a solution $x = a$, then the corresponding to it value of y is $y = f(a)$, i.e. the system has a solution $x = a$ and $y = f(a)$. If the single equation $f_2(x, f(x)) = 0$ has more than one solution, the corresponding value of y associated with each of them is obtained via $y = f(x)$.

We shall now illustrate these general concepts on a number of simple examples of systems of equations.

First, let us consider a system of two *linear* equations, i.e. algebraic equations that contain the unknown variables in the linear form:

$$\begin{cases} y + x = 5 \\ y - x = 10 \end{cases}.$$

The two equations are independent as it is not possible to obtain the second equation from the first via any permissible set of operations. Also, we shall easily find a unique solution of this system by the method of substitution, which also proves this point. Indeed, solving for y in the first equation, $y = 5 - x$, and substituting into the second, we obtain:

$$(5 - x) - x = 10 \quad \Longrightarrow \quad 2x = -5 \quad \Longrightarrow \quad x = -\frac{5}{2}.$$

Hence,

$$y = 5 - x = \frac{15}{2}.$$

Thus, the pair $(x = -5/2,\ y = 15/2)$ is the only solution of this system of equations.

Equivalently, we may simply sum both equations: by doing so, x cancels out and we immediately obtain $2y = 15$ leading to the same answer for y as before, More generally, it is always possible to eliminate one variable by carefully transforming equations in the linear system. Consider a general linear system of two equations:

$$\begin{cases} a_1 x + a_2 y = c \\ b_1 x + b_2 y = d \end{cases}.$$

To eliminate x, let us multiply the first equation by $b_1 \neq 0$ and the second by $a_1 \neq 0$ (if, for instance, $b_1 = 0$, the system is solved immediately since the second equation gives y straight away; hence, this case is trivial and can be disregarded; the same for the case of $a_1 = 0$):

$$\begin{cases} b_1 a_1 x + b_1 a_2 y = b_1 c \\ a_1 b_1 x + a_1 b_2 y = a_1 d \end{cases}. \tag{1.23}$$

Subtracting the second equation from the first (this is permissible as both sides of the second equation are equal, hence we subtract from both sides of the first equation the same number), we find that x cancels out:

$$(b_1 a_2 - a_1 b_2)\, y = b_1 c - a_1 d.$$

If $b_1a_2 - a_1b_2 \neq 0$, then y is immediately found; as y is known, substituting it into any of the original equations enables one to obtain x. If, however, $b_1a_2 - a_1b_2 = 0$, two cases are possible. Multiplying the first equation of (1.23) by b_1, we obtain in this case:

$$\begin{cases} b_1a_1x + b_1a_2y = b_1c \\ b_1x + b_2y = d \end{cases} \implies \begin{cases} b_1a_1x + a_1b_2y = b_1c \\ b_1x + b_2y = d \end{cases} \implies \begin{cases} b_1x + b_2y = \dfrac{b_1c}{a_1} \\ b_1x + b_2y = d \end{cases}.$$

If $d = b_1c/a_1$, then both equations are equivalent; we say that they are *linearly dependent*. In this case, one may only obtain a dependence of y on x (or x on y), i.e. there will be an infinite number of solutions as essentially we only have one algebraic equation: $y = t$ and $x = (d - b_2t)/b_1$, where $-\infty < t < \infty$ is an arbitrary parameter. If, however, $d \neq b_1c/a_1$, then both equations are incompatible and hence there are no solutions.

As an example of a linear system, consider

$$\begin{cases} 3x + 4y = 5 \\ 4x + 3y = 5 \end{cases}.$$

Multiplying the first equation by 4 and the second by 3, we obtain:

$$\begin{cases} 12x + 16y = 20 \\ 12x + 9y = 15 \end{cases}.$$

Subtracting the second equation from the first, we get $7y = 5$ yielding $y = 5/7$. Substituting this value into the second equation, we obtain the same value for x.

Let us now consider a system

$$\begin{cases} x + y = 5 \\ 2x + 2y = 10 \end{cases}.$$

Substituting $y = 5 - x$ from the first equation into the second, one obtains:

$$2x + 2(5 - x) = 10 \implies 2x - 2x = 0,$$

which is an identity valid for any x. This happened since our second equation is in fact equivalent to the first (it is obtained by multiplying the first by 2). Hence, our system of equations has an infinite number of solutions $x = t$ and $y = 5 - t$, where $-\infty < t < \infty$ is arbitrary. This is because in this case $c = 5$, $d = 10$, $a_1 = a_2 = 1$, and $b_1 = b_2 = 2$, so that $b_1a_2 - a_1b_2 = 2 - 2 = 0$ and $d = b_1c/a_1$. At the same time, the system

$$\begin{cases} x + y = 6 \\ 2x + 2y = 10 \end{cases}$$

has no solutions as being incompatible because $d \neq b_1c/a_1$.

Next consider a non-linear system of equations:

$$\begin{cases} y + 2x^2 = 10 \\ x + y = 1 \end{cases} .$$

We can solve for y in the second equation and substitute into the first:

$$(1 - x) + 2x^2 = 10 \quad \Longrightarrow \quad 2x^2 - x - 9 = 0 \quad \Longrightarrow \quad x_\pm = \frac{1}{4}\left(1 \pm \sqrt{73}\right) .$$

Two values of x are obtained. Correspondingly, for each of them there will be an appropriate value of $y_\pm = 1 - x_\pm = \frac{1}{4}\left(3 \mp \sqrt{73}\right)$. Hence, two solutions (x, y) have been found in this case: $\left(\frac{1}{4}\left(1 + \sqrt{73}\right), \frac{1}{4}\left(3 - \sqrt{73}\right)\right)$ and $\left(\frac{1}{4}\left(1 - \sqrt{73}\right), \frac{1}{4}\left(3 + \sqrt{73}\right)\right)$.

In some cases substitutions could be less trivial, as in the following example:

$$\begin{cases} x^2 y + xy^2 = 1 \\ 2x + 2y = 1 \end{cases} .$$

Solving for y from the second equation and substituting into the first result in a rather complicated equation that is cubic with respect to x. Instead of using this brute force method, we note that the following simple algebraic manipulations can be performed first:

$$\begin{cases} xy(x + y) = 1 \\ x + y = \frac{1}{2} \end{cases} .$$

Then, we substitute $x + y$ from the second equation into the first yielding a simpler system:

$$\begin{cases} xy = 2 \\ x + y = \frac{1}{2} \end{cases} ,$$

which can now be transformed into a quadratic equation by substituting $y = \frac{1}{2} - x$ from the second equation into the first:

$$x\left(\frac{1}{2} - x\right) = 2 \quad \Longrightarrow \quad 2x^2 - x + 4 = 0 \quad \Longrightarrow \quad x_\pm = \frac{1}{4}\left(1 \pm \sqrt{-31}\right) ,$$

i.e. we do not have any solutions within real numbers.

Problem 1.55 Solve

$$\begin{cases} x^2 y - x y^2 = 1 \\ x = 5 + y \end{cases}.$$

[Answer: $\left(\frac{1}{2} \left(-5 + \sqrt{\frac{129}{5}} \right), \frac{1}{2} \left(5 + \sqrt{\frac{129}{5}} \right) \right)$ *and*

$\left(\frac{1}{2} \left(-5 - \sqrt{\frac{129}{5}} \right), \frac{1}{2} \left(5 - \sqrt{\frac{129}{5}} \right) \right)$]

Problem 1.56 Solve equations

$$\begin{cases} x + y = 1 \\ x^3 + y^3 = 10 \end{cases}.$$

[Hint: *define   new   variables* $u = x + y$ *and* $v = xy$. Answer:
$\left(\frac{1}{2} \left(1 \pm \sqrt{13} \right), \frac{1}{2} \left(1 \mp \sqrt{13} \right) \right)$.]

Problem 1.57 Find real-valued solutions of the equations

$$\begin{cases} x + y = 1 \\ x^4 + y^4 = 10 \end{cases}.$$

[Hint: *define   new   variables* $u = x + y$ and $v = xy$. Answer:
$\left(\frac{1}{2} \left(1 \pm \sqrt{-3 + 2\sqrt{22}} \right), \frac{1}{2} \left(1 \mp \sqrt{-3 + 2\sqrt{22}} \right) \right)$.]

Problem 1.58 Solve

$$\begin{cases} x^2 y - x y^2 = 3 \\ x^3 - y^3 = 17 \end{cases}.$$

[Hint: *make use of formula* (1.5). Answer: $(1 + \sqrt{5/2}, -1 + \sqrt{5/2})$ *and* $(1 - \sqrt{5/2}, -1 - \sqrt{5/2})$.]

Problem 1.59 Solve

$$\begin{cases} x^4 + y^4 - x^2 y^2 = 2 \\ x^2 + y^2 = 2 \end{cases}.$$

[Answer: *all possible pairs* (x, y) *such that* $x^2 = 1 \pm 1/\sqrt{3}$ *and* $y^2 = 1 \mp 1/\sqrt{3}$.]

Problem 1.60 Solve

$$\begin{cases} -x + y + z = 1 \\ x - y + z = 1 \\ x + y - z = 1 \end{cases}.$$

[Answer: $(1, 1, 1)$.]

Problem 1.61 Solve

$$\begin{cases} x^2 + y^2 = 1 \\ x^2 - y^2 = -1 \end{cases}.$$

[Answer: $(0, \pm 1)$.]

Problem 1.62 Solve

$$\begin{cases} x^2 + y^2 = 1 \\ \frac{x+y}{x-y} + \frac{x-y}{x+y} = 3 \end{cases}.$$

[Answer: $(\pm\sqrt{5/6}, \pm 1/\sqrt{6})$ *and* $(\pm\sqrt{5/6}, \mp 1/\sqrt{6})$.]

There is a general trick in solving a specific type of a system of two algebraic equations containing homogeneous functions that we would like now to consider. A function $f(x, y)$ is called homogeneous of an integer degree n if it satisfies the following condition:

$$f(\lambda x, \lambda y) = \lambda^n f(x, y) . \tag{1.24}$$

For instance, the function $f(x, y) = x^2 + y^2$ is homogeneous of degree 2, while the function $f(x, y) = x^3 + 3x^2 y$ of degree 3. At the same time, the function $f(x, y) =$

$x^2 + xy^3$ is not homogeneous. Suppose we are given a set of two algebraic equations

$$\begin{cases} f_1(x, y) = a \\ f_2(x, y) = b \end{cases},$$ (1.25)

where a and b are constants. If the two functions are homogeneous of the same degree, say n, then this set of two equations can be manipulated into a single equation with respect to a single auxiliary variable, which may help in solving the whole set. Indeed, let us make the substitution $y = tx$. Since both functions are homogeneous, we can write:

$$f_1(x, y) = f_1(x, tx) = x^n f_1(1, t) \quad \text{and} \quad f_2(x, y) = f_2(x, tx) = x^n f_2(1, t),$$

so that equations become

$$\begin{cases} x^n f_1(1, t) = a \\ x^n f_2(1, t) = b \end{cases} \implies \begin{cases} bx^n f_1(1, t) = ab \\ ax^n f_2(1, t) = ab \end{cases},$$ (1.26)

and hence we can write:

$$bx^n f_1(1, t) = ax^n f_2(1, t) \implies x^n [bf_1(1, t) - af_2(1, t)] = 0.$$ (1.27)

Note that setting both left-hand sides of Eq. (1.26) to be equal to each other may result in acquiring spurious solutions, and hence one has to always check if the obtained solutions satisfy the original equations. The obtained Eq. (1.27) may give a solution $x = 0$ (and correspondingly the value of y from any of the original equations) and also an equation

$$bf_1(1, t) - af_2(1, t) = 0$$

for t. If the latter can be solved for t and checked to be a solution, then one of Eq. (1.26) is solved for x with the other pair y following immediately from $y = tx$.

Example 1.6 ▶ Solve the system of equations

$$\begin{cases} x^2 - xy + y^2 = 1 \\ y^2 - 2xy = 1 \end{cases}.$$ (1.28)

Solution. In this system, both functions $f_1(x, y) = x^2 - xy + y^2$ and $f_2(x, y) = y^2 - 2xy$ are homogeneous of degree 2, hence, we can use our method. Substituting $y = tx$ into both equations, we obtain:

$$\begin{cases} x^2 \left(1 - t + t^2\right) = 1 \\ x^2 \left(t^2 - 2t\right) = 1 \end{cases} \implies x^2 \left(1 - t + t^2\right) = x^2 \left(t^2 - 2t\right).$$

It is directly checked that $x = 0$ is a solution of our equations, as both original Eq. (1.28) yield $y^2 = 1$, i.e. they are consistent. Hence, the first pair of solutions is $(0, 1)$ and $(0, -1)$. Other solutions are obtained by solving the equation for the auxiliary parameter t, which gives

$$1 - t + t^2 = t^2 - 2t \implies t = -1 .$$

Correspondingly, $x^2 = 1/ \left(t^2 - 2t \right) = 1/3$ and hence we obtain two more solutions, $x_\pm = \pm 1/\sqrt{3}$, with the corresponding y values $y_\pm = tx_\pm = \mp 1/\sqrt{3}$. ◄

Problem 1.63 Solve Problem 1.59 using the above method.

Problem 1.64 Solve Problem 1.61 using the method of homogeneous functions.

1.3.7 Functional Algebraic Inequalities

Consider an inequality $f(x) < g(x)$. Here we would like to determine those values of x, which may form finite or infinite continuous intervals, for which this inequality is true. Similarly, one considers systems of inequalities

$$\begin{cases} f_1(x) < g_1(x) \\ f_2(x) < g_2(x) \\ \quad \dots \end{cases}$$

The particular sign is not that important, it could be either $>$, $<$, $\geq$ or $\leq$. Here we shall learn a few elementary rules in solving inequalities and give a few examples.

The consideration of a system of inequalities is identical to single inequalities; hence, we shall only give the rules for the latter. We shall formulate these as a set of simple theorems.

Theorem 1.1 *If an arbitrary function $h(x)$ is added to both sides of an inequality, the resulting one is equivalent to the initial one, i.e. it has the same solutions:*

$$f(x) < g(x) \iff f(x) + h(x) < g(x) + h(x) .$$

Proof: Indeed, consider $x = a$ that is a solution of the original inequality, i.e. it satisfies $f(a) < g(a)$. We know (Sect. 1.3.3) that one can add to both sides of the inequality an arbitrary number. Adding to both sides $h(a)$, we obtain

$$f(a) + h(a) < g(a) + h(a) , \tag{1.29}$$

i.e. $x = a$ satisfies this inequality as well. Reversely, assume that $x = a$ satisfies inequality (1.29). Taking $h(a)$ to the left-hand side (as $-h(a)$) and cancelling on $h(a)$, we obtain $f(a) < g(a)$, i.e. a satisfies the original inequality as well. **Q.E.D.**

As a consequence, it follows that one can take a term in an inequality and move it to the other side with an opposite sign.

Theorem 1.2 *If $f_1(x) < g_1(x)$ and $f_2(x) < g_2(x)$, then*

$$f_1(x) + f_2(x) < g_1(x) + g_2(x) ,$$

---

i.e. two inequalities of the same sign can be "summed up".

Proof: If $x = a$ satisfies both initial inequalities, then

$$f_1(a) < g_1(a) \implies f_1(a) + f_2(a) < g_1(a) + f_2(a) . \tag{1.30}$$

Similarly,

$$f_2(a) < g_2(a) \implies f_2(a) + g_1(a) < g_2(a) + g_1(a) . \tag{1.31}$$

The right-hand side of (1.30) is the same as the left-hand side of (1.31); at the same time, this expression, $g_1 + f_2$, is larger than $f_1 + f_2$ and, at the same time, smaller than $g_2 + g_1$ at $x = a$. Hence, $f_1 + f_2 < g_1 + g_2$ even more so, as required. **Q.E.D.**

Theorem 1.3 *If*

$$0 < f_1(x) < g_1(x) \quad and \quad 0 < f_2(x) < g_2(x) , \tag{1.32}$$

then

$$0 < f_1(x)f_2(x) < g_1(x)g_2(x) ,$$

---

i.e. two inequalities with both sides being positive can be "multiplied".

Proof: Consider $x = a$ that satisfies both inequalities (1.32). Since $f_2(a) > 0$, one can multiply both sides of the first inequality in (1.32) by it:

$$0 < f_1(a) f_2(a) < g_1(a) f_2(a) \ .$$

Similarly,

$$0 < f_2(a) < g_2(a) \quad \Longrightarrow \quad 0 < g_1(a) f_2(a) < g_1(a) g_2(a) \ .$$

Combining both inequalities, we obtain

$$0 < f_1(a) f_2(a) < g_1(a) f_2(a) < g_1(a) g_2(a) \ ,$$

as required. **Q.E.D.**

Theorem 1.4 *If $f(x) < g(x)$ and $h(x) > 0$, then one can multiply both sides of the inequality by this function: $f(x)h(x) < g(x)h(x)$. If, however, $h(x) < 0$ for some values of x, then the inequality changes sign upon multiplication with $h(x)$, i.e. in this case $f(x)h(x) > g(x)h(x)$.*

We leave to the reader to prove this theorem. As a consequence,

$$f(x) < g(x) \quad \Longrightarrow \quad -f(x) > -g(x) \ .$$

Problem 1.65 Prove:

$$0 < f(x) < g(x) \quad \Longrightarrow \quad 0 < \frac{1}{g(x)} < \frac{1}{f(x)} \ .$$

Let us now consider a few examples.
Find the values of x that satisfy

$$2x^2 - 4x + 11 > 0 \ .$$

The simplest method is based on finding the roots of this square polynomial: $x_\pm = 1 \pm \frac{3}{2}\sqrt{-2}$, i.e. there are no real roots, i.e. the parabola does not cross the x-axis and is for all values of x positive. Hence, $-\infty < x < \infty$. The same result can also be obtained directly by manipulating this expression into the full square:

$$2x^2 - 4x + 11 = 2\left(x^2 - 2x + 1\right) - 2 + 11 = 2(x - 1)^2 + 9 \ ,$$

which for all x is obviously positive.

Let us now consider the inequality

$$x^2 + 2x - 3 > 0 .$$

The roots of this square polynomial are $x_+ = 1$ and $x_- = -3$. Therefore, writing the polynomial in the factorised form, Eq. (1.19), we have an equivalent inequality:

$$(x - 1)(x + 3) > 0 ,$$

which is satisfied in two cases: both brackets are either positive or negative. In the first case

$$\begin{cases} x - 1 > 0 \\ x + 3 > 0 \end{cases} \implies \begin{cases} x > 1 \\ x > -3 \end{cases} \implies x > 1 ,$$

while in the second,

$$\begin{cases} x - 1 < 0 \\ x + 3 < 0 \end{cases} \implies \begin{cases} x < 1 \\ x < -3 \end{cases} \implies x < -3 .$$

Hence, the inequality is satisfied by the following semi-infinite intervals: $-\infty < x < -3$ and $1 < x < \infty$.

Problem 1.66 Find values of x that satisfy

$$2(x + 1)(x - 1)(x - 5) \leq 0 .$$

[Answer: $-\infty < x \leq -1$ *and* $1 \leq x \leq 5$.]

Problem 1.67 Find x satisfying the following system of inequalities:

$$\begin{cases} x^2 > 1 \\ 2x + 5 \leq 3x - 6 \end{cases} .$$

[Answer: $x \geq 11$.]

Problem 1.68 Solve the inequality

$$\frac{x + 1}{x - 1} \geq 0 .$$

[Answer: $-\infty < x \leq -1$ *and* $1 \leq x < \infty$.]

When solving the following inequality,

$$\frac{x+1}{x-1} > 2 ,$$

one have to be careful. First of all, we need to get rid of the denominator. This can be done by multiplying both sides of the inequality by $(x-1)$. However, depending on whether $(x-1)$ is positive or negative, the sign of the inequality may change. Hence, two cases need to be considered. If $x - 1 > 0$ (or $x > 1$), then $x + 1 > 2(x - 1)$ or $x < 3$. Both cases, $x > 1$ and $x < 3$, result in $1 < x < 3$, a finite interval. If, however, $x - 1 < 0$, i.e. $x < 1$, then we have $x + 1 < 2(x - 1)$, which yields $x > 3$ that contradicts the previous inequation of $x < 1$; hence, this case does not give any new solutions. We are left with the finite interval between 1 and 3 for x.

This inequality can also be solved by taking 2 from the right into the left-hand side and factorising:

$$\frac{x+1}{x-1} > 2 \implies \frac{x+1}{x-1} - 2 = \frac{-x+3}{x-1} > 0 ,$$

which results in considering the same cases, leading to an identical solution.

> **Problem 1.69** Find values of x that satisfy
>
> $$\frac{2x}{x+5} \leq \frac{1}{3} .$$
>
> [Answer: $-5 \leq x \leq 1$.]

1.4   Polynomials

1.4.1   Division of Polynomials

Next, let us discuss a division of two polynomials. Before we discuss this, however, it is instructive to recall division of two integer numbers N and M. Generally, we write $N - r = Md$ or $N = Md + r$, where $0 < r < M$ is a residue and d is the quotient.

Similarly with polynomials: if we divide $P_n(x)$ by $Q_m(x)$ with $m < n$, then generally one can write:

$$P_n(x) = Q_m(x)D_l(x) + R_k(x) \quad \text{or} \quad \frac{P_n(x)}{Q_m(x)} = D_l(x) + \frac{R_k(x)}{Q_m(x)} , \qquad (1.33)$$

where $R_k(x)$ is a polynomial of the order $k < m$ and the quotient $D_l(x)$ is also a polynomial, its order l must be such that multiplication of two polynomials $D_l(x)$

and $Q_m(x)$ would give a polynomial of order n, i.e. $l + m = n$. This is because we assumed that $k < m$, but the largest power n of $P_n(x)$ has to be provided by the right-hand side.

Consider as an example a division of the polynomial $P_4(x) = x^4 + x^3 + x^2 + x + 5$ by $Q_2(x) = 2x^2 + 1$. It can be verified by direct multiplication of the right-hand side that

$$x^4 + x^3 + x^2 + x + 5 = \underbrace{(2x^2 + 1)}_{Q_2(x)} \underbrace{\left(\frac{1}{2}x^2 + \frac{1}{2}x + \frac{1}{4}\right)}_{D_3(x)} + \underbrace{\left(\frac{1}{2}x + \frac{19}{4}\right)}_{R_1(x)}.$$

Here the quotient is $D_3(x) = \frac{1}{2}x^2 + \frac{1}{2}x + \frac{1}{4}$ and the reminder $R_1(x) = \frac{1}{2}x + \frac{19}{4}$.

The most convenient general approach to tackle the division of polynomials is the so-called long division method. Suppose we would like to divide the polynomial

$$P_n(x) = a_n x^n + a_{n-1} x^{n-1} + \cdots + a_0$$

by the polynomial ($m < n$)

$$Q_m(x) = b_m x^m + b_{m-1} x^{m-1} + \cdots + b_0 .$$

The idea is to multiply the second polynomial by $a_n x^n / b_m x^m$ and subtract from the first; this way the highest power term in $P_n(x)$ will be eliminated:

$$P_n(x) - \frac{a_n x^n}{b_m x^m} Q_m(x) = a'_{n-1} x^{n-1} + a'_{n-2} x^{n-2} + \cdots + a'_0 ,$$

where $a'_{n-1}, a'_{n-2}, a'_{n-3}$, etc. are the new coefficients. The obtained polynomial is of one power less than the original one. At the next step, we multiply $Q_m(x)$ by $a'_{n-1} x^{n-1} / b_m x^m$ and subtract from the polynomial obtained at the previous step to eliminate the highest power term in it; this yields a polynomial of power $n - 2$:

$$\left[P_n(x) - \frac{a_n x^n}{b_m x^m} Q_m(x) \right] - \frac{a'_{n-1} x^{n-1}}{b_m x^m} Q_m(x) = a''_{n-2} x^{n-2} + a''_{n-3} x^{n-3} + \cdots + a''_0 ,$$

where $a''_{n-2}, a''_{n-3}, a''_{n-4}$, etc. are the new coefficients. This process is continued until the final polynomial is of the order less than m, the order of the divisor:

$$P_n(x) - \frac{a_n x^n}{b_m x^m} Q_m(x) - \frac{a'_{n-1} x^{n-1}}{b_m x^m} Q_m(x) - \cdots = c_{m-1} x^{m-1} + \cdots + c_0 .$$

The division is finished. Indeed, let us rearrange:

$$P_n(x) = Q_m(x) \left(\frac{a_n}{b_m} x^{n-m} + \frac{a'_{n-1}}{b_m} x^{n-m-1} + \cdots \right) + R_{m-1}(x) ,$$

where $R_{m-1}(x) = c_{m-1} x^{m-1} + \cdots + c_0$ is the residue and the expression in the brackets in the right-hand side is the quotient, $D_{n-m}(x)$.

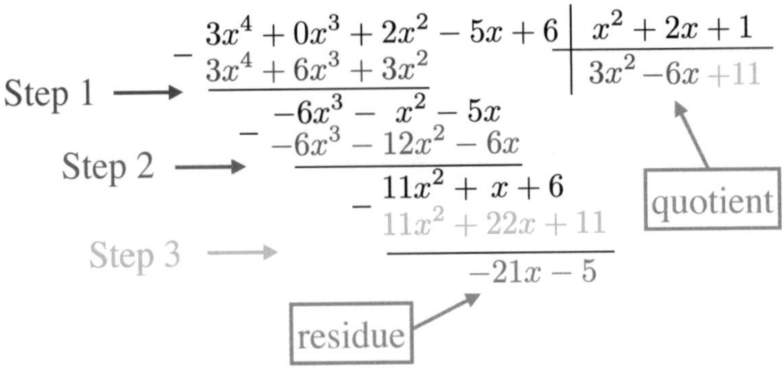

$$
\begin{array}{r|l}
3x^4 + 0x^3 + 2x^2 - 5x + 6 & x^2 + 2x + 1 \\
\underline{-\ 3x^4 + 6x^3 + 3x^2} & 3x^2 - 6x + 11 \\
-6x^3 - x^2 - 5x & \\
\underline{-\ 6x^3 - 12x^2 - 6x} & \\
11x^2 + x + 6 & \\
\underline{-\ 11x^2 + 22x + 11} & \\
-21x - 5 &
\end{array}
$$

Step 1 $\longrightarrow$

Step 2 $\longrightarrow$

Step 3 $\longrightarrow$

quotient

residue

Fig. 1.3 Long division example

Example 1.7 ▶ Divide $P_4(x) = 3x^4 + 2x^2 - 5x + 6$ by $Q_2(x) = x^2 + 2x + 1$.

Solution. At the first step, we multiply Q_2 by $3x^2$ and subtract from P_4:

$$P_4 - 3x^2 Q_2 = -6x^3 - x^2 - 5x + 6 .$$

At the next step, we subtract $-6x Q_2$, which gives:

$$P_4 - 3x^2 Q_2 + 6x Q_2 = 11x^2 + x + 6 .$$

Finally, at the last step, we subtract $11 Q_2$ that finally yields

$$P_4 - 3x^2 Q_2 + 6x Q_2 - 11 Q_2 = -21x - 5 .$$

Therefore, one can write:

$$P_4(x) = Q_2(x) \left(3x^2 - 6x + 11 \right) + (-21x - 5) ,$$

with the quotient being $3x^2 - 6x + 11$ and the residue $-21x - 5$. ◀

The discussed method can be presented in a much more convenient way. Indeed, let us return back to our example and look at Fig. 1.3. At the top line in the left we write the polynomial that we would like to divide (in black); we write all terms including the ones that have zero coefficients (as is the case for the x^3 term). At the right corner we write the divisor (also in black). Next, we check the highest power term of the original polynomial and of the divisor. It is seen that one has to multiply the divisor by $3x^2$ to match the highest power term in the original polynomial. Hence, we write $3x^2$ (shown in blue) under the horizontal line which we drew under the divisor and multiply the divisor by $3x^2$. The result of this multiplication we write under the original polynomial, making sure that the terms of the same power are properly aligned (also shown in blue). At that point the first step of the long division procedure has been done.

At the second step we subtract the blue polynomial $3x^4 + 6x^3 + 3x^2$ from the original polynomial, and also attach the next term of the original polynomial, $-5x$, to the result. The resulting expression $-6x^3 - x^2 - 5x$ is shown in black; it effectively replaces now our original polynomial. At that point we think of an expression that would eliminate the highest power term, $-6x^3$, in it when multiplied by the divisor. Obviously, this is $-6x$. We write it next to $3x^2$ (shown in red) under the divisor on the right, and multiply the divisor by it. The result, shown also in red, is written under the expression we are left with after the first step, keeping terms of the same power aligned. This finalises the second step of the procedure.

At the beginning of the third step, we subtract the red expression from the black one above it, attaching the last term (which is just 6) from the original polynomial; we get $11x^2 + x + 6$. We need to multiply the divisor by 11 to eliminate the $11x^2$ term of the last expression. Hence, we write 11 under the divisor (shown in green) and the result of the multiplication (also in green) under the result of the last subtraction. This finalises the third step.

Finally, at the last step we subtract the green expression from the black one just above it. The resulting expression, $-21x - 5$, shown in brown, is the residue; the whole expression that appears under the divisor, $3x^2 - 6x + 11$ (multicoloured), is the quotient. We have finished.

Problem 1.70 Divide $6x^4 + 39x^3 + 91x^2 + 89x + 30$ by $x + 2$. [Answer: the quotient is $6x^3 + 27x^2 + 37x + 15$, no residue.]

Problem 1.71 Divide $P_5(x) = 4 + 249x - 127x^2 - 3x^3 + 69x^4 + 3x^5$ by $Q_3(x) = 3x^3 - 6x + 11$. [Answer: the quotient is $x^2 + 23x + 1$ and the residue is $2x - 7$.]

Problem 1.72 Divide $-1 + x - 3x^2 + x^3 + 2x^4$ by $x - 1$. [Answer: the quotient is $2x^3 + 3x^2 + 1$ and there is no residue.]

The last problem illustrates another interesting method of transformation of polynomials called factorisation:

$$-1 + x - 3x^2 + x^3 + 2x^4 = (x - 1)\left(2x^3 + 3x^2 + 1\right) .$$

In that respect, we shall indicate one important point here.

Consider division of $P_n(x)$ with $n \geq 2$, by the simplest first order polynomial, $x - a$:

$$P_n(x) = (x - a)Q_{n-1}(x) + b ,$$

where b must then be a zero-order polynomial, i.e. a constant. This constant can easily be seen to be $P_n(a)$. Hence, we can write:

$$P_n(x) = (x - a)Q_{n-1}(x) + P_n(a) \,, \tag{1.34}$$

which is the so-called little Bèzout's theorem.[6] In particular, let the point $x = a$ be a root of the polynomial, i.e. the point at which the polynomial is zero: $P_n(a) = 0$. Then, the theorem states that in this case

$$P_n(x) = (x - a)Q_{n-1}(x) \,, \tag{1.35}$$

i.e. the polynomial is divisible by $x - a$, there is no residue. This means that $P_n(x)$ can be factorised using its root.

Next, if in turn $Q_{n-1}(x)$ has a root $x = b$, then similarly, we can factorise it as well, and write

$$P_n(x) = (x - a)(x - b)B_{n-2}(x) \,, \tag{1.36}$$

where $B_{n-2}(x)$ is another polynomial. This process can be continued until at the last step the final polynomial is a zero-order polynomial, i.e. a constant. Hence, the polynomial of order n can be represented as a product of differences $x - x_i$, where x_i are roots of the polynomial:

$$P_n(x) = a_n (x - x_1) (x - x_2) \cdots (x - x_n) \,. \tag{1.37}$$

We have explicitly written here the constant to be obtained at the last step as being a_n, the coefficient to the highest power of x in $P_n(x)$, as this is the only way to get correctly such a coefficient after multiplying all the brackets in the right-hand side of Eq. (1.37). Moreover, the polynomial cannot have more roots as if we assume that there exists yet another root x_{n+1} that is not equal to any of the roots $x_1, \ldots, x_n$, then none of the differences $x_{n+1} - x_1, x_{n+1} - x_2$, etc. in the expansion (1.37) would be equal to zero and hence $P_n(x_{n+1}) \neq 0$. This contradicts our assumption proving it wrong.

It may seem that, as follows from this argument, we have proven that any polynomial of order n should have n roots (although there could be repeated roots). Actually, the situation is not so trivial as we "assumed" that any polynomial has a root. The rigorous proof of this fact called the fundamental theorem of algebra is not trivial and will be given in Vol. II of the course. For the moment, we shall not bother ourselves with this and assume that any polynomial has at least one root (which could be complex[7]). From this, following the logic of the above argument, we can indeed conclude that any nth-order polynomial has exactly n (possibly complex)

---

[6] Named after Étienne Bèzout.
[7] Complex numbers will be introduced in Sect. 1.11.

roots (including all repeated ones). If the root x_k is repeated n_k times, then one can write from Eq. (1.37):

$$P_n(x) = a_n (x - x_1)^{n_1} (x - x_2)^{n_2} \cdots (x - x_l)^{n_l} \, ,$$

where $x_1, \ldots, x_l$ are all distinct roots, and $n_1 + \cdots + n_l = n$.

1.4.2 Finding Roots of Polynomials with Integer Coefficients

Often one needs to find roots of a polynomial. As we have learned earlier in Sect. 1.3.4, it is possible to express the roots of a quadratic polynomial via a general expression containing radicals. We shall also see in Sect. 1.11.1 that this is also possible to do for a general cubic equation, i.e. to find the roots of a general third-order polynomial. In fact, there is only one other case of the quartic equation for which a general solution is known, we shall also consider this case in Sect. 3.4.6. However, it is not possible to express the solution of a general polynomial of order more than four in terms of radicals; such an expression simply does not exist. Therefore, a progress can be made for specific classes of polynomials which we shall consider now.

One such method deals with the polynomial equation

$$P_n(x) = a_n x^n + a_{n-1} x^{n-1} + \cdots + a_1 x + a_0 = 0 \, , \tag{1.38}$$

containing exclusively integer coefficients. Here we assume that both a_n and a_0 are not equal to zero as in these cases we still need to solve an analogous problem, but for a polynomial of a lower order. So, without loss of generality, these cases can be excluded.

Theorem 1.5 *Consider an nth-order algebraic polynomial equation (1.38) with all coefficients a_0, a_1, etc. being integers. If p/q is a rational root of the equation (p and q are both integers and p is not divisible by q), then a_n must be divisible by q and a_0 must be divisible by p.*

Proof: Indeed, if p/q is the root of our equation, we can write:

$$a_n \left(\frac{p}{q}\right)^n + a_{n-1} \left(\frac{p}{q}\right)^{n-1} + \cdots + a_1 \frac{p}{q} + a_0 = 0 \, .$$

This equation can be rewritten in two different ways: if we multiply both sides of it by q^n / p and take all terms, apart from the one with a_0, to the other side,

$$\frac{a_0 q^n}{p} = -a_n p^{n-1} - a_{n-1} p^{n-2} q - \cdots - a_1 q^{n-1} \tag{1.39}$$

and, similarly, if we multiply both sides of it by q^{n-1} and leave in the left-hand side only the term with a_n:

$$\frac{a_n p^n}{q} = -a_{n-1}p^{n-1} - a_{n-2}p^{n-2}q - \cdots - a_0 q^{n-1}\,. \tag{1.40}$$

Let us first consider Eq. (1.39). In the right-hand side of it we obviously have an integer number; hence, the number $a_0 q^n/p$ in its left-hand side must also be an integer. However, we know that p is not divisible by q, and therefore p cannot contain in its representation as a product of all its divisors the number q in any power. That means that q^n/p cannot be an integer. This simply means that a_0/p must be an integer, i.e. a_0 must be divisible by p. The first part of the theorem is proven.

Similarly we prove its second part using Eq. (1.40): since the right-hand side is an integer, so be it for the left-hand side, $a_n p^n/q$. But p is not divisible by q, so a_n must then be. **Q.E.D.**

This theorem can be extremely useful in finding roots of polynomial equations of high orders: we simply need to list all possible divisors p of the free term a_0, and all divisors q of the coefficient a_n to the largest power of x, including both possible signs, and then consider their all possible ratios p/q. Of course, we are not guaranteed that this way we shall find solutions as the theorem contains an important "if", that "if p/q is a root". Still, if there is at least one rational root p/q of the polynomial, it is guaranteed to be such that a_0 is divisible by p and a_n by q. Once at least one root is determined, the polynomial can be factorised, and to determine the other roots a polynomial of a lower order will need to be analysed.

It is instructive to illustrate this method on a few examples.

Example 1.8 ▶ Find roots of the polynomial $P_3(x) = x^3 - 3x^2 - x + 3$.

Solution. Here $a_0 = 3$ and $a_3 = 1$. Hence, possible values of p are ± 1 and ± 3, and possible values of q are ± 1. Hence, possible rational numbers to try are ± 1 and ± 3. We do not know if any of these would be roots, but at least we can try to substitute each of them one by one and find out. Simple calculations yields that $P_3(1) = P_3(-1) = P_3(3) = 0$, but $P_3(-3) = -48 \neq 0$, i.e. we have found three roots of this equation: $+1, -1$ and $+3$. Therefore, the polynomial can also be written in the following factorised form:

$$x^3 - 3x^2 - x + 3 = (x + 1)(x - 1)(x - 3)\,. \blacktriangleleft$$

Example 1.9 ▶ Find roots of the polynomial $P_4(x) = 4x^4 + 16x^3 - 31x^2 - 109x + 60$.

Solution. Here $a_0 = 60$ and $a_4 = 4$. Possible p values are: $\pm 1, \pm 2, \pm 3, \pm 4, \pm 5$, $\pm 6, \pm 10, \pm 12, \pm 15, \pm 20, \pm 30$ and ± 60. Possible values of q are $\pm 1, \pm 2$ and ± 4.

Dividing all possible p by all possible q values, all possible rational numbers p/q are obtained. We do not write them all here as there are too many of them. And it is rather cumbersome to try them all. Instead, one may stop at finding one root and then reduce the order of the polynomial by division. Indeed, suppose we tried -3 and found out that it is indeed a root as $P_4(-3) = 0$. This means that the polynomial should be divisible by $x + 3$, so we can write $P_4(x) = (x + 3)Q_3(x)$, where $Q_3(x)$ is to be determined. Performing division, we obtain that $Q_3(x) = 4x^3 + 4x^2 - 43x + 20$. The same analysis is then performed for this polynomial. In this case, p could be ± 1, ± 2, ± 4, ± 5, ± 10 and ± 20, while possible values of q are still ± 1, ± 2 and ± 4. The number of candidates for the roots p/q is greatly reduced from the previous case, but is still significant. Nevertheless, we may try $1/2$ and discover that it is a root, $Q_3(1/2) = 0$. To reduce the problem further, we divide $Q_3(x)$ by $2x - 1$, which gives $Q_3(x) = (2x - 1)(-20 + 3x + 2x^2)$, i.e. the remaining roots are contained in the quadratic equation $-20 + 3x + 2x^2 = 0$ that gives $5/2$ and -4. Hence, there are four roots altogether, -4, -3, $1/2$ and $5/2$, and the polynomial is factorised as follows:

$$4x^4 + 16x^3 - 31x^2 - 109x + 60 = (x + 3)(x + 4)(2x - 1)(2x - 5) . \blacktriangleleft$$

The last example shows that in some cases the described method becomes really time-consuming since one has to go through a long list of possible root candidates. Fortunately, there exists one more statement that one may employ to simplify this procedure.

Theorem 1.6 *Consider an nth-order algebraic polynomial equation (1.38) with all coefficients a_0, a_1, etc. being integers. If p/q is a rational root of the equation (p and q are both integers and p is not divisible by q), then $P_n(k)$ must be divisible by $p - kq$ for any integer k.*

Proof: According to Bèzout's theorem (1.34), one can write, for any a,

$$P_n(x) = (x - a) Q_{n-1}(x) + P_n(a) .$$

Let us substitute $x = p/q$ into this equation and use the fact that it is its root, $P_n(p/q) = 0$. Hence,

$$\left(\frac{p}{q} - a \right) Q_{n-1} \left(\frac{p}{q} \right) + P_n(a) = 0$$

$$\implies P_n(a) = -\frac{p - aq}{q} Q_{n-1}\left(\frac{p}{q}\right)$$

$$= -\frac{p - aq}{q} \underbrace{\left[b_{n-1}\left(\frac{p}{q}\right)^{n-1} + b_{n-2}\left(\frac{p}{q}\right)^{n-2} + \cdots + b_1\frac{p}{q} + b_0\right]}_{Q_{n-1}(p/q)}.$$

Multiplying and dividing the right-hand side by q^{n-1}, we finally obtain:

$$P_n(a) = -\frac{p - aq}{q^n}\left[b_{n-1}p^{n-1} + b_{n-2}p^{n-2}q + \cdots + b_1 pq^{n-2} + b_0 q^{n-1}\right]$$

or

$$-\frac{P_n(a)q^n}{p - aq} = b_{n-1}p^{n-1} + b_{n-2}p^{n-2}q + \cdots + b_1 pq^{n-2} + b_0 q^{n-1}.$$

Now, let us consider $a = k$ being an integer k. $Q_{n-1}(x)$ is a polynomial with integer coefficients b_0, b_1, etc., since it was obtained by division of $P_n(x)$ by $x - k$, both containing integer coefficients. Therefore, the polynomial in the right-hand side in the expression above must be an integer number. Hence, the expression in the left-hand side must also be an integer. We know that $P_n(k)$ is an integer. However, the fraction $q^n/(p - kq)$ cannot be an integer since p is not divisible by q. Therefore, $P_n(k)$ must be divisible by $p - kq$. **Q.E.D.**

Let us illustrate on our previous Example 1.9 how this theorem may help to seep through all the possibilities.

Example 1.10 ▶ Find roots of the polynomial $P_4(x) = 4x^4 + 16x^3 - 31x^2 - 109x + 60$.

Solution. We know that all possible roots p/q, if exist, must be composed of p from $\pm 1, \pm 2, \pm 3, \pm 4, \pm 5, \pm 6, \pm 10, \pm 12, \pm 15, \pm 20, \pm 30$ and ± 60, and q from $\pm 1, \pm 2$ and ± 4. Let us test a possible root candidate $-5/2$ with $p = 5$ and $q = -2$. Consider $k = 1$, in which case $P_n(1) = -60$ and $p - kq = p - q = 7$; as 60 is not divisible by 7, this combination of p and q cannot be a root and must be excluded. If we, however, consider $p = 5$ and $q = +2$ and the same value of $k = 1$, then $p - kq = 3$ and 60 is divisible by it; hence, the candidate $p/q = 5/2$ cannot be excluded and must be tried. In fact, it is a root. ◀

The choice of the integer k is arbitrary; one may try $k = 1$, then $k = 2$, etc. to eliminate some candidates. Hence, the method of finding roots of a polynomial equation $P_n(x) = 0$ with integer coefficients could be as follows. First, find all possible divisors p of a_0 and q of a_n. Second, choose a value of k and calculate $P_n(k)$. Check various combinations of p and q and choose only those for which $P_n(k)$ is divisible by $p - kq$. Once various combinations p and q are eliminated, try another value of k to eliminate some of the remaining ones. In the end, verify if any combinations that are left are roots of the equation. If one root $x_1 = p/q$ is found, you may divide the polynomial on $qx - p$ to lower its order and repeat the analysis. Alternatively, you can try other remaining combinations to see if more roots $x_2, x_3, \ldots, x_m$ are found and then divide the polynomial on a product $G_m(x) = \gamma \, (x - x_1) \, (x - x_2) \ldots (x - x_m)$, where γ is an integer that would ensure that the polynomial $G_m(x)$ contains all coefficients being integers. The procedure is repeated for the resulting polynomial of lower order if required. Once you are left with a square polynomial, its roots can be determined directly using Eq. (1.18).

Problem 1.73 Find roots of the equation

$$4x^4 - 20x^3 + 26x^2 - 4x - 6 = 0 .$$

[Answer: $1, 3, 1/2 \pm \sqrt{3}/2$.]

Problem 1.74 Find roots of the equation

$$36x^4 + 96x^3 - 61x^2 - 74x + 3 = 0 .$$

[Answer: $1, -3, -1/3 \pm \sqrt{5}/6$.]

Problem 1.75 Show that the polynomial $-3 - 4x - 8x^2 - x^3 + 6x^4$ can be factorised as follows:

$$- 3 - 4x - 8x^2 - x^3 + 6x^4 = (3x^2 + x + 1)(2x - 3)(x + 1) . \quad (1.41)$$

[Hint: *determine the roots* -1 *and* $-3/2$ *and then divide the polynomial by the appropriate quadratic expression.*]

1.4.3 Vieta's Formulae

The roots of a polynomial can be related to its coefficients. These relationships can be established by opening the brackets in Eq. (1.37) and comparing the coefficients to the same powers of x with those in the polynomial $a_0 + a_1 x + \cdots + a_n x^n$. Even though this seems a reasonable approach, its legitimacy will be proven at the end of Sect. 1.4.4.

In order to consider the general case of arbitrary n, it is instructive to start from simple cases. Consider first the case of $n = 2$:

$$a_2 \, (x - x_1) \, (x - x_2) = a_2 \left[x^2 - (x_1 + x_2) \, x + x_1 x_2 \right] .$$

Comparing coefficients to the same powers of x in this expression with those in the polynomial $a_2x^2 + a_1x + a_0$, we immediately obtain:

$$a_2x_1x_2 = a_0 \quad \text{and} \quad a_2(x_1 + x_2) = -a_1 . \tag{1.42}$$

Problem 1.76 Consider a third-order polynomial $a_3x^3 + a_2x^2 + a_1x + a_0$ with roots x_1, x_2 and x_3. Show that the roots and the coefficients are related via:

$$a_3x_1x_2x_3 = -a_0 , \quad a_3(x_1x_2 + x_2x_3 + x_1x_3) = a_1 \quad \text{and}$$
$$a_3(x_1 + x_2 + x_3) = -a_2 . \tag{1.43}$$

Problem 1.77 Consider a fourth-order polynomial $a_4x^4 + a_3x^3 + a_2x^2 + a_1x + a_0$ with roots x_1, x_2, x_3 and x_4. Show that the roots and the coefficients are related via:

$$a_4x_1x_2x_3x_4 = a_0 , \quad a_4(x_1x_2x_3 + x_1x_2x_4 + x_1x_3x_4 + x_2x_3x_4) = -a_1 ,$$
$$a_4(x_1x_2 + x_1x_3 + x_1x_4 + x_2x_3 + x_2x_4 + x_3x_4) = a_2 \quad \text{and}$$
$$a_4(x_1 + x_2 + x_3 + x_4) = -a_3 . \tag{1.44}$$

Note that all coefficients are fully symmetric with respect to an arbitrary permutation of the roots. Moreover, one can see that each coefficient has all possible combinations of roots indices. For instance, consider the quartic case. a_1 relates to a sum (with the minus sign) of four terms corresponding[8] to $4!/3! = 4$ possible combinations of choosing three different indices from a pool of four, a_2 contains the sum of six terms due to $4!/(2!2!) = 6$ combinations, while a_3 contains $4!/3! = 4$ terms and a_0 contains only 1. Here $n! = 1 \cdot 2 \cdot 3 \ldots \cdot n$ is a short-hand notation for a product of all integers between 1 and n, called n-*factorial*. Note that the factorial of 0 is *defined* as 1, i.e. $0! = 1$.

It is seen that a_0 coefficient seems to be basically related to the product of all roots, a_{n-1} to the sum of all roots, a_{n-2}—to the sum of products of all possible pairs of different roots, a_{n-3}—to the sum of all possible triples of them, and so on. In order to formulate a general rule which emerges from the considered above particular cases, it is convenient first to introduce a simple notation for such sums of products of different roots. Let us denote by

$$[k]_n = \sum_{1 \le i_1 < i_2 < \cdots < i_k \le n} \underbrace{x_{i_1}x_{i_2} \cdots x_{i_k}}_{k}$$

---

[8] We shall discuss in more detail the numbers of such combinations in Sect. 1.16, where we shall also relate them to the so-called binomial coefficients to be considered in Sect. 1.13.

the sum of products of k different roots of the polynomial of order n. We shall also set $[0]_n = 1$. The quantity $[k]_n$ satisfies an obvious recurrence relationship for any $0 \le k < n$,

$$[k+1]_{n+1} = [k+1]_n + x_{n+1}[k]_n , \tag{1.45}$$

while for $k = n$:

$$[n+1]_{n+1} = (x_1 x_2 \cdots x_n) x_{n+1} = [n]_n x_{n+1} . \tag{1.46}$$

This might be easier to see by example. For instance, formulae (1.44) can compactly be rewritten as

$$a_4[4]_4 = a_0 , \quad a_4[3]_4 = -a_1 , \quad a_4[2]_4 = a_2 \quad \text{and} \quad a_4[1]_4 = -a_3 .$$

Of course, there is only one possible product of the roots in $[4]_4$, four possibilities in $[3]_4$ and so on. Also,

$$[3]_4 = x_1 x_2 x_3 + x_1 x_2 x_4 + x_1 x_3 x_4 + x_2 x_3 x_4$$

$$= \underbrace{x_1 x_2 x_3}_{[3]_3} + x_4 \underbrace{(x_1 x_2 + x_2 x_3 + x_1 x_3)}_{[2]_3} = [3]_3 + x_4 [2]_3 ,$$

as required (here $k = 2$ and $n = 3$ in formula (1.45)).

It seems that we are able to formulate now the general rule for the nth-order polynomial in a very compact way as follows:

$$
\begin{cases}
a_n[1]_n &= \quad -a_{n-1} \\
a_n[2]_n &= \quad a_{n-2} \\
a_n[3]_n &= \quad -a_{n-3} \\
\cdots \quad \cdots & \quad \cdots \\
a_n[k]_n &= \quad (-1)^k a_{n-k} \\
\cdots \quad \cdots & \quad \cdots \\
a_n[n]_n &= \quad (-1)^n a_0
\end{cases}
\tag{1.47}
$$

We shall prove this result using induction. So, it has been proven for n between 2 and 4 (the case of $n = 1$ is trivial, but the general result is valid for it as well). Next, we assume that it is valid for n, Eq. (1.47), i.e. an nth-order polynomial can be written via its roots as (note that we have changed the order of the terms in the polynomial)

$$P_n(x) = \sum_{k=0}^{n} a_{n-k} x^{n-k} = a_n \sum_{k=0}^{n} (-1)^k [k]_n x^{n-k} , \tag{1.48}$$

and consider the case of a $(n+1)$th-order polynomial with the coefficients b_k:

$$P_{n+1}(x) = \sum_{k=0}^{n+1} b_k x^k = b_{n+1} (x - x_{n+1}) (x - x_n) (x - x_{n-1}) \cdots (x - x_1) .$$

This polynomial has one extra root x_{n+1}, all other of its roots are the same as in the polynomial (1.48); however, its coefficients b_k may all be different. Therefore, we can write:

$$P_{n+1}(x) = \frac{b_{n+1}}{a_n} (x - x_{n+1}) \left[a_n (x - x_n)(x - x_{n-1}) \cdots (x - x_1) \right]$$

$$= \frac{b_{n+1}}{a_n} (x - x_{n+1}) P_n(x) = b_{n+1}(x - x_{n+1}) \sum_{k=0}^{n} (-1)^k [k]_n x^{n-k}$$

$$= b_{n+1} \left[\sum_{k=0}^{n} (-1)^k [k]_n x^{n-k+1} - \sum_{k=0}^{n} (-1)^k [k]_n x_{n+1} x^{n-k} \right].$$

We shall next separate out the $k = 0$ term in the first sum and the $k = n$ term in the second:

$$= b_{n+1} \left[x^{n+1} + \sum_{k=1}^{n} (-1)^k [k]_n x^{n-k+1} \right.$$

$$\left. - \sum_{k=0}^{n-1} (-1)^k [k]_n x_{n+1} x^{n-k} + (-1)^{n+1} [n]_n x_{n+1} x^0 \right].$$

In the first sum we shall change the summation index from k to $l = k - 1$, and will also note the recurrence relation (1.46):

$$= b_{n+1} \left[x^{n+1} + \sum_{l=0}^{n-1} (-1)^{l+1} [l + 1]_n x^{n-l} \right.$$

$$\left. + \sum_{k=0}^{n-1} (-1)^{k+1} [k]_n x_{n+1} x^{n-k} + (-1)^{n+1} [n + 1]_{n+1} x^0 \right].$$

The two sums can be now combined as both have the summation index running from 0 to $n - 1$; this is most easily done by replacing back l with k (of course, the actual symbol used as a summation index is irrelevant):

$$= b_{n+1} \left[x^{n+1} + \sum_{k=0}^{n-1} (-1)^{k+1} ([k + 1]_n + [k]_n x_{n+1}) x^{n-k} + (-1)^{n+1} [n + 1]_{n+1} x^0 \right]$$

$$= b_{n+1} \left[x^{n+1} + \sum_{k=0}^{n-1} (-1)^{k+1} [k + 1]_{n+1} x^{n-k} + (-1)^{n+1} [n + 1]_{n+1} x^0 \right],$$

where at the last step the recurrence relation (1.45) has been used. Finally, replacing the summation index k with $l = k + 1$, and joining the first and the last terms inside the square brackets into the sum, we finally obtain:

$$= b_{n+1} \left[x^{n+1} + \sum_{l=1}^{n} (-1)^l [l]_{n+1} x^{n+1-l} + (-1)^{n+1} [n+1]_{n+1} x^0 \right]$$

$$= b_{n+1} \sum_{l=0}^{n+1} (-1)^l [l]_{n+1} x^{n+1-l} ,$$

which is the required relation (compare with Eq. (1.48) written for n).

Relationships (1.47) and (1.48) represent the so-called Vieta's formulae.[9]

> **Problem 1.78** Construct a polynomial that has the following roots: $1, -1$ (repeated twice) and 3. $[P_4(x) = 3 + 2x - 4x^2 - 2x^3 + x^4.]$

1.4.4 Factorisation of Polynomials: Method of Undetermined Coefficients

Let us now discuss how one can factorise a polynomial using the method of undetermined coefficients. That could be useful in solving algebraic equations of the form $P_n(x) = 0$, $P_n(x)/P_m(x) = 0$ or simplifying the latter rational polynomial expression. One way of factorising polynomials is related to finding its roots and then representing it as a product via its roots, Eq. (1.37), or partially via its roots, as in Eq. (1.41). Here we shall consider a different method called the method of undetermined coefficients. The idea is to write the polynomial as a product of two (or more) polynomials of lower degrees with unknown coefficients and determine them by multiplying the polynomials and comparing the coefficients to the same powers of x.

Example 1.11 ▶ It is known that the polynomial $4x^4 + 8x^3 + 5x^2 + x - 3$ has a root $1/2$. Factorise it.

Solution. Since we know one root, we can write:

$$4x^4 + 8x^3 + 5x^2 + x - 3 = (2x - 1)\left(px^3 + qx^2 + rx + s\right) ,$$

---

[9] Named after François Viète (Franciscus Vieta).

where p, q, r and s are the coefficients to be determined. We cannot be more explicit in the right-hand side.

Multiplying the two polynomials in the right-hand side, we obtain:

$$4x^4 + 8x^3 + 5x^2 + x - 3 = 2px^4 + (2q - p) x^3 + (2r - q) x^2 + (2s - r) x - s .$$
$$(1.49)$$

It seems obvious that the coefficients on both sides to the same powers of x must be the same. Therefore, we can compare them. Below we shall write x in a particular power and then the equation for the corresponding coefficient; the equations are solved one after another from the top to the bottom:

$$
\begin{aligned}
x^0 &: \quad -s = -3 \quad \Longrightarrow \quad\quad s = 3 \\
x^1 &: \quad -r + 2s = 1 \Longrightarrow r = 2s - 1 = 5 \\
x^2 &: \quad -q + 2r = 5 \Longrightarrow q = 2r - 5 = 5 \\
x^3 &: \quad -p + 2q = 8 \Longrightarrow p = 2q - 8 = 2 \\
x^4 &: \quad\quad 2p = 4 \quad \Longrightarrow \quad\quad p = 2.
\end{aligned}
$$

It can be seen that the last two equations are consistent with each other. Hence, we are done:

$$4x^4 + 8x^3 + 5x^2 + x - 3 = (2x - 1) \left(2x^3 + 5x^2 + 5x + 3\right) .$$

At the next step, we have to factorise the obtained third-order polynomial $2x^3 + 5x^2 + 5x + 3$. Using the methods of the previous subsection, we can determine its root $-3/2$. Hence, we can write:

$$2x^3 + 5x^2 + 5x + 3 = (2x + 3) \left(px^2 + qx + r\right) ,$$

where p, q and r are new undetermined coefficients. Using the same method as in the previous case, we write:

$$2x^3 + 5x^2 + 5x + 3 = 2px^3 + (3p + 2q)x^2 + (3q + 2r)x + 3r ,$$
$$(1.50)$$

and hence:

$$
\begin{aligned}
x^0 &: \quad\quad 3r = 3 \quad \Longrightarrow r = 1 \\
x^1 &: \quad 2r + 3q = 5 \Longrightarrow q = 1 \\
x^2 &: \quad 2q + 3p = 5 \Longrightarrow p = 1 \\
x^3 &: \quad\quad 2p = 2 \quad \Longrightarrow p = 1
\end{aligned}
$$

and again the last two equations are fully consistent. Therefore,

$$2x^3 + 5x^2 + 5x + 3 = (2x + 3) \left(x^2 + x + 1\right) .$$

So, the final factorisation of the original polynomial reads:

$$4x^4 + 8x^3 + 5x^2 + x - 3 = (2x - 1) (2x + 3) \left(x^2 + x + 1\right) . \quad \blacktriangleleft$$

Problem 1.79 Using the method of undetermined coefficients, factorise the polynomial as shown below:

$$4x^4 + 8x^3 + 5x^2 + x - 3 = \left(x^2 + x + 1\right)\left(px^2 + qx + r\right) .$$

[Answer: $p = 4$, $q = -4$ and $r = -3$.]

Problem 1.80 Using the method of undetermined coefficients, find one factor in the product in the right-hand side:

$$4x^4 + 16x^3 - 31x^2 - 109x + 60 = (x + 3)\left(px^3 + qx^2 + rx + s\right) .$$

[Answer: $p = 4$, $q = 4$, $r = -43$ and $s = 20$.]

Problem 1.81 Simplify the following rational function:

$$f(x) = \frac{-2 - x + 4x^2 + 4x^3 + x^4}{2 + 5x + 6x^2 + 4x^3 + x^4} .$$

[Answer: $f(x) = (x^2 + x - 1)/(x^2 + x + 1)$.]

In our reasoning above we have silently assumed that if two polynomials are equal, then their coefficients must also be equal. Even though this may sound trivial, it requires a rigorous proof. To this end, we shall state and prove first a simple theorem.

Theorem 1.7 *If a polynomial $P_n(x)$ of order n equals to zero at $n + 1$ values $x_1, x_2, \ldots, x_{n+1}$ of x, it is a null polynomial, i.e. all of its coefficients are equal to zero.*

Proof: Indeed, let us assume that $P_n(x)$ is not a zero polynomial, and that it is a polynomial at least of the order $m \leq n$, i.e. its non-zero highest power coefficient $a_m \neq 0$. On the other hand, since the polynomial is equal to zero at $n + 1$ values of x, these must be its $n + 1$ roots, and hence it can be written in a factorised form:

$$P_n(x) = (x - x_1)(x - x_2) \cdots (x - x_{n+1}) f(x) ,$$

where $f(x)$ is some polynomial of a lower order. One can see that in the right-hand side we have a polynomial of order definitely larger than or equal to $n + 1$, which contradicts our assumption that in the left-hand side there is a polynomial of the order not greater than n. We have arrived at a contradiction, which can only be resolved if all coefficients of our polynomial are equal to zero, i.e. it is a null polynomial. **Q.E.D.**

This theorem has an important corollary: if two polynomials $P_n(x)$ and $Q_m(x)$ (with $m \le n$) are equal at $n + 1$ points, then they are identical. This statement follows from the previous theorem if we consider a difference $P_n(x) - Q_m(x)$. Note that this statement is closely related to a more profound notion of linearly independent functions which we shall consider in the second volume of this course, Chap. 1.

Hence, it is fully legitimate to compare the coefficients to the same power of x on both sides of the equation as we did while analysing Eqs. (1.49) or (1.50).

1.4.5   Multiplication of Polynomials

A product of two polynomials

$$A_n(x) = a_0 + a_1 x + a_2 x^2 + \cdots + a_n x^n = \sum_{i=0}^{n} a_i x^i$$

and

$$B_m(x) = b_0 + b_1 x + b_2 x^2 + \cdots + b_m x^m = \sum_{i=0}^{m} a_i x^i$$

is also a polynomial

$$C_{n+m}(x) = c_0 + c_1 x + c_2 x^2 + \cdots + c_{n+m} x^{n+m} = \sum_{i=0}^{n+m} c_i x^i .$$

Indeed, multiplying any two terms of $A_n(x)$ with $B_m(x)$, we get a certain power of x which is between 0 and $n + m$, i.e. it must be a polynomial of the order $n + m$. For instance,

$$\left(1 + x + x^2\right)\left(-1 + 3x + 2x^4\right) = -1 + 2x + 2x^2 + 3x^3 + 2x^4 + 2x^5 + 2x^6 ,$$
$$(1.51)$$

i.e. a polynomial of the order $2 + 4 = 6$.

Instead of multiplying the polynomials directly, which is tedious, one may use simple formulae for the coefficients c_i ($i = 0, \ldots, n + m$) of the resulting polynomial that we shall derive here. At the same time, we will have a useful exercise in working with the summation symbols. Let us multiply $A_n(x)$ and $B_m(x)$:

$$A_n(x) B_m(x) = \left(\sum_{i=0}^{n} a_i x^i\right)\left(\sum_{j=0}^{m} b_j x^j\right) = \sum_{i=0}^{n}\sum_{j=0}^{m} a_i b_j x^{i+j} . \qquad (1.52)$$

Note that above we have used a different summation index j in the second sum to avoid confusion, since the index i was already used in the first sum. As we work with both sums at the same time in the same mathematical expression, it is important to use different summation indices as otherwise the expression would have little sense as it is not clear which sum the index should belong to. Note that using the same summation index in separate expressions (as we did above) is absolutely fine.

In Eq. (1.52), we multiplied the individual terms $a_i x^i$ and $b_j x^j$ with each other (as we would if using direct multiplication) yielding the term $a_i b_j x^{i+j}$. So, the result of the multiplication in Eq. (1.52) is a sum of terms like this one. It is easy to see that the same power $i + j$ of x may be obtained by multiplying different terms. For instance, the zero power can only be obtained by multiplying the free terms of the two polynomials,

$$c_0 = a_0 b_0 \,, \tag{1.53}$$

the first power $i + j = 1$ can be obtained only by multiplying the terms with the indices (i, j) equal to $(0, 1)$ and $(1, 0)$, i.e.

$$c_1 = a_0 b_1 + a_1 b_0 \,, \tag{1.54}$$

while the second power $i + j = 2$ is obtained by three combinations $(0, 2)$, $(1, 1)$ and $(2, 0)$:

$$c_2 = a_0 b_2 + a_1 b_1 + a_2 b_0 \,. \tag{1.55}$$

This process can be continued, until we arrive at the last coefficient, for which we have a single combination of indices (n, m), i.e.

$$c_{n+m} = a_n b_m \,. \tag{1.56}$$

Let us try to work directly with the double lattice sums in Eq. (1.52) and derive a general coefficient c_k of the product polynomial. To this end, we notice that only those values of the indices (i, j) would contribute to c_k for which $i + j = k$. Then, the original expression of Eq. (1.52) can be replaced as follows:

$$A_n(x) B_m(x) = \sum_{k=0}^{n+m} \left(\sum_{i+j=k} a_i b_j \right) x^k = \sum_{k=0}^{n+m} c_k x^k \,, \tag{1.57}$$

where the expression within the round brackets is a single sum over all possible values of i and j such that their sum is equal to k. It is easy to see that we can list all such terms by summing over i but then choosing j as $k - i$. This yields:

$$c_k = \sum_{i+j=k} a_i b_j = \sum_{i=\max(0, k-m)}^{k} a_i b_{k-i} \,, \tag{1.58}$$

which is the desired general expression. The lowest value of index i in the summation above is the maximum value between two integers, 0 and $k - m$. Indeed, for all values of $k \leq m$, we should start from a_0; however, as the value of k increases, the first possible value of i should be $i = k - m > 0$. This is clearly illustrated on an example.

Let us multiply the polynomials in Eq. (1.51) using this general method. We only need to calculate the c_k coefficients:

$$
\begin{array}{llll}
c_0 : & a_0 b_0 & = & 1 \cdot (-1) & = -1 \\
c_1 : & a_0 b_1 + a_1 b_0 & = & 1 \cdot 3 + 1 \cdot (-1) & = +2 \\
c_2 : & a_0 b_2 + a_1 b_1 + a_2 b_0 & = & 1 \cdot 0 + 1 \cdot 3 + 1 \cdot (-1) & = +2 \\
c_3 : & a_0 b_3 + a_1 b_2 + a_2 b_1 & = & 1 \cdot 0 + 1 \cdot 0 + 1 \cdot 3 & = +3 \\
c_4 : & a_0 b_4 + a_1 b_3 + a_2 b_2 & = & 1 \cdot 2 + 1 \cdot 0 + 1 \cdot 0 & = +2 \\
c_5 : & a_1 b_4 + a_2 b_3 & = & 1 \cdot 2 + 1 \cdot 0 & = +2 \\
c_6 : & a_2 b_4 & = & 1 \cdot 2 & = +2
\end{array}
$$

Problem 1.82 Multiply two polynomials, $1 + x^2 + x^4$ and $2 + x^4 + x^5 - x^6$. [Answer: $2 + 2x^2 + 3x^4 + x^5 + x^7 + x^9 - x^{10}$.]

1.5   Elementary Geometry

1.5.1   Circle, Angles, Lines, Intersections, Polygons

Here we shall revisit the main ideas of the elementary geometry. We shall start from a circle on the $x - y$-plane. The simplest circle of radius R is the one centred at the origin. Every point (x, y) on the circle satisfies the equation $x^2 + y^2 = R^2$. Such a circle of unit radius is shown in Fig. 1.4a.

Next, we shall consider angles. We have already mentioned above that two lines can be "perpendicular", meaning they make 90 degrees (denoted 90°) with each other. This angle corresponds to the most symmetrical position of two lines crossing each other when they make four such angles with each other. Here we shall introduce general angles properly.

If we draw a circle of unit radius, see Fig. 1.4a, then the positive angle $\theta = \angle AOB$ formed by two lines AO and BO crossing at point O is defined by the length L_{AB} of the arc AB relative to the length L_{circle} of the whole circle. Here the point A is obtained from the point B by moving around the circle in the anticlockwise direction. The whole circle corresponds to the angle of 360°, half of the circle to 180° and a quarter to 90°, as shown in Fig. 1.4b. Alongside degrees, *radians* are also frequently used: 360° are equivalent to 2π radians, where $\pi = 3.1415\ldots$ is an irrational number. Therefore, generally the angle θ in Fig. 1.4a is given by the formula:

$$
\theta = \angle AOB = 360 \frac{L_{AB}}{L_{circle}} \text{ (degrees)} \quad \text{or} \quad 2\pi \frac{L_{AB}}{L_{circle}} \text{ (radians)}.
$$

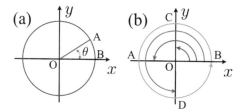

Fig. 1.4 The angle $\theta = \angle AOB$ in (**a**) is formed by lines AO and BO crossing at the point O in the centre of the unit radius circle. Specific angles of $90° = \pi/2$ (blue), $180° = \pi$ (red), $270° = 3\pi/2$ (magenta) and $360° = 2\pi$ (orange) are shown in (**b**)

Fig. 1.5 a When a transversal line T crosses two parallel lines L_1 and L_2, eight angles are formed, four at each crossing. **b** The sum of the angles of any triangle is $180° \equiv \pi$

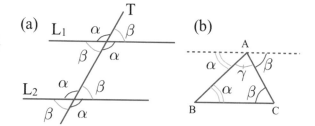

The angle of $180° = \pi$ corresponds to the half of the circle, while quarter of the circle gives $90° = \pi/2$. The latter angle is sometimes called the right angle. One also introduces *negative angles* which correspond to going around the unit circle in the clockwise direction starting from the positive part of the x-axis. This is a fully equivalent representation. Then for instance $\angle BOD$ in Fig. 1.4b is $-90°$ (or $-\pi/2$). Hence the angles $-90°$ and $270°$ correspond to the same point D on the unit circle and hence are equivalent.

Consider now a line L_1 in Fig. 1.5a. If we cross it with another line T at some angle, then at the crossing there will be four angles, but at most two will be different. This is because the opposite angles are always the same. Moreover, it can also be seen that the supplementary angles (defined as two angles on the same side of a straight line) on any side sum up to $180°$ (or π): $\alpha + \beta = \pi$. Exactly the same angles will be formed at the crossing with another line L_2 parallel to the first one: if we slide L_1 along T keeping it parallel all the way (i.e. keeping the same angles α and β), then we shall arrive at L_2 and the crossing will be perfectly identical; this follows from the fact that the lines are parallel: we call lines parallel if they never cross. This discussion gives us a useful relationship between various angles formed at the crossing of two parallel lines with a third (called transversal) line, as shown in Fig. 1.5a.

We shall be using a lot of various 2D figures, and among them the *triangle* and *parallelogram* play an essential role. Consider first a general triangle $\triangle ABC$ shown in Fig. 1.5b.

Fig. 1.6 a A right triangle
ABC with sides a, b and c
and two acute angles α and
β. **b** An isosceles triangle
with two equal sides
$AC = AB$ and two equal
angles
$\angle ACB = \angle ABC = \alpha$. **c** A
general triangle ABC with
obtuse angle γ

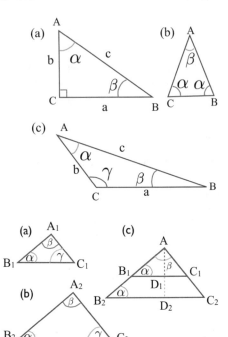

Fig. 1.7 Triangles in **a** and **b**
are *similar*; the heights of the
triangles in (c), AD_1 and
AD_2, differ by the same
factor as the sides

Problem 1.83 Prove using Fig. 1.5b that the sum of three (internal) angles of
a general triangle is 180°, i.e. $\alpha + \beta + \gamma = \pi$. Note that the dashed line is
drawn parallel to BC.

The triangle in Fig. 1.6a is called a right triangle as one of its angles is 90° (or
$\pi/2$). Then, the sum of its other two angles is $\pi/2$ as well. A general triangle,
Fig. 1.6c would have its three angles α, β and γ all different, and its three sides
AB, BC and AC different as well. If a triangle has all three sides equal to each
other, $AB = BC = AC$, the three angles are equal as well simply by symmetry:
$\alpha = \gamma = \beta$. It is called equilateral. If only two sides are the same, e.g. $AB = AC$ as
in Fig. 1.6b, then the angles which are opposite to these sides are the same, again by
symmetry. This is an isosceles triangle. If one draws a perpendicular from its unique
angle to the opposite side, it will divide the angle in equal halves and the opposite
side into two equal lines.

In practice very often it is necessary to consider *similar triangles*, shown in
Fig. 1.7a and b; these are the ones which are scaled up or down with respect to
each other; these have exactly the same angles, and their sides are scaled by the same
factor: $A_1B_1/A_2B_2 = A_1C_1/A_2C_2 = B_1C_1/B_2C_2$.

Fig. 1.8 n-side convex polygon

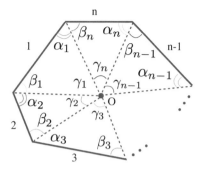

Problem 1.84 Show that the heights of the two triangles, AD_1 and AD_2, see Fig. 1.7c, obtained by drawing perpendiculars from A to the opposite side, are scaled by exactly the same factor as their sides. [Hint: *compare, e.g. triangles AB_1D_1 and AB_2D_2.*]

Consider now a convex polygon consisting of n sides, shown in Fig. 1.8. Let us show that the sum of its internal angles is equal to $\pi (n-2)$. Indeed, let us choose an arbitrary point O inside the polygon and connect it with every vertex by a dashed line as shown. This way we form n triangles. The sum of the angles of each triangle is π, e.g. for the ith triangle ($i = 1, 2, \ldots, n$) we can write $\alpha_i + \beta_i + \gamma_i = \pi$. Here the sum of all alpha and beta angles gives the sum of all internal angles of the polygon, the quantity we are after, while all gamma angles would sum up to 2π (the complete circle). Hence, we can write:

$$(\alpha_1 + \beta_1 + \gamma_1) + (\alpha_2 + \beta_2 + \gamma_2) + \cdots + (\alpha_n + \beta_n + \gamma_n) = n\pi ,$$

or, rearranging:

$$(\alpha_1 + \beta_1) + (\alpha_2 + \beta_2) + \cdots + (\alpha_n + \beta_n) = n\pi - \underbrace{(\gamma_1 + \gamma_2 + \cdots + \gamma_n)}_{2\pi} = \pi (n-2) ,$$

yielding the required result. This formula gives for the sum of internal angles of the triangle ($n = 3$) the correct value of π, for a quadrilateral ($n = 4$) the value of 2π, for a pentagon ($n = 5$) 3π, for a hexagon ($n = 6$) 4π and so on.

1.5.2 Areas of Simple Plane Figures

In applications it is frequently needed to calculate areas of various objects, and we shall come across this type of problem very often in this course. Let us consider some elementary planar objects here. A square has four sides of the same length a and all four of its angles are the right angles, Fig. 1.9a. Its area is $S_{square} = a^2$. The square is generalised by a rectangle in which case the two adjacent sides are not the same and

Fig. 1.9 **a** Square $ABCD$; **b** parallelogram $ABCD$; **c** to illustrate that the parallelogram $ABCD$ consists of two identical triangles; **d** trapezoid $ABCD$

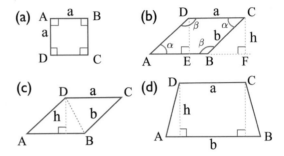

equal to a and b, while the four angles are still of $90°$; its area $S_{rectang} = ab$. Next, we bend the rectangle by pressing on one of its sides and obtain a parallelogram, Fig. 1.9b. Its opposite sides are of the same length, but the four angles are in general different from $90°$, although $\alpha + \beta = 180°$.

Problem 1.85 Show using Fig. 1.9b that the area of the parallelogram $S_{parallelogram} = ah$, where h is the height shown in the figure.

Problem 1.86 Show using Fig. 1.9c that the area of a triangle $\triangle ABD$ is equal to

$$S_\triangle = \frac{1}{2}ah \,, \tag{1.59}$$

where h is the height (the perpendicular drawn to the side a).

Problem 1.87 Show that the area of the trapezoid shown in Fig. 1.9d that has the height h and the parallel upper and lower sides a and b, respectively, is

$$S_{trapezoid} = \frac{1}{2}h(a+b) \,. \tag{1.60}$$

1.6   Trigonometric Functions

The other important class of functions are *trigonometric* ones. But first, we have to briefly return to the right triangles, i.e. the ones which have one right angle (so that the sum of the other two is exactly $90°$), see Fig. 1.6a. The sides a, b and c of the triangle $\triangle ABC$ are related by the famous Pythagorean (or Pythagoras) theorem stating that $a^2 + b^2 = c^2$. There are many proofs of this theorem, one of the simplest is based on the construction of four identical triangles depicted in Fig. 1.10. They form two squares: the outer of side c and the inner of side $d = a - b$.

Fig. 1.10 To the proof of the Pythagoras theorem

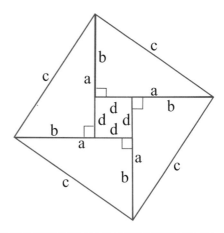

Problem 1.88 Prove the Pythagorean theorem using the construction in Fig. 1.10 made of four identical right triangles of Fig. 1.6a. [Hint: *establish first, using properties of the triangles, that the outer and inner figures are exact squares, and then consider the area of the outer square as composed of the areas of four right triangles and one inner square.*]

With the help of the Pythagoras theorem, we are able to calculate the distance between two arbitrary points $A(x_1, y_1)$ and $B(x_2, y_2)$ on the $x-y$-plane, see Fig. 1.11a. Indeed, $d = AB$ is the hypothenuse of the right triangle $\triangle ABC$. Hence,

$$d = \sqrt{AC^2 + BC^2} = \sqrt{(x_2 - x_1)^2 + (y_2 - y_1)^2} . \tag{1.61}$$

Problem 1.89 Using Fig. 1.11b and the Pythagoras theorem twice, show that the distance between two points $A(x_1, y_1, z_1)$ and $B(x_2, y_2, z_2)$ in the 3D space is given by the following formula:

$$d = \sqrt{(x_2 - x_1)^2 + (y_2 - y_1)^2 + (z_1 - z_2)^2} . \tag{1.62}$$

We are now in the position to define the necessary trigonometric functions. The *sine* function of angle α is defined from the right triangle of Fig. 1.6a as the ratio of the opposite side of the triangle a to its hypotenuse c:

$$\sin \alpha = \frac{a}{c} = \frac{a}{\sqrt{a^2 + b^2}} . \tag{1.63}$$

Similarly, $\sin \beta$ for the other acute angle is defined as b/c, where b is the length of the side opposite to the angle β. The *cosine* function is defined as the ratio of the

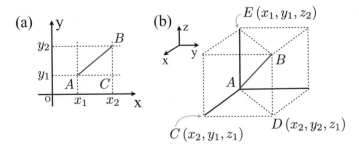

Fig. 1.11 To the calculation of the distance d between two points A and B **a** on the $x - y$-plane and **b** in 3D space

adjacent (with respect to the angle) side of the triangle to its hypotenuse:

$$\cos \alpha = \frac{b}{c} = \frac{b}{\sqrt{a^2 + b^2}} \ . \tag{1.64}$$

From these two definitions, we immediately have our first trigonometric identity:

$$\sin^2 \alpha + \cos^2 \alpha = \frac{a^2}{a^2 + b^2} + \frac{b^2}{a^2 + b^2} = 1 \ , \tag{1.65}$$

which follows from the Pythagorean theorem. By looking at our triangle, we can also establish the following identities:

$$\cos \alpha = \frac{b}{c} = \sin \beta = \sin \left(\frac{\pi}{2} - \alpha \right) \quad \text{and} \quad \sin \alpha = \frac{a}{c} = \cos \beta = \cos \left(\frac{\pi}{2} - \alpha \right) \ .$$

Angles in a right triangle are limited to $90°$ (or $\pi/2$), and hence our definitions of the sine and cosine functions. However, these are generalised to any angle by means of the unit radius circle in the $x - y$-plane depicted in Fig. 1.12. If we consider a radius OP at an angle α with the x-axis, Fig. 1.12a, then for angles $0 \le \alpha \le \pi/2$ the value of $\sin \alpha$ corresponds to the projection OA of the radius on the y-axis; correspondingly, $\cos \alpha$ will be the projection on the x-axis, OB. The same holds for any other angle. For instance, the radius OP in Fig. 1.12b corresponds to the angle $\pi/2 < \alpha < \pi$. In this case $\cos \alpha$ is defined as $-$OD and is negative (OD itself is a positive length) since the point D is on the negative part of the x-axis; at the same time, $\sin \alpha = $ OA and is still positive. By rotating the radius around the circle the two projections change sign, and so do the two functions, sine and cosine. In fact, sine and cosine functions for arbitrary angle α can be defined as the projection on the y- and x-axes, respectively, of the line drawn from the origin of the circle of unit radius and making angle α with the x-axis.

The sine and cosine functions of different angles are related to each other. Firstly, by rotating the radius by 2π we arrive at the same point on the circle, so that both

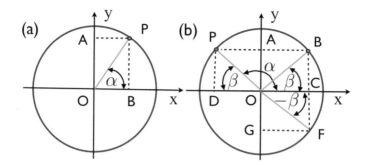

Fig. 1.12 To the definition of the trigonometric functions for arbitrary angles

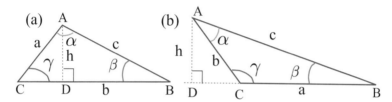

Fig. 1.13 To the proof of the cosine theorem for acute (**a**) and obtuse (**b**) angle γ

functions remain the same for angles α and $\alpha + 2\pi$; we say that the functions are *periodic* with the period 2π:

$$\sin(\alpha + 2\pi) = \sin\alpha \quad \text{and} \quad \cos(\alpha + 2\pi) = \cos\alpha . \tag{1.66}$$

Secondly, consider points P and B in Fig. 1.12b. We have:

$$\sin\alpha = \sin(\pi - \beta) = PD = BC = \sin\beta ,$$
$$\cos\alpha = \cos(\pi - \beta) = -OD = -OC = -\cos\beta . \tag{1.67}$$

If *negative angles* are used that correspond to rotating the radius in the clockwise direction, the same definitions for the sine and cosine functions apply, see Fig. 1.12b:

$$\sin(-\beta) = -OG = -OA = -\sin\beta \tag{1.68}$$

and

$$\cos(-\beta) = OC = \cos\beta . \tag{1.69}$$

A function $f(x)$ is called *even* with respect to its argument x, if $f(-x) = f(x)$ for every value of the x, while it is called *odd* if $f(-x) = -f(x)$. It follows from these definitions that the sine is an odd function, while cosine is an even one.

Now we shall consider a general triangle to derive the cosine theorem which generalises the Pythagoras theorem to arbitrary triangles. To this end, consider first triangle $\triangle ABC$ in Fig. 1.13a with an acute angle γ. By the Pythagoras theorem,

$$h^2 + (b - r)^2 = c^2 ,$$

where $b = CB$, $r = CD = a \cos \gamma$ and $h = a \sin \gamma$. Substituting h and r into the above equation, we obtain:

$$a^2 \sin^2 \gamma + (b - a \cos \gamma)^2 = c^2 .$$

Opening the brackets and using identity (1.65), we arrive at the cosine theorem, stating that

$$c^2 = a^2 + b^2 - 2ab \cos \gamma . \tag{1.70}$$

It expresses a side of a triangle via its other two sides and the angle between them.

Problem 1.90 Repeat the proof for the triangle in Fig. 1.13b in which angle γ is obtuse.

Problem 1.91 Prove that the area of a triangle, Eq. (1.59), can be written exclusively via its sides a, b and c as (Heron's formula[10]):

$$S_\triangle = \sqrt{p (p - a) (p - b) (p - c)}$$

$$= \frac{1}{4}\sqrt{(a + b + c)(-a + b + c)(a - b + c)(a + b - c)} , \tag{1.71}$$

where $2p = a + b + c$ is the perimeter. [Hint: *first express the height h in Fig. 1.13a via the sides of the triangle by writing $h = c \sin \beta$; express the sine via the cosine and use the cosine theorem. Then factorise the expression under the square root.*]

Problem 1.92 Let h_a, h_b and h_c be the heights of a triangle that are drawn perpendicularly to its sides a, b and c, respectively. Show that the area of the triangle $S_\triangle$ satisfies an identity:

$$S_\triangle = \frac{1}{\sqrt{\left(\frac{1}{h_a} + \frac{1}{h_b} + \frac{1}{h_c}\right)\left(-\frac{1}{h_a} + \frac{1}{h_b} + \frac{1}{h_c}\right)\left(\frac{1}{h_a} - \frac{1}{h_b} + \frac{1}{h_c}\right)\left(\frac{1}{h_a} + \frac{1}{h_b} - \frac{1}{h_c}\right)}} . \tag{1.72}$$

[Hint: *use $S_\triangle = \frac{1}{2}ah_a$ (and similarly for the other two sides) in Heron's formula.*]

---

[10] Named after Hero (Heron) of Alexandria.

Problem 1.93 Express sides of a triangle via its heights.

Problem 1.94 Using the same triangle in Fig. 1.13a, prove the *sine theorem* stating that the ratio of the sine of an angle in a triangle to the length of the opposite side is the same for all three angles (sides):

$$\frac{\sin \beta}{b} = \frac{\sin \gamma}{c} = \frac{\sin \alpha}{a} \,. \tag{1.73}$$

[Hint: *expressing the height h using either β or γ will immediately give the first part of the above identity; the other part is proven similarly by introducing the appropriate height h′.*]

Problem 1.95 *Repeat this derivation for the triangle in Fig. 1.13b with the obtuse angle γ.*

Two other frequently used functions are $\tan \alpha = \sin \alpha / \cos \alpha$ and $\cot \alpha = \cos \alpha / \sin \alpha = 1 / \tan \alpha$. All their properties follow from their definition and the properties of the sine and cosine functions.

By looking at the right triangle in Fig. 1.6a, we see that

$$\tan \alpha = \frac{\sin \alpha}{\cos \alpha} = \frac{a/c}{b/c} = \frac{a}{b} \,, \tag{1.74}$$

i.e. it is equal to the ratio of the opposite and adjacent sides of the triangle. For $\cot \alpha$ this is the other way round.

Sine and cosine of some angles such as $30° = \pi/6$, $60° = \pi/3$, $45° = \pi/4$ and some others can be calculated using special geometric considerations. Consider an equilateral $\triangle ABC$ shown in Fig. 1.14a; it has all three angles equal: $\alpha = 60°$. Next, we draw the height from the corner A that will split the side CB into two equal parts. It will also split the angle $\angle CAB$ into two equal angles $\gamma = 30°$.

Problem 1.96 Show using Fig. 1.14a, where an equilateral triangle $\triangle ABC$ is drawn, that $\sin 30° = \cos 60° = 1/2$. Correspondingly, show using the Pythagoras theorem that $\sin 60° = \cos 30° = \sqrt{3}/2$.

Fig. 1.14 To the proof of **a** sine and cosine of 30° and 60° and **b** sin $(\pi/4)$

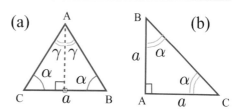

Fig. 1.15 Illustration for the proof of double angle trigonometric identities

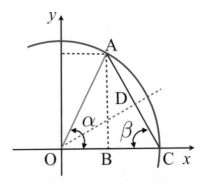

Problem 1.97 Similarly, show using Fig. 1.14b, where an isosceles right triangle $\triangle ABC$ is depicted, that $\sin 45° = \cos 45° = 1/\sqrt{2}$.

Next, we shall consider two important identities for the sine and cosine functions that we shall soon need, the so-called double angle formulae.

Consider a circle of unit radius as shown in Fig. 1.15. The triangle $\triangle AOC$ is an isosceles triangle since it has two of its sides, OC and OA, equal (in fact, equal to 1), hence two angles $\angle OAC = \beta$ and $\angle OCA$ are also equal, and since all three angles of a triangle sum up to π, we have $2\beta + \alpha = \pi$. Also, the height of the triangle $AB = \sin \alpha$. Then, draw the bisector of the $\angle AOC$ which crosses AC at the point D; then, $AC = 2DC = 2 \sin \frac{\alpha}{2}$. On the other hand,

$$AC = \frac{AB}{\sin \beta} = \frac{\sin \alpha}{\sin \beta} = \frac{\sin \alpha}{\sin \left(\frac{\pi}{2} - \frac{\alpha}{2} \right)} = \frac{\sin \alpha}{\cos \frac{\alpha}{2}} ,$$

which must be equal to $2 \sin \frac{\alpha}{2}$. Hence, $\sin \alpha = 2 \sin \frac{\alpha}{2} \cos \frac{\alpha}{2}$ or written in a more conventional form:

$$\sin (2\alpha) = 2 \sin \alpha \cos \alpha . \tag{1.75}$$

Similarly, looking again at the same figure, we can write:

$$OB = \cos \alpha = 1 - BC = 1 - AC \cos \beta = 1 - \frac{\sin \alpha}{\cos \frac{\alpha}{2}} \cos \left(\frac{\pi}{2} - \frac{\alpha}{2} \right)$$

$$= 1 - \frac{\sin \alpha}{\cos \frac{\alpha}{2}} \sin \frac{\alpha}{2} = 1 - 2 \sin^2 \frac{\alpha}{2} ,$$

where we decomposed $\sin \alpha$ using (1.75). Written in a more conventional form:

$$\cos (2\alpha) = 1 - 2 \sin^2 \alpha = \cos^2 \alpha - \sin^2 \alpha = -1 + 2 \cos^2 \alpha , \tag{1.76}$$

where we have also used Eq. (1.65). The derived formulae enable one to express the sine and cosine functions of the angle α via sine (cosine) functions of either double or half angle, 2α or $\alpha/2$.

Problem 1.98 Prove the following identities:

$$1 + \sin x = \left(\cos\frac{x}{2} + \sin\frac{x}{2}\right)^2 ;$$

$$\frac{1 + \sin x}{\cos x} = \frac{\cos x}{1 - \sin x} = \frac{\cos\frac{x}{2} + \sin\frac{x}{2}}{\cos\frac{x}{2} - \sin\frac{x}{2}} = \frac{1 + \tan\frac{x}{2}}{1 - \tan\frac{x}{2}} .$$

1.7   Golden Ratio and Golden Triangle. Fibonacci Numbers

The number ϕ satisfying the following identity:

$$\phi = \frac{a + b}{a} = \frac{a}{b}$$

with $a > b$ is called golden ratio. It is an irrational number which can easily be calculated. Indeed, from the above we have

$$\phi = \frac{a + b}{a} = 1 + \frac{b}{a} = 1 + \frac{1}{\phi} \implies \phi^2 = \phi + 1, \qquad (1.77)$$

which is a quadratic equation with respect to ϕ whose solutions are $\left(1 \pm \sqrt{5}\right)/2$. The root with the plus is $\phi = 1.61803\ldots$ and is what is called the golden ratio; the other root (with the minus) is denoted $\psi = -0.618033\ldots$ and is called (minus) reciprocal of the golden ratio. The two roots are related to each other as $1/\phi = -\psi$.

Problem 1.99 Use the appropriate Vieta's formula (Sect. 1.4.3) to prove that $\phi\psi = -1$.

Problem 1.100 Use the definition of ϕ in Eq. (1.77) to show that ϕ can be written either as a continued square root

$$\phi = \sqrt{1 + \sqrt{1 + \sqrt{1 + \sqrt{1 + \cdots}}}}$$

or a continued fraction

$$\phi = 1 + \cfrac{1}{1 + \cfrac{1}{1 + \cfrac{1}{1 + \cdots}}} .$$

One can calculate ϕ to any given precision by using these formulae. For instance,

$$\sqrt{1 + \sqrt{1 + \sqrt{1 + \sqrt{1 + 1}}}} = 1.6118\ldots,$$

$$1 + \cfrac{1}{1 + \cfrac{1}{1 + \frac{1}{1+1}}} = 1.6.$$

The golden ratio has been known since ancient times fascinating mathematicians, architects and artists in equal measure. Many claim to have found manifestations of the golden ratio in nature as well. We shall only consider here a remarkable geometrical object called the golden triangle.

Consider an isosceles triangle $\triangle ABC$, see Fig. 1.16, with its sides $a = AC$ and $b = AB = BC$. Let its angles by the bottom side be $\angle BAC = \angle BCA = 2\alpha$. The sides of this triangle must be related to each other such that $b/a = \phi$. As we shall see in a minute, this is indeed the case. Hence, when one draws the line AD that bisects the angle $\angle BAC$, the obtained smaller triangle $\triangle ADC$ is also an isosceles triangle with $AD = AC = a$.

Indeed, since the smaller triangle is an isosceles one by construction, $\angle ACD = \angle CDA = 2\alpha$. Since the sum of all angles of the small triangle is π, we have $5\alpha = \pi$ or $\alpha = \pi/5 = 36°$. Correspondingly, $\angle ABC = \alpha$, so that the sum of all angles of the original (bigger) triangle would also be $5\alpha = \pi$. As a consequence of this, $\triangle ADB$ is also an isosceles triangle with $AD = BD = a$.

Once we know the angles, we can look at the ratio of the sides of these triangles. We shall start from the bigger triangle. Let the ratio of its sides, b/a, be x. Of course, $x > 0$. Then, consider the triangle $\triangle ADB$ with the perpendicular FD cutting the opposite side AB in half. From the right triangle $\triangle AFD$ we have $\cos(\alpha) = AF/AD = (b/2)/a = x/2$. Similarly, we can drop the perpendicular BE onto the bottom side of the original triangle cutting AC in half. Since $\triangle AEB$ is the right triangle, $\cos(2\alpha) = AE/AB = 1/2x$.

Fig. 1.16 Golden triangle $\triangle ABC$. The line BE is perpendicular to the bottom side of the large triangle and cuts it in half. Similarly, FD is perpendicular to AB cutting it in half

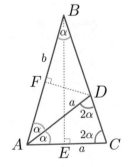

The cosine of the double angle can be related to that of the single one via Eq. (1.76). Therefore, we obtain the following algebraic equation for the ratio x:

$$\frac{1}{2x} = 2\left(\frac{x}{2}\right)^2 - 1 \implies x^3 - 2x - 1 = 0.$$

This cubic equation can easily be solved by splitting $2x$ as $x + x$ and then factorising:

$$x^3 - x - (x + 1) = x\left(x^2 - 1\right) - (x + 1)$$

$$= x(x - 1)(x + 1) - (x + 1) = (x + 1)\left(x^2 - x - 1\right) = 0.$$

The first root, $x = -1$, is negative and has to be rejected. The other two roots are given by the golden ratio Eq. (1.77) from which we can only accept the positive root $x = \phi$ as the other one, ψ, is negative. Hence, this remarkable triangle must have the ratio of its long side to the short one equal exactly to the golden ratio.

The smaller triangle has the same angles as the bigger one and hence is similar, which means that $AD/DC = \phi$ as well. This can also be seen explicitly. Indeed, $DC = b - BD = b - a$, so that

$$\frac{AC}{DC} = \frac{a}{b - a} = \frac{1}{\phi - 1},$$

which is ϕ according to Eq. (1.77) (prove this!).

The golden triangle enables one to calculate the cosine and sine of $36°$ and $72°$.

Problem 1.101 Show that

$$\sin\left(36°\right) = \sqrt{\frac{5 - \sqrt{5}}{8}} = \sqrt{\frac{2\phi - 1}{4\phi}},$$

$$\cos\left(36°\right) = \sqrt{\frac{3 + \sqrt{5}}{8}} = \sqrt{\frac{2\phi + 1}{4\phi}},$$

$$\sin\left(72°\right) = \sqrt{\frac{5 + \sqrt{5}}{8}} = \sqrt{\frac{4\phi + 3}{4(\phi + 1)}},$$

$$\cos\left(72°\right) = \frac{1}{1 + \sqrt{5}} = \frac{1}{2\phi}.$$

The golden ratio is closely related to a special sequence of integer numbers called Fibonacci[11] sequence. This sequence of numbers, usually denoted as F_k (where $k = 0, 1, 2, 3, \ldots$ are integer indices), starts from $F_0 = 0$ and $F_1 = 1$, and each consecutive number is constructed as a sum of two preceding ones: $F_2 = F_1 + F_0 = 1$, $F_3 = F_2 + F_1 = 2$ and so on. Generally, Fibonacci numbers satisfy:

$$F_k = F_{k-1} + F_{k-2}, \quad k = 2, 3, 4, \ldots . \tag{1.78}$$

Below are the first 13 numbers:

$$0, 1, 1, 2, 3, 5, 8, 13, 21, 34, 55, 89, 144, \ldots .$$

The Fibonacci numbers were originally defined for positive (and zero) indices. However, they can formally be extended to negative indices as well (these are called negaFibonacci numbers) by continuing their definition (1.78) into negative indices. This can formally be done by rewriting this definition as $F_{k-2} = F_k - F_{k-1}$, expressing the number F_{k-2} via the two that follow in the direction of increasing indices. Then, $F_{-1} = F_1 - F_0 = 1$, $F_{-2} = F_0 - F_{-1} = -1$, $F_{-3} = F_{-1} - F_{-2} = 2$ and so on. It can be noticed that $F_{-1} = F_1$, $F_{-2} = -F_2$, $F_{-3} = F_3$. In fact, a general statement is true.

> **Problem 1.102** Prove using induction that generally $F_{-k} = (-1)^{k+1} F_k$ for any $k \geq 0$.

There exists a closed formula for a general Fibonacci number, F_k, called Binet's formula[12]:

$$F_k = \frac{\phi^k - \psi^k}{\phi - \psi} . \tag{1.79}$$

Indeed, because ϕ and ψ both satisfy their generating quadratic equation (1.77), we immediately can write that their powers satisfy the Fibonacci rule as well:

$$\phi^k = \phi^{k-1} + \phi^{k-2} \quad \text{and} \quad \psi^k = \psi^{k-1} + \psi^{k-2} .$$

Correspondingly, their arbitrary linear combination $\mathcal{F}_k = \alpha \phi^k + \beta \psi^k$ would satisfy the Fibonacci rule:

$$\mathcal{F}_k = \alpha \left(\phi^{k-1} + \phi^{k-2} \right) + \beta \left(\psi^{k-1} + \psi^{k-2} \right)$$

$$= \left(\alpha \phi^{k-1} + \beta \psi^{k-1} \right) + \left(\alpha \phi^{k-2} + \beta \psi^{k-2} \right) = \mathcal{F}_{k-1} + \mathcal{F}_{k-2} .$$

---

[11] Named after Leonardo of Pisa, also known as Fibonacci.
[12] Named after Jacques Philippe Marie Binet.

Therefore, the numbers $\mathcal{F}_k$ would form the Fibonacci sequence F_k if they start from $\mathcal{F}_0 = 0$ and $\mathcal{F}_1 = 1$. Solving equations

$$\begin{cases} 0 = \ \alpha + \beta \\ 1 = \alpha\phi + \beta\psi \end{cases}$$

gives $\alpha = -\beta = 1/(\phi - \psi)$. Using these values in $F_k = \alpha\phi^k + \beta\psi^k$, we obtain Eq. (1.79).

Problem 1.103 Consider a generalisation of the integer Fibonacci sequence that is generated by the recurrence relation $F_k = a F_{k-1} + b F_{k-2}$, where a and b are some integers, and $F_0 = 0$ and $F_1 = 1$ as in the Fibonacci sequence. Using the method outlined above, show that the same Binet formula (1.79) is valid for F_k, in which ϕ and ψ are two roots of the quadratic equation $x^2 = ax + b$. Let us assume for simplicity that $b + a^2/4 > 0$ which guarantees that the two roots of this quadratic equations are real and different.

Problem 1.104 Using the results of the previous problem, show that the sequence $G_k = 2G_{k-1} + G_{k-2}$ with $G_0 = 0$ and $G_1 = 1$ (these are: $0, 1, 2, 5, 12, 29, 70, 169$, etc.) can be generated by Binet's formula with $\phi = 1 + \sqrt{2}$ and $\psi = 1 - \sqrt{2}$.

1.8   Essential Smooth 2D Curves

Here we shall describe several essential 2D curves and the corresponding equations $y = y(x)$ for them. Some other famous curves will be considered in Sect. 1.19.3.

We shall start by considering equations for a general circle and of an ellipse. Coordinates (x, y) of points on a circle of radius R centred in the $x - y$-plane at the point (x_0, y_0) can be found in the following way, see Fig. 1.17. Let α be the angle the line PA connecting the centre of the circle and the point $A(x, y)$ on it makes with the dashed horizontal axis. Obviously,

$$x - x_0 = R \cos\alpha \quad \text{and} \quad y - y_0 = R \sin\alpha .$$

Therefore,

$$\left(\frac{x - x_0}{R}\right)^2 + \left(\frac{y - y_0}{R}\right)^2 = \cos^2\alpha + \sin^2\alpha = 1 ,$$

so that the points (x, y) lying on the circle satisfy the equation:

$$(x - x_0)^2 + (y - y_0)^2 = R^2 . \tag{1.80}$$

Fig. 1.17 A circle of radius
R centred at the point
$P\,(x_0,\,y_0)$

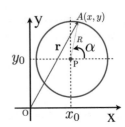

Fig. 1.18 An ellipse with the
centre of symmetry at the
centre of the coordinate
system

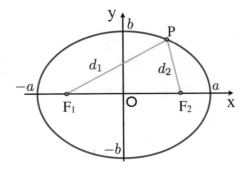

The circle of radius R centred at the origin is given by $x^2 + y^2 = R^2$. In this equation, both coordinates enter equally.

The simplest generalisation of the circle is an *ellipse*. It is obtained by stretching (squeezing) the circle along either x- or y-direction and the points lying on it satisfy the following equation:

$$\left(\frac{x}{a}\right)^2 + \left(\frac{y}{b}\right)^2 = 1 , \tag{1.81}$$

where a and b are two positive constants and $(x,\,y)$ are coordinates of arbitrary points on the curve. This ellipse has the centre of the 2D coordinates system O as its symmetry point, see Fig. 1.18. There are two axes: the one drawn between points with coordinates $(-a, 0)$ and $(a, 0)$ along the x-axis and the other between points $(0, -b)$ and $(0, b)$ along the y-axis. The longer of the two axes is called the *major axis* of the ellipse, while the smaller one the *minor*. If $a = b$, the ellipse coincides with a circle of radius $R = a = b$. The *vertex* points $(\pm a, 0)$ and $(0, \pm b)$ correspond to the points on the curve of the ellipse which have, respectively, the largest and the smallest distance from the centre (specifically for the ellipse drawn in Fig. 1.18). The distance to the centre O from any other point on the curve of the ellipse lies between b and a values ($a > b$ for the ellipse in the figure). One also defines *eccentricity* of an ellipse,

$$e = \sqrt{1 - \left(\frac{b}{a}\right)^2} , \tag{1.82}$$

where it is assumed that b and a are halves of the minor and major axes, respectively ($a > b$).

An ellipse has two special points called *focal points* (or *foci*) shown in the figure as $F_1(-f, 0)$ and $F_2(f, 0)$, where $f = \sqrt{a^2 - b^2}$. A peculiar property of the ellipse is that the sum of distances $F_1 P + F_2 P = d_1 + d_2$ from the focal points to any point P on the ellipse is a constant quantity equal to $2a$ (that value can be obtained immediately by taking the point P to be one of the vertices on the major axis in which case $d_1 + d_2$ becomes clearly equal to the full length of the major axis). Let us show that points that satisfy this condition do indeed form an ellipse. We have:

$$d_1 = \sqrt{(x + f)^2 + y^2}$$

is the distance between points $P(x, y)$ and $F_1(-f, 0)$, while

$$d_2 = \sqrt{(x - f)^2 + y^2}$$

is the distance between points $P(x, y)$ and $F_2(f, 0)$. Hence, the equation $d_1 + d_2 = 2a$ becomes

$$\sqrt{(x + f)^2 + y^2} + \sqrt{(x - f)^2 + y^2} = 2a .$$

Let us take the second root in the left-hand side to the right-hand side,

$$\sqrt{(x + f)^2 + y^2} = 2a - \sqrt{(x - f)^2 + y^2} ,$$

and then square both sides of the equation and simplify:

$$(x + f)^2 + y^2 = 4a^2 - 4a\sqrt{(x - f)^2 + y^2} + (x - f)^2 + y^2$$

$$\Rightarrow \quad a^2 - fx = a\sqrt{(x - f)^2 + y^2} .$$

Taking the square of both sides again and further simplifying, we obtain that x and y do indeed satisfy Eq. (1.81) of the ellipse.

Problem 1.105 Show that equation of the ellipse whose centre is located at the point $O(x_0, y_0)$ is

$$\left(\frac{x - x_0}{a}\right)^2 + \left(\frac{y - y_0}{b}\right)^2 = 1 . \tag{1.83}$$

Problem 1.106 The ellipse of Eq. (1.81) is rotated by $90°$ about the origin. Write its equation.

Problem 1.107 In this problem, we assume that the equation of the ellipse (1.81) is given and will find the foci. Let the foci have coordinates $(\pm f, 0)$. Using the equation of the ellipse, show that for any point (x, y) on the ellipse the distance $D = d_1 + d_2 = 2a$ is constant if and only if $f^2 = a^2 - b^2$. Here the proof consists of two parts: (i) assume that $D = 2a$ for any point on the ellipse, and then find f; (ii) given $f = \sqrt{a^2 - b^2}$, show that the value of D is constant and equal to $2a$. [Hint: *in the first case write the condition* $D = d_1 + d_2 = 2a$ *explicitly, replace* y *with a proper expression with* x *using Eq. (1.81); then perform a sequence of squares similar to the method used above; in the second case, construct* $d_1 + d_2$ *with the given value of* f *and with* y *expressed via* x *as before, and then use identity* (1.14).]

Another smooth curve which we shall consider is called hyperbola. If in the case of an ellipse the sum of distances from each point on the ellipse to the two focal points remains constant, the hyperbola is defined such that an absolute value of the difference of such distances remains constant, a somewhat similar idea. Consider a derivation of the equation of the canonical hyperbola that is symmetric with respect to both x- and y-axes. It consists of two separate curves and is shown in Fig. 1.19 (in blue). The focal points $F_1(-c, 0)$ and $F_2(c, 0)$ are arranged symmetrically on the x-axis with respect to the centre point O; this point is at the distance a from both hyperbolas which are run at $x \geq a$ (the right hyperbola) and $x \leq -a$ (the left one).

If $P(x, y)$ is a general point on either of the two curves, then, according to the definition, we can write:

$$|d_1 - d_2| = 2a .$$

The value of $2a$ in the right-hand side can easily be obtained by taking the point P as any of the two closest points to the centre, either $(a, 0)$ or $(-a, 0)$. Since $d_1 = \sqrt{(x + c)^2 + y^2}$ and $d_2 = \sqrt{(x - c)^2 + y^2}$, we can also rewrite the above equation

Fig. 1.19 Hyperbola. F_1 and F_2 are two focal points, O is the centre, and the dashed red lines are its two asymptotes

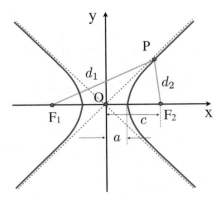

as

$$\left| \sqrt{(x+c)^2 + y^2} - \sqrt{(x-c)^2 + y^2} \right| = 2a \ .$$

Squaring both sides once, we get:

$$(x+c)^2 + y^2 + (x-c)^2 + y^2 - 2\sqrt{[(x+c)^2 + y^2][(x-c)^2 + y^2]} = 4a^2$$

$$\implies \left(x^2 + y^2 + c^2\right) - \sqrt{[(x+c)^2 + y^2][(x-c)^2 + y^2]} = 2a^2$$

$$\implies \left(x^2 + y^2 + c^2\right) - 2a^2 = \sqrt{\left(x^2 + y^2 + c^2 + 2xc\right)\left(x^2 + y^2 + c^2 - 2xc\right)}$$

$$\implies \left(x^2 + y^2 + c^2\right) - 2a^2 = \sqrt{\left(x^2 + y^2 + c^2\right)^2 - 4x^2c^2} \ .$$

Taking square of both sides again, we obtain:

$$-a^2 \left(x^2 + y^2 + c^2\right) + a^4 = -x^2 c^2$$

$$\implies \left(c^2 - a^2\right)x^2 - a^2 y^2 = a^2 \left(c^2 - a^2\right) \ .$$

Introducing $b^2 = c^2 - a^2 > 0$ and dividing both sides by $a^2 b^2$, we finally get:

$$\left(\frac{x}{a}\right)^2 - \left(\frac{y}{b}\right)^2 = 1 \ , \tag{1.84}$$

which is the required equation. As one can see, it differs from the equation for the ellipse (1.81) by the minus sign in the left-hand side. The curve for $x \geq a$ is obtained by solving the equation with respect to x and taking the plus sign, $x = \frac{a}{b}\sqrt{b^2 + y^2}$, while the left curve in Fig. 1.19 corresponding to negative x values, $x \leq -a$, reads $x = -\frac{a}{b}\sqrt{b^2 + y^2}$.

The shape of a hyperbola is characterised by its eccentricity

$$e = \sqrt{1 + \frac{b^2}{a^2}} \ . \tag{1.85}$$

It is possible to introduce two straight lines that contain both hyperbola curves between them; moreover, the hyperbolas approach these lines closer and closer as x approaches $\pm\infty$. These lines are called asymptotes of the hyperbola; we shall consider the notion of asymptotes more closely (and rigorously) later on in Sect. 3.11, so our consideration here will mostly be intuitive. Let us find the straight line $y = \frac{b}{a}x$

and show that for all $x \geq a$ it lies above the hyperbola $y = \frac{b}{a}\sqrt{x^2 - a^2}$. Indeed, their difference

$$\Delta y(x) = \frac{b}{a}x - \frac{b}{a}\sqrt{x^2 - a^2} = \frac{b}{a}x\left(1 - \sqrt{1 - \frac{a^2}{x^2}}\right)$$

is positive as the expression under the square root is less than one. Also as x becomes larger and larger, a^2/x^2 becomes smaller and smaller, and hence the square root approaches the value of 1 for very large x, leading the expression in the brackets to tend to zero. Similarly, the same asymptote runs below the bottom left part of the hyperbola, while the asymptote $y = -\frac{b}{a}x$ runs above the left upper part and below the right bottom parts of the hyperbola.

Problem 1.108 Consider two points $x_2 > x_1 \geq a$ on the hyperbola $y = \frac{b}{a}\sqrt{x^2 - a^2}$. Prove that $\Delta y(x_2) < \Delta y(x_1)$, i.e. as x increases, the hyperbola approaches closer to the asymptote $y = \frac{b}{a}x$. [Hint: *assume the opposite, i.e. that $\Delta y(x_2) \geq \Delta y(x_1)$; rearrange it such that both sides were positive and square; rearrange and then square again; you should arrive at an inequality that is definitely wrong; we arrive at a contradiction proving that the opposite inequality sign must be used.*]

1.9   Simple Determinants

In many instances, it appears to be extremely convenient to use special objects called *determinants*.

They are properly defined in linear algebra using matrices in Vol. II; however, it is quite useful for us to give a formal definition of the 2×2 and 3×3 determinants already here; we shall do it quite formally without referring to matrices, so that determinants can be used in this and the following chapters as a convenient notation.

A 2×2 determinant is a scalar (a number) composed from four numbers a_{11}, a_{12}, a_{21} and a_{22} (which are distinguished by using unique double indices for convenience) written on a 2×2 grid between two vertical lines and its value is calculated as follows:

$$\begin{vmatrix} a_{11} & a_{12} \\ a_{21} & a_{22} \end{vmatrix} = a_{11}a_{22} - a_{12}a_{21} . \tag{1.86}$$

A 3×3 determinant is composed of nine numbers arranged in three rows and three columns, and it is calculated as shown below:

$$\begin{vmatrix} a_{11} & a_{12} & a_{13} \\ a_{21} & a_{22} & a_{23} \\ a_{31} & a_{32} & a_{33} \end{vmatrix} = a_{11}\begin{vmatrix} a_{22} & a_{23} \\ a_{32} & a_{33} \end{vmatrix} - a_{12}\begin{vmatrix} a_{21} & a_{23} \\ a_{31} & a_{33} \end{vmatrix} + a_{13}\begin{vmatrix} a_{21} & a_{22} \\ a_{31} & a_{32} \end{vmatrix}$$

$$= a_{11}(a_{22}a_{33} - a_{23}a_{32}) - a_{12}(a_{21}a_{33} - a_{23}a_{31}) + a_{13}(a_{21}a_{32} - a_{22}a_{31}) \ .$$
$$\tag{1.87}$$

It is seen that the 3×3 determinant is expressed via three 2×2 determinants as shown. Note how the elements of the determinants are denoted: they have a double index, e.g. a_{13} has the left index 1 and the right index 3 in the 3×3 determinant. The left index indicates the row in which the given element is located, while the right index indicates the column. Indeed, the element a_{13} can be found at the end of the first row in the 3×3 determinant. We note that these notations come from matrices.

When defining the value of the determinant in (1.87), we "expanded" it along the first row as the elements of the first row (a_{11}, a_{12} and a_{13}) appear before the 2×2 determinants in the expansion. Notice that these 2×2 determinants can be obtained by removing all elements of the row and column which intersect at the particular element which appears as a pre-factor to them. Indeed, the determinant $\begin{vmatrix} a_{22} & a_{23} \\ a_{32} & a_{33} \end{vmatrix}$ is obtained by crossing out the first row and the first column in the original 3×3 determinant as these intersect exactly at the a_{11} element. The same is true for the other two as well as can easily be checked. It can be shown that the determinant can be expanded in a similar manner with respect to any row or any column although care is needed in choosing appropriate signs for each term.

An important property of the determinants that we shall need in the following is that a determinant is equal to zero if any one of its rows (or columns) is a linear combination of the other rows (columns). Let us consider in more detail what this means. Starting from the 2×2 case, if the second row is linearly dependent on the first, that means that the elements of the second row are related to the corresponding elements of the first row via the same scaling factor λ, i.e. $a_{21} = \lambda a_{11}$ and $a_{22} = \lambda a_{12}$. In this case, using the definition (1.86) of the determinant, we have:

$$\begin{vmatrix} a_{11} & a_{12} \\ a_{21} & a_{22} \end{vmatrix} = \begin{vmatrix} a_{11} & a_{12} \\ \lambda a_{11} & \lambda a_{12} \end{vmatrix} = \lambda a_{11} a_{12} - \lambda a_{12} a_{11} = 0 \ ,$$

as required. In the 3×3 case, the linear dependence is defined as follows: each element of, say, the first row is expressed via the corresponding elements of the other two rows using the same linear factors λ and μ:

$$a_{11} = \lambda a_{21} + \mu a_{31} , \quad a_{12} = \lambda a_{22} + \mu a_{32} \quad \text{and} \quad a_{13} = \lambda a_{23} + \mu a_{33} \ . \tag{1.88}$$

You can see that elements of the first row are indeed expressed as a linear combination of the elements from the other two rows taken from *the same column*. [Jumping a little bit ahead, what we have just formulated can also be recast in a simple form if each row is associated with a vector; then the linear combination expressed above would simply mean that the vector given by the first row is a linear combination of the vectors (with the coefficients λ and μ) corresponding to the second and the third rows.] Now, let us see explicitly that if we have such a linear combination (1.88), then the determinant is equal to zero. This can be easily established from the definition of the determinant.

Problem 1.109 Show using the definition of the 3×3 determinant that if the elements of the first row are related to those of the second and the third rows as given in Eq. (1.88), then the determinant is zero:

$$\begin{vmatrix} a_{11} \ a_{12} \ a_{13} \\ a_{21} \ a_{22} \ a_{23} \\ a_{31} \ a_{32} \ a_{33} \end{vmatrix} = \begin{vmatrix} \lambda a_{21} + \mu a_{31} & \lambda a_{22} + \mu a_{32} & \lambda a_{23} + \mu a_{33} \\ a_{21} & a_{22} & a_{23} \\ a_{31} & a_{32} & a_{33} \end{vmatrix} = 0$$

for any values of λ and μ.

The other important property which we shall also need rather soon is that the determinant changes sign if two rows (or columns) are swapped around (permuted). Again, this is most easily checked by a direct calculation: use definition (1.87) where say elements a_{1i} of the first row are replaced by the elements a_{2i} of the second, and *vice versa*:

$$\begin{vmatrix} a_{21} \ a_{22} \ a_{23} \\ a_{11} \ a_{12} \ a_{13} \\ a_{31} \ a_{32} \ a_{33} \end{vmatrix} = a_{21} \left(a_{12} a_{33} - a_{13} a_{32} \right) - a_{22} \left(a_{11} a_{33} - a_{13} a_{31} \right) + a_{23} \left(a_{11} a_{32} - a_{12} a_{31} \right) .$$

Rearranging the terms in the right-hand side, we obtain:

$$\begin{vmatrix} a_{21} \ a_{22} \ a_{23} \\ a_{11} \ a_{12} \ a_{13} \\ a_{31} \ a_{32} \ a_{33} \end{vmatrix} = -a_{11} \left(a_{22} a_{33} - a_{23} a_{32} \right) + a_{12} \left(a_{21} a_{33} - a_{23} a_{31} \right)$$

$$-a_{13} \left(a_{21} a_{32} - a_{22} a_{31} \right) = - \begin{vmatrix} a_{11} \ a_{12} \ a_{13} \\ a_{21} \ a_{22} \ a_{23} \\ a_{31} \ a_{32} \ a_{33} \end{vmatrix} ,$$

i.e. the minus of the original determinant. The same happens if any two rows or columns are interchanged. If two interchanges take place, the determinant does not change its sign.

Problem 1.110 Show that

$$|A| = \begin{vmatrix} 1 \ 0 \ 2 \\ 3 \ -1 \ 0 \\ 0 \ 5 \ 1 \end{vmatrix} = 29 , \quad |B| = \begin{vmatrix} 1 \ 1 \ 0 \\ 0 \ 2 \ 1 \\ 3 \ -1 \ 0 \end{vmatrix} = 4 .$$

Problem 1.111 Prove that if a determinant has two identical rows (or columns), it is equal to zero.

Problem 1.112 Find all values of x satisfying the following equations:

$$(a) \quad \begin{vmatrix} x & 0 & 0 \\ 0 & 1 & 2 \\ 1 & x & 0 \end{vmatrix} + 2 \begin{vmatrix} 0 & x^2 & 0 \\ 1 & 1 & 1 \\ 0 & 0 & -2x \end{vmatrix} = 0 ;$$

$$(b) \quad \begin{vmatrix} x & 0 & 1 \\ \alpha & x & 0 \\ 2 & 0 & x \end{vmatrix} = 0 .$$

[Answer: (a) $x = 0$, $1/2$; (b) 0, $\pm\sqrt{2}$.]

1.10 Vectors

1.10.1 Three-Dimensional Space

As the 2D space is a particular case of the 3D space, it is sufficient to discuss the 3D space, and this is what we shall do here.

It is convenient to define objects, called *vectors*, belonging to the space denoted $\mathbb{R}^3$, such that they can undergo summation and subtraction operations resulting in the same type of objects (i.e. vectors as well) belonging to the same space $\mathbb{R}^3$. Vectors are defined by two points in space, A and B, and a direction between them, and are sometimes denoted by a directed arrow above the letters AB as $\overrightarrow{AB}$. This notation we shall be rarely using; however, instead, another notation will be frequently employed whereby a vector is shown by a bold letter, e.g. $\mathbf{a}$. If Cartesian coordinates of points A and B are (x_A, y_A, z_A) and (x_B, y_B, z_B), respectively, and the vector $\mathbf{a} = \overrightarrow{AB}$ is directed from A to B, then it is said that it has Cartesian coordinates $x_B - x_A, y_B - y_A$ and $z_B - z_A$. Then the vector is denoted simply as $(x_B - x_A, y_B - y_A, z_B - z_A)$ with all its coordinates listed inside the round brackets. It is seen that a vector is specified by three real numbers (its coordinates), hence the notation $\mathbb{R}^3$ for the space. The length of the vector (or its magnitude) is the distance between two points, A and B, and it is denoted $\left|\overrightarrow{AB}\right|$ or $|\mathbf{a}|$. Often, the length of a vector is written using the same non-bold letter, e.g. $|\mathbf{a}| = a$. Two vectors $\mathbf{a}$ and $\mathbf{b}$ are said to be identical, $\mathbf{a} = \mathbf{b}$, if they have the same length and direction. This means that a vector can be translated in space to any position and remain the same. In particular, any vector can be translated to the position in which its starting point is at the centre of the coordinate system (that has zero coordinates) as shown in Fig. 1.1c. In this case, the ending point of the vector P has coordinates $x_P - x_O = x$, $y_P - y_O = y$ and $z_P - z_O = z$ of the point P. It is easy to see from the same figure that the length of the vector,

$$\left|\overrightarrow{OP}\right| = |\mathbf{a}| = \sqrt{x^2 + y^2 + z^2} = \sqrt{(x_P - x_O)^2 + (y_P - y_O)^2 + (z_P - z_O)^2},$$
$$(1.89)$$

corresponds to a diagonal of the cuboid with sides x, y and z. The formula above is obtained by using the Pythagoras theorem twice: first, we find the diagonal $OB = \sqrt{x^2 + y^2}$ of the right triangle $\triangle OAB$, then we obtain $OP = \sqrt{OB^2 + z^2} = \sqrt{x^2 + y^2 + z^2}$ as the diagonal of the other right triangle $\triangle OPB$, see Fig. 1.1c (also compare with Problem 1.89).

A vector of unit length is called *unit vector*. A null vector, $\mathbf{0}$, has all its coordinates equal to zero and hence is of the zero length. As was mentioned, formula (1.89) also naturally gives the distance OP between points O and P (cf. Eq. (1.62)).

Two vectors are *collinear* if they have the same or opposite directions; in other words, if translated such that they begin at the centre of the coordinate system, they lie on the same line. Two vectors specified explicitly via their start and end points are *coplanar*, if there is a plane such that both vectors lie in that plane (no translation of the vectors is to be applied).

Now we are in a position to define operations on vectors. A vector $\mathbf{a}$, if multiplied by a scalar (a real number) c, results in another vector $\mathbf{b} = c\mathbf{a}$ which is collinear with $\mathbf{a}$ and has the length $|\mathbf{b}| = c|\mathbf{a}|$. Obviously, coordinates of $\mathbf{b}$ are obtained by multiplying the coordinates of $\mathbf{a}$ by the same factor c. The vector $\mathbf{b}$ has the same direction as $\mathbf{a}$ if $c > 0$, and opposite to that of $\mathbf{a}$ if $c < 0$, and is the null vector if $c = 0$. Dividing a vector $\mathbf{a}$ by its length $|\mathbf{a}|$ results in a unit vector (of unit length): $\mathbf{e} = \mathbf{a}/|\mathbf{a}|$ which has the same direction as $\mathbf{a}$.

The sum of two vectors $\mathbf{a}$ and $\mathbf{b}$ is defined as follows: translate the vector $\mathbf{a}$ into the vector $\overrightarrow{AB}$, vector $\mathbf{b}$ into $\overrightarrow{BC}$, then the sum $\mathbf{c} = \mathbf{a} + \mathbf{b}$ is defined as the vector $\overrightarrow{AC}$, see Fig. 1.20a. It is easy to see using the geometrical representation of the vectors in the 3D Cartesian system that Cartesian coordinates of the vector $\mathbf{c}$ are obtained by summing up the corresponding coordinates of $\mathbf{a}$ and $\mathbf{b}$. Similarly, one defines a difference $\mathbf{b} = \mathbf{c} - \mathbf{a}$ of two vectors $\mathbf{c}$ and $\mathbf{a}$ as a vector $\mathbf{b}$ for which $\mathbf{c} = \mathbf{a} + \mathbf{b}$, also see Fig. 1.20a. Correspondingly, one can also define a general linear operation on

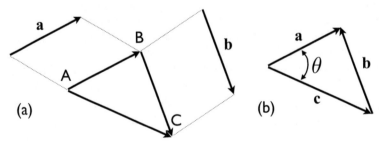

Fig. 1.20 Sum of two vectors $\mathbf{a}$ and $\mathbf{b}$

two or any number n of vectors:

$$\mathbf{b} = c_1\mathbf{a}_1 + c_2\mathbf{a}_2 + \cdots + c_n\mathbf{a}_n ,$$

where c_1, c_2, etc. are real numbers. Note that any such *linear combination* of vectors results in a vector in the same space $\mathbb{R}^3$. The summation operation of vectors is both commutative and associative, similar to real numbers; this is because by summing vectors we actually perform summation of their coordinates which are real numbers, and for them this operation is both commutative and associative.

A *scalar* or *dot product* of two vectors $\mathbf{a} = (a_1, a_2, a_3)$ and $\mathbf{b} = (b_1, b_2, b_3)$ is defined as a scalar (a real number)

$$\mathbf{a} \cdot \mathbf{b} = (\mathbf{a}, \mathbf{b}) = a_1 b_1 + a_2 b_2 + a_3 b_3 . \qquad (1.90)$$

Both notations for the dot product (as $\mathbf{a} \cdot \mathbf{b}$ or $(\mathbf{a}, \mathbf{b})$) will be frequently used. The dot product of the vector with itself gives square of its length: $(\mathbf{a}, \mathbf{a}) = \mathbf{a}^2 = a^2$. Hence, the length of a vector $\mathbf{a}$ can be expressed via its dot product with itself:

$$|\mathbf{a}| = \sqrt{\mathbf{a} \cdot \mathbf{a}} = \sqrt{(\mathbf{a}, \mathbf{a})} = \sqrt{a_1^2 + a_2^2 + a_3^2} ,$$

which directly follows from the definitions of the two.

The dot product is *commutative*, $(\mathbf{a}, \mathbf{b}) = (\mathbf{b}, \mathbf{a})$, and *distributive*, $(\mathbf{c}, \mathbf{a} + \mathbf{b}) = (\mathbf{c}, \mathbf{a}) + (\mathbf{c}, \mathbf{b})$, which can easily be checked from their definitions, e.g. for the latter:

$$\begin{aligned}
(\mathbf{c}, \mathbf{a} + \mathbf{b}) &= c_1(a_1 + b_1) + c_2(a_2 + b_2) + c_3(a_3 + b_3) \\
&= (c_1 a_1 + c_2 a_2 + c_3 a_3) + (c_1 b_1 + c_2 b_2 + c_3 b_3) \\
&= (\mathbf{c}, \mathbf{a}) + (\mathbf{c}, \mathbf{b}) .
\end{aligned}$$

Two vectors $\mathbf{a}$ and $\mathbf{b}$ are said to be *orthogonal,* if their dot product is zero. It is easy to see that this makes perfect sense since the orthogonal vectors make an angle of $90°$ with each other. Indeed, consider two vectors $\mathbf{a}$ and $\mathbf{c}$ with the angle θ between them, Fig. 1.20b. Using distributivity and commutativity of the dot product, we can write:

$$\begin{aligned}
(\mathbf{a} - \mathbf{c})^2 &= (\mathbf{a} - \mathbf{c}, \mathbf{a} - \mathbf{c}) \\
&= (\mathbf{a}, \mathbf{a}) - 2(\mathbf{a}, \mathbf{c}) + (\mathbf{c}, \mathbf{c}) \\
&= |\mathbf{a}|^2 - 2(\mathbf{a}, \mathbf{c}) + |\mathbf{c}|^2 \\
&= a^2 + c^2 - 2(\mathbf{a}, \mathbf{c}) ,
\end{aligned}$$

where $a = |\mathbf{a}|$ and $c = |\mathbf{c}|$. On the other hand, $\mathbf{a} - \mathbf{c} = \mathbf{b}$, and $|\mathbf{a} - \mathbf{c}| = |\mathbf{b}| = b$ is the length of the side of the triangle in Fig. 1.20b which is opposite to the angle θ. Therefore, from the well-known cosine theorem of Eq. (1.70),

$$b^2 = |\mathbf{a} - \mathbf{c}|^2 = a^2 + c^2 - 2ac \cos\theta .$$

Comparing this with our previous result, we obtain another definition of the dot product in the 3D space as

$$(\mathbf{a}, \mathbf{c}) = |\mathbf{a}| \, |\mathbf{c}| \cos\theta \; . \tag{1.91}$$

Therefore, if the dot product of two vectors of non-zero lengths is equal to zero, the two vectors make the right angle with each other, i.e. they are orthogonal.

For instance, if $\mathbf{u} = (0, 1, 3)$, $\mathbf{v} = (-1, 1, 2)$, then $\mathbf{u} + \mathbf{v} = (-1, 2, 5)$, $\mathbf{u} - \mathbf{v} = (1, 0, 1)$, and

$$(\mathbf{u}, \mathbf{v}) = 0 \cdot (-1) + 1 \cdot 1 + 3 \cdot 2 = 7 \; .$$

It is convenient to introduce three special unit vectors, frequently called *unit base vectors* of the Cartesian system,

$$\mathbf{i} = (1, 0, 0) \; , \quad \mathbf{j} = (0, 1, 0) \; , \quad \mathbf{k} = (0, 0, 1) \; , \tag{1.92}$$

which run along the Cartesian x, y and z axes. These unit vectors are orthogonal, $(\mathbf{i}, \mathbf{j}) = (\mathbf{i}, \mathbf{k}) = (\mathbf{j}, \mathbf{k}) = 0$, and all have the unit length. These are said to be *orthonormal*. Then, an arbitrary vector $\mathbf{a} = (a_1, a_2, a_3)$ can always be written as a linear combination of these unit base vectors:

$$\mathbf{a} = a_1(1, 0, 0) + a_2(0, 1, 0) + a_3(0, 0, 1) = a_1\mathbf{i} + a_2\mathbf{j} + a_3\mathbf{k} \; . \tag{1.93}$$

There also exists another useful multiplication operation between two vectors which is called a *vector* (or *cross*) *product*. If $\mathbf{a} = (a_1, a_2, a_3)$ and $\mathbf{b} = (b_1, b_2, b_3)$, then one defines the third vector $\mathbf{c} = \mathbf{a} \times \mathbf{b}$ called their vector product as a vector of length

$$|\mathbf{c}| = |\mathbf{a} \times \mathbf{b}| = |\mathbf{a}| \, |\mathbf{b}| \sin\theta = ab \sin\theta \tag{1.94}$$

and directed as shown in Fig. 1.21. Note that the vector $\mathbf{c}$ is perpendicular to both vectors $\mathbf{a}$ and $\mathbf{b}$.

In order to derive an explicit expression for the components of the vector product, $\mathbf{c} = \mathbf{a} \times \mathbf{b} = (c_1, c_2, c_3)$, let us first make use of the conditions that the vector $\mathbf{c}$ is perpendicular to both $\mathbf{a}$ and $\mathbf{b}$:

$$\mathbf{c} \perp \mathbf{a} \;\; \Rightarrow \;\; (\mathbf{c}, \mathbf{a}) = 0 \;\; \Rightarrow \;\; a_1c_1 + a_2c_2 + a_3c_3 = 0 \; , \tag{1.95}$$

Fig. 1.21 **a** Definition of the vector product of two vectors **a** and **b**. **b** The right-hand rule helps to remember the direction of the cross product **c** of two vectors **a** and **b**

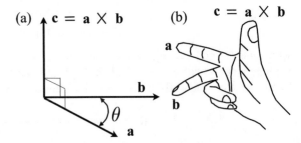

$$\mathbf{c} \perp \mathbf{b} \quad \Rightarrow \quad (\mathbf{c}, \mathbf{b}) = 0 \quad \Rightarrow \quad b_1 c_1 + b_2 c_2 + b_3 c_3 = 0 . \tag{1.96}$$

Multiplying the first equation by b_1 and the second by a_1 and subtracting from each other, one can relate c_2 to c_3:

$$c_2 = \frac{b_1 a_3 - b_3 a_1}{b_2 a_1 - b_1 a_2} c_3 . \tag{1.97}$$

Here we assumed that $b_2 a_1 - a_2 b_1 \neq 0$. Substituting the above expression back into any of the two Eqs. (1.95) or (1.96), we also obtain a formula relating c_1 to c_3:

$$c_1 = \frac{b_3 a_2 - b_2 a_3}{b_2 a_1 - b_1 a_2} c_3 . \tag{1.98}$$

The obtained components of $\mathbf{c} = (c_1, c_2, c_3)$ make it perpendicular to both $\mathbf{a}$ and $\mathbf{b}$ for any value of c_3; choosing the latter fixes the length of the vector $\mathbf{c}$. So, we must choose it in such a way as to comply with the definition Eq. (1.94). Let us then consider the square of the length and make use of the fact that the dot product of $\mathbf{a}$ and $\mathbf{b}$ gives us access to the cosine of the angle θ between the two vectors, Eq. (1.91):

$$\begin{aligned}
c^2 &= a^2 b^2 \sin^2 \theta = a^2 b^2 \left(1 - \cos^2 \theta\right) \\
&= a^2 b^2 \left[1 - \left(\frac{\mathbf{a} \cdot \mathbf{b}}{ab}\right)^2\right] = a^2 b^2 - (\mathbf{a} \cdot \mathbf{b})^2 \\
&= \left(a_1^2 + a_2^2 + a_3^2\right)\left(b_1^2 + b_2^2 + b_3^2\right) - (a_1 b_1 + a_2 b_2 + a_3 b_3)^2 \\
&= (a_2 b_1 - a_1 b_2)^2 + (a_3 b_1 - a_1 b_3)^2 + (a_2 b_3 - a_3 b_2)^2 .
\end{aligned} \tag{1.99}$$

This should be equal to

$$\begin{aligned}
c^2 &= c_1^2 + c_2^2 + c_3^2 = c_3^2 \left[\left(\frac{b_3 a_2 - b_2 a_3}{b_2 a_1 - b_1 a_2}\right)^2 + \left(\frac{b_1 a_3 - b_3 a_1}{b_2 a_1 - b_1 a_2}\right)^2 + 1\right] \\
&= \gamma \left[(a_2 b_1 - a_1 b_2)^2 + (a_3 b_1 - a_1 b_3)^2 + (a_2 b_3 - a_3 b_2)^2\right],
\end{aligned}$$

where

$$\gamma = \left(\frac{c_3}{b_2 a_1 - b_1 a_2}\right)^2$$

and we have made use of Eqs. (1.97) and (1.98) for c_1 and c_2. Comparing (1.99) with the result above, we see that γ must be equal to one, i.e. c_3 must be equal to $\pm (b_2 a_1 - b_1 a_2)$. To fix the sign here, consider a particular case when $\mathbf{a}$ is the unit vector directed along the x-axis, $\mathbf{a} = \mathbf{i} = (1, 0, 0)$, and $\mathbf{b} = \mathbf{j} = (0, 1, 0)$ is the unit vector along the y-axis. According to Fig. 1.21, the vector product of the two should be directed along the z-axis, i.e. c_3 must be positive. Therefore, in

$$c_3 = \pm (b_2 a_1 - b_1 a_2) = \pm (1 \cdot 1 - 0 \cdot 0) = \pm 1$$

the plus sign must be chosen. We then have $c_3 = b_2 a_1 - b_1 a_2$, which, by virtue of Eqs. (1.97) and (1.98), enables us to work out the other two components of $\mathbf{c}$ as well:

$$c_1 = b_3 a_2 - b_2 a_3 \text{ and } c_2 = b_1 a_3 - b_3 a_1 \,.$$

Hence, we can finally explicitly write the vector (cross) product of $\mathbf{a}$ and $\mathbf{b}$ as

$$\mathbf{a} \times \mathbf{b} = (a_2 b_3 - a_3 b_2)\mathbf{i} + (a_3 b_1 - a_1 b_3)\mathbf{j} + (a_1 b_2 - a_2 b_1)\mathbf{k} \,. \qquad (1.100)$$

The vector product of two vectors can most easily be remembered if written via a *determinant* which we introduced in Sect. 1.9; specifically, compare Eqs. (1.87) and (1.100). Hence, one can write:

$$\mathbf{a} \times \mathbf{b} = \begin{vmatrix} \mathbf{i} & \mathbf{j} & \mathbf{k} \\ a_1 & a_2 & a_3 \\ b_1 & b_2 & b_3 \end{vmatrix} \,. \qquad (1.101)$$

For instance, if $\mathbf{u} = (0, 1, 3)$, $\mathbf{v} = (-1, 1, 2)$, then

$$\begin{aligned} \mathbf{u} \times \mathbf{v} &= (1 \cdot 2 - 3 \cdot 1)\mathbf{i} + (3 \cdot (-1) - 0 \cdot 2)\mathbf{j} + (0 \cdot 1 - 1 \cdot (-1))\mathbf{k} \\ &= (-1, -3, 1) \,. \end{aligned}$$

Problem 1.113 Prove that formula (1.100) has perfect sense also in the case of $b_2 a_1 - a_2 b_1 = 0$, i.e. check explicitly that in this case the vector $\mathbf{a} \times \mathbf{b}$ given by Eq. (1.100) is automatically perpendicular to both vectors $\mathbf{a}$ and $\mathbf{b}$, and that its length corresponds to the definition Eq. (1.94). In the same vein, verify that formula (1.100) is also valid in the cases of a_1 and/or b_1 equal to zero (the derivation performed above silently assumed that they are not).

Problem 1.114 If $\mathbf{a} = (1, 0, 1)$ and $\mathbf{b} = (-1, 1, 0)$, calculate: $(\mathbf{a}, \mathbf{b})$ and $\mathbf{a} \times \mathbf{b}$.

Problem 1.115 For $\mathbf{a} = (x, x^2, x^3)$ and $\mathbf{b} = (x^{-1}, x^{-2}, x^{-3})$, calculate: $x^2\mathbf{a}$, $(\mathbf{a}, \mathbf{b})$, $\mathbf{a} \times \mathbf{b}$, as well as $|\mathbf{a}|$ and $|\mathbf{b}|$. [Answers: (x^3, x^4, x^5); 3; $(x^{-1} - x, -x^{-2} + x^2, x^{-1} - x)$; $x\sqrt{1 + x^2 + x^4}$ and $x^{-3}\sqrt{1 + x^2 + x^4}$.]

Problem 1.116 A boat is initially positioned at point A at the bottom bank of a river, Fig. 1.22. The river flows to the right with velocity **v**, while the boat engine can provide velocity **u** (assume that $|\mathbf{u}| > |\mathbf{v}|$). In the first journey, the boat goes directly perpendicular to the river flow to point B at the other bank and then back to the original point A, while in the second journey it goes from A to point C at the same bank and then back. Calculate (a) the direction the boat must take for the first journey to travel perpendicular to the river flow, and (b) the times of travel, $t_\perp$ and $t_\|$, in both cases. [Answer: *(a) the angle between* **u** *and* **v** *is* $\pi/2 + \alpha$, *where* $\sin \alpha = v/u$; *(b)* $t_\perp = 2d/\sqrt{u^2 - v^2}$ *and* $t_\| = 2du/(u^2 - v^2)$.]

Another mathematical expression for the vector product, very useful in practical analytical work, is based on the so-called Levi-Civita symbol ϵ_{ijk} defined as follows: if its indices i, j, k are all different and form a cyclic order (123, 231 or 312), then it is equal to $+1$, if the order is non-cyclic, then it is equal to -1; finally, it is equal to 0 if at least two indices are the same. With this definition, as can easily be checked, the component i of the vector product of **a** and **b** can be written as a double sum over all their components with the Levi-Civita symbol:

$$c_i = [\mathbf{a} \times \mathbf{b}]_i = \sum_{j=1}^{3} \sum_{k=1}^{3} \epsilon_{ijk} a_j b_k = \sum_{j,k=1}^{3} \epsilon_{ijk} a_j b_k . \tag{1.102}$$

In the last passage here a single sum symbol was used to simplify the notations; this is what we shall be frequently doing.

If two vectors are collinear (parallel), then the angle between them $\theta = 0$, hence $\sin \theta = 0$ and thus their vector product is zero. In particular, $\mathbf{a} \times \mathbf{a} = 0$.

It also follows from its definition that the vector product is *anti-commutative*,

$$\mathbf{a} \times \mathbf{b} = -\mathbf{a} \times \mathbf{b}$$

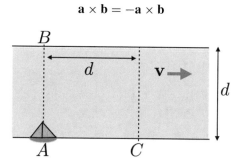

Fig. 1.22 A boat of Problem 1.116 starts at the bottom bank at point A and performs two different return journeys: from A to point B at the opposite bank and back, and between A and C along the same bank. In both cases, the total distance the boat travels is $2d$. The river flows to the right with velocity **v**

(the same length, but opposite direction), and non-associative,

$$(\mathbf{a} \times \mathbf{b}) \times \mathbf{c} \neq \mathbf{a} \times (\mathbf{b} \times \mathbf{c}) \ .$$

However, the vector product is *distributive*:

$$\mathbf{a} \times (\mathbf{b} + \mathbf{c}) = \mathbf{a} \times \mathbf{b} + \mathbf{a} \times \mathbf{c} \ .$$

Considering all vector products between the three unit base vectors (either directly using the definition based on Fig. 1.21 or by means of the coordinate representation Eqs. (1.100) or (1.101)), we can establish the following simple identities:

$$\mathbf{i} \times \mathbf{j} = \mathbf{k}, \ \ \mathbf{i} \times \mathbf{k} = -\mathbf{j} \ \text{and} \ \mathbf{j} \times \mathbf{k} = \mathbf{i} \ .$$

Of course, the vector multiplication of either of the three vectors with themselves gives zero.

Problem 1.117 Using Eq. (1.100), prove that the vector product is anti-commutative.

Problem 1.118 Using Eq. (1.100), prove that the vector product is distributive. [Hint: *the algebra could be somewhat simplified if vector* **a** *is chosen along the x-axis and vector* **b** *in the* (x, y)-*plane; of course, the result should not depend on the orientation of the Cartesian axes!*]

Problem 1.119 Assuming the distributivity of the vector product, derive formula (1.100) by cross-multiplying vectors $\mathbf{a} = a_1 \mathbf{i} + a_2 \mathbf{j} + a_3 \mathbf{k}$ and $\mathbf{b} = b_1 \mathbf{i} + b_2 \mathbf{j} + b_3 \mathbf{k}$ directly and using the given above expressions for the cross products of the unit base vectors.

Problem 1.120 Using the explicit expression for the vector product given above, prove the following identities:

$$[\mathbf{a} \times \mathbf{b}] \cdot [\mathbf{c} \times \mathbf{d}] = (\mathbf{a} \cdot \mathbf{c})(\mathbf{b} \cdot \mathbf{d}) - (\mathbf{a} \cdot \mathbf{d})(\mathbf{b} \cdot \mathbf{c}) \ , \tag{1.103}$$

$$\mathbf{a} \times [\mathbf{b} \times \mathbf{c}] = (\mathbf{a} \cdot \mathbf{c})\mathbf{b} - (\mathbf{a} \cdot \mathbf{b})\mathbf{c} \ , \tag{1.104}$$

$$[\mathbf{a} \times \mathbf{b}] \times \mathbf{c} = (\mathbf{a} \cdot \mathbf{c})\mathbf{b} - (\mathbf{c} \cdot \mathbf{b})\mathbf{a} \ , \tag{1.105}$$

and hence, correspondingly, that

$$\mathbf{a} \times [\mathbf{b} \times \mathbf{c}] + \mathbf{b} \times [\mathbf{c} \times \mathbf{a}] + \mathbf{c} \times [\mathbf{a} \times \mathbf{b}] = 0 \,. \qquad (1.106)$$

Problem 1.121 Show that the area of a parallelogram defined by two vectors **a** and **b**, as shown in Fig. 1.23a, is given by the absolute value of their vector product:

$$S_{paralellogram} = |\mathbf{a} \times \mathbf{b}| \,. \qquad (1.107)$$

It is also possible to mix the dot and vector products into the so-called *mixed* or *triple product*, denoted by putting the three vectors within square brackets, e.g. $[\mathbf{c}, \mathbf{a}, \mathbf{b}]$. Indeed, a vector product of two vectors is a vector which can be dot-multiplied with a third vector giving a scalar:

$$
\begin{aligned}
[\mathbf{c}, \mathbf{a}, \mathbf{b}] &= \mathbf{c} \cdot [\mathbf{a} \times \mathbf{b}] \\
&= (c_1 \mathbf{i} + c_2 \mathbf{j} + c_3 \mathbf{k}) \cdot [(a_2 b_3 - a_3 b_2)\mathbf{i} + (a_3 b_1 - a_1 b_3)\mathbf{j} + (a_1 b_2 - a_2 b_1)\mathbf{k}] \\
&= (a_2 b_3 - a_3 b_2)c_1 + (a_3 b_1 - a_1 b_3)c_2 + (a_1 b_2 - a_2 b_1)c_3 \\
&= \begin{vmatrix} c_1 & c_2 & c_3 \\ a_1 & a_2 & a_3 \\ b_1 & b_2 & b_3 \end{vmatrix} \,.
\end{aligned}
$$

$$(1.108)$$

The last equality trivially follows from formula (1.101) as we effectively replace **i** with c_1, **j** with c_2 and **k** with c_3 in Eq. (1.100); this is exactly what we have done in Eq. (1.108). The triple product has some important symmetry properties listed below which follow from its determinant representation and the fact, discussed earlier in Sect. 1.9, that a determinant changes sign if two of its rows are exchanged:

$$[\mathbf{c}, \mathbf{a}, \mathbf{b}] = [\mathbf{a}, \mathbf{b}, \mathbf{c}] = [\mathbf{b}, \mathbf{c}, \mathbf{a}] = -[\mathbf{c}, \mathbf{b}, \mathbf{a}] = -[\mathbf{b}, \mathbf{a}, \mathbf{c}] = -[\mathbf{a}, \mathbf{c}, \mathbf{b}] \,, \quad (1.109)$$

Fig. 1.23 Illustrations for the calculation of the area **a** of a parallelogram defined by two vectors **a** and **b**, and **b** of the volume of a parallelepiped defined by three non-planar vectors **a**, **b** and **c**

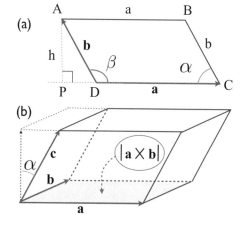

which show that the vectors in the triple product can be "rotated" in a cyclic order $\mathbf{abc} \to \mathbf{bca} \to \mathbf{cab} \to \mathbf{abc} \to \dots$ and this does not affect the sign of the product; if the order is destroyed, a minus sign appears. If two permutations are performed that both are opposite to the cyclic order, e.g. $\mathbf{abc} \to \mathbf{bac} \to \mathbf{bca}$, then no change of the sign occurs.

Problem 1.122 Show that

$$[\mathbf{a}, \mathbf{a}, \mathbf{b}] = [\mathbf{a}, \mathbf{b}, \mathbf{a}] = [\mathbf{b}, \mathbf{a}, \mathbf{a}] = 0 . \qquad (1.110)$$

[Hint: *use the explicit expression* (1.108) *for the triple product preceding the determinant.*[13]]

Problem 1.123 Using Fig. 1.23b show that the volume of a parallelepiped formed by three non-planar vectors $\mathbf{a}$, $\mathbf{b}$ and $\mathbf{c}$ is given by the absolute value of their triple product:

$$V_{parallelepiped} = |[\mathbf{a}, \mathbf{b}, \mathbf{c}]| = |[\mathbf{a} \times \mathbf{b}] \cdot \mathbf{c}| . \qquad (1.111)$$

The immediate consequence of the definition of the triple product is that it is equal to zero if the three vectors lie in the same plane, i.e. are *coplanar*. Indeed, in this case, one of the vectors must be a linear combination of the other two and hence the volume of the parallelepiped constructed from these three vectors will obviously be zero: in the corresponding determinant of the triple product one of its rows will be a linear combination of the other two.

The Cartesian basis introduced above, $\{\mathbf{i}, \mathbf{j}, \mathbf{k}\}$, is not the only one possible; one can define any three non-collinear and non-planar vectors $\{\mathbf{a}_1, \mathbf{a}_2, \mathbf{a}_3\}$ to serve as a new basis and then any vector $\mathbf{x}$ from $\mathbb{R}^3$ can then be expanded with respect to them:

$$\mathbf{x} = g_1\mathbf{a}_1 + g_2\mathbf{a}_2 + g_3\mathbf{a}_3 . \qquad (1.112)$$

This is proven in the following way. Let us write the new basis via their Cartesian coordinates: $\mathbf{a}_1 = (a_{11}, a_{12}, a_{13})$ and similarly for $\mathbf{a}_2$ and $\mathbf{a}_3$. Here a_{ij} corresponds to the jth Cartesian component ($j = 1, 2, 3$) of the vector $\mathbf{a}_i$ ($i = 1, 2, 3$). Then the Cartesian components $\{x_j\}$ of $\mathbf{x}$ can be expressed via $\{a_{ij}\}$ as follows:

$$x_1 = g_1a_{11} + g_2a_{21} + g_3a_{31} ,$$

$$x_2 = g_1a_{12} + g_2a_{22} + g_3a_{32} ,$$

---

[13] The required statement also follows immediately from the determinant formula for the triple product since the determinant is equal to zero if any two of its rows are identical.

$$x_3 = g_1 a_{13} + g_2 a_{23} + g_3 a_{33} . \tag{1.113}$$

One can see that the "coordinates" $\{g_1, g_2, g_3\}$ in the new basis define uniquely the Cartesian components $\{x_i\}$ of the vector $\mathbf{x}$; inversely, given the Cartesian components of $\mathbf{x} = (x_1, x_2, x_3)$, one determines uniquely the coordinates $\{g_1, g_2, g_3\}$ in the new basis by solving the system of three algebraic equations (1.113). It is shown in linear algebra (see Vol. II, Chap. 1) that for the unique solution of these equations to exist it is necessary that the following condition be satisfied:

$$\begin{vmatrix} a_{11} & a_{21} & a_{31} \\ a_{12} & a_{22} & a_{32} \\ a_{13} & a_{23} & a_{33} \end{vmatrix} \neq 0 , \tag{1.114}$$

i.e. the determinant composed of Cartesian components of the three basis vectors be non-zero. It must be clear that this condition is satisfied as long as the three vectors are non-collinear and non-coplanar (for planar vectors one must be a linear combination of the other two, e.g. $\mathbf{a}_3 = \lambda \mathbf{a}_1 + \mu \mathbf{a}_2$ with λ and μ being real numbers, and then in this case the third column is a linear combination of the first two in the determinant, which would force it to be zero in contradiction with Eq. (1.114)).

Example 1.12 ▶ Expand vector $\mathbf{x} = (2, 1, -1)$ in terms of three vectors $\mathbf{a}_1 = (1, 0, 0)$, $\mathbf{a}_2 = (1, 1, 0)$ and $\mathbf{a}_3 = (1, 1, 1)$.

Solution. Solving the system of linear equations (1.113),

$$\begin{cases} g_1 + g_2 + g_3 = 2 \\ g_2 + g_3 = 1 \\ g_3 = -1 \end{cases} ,$$

we obtain $g_3 = -1$, then $g_2 = 2$ and $g_1 = 1$, i.e. the expansion sought for is $\mathbf{x} = \mathbf{a}_1 + 2\mathbf{a}_2 - \mathbf{a}_3$, which can now be checked by direct calculation. ◀

Alongside a direct space basis $\{\mathbf{a}_1, \mathbf{a}_2, \mathbf{a}_3\}$, it is sometimes useful to introduce another special basis, $\{\mathbf{b}_1, \mathbf{b}_2, \mathbf{b}_3\}$, called *reciprocal* or *bi-orthogonal* basis. These basis vectors are defined by the following relations:

$$\mathbf{b}_1 = \frac{1}{v_c} [\mathbf{a}_2 \times \mathbf{a}_3] , \quad \mathbf{b}_2 = \frac{1}{v_c} [\mathbf{a}_3 \times \mathbf{a}_1] , \quad \mathbf{b}_3 = \frac{1}{v_c} [\mathbf{a}_1 \times \mathbf{a}_2] , \tag{1.115}$$

where $v_c = [\mathbf{a}_1, \mathbf{a}_2, \mathbf{a}_3]$ is the volume of the parallelepiped (assumed to be positive) formed by the three original basis vectors $\{\mathbf{a}_1, \mathbf{a}_2, \mathbf{a}_3\}$. Notice that the index of the reciprocal vector in any formula in Eq. (1.115) and the indices of the two direct space vectors form an ordered triple. It is easy to see using properties (1.109) and (1.110) of the triple product that the new basis is related to the old one via:

$$\mathbf{a}_i \cdot \mathbf{b}_j = \begin{cases} 0 , & \text{if } i \neq j \\ 1 , & \text{if } i = j \end{cases} = \delta_{ij} , \tag{1.116}$$

where we have introduced the *Kronecker symbol* which is equal to zero if its two indices are different, otherwise it is equal to one.[14]

The volumes of the parallelepipeds $v_c = [a_1, a_2, a_3]$ and $v_g = [b_1, b_2, b_3]$, formed by the two sets of vectors, are inversely related. Indeed,

$$v_g = b_1 \cdot [b_2 \times b_3] = \frac{1}{v_c}[a_2 \times a_3] \cdot [b_2 \times b_3]$$

$$= \frac{1}{v_c}\{(a_2 \cdot b_2)(a_3 \cdot b_3) - (a_2 \cdot b_3)(a_3 \cdot b_2)\} = \frac{1}{v_c}, \qquad (1.117)$$

where we made use of (1.103) and then (1.116) for the four dot products of direct and reciprocal vectors. Moreover, it is easy to see that direct basis vectors are reciprocal to the reciprocal ones; in other words, the two sets of vectors are mutually reciprocal. This is proven in the following problem.

Problem 1.124 Show that direct basis vectors can be written via the reciprocal vectors similar to Eq. (1.115), i.e. that

$$a_1 = \frac{1}{v_g}[b_2 \times b_3] \ , \ a_2 = \frac{1}{v_g}[b_3 \times b_1] \ , \ a_3 = \frac{1}{v_g}[b_1 \times b_2] \ .$$

[Hint: *in calculating vector products of reciprocal vectors replace only one of them via direct ones using Eq. (1.115), and then make use of either Eq. (1.104) or (1.105) for the double vector product.*]

Problem 1.125 The crystal lattice is characterised by three lattice vectors a_1, a_2 and a_3, which form its primitive unit cell. By repeating this cell in all three directions the whole periodic crystal lattice is constructed. Let a, b and c be the lengths of the three vectors, and α, β and γ the angles between the first and the third, the second and the third, and the first and the second vectors, respectively. Assuming that a_1 lies along the x-axes, a_2 in the $x - y$-plane, work out the Cartesian components of all three lattice vectors.

Problem 1.126 Consider a set of three non-coplanar lattice vectors a_1, a_2 and a_3 that form a parallelepiped with volume $v_c = [a_1, a_2, a_3]$. Instead of these vectors, one may also consider a different set of lattice vectors that are specified by a linear combination of the former ones with integer coefficients, $A_i = \sum_{j=1}^{3} T_{ij}a_j$, where T_{ij} are integer coefficients (they form a 3×3 matrix).

---

[14] In solid-state physics and crystallography, the reciprocal lattice is defined with an additional factor of 2π, so that instead of (1.116) we have $a_i \cdot b_j = 2\pi\delta_{ij}$.

Show that the volume of the parallelepiped formed by the new vectors $V_c = [\mathbf{A}_1, \mathbf{A}_2, \mathbf{A}_3]$ is related to the one of the former set via $V_c = v_c \det \mathbf{T}$, where $\det \mathbf{T}$ is the 3×3 determinant formed by the coefficients T_{ij}.

Direct and reciprocal vectors are frequently used in crystallography and solid-state physics to build the crystal lattices in both spaces.

1.10.2 N-Dimensional Space

Most of the material we considered above is generalised to abstract spaces of arbitrary N dimensions ($N \geq 1$). In this space, any point X is specified by N coordinates x_1, $x_2, \ldots, x_N$, i.e. $X = (x_1, \ldots, x_N)$. The "distance" d_{XY} between two points $X = (x_1, \ldots, x_N)$ and $Y = (y_1, \ldots, y_N)$ is defined as a direct generalisation of the result (1.89) obtained for the 3D space:

$$d_{XY} = \sqrt{\sum_{i=1}^{N} (x_i - y_i)^2} . \tag{1.118}$$

We can also define an N-dimensional vector $\overrightarrow{XY} = \mathbf{a} = (y_1 - x_1, y_2 - x_2, \ldots, y_N - x_N)$ connecting the points X and Y, so that d_{XY} would serve as its length. Similar to the 3D case, vectors can be added or subtracted from each other or multiplied by a number; each such operation results in a vector in the same space: if there are two vectors $\mathbf{a} = (a_1, \ldots, a_N)$ and $\mathbf{b} = (b_1, \ldots, b_N)$, then we define

$$\mathbf{a} \pm \mathbf{b} = \mathbf{c} = (c_1, \ldots, c_N) \quad \text{with} \quad c_i = a_i \pm b_i , \ i = 1, \ldots, N ,$$

and

$$\alpha \mathbf{a} = \mathbf{g} \quad \text{with} \quad g_i = \alpha a_i , \ i = 1, \ldots, N ,$$

where α is a number from $\mathbb{R}$. Correspondingly, we can also define a dot (or scalar) product of two vectors $\mathbf{a} = (a_1, \ldots, a_N)$ and $\mathbf{b} = (b_1, \ldots, b_N)$ as a direct generalisation of Eq. (1.90):

$$\mathbf{a} \cdot \mathbf{b} = (\mathbf{a}, \mathbf{b}) = \sum_{i=1}^{N} a_i b_i = a_1 b_1 + a_2 b_2 + \cdots + a_N b_N . \tag{1.119}$$

As was the case for the 3D space, the square of a vector length is given by the dot product of the vector with itself: $|\mathbf{a}|^2 = \mathbf{a} \cdot \mathbf{a}$. All properties of the dot product are directly transferred to the N-dimensional space. The distance between two points X and Y is then given by

$$d_{XY} = \sqrt{(\mathbf{x} - \mathbf{y})^2} = \sqrt{(\mathbf{x} - \mathbf{y}) \cdot (\mathbf{x} - \mathbf{y})} = |\mathbf{x} - \mathbf{y}| .$$

Next, we introduce N basis vectors (or unit base vectors) of the space as $\mathbf{e}_1 = (1, 0, 0, \ldots, 0), \mathbf{e}_2 = (0, 1, 0, \ldots, 0)$, etc., $\mathbf{e}_N = (0, 0, \ldots, 0, 1)$. Each of these vectors consists of $N - 1$ zeros and a single number one; the latter is positioned exactly at the ith place in $\mathbf{e}_i$. Each such vector is obviously of unit length and they are orthogonal, i.e. $\mathbf{e}_i \cdot \mathbf{e}_j = \delta_{ij}$. Then, any vector $\mathbf{a} = (a_1, \ldots, a_N)$ in such a space can be expanded via the basis vectors as

$$\mathbf{a} = a_1\mathbf{e}_1 + a_2\mathbf{e}_2 + \cdots + a_N\mathbf{e}_N = \sum_{i=1}^{N} a_i\mathbf{e}_i .$$

So far we have been using only vector-rows, e.g. $\mathbf{x} = (x_1, \ldots, x_N)$, when the coordinates of the vector are listed along a line (a row). In some cases, vector-columns

$$\mathbf{x} = \begin{pmatrix} x_1 \\ \cdots \\ x_N \end{pmatrix}$$

may be more useful. We shall see in Vol. II, when studying linear algebra, that a vector-column is a more natural and hence more appropriate representation of a vector.

1.10.3 My Father's Number Pyramid

My father Nohim Icikovich Kantorovich, purely empirically, some time around 1970–75, discovered a beautiful arithmetic problem: he found rules whereby one can mentally construct sets of numbers that satisfy interesting sequences, examples of which are shown in Fig. 1.24. In each identity there are sums of squares of three integers in the left- and right-hand sides.

Consider first the upper panel (a). One can easily check that the first identity containing squares of single-digit numbers is true. What is remarkable, however, is that, as shown by other lines in (a), one can form from these digits squares of double, triple, etc. digit numbers and the equalities would still remain true as can also be checked explicitly. What is astonishing is that this sequence can be continued indefinitely and the equality can still be guaranteed to be true! But there is more to it! The sequence $(4, 8, 3; 6, 2, 7)$ used in (a) is not unique. Another sequence, $(1, 6, 8; 9, 4, 2)$, also exists and one can build similarly the "pyramid" of equalities from it as shown in panel (b).

But this is where all the fun begins! Let us have a look at panel (c). There we start from the second sequence and form a single-digit equality. But then we construct a double-digit equality using the first sequence of single digits: we attach 4 to 1 of the first number in the left-hand side forming 41, similarly we form 86 by attaching 8 to 6, and so on. It is easy to check that the equality still holds! We shall say that the second equality in (c) is obtained by "attaching" the set $(4, 8, 3; 6, 2, 7)$ to the set

(a)
$$4^2 + 8^2 + 3^2 = 6^2 + 2^2 + 7^2$$
$$44^2 + 88^2 + 33^2 = 66^2 + 22^2 + 77^2$$
$$444^2 + 888^2 + 333^2 = 666^2 + 222^2 + 777^2$$
$$4444^2 + 8888^2 + 3333^2 = 6666^2 + 2222^2 + 7777^2$$
$$44444^2 + 88888^2 + 33333^2 = 66666^2 + 22222^2 + 77777^2$$

(b)
$$1^2 + 6^2 + 8^2 = 9^2 + 4^2 + 2^2$$
$$11^2 + 66^2 + 88^2 = 99^2 + 44^2 + 22^2$$
$$111^2 + 666^2 + 888^2 = 999^2 + 444^2 + 222^2$$
$$1111^2 + 6666^2 + 8888^2 = 9999^2 + 4444 + 2222^2$$
$$11111^2 + 66666^2 + 88888^2 = 99999^2 + 44444^2 + 22222^2$$

(c)
$$1^2 + 6^2 + 8^2 = 9^2 + 4^2 + 2^2$$
$$41^2 + 86^2 + 38^2 = 69^2 + 24^2 + 72^2$$
$$141^2 + 686^2 + 838^2 = 969^2 + 424^2 + 272^2$$
$$4141^2 + 8686^2 + 3838^2 = 6969^2 + 2424^2 + 7272^2$$
$$14141^2 + 68686^2 + 83838^2 = 96969^2 + 42424^2 + 27272^2$$
$$1441^2 + 6886^2 + 8338^2 = 9669^2 + 4224^2 + 2772^2$$
$$14641^2 + 68786^2 + 83238^2 = 96469^2 + 42324^2 + 27872^2$$

Fig. 1.24 Sequences of equalities containing a sum of squares of integers on both sides

$(1, 6, 8; 9, 4, 2)$ from the left. In fact, if you attach similarly that set from the right instead, the equality would still hold:

$$14^2 + 68^2 + 83^2 = 96^2 + 42^2 + 27^2.$$

What we did next in panel (c) was that we attached from the left the second set again forming an equality with three-digit numbers. You can easily check that the equality is still true. Then we continued this process up to the fifth equality in (c) attaching the first and then the second sequence. Again, the equalities remain to be true, and in fact this process of attaching the two sets can be continued indefinitely; in fact, each of the sets can be attached in any order!

Once this is true, it should come with no surprise that one can "remove" digits at the same position from each number in any equality without destroying its correctness, and this is what has been done in the sixth equality in (c) where we removed the third digit from each number. In the final seventh equality in (c) we have "inserted" yet a third set of digits, $(6, 7, 2; 4, 3, 8)$, at the position of the third digit in each number. Remarkably, the equality still holds. This other set of digits can be inserted at any position: from the left, right or inside.

Summarising, there are several sets of six numbers: the first three to be used in the left-hand side and the other three in the right. These sets can be used in constructing identities containing sums of squares of integers in both sides of the equalities. Each number is constructed by "attaching" a single-digit set (either from the left, right or inside) to the numbers in the current equality. One can attach each time the same or a different single-digit set and this can be done indefinitely. Of course, one can also remove a set from the given identity as this would simply correspond to a different order in which sets are attached. My father found a simple rule that enables one to construct (likely almost all) these single-digit sets, and asserted that one should be able to attach these sets in any order constructing beautiful pyramids of squares of numbers that can run to very large numbers indeed!

Is it not something worth understanding? Below we shall give a detailed explanation of this remarkable phenomenon with integer numbers (then a few extensions will also be considered) and will establish a sufficient condition that would allow one to build many such single-digit sets. We shall perform our analysis in the general case of $n \geq 2$ numbers in each side of the equalities as it appears that it is not restricted only to the case of $n = 3$ considered above.

We shall start from definitions. It is convenient to use the language of vectors here albeit with some minor (but important) modification. Let us define vectors $\mathbf{X}$ in a $(2n)$-dimensional space which will be written via their components as $\mathbf{X} = (x_1, x_2, \ldots, x_n; y_1, y_2, \ldots, y_n)$. As you can see, we have divided the vector into two equal parts and use x_i for the components in the first part and y_i in the second ($i = 1, \ldots, n$). The components x_i and y_i of the vectors are formed only by single-digit integer numbers between 0 and 9.

Then we shall define a dot product of two vectors $\mathbf{X}^{(a)}$ and $\mathbf{X}^{(b)}$ as follows:

$$\mathbf{X}^{(a)} \cdot \mathbf{X}^{(b)} = \sum_{i=1}^{n} \left(x_i^{(a)} x_i^{(b)} - y_i^{(a)} y_i^{(b)} \right).$$

This is very close to the usual dot product (1.119) of two vectors; the only difference is that there is the minus sign by the product of the y-components. In what follows, different vectors will be distinguished by their superscript in the round brackets as was done above.

Next, it is convenient to introduce some additional notations. By attaching to each other different vectors $\mathbf{X}^{(1)}, \mathbf{X}^{(2)}, \ldots, \mathbf{X}^{(k)}$, we shall construct a set of k-digital numbers. Each of those numbers will be denoted as

$$\left[x_i^{(k)} x_i^{(k-1)} \cdots x_i^{(2)} x_i^{(1)} \right] = x_i^{(k)} 10^{k-1} + x_i^{(k-1)} 10^{k-2} + \cdots + x_i^{(2)} 10^1 + x_i^{(1)} 10^0,$$

if it is at the ith position in the left part of the equality, i.e. composed of x-components, and

$$\left[y_i^{(k)} y_i^{(k-1)} \cdots y_i^{(2)} y_i^{(1)} \right] = y_i^{(k)} 10^{k-1} + y_i^{(k-1)} 10^{k-2} + \cdots + y_i^{(2)} 10^1 + y_i^{(1)} 10^0,$$

if of y-components. These are just usual representations of the integer numbers written in the decimal form. For instance, [30562] is the same as the number 30562. However, it is convenient to enclose all digits into square brackets for the sake of the notations. The superscripts by the x and y digits above indicate which vectors $\mathbf{X}^{(a)}$ were used to construct them. There are $2n$ such multi-digit numbers after k vectors

$\mathbf{X}^{(1)}$, $\mathbf{X}^{(2)}$, ..., $\mathbf{X}^{(k)}$ which were sequentially attached one after another. We shall now collect all these numbers into a vector

$$\left[\mathbf{X}^{(k)}\cdots\mathbf{X}^{(1)}\right]$$

$$= \left(\left[x_1^{(k)}\cdots x_1^{(1)}\right],\ldots,\left[x_n^{(k)}\cdots x_n^{(1)}\right]; \left[y_1^{(k)}\cdots y_1^{(1)}\right],\ldots,\left[y_n^{(k)}\cdots y_n^{(1)}\right]\right)$$

and can define a dot product of any two such vectors as

$$\left[\mathbf{X}^{(k)}\cdots\mathbf{X}^{(1)}\right]\cdot\left[\mathbf{X}^{(k')}\cdots\mathbf{X}^{(1')}\right]$$

$$= \sum_{i=1}^{n}\left\{\left[x_i^{(k)}\cdots x_i^{(1)}\right]\cdot\left[x_i^{(k')}\cdots x_i^{(1')}\right] - \left[y_i^{(k)}\cdots y_i^{(1)}\right]\cdot\left[y_i^{(k')}\cdots y_i^{(1')}\right]\right\}.$$

When a vector $\left[\mathbf{X}^{(k)}\cdots\mathbf{X}^{(1)}\right]$ is multiplied with itself, this will be simply denoted as its square, $\left[\mathbf{X}^{(k)}\cdots\mathbf{X}^{(1)}\right]^2$.

For example, consider vectors $\mathbf{X}^{(1)} = (4, 8, 3; 6, 2, 7)$ and $\mathbf{X}^{(2)} = (1, 6, 8; 9, 4, 2)$ corresponding to $n = 3$. By attaching the second to the first, we form a vector of two-digit numbers $\left[\mathbf{X}^{(2)}\mathbf{X}^{(1)}\right] = (14, 68, 83; 96, 42, 27)$. Its dot product with itself is

$$\left[\mathbf{X}^{(2)}\mathbf{X}^{(1)}\right]\cdot\left[\mathbf{X}^{(2)}\mathbf{X}^{(1)}\right] = \left[\mathbf{X}^{(2)}\mathbf{X}^{(1)}\right]^2$$
$$= 14^2 + 68^2 + 83^2 - 96^2 - 42^2 - 27^2$$
$$= 0.$$

If we use yet another vector $\mathbf{X}^{(3)} = (6, 7, 2; 4, 3, 8)$, then we can define, e.g. a three-digit vector $\left[\mathbf{X}^{(2)}\mathbf{X}^{(1)}\mathbf{X}^{(3)}\right] = (146, 687, 832; 964, 423, 278)$ and construct its dot product with $\left[\mathbf{X}^{(2)}\mathbf{X}^{(1)}\right]$ as

$$\left[\mathbf{X}^{(2)}\mathbf{X}^{(1)}\mathbf{X}^{(3)}\right]\cdot\left[\mathbf{X}^{(2)}\mathbf{X}^{(1)}\right] = 146\cdot 14 + 687\cdot 68 + 832\cdot 83 - 964\cdot 96 - 423\cdot 42 - 278\cdot 27$$
$$= 0.$$

We shall see that the fact that these dot products are all equal to zero is not accidental. Now we are ready to prove the following theorem.

Theorem 1.8 *In order to be able to "attach" single-digit vectors (as defined above), in any order, to each other in building the number pyramid, it is necessary and sufficient for these vectors to be mutually orthogonal and of zero "length", i.e. for any two vectors $\mathbf{X}^{(a)}$ and $\mathbf{X}^{(b)}$ we should have $\mathbf{X}^{(a)}\cdot\mathbf{X}^{(b)} = 0$ (including the case of $a = b$).*

Proof: Let us first prove the sufficient condition: we assume that the vectors are mutually orthogonal and of zero length and then show that one can easily build the number pyramids by means of the attachment operation. First of all, we note that the very first equality of a number pyramid is basically the condition of the length of a vector being zero:

$$\mathbf{X}^{(1)} \cdot \mathbf{X}^{(1)} = [\mathbf{X}^{(1)}]^2 = \sum_{i=1}^{n} \left(x_i^{(1)} x_i^{(1)} - y_i^{(1)} y_i^{(1)} \right) = 0$$

$$\implies \sum_{i=1}^{n} \left(x_i^{(1)} \right)^2 = \sum_{i=1}^{n} \left(y_i^{(1)} \right)^2 .$$

Next, let us consider attaching to the dot product above another vector $\mathbf{X}^{(2)}$ to build an expression containing squares of two-digit numbers. By doing so, we construct an expression:

$$[\mathbf{X}^{(2)} \mathbf{X}^{(1)}]^2 = \sum_{i=1}^{n} \left\{ \left[x_i^{(2)} x_i^{(1)} \right]^2 - \left[y_i^{(2)} y_i^{(1)} \right]^2 \right\} . \tag{1.120}$$

Expanding the two-digit numbers explicitly, we can write:

$$[\mathbf{X}^{(2)} \mathbf{X}^{(1)}]^2 = \sum_{i=1}^{n} \left\{ \left(10 x_i^{(2)} + x_i^{(1)} \right)^2 - \left(10 y_i^{(2)} + y_i^{(1)} \right)^2 \right\}$$

$$= 100 \sum_{i=1}^{n} \left\{ \left(x_i^{(2)} \right)^2 - \left(y_i^{(2)} \right)^2 \right\} + \sum_{i=1}^{n} \left\{ \left(x_i^{(1)} \right)^2 - \left(y_i^{(1)} \right)^2 \right\} + 2 \cdot 10 \sum_{i=1}^{n} \left\{ x_i^{(1)} x_i^{(2)} - y_i^{(1)} y_i^{(2)} \right\}$$

$$= 100 \left[\mathbf{X}^{(2)} \right]^2 + \left[\mathbf{X}^{(1)} \right]^2 + 20 \mathbf{X}^{(1)} \cdot \mathbf{X}^{(2)} . \tag{1.121}$$

Since the vectors are mutually orthogonal and of zero length, this expression is equal to zero. It is next proven by induction that for any order k we have $[\mathbf{X}^{(k)} \mathbf{X}^{(k-1)} \cdots \mathbf{X}^{(1)}]^2 = 0$. Indeed, assuming that this is valid for some k, let us consider the case of $k + 1$:

$$\left[\mathbf{X}^{(k+1)} \mathbf{X}^{(k)} \mathbf{X}^{(k-1)} \cdots \mathbf{X}^{(1)} \right]^2 = \sum_{i=1}^{n} \left\{ \left(\sum_{j=1}^{k+1} 10^{j-1} x_i^{(j)} \right)^2 - \left(\sum_{j=1}^{k+1} 10^{j-1} y_i^{(j)} \right)^2 \right\}$$

$$= \sum_{i=1}^{n} \left\{ \left(10^k x_i^{(k+1)} + \sum_{j=1}^{k} 10^{j-1} x_i^{(j)} \right)^2 - \left(10^k y_i^{(k+1)} + \sum_{j=1}^{k} 10^{j-1} y_i^{(j)} \right)^2 \right\} .$$

Opening the brackets, the terms containing the first part of the numbers (the x-part) can be worked into

$$\sum_{i=1}^{n} \left(10^k x_i^{(k+1)} + \sum_{j=1}^{k} 10^{j-1} x_i^{(j)} \right)^2$$

$$= \sum_{i=1}^{n} \left[\left(\sum_{j=1}^{k} 10^{j-1} x_i^{(j)} \right)^2 + 10^{2k} \left(x_i^{(k+1)} \right)^2 + 2 \cdot 10^k x_i^{(k+1)} \left(\sum_{j=1}^{k} 10^{j-1} x_i^{(j)} \right) \right]$$

and similarly for the y-part, so that

$$\left[\mathbf{X}^{(k+1)} \mathbf{X}^{(k)} \mathbf{X}^{(k-1)} \cdots \mathbf{X}^{(1)} \right]^2$$

$$= \left[\mathbf{X}^{(k)} \mathbf{X}^{(k-1)} \cdots \mathbf{X}^{(1)} \right]^2 + 10^{2k} \left[\mathbf{X}^{(k+1)} \right]^2 + 2 \sum_{j=1}^{k} 10^{k+j-1} \mathbf{X}^{(k+1)} \cdot \mathbf{X}^{(j)}.$$

The first term in the right-hand side is zero due to our assumption, the other two are zero because the single-digit vectors are of zero length and mutually orthogonal. So, the whole expression is zero. Since the case of $k = 2$ was considered explicitly above, the required statement has been fully proven. So, if the vectors are mutually orthogonal and of zero length, then for any k we have $\left[\mathbf{X}^{(k)} \mathbf{X}^{(k-1)} \cdots \mathbf{X}^{(1)} \right]^2 = 0$. This finalises the proof of sufficiency.

Let us now provide the second part of the proof, the necessary condition. We are given that

$$\left[\mathbf{X}^{(k)} \mathbf{X}^{(k-1)} \cdots \mathbf{X}^{(1)} \right]^2 = 0$$

for any order $k = 1, 2, 3, \ldots$ and for any vectors, and we have to prove that the vectors must be of zero length and mutually orthogonal. This part is actually very simple. We start from $k = 1$ which leads directly to the condition of zero length. By considering the case of $k = 2$ we arrive at orthogonality of any two vectors, see Eq. (1.121), since the vectors are of zero length. **Q.E.D.**

So, the trick for building number pyramids is to be able to construct single-digit vectors that satisfy the required conditions of zero length and orthogonality. My father, purely empirically, found a simple condition that allows constructing not one but many such vectors for any $n \geq 2$. The conditions he found are sufficient and may not exhaust all possible vectors. Still they provide a rather extended set of them. The first condition is that, for any single-digit vector, we must have $x_i + y_i = N$ for any $i = 1, \ldots, n$. To work out the second condition, let us assume that the sum of the digits in the x-part of a vector and in the y-part of it is the same and equal to M:

$$\sum_{i=1}^{n} x_i = \sum_{i=1}^{n} y_i = M \implies \sum_{i=1}^{n} (x_i - y_i) = \sum_{i=1}^{n} (2x_i - N) = 0,$$

from which it immediately follows that $2M - nN = 0$ or $N = 2M/n$, which is the second condition. Hence, only one number, e.g. M, need to be specified to define all vectors, and this number must be such that $2M$ is divisible by n as N must be integer.

Now, let us consider different vectors constructed by these two conditions, i.e. by choosing a particular value of M (for a fixed n). Then, it is shown in the following problem that vectors specified by the two conditions are of zero length and mutually orthogonal.

Problem 1.127 Consider all single-digit vectors $\mathbf{X} = (x_1, \ldots, x_n; y_1, \ldots, y_n)$ constructed in such a way that $x_i + y_i = N$ and $\sum_{i=1}^{n} x_i = M$, where $N = 2M/n$. Show that $\sum_{i=1}^{n} y_i = M$, and that $[\mathbf{X}]^2 = 0$. Moreover, show that for any two distinct vectors $\mathbf{X}$ and $\mathbf{X}'$ that are constructed using the same number M we have $\mathbf{X} \cdot \mathbf{X}' = 0$, i.e. the vectors are of zero length and mutually orthogonal.

The examples given in Fig. 1.24 correspond to a collection of three vectors $(4, 8, 3; 6, 2, 7)$, $(1, 6, 8; 9, 4, 2)$ and $(6, 7, 2; 4, 3, 8)$ that are built using $M = 15$ and correspondingly $N = 10$. One can write a simple code and find that in the $n = 3$ case for $M = 15$ there are altogether 61 vectors possible (accepting 0 as well as a digit). If $M = 9$, then 37 vectors can be constructed. Similarly one can consider cases of larger n.

You can show off in front of your friends by demonstrating this problem. Let us take the case of $n = 3$. Ask your friends to suggest two numbers between 0 and 9. These might be your x_1 and x_2, for instance. Then, think of an appropriate value of M, work out N and hence calculate the remaining four numbers x_3, y_1, y_2 and y_3; write the first sum of squares made by these six numbers. Then ask your friends again to give you two numbers, work out the remaining four and attach them to the previous equality forming a sum of double-digit numbers. Then continue this process, repeating the same attachment from time to time to increase the numbers. Everybody would think that you can square big numbers and sum them up. In fact, you would simply use a very clever maths discovered by my father! Do not disclose your secret, keep it to yourself!

When I showed the above to my brother Ephim, he did not spend more than 10 min to suggest two possible extensions to this entertaining numerical problem. The first one relates to the fact, already mentioned, that the vectors constructed in the way proposed by my father are not exhaustive. For instance, the zero vector $(0, \ldots, 0; 0, \ldots, 0)$ can always be added to the set as it is automatically mutually orthogonal to any other and is of zero length. For instance, the last line in Fig. 1.24c may look either like

$$146410^2 + 687860^2 + 832380^2 = 964690^2 + 423240^2 + 278720^2$$

or

$$1406410^2 + 6807860^2 + 8302380^2 = 9604690^2 + 4203240^2 + 2708720^2.$$

The second observation of my brother is more intriguing: it must also be possible to extend these number pyramids to *real numbers* since the mantissa part of the real numbers in the decimal form can be written using 10 in negative powers, and hence the given above proof would still be valid. Or, looking at this from a different perspective, one can just divide both sides of any equality with integer numbers by 10^{2m} with some integer m. Basically, the last line above can be rewritten by placing the decimal point somewhere inside, e.g. between the fourth and fifth digits:

$$140.6410^2 + 680.7860^2 + 830.2380^2 = 960.4690^2 + 420.3240^2 + 270.8720^2.$$

Quite fascinating, isn't it?

Still, a few more extensions are possible. Indeed, we have talked above about taking the square of the multi-digit vector. In fact, it must be clear from the above proof that a dot product of two such different vectors must also be zero:

$$\left[\mathbf{X}^{(k)} \cdots \mathbf{X}^{(1)}\right] \cdot \left[\mathbf{X}^{(k')} \cdots \mathbf{X}^{(1')}\right] = 0.$$

For instance, using again the mentioned above single-digit vectors, we can write:

$$146.4101 \cdot 14.6 + 687.8606 \cdot 68.7 + 832.3808 \cdot 83.2 = 964.6909 \cdot 96.4 + 423.2404 \cdot 42.3 + 278.7202 \cdot 27.8.$$

Starting from this line, if desired, I can remove some single-digit vectors and/or attach some, the equality will hold. So, it is possible to construct another type of the number pyramid using a product of different numbers instead of their square. Because of this, the number pyramid can be extended to a product (or square) of complex numbers as well. The complex numbers are introduced in the next section.

1.11 Introduction to Complex Numbers

1.11.1 Cardano's Formula

We know that a square of a real number is always positive. Therefore, a square root of a negative number does not make sense within the manifold of real numbers. This fact causes some problems in solving algebraic equations of power more than one. Indeed, well back in sixteenth century it was found that solutions of some third-order cubic equations may contain square roots of negative numbers as a part of the formula for the roots. At the same time, the roots are all real. Hence, complex numbers were introduced to serve as an intermediate device that need to be dealt with in order to continue to the final answer. An extension of real numbers to a form which allows representation of a square root of negative numbers was first introduced by a sixteenth-century Italian mathematician Gerolamo Cardano and is called *complex numbers*.[15]

To illustrate better what has been said above and hence justify in detail the necessity of introducing complex numbers, it is indeed instructive to repeat the root passed by Italian mathematicians of the sixteenth century and consider solutions of the so-called depressed cubic equation

$$x^3 + px + q = 0 \, . \tag{1.122}$$

This is the cubic equation without the x^2 term.

Problem 1.128 Consider the most general cubic equation

$$x^3 + ax^2 + bx + c = 0 \, . \tag{1.123}$$

Show that by the transformation $x = y - a/3$ this equation turns into the depressed form with respect to the new variable y, with $p = b - a^2/3$ and $q = c - ba/3 + 2a^3/27$.

The first solution of the depressed equation was probably given by Italian mathematician Niccolo Tartaglia in the sixteenth century although earlier attempts also existed. However, it was published by Gerolamo Cardano in 1545 and bears his name.[16]

To solve the cubic equation, we express x as a sum of two other unknowns, $x = a + b$, which yields upon substitution into the depressed form of an equation:

$$(a + b)^3 + p(a + b) + q = 0 \, .$$

Opening the brackets and rearranging, we get:

$$a^3 + b^3 + (p + 3ab)(a + b) + q = 0 \, .$$

Since there is only one unknown x that was replaced by two, we are able to impose an extra condition on a and b. It is convenient to require that $p + 3ab = 0$, which leads to two equations with respect to the two variables:

$$\begin{cases} a^3 + b^3 = -q \\ 3ab = -p \end{cases} \implies \begin{cases} a^3 + b^3 = -q \\ a^3 b^3 = -p^3/27 \end{cases} , \tag{1.124}$$

---

[15] These were developed further by Rafael Bombelli and William Rowan Hamilton and many others.
[16] Quartic equation was solved by Lodovico de Ferrari, a pupil of Cardano, but Cardano seemed to have stolen the solution and published it in 1545. It was later proven (incompletely) in 1799 by Italian mathematician Paolo Ruffini and then completely in 1824 by Norwegian mathematician Niels Henrik Abel that a solution of general quintic and higher order equations is impossible to write in radicals. So it was rigorously established that a general solution in radicals exists only for up to the quartic equation.

where in the last passage we cubed both sides of the second equation. The obtained two equations can actually be transformed into an equation that we can solve. The equations for a^3 and b^3 look exactly the same due to symmetry. Let us obtain an equation for a^3. Expressing $b^3 = -q - a^3$ from the first equation and substituting into the second, one obtains:

$$a^6 + qa^3 - \frac{p^3}{27} = 0 .$$

This is a quadratic equation with respect to a^3. Its solutions, see Eq. (1.18), are:

$$a^3 = -\frac{q}{2} \pm \sqrt{\frac{q^2}{4} + \frac{p^3}{27}} = -Q \pm \sqrt{Q^2 + P} ,$$

where $Q = q/2$ and $P = p^3/27$ were introduced to simplify the notations. There are two solutions, each given by the plus or minus sign. Similarly, we obtain exactly the same expression for b^3, also two solutions. Which ones should we take? Let us choose the plus sign for the a^3 and consider b^3 from the second equation in (1.124):

$$b^3 = -\frac{P}{a^3} = -\frac{P}{-Q + \sqrt{Q^2 + P}} .$$

Next, we multiply the denominator and numerator by the same expression $-Q - \sqrt{Q^2 + P}$, and make use of the identity (1.4) for the difference of the squares of two numbers in the denominator:

$$b^3 = -\frac{P\left(-Q - \sqrt{Q^2 + P}\right)}{\left(-Q + \sqrt{Q^2 + P}\right)\left(-Q - \sqrt{Q^2 + P}\right)} = -\frac{P\left(-Q - \sqrt{Q^2 + P}\right)}{Q^2 - \left(Q^2 + P\right)}$$

$$= -Q - \sqrt{Q^2 + P} .$$

Hence, choosing plus in the solution for a^3, we have to choose minus in the solution for b^3, and *vice versa*. Without loss of generality, due to inherent symmetry between a and b, we can finally write expressions for them:

$$a^3 = -Q + \sqrt{Q^2 + P} \quad \text{and} \quad b^3 = -Q - \sqrt{Q^2 + P} ,$$

and therefore

$$a = \sqrt[3]{-Q + \sqrt{Q^2 + P}} \quad \text{and} \quad b = \sqrt[3]{-Q - \sqrt{Q^2 + P}} ,$$

leading immediately to the final solution called Cardano's formula:

$$x = \sqrt[3]{-\frac{q}{2} + \sqrt{\frac{q^2}{4} + \frac{p^3}{27}}} + \sqrt[3]{-\frac{q}{2} - \sqrt{\frac{q^2}{4} + \frac{p^3}{27}}} , \qquad (1.125)$$

where we have replaced back the original expressions for Q and P. We shall see in Sect. 3.4.5 that each cubic root gives up to three values; however, overall there will only be up to three different solutions for x possible, when we choose proper combinations of these values.

At this point, we are not interested in finding all solutions; this would require learning a bit more about complex numbers. Instead, we would simply like to understand whether this formula gives real solutions exclusively via operating with real numbers $\mathbb{R}$. It is easy to construct an example in which the solutions are all real, but the expression under the square root in Eq. (1.125) is negative and hence it appears that Cardano's formula cannot be used!

Indeed, consider the equation

$$x^3 - 19x - 30 = 0 ,$$

corresponding to $p = -19$ and $q = -30$ of the depressed equation (1.122). For these values, however, the expression under the square root $q^2/4 + p^3/27$ is negative:

$$x = \sqrt[3]{15 + \sqrt{225 - \frac{6859}{27}}} + \sqrt[3]{15 - \sqrt{225 - \frac{6859}{27}}}$$

$$\simeq \sqrt[3]{15 + \sqrt{-29.037}} + \sqrt[3]{15 - \sqrt{-29.037}} ,$$

that is, we have to take the square root of a negative number! At the same time, the three roots of this equation are all real: these are 5, -2 and -3, as can easily be verified. Hence, it appears that one need to take the square root of a negative number to arrive at a real root of the equation and there is no way to avoid this!

This simple fact triggered mathematicians to define square roots of negative numbers in a systematic way and eventually develop consistent theory of complex numbers. Today complex numbers are widely used in solving various problems in natural sciences.

1.11.2   Complex Numbers

These are defined by a pair of two real numbers x (which is called a real part) and y (a complex part) and a special object i as follows: $z = x + iy$. The i is defined in such a way that its square is equal to -1, i.e. $i^2 = -1$ or $i = \sqrt{-1}$. It is also a complex number with zero real part and the complex part equal to one.

Two complex numbers are identical if and only if their corresponding real and imaginary parts coincide.

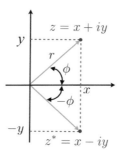

Fig. 1.25 A complex number $z = x + iy$ is depicted as a point (x, y) with coordinates x and y on the 2D plane, the so-called complex plane. Alternatively the vector from the origin of the coordinate system to the point is also frequently used. The vector has the length r and makes the angle ϕ with the x-axis. The complex conjugate number $(z)^* = z^* = x - iy$ is depicted by the point reflected upon the x-axis. It has the same r but its phase is $-\phi$

A complex number is conveniently shown as a point (x, y) on a 2D plane, called the *complex plane*, with its real and imaginary parts x and y being the point's coordinates, Fig. 1.25. The distance $r = \sqrt{x^2 + y^2}$ from the point to the origin of the coordinate system is called the *absolute value* $|z|$ of the complex number z, while the angle ϕ (see the figure) the *argument* or *phase*. Of course, $x = r \cos \phi$ and $y = r \sin \phi$, i.e. the complex number can also be written in its trigonometric form as

$$z = x + iy = r \left(\cos \phi + i \sin \phi \right) . \tag{1.126}$$

To the phase of a complex number a multiple of 2π may be added without any change of the complex number itself, i.e. the phases ϕ and $\phi + 2\pi k$ with $k = \pm 1, \pm 2, \dots$ result in the same complex number. The phase can be obtained from the values of $y = r \sin \phi$ and $x = r \cos \phi$ via

$$\frac{y}{x} = \frac{r \sin \phi}{r \cos \phi} = \tan \phi , \tag{1.127}$$

i.e. by solving the above equation with respect to the angle ϕ. We shall learn in Sect. 2.3.9 that there is a function which does the job, called arctangent, but for now we shall just limit ourselves with this result.

Problem 1.129 Determine the absolute value r and the phase ϕ of the following complex numbers: $1 + i$; $1 - i$; $\sqrt{3} + i$. [Answers: $r = \sqrt{2}, \phi = \pi/4$; $r = \sqrt{2}, \phi = -\pi/4$; $r = 2, \phi = \pi/6$.]

The vector from the coordinate origin to the point (x, y) on the complex plane can also be used to represent the complex number $z = x + iy$ on the plane, Fig. 1.25. In that case r is the vector length and ϕ is the angle the vector makes with the x-axis. Two identical complex numbers correspond to the same point on the complex plane.

If $y = 0$, the numbers are positioned on the x-axis and are purely real, and the phase is $\phi = \pi n$ with any positive and negative integer n, including zero: even n corresponds to positive, while odd n corresponds to the negative direction of the x-axis. If the complex number is purely imaginary, then the phase is given generally as $\phi = \pi/2 + \pi n$. Any arithmetic operation with any two real numbers would result in a number on the same axis, i.e. within the manifold of real numbers as we discussed at length above. However, taking a square root of a negative number takes the result immediately out of that axis, e.g. $\sqrt{-9} = \sqrt{9(-1)} = \sqrt{9}\sqrt{-1} = 3i$.

The addition and subtraction (two of the four elementary operations) on complex numbers are defined in the following way:

$$(x_1 + iy_1) + (x_2 + iy_2) = x_3 + iy_3 = (x_1 + x_2) + i(y_1 + y_2) \quad \text{(addition)}, \tag{1.128}$$

$$(x_1 + iy_1) - (x_2 + iy_2) = x_3 + iy_3 = (x_1 - x_2) + i(y_1 - y_2) \quad \text{(subtraction)}. \tag{1.129}$$

These definitions ensure that the sum (difference) of two numbers are equivalent to a sum (difference) of two vectors on the complex plane corresponding to them. It is also seen that the sum (difference) is commutative ($z_1 + z_2 = z_2 + z_1$ and $z_1 - z_2 = -z_2 + z_1$) and associative, e.g. $(z_1 + z_2) + z_3 = z_1 + (z_2 + z_3)$.

When defining the product of two numbers, $z_3 = z_1 z_2$, a number of conditions are to be satisfied: (i) the product must be equivalent to the product of real numbers if the imaginary parts of z_1 and z_2 are both zeros, i.e. $x_3 = x_1 x_2$ and $y_3 = 0$; (ii) the product should be a complex number, i.e. it must have the form $x_3 + iy_3$; (iii) algebraic rules must be respected. These conditions are satisfied if we treat a product of two complex numbers using usual algebraic rules; in fact, the form $x + iy$ already contains such a product (with i being multiplied by y, i.e. i is a number in its own right). Hence we *define* the product as follows:

$$(x_1 + iy_1)(x_2 + iy_2) = x_1 x_2 + iy_1 x_2 + ix_1 y_2 + i^2 y_1 y_2$$

$$= \left(x_1 x_2 + i^2 y_1 y_2\right) + i(x_1 y_2 + y_1 x_2) .$$

Since $i^2 = -1$ by definition, we finally obtain:

$$(x_1 + iy_1)(x_2 + iy_2) = (x_1 x_2 - y_1 y_2) + i(x_1 y_2 + y_1 x_2) \quad \text{(product)}. \tag{1.130}$$

Problem 1.130 Check explicitly that the product is both commutative ($z_1 z_2 = z_2 z_1$) and associative, i.e. $(z_1 z_2) z_3 = z_1 (z_2 z_3)$.

It is easy to see from Eq. (1.130) that if $y_1 = y_2 = 0$, then the imaginary part of the product is also equal to zero, i.e. we remain within the real numbers and that indeed we get $x_1 x_2$. Also, the product clearly has the form of a complex number.

Finally, the product is commutative, i.e. $z_1 z_2 = z_2 z_1$, and associative, $(z_1 z_2) z_3 = z_1 (z_2 z_3)$, which guarantee (together with the analogous properties of the sum and difference) that the algebraic rules, however complex, will indeed be respected when manipulating complex numbers.

Next, we consider division of complex numbers. This is defined as an operation inverse to the product, i.e. $z_1 / z_2 = z_3$ is understood in the sense that $z_1 = z_2 z_3$. It is shown in the next problem that

$$\frac{z_1}{z_2} = \frac{x_1 + i y_1}{x_2 + i y_2} = \frac{x_1 x_2 + y_1 y_2}{x_2^2 + y_2^2} + i \frac{y_1 x_2 - x_1 y_2}{x_2^2 + y_2^2} \quad \text{(division)}. \qquad (1.131)$$

Problem 1.131 Using this definition of the division and explicit definition of the product Eq. (1.130) prove the above formula for the division. [*Hint: write* $z_1 = z_2 z = z_2 (x + iy)$ *and solve for x and y using the fact that two complex numbers are equal if and only if their respective real and imaginary parts are equal.*]

We shall see later on in Sect. 3.4 that when we multiply two complex numbers, their absolute values multiply and the phases sum up; when we divide two numbers, their absolute values are divided and the phases are subtracted.

Finally, a special operation characteristic only to complex numbers is also introduced whereby the signs to all i in an expression are inverted. This is called *complex conjugation* and is denoted by the superscript *, i.e. $(5 - i6)^* = 5 + i6$. The complex conjugate number z^* corresponds to the reflection of the position of the number z on the complex plane with respect to the x-axis, Fig. 1.25.

Problem 1.132 Prove the following identities:

$$z + z^* = 2x \quad \text{and} \quad z - z^* = 2iy .$$

Problem 1.133 Prove the following inequality:

$$|z| = |x + iy| \le |x| + |y| . \qquad (1.132)$$

Problem 1.134 The square of the absolute value of a complex number r^2 is very often denoted $|z|^2$ and called the square of the module of the complex number z. Prove:

$$|z|^2 = x^2 + y^2 = z z^* . \qquad (1.133)$$

Another (much simpler) way of proving the result for the division (1.131) can be done using complex conjugation: multiply and divide the number z_1/z_2 by z_2^* and make use of Eq. (1.133):

$$\frac{x_1 + iy_1}{x_2 + iy_2} = \frac{x_1 + iy_1}{x_2 + iy_2} \cdot \frac{x_2 - iy_2}{x_2 - iy_2} = \frac{(x_1 + iy_1)(x_2 - iy_2)}{x_2^2 + y_2^2}$$

$$= \frac{x_1 x_2 + y_1 y_2}{x_2^2 + y_2^2} + i\frac{y_1 x_2 - x_1 y_2}{x_2^2 + y_2^2} .$$

Problem 1.135 Calculate the following complex numbers (i.e. present them in the form $a + ib$):

$$(1+i)(3-2i) , \quad (1+i)(3-2i)^* , \quad \frac{1+i}{3-2i} , \quad (1+i) - (3-2i) .$$

[Answers: $5 + i$; $1 + 5i$; $(1 + 5i)/13$; $-2 + 3i$.]

Problem 1.136 Work out the absolute values and the phases of the numbers: $2 + 2i, 2 - 2i, i, 1$ and -1. [Answers: $2\sqrt{2}, \pi/4$; $2\sqrt{2}, -\pi/4$; $1, \pi/2$; $1, 0$; $1, \pi$.]

1.11.3 Square Root of a Complex Number

Above we have introduced main algebraic operations on complex numbers, such as summation, subtraction, multiplication and division. These operations enable one to introduce their integer powers as, for instance,

$$(x + iy)^2 = (x + iy)(x + iy) = (x^2 - y^2) + i2xy,$$

i.e. the real part of the square is $x^2 - y^2$, while the imaginary part is $2xy$. Similarly, higher powers are also defined.

Problem 1.137 Show that

$$(x + iy)^4 = (x^4 - 6x^2y^2 + y^4) + i4xy(x^2 - y^2) .$$

More complex operations can also be defined such as an arbitrary power of a complex number. We shall postpone this to Sect. 3.4. Here we shall only consider the square root of a complex number.

Given the number $z = x + iy$, its square root $\sqrt{z} = a + ib$ is defined such that $(a + ib)^2 = x + iy$. Opening the square and comparing the real and imaginary parts on both sides, we obtain two equations for determining a and b:

$$\begin{cases} a^2 - b^2 = x \\ 2ab = y \end{cases} \implies \begin{cases} a^2 - b^2 = x \\ b = y/2a \end{cases} \implies a^2 - \left(\frac{y}{2a}\right)^2 = x .$$

The obtained equation is quadratic with respect to $c = a^2$:

$$4c^2 - 4xc - y^2 = 0 \implies c = \frac{1}{2}\left(x \pm \sqrt{x^2 + y^2}\right) .$$

As $a = \pm\sqrt{c}$ and is real, c must be positive. Hence, we have to reject the minus sign above and adopt a single value of c with the plus sign only. This yields two possible values for a:

$$a = \pm\sqrt{\frac{1}{2}\left(x + \sqrt{x^2 + y^2}\right)} \implies b = \frac{y}{2a} = \pm\frac{y}{\sqrt{2\left(x + \sqrt{x^2 + y^2}\right)}} .$$

Therefore, we obtain:

$$\sqrt{x + iy} = \pm\left(\sqrt{\frac{1}{2}\left(x + \sqrt{x^2 + y^2}\right)} + i\frac{y}{\sqrt{2\left(x + \sqrt{x^2 + y^2}\right)}}\right) . \qquad (1.134)$$

This formula is applicable to all cases apart from the one in which we calculate a square root of a negative real number. Indeed, if $x = -|x| < 0$ and $y = 0$, then $\sqrt{x^2 + y^2} = \sqrt{x^2} = |x|$ (recall that when calculating c in the derivation above, we adopted the positive value of the square root), and then the denominator of the imaginary part becomes $\sqrt{2(x + |x|)} = 0$. Hence, we have $0/0$ in formula (1.134) in the imaginary part.

Therefore, this case needs to be considered separately. We have:

$$\begin{cases} a^2 - b^2 = x \\ 2ab = 0 \end{cases} .$$

Because of the second equation, two cases are possible, either $a = 0$ or $b = 0$. If $b = 0$, then from the first equation we get $a^2 = x < 0$, which does not have solutions in real numbers (note that a and b must be real!), so this case need to be rejected. If

$a = 0$, then the first equation gives $-b^2 = x$ or $b^2 = -x = |x|$, so that $b = \pm\sqrt{|x|}$. Therefore, in this particular case, $\sqrt{-|x|} = \pm i\sqrt{|x|}$.

If $y = 0$, but $x > 0$, then from Eq. (1.134) we immediately obtain

$$\sqrt{x + i0} = \pm\sqrt{\frac{1}{2}\left(x + \sqrt{x^2}\right)} = \pm\sqrt{\frac{1}{2}2x} = \pm\sqrt{x}\,,$$

i.e. what we would have expected when taking the square root of a real positive number.

Problem 1.138 Show that formula (1.134) can also be rewritten as ($y \neq 0$)

$$\sqrt{x + iy} = \pm\left(\sqrt{\frac{1}{2}\left(x + \sqrt{x^2 + y^2}\right)} + i\frac{y}{|y|}\sqrt{\frac{1}{2}\left(-x + \sqrt{x^2 + y^2}\right)}\right).$$
$$(1.135)$$

Problem 1.139 Using Eq. (1.134) or (1.135), prove the following identities:

$$\sqrt{1 + i} = \pm\left(\sqrt{\frac{1}{2}\left(1 + \sqrt{2}\right)} + i\sqrt{\frac{1}{2}\left(\sqrt{2} - 1\right)}\right);$$

$$\sqrt{\frac{1}{2} - 5i} = \pm\left(\frac{1}{2}\sqrt{1 + \sqrt{101}} - \frac{i}{2}\sqrt{\sqrt{101} - 1}\right);$$

$$\sqrt{2 + i\sqrt{2}} = \pm\left(\sqrt{\frac{1}{2}\left(2 + \sqrt{6}\right)} + i\sqrt{\frac{1}{2}\left(-2 + \sqrt{6}\right)}\right).$$

Using complex numbers, one can now solve any quadratic equation using Eq. (1.18). For example, consider

$$x^2 + 2x + 2 = 0 \quad \Longrightarrow \quad x_\pm = -1 \pm \sqrt{1 - 2} = -1 \pm i\,.$$

Problem 1.140 Solve $x^2 + 2x + 4 = 0$. [Answer: $-1 \pm i\sqrt{3}$.]

Problem 1.141 Solve $x^2 + 2ix + 4 = 0$. [Answer: $i\left(-1 \pm \sqrt{5}\right)$.]

Problem 1.142 Solve $x^2 + (1 - i)x + 4 + 7i = 0$. [Answer: $1 - 2i$ *and* $-2 + 3i$.]

Formulae (1.134) and (1.135) for the root $\sqrt{z}$ of the complex number z are quite cumbersome; also, they do not work for a special case of the real part of z being negative and the imaginary part being zero. There is another, much more convenient (and more powerful, see Sect. 3.4), formula based on the trigonometric form (1.126) of the complex number. Consider

$$z = r \left(\cos \psi + i \sin \psi \right) ,$$

and the root of it,

$$\sqrt{z} = R \left(\cos \varphi + i \sin \varphi \right) .$$

We should calculate the absolute value, R, and the phase, φ, of the square root. From definition,

$$R^2 \left(\cos \varphi + i \sin \varphi \right)^2 = r \left(\cos \psi + i \sin \psi \right) ,$$

i.e. equating the real and imaginary parts on both sides, we get:

$$\begin{cases} R^2 \left(\cos^2 \varphi - \sin^2 \varphi \right) = r \cos \psi \\ 2R^2 \cos \varphi \sin \varphi = r \sin \psi \end{cases} .$$

Using Eqs. (1.75) and (1.76), these equations can be manipulated into:

$$\begin{cases} R^2 \cos (2\varphi) = r \cos \psi \\ R^2 \sin (2\varphi) = r \sin \psi \end{cases} . \tag{1.136}$$

Square both parts of each equation and add them together:

$$R^4 \left(\cos^2 (2\varphi) + \sin^2 (2\varphi) \right) = r^2 \left(\cos^2 \psi + \sin^2 \psi \right)$$
$$\Rightarrow R^4 = r^2 \Rightarrow R = \sqrt{r} .$$

So, the absolute value of the square root of a complex number is the square root of its absolute value. Then, Eq. (1.136) takes a simpler form:

$$\begin{cases} \cos (2\varphi) = \cos \psi \\ \sin (2\varphi) = \sin \psi \end{cases} , \tag{1.137}$$

from which we deduce that $2\varphi = \psi + 2\pi n$, where $n = 0, \pm 1, \pm 2, \ldots$. Only two values of n lead to different values of the phase φ, which are $n = 0$, yielding $\varphi = \frac{\psi}{2}$, and $n = 1$, giving $\varphi = \frac{\psi}{2} + \pi$; all others result in equivalent phases. Hence, we arrive at two solutions:

$$\left(\sqrt{z}\right)_1 = \sqrt{r}\left(\cos\frac{\psi}{2} + i\sin\frac{\psi}{2}\right) \tag{1.138}$$

and

$$\left(\sqrt{z}\right)_2 = \sqrt{r}\left[\cos\left(\frac{\psi}{2}+\pi\right) + i\sin\left(\frac{\psi}{2}+\pi\right)\right] = -\sqrt{r}\left(\cos\frac{\psi}{2} + i\sin\frac{\psi}{2}\right). \tag{1.139}$$

The obtained equations do not have limitations and work in all cases. Indeed, consider positive real numbers r, for which the phase $\psi = 0$. In this case $\left(\sqrt{z}\right)_{1,2} = \pm\sqrt{r}$. For negative real numbers $\psi = \pi$ and hence $\left(\sqrt{z}\right)_{1,2} = \pm i\sqrt{r}$.

Problem 1.143 Repeat the calculation of Problem 1.139 using this time the trigonometric form of the square root. Check the following identities:

$$\sqrt{1+i} = \pm\sqrt{2}\left(\cos\frac{\pi}{8} + i\sin\frac{\pi}{8}\right) ;$$

$$\sqrt{\frac{1}{2}-5i} = \pm\frac{101^{1/4}}{\sqrt{2}}\left(\cos\frac{\varphi}{2} - i\sin\frac{\varphi}{2}\right), \quad \tan\varphi = 10 ;$$

$$\sqrt{2+i\sqrt{2}} = \pm 6^{1/4}\left(\cos\frac{\varphi}{2} + i\sin\frac{\varphi}{2}\right), \quad \tan\varphi = \frac{1}{\sqrt{2}} .$$

1.11.4 Polynomials with Complex Coefficients

The theorem proven for real polynomials can now be straightforwardly extended for polynomials with complex coefficients. Namely, the little Bèzout's theorem is also valid, Eq. (1.34), and its consequence, Eq. (1.35), for a single root. Hence, in the same way as in Sect. 1.4.1, we can prove that a general polynomial $P_n(x)$ has exactly n complex roots and can be written via its roots using Eq. (1.37) or (1.48).

Problem 1.144 Prove identities:

$$x^2 + 1 = (x+i)(x-i) ;$$

$$x^4 + 1 = \left(x - \frac{1+i}{\sqrt{2}}\right)\left(x - \frac{1-i}{\sqrt{2}}\right)\left(x + \frac{1+i}{\sqrt{2}}\right)\left(x + \frac{1-i}{\sqrt{2}}\right).$$

Problem 1.145 Obtain the polynomial that has three roots: $x_1 = i$, $x_2 = -i$ and $x_3 = \frac{1}{2} + i$. [Answer: $-(1 + 2i) + 2x - (1 + 2i)x^2 + 2x^3$.]

1.11.5 Factorisation of a Polynomial with Real Coefficients

We shall now consider a polynomial

$$P_n(x) = a_0 + a_1 x + \cdots + a_n x^n$$

of order n with real coefficients. Assume that this polynomial has a complex root $z = a + ib$ (with $b > 0$), i.e. $P_n(a + ib) = 0$. Let us take the complex conjugate of this expression: $P_n(a - ib) = 0$. This proves that the complex number $z^* = a - ib$ must also be a root of the polynomial. This proves an important statement: in a polynomial with real coefficients complex roots (if exist) always appear in complex conjugate pairs.

Consider now a product representation of a polynomial via its roots, Eq. (1.37). Specifically, let us consider the product of the differences related to two complex conjugate roots z and z^*:

$$(x - z)\left(x - z^*\right) = (x - z)(x - z)^* = ZZ^* = X^2 + Y^2 \, ,$$

where $Z = x - z = X + iY$ with the real and imaginary parts being $X = x - a$ and $Y = -b$. Hence,

$$(x - z)\left(x - z^*\right) = (x - a)^2 + b^2 = x^2 - 2ax + \left(b^2 + a^2\right)$$

is a square polynomial. It does not have real roots. Indeed, using Eq. (1.18), we get:

$$x_\pm = a \pm \sqrt{a^2 - \left(b^2 + a^2\right)} = a \pm \sqrt{-b^2} = a \pm ib \, ,$$

as required. Therefore, any polynomial with real coefficients can be written as a product of factors $x - x_k$ for each real root and of factors $x^2 - 2ax + \left(b^2 + a^2\right)$ due to each pair $a \pm ib$ of complex conjugate roots.

1.12 Summation of Finite Series

It is often necessary to sum a finite set of terms[17]

$$S_n = a_0 + a_1 + a_2 + \cdots + a_n = \sum_{i=0}^{n} a_i \, , \tag{1.140}$$

---

[17] In this subsection, i is the summation index, *not* the complex $i = \sqrt{-1}$.

containing $n + 1$ terms constructed via a certain rule, i.e. each term a_i is constructed from its number i (the index). Here we shall consider several examples of such finite sums.

In a *geometric progression* $a_i = a_0 q^i$, i.e. the numbers a_i form a finite *sequence* $a_0, a_0 q, a_0 q^2, \ldots, a_0 q^n$. This means that each next term is obtained from the previous one via a multiplication with q:

$$a_{i+1} = a_i q , \text{ where } i = 0, 1, 2, \ldots . \tag{1.141}$$

This is called a *recurrence relation*. We would like to calculate the sum

$$S_n = \sum_{i=0}^{n} a_i = a_0 + a_0 q + a_0 q^2 + \cdots + a_0 q^n , \tag{1.142}$$

containing $n + 1$ terms. The trick here is to notice that if we multiply S_n by q, we can manipulate the result into an expression containing S_n again:

$$S_n q = a_0 q + a_0 q^2 + a_0 q^3 + \cdots + a_0 q^{n+1}$$

$$= \left(a_0 + a_0 q + a_0 q^2 + \cdots + a_0 q^n \right) - a_0 + a_0 q^{n+1}$$

$$= S_n + a_0 \left(q^{n+1} - 1 \right) .$$

Solving this equation with respect to S_n, we finally obtain:

$$S_n = a_0 \frac{q^{n+1} - 1}{q - 1} , \tag{1.143}$$

which is the desired result. Note that this formula is valid for both $q > 1$ and $q < 1$, but not for $q = 1$ as in this case we have a singularity (division of zero by zero). In the latter case, however, all terms in the sum (1.142) are identical and $S_n = (n + 1)a_0$ which is a trivial result. We shall see later on in Sect. 3.10 that the formula (1.143), in spite of the singularity, can in fact still be used for the case of $q = 1$, i.e. it is actually quite general, but we have to learn limits before we can establish that and hence have to postpone the necessary discussion.

Although we have derived the sum of the geometric progression from its definition, it is also instructive to illustrate using this example yet again how the induction principle we discussed in Sect. 1.1 works. So, let us suppose that we *are given* the result (1.143) for the sum of the progression (1.142), and now we would like to *prove* that it is correct for any n. First of all, we check that the formula (1.143) works for the lowest value of $n = 0$ and, indeed, we get

$$S_0 = a_0 \frac{q^1 - 1}{q - 1} = a_0 ,$$

as expected. Then we *assume* that (1.143) is valid for some general n, and hence we consider

$$S_{n+1} = S_n + a_0 q^{n+1} = a_0 \frac{q^{n+1} - 1}{q - 1} + a_0 q^{n+1}$$

$$= a_0 \frac{q^{n+1} - 1 + q^{n+2} - q^{n+1}}{q - 1} = a_0 \frac{q^{n+2} - 1}{q - 1} \, ,$$

which does have the correct form as it can be obtained from (1.143) by the substitution $n \to n + 1$, as required. According to the principle of mathematical induction, the tried formula is proven to be correct.

Example 1.13 ► Sum up the finite series

$$3^2 + 3 + 1 + \frac{1}{3} + \frac{1}{3^2} + \cdots + \frac{1}{3^{10}} \, .$$

Solution. This is a geometric progression with $a_0 = 3^2 = 9$ and $q = 1/3$, containing $n + 1 = 13$ terms. Using our general result (1.143), we thus get

$$S_{12} = 9 \frac{(1/3)^{13} - 1}{1/3 - 1} = \frac{27}{2} \left(1 - \frac{1}{3^{13}} \right) . \, ◄$$

Problem 1.146 Sum up the finite numerical series:

(a) $1 + 2 + 4 + 8 + \cdots + 1024$; (b) $3^2 - 3 + 1 - \frac{1}{3} + \frac{1}{3^2} - \cdots + \frac{1}{3^{10}}$.

[Answers: $2^{11} - 1$; $\left(3^3 + 3^{-10} \right) / 4$.]

Problem 1.147 Show that

$$\sum_{n=1}^{N} \left(a^n + \frac{1}{a^n} \right)^2 = 2N + \frac{\left(a^{2N+2} + 1 \right) \left(a^{2N} - 1 \right)}{a^{2N} \left(a^2 - 1 \right)} \, .$$

[Hint: *open the brackets first.*]

In an *arithmetic progression* $a_i = a_0 + ir$, i.e. each term is obtained from the previous one by adding a constant r:

$$a_{i+1} = a_i + r \, , \quad \text{where } i = 0, 1, 2, \ldots , \tag{1.144}$$

and we would like to calculate the sum:

$$S_n = \sum_{i=0}^{n} a_i = a_0 + [a_0 + r] + [a_0 + 2r] + \cdots + [a_0 + (n-1)r] + [a_0 + nr] .$$

(1.145)

The trick here which will enable us to derive a simple expression for the sum S_n is to notice that each term in the series can also be written via the last term: $a_k = a_0 + kr = a_n - (n-k)r$. Therefore, the sum (1.145) can alternatively be written as:

$$S_n = [a_n - nr] + [a_n - (n-1)r] + [a_n - (n-2)r] + \cdots + [a_n - r] + a_n .$$

(1.146)

Here we have exactly the same terms containing r as in (1.145), but in the reverse order and with the minus sign each. Therefore, all these terms cancel out if we sum the two equalities (1.145) and (1.146):

$$2S_n = (n+1)a_0 + (n+1)a_n \;\Rightarrow\; S_n = \frac{a_0 + a_n}{2}(n+1) ,$$

(1.147)

where the factor of $(n+1)$ appeared since we have both a_0 and a_n that number of times in the two sums above. The final result is very simple: you take the average of the first and the last terms which is to be multiplied by the total number of terms in the finite series (which is $n+1$).

Problem 1.148 Calculate the sum of all integers from 10 to 1000. [Answer: 500455.]

Many other finite series are known, some of them are given in the problems below.

Problem 1.149 Use the mathematical induction to prove the formulae given below:

$$K_2(n) = \sum_{i=1}^{n} i^2 = 1 + 2^2 + 3^2 + \cdots + n^2 = \frac{1}{6}n(n+1)(2n+1) ;$$ (1.148)

$$K_3(n) = \sum_{i=1}^{n} i^3 = 1 + 2^3 + 3^3 + \cdots + n^3 = \frac{1}{4}n^2(n+1)^2 ;$$ (1.149)

$$\sum_{i=2}^{n} \frac{1}{i^2 - 1} = \frac{3}{4} - \frac{2n+1}{2n(n+1)} ;$$ (1.150)

$$\sum_{i=0}^{n}(a_0 + ir)q^i = a_0\frac{1 - q^{n+1}}{1 - q} + \frac{rq}{(1 - q)^2}\left[1 - (n + 1)q^n + nq^{n+1}\right].$$

(1.151)

The latter series is a combination of geometric and arithmetic progressions.

Problem 1.150 Prove that

$$\sum_{i=1}^{n}\frac{1}{i(i + 1)} = 1 - \frac{1}{n + 1}.$$

(1.152)

[Hint: *make use of the fact that* $\frac{1}{i(i+1)} = \frac{1}{i} - \frac{1}{i+1}$.]

Problem 1.151 Prove the following identities using the previous results:

$$\sum_{i=1}^{n}(2i)^2 = 2^2 + 4^2 + 6^2 + \cdots + (2n)^2 = \frac{2}{3}n(n + 1)(2n + 1);$$

$$\sum_{i=0}^{n}(2i + 1)^2 = 1^2 + 3^2 + 5^2 + \cdots + (2n + 1)^2 = \frac{1}{3}(n + 1)(2n + 1)(2n + 3).$$

Problem 1.152 The following steps enable one to derive the sum in Eq. (1.148). Consider the difference

$$(i + 1)^3 - i^3 = 3i^2 + 3i + 1.$$

Write this equation for $i = 1, 2, 3, \ldots, n$ and then sum all equations up.

Problem 1.153 To derive Eq. (1.149), use a similar method by considering the difference $(i + 1)^4 - i^4$.

Problem 1.154 Prove the following formulae (instead of the induction, conceive a method based on the knowledge of the sums in Eqs. (1.148) and (1.149)):

Fig. 1.26 Spherical balls arranged into a pyramid. Each layer consists of a square arrangement of the balls

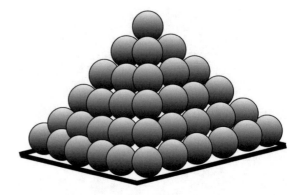

$$(a)\ 1 \cdot 2 + 2 \cdot 3 + \cdots + n(n+1) = \frac{1}{3}n(n+1)(n+2)\ ;$$

$$(b)\ 1 \cdot 2 \cdot 3 + 2 \cdot 3 \cdot 4 + \cdots + n(n+1)(n+2) = \frac{1}{4}n(n+1)(n+2)(n+3)\ .$$

Problem 1.155 Consider packing of balls into a pyramid as shown in Fig. 1.26. Each layer is formed by a square arrangement of the balls. If the bottom layer contains n^2 balls, show that the total number of balls is $\frac{1}{6}n(n+1)(2n+1)$.

Problem 1.156 Similarly to the previous problem, consider a pyramid of spherical balls, but this time each layer having their triangular arrangement. Show that if the bottom layer has n balls on one side, the total number of balls will be $\frac{1}{6}n(n+1)(n+2)$.

Problem 1.157 Consider a hexagonal arrangement of spherical balls with n balls on each side of a hexagon. Show that the total number of such balls in the hexagon is $3n(n-1)+1$.

1.13   Binomial Formula

There are numerous examples when one needs to expand an expression like $(a+b)^n$ with an integer power $n \geq 1$. We can easily multiply the brackets for small values of n to get:

$$(a+b)^1 = a + b = a^1b^0 + a^0b^1\ ,$$

Fig. 1.27 Pascal's triangle

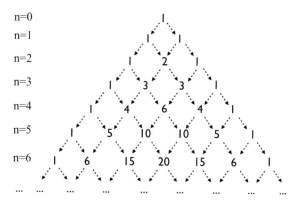

$$(a + b)^2 = a^2 + 2ab + b^2 = a^2b^0 + 2a^1b^1 + a^0b^2 \,,$$

$$(a + b)^3 = a^3 + 3a^2b + 3ab^2 + b^3 = a^3b^0 + 3a^2b^1 + 3a^1b^2 + a^0b^3 \,,$$

$$(a + b)^4 = a^4 + 4a^3b + 6a^2b^2 + 4ab^3 + b^4 = a^4b^0 + 4a^3b^1 + 6a^2b^2 + 4a^1b^3 + a^0b^4 \,.$$

This can be continued, but the expressions get much longer. We can notice, however, some general rules in expanding the $(a + b)^n$ from the particular cases considered above. Indeed, the expansions contain all possible products $a^i b^{n-i}$, starting from $i = 0$ and ending with $i = n$, such that the sum of powers is always $i + (n - i) = n$. The coefficients to these products form sequences: $(1, 1)$ for $n = 1$, $(1, 2, 1)$ for $n = 2$, $(1, 3, 3, 1)$ for $n = 3$, $(1, 4, 6, 4, 1)$ for $n = 4$, and so on. One may notice that a set of coefficients for the power $n + 1$ can actually be obtained from the coefficients corresponding to the previous power n. This elegant rule known as Pascal's triangle[18] is illustrated in Fig. 1.27. Consider, for instance, the next to $n = 4$ set of coefficients, i.e. for $n = 5$. There will be six terms in the expansion with coefficients $(1, 5, 10, 10, 5, 1)$, and they are constructed following two simple rules: (i) the first and the last coefficients (for the terms $a^5b^0 = a^5$ and $a^0b^5 = b^5$) are 1; (ii) any other coefficient is obtained as a sum of the two coefficients directly above it from the previous line (corresponding to $n = 4$ in our case).

Although this construction is elegant, it can only be useful for small values of n; moreover, it is not useful at all if one has to manipulate the binomial expressions $(a + b)^n$ analytically for a general integer n. This problem is solved by the *binomial formula* (or theorem)[19]:

$$(a + b)^n = \sum_{i=0}^{n} \binom{n}{i} a^i b^{n-i} \,, \tag{1.153}$$

---

[18] After Blaise Pascal.

[19] The formula and the triangle were described by Blaise Pascal, but apparently were known before him; in particular, a Persian mathematician and engineer Abu Bakr ibn Muhammad ibn al Husayn al-Karaji (or al-Karkhi) wrote about them in the tenth century as well; he has also proved the formula by introducing the principle of mathematical induction.

where

$$\binom{n}{i} = \frac{n!}{i!(n-i)!} \qquad (1.154)$$

are called *binomial coefficients*. Recall that $n!$ represents a product of integers between 1 and n, and that by definition $0! = 1$.

Problem 1.158 Show that

$$\frac{1}{2}\left(1+\frac{1}{2}\right)\left(2+\frac{1}{2}\right)\ldots\left(n-\frac{1}{2}\right)\left(n+\frac{1}{2}\right) = \frac{(2n+1)!}{2^{2n+1}n!} ,$$

where n is a positive integer.

The formula (1.153) can also formally be written as follows:

$$(a+b)^n = \sum_{k_1+k_2=n} \binom{n}{k_1,k_2} a^{k_1} b^{k_2} , \qquad (1.155)$$

where $\binom{n}{k_1,k_2} = \binom{n}{k_1}$. The sum in this formula is actually a double sum as it is taken over all possible indices k_1 and k_2 between 0 and n such that $k_1 + k_2 = n$. It is easy to see that this formula is equivalent to the previous one, Eq. (1.153): if we take $k_1 \equiv i$, then $k_2 = n - i$ has a single value and the sum in (1.155) becomes a single sum over i between 0 and n.

Let us establish some essential properties of these coefficients:

$$\binom{n}{0} = \frac{n!}{0!n!} = 1 , \quad \binom{n}{n} = \frac{n!}{n!0!} = 1 , \qquad (1.156)$$

i.e. the first and the last coefficients (which are by the highest powers of a and b) are, for any n, equal to 1. This is the first property of Pascal's triangle. The second property of Pascal's triangle is established by the following identity:

$$\begin{aligned}
\binom{n}{i-1} + \binom{n}{i} &= \frac{n!}{(i-1)!(n-i+1)!} + \frac{n!}{i!(n-i)!} \\
&= \frac{n!}{(i-1)!(n-i)!}\left[\frac{1}{n-i+1} + \frac{1}{i}\right] \\
&= \frac{n!}{(i-1)!(n-i)!}\frac{n+1}{i(n-i+1)} \\
&= \frac{(n+1)!}{i!(n+1-i)!} = \binom{n+1}{i} ,
\end{aligned} \qquad (1.157)$$

where we have made use of the fact that $m! = (m - 1)!m$ for any positive integer m. We can now indeed see that two binomial coefficients by the $(i - 1)$th and ith terms of the expansion of $(a + b)^n$ give rise to the ith coefficient of the $(a + b)^{n+1}$ expansion as stipulated by Pascal's triangle.

Problem 1.159 Show by direct calculation that

$$\binom{n}{1} = \binom{n}{n-1} = n \; ; \quad \binom{n}{2} = \binom{n}{n-2} = \frac{1}{2} n(n - 1) \; ;$$

$$\binom{n}{3} = \binom{n}{n-3} = \frac{1}{6} n(n - 1)(n - 2) \; ;$$

$$\binom{n}{4} = \binom{n}{n-4} = \frac{1}{24} n(n - 1)(n - 2)(n - 3) \; .$$

Problem 1.160 Prove the symmetry property:

$$\binom{n}{k} = \binom{n}{n-k} \; , \tag{1.158}$$

which basically means that the binomial coefficients are the same in the binomial formula whether one counts from the beginning to the end or *vice versa*. Equation (1.156) is a particular case of this symmetry written for the first and the last coefficients.

Problem 1.161 Prove an identity:

$$\binom{n}{k}\binom{n-k}{m-k} = \binom{m}{k}\binom{n}{m} \; .$$

These properties do not yet prove the binomial formula. However, before proving it, let us first see that it actually works. Consider the case of $n = 5$. Since the sum runs from $i = 0$ up to $i = 5$, there will be six terms with the coefficients:

$$\binom{5}{0} = 1 \; , \quad \binom{5}{1} = \frac{5!}{1!4!} = 5 \; , \quad \binom{5}{2} = \frac{5!}{2!3!} = 10 \; ,$$

$$\binom{5}{3} = \frac{5!}{3!2!} = 10 \; , \quad \binom{5}{4} = \frac{5!}{4!1!} = 5 \; , \quad \binom{5}{0} = 1 \; ,$$

i.e. the same results as we have got with Pascal's triangle.

Now we shall prove the formula using induction. First of all, we check that it works for $n = 1$. Indeed, since $\binom{1}{0} = \binom{1}{1} = 1$, we immediately see that $(a+b)^1 = 1 \cdot a^1 b^0 + 1 \cdot a^0 b^1 = a + b$, as required. Next, we *assume* that formula (1.153) is valid for a general n, and we then consider the expansion for the next value of $(n+1)$:

$$(a+b)^{n+1} = (a+b)(a+b)^n = (a+b) \sum_{i=0}^{n} \binom{n}{i} a^i b^{n-i}$$

$$= \sum_{i=0}^{n} \binom{n}{i} a^{i+1} b^{n-i} + \sum_{i=0}^{n} \binom{n}{i} a^i b^{n+1-i} .$$

In the first sum we separate out the term with $i = n$, while in the second the term with $i = 0$:

$$(a+b)^{n+1} = \binom{n}{n} a^{n+1} b^0 + \sum_{i=0}^{n-1} \binom{n}{i} a^{i+1} b^{n-i} + \sum_{i=1}^{n} \binom{n}{i} a^i b^{n+1-i} + \binom{n}{0} a^0 b^{n+1} .$$

Note that now the summation index i ends at $(n-1)$ in the first sum and starts from $i = 1$ in the second. Also note that, according to properties (1.156), the first and the last binomial coefficients in the last expression are both equal to one. We shall now change the summation index in the first sum to make it look similar to the second: instead of i running between 0 and $(n-1)$ we introduce $k = i + 1$ changing between 1 and n:

$$\sum_{i=0}^{n-1} \binom{n}{i} a^{i+1} b^{n-i} = \sum_{k=1}^{n} \binom{n}{k-1} a^k b^{n-k+1} = \sum_{i=1}^{n} \binom{n}{i-1} a^i b^{n-i+1} ,$$

where in the last passage we replaced k back with i for convenience.[20] Now the two sums look really similar and we can combine all our manipulations into:

$$(a+b)^{n+1} = a^{n+1} b^0 + \sum_{i=1}^{n} \left[\binom{n}{i-1} + \binom{n}{i} \right] a^i b^{n-i+1} + a^0 b^{n+1} . \quad (1.159)$$

Since $\binom{n+1}{0} = \binom{n+1}{n+1} = 1$ according to Eq. (1.156) and the sum of two binomial coefficients in the square brackets is $\binom{n+1}{i}$ because of Eq. (1.157), we see that we managed to manipulate $(a+b)^{n+1}$ in (1.159) exactly into the right-hand side of (1.153) with $(n+1)$ instead of n. This completes the proof by induction of the binomial formula.

---

[20] After all, the summation index is a dummy index which is only used to indicate how the summation is made; any letter can be used for it.

Example 1.14 ▶ Calculate the following sum of the binomial coefficients:

$$\sum_{i=0}^{n} \binom{n}{i} = 1 + \binom{n}{1} + \binom{n}{2} + \cdots + \binom{n}{n-1} + 1 \, .$$

Solution. Consider the binomial expansion of order n for $a = b = 1$:

$$(1+1)^n = 2^n \equiv 1 + \binom{n}{1} + \binom{n}{2} + \cdots + \binom{n}{n-1} + 1 \, , \qquad (1.160)$$

which shows that the sum is equal to 2^n. ◀

Problem 1.162 Evaluate the expression $(1+0.1)^5$. [Answer: 1.61051.]

Problem 1.163 Evaluate the expression $(1+0.01)^5$ to the fourth decimal digit. [Hint: *estimate first every term in the binomial expansion to decide which ones are necessary to include within the required precision.* Answer: $\simeq 1.051$.]

Problem 1.164 Calculate $\left(1+\sqrt{5}\right)^5$. [Answer: $176 + 80\sqrt{5}$.]

Problem 1.165 Calculate $\left(1+\sqrt{2}\right)^7$. [Answer: $239 + 169\sqrt{2}$.]

Problem 1.166 Prove the formulae:

$$\binom{n}{2} + \binom{n}{4} + \binom{n}{6} + \cdots + \binom{n}{n-1} = 2^{n-1} - 1 \, , \quad \text{if } n = \text{odd,}$$
$$(1.161)$$

$$\binom{n}{2} + \binom{n}{4} + \binom{n}{6} + \cdots + \binom{n}{n-2} = 2^{n-1} - 2 \, , \quad \text{if } n = \text{even.}$$
$$(1.162)$$

[Hint: *consider the expansion for* $(1+1)^{n-1}$ *and make use of Eq.* (1.157).]

Problem 1.167 Similarly, prove:

$$\binom{n}{1} + \binom{n}{3} + \binom{n}{5} + \cdots + \binom{n}{n-2} = 2^{n-1} - 1 \, , \quad \text{if } n = \text{odd,}$$
$$(1.163)$$

$$\binom{n}{1} + \binom{n}{3} + \binom{n}{5} + \cdots + \binom{n}{n-1} = 2^{n-1}, \quad \text{if } n = \text{even}. \quad (1.164)$$

Problem 1.168 Prove that

$$\sum_{k=0}^{n} \binom{n}{k}^2 = \sum_{k=0}^{n} \binom{n}{k}\binom{n}{n-k} = \binom{2n}{n}. \quad (1.165)$$

[Hint: *expand* $(1 + x)^n (1 + x)^n$ *in two different ways and compare the coefficients by* x^n. *Note the symmetry property* (1.158).]

Many other identities involving the binomial coefficients can be established in a similar fashion by giving a and b in $(a + b)^n$ special values.

1.14   *Summae Potestatum* and Bernoulli Numbers

In Sect. 1.12, we have come across a problem of finding a finite sum of p-powers of the first n integers,

$$K_p(n) = \sum_{i=1}^{n} i^p = 1^p + 2^p + \cdots + n^p. \quad (1.166)$$

Indeed, in Eqs. (1.148) and (1.149), we obtained closed results for the sum of squares, $K_2(n)$, and cubes, $K_3(n)$, of the integers, respectively. The sum of the first n integer numbers also belongs to the same class of the finite sums corresponding to $p = 1$,

$$K_1(n) = 1^1 + 2^1 + \cdots + n^1 = \sum_{i=1}^{n} i = \frac{1}{2}(1 + n)n, \quad (1.167)$$

it is easily summed up representing an arithmetic progression, see Eq. (1.147).

It is seen from these three examples that an expression for $K_p(n)$ seems to be a polynomial of n of the order $p + 1$. A general problem of calculating the sum of p-powers of the integers was considered by Jacob Bernoulli. In his book "Ars Conjectandi" that was published posthumously in 1713 under the title *Summae Potestatum* Bernoulli gave a general form of such a polynomial in n for a general value of p:

$$K_p(n) = \frac{1}{p+1} \sum_{j=0}^{p} (-1)^j B_j \binom{p+1}{j} n^{p+1-j}, \tag{1.168}$$

where B_j are universal numbers (they are the same for any value of p, which is non-trivial!) that are nowadays called by Bernoulli's name.[21] The first few Bernoulli numbers are:

$$B_0 = 1, \ B_1 = -\frac{1}{2}, \ B_2 = \frac{1}{6}, \ B_3 = 0, \ B_4 = -\frac{1}{30}, \ B_5 = 0, \ B_6 = \frac{1}{42}, \ B_7 = 0, \ B_8 = -\frac{1}{30}. \tag{1.169}$$

Problem 1.169 Using Bernoulli formula (1.168), confirm Eqs. (1.167), (1.148) and (1.149) for $p = 1, 2, 3$. Then obtain the following results:

$$K_4(n) = \frac{1}{30} n(n + 1)(2n + 1) \left(3n^2 + 3n - 1\right) ; \tag{1.170}$$

$$K_5(n) = \frac{1}{12} n^2 (n + 1)^2 (2n^2 + 2n - 1) . \tag{1.171}$$

We shall prove formula (1.168) in Sect. 7.3.4. In the following, we shall develop a simple recursive procedure for calculating $K_p(n)$ that is not based on Bernoulli numbers, but still enables one to calculate such sums progressively from smaller to larger values of the power p. To this end, consider the following expression:

$$\sum_{i=1}^{n} (i + 1)^{p+1} - \sum_{i=1}^{n} i^{p+1} = \left[2^{p+1} + \cdots + (n + 1)^{p+1}\right] - \left[1^{p+1} + \cdots + n^{p+1}\right]$$

$$= (n + 1)^{p+1} - 1 .$$

On the other hand, using the binomial formula for $(i + 1)^{p+1}$, the same expression in the left-hand side can be manipulated as follows:

---

[21] Actually, Bernoulli started his summation from $j = 2$ separating out the $j = 0$ and $j = 1$ terms, basically pre-defining the first two numbers as $B_0 = 1$ and $B_1 = -1/2$.

$$\sum_{i=1}^{n}\left[(i+1)^{p+1}-i^{p+1}\right]=\sum_{i=1}^{n}\left[\sum_{q=0}^{p+1}\binom{p+1}{q}i^q-i^{p+1}\right]=\sum_{i=1}^{n}\sum_{q=0}^{p}\binom{p+1}{q}i^q$$

$$=\sum_{q=0}^{p}\binom{p+1}{q}\sum_{i=1}^{n}i^q=\sum_{q=0}^{p}\binom{p+1}{q}K_q(n)$$

$$=\underbrace{\binom{p+1}{0}}_{1}\underbrace{K_0(n)}_{n}+\sum_{q=1}^{p-1}\binom{p+1}{q}K_q(n)+\underbrace{\binom{p+1}{p}}_{p+1}K_p(n)$$

$$=n+\sum_{q=1}^{p-1}\binom{p+1}{q}K_q(n)+(p+1)K_p(n).$$

Comparing both expressions, we obtain the following final result:

$$K_p(n)=\frac{1}{p+1}\left\{(n+1)\left[(n+1)^p-1\right]-\sum_{q=1}^{p-1}\binom{p+1}{q}K_q(n)\right\}.\qquad(1.172)$$

This is a recurrence relation that enables one to derive a particular sum $K_p(n)$ of integers in powers p given all the analogous sums with smaller powers q.

It is a simple exercise to re-derive now our formulae (1.148) and (1.149) using the recurrence relation (1.172). Indeed,

$$K_2(p)=\frac{1}{3}\left\{(n+1)\left[(n+1)^2-1\right]-\binom{3}{1}K_1(n)\right\}=\frac{1}{6}n(n+1)(2n+1),$$

$$K_3(p)=\frac{1}{4}\left\{(n+1)\left[(n+1)^3-1\right]-\binom{4}{1}K_1(n)-\binom{4}{2}K_2(n)\right\}=\frac{1}{4}n^2(n+1)^2,$$

as required.

Problem 1.170 Re-derive $K_4(n)$ and $K_5(n)$ given in Eqs. (1.170) and (1.171) using the recursive formula (1.172).

Problem 1.171 Prove that

$$1\cdot2\cdot3\cdot4+2\cdot3\cdot4\cdot5+3\cdot4\cdot5\cdot6+\cdots+n(n+1)(n+2)(n+3)$$

$$=\frac{1}{5}n(n+1)(n+2)(n+3)(n+4).$$

[Hint: *first, write a general kth term of the sum, a_k, via its number k and then relate the sum $\sum_k a_k$ to the sums (1.166), there will be a few values of p which you would need.*]

1.15 Prime Numbers

Throughout history of mathematics scientists and philosophers studied the so-called *prime numbers*, i.e. integers greater than 1 that are not divisible by any integer between 1 and themselves. For instance, 4 is divisible by 2 (apart from 1 and 4), so it is not a prime number, while 3 can only be divisible by itself and 1, and so it is. The following numbers are prime numbers: 2, 3, 5, 7, 11, 13, 17, 19, 23 and so on. Any non-prime number can be written as a product of prime numbers, e.g. $8 = 2 \cdot 2 \cdot 2, 22 = 2 \cdot 11$, and $48 = 2 \cdot 2 \cdot 2 \cdot 2 \cdot 3$. Non-prime numbers are therefore divisible by the primes present in their representation, and are also divisible by their various products. For instance, 48 is divisible by 2, 3, 4, 6, 8, 12, 16 and 24.

Problem 1.172 Express the following numbers as products of the appropriate prime numbers: 128, 524 and 1035.

There is no general formula to give prime numbers in a systematic way. Pierre de Fermat attempted to suggest such a formula by proposing that numbers $2^{2^n} + 1$ for any $n = 0, 1, 2, 3, 4, \ldots$ are prime numbers. Indeed, with the listed values of n we get 3, 5, 17, 257 and 65537 that are all prime numbers. However, Leonhard Euler proved him wrong by offering an example that contradicts Fermat's proposal: for $n = 5$ the number $2^{2^5} + 1 = 4294967297$ is not prime as it is divisible by 641.

Other proposals also exist. For instance, there is a formula that surprisingly gives a lot of prime numbers: $n^2 + n + 41$. Indeed, for all values of n between 1 and 39 inclusive we obtain only prime numbers: 43, 47, 53, 61, etc., 1601. However, $n = 40$ gives the number 1681 that is not prime (it is divisible by 41), $n = 41$ gives 1763 that is also not prime (it is divisible by 41 as well), then $n = 42, 43$ give prime numbers 1847 and 1933, but $n = 44$ yields 2021 that is divisible by 43 and so on. Hence, this formula does give many prime numbers, but also non-prime numbers as well.

We shall now illustrate the usefulness of formula (1.12) on one peculiar example. Let us show that if $m = 2^n - 1$ is a prime number (n—integer), then n is also a prime number. We shall prove this statement by contradiction. Let us assume that n is not a prime number; hence it can be written as a product of two integers, $n = kl$, both of which are between 1 and n. Using Eq. (1.12), we can then write:

$$2^n - 1 = 2^{kl} - 1 = \left(2^k\right)^l - 1 = \left(2^k - 1\right)\left(2^{k(l-1)} + 2^{k(l-2)} + \cdots + 1\right).$$

Note that $2^k - 1$ must be larger than 1 as k is larger than 1. It is seen then that the number $2^n - 1$ is represented as a product of two integer numbers and hence is not a prime number, which contradicts the initial knowledge about that number. Hence, our assumption must be wrong, and n must also be a prime number. Note that in proving the required statement it was essential that n can be represented as a product of two integers $k > 1$ and $l > 1$; if we simply used formula (1.12) directly with n, we would have got

$$2^n - 1 = (2 - 1)(2^{n-1} + 2^{n-2} + \cdots + 1) = 2^{n-1} + 2^{n-2} + \cdots + 1,$$

i.e. the number $2^n - 1$ is simply equal to a sum of integers, but not as a product of two integers.

There is also one remarkable theorem related to prime numbers:

Theorem 1.9 (Fermat's little theorem) *Let p be a prime number. Then, if an integer number a is not divisible by p, then $N = a^{p-1} - 1$ is.*

Proof: Since a is not divisible by p, the numbers $2a, 3a, \ldots, (p-1)a$ will also be not divisible by p, i.e. we can write:

$$\begin{cases} a = d_1 p + r_1 \\ 2a = d_2 p + r_2 \\ 3a = d_3 p + r_3 \\ \cdots \\ (p-1)a = d_{p-1} p + r_{p-1} \end{cases}, \qquad (1.173)$$

where $d_1, \ldots, d_{p-1}$ are corresponding quotients and $r_1, \ldots, r_{p-1}$ the residues. It is easy to see that all residues are different. Indeed, assume that two of them coincide, e.g. $r_i = r_j$ for $j > i$. Then, subtracting $ia = d_i p + r_i$ from $ja = d_j p + r_j$, we obtain:

$$(j - i)a = (d_j - d_i) p.$$

Here $k = j - i$ is a positive integer between 1 and $p - 2$, and the above result shows that ka must be divisible by p. However, this contradicts the fact that a is not divisible by p, making our assumption that the two residues r_i and r_j are equal, wrong. Since all the residues are different integers, there are $p - 1$ of them and all of then must be smaller than p, we conclude that all of them must be different integer numbers between 1 and $p - 1$ (not necessarily in a specific order though). In other words, their product

$$r_1 r_2 \cdots r_{p-1} = 1 \cdot 2 \cdot 3 \cdot \ldots \cdot (p - 1).$$

Next, let us multiply left-hand sides of all Eq. (1.173) with their right-hand sides. In the left-hand side we shall simply have

$$\text{LHS} = 1 \cdot 2 \cdot 3 \cdot \ldots \cdot (p-1) a^{p-1} \,,$$

while in the right-hand side,

$$\text{RHS} = (d_1 p + r_1)(d_2 p + r_2) \cdots (d_{p-1} p + r_{p-1}) \,.$$

It is easy to convince yourself after performing multiplications in the right-hand side, that, apart from the single term containing the product of all residues, $r_1 r_2 \cdots r_{p-1}$, all other terms would contain at least one p as a multiplier, i.e.

$$\text{RHS} = Mp + r_1 r_2 \cdots r_{p-1} = Mp + 1 \cdot 2 \cdot 3 \cdot \ldots \cdot (p-1) \,,$$

where M is some integer. Therefore, equating both sides, we get:

$$1 \cdot 2 \cdot 3 \cdot \ldots \cdot (p-1) a^{p-1} = Mp + 1 \cdot 2 \cdot 3 \cdot \ldots \cdot (p-1) \,,$$

or

$$1 \cdot 2 \cdot 3 \cdot \ldots \cdot (p-1) \left(a^{p-1} - 1 \right) = Mp \,.$$

By looking at the right-hand side, it is clear that the number in the left-hand side must be divisible by p. However, the product $1 \cdot 2 \cdot 3 \cdot \ldots \cdot (p-1)$ is obviously not divisible by p, so then $N = a^{p-1} - 1$ must be. **Q.E.D.**

Let us illustrate this theorem by an example. Consider a prime number $p = 7$ and a number $a = 13$ that is not divisible by 7. Then, the number $N = 13^{7-1} - 1 = 13^6 - 1 = 4826808$ is divisible by 7, the quotient is 689544.

Problem 1.173 Let p be a prime number. Then prove that for any integer a the number $a^p - a$ is divisible by p.

Problem 1.174 Prove that if $a^p - b^p$ is divisible by p, then it is also divisible by p^2 if p is a prime number. [Hint: *first, represent $a^p - b^p$ as a sum of three terms ones of which is $a - b$. Then, using Fermat's little theorem, prove that $a - b$ is divisible by p, i.e. $a = kp + b$, with some integer k. Then prove that every term in the expansion of $(kp + b)^p - b^p$ contains at least p^2.*]

Let us consider a few examples to illustrate the results of the first problem. If we take $p = 7$ and $a = 5$, then $5^7 - 5 = 78120 = 7 \cdot 11160$, i.e. is indeed divisible by 7.

1.16   Combinatorics and Multinomial Theorem

The binomial coefficients we derived in the previous section have another significance related to the number of *combinations* in which different objects can be arranged with respect to each other when placed on a single line. This may seem only to deserve some attention of a mathematician as a purely abstract problem, which is entirely irrelevant to a physicist or engineer. Apparently, this is not the case: combinatorics, known as the science of counting different combinations of objects, has important applications in many sciences and hence deserves our full attention here.

There are several questions one may ask, the simplest of them being as follows: in how many combinations can one arrange n different objects? For instance, consider different numbers: there will be exactly two combinations, (ab) and (ba), in arranging two different numbers a and b; if there are three numbers, a, b and c; however, the number of combinations is six, namely, (abc), (acb), (bac), (bca), (cab) and (cba). It is easy to relate these combinations of three objects to the combinations of the two: in order to obtain all combinations of the three objects, we fix one object at the first "place" and perform all combinations of the other two objects; we know there are exactly two such combinations possible. For instance, choosing c at the first position, a and b are left giving us two possibilities: c (ab) and c (ba). Since there are three different objects, then each of them is to be placed in the first position, and each time we shall have only two combinations, which brings the total number of them to $3 \times 2 = 6$. As the next problem shows, there are $n!$ combinations of n different objects. For $n = 3$ we get exactly $3! = 6$ combinations.

Problem 1.175 Use the method of induction to prove that n distinct objects can be arranged exactly in $n!$ different combinations.

Suppose now that we have n identical boxes in which we would like to put $m < n$ *different* balls (e.g. of different colour). In how many combinations the balls can be allocated to boxes assuming that a box can accommodate only one ball? Let us start from the first ball. It can be placed in any of the n boxes, so there are n possibilities. If we take—for any of those possibilities—a second ball, there are only $n - 1$ empty boxes left for it, so we have $n - 1$ possibilities for the second ball, for every arrangement of the first ball, i.e. $n(n - 1)$ arrangements altogether. Taking a third ball would leave only $n - 2$ empty boxes for each of the previous arrangements, so the total number of arrangements is then $n(n - 1)(n - 2)$. Continuing this process, we have for m balls the following total number of possible arrangements:

$$A_m^n = n(n-1)(n-2)\cdots(n-m+1) = \frac{n!}{(n-m)!} . \tag{1.174}$$

The obtained solution can be interpreted in the following way: let us put our m balls into first m boxes, e.g. by counting boxes from 1 to m. By making $n!$ permutations of all boxes we shall have all possible arrangements of the balls. However, many arrangements will be the same since, for the given filled boxes, any permutation of the empty $n - m$ boxes would not give a different combination. Hence, $n!$ must be divided by $(n - m)!$ to count only different arrangements (remember that all boxes are identical).

Problem 1.176 There are 10 shelves each can handle a single book. In how many ways 5 books can be arranged on these shelves? [Answer: *30240.*]

What if now some of the balls are the same? In this case, some of the arrangements would be identical and hence need to be excluded. For instance, consider two white balls and three black ones, and let us try to calculate the total number of combinations in which we can arrange them. If the balls were all different, then there would have been $5! = 120$ such combinations. However, in our case, the number of combinations must be much smaller since we have to remove all combinations (or arrangements) in which we permute balls of the same colour as these would obviously not bring any new arrangements of black and white balls. Within the $5! = 120$ combinations, for *any* given arrangement of three black balls in five places, there will exactly be $3! = 6$ equivalent combinations of the black ones permuted between themselves; similarly, for any given arrangement of two white balls in five positions, there will be $2! = 2$ identical combinations associated with the permutation of the two white balls between themselves. Hence, this simple argument gives us $5!/(2! \cdot 3!) = 120/(2 \cdot 6) = 10$ different combinations altogether, all shown in Fig. 1.28. The same argument can be extended to n black and m white balls, in which case the total number of combinations will be $(n + m)!/(n! \cdot m!)$. But this is nothing but the binomial coefficient $\begin{pmatrix} n + m \\ n \end{pmatrix} = \begin{pmatrix} n + m \\ m \end{pmatrix}$, check out Eq. (1.154). This is not accidental: if we look closely at the formula (1.153), each term in the sum contains ith power of a and $(n - i)$th power of b; if we multiply i numbers a and $n - i$ numbers b, like

$$\underbrace{aaa\ldots}_{i}\underbrace{bbbb\ldots}_{n-i},$$

then obviously the result does not depend on the order in which we multiply the numbers, it will be still $a^i b^{n-i}$. However, the total number of combinations in which we arrange the numbers in the product is equal to the total number of arrangements of i numbers a and $n - i$ numbers b, i.e. it is $\begin{pmatrix} n \\ i \end{pmatrix}$, the binomial coefficient in Eq. (1.153).

Finally, we are now ready to generalise our result to a problem of arranging m balls into m boxes (no empty boxes to be left each time), when there are k_1 balls of

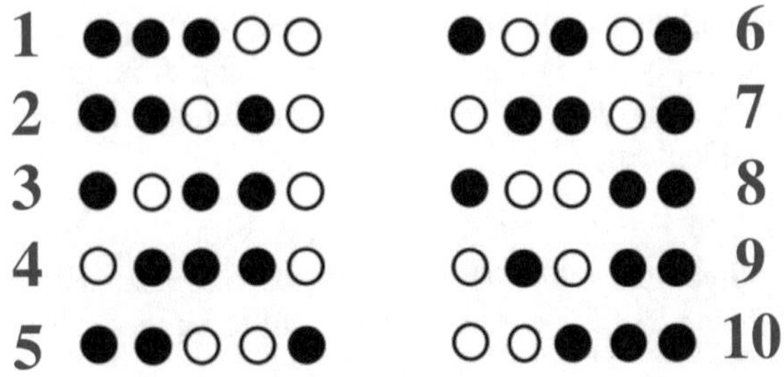

Fig. 1.28 There are 10 combinations in arranging three black and two white balls

$$
\begin{array}{ll}
1 \; \bullet\bullet\bullet\circ\circ & \bullet\circ\bullet\circ\bullet \; 6 \\
2 \; \bullet\bullet\circ\bullet\circ & \circ\bullet\bullet\circ\bullet \; 7 \\
3 \; \bullet\circ\bullet\bullet\circ & \bullet\circ\circ\bullet\bullet \; 8 \\
4 \; \circ\bullet\bullet\bullet\circ & \circ\bullet\circ\bullet\bullet \; 9 \\
5 \; \bullet\bullet\circ\circ\bullet & \circ\circ\bullet\bullet\bullet \; 10
\end{array}
$$

one colour, k_2 of another and so on, n colours altogether.[22] Since the total number of balls is $k_1 + k_2 + \cdots + k_n = m$, the number of different (unique) combinations will be

$$
\begin{pmatrix} m \\ k_1, k_2, \ldots, k_n \end{pmatrix} = \frac{(k_1 + k_2 + \cdots + k_n)!}{k_1! k_2! \ldots k_n!} = \frac{\left(\sum_{i=1}^{n} k_i\right)!}{\prod_{i=1}^{n} k_i!}, \tag{1.175}
$$

where the symbol $\prod$ means "a product of", i.e. in our case above this is simply the product of $k_1!$, $k_2!$, etc. The bracket notation is not accidental as will be clear in a moment; also, in the case of $n = 2$ the expression above is exactly the binomial coefficient in (1.155). This equation is readily understood: the total number of permutations of all m balls is $m!$. However, this number is to be reduced by the numbers $k_1!$, $k_2!$, and so on of permutations of all identically coloured balls, so we divide $m!$ by the product of $k_1!$, $k_2!$, etc. If the number of boxes n is larger than the total number of balls m, then this formula is easily generalised by introducing $k_0 = n - m$ "transparent" ("non-existent") balls which fill in all empty boxes. Then, the number of combinations becomes

$$
\begin{pmatrix} n \\ k_0, k_1, k_2, \ldots, k_n \end{pmatrix} = \frac{\left(\sum_{i=0}^{n} k\right)!}{\prod_{i=0}^{n} k_i!} = \frac{n!}{\prod_{i=0}^{n} k_i!}. \tag{1.176}
$$

Problem 1.177 The same 10 shelves. We have 2 green books, 1 red book and 2 blue ones. In how many ways these 5 books can be arranged on the shelves? [Answer: *7560.*]

---

[22] Of course, many other interpretations of the same result exist. For instance, in how many ways one can distribute m bins in n boxes so that there will be k_1 bins in the first box, k_2 in the second and so on, $k_1 + k_2 + \cdots + k_n = m$ altogether.

Problem 1.178 We have the same 10 shelves as in the previous problem, but this time we would like to place in all possible ways 5 identical books. In how many combinations this can be done? [Answer: *252.*]

If we now multiply the sum of n numbers by itself m times,

$$\underbrace{(a_1 + a_2 + \cdots + a_n)(a_1 + a_2 + \cdots + a_n) \cdots (a_1 + a_2 + \cdots + a_n)}_{m \text{ times}}$$

$$= (a_1 + a_2 + \cdots + a_n)^m ,$$

then upon opening the brackets and multiplying all terms, we shall have all possible combinations of the products of all individual numbers; there will be exactly m numbers in each such term, and some of the numbers may be the same or appear in different order.

Problem 1.179 As an example, consider $(a_1 + a_2 + a_3)^3$. First, expand the cube without combining identical terms such as $a_i a_j a_k$, $a_k a_j a_i$, etc. (where i, j, k are 1, 2 or 3); verify that there will be all possible combinations of the indices (i, j, k). Then, combine identical terms having the same powers for a_1, a_2 and a_3. Check that the number of identical terms (and hence the numerical pre-factor to each term) for the given indices (i, j, k) corresponds to the number of ways the indices can be permuted giving different sequences. Verify that this number is given by Eq. (1.175).

Returning to the general case of the expansion of $(a_1 + a_2 + \cdots + a_n)^m$, we can combine together all combinations which contain a_1 exactly k_1 times, a_2 exactly k_2 times and so on, as they give an identical contribution of $a_1^{k_1} a_2^{k_2} \ldots a_n^{k_n}$. Of course, $k_1 + k_2 + \cdots + k_n = m$ as there are exactly m numbers in each elementary term. How many such identical terms there will be? The answer is given by the same Eq. (1.175)! Therefore, combining all identical terms, we arrive at an expansion in which a numerical pre-factor (1.175) is assigned to each different term, i.e. we finally obtain:

$$(a_1 + a_2 + \cdots + a_n)^m = \sum_{k_1 + k_2 + \cdots + k_n = m} \binom{m}{k_1, k_2, \ldots, k_n} a_1^{k_1} a_2^{k_2} \ldots a_n^{k_n} ,$$

$$(1.177)$$

which is called the *multinomial expansion*. The sum here is actually a multiple sum and it is run over all possible indices k_1, k_2, etc. between 0 and m in such a way that their sum is equal to m.

Problem 1.180 Prove Eq. (1.177) using the method of induction. [Hint: *when considering the induction hypothesis for the case of $n + 1$, combine the last two numbers in the expression $(a_1 + a_2 + \cdots + a_n + a_{n+1})^m$ into a single number $b_n = a_n + a_{n+1}$, apply the hypothesis and use the binomial theorem (1.155).*]

Problem 1.181 Consider $2n$ objects. Show that the number of different pairs of objects which can be selected out of all objects is

$$\mathcal{N}_{pairs}(n) = \frac{(2n)!}{2^n n!} \,. \tag{1.178}$$

[Hint: *first, derive a recurrence formula $\mathcal{N}_{pairs}(n + 1) = \mathcal{N}_{pairs}(n)(2n + 1)$. Then, either use the method of induction or use the recurrence formula recursively. When selecting pairs, it might be helpful to connect each pair with a line.*]

Problem 1.182 There are 88 keys in the piano. How many different combinations of 5 different notes one can play? [Answer: *39175752.*]

Often in applications one has to consider multiple sums over k indices that change between 1 and n and which are all different. This requirement can be achieved if we constrain all the indices $j_1, \ldots, j_k$ by the following condition: $1 = j_1 < j_2 < \cdots < j_k = n$. It appears that the number of terms in such a sum (that is equivalent to summing up the unity) is equal to the binomial coefficient:

$$A_k(n) = \sum_{1 = j_1 < j_2 < \cdots < j_k = n} 1 = \binom{n}{k}. \tag{1.179}$$

Before we prove this, let us consider a few simple cases first. In the simplest case of $k = 1$, we get

$$A_1(n) = \sum_{j_1=1}^{n} 1 = n = \binom{n}{1}.$$

In the next case of $k = 2$ for each j_1 between 1 and $n - 1$ the index j_2 changes between $j_1 + 1$ and n. Hence, the total number of terms is

$$A_2(n) = \underbrace{(n - 1)}_{j_1=1} + \underbrace{(n - 2)}_{j_1=2} + \cdots + \underbrace{1}_{j_{n-1}=n-1}$$

$$= \frac{1}{2}[1 + (n - 1)](n - 1) = \frac{1}{2}n(n - 1) = \binom{n}{2},$$

as this is an arithmetic progression.

Let us now prove the general result. To this end, we rewrite the multiple sum as follows:

$$A_k(n) = \frac{1}{k!} B_k(n) , \quad \text{where} \quad B_k(n) = \left(\sum_{1 \le j_1, \ldots, j_k \le n} \right)' 1 .$$

Here we sum over all distinct values of the indices; in other words, each index is allowed to run from 1 to n, however, terms with at least two identical indices are excluded (which is indicated by the prime). There are $k!$ permutations of identical sets of indices in this sum. Note that in the original sum (1.179) we only have strictly one such set, hence the pre-factor $1/k!$ in front of the above expression.

To calculate the numbers $B_k(n)$, we shall establish a recurrence relation for them:

$$B_k(n) = \sum_{j_1=1}^{n} \left(\sum_{1 \le j_2, \ldots, j_k \le n} \right)'' 1 = \sum_{j_1=1}^{n} B_{k-1}(n-1) = n B_{k-1}(n-1) .$$

In the above expression, the summations within the brackets run over all indices $j_2, j_3, \ldots, j_n$ between 1 and n, which exclude any repeated ones and the index j_1 (indicated by the double prime); that is why in the right-hand side in $B_{k-1}(n-1)$ we reduced k by one as the number of indices is by one less, and we have also reduced n by one since the number of values between 1 and n is less by one as j_1 is to be excluded. Finally, $B_{k-1}(n-1)$ is the same for any j_1 which allowed calculating the j_1 sum as shown. Using the recurrence relation just obtained, we have:

$$B_k(n) = n B_{k-1}(n-1) = n(n-1) B_{k-2}(n-2) = n(n-1) \ldots B_1(n-k+1)$$

$$= n(n-1)(n-2) \cdots (n-k+1) = \frac{n!}{(n-k)!} ,$$

so that

$$A_k(n) = \frac{1}{k!} B_k(n) = \frac{n!}{k!(n-k)!} = \binom{n}{k} ,$$

as required.

1.17 Elements of Classical Probability Theory

The concept of probability plays fundamental role in physics, chemistry, financial science and many other disciplines. As we have already considered main ideas of combinatorics, we are quite ready to discuss probability.

1.17.1 Trials, Outcomes and Sets

It is convenient to start from terminology. Suppose, when we run a certain experiment, at each particular *trial* various *outcomes* are possible. For instance, when we pick a card from a pack, we may choose a different card at each trial, and any card may appear equally likely over many such trials. Then, the total number of all possible different outcomes is 52, the number of different cards in the pack (4 suits, 13 cards in each). We may then ask, for instance, how likely it is to pick up a spade; this is called an *event* as it consists of many possible outcomes (in this case, the number of cards of that suit that is 13).

Let us indicate by dots all possible outcomes U of a particular experiment as shown in Fig. 1.29a. These dots form a *complete set of all possible outcomes*. Giving a different example of a die, there are only six possible outcomes. If the die is not biased, all six outcomes are equally likely; if it is biased,[23] then different outcomes will be of a variable likelihood.

Consider a particular event A consisting of some particular outcomes, e.g. picking up cards with even numbers from the pack. In mathematical logic A is called a *set*. We can schematically show the event A by an ellipse (generally, by a closed curve) comprising all such outcomes as shown in Fig. 1.29b (these are called Venn[24] diagrams). Another event B (shown by a different ellipsoidal object on the same figure) consists of some outcomes of A, but also contains different outcomes. Event C has common outcomes with A and B, but there are also some distinct outcomes. The three events may and may not have common outcomes, which would either show or not as an overlap in their diagrams.

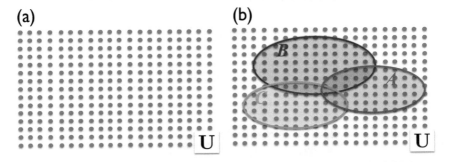

Fig. 1.29 a All possible outcomes in a certain experiment are shown by dots. Dots are of the same size indicating that in this particular case the likelihood of different outcomes is the same (all outcomes are equally likely). **b** Three different events A, B and C, each corresponding to a collection of various possible outcomes, are shown by ellipses (Venn diagrams). Note that some of the outcomes are common to several events

---

[23] For instance, if its centre of mass is shifted towards one of the faces; the opposite face would then show up more often.

[24] These are diagrams from mathematical logic introduced by John Venn.

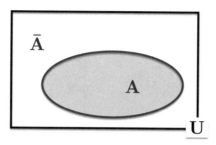

Fig. 1.30 Event A and its complement $\overline{A}$ give the complete set of all outcomes U. Note that we do not show by dots all possible outcomes here for simplicity; it is implied that all such outcomes are contained inside the black rectangular corresponding to their complete set U

For instance, consider again the card picking experiment. If A corresponds to picking up a spade (13 possible outcomes) and B picking up an ace (4 possible outcomes), then the outcome of the spade ace (just one outcome) is common to both; it will show as an overlap of objects A and B on the Venn diagram.

All outcomes which do not belong to A form another set of outcomes which we shall denote $\overline{A}$ and call the *complement* to A, see Fig. 1.30. All possible outcomes corresponding to two events A and B we shall call a *union* of them and will write as $A \cup B$, see Fig. 1.31a. Obviously, $A \cup \overline{A} = U$, i.e. the union of the outcomes from A and its complement is the complete set of all possible outcomes.

The outcomes common to two events A and B also form a set, which we shall call an *intersection* and denote $A \cap B$. On Venn diagrams an intersection A and B is the region where the two objects A and B overlap, Fig. 1.31b. There is also a special symbol $\varnothing$ used for the event with no outcomes (the empty event). Obviously, $A \cap \overline{A} = \varnothing$.

Problem 1.183 Operations on the sets introduced above satisfy certain important properties. Verify these using Venn diagrams:

$$A \cup B = B \cup A \text{ and } A \cap B = B \cap A \text{ (commutativity)}$$

$$(A \cap B) \cap C = A \cap (B \cap C) \text{ and } (A \cup B) \cup C = A \cup (B \cup C) \text{ (associativity)}$$

$$A \cap (B \cup C) = (A \cap B) \cup (A \cap C) \text{ and } A \cup (B \cap C) = (A \cup B) \cap (A \cup B) \text{ (distributivity)}$$

$$A \cap A = A \text{ and } A \cup A = A \text{ (idempotency)}.$$

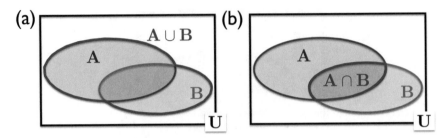

Fig. 1.31 Union (**a**) and intersection (**b**) of two events A and B, shown by the blue closed curves

> **Problem 1.184** Prove using Venn diagrams, or otherwise, the following identities between sets:
>
> $$A \cap \overline{(A)} = A; \quad \overline{A \cup B} = \overline{A} \cap \overline{B}; \quad \overline{A \cap B} = \overline{A} \cup \overline{B}.$$

1.17.2 Definition of Probability of a Random Event

Probability is a numerical measure, between zero and one, of a likelihood of a random event. If the occurrence of the event A in an experiment is considered certain, the probability $P(A)$ of it is equal to one, if the event cannot happen (the empty event), $P(A) = 0$. The larger the likelihood of the event to happen, the larger the probability. In practical terms, $P(A)$ is defined by running an experiment N times (N trials with $N \gg 1$), and counting the number N_A of attempts that gave the favourable outcome A; then $P(A) = N_A/N$. It is expected that if a different set of the same experiment is run, the ratio N_A/N in the new experimental run may not be the same as in the previous run, but should be close to it (provided the number of trials $N \gg 1$). The more attempts N is made, the closer the calculated probability will be to its correct value.

Consider a die. It has six sides with numbers from one to six on each of them. If we throw the die, one of these six numbers will show up. If we throw it again, we may get the same or a different number. If we run this experiment many times, we shall find that, on average, each number appears practically the same number of times. Obviously, each event is entirely random: we cannot say before throwing the die, which number will play each time (unless, of course, the die is biased, i.e. has some defect). Hence, the probability of getting a certain outcome, e.g. number 2, is 1/6.

Consider another experiment. Suppose, we have 10 black and 20 white balls in a box. What are the probabilities P_w and P_b of picking up a white or black ball from the box (we assume that after picking up a ball, it is placed back into the box)? Since there are two times more white balls than black ones, it seems plausible that when running this experiment many times, we would pick up white balls two times more often than the black one, on average. This means, that out of N trials, black balls will be picked up close to $N/3$ times and while balls close to $2N/3$ times, leading to probabilities $P_w = 2/3$ and $P_b = 1/3$. The more attempts we make, the closer the ratios of favourable to all trials will be to the probabilities of $2/3$ and $1/3$, respectively.

A simple way of calculating the probability of a particular event A exists if all possible outcomes in the given experiment,

$$A_1, A_2, \ldots, A_N, \tag{1.180}$$

are equally likely (of equal probability). These form the complete set of outcomes. Then, if all these events are equally likely, their probabilities are the same. Hence, to obtain the probability $P(A)$ of a certain event A, one needs to count the total number of successful (favourable) outcomes, N_A, and then the probability will simply be $P(A) = N_A/N$.

Consider a few examples. First, consider the probability of having a number larger than 4 when we throw a die. There are 6 outcomes possible altogether, all of the same probability (equal to $1/6$), they form the complete set of outcomes. However, there are only two favourable outcomes satisfying our condition of getting a number larger than 4, that is, when we get either 5 or 6. Hence, the required probability is $2/6 = 1/3$.

Now, we have two dices, and we throw them together, what is the probability to get the sum of the two numbers equal to 8? In this case, the total number of outcomes is $6 \times 6 = 36$, which corresponds to all possible combinations of numbers we may have when throwing two dices, i.e. the complete set of outcomes (1.180) is

$$\underbrace{(1, 1), (1, 2), \ldots, (6, 1), (2, 1), \ldots, (6, 1), \ldots, (6, 6)}_{36}.$$

However, there are only 5 favourable combinations, $(2, 6), (3, 5), (4, 4), (5, 3)$ and $(6, 2)$ that give the sum of 8. Therefore, the probability of getting this sum is $5/36 \approx 0.14$.

Problem 1.185 Calculate the probability of picking the sum of two numbers being between 8 and 10 (inclusive) when throwing two dices. [Answer: $1/3$.]

Problem 1.186 Consider a box, in which there are balls of 5 colours: 5 black, 6 white, 7 green, 8 red and 9 yellow. What is the probability of picking (a) 2 red balls; (b) 1 ball green and 1 black? [Answer: (a) $\binom{8}{2} / \binom{35}{2} = 4/85$; (b) $5 \times 7/ \binom{35}{2}$.]

Problem 1.187 Consider the same box of balls as in the previous problem. What is the probability of picking 4 balls that are (a) 1 black, 1 green, 1 red and 1 yellow; (b) all of different colours? [Answer: $(a)\ 5 \times 7 \times 8 \times 9 / \binom{35}{2}$; $(b)\ 11274 / \binom{35}{2}$.]

1.17.3 Main Theorems of Probability

Let us suppose that there are in total N possible outcomes (1.180) in a certain experiment, and all of them are equally likely. Consider two events, A and B. These give rise to the following four events:

$$C_1 = A \cap B, \; C_2 = A \cap \overline{B}, \; C_3 = \overline{A} \cap B, \; \text{and} \; C_4 = \overline{A} \cap \overline{B} \qquad (1.181)$$

that are shown in Fig. 1.32. It is easy to see that these four events that we generated from events A and B form the complete set U. Out of all N outcomes, there are N_1 outcomes corresponding to the event C_1, N_2 outcomes to C_2 and so on. Obviously, $N_1 + N_2 + N_3 + N_4 = N$. Hence, the probabilities of the four events (1.181) are:

$$P(A \cap B) = \frac{N_1}{N}, \; P\left(A \cap \overline{B}\right) = \frac{N_2}{N}, \; P\left(\overline{A} \cap B\right) = \frac{N_3}{N}, \; P\left(\overline{A} \cap \overline{B}\right) = \frac{N_4}{N}.$$
$$(1.182)$$

Theorem 1.10 (Addition Rule for Probabilities) *The probability of the union of A and B, i.e. of any outcome included either in A or B, is*

$$P(A \cup B) = P(A) + P(B) - P(A \cap B). \qquad (1.183)$$

Proof: The probability $P(A)$ of A corresponds to $N_1 + N_2$ favourable outcomes, since only in the first two events A is included, and is therefore given by

$$P(A) = \frac{N_1 + N_2}{N} = P(A \cap B) + P\left(A \cap \overline{B}\right), \qquad (1.184)$$

while similarly

$$P(B) = \frac{N_1 + N_3}{N} = P(A \cap B) + P\left(\overline{A} \cap B\right). \qquad (1.185)$$

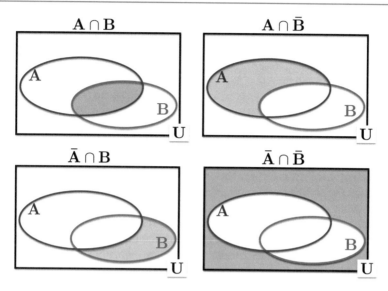

Fig. 1.32 Events (1.181) based on events A and B are shown in each case as grey. Their union forms the complete set U

Then, the probability of the union of A and B is

$$P(A \cup B) = \frac{N_1 + N_2 + N_3}{N}.$$

Using Eqs. (1.182), (1.184) and (1.185), it is easy to see that the addition theorem (1.183) follows immediately. **Q.E.D.**

It follows from Fig. 1.31a that the total area under the union $A \cup B$ is indeed equal to the sum of the areas of A and B minus their intersection as it is counted twice, i.e. the addition theorem trivially follows from the Venn diagrams.

Two events are called *mutually exclusive* or *independent* if the probability for them to happen together is impossible, i.e. if their intersection is the empty event. On the Venn diagram, two independent events A and B do not overlap. Hence, if the two events A and B are independent, then it follows from the addition theorem (1.183) that the probability of their union is simply the sum of their individual probabilities,

$$P(A \cup B) = P(A) + P(B), \tag{1.186}$$

since $P(A \cap B) = P(\varnothing) = 0$. The obtained statements are trivially generalised to any number of independent events. In particular, individual outcomes (1.180) of the complete set U of such outcomes are independent. Since the probability of U is one, we have that

$$P(A_1 \cup A_2 \cup \cdots \cup A_N) = \sum_{i=1}^{N} P(A_i) = P(U) = 1,$$

and hence for each individual outcome $P(A_i) = 1/N$, as they all are equally likely.

Problem 1.188 There are white and black balls in a box. It is known that on average the white balls are picked three times more often than the black ones (each time a ball is picked, it is placed back into the box). Determine the probabilities P_w and P_b for picking up a white and a black ball. Determine the numbers N_w and N_b of white and black balls in the box if the their total number is 12. [Answer: $P_w = 3/4$, $P_b = 1/4$; $N_w = 9$, $N_b = 3$.]

To formulate the next important theorem, we need to introduce the so-called conditional probability, $P(B|A)$. This is the probability of B given A has happened. For instance, the probability of getting number 6 when throwing a die is $1/6$. However, if we ask what is the probability of getting 6 if we know that an even number has been obtained, then it is equal to $1/3$, since there are only 3 even numbers possible.

If the two events A and B are independent, then obviously, $P(A|B) = P(A)$ and $P(B|A) = P(B)$. If the two events A and B are mutually exclusive, $P(A|B) = P(B|A) = 0$.

The idea of the conditional probability can be illustrated by Fig. 1.32 for the intersection $A \cap B$, where the intersection is shown in grey: the probability $P(A|B)$ corresponds to the ratio of the shaded area of A to the whole area of B: since we know that B has happened, the complete set of outcomes becomes the one comprised of B only, hence $P(A|B)$ is given by the number of favourable outcomes (the shaded area) to the total number of them (all those in B).

Theorem 1.11 (Product Rule for Probabilities) *The probability of the intersection of A and B, i.e. of any outcome that is included in both A and B (corresponds to the overlap of the Venn diagrams for A and B), is*

$$P(A \cap B) = P(A)P(B|A) = P(B)P(A|B). \qquad (1.187)$$

Proof: Consider again the complete set of events (1.181) and let us work out the conditional probability $P(B|A)$. The number of individual outcomes of A is $N_1 + N_2$, and among those, only N_1 outcomes are favourable to B, hence

$$P(B|A) = \frac{N_1}{N_1 + N_2} = \frac{N_1/N}{(N_1 + N_2)/N} = \frac{P(A \cap B)}{P(A)},$$

from which the first part of the desired result follows. Similarly,

$$P(A|B) = \frac{N_1}{N_1 + N_3} = \frac{N_1/N}{(N_1 + N_3)/N} = \frac{P(A \cap B)}{P(B)},$$

and the second part of the theorem is obtained. **Q.E.D.**

In the case of independent events A and B, we obtain from the first part of Eq. (1.187) that

$$P(A \cap B) = P(A)P(B), \tag{1.188}$$

since $P(B|A) = P(B)$, i.e. the probability of two events to happen is simply given as a product of their individual probabilities. Using this result in Eq. (1.187), we also obtain $P(A|B) = P(A)$, i.e. the independence of A and B is mutual, as has already been concluded above on intuitive grounds.

Problem 1.189 Consider three events A, B and C. Show that the addition theorem becomes in this case:

$$P(A \cup B \cup C) = P(A) + P(B) + P(C)$$
$$- P(A \cap B) - P(A \cap C) - P(B \cap C) + P(A \cap B \cap C).$$

Illustrate this result by an appropriate Venn diagram.

Problem 1.190 Generalise the product rule of probabilities for three events by choosing $D = A \cap B$:

$$P(A \cap B \cap C) = P(A)P(B|A)P(C|A \cap B).$$

By choosing $D = B \cap C$, show that

$$P(B \cap C|A) = P(B|A)P(C|A \cap B).$$

Consider some examples to illustrate the obtained results.

Consider a box with three black and five white balls. Two balls are taken one after another. What is the probability for the second ball to be white, if the first ball is known to have been white and it has not been placed back into the box? The first event, A, corresponds to the first ball (white) being taken from the box, its probability $P(A) = 5/(3 + 5) = 5/8$. The second picking of a ball from the box (event B) corresponds to the conditional probability $P(B|A)$: there are 3 black and 4 white balls left in the box, and therefore $P(B|A) = 4/7$. Correspondingly, the probability to pick up 2 white balls one after another is

$$P(A \cap B) = P(A)P(B|A) = \frac{5}{8} \times \frac{4}{7} = \frac{5}{14}.$$

This result can also be checked as follows. Picking one ball after another can be considered as picking a pair of balls. There are altogether $\binom{8}{2} = 28$ pairs possible to form out of 8 balls, and only $\binom{5}{2} = 10$ of them correspond to the pairs with both balls being white. Hence, the probability of picking a white pair is $P = 10/28 = 5/14$, the same result.

Consider a probability of picking from a pack of cards either a king or a queen. Here A corresponds to picking up a king, $P(A) = 4/52 = 1/13$. The same probability is for picking up a queen, $P(B) = 1/13$. Both events are independent (it is not possible to pick up both a king and a queen), so

$$P(A \cup B) = P(A) + P(B) = \frac{2}{13}.$$

Now consider a more complex example: what is the probability of picking up either a king or a spade? Picking up a king (event A) gives $P(A) = 4/52 = 1/13$, while picking up independently a spade (event B) is $P(B) = 13/52 = 1/4$. The two events are not independent anymore as one can also pick up the spade king. The probability of this is $P(A \cap B) = 1/52$. Hence,

$$P(A \cup B) = P(A) + P(B) - P(A \cap B) = \frac{1}{13} + \frac{1}{4} - \frac{1}{52} = \frac{16}{52} = \frac{4}{13}.$$

Again, this result can be checked by counting all possibilities. Out of 52 cards we have 13 spades and 4 kings; however, one of them is the spade king, which we have already counted, so the total number of favourable possibilities is only $13 + 3 = 16$, giving the probability $16/52 = 4/13$, which is the same.

Problem 1.191 There are two packs of cards. From each a card is taken. What is the probability that: (a) both cards are of suit hearts, (b) at least one is of this suit. Check your calculation also by counting explicitly all possibilities. [Answer: *(a)* 1/16; *(b)* 7/16.]

Problem 1.192 Two packs of cards are thoroughly mixed. Then two cards are taken one after another. What is the probability that: (a) both cards are of suit hearts, (b) at least one is of this suit. [Answer: *(a)* 25/412; *(b)* 181/412.]

Problem 1.193 Two dices are played at the same time n times. Find the probability of having both fives at least ones. Use the fact the required event is complementary to the one in which two fives never appeared. [Answer: $1 - (1 - p)^n$, *where* $p = 1/36$.]

It is instructive to check the answer of the above problem by counting all possibilities. Let $p = 1/36$ be the probability to have two fives and $q = 1 - p$ not to have them. Each time we play the dices we may have either one or another event with the probabilities p or q, respectively. Consider all possible sequences of these events after playing the dices n times in which the pair of fives appeared exactly once:

$$Y \underbrace{N \ldots N}_{n-1}, \quad NY \underbrace{N \ldots N}_{n-2}, \quad \ldots, \quad \underbrace{N \ldots N}_{n-1} Y,$$

where Y means that two fives appeared, and N that they did not. All these events have identical probabilities of pq^{n-1}, and there are $\binom{n}{1}$ of them, i.e. the total probability of these independent events is the sum of their individual probabilities amounting to $\binom{n}{1} pq^{n-1}$. There will also be events in which two times the fives were obtained, e.g. $YY \underbrace{N \ldots N}_{n-2}$, etc. There are $\binom{n}{2}$ such events, each of the probability $p^2 q^{n-2}$, giving $\binom{n}{2} p^2 q^{n-2}$ overall. Clearly, we can similarly obtain probabilities for the two fives to appear three, four, etc., n times. The total probability of all these independent events is then

$$P_1 = \sum_{k=1}^{n} \binom{n}{k} p^k q^{n-k} = \underbrace{(p+q)^n}_{1} - \binom{n}{0} p^0 q^n$$
$$= 1 - q^n = 1 - (1-p)^n,$$

as required.

Problem 1.194 Two cards are taken from a pack one after another. It is known that the first card was a queen. What is the probability of the second card to be a king if: (a) the first card was put back into the pack and (b) it was not. [Answer: *(a)* $1/169$; *(b)* $4/663$.]

Problem 1.195 Pairs of cards are picked from a pack and not placed back. What is the probability that a king and a queen are picked two times one after another. [Answer: $24/270725$.]

Problem 1.196 In this problem, we shall consider a diffusion of a particle on a one-dimensional lattice. Consider an infinite set of equidistant points along the x-axis. We shall number the points by integer numbers between $-\infty$ and ∞, see Fig. 1.33a. The particle can only diffuse to the nearest point, either to the right, with probability p, or to the left, with probability q. Obviously, $p + q = 1$. Assuming that initially the particle occupies the zero lattice site, consider all possible trajectories the particle can make in order to find itself after N steps at the lattice site with the number $n > 0$. Some such trajectories are shown in Fig. 1.33b. Correspondingly show that the probability for such an event is given, after counting all such trajectories, by the formula:

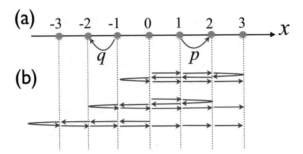

Fig. 1.33 Diffusion of a particle on a 1D lattice of equidistant lattice sites, Problem 1.196. **a** The particle can jump one site either to the right (with probability p) or to the left (with probability $q = 1 - p$). **b** Three possible trajectories are shown each consisting of $N = 11$ steps after which the particle eventually ends up at the site $n = 3$. There are $\binom{11}{7} = 330$ such trajectories possible

$$P_N(n) = \frac{N!}{\left(\frac{1}{2}(N-n)\right)!\left(\frac{1}{2}(N+n)\right)!} p^{(N+n)/2} q^{(N-n)/2}. \qquad (1.189)$$

Does the probability to find the particle on the lattice at all possible sites available to it after N steps is equal to one as it should? [Hint: *calculate the number of steps $n_\rightarrow$ to the right and $n_\leftarrow$ to the left and argue that the total number of such trajectories, with the given $n_\rightarrow$ and $n_\leftarrow$ is given by the binomial coefficient* $\binom{N}{n_\rightarrow}$. *Then consider whether each of these trajectories has an equal probability or not. To check the total probability, sum all probabilities* (1.189) *for all possible values of $n_\rightarrow$.*]

The probability theory considered so far was based on an assumption that all elementary outcomes in a particular experiment are of equal probabilities. This was essential for proving the addition and multiplication rules for the probabilities. To be able to study more general cases, a modern axiomatic probability theory was created in 1933 by A. N. Kolmogorov. The theory considered above is a particular case of this more general theory.

1.18   Some Important Inequalities

1.18.1 Cauchy–Bunyakovsky–Schwarz Inequality

There is one famous result in mathematics known as Cauchy–Bunyakovsky–Schwarz inequality. It states that for any real set of numbers x_i and y_i (where $i = 1, \ldots, n$)

$$\left(\sum_{i=1}^{n} x_i y_i\right)^2 \leq \left(\sum_{i=1}^{n} x_i^2\right)\left(\sum_{i=1}^{n} y_i^2\right) \quad \text{or} \quad \sum_{i=1}^{n} x_i y_i \leq \sqrt{\sum_{i=1}^{n} x_i^2}\sqrt{\sum_{i=1}^{n} y_i^2}.$$
$$(1.190)$$

To prove this, consider a function of some real variable t defined in the following way:

$$f(t) = \sum_{i=1}^{n} (tx_i + y_i)^2 = t^2 \underbrace{\left(\sum_{i=1}^{n} x_i^2\right)}_{a} + 2t \underbrace{\left(\sum_{i=1}^{n} x_i y_i\right)}_{b} + \underbrace{\left(\sum_{i=1}^{n} y_i^2\right)}_{c} \geq 0 \, .$$

This is a parabola with respect to t; it is non-negative, and hence should have either one or no roots. This means that the discriminant of this square polynomial $\mathcal{D} = 4b^2 - 4ac$ is either equal to zero (one root: the parabola just touches the t axis at a single point) or is negative (no roots: the parabola is lifted above the t axis), therefore

$$\mathcal{D} = 4\left(b^2 - ac\right) = 4\left[\left(\sum_{i=1}^{n} x_i y_i\right)^2 - \left(\sum_{i=1}^{n} x_i^2\right)\left(\sum_{i=1}^{n} y_i^2\right)\right] \leq 0 \, ,$$

which is exactly what we wanted to prove.

Using this rather famous inequality, we shall now prove another one related to lengths of vectors. Indeed, consider two vectors $\mathbf{x} = (x_1, x_2, x_3)$ and $\mathbf{y} = (y_1, y_2, y_3)$. Then, the length of their sum will be shown to be less than or equal to the sum of their individual lengths:

$$|\mathbf{x} + \mathbf{y}| \leq |\mathbf{x}| + |\mathbf{y}| \, . \tag{1.191}$$

Indeed,

$$|\mathbf{x} + \mathbf{y}| = \sqrt{\sum_{i=1}^{3} (x_i + y_i)^2} = \sqrt{\sum_i x_i^2 + \sum_i y_i^2 + 2\sum_i x_i y_i} \, .$$

Now we can apply the inequality (1.190) for the third term inside the square root, which gives:

$$|\mathbf{x} + \mathbf{y}| = \sqrt{\sum_i x_i^2 + \sum_i y_i^2 + 2\sum_i x_i y_i} \leq \sqrt{\sum_i x_i^2 + \sum_i y_i^2 + 2\sqrt{\sum_i x_i^2}\sqrt{\sum_i y_i^2}}$$

$$= \sqrt{\left(\sqrt{\sum_i x_i^2} + \sqrt{\sum_i y_i^2}\right)^2} = \sqrt{\sum_i x_i^2} + \sqrt{\sum_i y_i^2} = |\mathbf{x}| + |\mathbf{y}| \, ,$$

which is the desired result.[25]

---

[25] The same inequality is of course valid for vectors in a space of any dimension as we did not really use the fact that our vectors only have three components.

Problem 1.197 Prove similarly that for any two complex numbers z_1 and z_2, we have:

$$|z_1 + z_2| \leq |z_1| + |z_2| \ . \tag{1.192}$$

Problem 1.198 Using induction, generalise this result for a sum of n complex numbers:

$$\left| \sum_i z_i \right| \leq \sum_i |z_i| \ . \tag{1.193}$$

Written in components, this inequality can also be manipulated into the following one that is valid for n pairs (x_i, y_i) of real numbers:

$$\sqrt{\left(\sum_{i=1}^{n} x_i \right)^2 + \left(\sum_{i=1}^{n} y_i \right)^2} \leq \sum_{i=1}^{n} \sqrt{x_i^2 + y_i^2} \ .$$

Problem 1.199 Prove that for any two complex numbers z_1 and z_2:

$$|z_1 - z_2| \geq |z_1| - |z_2| \ . \tag{1.194}$$

[Hint: *make a substitution in Eq. (1.192).*]

Problem 1.200 Prove that one side of a triangle is always shorter than the sum of two other sides.

Of course, the same inequalities are valid for real numbers as well, as these are particular cases of their complex counterparts with zero imaginary parts:

$$\left| \sum_{i=1}^{n} x_i \right| \leq \sum_{i=1}^{n} |x_i| \ , \tag{1.195}$$

i.e. the absolute value of a sum of real numbers is always less than or equal to the sum of their absolute values.

However, it is instructive to prove the above inequality independently as well. Let us first consider the case of $n = 2$. Then, we can write:

$$|x_1 + x_2| = \sqrt{(x_1 + x_2)^2} = \sqrt{x_1^2 + x_2^2 + 2x_1 x_2}$$

$$= \sqrt{(|x_1| + |x_2|)^2 - 2|x_1 x_2| + 2x_1 x_2}$$

$$= \sqrt{(|x_1| + |x_2|)^2 + 2(x_1 x_2 - |x_1 x_2|)} \ .$$

Fig. 1.34 To the proof of inequality (1.196): the area of the △OAC is less than the area of the sector OAC which in turn is less than the area of the △ODC

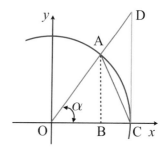

Two cases are possible: if $x_1 x_2 \geq 0$, then $x_1 x_2 - |x_1 x_2| = x_1 x_2 - x_1 x_2 = 0$; if, however, $x_1 x_2 < 0$, then $x_1 x_2 - |x_1 x_2| = x_1 x_2 + x_1 x_2 = 2 x_1 x_2 < 0$. Hence, $2 (x_1 x_2 - |x_1 x_2|) \leq 0$, and, therefore,

$$|x_1 + x_2| = \sqrt{(|x_1| + |x_2|)^2 + 2 (x_1 x_2 - |x_1 x_2|)} \leq \sqrt{(|x_1| + |x_2|)^2} = |x_1| + |x_2| \ ,$$

as required. Now we can consider the case of three numbers, for which we can use the inequality for two numbers proven above:

$$|x_1 + x_2 + x_3| = |x_1 + (x_2 + x_3)| \leq |x_1| + |x_2 + x_3| \leq |x_1| + |x_2| + |x_3| \ ,$$

and so on.

1.18.2 Angles Inequality

In the following we shall also need one very useful result that is used in various estimates and proofs:

$$\sin \alpha < \alpha < \tan \alpha \ . \tag{1.196}$$

To show that, consider Fig. 1.34 where we have a circle of unit radius $R = 1$ (blue), two triangles, △OAC and △ODC, as well as a sector OAC. Obviously, the area of the △OAC is strictly smaller than the area of the sector, which in turn is smaller than the area of the △ODC. Both triangles have the same unit base $OC = 1$, the height of △OAC is $AB = OA \sin \alpha = \sin \alpha$, while the height of the △ODC is $DC = OC \tan \alpha = \tan \alpha$. The areas of the two triangles are easy to calculate as $S_{\triangle OAC} = \frac{1}{2} OC \cdot AB = \frac{1}{2} \sin \alpha$ and $S_{\triangle ODC} = \frac{1}{2} OC \cdot DC = \frac{1}{2} \tan \alpha$. The area of the sector OAC of radius $R = 1$ can be calculated as a fraction $\alpha / 2\pi$ of the area $\pi R^2 = \pi$ of the whole circle[26]: $\left(\frac{\alpha}{2\pi} \right) \pi R^2 = \frac{1}{2} \alpha$. Comparing the three and cancelling out on the common factor of $1/2$, we arrive at (1.196).

---

[26] The area of a circle of radius R is πR^2, this will be established later on in Sect. 4.7.2.

Fig. 1.35 2D packing of circles: **a** loose packing; **b** compact packing

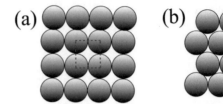

Problem 1.201 Prove the same result (1.196) by considering the lengths of AB and DC as well the circle arc length AC.[27]

Problem 1.202 Consider a 2D packing problem. In Fig. 1.35, two possible packings of identical circles are shown, one is loose and another—compact. Show that the so-called packing ratio, which is a fraction of the area occupied by the circles, is $\pi/4 \approx 0.7854$ for the loose and $\pi/2\sqrt{3} \approx 0.9069$ for the compact arrangement of the circles. [Hint: *consider areas shown by the dashed lines and work out the fraction of the area occupied by the circles within the identified shapes.*]

1.18.3 Four Averages of Positive Numbers

We shall finish this section by considering four types of averages of n positive numbers a_i, $i = 1, \ldots, n$. The simplest ones are the *arithmetic average*

$$M_{arithm} = \frac{1}{n}(a_1 + a_2 + \cdots + a_n) = \frac{1}{n}\sum_{i=1}^{n} a_i \,, \qquad (1.197)$$

and the so-called *geometric average*

$$M_{geom} = \sqrt[n]{a_1 a_2 \cdots a_n} = \sqrt[n]{\prod_{i=1}^{n} a_i} \,. \qquad (1.198)$$

Then, one can also define the *harmonic average*

$$M_{harm} = \frac{n}{\sum_{i=1}^{n} \frac{1}{a_i}} \qquad (1.199)$$

and the *quadratic average*

---

[27] The length of the circle circumference is $2\pi R$, where R is the circle radius (see Sect. 4.7.1).

$$M_{quadr} = \sqrt{\frac{1}{n} \sum_{i=1}^{n} a_i^2}. \tag{1.200}$$

The defined quantities do correspond to averages in a sense that their values are always larger or equal the smallest element in the set, and smaller or equal its largest element. In other words, if we order the elements in the set,

$$0 \le a_1 \le a_2 \le \cdots \le a_n,$$

then for any of the averages we have $a_1 \le M \le a_n$. Let us prove this statement specifically for the harmonic average. Since $a_i \ge a_1$, then $1/a_i \le 1/a_1$ and hence

$$\frac{1}{\frac{1}{a_i} + \Delta} \ge \frac{1}{\frac{1}{a_1} + \Delta}$$

for any $\Delta \ge 0$. Therefore,

$$M_{harm} = \frac{n}{\sum_{i=1}^{n} \frac{1}{a_i}}$$

$$= \frac{n}{\frac{1}{a_2} + \left(\frac{1}{a_1} + \frac{1}{a_3} + \cdots + \frac{1}{a_n}\right)} \ge \frac{n}{\frac{1}{a_1} + \left(\frac{1}{a_1} + \frac{1}{a_3} + \cdots + \frac{1}{a_n}\right)}$$

$$= \frac{n}{\frac{1}{a_3} + \left(\frac{2}{a_1} + \frac{1}{a_4} + \cdots + \frac{1}{a_n}\right)} \ge \frac{n}{\frac{1}{a_1} + \left(\frac{2}{a_1} + \frac{1}{a_4} + \cdots + \frac{1}{a_n}\right)}$$

$$= \frac{n}{\frac{1}{a_4} + \left(\frac{3}{a_1} + \frac{1}{a_5} + \cdots + \frac{1}{a_n}\right)}$$

$$\ge \cdots \ge \frac{n}{\frac{1}{a_n} + \frac{n-1}{a_1}} \ge \frac{n}{\frac{1}{a_1} + \frac{n-1}{a_1}} = a_1.$$

Problem 1.203 Prove similarly that $M_{harm} \le a_n$.

Problem 1.204 Prove that $a_1 \le M_{arithm} \le a_n$.

Problem 1.205 Prove that $a_1 \le M_{geom} \le a_n$.

Problem 1.206 Prove that $a_1 \le M_{quadr} \le a_n$.

Hence, all four averages lie somewhere between the values of the largest and smallest elements of the set and hence do deserve being called "averages".

Next we shall prove that these averages give somewhat different characterisation of the elements in the set since they satisfy the following inequalities:

$$M_{harm} \leq M_{geom} \leq M_{arithm} \leq M_{quadr} \, . \tag{1.201}$$

So, the harmonic averages are the smallest of the four, while the quadratic—the largest; the other two lie somewhere in between.

We start by proving the inequality $M_{geom} \leq M_{arithm}$, i.e.

$$\frac{1}{n}(a_1 + a_2 + \cdots + a_n) \geq \sqrt[n]{a_1 a_2 \cdots a_n} \tag{1.202}$$

The proof is so ingenious that it would be a shame not to reproduce it here.[28] We begin with the case of $n = 2$. This was considered in Problem 1.18.

To consider the general case of any n, we shall prove next that Eq. (1.202) is valid for any n that can be represented as an integer power of 2, i.e. $n = 2^l$. This can be done by induction. Indeed, the case of $n = 2^1$ was already proven in the above problem. Assume next that it is also valid for some $m = 2^l$, and let us prove that it is valid for $n = 2m = 2^{l+1}$. In this case, one can split all numbers $a_1, \ldots, a_{2m}$ into m pairs and then apply the assumed inequality for m numbers:

$$\frac{1}{2m}(a_1 + a_2 + \cdots + a_{2m}) = \frac{1}{m}\left(\frac{a_1 + a_2}{2} + \frac{a_3 + a_4}{2} + \cdots + \frac{a_{2m-1} + a_{2m}}{2}\right)$$

$$\geq \sqrt[m]{\frac{a_1 + a_2}{2} \frac{a_3 + a_4}{2} \cdots \frac{a_{2m-1} + a_{2m}}{2}} \, .$$

Next, for every average of a pair under the root we can apply inequality (1.15):

$$\sqrt[m]{\frac{a_1 + a_2}{2} \frac{a_3 + a_4}{2} \cdots \frac{a_{2m-1} + a_{2m}}{2}} \geq \sqrt[m]{\sqrt{a_1 a_2}\sqrt{a_3 a_4} \cdots \sqrt{a_{2m-1} a_{2m}}}$$

$$= \sqrt[2m]{a_1 a_2 \cdots a_{2m}} \, ,$$

as required.

Next, we should consider an arbitrary n. For any given n, we can always find a natural k such that $n + k = 2^p$. For instance, with $n = 9$, the smallest possible k would be $k = 7$, so that $n + k = 16 = 2^4$; however, the k does not need to be the smallest possible, but could be chosen as such if desired. Next, let us add to

---

[28] This inequality was first proven by nineteenth-century French mathematician Augustin-Louis Cauchy.

our set of numbers additional positive numbers $a_{n+1}, \ldots, a_{n+k}$. Since $n + k$ can be represented as a power of 2, we can use our previous result and write:

$$\frac{1}{n+k} (a_1 + \cdots + a_n + a_{n+1} + \cdots + a_{n+k}) \geq \sqrt[n+k]{a_1 a_2 \cdots a_n a_{n+1} a_{n+2} \cdots a_{n+k}} .$$

$$(1.203)$$

The additional numbers that we have artificially attached to our set can be chosen as we wish. Let us choose them all equal and as follows:

$$a_{n+1} = a_{n+2} = \cdots = a_{n+k} = \frac{1}{n} (a_1 + a_2 + \cdots + a_n) .$$

Then the left-hand side of inequality (1.203) takes the form:

$$\text{LHS} = \frac{1}{n+k} \left(a_1 + \cdots + a_n + k \frac{a_1 + \cdots + a_n}{n} \right) = \frac{a_1 + \cdots + a_n}{n} ,$$

while its right-hand side becomes

$$\sqrt[n+k]{a_1 a_2 \cdots a_n \left(\frac{a_1 + \cdots + a_n}{n} \right)^k} ,$$

so that we can write:

$$\frac{a_1 + \cdots + a_n}{n} \geq \sqrt[n+k]{a_1 a_2 \cdots a_n \left(\frac{a_1 + \cdots + a_n}{n} \right)^k} .$$

Next, we raise both sides into the $(n + k)$th power[29]:

$$\left(\frac{a_1 + \cdots + a_n}{n} \right)^{n+k} \geq a_1 a_2 \cdots a_n \left(\frac{a_1 + \cdots + a_n}{n} \right)^k ,$$

or

$$\left(\frac{a_1 + \cdots + a_n}{n} \right)^n \geq a_1 a_2 \cdots a_n ,$$

which after taking the nth root of both sides goes into the desired result we are after.

What is left to see is at which conditions the equal sign is reached. Clearly, if all numbers are identical and equal to a, then we shall get the equal sign as on both sides we simply get a. Let us prove that the equal sign is never reached if at least one of the numbers is different. Suppose, $a_1 \neq a_2$, then

---

[29] As both sides are positive, this is permissible.

$$\frac{a_1 + \cdots + a_n}{n} = \frac{\frac{a_1+a_2}{2} + \frac{a_1+a_2}{2} + a_3 + \cdots + a_n}{n}$$

$$\geq \sqrt[n]{\frac{a_1 + a_2}{2} \frac{a_1 + a_2}{2} a_3 a_4 \cdots a_n} \,,$$

where we have made use of our inequality (1.202). Since a_1 is strictly not equal to a_2, we have from Eq. (1.15) that

$$\frac{a_1 + a_2}{2} > \sqrt{a_1 a_2}$$

with the "greater than" sign. Therefore,

$$\frac{a_1 + \cdots + a_n}{n} \geq \sqrt[n]{\frac{a_1 + a_2}{2} \frac{a_1 + a_2}{2} a_3 a_4 \cdots a_n}$$

$$> \sqrt[n]{a_1 a_2 a_3 a_4 \cdots a_n} \,,$$

with the strictly "greater than" sign. The prove is concluded.

Various inequalities can be constructed by choosing n numbers in Eq. (1.202) in different ways.

Next, we shall prove that $M_{harm} \leq M_{geom}$. This is suggested as a problem for the reader.

Problem 1.207 By applying inequality (1.202) to $b_i = 1/a_i$, prove that $M_{harm} \leq M_{geom}$.

To finish proving Eq. (1.201), it is only remained to demonstrate that $M_{arithm} \leq M_{quadr}$, i.e. that

$$\frac{1}{n} \sum_{k=1}^{n} a_k \leq \sqrt{\frac{1}{n} \sum_{k=1}^{n} a_k^2} \,. \tag{1.204}$$

It is easy to see (by squaring both sides) that what we actually need to prove is:

$$\left(\sum_{k=1}^{n} a_k \right)^2 \leq n \sum_{k=1}^{n} a_k^2 \,. \tag{1.205}$$

We shall prove this by induction. In the case of $n = 1$ we have an obvious equality. The case of $n = 2$ was proven in Problem 1.19.

Next, we assume that the inequality (1.205) is valid for some n, and prove that it is also valid for the next value of n, i.e.

$$\left(\sum_{k=1}^{n+1} a_k \right)^2 \leq (n + 1) \sum_{k=1}^{n+1} a_k^2 \,.$$

Consider the difference of the left- and right-hand sides:

$$(n+1)\sum_{k=1}^{n+1} a_k^2 - \left(\sum_{k=1}^{n+1} a_k\right)^2 = (n+1)\sum_{k=1}^{n+1} a_k^2 - \left[\left(\sum_{k=1}^{n} a_k\right) + a_{n+1}\right]^2$$

$$= (n+1)\sum_{k=1}^{n+1} a_k^2 - \left[\left(\sum_{k=1}^{n} a_k\right)^2 + a_{n+1}^2 + 2a_{n+1}\left(\sum_{k=1}^{n} a_k\right)\right].$$

Due to our assumption (1.205), we can estimate the above expression as

$$\geq (n+1)\sum_{k=1}^{n+1} a_k^2 - \left[n\sum_{k=1}^{n} a_k^2 + a_{n+1}^2 + 2a_{n+1}\left(\sum_{k=1}^{n} a_k\right)\right]$$

$$= na_{n+1}^2 + \sum_{k=1}^{n} a_k^2 - 2a_{n+1}\left(\sum_{k=1}^{n} a_k\right)$$

$$= \sum_{k=1}^{n} \left(a_{n+1}^2 + a_k^2 - 2a_{n+1}a_k\right) = \sum_{k=1}^{n} (a_{n+1} - a_k)^2 \geq 0\,,$$

as required. This fully proves the inequality (1.201).

There is a curious illustration of the usefulness of the inequality $M_{geom} \leq M_{arithm}$. We shall prove that of all triangles with the fixed perimeter $2p$, the equilateral triangle (with equal sides $a = b = c = 2p/3$) has the largest area. To prove this, let us consider a triangle with sides a, b and c. Its area is given by Heron's formula (1.71):

$$S_\triangle = \sqrt{p(p-a)(p-b)(p-c)}\,.$$

The idea is to apply the inequality $M_{geom} \leq M_{arithm}$ to the product of terms under the square root:

$$(p-a)(p-b)(p-c) \leq \left[\frac{(p-a)+(p-b)+(p-c)}{3}\right]^3 = \left[\frac{p}{3}\right]^3 = \frac{p^3}{27}\,,$$

so that

$$S_\triangle \leq \sqrt{\frac{p^4}{27}} = \frac{p^2}{3\sqrt{3}}\,.$$

So, the area of the triangle cannot exceed the value in the right-hand side. In fact, the equilateral triangle has its area equal precisely this value:

$$S_\triangle = \sqrt{p(p-a)(p-b)(p-c)} = \sqrt{p\left(p-\frac{2p}{3}\right)^3} = \frac{p^2}{3\sqrt{3}}\,.$$

This proves the above made statement.

Problem 1.208 Prove the following inequality:

$$\sqrt[n]{n!} \le \frac{n+1}{2}\,.$$

[Hint: *use Eq. (1.202) choosing the numbers $a_1, \ldots, a_n$ such that the product under the root is reproduced.*]

Problem 1.209 Prove that for any integer n

$$\left(1 + \frac{1}{n+1}\right)^{n+1} > \left(1 + \frac{1}{n}\right)^{n}\,.$$

Note that here we have strictly the "greater than" sign. To prove this inequality, choose $n+1$ numbers $a_1 = a_2 = \cdots = 1 + \frac{1}{n}$ and $a_{n+1} = 1$, and apply inequality (1.202).

Problem 1.210 Using induction, prove that for any $a > -1$,

$$(1 + a)^n \ge 1 + an\,. \qquad (1.206)$$

The condition $a > -1$ ensures that $1 + a$ in the left-hand side is positive. This is called Bernoulli's inequality.[30]

Problem 1.211 Prove that

$$\sum_{k=1}^{n} \frac{1}{n+k} > \frac{1}{2}\,.$$

[Hint: *use the fact that $1/\left(1 + \frac{k}{n}\right) \ge 1/2$.*]

1.19   Lines, Planes and Spheres

We shall be often dealing with lines and surfaces in the course, and it is convenient at this stage to summarise a number of useful results here. We shall start from lines, then turn to surfaces, such as planes and spheres, after which we shall consider various geometrical problems related to all of them.

---

[30] Named after Jacob Bernoulli.

Fig. 1.36 An illustration to
the derivation of the
equations for a line

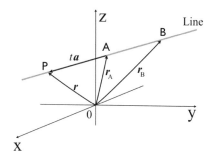

1.19.1 Straight Lines

A straight line in the 3D space can be specified by any two points $A(x_A, y_A, z_A)$
and $B(x_B, y_B, z_B)$ on the line, or alternatively, by vectors $\mathbf{r}_A = (x_A, y_A, z_A)$ and
$\mathbf{r}_B = (x_B, y_B, z_B)$. If the vector $\mathbf{r} = (x, y, z)$ corresponds to an arbitrary point P on
the line, see Fig. 1.36, then we can construct a vector $\mathbf{a} = \overrightarrow{AB} = \mathbf{r}_B - \mathbf{r}_A$ along the
line, so that any point P on the line characterised by the vector $\mathbf{r}$ can be obtained by
adding to $\mathbf{r}_A$ the scaled vector $\mathbf{a}$:

$$\mathbf{r} = \mathbf{r}_A + t\mathbf{a} = \mathbf{r}_A + t(\mathbf{r}_B - \mathbf{r}_A) \; , \qquad (1.207)$$

where $-\infty < t < \infty$ is a scalar parameter. This is a parametric equation of the line,
specified either via two points A and B or via a vector $\mathbf{a}$ (it is frequently convenient
for it to be a unit vector) and a point A on the line.

Problem 1.212 The vector equation (1.207) is equivalent to three scalar ones
for the three Cartesian components of $\mathbf{r}$. Show that the equation of a line can
also alternatively be written as follows:

$$\frac{x - x_A}{x_B - x_A} = \frac{y - y_A}{y_B - y_A} = \frac{z - z_A}{z_B - z_A} . \qquad (1.208)$$

Problem 1.213 Inversely, show that given Eq. (1.208) for a line, its parametric
equation form is

$$\mathbf{r} = \begin{pmatrix} x_A \\ y_A \\ z_A \end{pmatrix} + t \begin{pmatrix} x_B - x_A \\ y_B - y_A \\ z_B - z_A \end{pmatrix} , \qquad (1.209)$$

which is Eq. (1.207).

Problem 1.214 Show that the line $\mathbf{r} = \mathbf{r}_A + t\mathbf{a}$ can also be written via the cross product:

$$(\mathbf{r} - \mathbf{r}_A) \times \mathbf{a} = 0 . \tag{1.210}$$

Two lines are parallel (collinear) if the vectors $\mathbf{a}_1$ and $\mathbf{a}_2$ in the corresponding parametric equations are parallel ($\mathbf{a}_1 \parallel \mathbf{a}_2$) , i.e. if $\mathbf{a}_1 \times \mathbf{a}_2 = 0$; two lines are perpendicular if these vectors are orthogonal, $\mathbf{a}_1 \cdot \mathbf{a}_2 = 0$ (or $\mathbf{a}_1 \perp \mathbf{a}_2$). Note, however, that in the second case the lines may not necessarily cross. A simple condition for two lines to cross (even if they are not perpendicular) is considered later on in this section.

Now consider specifically the $x - y$-plane.

Problem 1.215 Show from the result above that in the $x - y$-plane ($z = 0$, a 2D space), the equation of a line can be formally written as

$$Ax + By = C . \tag{1.211}$$

Note that this form is more general than the familiar $y = kx + c$ (with k and c being constants) since it contains also lines parallel to the y-axis ($Ax = C$) which are not contained in the other form. Give a geometrical interpretation of the constants k and b of the line equation $y = kx + b$.

Problem 1.216 Show that the line $y = kx + b$ can be written in the parametric form, e.g. as

$$\mathbf{r} = \begin{pmatrix} x \\ y \end{pmatrix} = \begin{pmatrix} 0 \\ b \end{pmatrix} + \frac{t}{\sqrt{1 + k^2}} \begin{pmatrix} 1 \\ k \end{pmatrix} . \tag{1.212}$$

Problem 1.217 Two lines $y = k_1 x + b_1$ and $y = k_2 x + b_2$ are specified on the $x - y$-plane. Show using Eq. (1.212) that the two lines are parallel if $k_1 = k_2$ and perpendicular if $k_2 = -1/k_1$. Verify that these results make sense from the geometrical point of view. [Hint: *in the first case write the line vectors $\mathbf{a}_1$ and $\mathbf{a}_2$ (defining the directions of the two lines) in 3D with zero z-component.*]

1.19.2   Polar and Spherical Coordinates

In 2D space *polar coordinates* are frequently used instead of the Cartesian ones. In polar coordinates the length $r = |\mathbf{r}|$ of the vector $\mathbf{r}$ and its angle ϕ with the x-

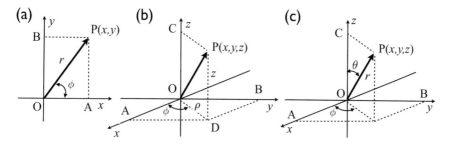

Fig. 1.37 Definitions of polar (**a**), cylindrical (**b**) and spherical (**c**) coordinate systems. The components of the vector **r** are its projections $x = OA$, $y = OB$ and $z = OC$ on the Cartesian axes

axis ($0 \leq \phi \leq 2\pi$) are used instead of the Cartesian coordinates (x, y) as shown in Fig. 1.37a. As it follows from the figure, the relationship between the two types of coordinates is provided by the following *connection relations*:

$$x = r \cos \phi , \quad y = r \sin \phi . \tag{1.213}$$

In 3D space polar coordinates are generalised to *cylindrical* coordinates by adding the z-coordinate as shown in Fig. 1.37b; in this case (x, y, z) are replaced by (r, ϕ, z) as follows:

$$x = \rho \cos \phi , \quad y = \rho \sin \phi , \quad z = z . \tag{1.214}$$

Note that ρ here is the length of the projection OD of the vector **r** on the $x - y$-plane.

Finally, we shall also introduce another very useful coordinate system—*spherical*. It is defined in Fig. 1.37c. Here the coordinates used are (r, θ, ϕ), where this time r is the length OP of the vector **r**, θ is the angle it makes with the z-axis ($0 \leq \theta \leq \pi$) and ϕ is the same angle as in the two previous systems ($0 \leq \phi \leq 2\pi$). The connection relations in this case are:

$$x = r \sin \theta \cos \phi , \quad y = r \sin \theta \sin \phi , \quad z = r \cos \theta , \tag{1.215}$$

since $CP = r \sin \theta$ and hence $x = OA = CP \cos \phi$, $y = OB = CP \sin \phi$ and $z = OC = r \cos \theta$.

The special coordinate systems introduced above are frequently used in the calculus to simplify calculations, and we shall come across many such examples throughout the course. Polar coordinates are useful in 2D calculations related to objects having circular symmetry; cylindrical coordinates—for 3D objects with cylindrical symmetry (with the axis of the cylinder along the z-axis), while the spherical coordinate system is convenient for the 3D calculations with objects having spherical symmetry. Many other so-called *curvilinear coordinates* also exist.

1.19.3 Curved Lines

Here we shall briefly mention an application of polar coordinates for specifying curved lines in the 2D space. The idea is based on writing r as a function of ϕ, e.g. $r = r(\phi)$. It is essential to understand that this is simply a prescription for writing the Cartesian coordinates,

$$x(\phi) = r(\phi)\cos\phi \quad \text{and} \quad y(\phi) = r(\phi)\sin\phi , \tag{1.216}$$

as parametric equations of curved lines via the angle ϕ serving as a single parameter. Because of that, $r = r(\phi)$ may be even negative, i.e. only its absolute value corresponds to the distance of the (x, y)-point to the centre O of the coordinate system.

As an example, consider a circle of radius R centred at the origin. Its equation is given simply by $r = R$, i.e. r does not depend on the angle ϕ. In that case from (1.216) we get (by taking the square of both sides of the two equations and summing them up) that the equation of the circle is the familiar $x^2 + y^2 = R^2$.

Problem 1.218 Show that the curve specified via $r = 2R\sin\phi$ corresponds to a circle of radius R centred at $(0, R)$. [Hint: *solve for ϕ from one of the equations for x or y, and derive an equation $f(x, y) = 0$ for the curve.*]

Problem 1.219 Various famous curves are shown in Fig. 1.38. Convince yourself that these geometrical figures correspond to the equations given. In the case of the three-petal rose (a) verify the range of the angle ϕ necessary to plot a single petal.[31]

Problem 1.220 Consider a family of curves specified by $r = a\cos(k\phi)$ and sketch them. How does the number of petals depend on the value of k?

A sphere, ellipse, hyperbola and parabola can all be written in a unified way in polar coordinates as

$$\frac{1}{r} = A(1 + e\cos\phi) , \tag{1.217}$$

where A is a positive constant and e is the eccentricity (also positive), Sect. 1.8. Indeed, if $e = 0$ we obtain $r = 1/A$ being a constant and hence this case corresponds to a circle.

---

[31] This and other related curves were studied by an Italian mathematician Guido Grandi.

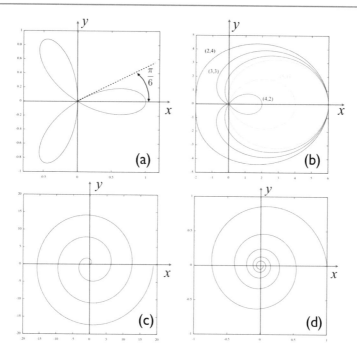

Fig. 1.38 Examples of various famous curves specified in polar coordinates: **a** a three-leaf rose (or rhodonea curve) specified as $r = a\cos(3\phi)$ (plotted for $a = 1$); **b** snail curves (limacons of Pascal) specified via $r = b + a\cos\phi$ and shown for various pairs of (a, b); note how the curves continuously change their shape when going from $a < b$ to $a > b$ across $a = b$; **c** Archimedean spiral $r = a\phi$ (plotted for $a = 1$); **d** spiral $r = ae^{-\alpha\phi}$ (plotted for $a = 1$ and $\alpha = 0.1$). In the latter two cases only a finite number of revolutions are depicted

Problem 1.221 Show that in the case of $e = 1$ we obtain a 90° rotated parabola (with the symmetry axis along the x-axis)

$$x = \frac{1}{2A} - \frac{A}{2}y^2 .$$

Problem 1.222 Show that in the case of $e < 1$ we can rearrange Eq. (1.217) into the form of an ellipse, Eq. (1.83), with $x_0 \neq 0$ and $y_0 = 0$ (i.e. it is x-shifted) and

$$\frac{1}{a} = A(1 - e^2) \quad \text{and} \quad \frac{1}{b} = A\sqrt{1 - e^2} ,$$

while in the case of $e > 1$ we similarly obtain Eq. (1.84) of a x-shifted hyperbola with

$$\frac{1}{a} = A(e^2 - 1) \quad \text{and} \quad \frac{1}{b} = A\sqrt{e^2 - 1} .$$

It is easy to see that in both cases e is indeed the corresponding eccentricity, Eqs. (1.82) and (1.85), respectively.

1.19.4 Planes

A plane can be specified by either three points (A, B, C) or by a normal vector $\mathbf{n}$ and a single point A, see Fig. 1.39. The two ways are closely related as the normal can be constructed from the three points by performing a vector product $\mathbf{n} \propto \overrightarrow{CB} \times \overrightarrow{CA} = (\mathbf{r}_B - \mathbf{r}_C) \times (\mathbf{r}_A - \mathbf{r}_C)$. So, it is sufficient to consider the specification of the plane via a normal and a point, say the point A. If we take an arbitrary point P on the plane corresponding to the vector $\mathbf{r}$, then the vector $\overrightarrow{AP} = \mathbf{r} - \mathbf{r}_A$ lies within the plane, and hence is orthogonal to the normal:

$$\mathbf{n} \cdot (\mathbf{r} - \mathbf{r}_A) = 0 \implies \mathbf{r} \cdot \mathbf{n} = \mathbf{r}_A \cdot \mathbf{n} \text{ or } \mathbf{r} \cdot \mathbf{n} = h , \tag{1.218}$$

where $h = \mathbf{r} \cdot \mathbf{n}_A$. Either form serves as the desired equation of the plane. One can see that the point $\mathbf{r}_A$ lies on the plane (as $\mathbf{r} = \mathbf{r}_A$ also satisfies the equation), and that the normal does not need to be a unit vector. In addition, either direction of the normal can be taken "up" or "down".

Another form of the equation for the plane follows immediately from (1.218) if we write explicitly the dot product in the left-hand side:

$$ax + by + cz = h . \tag{1.219}$$

It follows from this discussion that the coefficients (a, b, c) in this equation form components of the normal to the surface. Note that if $h \neq 0$, one may divide both sides of the equation by it, resulting in a simpler form $a'x + b'y + c'z = 1$; however, for some surfaces $h = 0$ (e.g. the $x - y$-plane has the equation $z = 0$), so that the form above is the most general. Let us mention some particular cases which might be of interest: (i) $h = 0$ corresponds to a plane passing through the centre of the coordinate system; (ii) $a = 0$ corresponds to a plane perpendicular to the x-axis; (iii) $a = b = 0$ corresponds to a plane $cz = h$ which is parallel to the $x - y$-plane.

Fig. 1.39 An illustration to the derivation of the equations for a plane

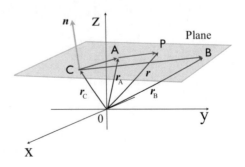

Problem 1.223 Determine one of the unit normals to the plane specified by the following equations: (a) $x + y + z = -1$; (b) $3x + 4y = -1$; (c) $3z = 8$; (d) $y = 0$. [Answer: *(a)* $\mathbf{n} = \frac{1}{\sqrt{3}}(1, 1, 1)$; *(b)* $\mathbf{n} = \frac{1}{5}(3, 4, 0)$; *(c)* $\mathbf{n} = (0, 0, 1)$; *(d)* $\mathbf{n} = (0, 1, 0)$. *Note that opposite directions of the normal vectors are also acceptable.*]

We shall now derive a parametric equation of a plane. In Eq. (1.207) for a line we only had one parameter t; this is because a line is equivalent to a 1D space. In the case of a plane there must be two such parameters as the plane is equivalent to a 2D space. One may say that a line has one "degree of freedom", while a plane has two. Now, consider two vectors $\mathbf{a}_1$ and $\mathbf{a}_2$ lying in the plane (it is usually convenient to choose them of unit length), and the vector $\mathbf{r}_A$ pointing to a point A on the plane. Then, evidently, any point P on the plane would have the vector

$$\mathbf{r} = \mathbf{r}_A + t_1\mathbf{a}_1 + t_2\mathbf{a}_2 \tag{1.220}$$

pointing to it; here t_1 and t_2 are the two parameters, each of which may take any values from $\mathbb{R}$.

1.19.5 Circle and Sphere

Using vector notations, it is very easy to rewrite Eq. (1.80) of a circle, $(x - x_0)^2 + (y - y_0)^2 = R^2$, in a much simpler way. Indeed, what we actually have in the left-hand side is the distance squared of the two points, (x, y) and (x_0, y_0), on the $x - y$-plane. Hence, the equation of a circle can simply be rewritten as

$$|\mathbf{r} - \mathbf{r}_0| = R , \tag{1.221}$$

where $\mathbf{r}_0 = (x_0, y_0)$ is the centre of the circle. This simple formula states that any point $\mathbf{r}$ on the circle stays at the distance R from its centre.

This result can easily be generalised to a sphere, which is the 3D analogue of a circle. Indeed, in the case of the sphere of radius R centred at the point $\mathbf{r}_0 = (x_0, y_0, z_0)$ any point $\mathbf{r}$ on the sphere is separated from its centre $\mathbf{r}_0$ by the same distance R. Hence, Eq. (1.221) is formally valid for the sphere as well, but need to be understood as written in the 3D space. Writing down this equation explicitly, after squaring both sides, we obtain:

$$(x - x_0)^2 + (y - y_0)^2 + (z - z_0)^2 = R^2 , \tag{1.222}$$

which is an equivalent equation of a sphere.

Problem 1.224 Find the radius and the central point of the sphere $x^2 + 2x + y^2 + z^2 - 4z - 1 = 0$. [Answer: $\mathbf{r}_0 = (-1, 0, 2)$ *and* $R = \sqrt{6}$.]

1.19.6 Typical Problems for Lines, Planes and Spheres

Let us now consider a number of typical problems which one comes across in applications related to lines, planes and spheres.

Distance from a point to a line or plane. Let us take a point P specified by the vector $\mathbf{r}_P$ and a line $\mathbf{r} = \mathbf{r}_A + t\mathbf{a}$ specified via the point A and the vector $\mathbf{a}$. What we would like to calculate is the distance between them; this will be the length of the perpendicular drawn from the point P to the line as shown in Fig. 1.40. First, consider the case of the acute angle made by the vectors $\overrightarrow{AP}$ and $\mathbf{a}$, i.e. when the dot product $\overrightarrow{AP} \cdot \mathbf{a} > 0$, Fig. 1.40a. Then the angle ϕ can be obtained from the dot product of $\mathbf{a}$ and $\overrightarrow{AP}$, i.e. $\overrightarrow{AP} \cdot \mathbf{a} = |\mathbf{a}| \left|\overrightarrow{AP}\right| \cos \phi$. Calculating the dot product using components of the vectors $\mathbf{a}$ and $\overrightarrow{AP}$, the angle ϕ is evaluated. Then, the required distance

$$BP = \left|\overrightarrow{AP}\right| \sin \phi = |\mathbf{r}_P - \mathbf{r}_A| \sqrt{1 - \cos^2 \phi}$$

$$= |\mathbf{r}_P - \mathbf{r}_A| \sqrt{1 - \left(\frac{(\mathbf{r}_P - \mathbf{r}_A) \cdot \mathbf{a}}{|\mathbf{r}_P - \mathbf{r}_A| \, |\mathbf{a}|}\right)^2} . \tag{1.223}$$

If the dot product $\overrightarrow{AP} \cdot \mathbf{a} < 0$, then the angle ϕ calculated via the dot product $\overrightarrow{AP} \cdot \mathbf{a} = |\mathbf{a}| \left|\overrightarrow{AP}\right| \cos \phi$ will be between $\pi/2$ and π, Fig. 1.40b, and hence the complementary angle is to be used instead: $BP = \left|\overrightarrow{AP}\right| \sin (\pi - \phi) = \left|\overrightarrow{AP}\right| \sin \phi$. However, this does not change the final result (1.223).

Let us now find the point B on the line $\mathbf{r} = \mathbf{r}_A + t\mathbf{a}$ that is the closest to the point P, see the same Fig. 1.40a, where we again assumed that the dot product $\mathbf{a} \cdot \overrightarrow{AP} > 0$. In this case $\overrightarrow{AP} \cdot \mathbf{a} = \left|\overrightarrow{AP}\right| |\mathbf{a}| \cos \phi$ with the angle $0 < \phi \le \pi/2$, so that

$$\cos \phi = \frac{\overrightarrow{AP} \cdot \mathbf{a}}{\left|\overrightarrow{AP}\right| |\mathbf{a}|} \quad \Longrightarrow \quad AB = \left|\overrightarrow{AP}\right| \cos \phi = \frac{\overrightarrow{AP} \cdot \mathbf{a}}{|\mathbf{a}|} . \tag{1.224}$$

Fig. 1.40 To the calculation of the distance PB between the point P and the line λ. The line direction $\mathbf{a}$ makes either an acute (**a**) or obtuse (**b**) angle to the vector $\overrightarrow{AP}$

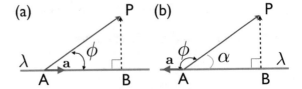

On the other hand, from the line equation, we can write $\mathbf{r}_B - \mathbf{r}_A = t_B \mathbf{a}$, where t_B is the value of the parameter t giving the point B; its value follows immediately from the last equation as

$$t_B = \frac{|\mathbf{r}_B - \mathbf{r}_A|}{|\mathbf{a}|} = \frac{AB}{|\mathbf{a}|} = \frac{\overrightarrow{AP} \cdot \mathbf{a}}{|\mathbf{a}|^2} . \tag{1.225}$$

Since the dot product $\mathbf{a} \cdot \overrightarrow{AP} > 0$, the point P is further away from the point A along the direction of $\mathbf{a}$, i.e. the parameter $t_B > 0$. Therefore, we obtain the vector of the point B as

$$\mathbf{r}_B = \mathbf{r}_A + \frac{\overrightarrow{AP} \cdot \mathbf{a}}{|\mathbf{a}|^2}\mathbf{a} = \mathbf{r}_A + \frac{(\mathbf{r}_P - \mathbf{r}_A) \cdot \mathbf{a}}{|\mathbf{a}|^2}\mathbf{a} \tag{1.226}$$
$$= \mathbf{r}_A + [(\mathbf{r}_P - \mathbf{r}_A) \cdot \widehat{\mathbf{a}}]\widehat{\mathbf{a}} ,$$

where $\widehat{\mathbf{a}} = \mathbf{a}/|\mathbf{a}|$ is the unit vector in the direction of $\mathbf{a}$. If the direction of $\overrightarrow{AB}$ is opposite to that of $\mathbf{a}$, Fig. 1.40b, the same result is obtained. Indeed, the dot product $\overrightarrow{AP} \cdot \mathbf{a} = \left|\overrightarrow{AP}\right| |\mathbf{a}| \cos\phi$ is in this case negative with the angle $\pi/2 \leq \phi < \pi$, and hence the distance AB comes out negative, leading to a negative value of the parameter t_B in Eq. (1.225). This is to be expected as this time the point B is before the point A. Using this value of t_B in the line equation, we again arrive at the final result (1.226), which therefore stands in both cases.

The obtained expression can also be derived directly from Fig. 1.40. Indeed, the distance $AB = (\mathbf{r}_P - \mathbf{r}_A) \cdot \widehat{\mathbf{a}}$, hence the vector $\overrightarrow{AB} = AB\,\widehat{\mathbf{a}} = \mathbf{r}_B - \mathbf{r}_A$.

The above result can be verified by determining again the distance BP between point P and the line. Indeed, it can now be calculated directly as $BP = |\mathbf{r}_P - \mathbf{r}_B|$.

Problem 1.225 Prove by calculating directly the distance $BP = |\mathbf{r}_P - \mathbf{r}_B|$ from Eq. (1.226) that the previously derived formula (1.223) is reproduced. [Hint: *calculate the vector* $\mathbf{r}_P - \mathbf{r}_B$ *and work out explicitly its square.*]

Example 1.15 ▶ Consider a simple 2D example of the line $\mathbf{r} = (1, 3) + t(1, 1)$ and point $P(1, 1)$. Obtain the point $\mathbf{r}_B$ on the line closest to the point P.

Solution. Here $\mathbf{a} = (1, 1), \mathbf{r}_A = (1, 3)$ and $\mathbf{r}_P - \mathbf{r}_A = (0, -2)$. Hence, the dot product $(\mathbf{r}_P - \mathbf{r}_A) \cdot \mathbf{a} = -2$, and therefore

$$\mathbf{r}_B = \begin{pmatrix} 1 \\ 3 \end{pmatrix} + \frac{-2}{\left(\sqrt{2}\right)^2} \begin{pmatrix} 1 \\ 1 \end{pmatrix} = \begin{pmatrix} 1 \\ 3 \end{pmatrix} - \begin{pmatrix} 1 \\ 1 \end{pmatrix} = \begin{pmatrix} 0 \\ 2 \end{pmatrix} ,$$

the desired result that can easily be checked by a simple drawing. ◀

Problem 1.226 Calculate the distance between the line $\mathbf{r} = (1,1,1) + t(0,1,0)$ and the centre of the coordinate system; determine the point on the line which is the closest to the centre. Check the calculation of the distance using Eq. (1.223). [Answer: $\sqrt{2}$, $(1,0,1)$.]

Next, let us consider the 2D case where the line is given by the equation $y = kx + b$. Let point A be $(0, b)$ (i.e. it is the point at which the line crosses the y-axis). The point P has coordinates (x_P, y_P). Then, by considering two points $A = (0, b)$ and $(x_P, kx_P + b)$ on the line, we can calculate its direction vector as

$$\mathbf{a} = (x_P - 0)\mathbf{i} + [(kx_P + b) - b]\mathbf{j} = x_P\mathbf{i} + x_P k\mathbf{j} .$$

The vector $\overrightarrow{AP} = (x_P, y_P - b)$. Therefore,

$$\cos\phi = \frac{\overrightarrow{AP} \cdot \mathbf{a}}{\left|\overrightarrow{AP}\right| |\mathbf{a}|} = \frac{x_P^2 + kx_P(y_P - b)}{\sqrt{x_P^2 + (y_P - b)^2} x_P \sqrt{1 + k^2}} = \frac{x_P + k(y_P - b)}{\sqrt{x_P^2 + (y_P - b)^2}\sqrt{1 + k^2}}$$

and hence

$$\sin^2\phi = 1 - \cos^2\phi = 1 - \frac{x_P^2 + 2kx_P(y_P - b) + k^2(y_P - b)^2}{\left[x_P^2 + (y_P - b)^2\right]\left(1 + k^2\right)}$$

$$= \frac{(kx_P - y_P + b)^2}{\left[x_P^2 + (y_P - b)^2\right]\left(1 + k^2\right)} = \frac{(kx_P - y_P + b)^2}{\left(1 + k^2\right)\left|\overrightarrow{AP}\right|^2} ,$$

so that the required distance is

$$d = \left|\overrightarrow{AP}\right| \sin\phi = \frac{|kx_P - y_P + b|}{\sqrt{1 + k^2}} . \tag{1.227}$$

Now we shall turn our attention to a distance between a point and a plane.

Problem 1.227 It is easy to see that the distance between a point $P(x_P, y_P, z_P)$ and a plane $ax + by + cz = h$ is (see Fig. 1.41a):

$$d = BP = \left|\mathbf{n} \cdot \overrightarrow{AP}\right| = |\mathbf{n} \cdot (\mathbf{r}_P - \mathbf{r}_A)| , \tag{1.228}$$

where $\mathbf{n}$ is the unit normal to the plane and the absolute value of the dot product is required to ensure the positivity of d in case the normal $\mathbf{n}$ is chosen to make an obtuse angle with the vector $\overrightarrow{AP}$. Here A is an arbitrary point on the plane, the result must not depend on its choice. Indeed, manipulate this expression into

$$d = \frac{|ax_P + by_P + cz_P - h|}{\sqrt{a^2 + b^2 + c^2}}. \tag{1.229}$$

Crossing of two lines. The necessary conditions for crossing of two lines can be established as follows. Suppose, we are given two lines $\mathbf{r} = \mathbf{r}_1 + t_1\mathbf{a}_1$ and $\mathbf{r} = \mathbf{r}_2 + t_2\mathbf{a}_2$, where t_1 and t_2 are two parameters. The lines will cross if and only if there exists a plane to which they both belong. This means that the vectors $\mathbf{a}_1$, $\mathbf{a}_2$ and $\mathbf{r}_1 - \mathbf{r}_2$ should lie in this plane. This in turn means that the vector $\mathbf{r}_1 - \mathbf{r}_2$ can be represented as a linear combination $s_1\mathbf{a}_1 + s_2\mathbf{a}_2$ of the vectors $\mathbf{a}_1$ and $\mathbf{a}_2$ with some numbers s_1 and s_2 since any vector within the plane can be represented like this unless the two vectors $\mathbf{a}_1$, $\mathbf{a}_2$ are parallel which we assume is not the case.[32] There is also an algebraic way to show precisely that: since two lines cross, the point $\mathbf{r}$ of crossing will be common to both of them, therefore $\mathbf{r}_1 + t_1\mathbf{a}_1 = \mathbf{r}_2 + t_2\mathbf{a}_2$ and hence $\mathbf{r}_1 - \mathbf{r}_2 = -t_1\mathbf{a}_1 + t_2\mathbf{a}_2$, i.e. the desired linear combination. The simplest way of writing this condition down is by using the triple product:

$$[\mathbf{r}_1 - \mathbf{r}_2, \mathbf{a}_1, \mathbf{a}_2] = \begin{vmatrix} x_1 - x_2 & y_1 - y_2 & z_1 - z_2 \\ (\mathbf{a}_1)_x & (\mathbf{a}_1)_y & (\mathbf{a}_1)_z \\ (\mathbf{a}_2)_x & (\mathbf{a}_2)_y & (\mathbf{a}_2)_z \end{vmatrix} = 0, \tag{1.230}$$

i.e. the determinant composed of the components of the three vectors is zero. Indeed, as we know from the properties of the determinants mentioned above, it is equal to zero if any of its rows (or columns) is a linear combination of any other rows (columns). Saying it differently, if the three vectors lie in the same plane, their triple product is zero (see Sect. 1.10), which immediately brings us to the above condition, cf. Eq. (1.108). The above condition must be supplemented by the condition for the two lines not be parallel, $\mathbf{a}_1 \times \mathbf{a}_2 \neq 0$, since, if they are, the condition (1.230) is satisfied automatically while the lines never cross.

Note that the condition $\mathbf{a}_1 \times \mathbf{a}_2 = 0$ for two lines be parallel and hence not to cross is not a necessary condition for avoided crossing as lines not satisfying this condition may still not cross as we shall see below.

Fig. 1.41 a To the calculation of the distance PB between the point P and the plane σ; **c** finding the points A of intersection of the line λ with the plane σ

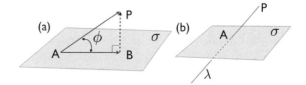

---

[32] Note that the case of $\mathbf{a}_1 \parallel \mathbf{a}_2$ is not interesting here as in this case the lines do not cross—unless they are the same, of course, which is of even less interest!

Example 1.16 ▶ A line is given by the equation $\mathbf{r} = (1, 0, 0) + t_1(1, 0, 0)$. There is also a family of lines specified via $\mathbf{r} = (0, 1, 1) + t_2(0, \cos\varphi, \sin\varphi)$; each line in the family corresponds to a particular value of the angle φ. Determine which line in the family (i.e. at which value of the angle φ) crosses the first line, and find the crossing point.

Solution. We have to solve Eq. (1.230) with $\mathbf{r}_1 - \mathbf{r}_2 = (1, -1, -1)$, $\mathbf{a}_1 = (1, 0, 0)$ and $\mathbf{a}_2 = (0, \cos\varphi, \sin\varphi)$; hence, building up the determinant and calculating it, we get:

$$\begin{vmatrix} 1 & -1 & -1 \\ 1 & 0 & 0 \\ 0 & \cos\varphi & \sin\varphi \end{vmatrix} = 1 \cdot (0 \cdot \sin\varphi - 0 \cdot \cos\varphi) - (-1) \cdot (\sin\varphi - 0 \cdot 0) - 1 \cdot (\cos\varphi - 0 \cdot 0)$$

$$= \sin\varphi - \cos\varphi = 0$$

$$\Longrightarrow \quad \sin\varphi = \cos\varphi,$$

i.e. $\tan\varphi = 1$ and $\varphi = \pi/4$, so the equation of the second line is $\mathbf{r} = (0, 1, 1) + t_2\left(0, 1/\sqrt{2}, 1/\sqrt{2}\right)$. Let us also check that the two lines are not parallel:

$$\mathbf{a}_1 \times \mathbf{a}_2 = \begin{vmatrix} \mathbf{i} & \mathbf{j} & \mathbf{k} \\ 1 & 0 & 0 \\ 0 & 1/\sqrt{2} & -1/\sqrt{2} \end{vmatrix} = \frac{1}{\sqrt{2}}\mathbf{j} + \frac{1}{\sqrt{2}}\mathbf{k} \neq 0.$$

The point of crossing is then obtained by solving the vector equation

$$\mathbf{r} = (1, 0, 0) + t_1(1, 0, 0) = (0, 1, 1) + t_2\left(0, \frac{1}{\sqrt{2}}, \frac{1}{\sqrt{2}}\right)$$

$$\Longrightarrow \quad \begin{cases} 1 + t_1 = 0 \\ -1 - t_2\frac{1}{\sqrt{2}} = 0 \\ -1 - t_2\frac{1}{\sqrt{2}} = 0 \end{cases}$$

with respect to t_1 and t_2 in x-, y- and z-components. There are three equations to solve for only two variables; however, because of the condition of crossing (1.230) we just used, one of the equations should be equivalent to the other two, so we just have to consider two independent equations. In our case these are $1 + t_1 = 0$ for the x-component and $0 = 1 + t_2/\sqrt{2}$ for the y (the z-component equation, $0 = 1 + t_2/\sqrt{2}$, is in this case the same as the second), which give $t_1 = -1$ and $t_2 = -\sqrt{2}$, and hence the point of crossing $\mathbf{r} = (1, 0, 0) + t_1(1, 0, 0) = (0, 0, 0)$ is the centre of the coordinate system (of course, the same result is obtained by using the equation of the second line and the value of t_2). This problem and its solution can easily be verified by plotting the two lines in the Cartesian frame. ◀

Problem 1.228 Verify if the two lines $\mathbf{r} = (1, 1, 1) + t(1, 1, 1)$ and $\mathbf{r} = (1, 0, 0) + t\,(a, b, b)$ cross. Then determine the point of crossing and the angle at which they cross. [Answer: $(1, 1, 1)b/(b - a)$, the angle ϕ is determined from $\cos\phi = (a + 2b)/\sqrt{3(a^2 + 2b^2)}$.]

Crossing point of a line and a plane. Consider a line λ, specified by the equation $\mathbf{r} = \mathbf{r}_B + t\mathbf{a}$, which crosses the plane σ given by the equation $\mathbf{n} \cdot \mathbf{r} = \mathbf{n} \cdot \mathbf{r}_C = h$. The line will definitely cross the plane at some point A, Fig. 1.41b, if the vector $\mathbf{a}$ is not perpendicular to the line normal: $\mathbf{a} \cdot \mathbf{n} \neq 0$. To determine the crossing point $\mathbf{r}$, we substitute the $\mathbf{r}$ from the line equation into the plane equation to get the value of t, and hence determine the crossing point:

$$t = \frac{\mathbf{n} \cdot (\mathbf{r}_C - \mathbf{r}_B)}{\mathbf{n} \cdot \mathbf{a}},$$

giving

$$\mathbf{r}_{crossing} = \mathbf{r}_B + \frac{\mathbf{n} \cdot (\mathbf{r}_C - \mathbf{r}_B)}{\mathbf{n} \cdot \mathbf{a}}\mathbf{a} = \mathbf{r}_B + \frac{h - \mathbf{n} \cdot \mathbf{r}_B}{\mathbf{n} \cdot \mathbf{a}}\mathbf{a}, \qquad (1.231)$$

where $\mathbf{r}_B$ is a point on the line, while $\mathbf{r}_C$ is a point on the plane. Calculating the dot product $\mathbf{a} \cdot \mathbf{n}$, one can also determine the angle ϕ at which the line crosses the plane (with respect to the normal to the plane). The above formula has a singularity when $\mathbf{n} \cdot \mathbf{a} = 0$ as this case corresponds to the line being parallel to the plane, so that they never cross.

Example 1.17 ▶ Determine at which point the line $\mathbf{r} = (1, 0, 0) + t(1, 0, 0)$ crosses the plane $-x + y + z = 1$, and at which angle.

Solution. First of all, we check that the line is not perpendicular to the normal of the plane $\mathbf{n} = (-1, 1, 1)$ (this is not the unit normal): indeed, $\mathbf{a} = (1, 0, 0)$ and $\mathbf{a} \cdot \mathbf{n} = -1 \neq 0$. Hence, the line definitely crosses the plane. Then, we use Eq. (1.231). We have: $\mathbf{n} \cdot \mathbf{r}_C = 1$ and $\mathbf{n} \cdot \mathbf{r}_B = -1$. Therefore,

$$\mathbf{r}_{crossing} = \begin{pmatrix} 1 \\ 0 \\ 0 \end{pmatrix} + \frac{1 - (-1)}{-1}\begin{pmatrix} 1 \\ 0 \\ 0 \end{pmatrix} = \begin{pmatrix} -1 \\ 0 \\ 0 \end{pmatrix}.$$

To determine the angle, we calculate the dot product $\mathbf{n} \cdot \mathbf{a} = -1$. Since $|\mathbf{n}| = \sqrt{3}$ and $|\mathbf{a}| = 1$, then $\cos\phi = -1/\left(1 \cdot \sqrt{3}\right) = -1/\sqrt{3}$. From this equation one can determine the angle. Note that it lies between $90°$ and $180°$ due to the chosen direction of the vector $\mathbf{a}$ of the line: if we choose its opposite direction (which we can always do), then $\cos\phi$ would be positive and the angle ϕ will be below $90°$. ◀

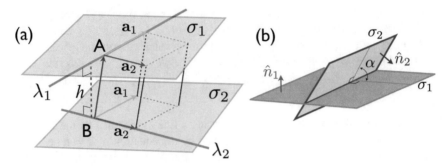

Fig. 1.42 a Finding the distance h between two non-parallel and non-crossing lines λ_1 and λ_2; **b** finding the angle α between two planes

Problem 1.229 At which point the line $\mathbf{r} = (1, 0, 0) + t(1, 0, 0)$ crosses the plane $3x + 4y - z = 3$. [Answer: $(1, 0, 0)$.]

Problem 1.230 Determine if the line

$$\frac{x - 2}{2} = \frac{y - 3}{3} = \frac{z - 4}{4}$$

crosses the $x - y$-plane, and if it does, at which point. [Answer: *yes*, $(0, 0, 0)$.]

Problem 1.231 Determine if the line $x/3 = -y/3 = z$ crosses the plane $x + y = 3$. [Answer: *no*.]

Distance between two non-parallel and non-planar lines. Consider two lines, λ_1 and λ_2, which are not parallel (their vectors $\mathbf{a}_1$ and $\mathbf{a}_2$ are not parallel, $\mathbf{a}_1 \times \mathbf{a}_2 \neq 0$) and do not cross, Fig. 1.42a. We would like to calculate the distance h between the two lines, i.e. to find a point on one line and also one on the other such that the distance between them is the smallest possible. It is easy to see that the line connecting these two points will be perpendicular to both lines. Indeed, let the equations of the two lines be $\mathbf{r} = \mathbf{r}_A + t_1\mathbf{a}_1$ and $\mathbf{r} = \mathbf{r}_B + t_2\mathbf{a}_2$. Let us first draw the vector $\mathbf{a}_2$ from the point A on the first line, and the vector $\mathbf{a}_1$ from the point B on the second, see Fig. 1.42a. Two vectors, $\mathbf{a}_1$ and $\mathbf{a}_2$, drawn from the point A form a plane σ_1 containing the line λ_1 with the normal $\mathbf{n}_1 = \mathbf{a}_1 \times \mathbf{a}_2$, whereas the same pair of vectors drawn from the point B forms another plane σ_2 containing the other line λ_2 and with the same normal $\mathbf{n}_2 = \mathbf{n}_1$. Since the two normal vectors are collinear, the two planes are parallel. Hence, the two lines lie in parallel planes. The distance between these planes, which is given by the perpendicular line drawn from one plane to the other, is exactly what is needed—the minimal distance between the points of the two lines, it is shown as a blue dashed line in Fig. 1.42a.

One way of obtaining the distance h is to calculate the distance between two arbitrary points on the two lines (specified by the parameters t_1 and t_2) and then minimise it with respect to them. For this task to be accomplished, we however require a derivative which we have not yet considered. So, we shall use another purely geometrical method. The vectors $\overrightarrow{BA} = \mathbf{r}_A - \mathbf{r}_B$, $\mathbf{a}_1$ and $\mathbf{a}_2$ form a parallelepiped shown in the figure. The volume of it is determined by the absolute value of the triple product of these three vectors. On the other hand, its volume is also given by the area of the parallelogram at the base of the parallelepiped and formed by $\mathbf{a}_1$ and $\mathbf{a}_2$ (which is equal to the absolute value of the vector product $|\mathbf{a}_1 \times \mathbf{a}_2|$, see Eq. (1.107)), multiplied by its height which is h, see Problem 1.123. Hence,

$$h = \frac{|[\mathbf{r}_B - \mathbf{r}_A, \mathbf{a}_1, \mathbf{a}_2]|}{|\mathbf{a}_1 \times \mathbf{a}_2|} = \frac{|(\mathbf{r}_B - \mathbf{r}_A) \cdot [\mathbf{a}_1 \times \mathbf{a}_2]|}{|\mathbf{a}_1 \times \mathbf{a}_2|} , \qquad (1.232)$$

which is the desired result.

Problem 1.232 Show that Eq. (1.232) is consistent with the condition (1.230) for the lines to cross.

Problem 1.233 Calculate the distance between lines $\mathbf{r} = (1, 1, 1) + t(1, 1, 1)$ and $\mathbf{r} = (0, 1, 0) + t(-1, 2, 3)$. [Answer: $h = 4/\sqrt{26}$.]

Problem 1.234 Consider a line $\mathbf{r} = (1, 1, 1) + t(-1, 1, 0)$. Next, consider a family of lines $\mathbf{r} = (-1, 1 - 1) + t(\cos\phi, \sin\phi, 0)$; each line in this family is specified by a specific parameter $0 \le \phi \le 2\pi$. Show that the first line does not cross any line from that family.

Distance between two parallel planes. We know that two planes are parallel if their normal vectors $\mathbf{n}_1$ and $\mathbf{n}_2$ are collinear, i.e. $\mathbf{n}_1 \times \mathbf{n}_2 = 0$. Consider two planes σ_1 and σ_2 with collinear normal vectors $\mathbf{n}_1$ and $\mathbf{n}_2$, given by the equations $\mathbf{n}_1 \cdot \mathbf{r} = \mathbf{n}_1 \cdot \mathbf{r}_A$ and $\mathbf{n}_2 \cdot \mathbf{r} = \mathbf{n}_2 \cdot \mathbf{r}_B$.

Problem 1.235 Show, using the same method as we have already used above for calculating the distance between a point and a line (or a plane), cf. Figs. 1.40 and 1.41a, that the distance between two parallel planes can be calculated via

$$h = \frac{|(\mathbf{r}_B - \mathbf{r}_A) \cdot \mathbf{n}_1|}{|\mathbf{n}_1|} = \frac{|\mathbf{r}_B \cdot \mathbf{n} - \mathbf{r}_A \cdot \mathbf{n}_1|}{|\mathbf{n}_1|} . \qquad (1.233)$$

Problem 1.236 Show that the distance between two planes $ax + by + cz = h_1$ and $ax + by + cz = h_2$ can be written as

$$h = \frac{|h_1 - h_2|}{\sqrt{a^2 + b^2 + c^2}}. \tag{1.234}$$

Crossing of two planes. Orientation of a plane is determined by its normal vector. Two planes are parallel if their corresponding normal vectors are collinear, i.e. their vector product is equal to zero. Two planes are perpendicular, if the dot product of their normal vectors is equal to zero. In general, one can define an angle α between two planes as shown in Fig. 1.42b. Two planes are parallel if their normal vectors $\mathbf{n}_1$ and $\mathbf{n}_2$ are parallel, i.e. if $\mathbf{n}_1 \times \mathbf{n}_2 = 0$, then the angle $\alpha = 0$. Two planes are perpendicular to each other if their normals are orthogonal, i.e. if $\mathbf{n}_1 \cdot \mathbf{n}_2 = 0$; in this case, the angle $\alpha = \pi/2$.

Problem 1.237 Show that the angle α between two crossing planes, as shown in Fig. 1.42b, is the same as the angle between their normal vectors. Correspondingly, demonstrate that the angle α between two planes given by equations $a_1 x + b_1 y + c_1 z = d_1$ and $a_2 x + b_2 y + c_2 z = d_2$ is given by:

$$\cos \alpha = \frac{a_1 a_2 + b_1 b_2 + c_1 c_2}{\sqrt{a_1^2 + b_1^2 + c_1^2}\sqrt{a_2^2 + b_2^2 + c_2^2}} \tag{1.235}$$

(either α or $\pi - \alpha$ can be chosen). In particular, verify the above criteria for the two planes to be parallel or perpendicular to each other.

Unless two planes are parallel, they intersect at a line. Let us derive the equation of that line. This is most easily accomplished by noticing that a vector $\mathbf{a}$ along the line in Eq. (1.207) is perpendicular to both normal vectors of the planes $\mathbf{n}_1$ and $\mathbf{n}_2$, i.e. we can choose a vector $\mathbf{a}$ along the line as $\mathbf{a} = \mathbf{n}_1 \times \mathbf{n}_2$. The only other unknown to find is a point A with vector $\mathbf{r}_A$ lying on that crossing, as after that the parametric equation of the line is completely determined. To find a point on the line, we should solve for a point $\mathbf{r}$ that simultaneously satisfies equations for both planes taken in either of the forms discussed above; however, the simplest are the first two, i.e. either (1.218) or (1.219). This step will result in two algebraic equations for three unknowns x, y and z, so that one can only express any two of them via the third; we can choose that third coordinate arbitrarily (our particular choice will result in a certain position of the point A along the line) and then would be able to determine the other two. Alternatively, the chosen coordinate may serve as a parameter t for the line equation.

Example 1.18 ▶ Determine the parametric equation of the line at which two surfaces $x + y + z = 1$ and $x = 0$ cross.

Solution. The normal of the first surface $\mathbf{n}_1 = (1, 1, 1)$, while $\mathbf{n}_2 = (1, 0, 0)$ for the second. Therefore, the vector of the line

$$\mathbf{a} = \begin{vmatrix} \mathbf{i} & \mathbf{j} & \mathbf{k} \\ 1 & 1 & 1 \\ 1 & 0 & 0 \end{vmatrix} = \mathbf{j} - \mathbf{k} = (0, 1, -1) .$$

Considering the two equations simultaneously, we get $y + z = 1$. Taking $y = 0$, we get $z = 1$ and hence we now have the point $(0, 0, 1)$ on the crossing line. The parametric equation, written using vector-columns, is thus

$$\mathbf{r} = \begin{pmatrix} 0 \\ 0 \\ 1 \end{pmatrix} + t \begin{pmatrix} 0 \\ 1 \\ -1 \end{pmatrix} = \begin{pmatrix} 0 \\ t \\ 1 - t \end{pmatrix} .$$

It is easy to see that any point on this line (i.e. for any value of the t) satisfies the equations for both planes simultaneously.

Alternatively, consider y as the parameter t. Then, we have $x = 0$, $y = t$ and $z = 1 - y = 1 - t$. We see that exactly the same result is obtained. ◀

Problem 1.238 Prove that this method will always give the correct solution, i.e. points on the line constructed in this manner will satisfy Eq. (1.218) for both planes.

Problem 1.239 Determine the equation of the line at which two planes $2x - y + z = 2$ and $-x + y + z = 1$ cross. [Answer: $\mathbf{r} = (3, 4, 0) + t(-2, -3, 1)$.]

Problem 1.240 Determine the line at which planes $x + y + z = 1$ and $y + z = 0$ cross. [Answer: $\mathbf{r} = (1, 0, 0) + t(0, -1, 1)$.]

Crossing of line and sphere. Consider a sphere of radius R and centred at point $\mathbf{r}_O = (x_O, y_O, z_O)$. Its equation is given by either of Eqs. (1.221) or (1.222). Let us determine if the line λ_1 given by the equation $\mathbf{r} = \mathbf{r}_A + t\mathbf{a}$ crosses the sphere, see Fig. 1.43. Note that in the figure we drew the system within the plane passing through the line and the centre O of the sphere; obviously, this is always possible.

Fig. 1.43 Lines λ_1, λ_2 and λ_3 and a sphere of radius R centred at point O. The first line does not have common points with the sphere, the second line touches the sphere at the single point D, while the third line crosses it at two points, F and G

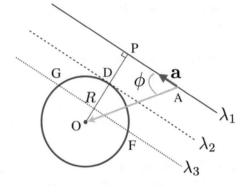

Problem 1.241 Show that the normal to this plane is given by $\mathbf{n} = (\mathbf{r}_O - \mathbf{r}_A) \times \mathbf{a}$, where $\mathbf{r}_A$ is any point on the line.

The problem is actually very simple as we know how to determine the distance between a point and a line. We also know that the shortest distance between a line and a point is the perpendicular drawn from the point onto the line. Therefore, to check if the line crosses the sphere, we need to calculate the distance between the line and the centre of the sphere. Applying Eq. (1.223), we have (see also Fig. 1.43, where the angle $\angle PAO = \phi$ is acute):

$$\overrightarrow{AO} \cdot \mathbf{a} = (\mathbf{r}_O - \mathbf{r}_A) \cdot \mathbf{a} = |\mathbf{r}_O - \mathbf{r}_A| \, |\mathbf{a}| \cos \phi$$

$$\implies \quad \sin \phi = \sqrt{1 - \left[\frac{(\mathbf{r}_O - \mathbf{r}_A) \cdot \mathbf{a}}{|\mathbf{r}_O - \mathbf{r}_A| \, |\mathbf{a}|} \right]^2} \, ,$$

so that the required distance

$$d = OP = |\mathbf{r}_O - \mathbf{r}_A| \sqrt{1 - \left[\frac{(\mathbf{r}_O - \mathbf{r}_A) \cdot \mathbf{a}}{|\mathbf{r}_O - \mathbf{r}_A| \, |\mathbf{a}|} \right]^2}$$

$$= \sqrt{|\mathbf{r}_O - \mathbf{r}_A|^2 - \left[\frac{(\mathbf{r}_O - \mathbf{r}_A) \cdot \mathbf{a}}{|\mathbf{a}|} \right]^2} \, .$$

$$(1.236)$$

The same formula is obtained when the angle between $\mathbf{a}$ and $\overrightarrow{AO}$ is obtuse (check it!).

Using this result, we can easily verify if the given line crosses the sphere: if $d > R$, the line does not cross; if $d = R$, it crosses it at a single point D, Fig. 1.43, and, finally, if $0 \leq d < R$, then it crosses it at two points. Note that at the touch point D the line will be perpendicular to the line going to the sphere centre. Point D can then be determined using Eq. (1.226):

$$\mathbf{r}_D = \mathbf{r}_A + \frac{(\mathbf{r}_O - \mathbf{r}_A) \cdot \mathbf{a}}{|\mathbf{a}|^2} \mathbf{a} . \tag{1.237}$$

To determine two crossing points for the case $0 \le d < R$, it is necessary to solve simultaneously equations of the line and sphere. However, a simpler method exists.

Indeed, all three cases considered above can be investigated by substituting the line equation into the equation of the sphere and then trying to find real solutions of the line parameter t. We have:

$$\begin{cases} \mathbf{r} = \mathbf{r}_A + t\mathbf{a} \\ |\mathbf{r} - \mathbf{r}_O| = R \end{cases} \implies |(\mathbf{r}_A - \mathbf{r}_O) + t\mathbf{a}| = R .$$

Denoting $\mathbf{g} = \mathbf{r}_A - \mathbf{r}_O$, we obtain, upon squaring both sides, a quadratic equation for t:

$$a^2 t^2 + 2 (\mathbf{a} \cdot \mathbf{g}) t + (g^2 - R^2) = 0 ,$$

where $g = |\mathbf{g}|$ and $a = |\mathbf{a}|$, whose solutions, see Eq. (1.18), are:

$$t_{\pm} = \frac{1}{a^2} \left(-\mathbf{a} \cdot \mathbf{g} \pm \sqrt{\mathcal{D}} \right) , \quad \text{where} \quad \mathcal{D} = (\mathbf{a} \cdot \mathbf{g})^2 - a^2 (g^2 - R^2) . \tag{1.238}$$

If $\mathcal{D} < 0$, then no real values of t exist that satisfy both equations; hence, the line and the sphere do not cross. If $\mathcal{D} = 0$, one real value of t exists, and hence the line only touches the sphere. If, finally, $\mathcal{D} > 0$, then two values of t exist; using those, we can find two points $\mathbf{r}_G = \mathbf{r}_A + t_+ \mathbf{a}$ and $\mathbf{r}_F = \mathbf{r}_A + t_- \mathbf{a}$ at which the line crosses the sphere, see Fig. 1.43.

Problem 1.242 Using explicitly the condition $\mathcal{D} < 0$, prove that $d > R$, where d is given by Eq. (1.236).

Problem 1.243 Using explicitly the condition $\mathcal{D} = 0$, show that the crossing point coincides with the one given by Eq. (1.237).

Problem 1.244 Determine the range of the angles α at which the line $\mathbf{r} = t(0, \cos \alpha, \sin \alpha)$ crosses the sphere $x^2 + (y - 1)^2 + z^2 = R^2$. [Answer: $|\sin \alpha| \le R$, i.e. $-\alpha_0 \le \alpha \le \alpha_0$, where $\sin \alpha_0 = R$.]

Problem 1.245 Determine the crossing points of the line and the sphere from the previous problem. [Answer: $\mathbf{r} = \left(\cos \alpha \pm \sqrt{R^2 - \sin^2 \alpha} \right) (0, \cos \alpha, \sin \alpha).$]

Fig. 1.44 A plane
$z = z_0 < R$ crosses a sphere
of radius R and centred at
the origin

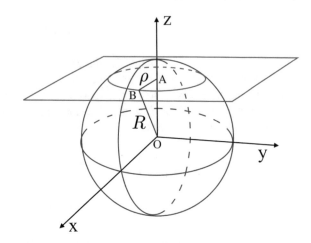

Crossing of planes and spheres. Consider now a sphere $|\mathbf{r} - \mathbf{r}_O| = R$ of radius R centred at point $\mathbf{r}_O$, and a plane $\mathbf{r} \cdot \mathbf{n} = \mathbf{r}_A \cdot \mathbf{n} = h$ with the normal $\mathbf{n}$ and passing through the point $\mathbf{r}_A$. The first question one should ask is what is the shape of the cross section? It is easy to see that this is a circle. Indeed, from symmetry considerations it is clear that the result cannot depend on the actual position of the sphere and of the plane. Therefore, it suffices to consider a sphere centred at the origin and a plane parallel to the $x - y$-plane and crossing the z-axis at the height $OA < R$, see Fig. 1.44.

The equation of the plane is $z = h$ (where $0 \le h \le R$), where h is the distance from the plane to the centre of the sphere, while $x^2 + y^2 + z^2 = R^2$ is the equation of the sphere. Substituting $z = h$ into it, we immediately obtain:

$$x^2 + y^2 = R^2 - h^2 = \rho^2 .$$

This is the equation of a circle in the $x - y$-plane (obviously, the circle is parallel to the $x - y$-plane) of radius $\rho = \sqrt{R^2 - h^2}$. This result also immediately follows from $\triangle OAB$ in which $OA = h$ and $OB = R$. It is also obvious that $\rho = 0$ corresponds to the situation in which the plane touches the sphere, so that $h = R$. For $h > R$ the radius of the circle ρ is not real; the plane does not cross the sphere.

Generally, given the plane $\mathbf{n} \cdot \mathbf{r} = \mathbf{n} \cdot \mathbf{r}_A = h$ and the sphere of radius R centred at point $\mathbf{r}_O$, the distance d between the plane and the centre of the sphere is determined by (cf. Problem 1.227)

$$d = \frac{|\mathbf{n} \cdot (\mathbf{r}_A - \mathbf{r}_O)|}{|\mathbf{n}|} = \frac{|h - \mathbf{n} \cdot \mathbf{r}_O|}{|\mathbf{n}|} .$$

Hence, if $d > R$, then the plane does not cross the sphere; when $d = R$, the plane touches the sphere at a single point that is obtained by $\mathbf{r}_O \pm R\,\mathbf{n}$ (obviously, there are two parallel planes of the same normal $\mathbf{n}$ that can touch the sphere), and if $d < R$, then the plane crosses the sphere with a circle of radius $\rho = \sqrt{R^2 - d^2}$ and centred at $\mathbf{r}_O \pm d\,\mathbf{n}$ in the cross section.

Problem 1.246 Determine whether the plane $x + y + z = 1$ crosses the sphere $(x - 1)^2 + (y + 1)^2 + z^2 = 1$, and if it does, what is the distance from the plane to the sphere centre and what is the radius of the circle in the cross section. [Answer: *yes, $h = 1/\sqrt{3}$ and $\rho = \sqrt{2/3}$.*]

Problem 1.247 Determine the centre O of the sphere of radius R that touches the plane $4x + 3y = 1$ at the point $P(1, 1, 1)$. [Answer: $\mathbf{r}_O = (1, 1, 1) \pm R\,(4/5, 3/5, 0)$.]

Problem 1.248 Consider two planes specified via

$$\mathbf{r} \cdot \begin{pmatrix} 0 \\ 1 \\ 1 \end{pmatrix} = 2 \text{ and } \mathbf{r} \cdot \begin{pmatrix} 1 \\ 0 \\ 1 \end{pmatrix} = 2,$$

and a sphere of radius R centred at the point $\mathbf{r}_O = (1, 1, 1)$. Determine the line at which these planes intersect, and then show that for any R there are exactly two common points these three figures share; find these points. [Answer: $(1 \pm R/\sqrt{3}, 1 \pm R/\sqrt{3}, 1 \mp R/\sqrt{3})$.]

Functions

2

2.1 Definition and Main Types of Functions

As was mentioned already in Sect. 1.3.4, a function $y = f(x)$ establishes a correspondence between values of x from $\mathbb{R}$ and values of y also from $\mathbb{R}$. It is not an essential condition for the function to exist that the values of x and/or y form continuous intervals. For instance, consider a factorial function $y = n! = 1 \cdot 2 \cdot 3 \cdot 4 \cdots \cdot n$. It makes a correspondence between all integer numbers (serving as the values of x) and some specific set of integer numbers $(1, 2, 6, 24,$ etc.) serving as y. It is essential, however, that to a single value of x there is always one and only one value of y. Note that this condition may not necessarily go in the reverse direction as the same values of y may correspond to different values of $x_1 \neq x_2$. Indeed, consider, e.g. the function $y = x^2$ for which two values of $x = \pm 1$ result in the same value of $y = 1$.

Four arithmetic operations defined between real numbers (addition, subtraction, multiplication and division) can be applied to any two functions $f(x)$ and $g(x)$ to define new functions: $f(x) \pm g(x)$, $f(x)g(x)$ and $f(x)/g(x)$ (in the latter case, the resulting function is not defined for the values of x where $g(x) = 0$). In addition, we can define a *composition* of two functions $f(x)$ and $g(x)$ via the following chain of operations: $y = g(w)$ and $w = f(x)$, i.e. $y = g(f(x))$. For instance, $y = \sin^2(x)$ is a composition of the functions $y = x^2$ and $y = \sin(x)$. Obviously, one can also define a composition of more than two functions, e.g. $y = g(f(d(x)))$.

Functions can be specified in various ways, the most useful one of which is analytic using either a single expression (e.g. $y = x^3 + 1$) or several as is the case for the *Heaviside* or *unit step function*:

$$H(x) = \begin{cases} 1, & \text{if } x \geq 0 \\ 0, & \text{if } x < 0 \end{cases}, \tag{2.1}$$

© The Author(s), under exclusive license to Springer Nature Switzerland AG 2022
L. Kantorovich, *Mathematics for Natural Scientists*, Undergraduate Lecture Notes
in Physics, https://doi.org/10.1007/978-3-030-91222-2_2

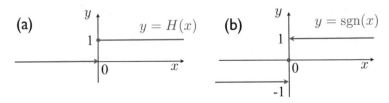

Fig. 2.1 Heaviside (**a**) and sign (**b**) functions, see Eqs. (2.1) and (2.2). The arrows at $x = 0$ at the end of some lines indicate that this point is excluded, while the fat dot—included

where we used $H(0) = 1$. It is shown in Fig. 2.1a. Other definitions of $H(x)$ at $x = 0$ are also possible. Another useful example of such a definition is the *sign function* which is defined similarly as:

$$\text{sgn}(x) = \begin{cases} -1, & \text{if } x < 0 \\ 0, & \text{if } x = 0, \\ 1, & \text{if } x > 0 \end{cases} \tag{2.2}$$

shown in Fig. 2.1b. Moreover, the two functions are related for all $x \neq 0$ as $\text{sgn}(x) = 2H(x) - 1$. However, if $H(x)$ was defined at $x = 0$ as $H(0) = 1/2$, then the mentioned relationship would hold even for $x = 0$. This redefinition of $H(x)$ at $x = 0$ is sometimes useful. Another possibility is to define $H(0) = 0$.

It can easily be checked that the function $H(x - a)$ makes a unit jump at $x = a$, while the function $AH(x - a)$ jumps by A at $x = a$.

Problem 2.1 Prove that

$$\text{sgn}(x) = 2H(x) - 1 = H(x) - H(-x).$$

Problem 2.2 Show that the following function can be defined ($a < b$):

$$W_{ab}(x) = H(x - a) - H(x - b) = \begin{cases} 0, & \text{if } x < a \\ 1, & \text{if } a \leq x < b. \\ 0 & \text{if } x \geq b \end{cases}$$

This function makes a unit step between $x = a$ and $x = b$ and is called the rectangular window function.

Problem 2.3 Consider a composite function

$$f(x) = \begin{cases} f_1(x), & \text{if } -\infty < x < a \\ f_2(x), & \text{if } a < x \le b \\ f_3(x), & \text{if } x > b \end{cases}.$$

Check that it can also be written as

$$f(x) = f_1(x)H(a-x) + f_2(x)W_{ab}(x) + f_3(x)H(x-b)$$

if we set $H(0) = 0$ for the Heaviside function. Note that the function $f(x)$ is not defined at $x = a$.

We shall frequently be using graphical representations of functions where the points x and $y = f(x)$ are displayed on the $x - y$ plane. This could be either a collection of points (as is the case, e.g. for the factorial function) or a continuous line (as e.g. for the parabola $y = x^2$). However, when discussing functions, we shall mostly be dealing with continuous functions, which are defined on a continuous interval of the values of x in $\mathbb{R}$.

Now we shall talk about the classifications of functions. A function can be *limited from above* on the given interval of x if for all values of the x within it there exist a real M such that $f(x) \le M$. For instance, $y = -x^2$ is limited from above, $y < M = 0$, for all x in $\mathbb{R}$, i.e. for any $-\infty < x < \infty$. Similarly, if there exist such real m that for all values of x we have $f(x) \ge m$, then the function $y = f(x)$ is *limited from below*, as is the case e.g. for $y = x^2$ ($m = 0$ in this case). A function can also be limited both from above and below if there exist such positive M that $|f(x)| \le M$. An obvious example is $y = \sin(x)$ since $-1 \le \sin(x) \le 1$, i.e. for the sine function $M = 1$. On the other hand, the function $y = 1/x^2$ is not limited in the interval $0 < x < \infty$ since it is impossible to find a single constant $M > 0$ such that $1/x^2 \le M$ for all positive x: whatever value of M we take, there always be values of x such that $1/x^2 > M$ in the interval $0 < x < 1/\sqrt{M}$.

A function $y = f(x)$ is *strictly increasing monotonically* within an interval, if for any $x_1 < x_2$ follows $f(x_1) < f(x_2)$, while the function is *strictly decreasing monotonically* if $f(x_1) > f(x_2)$ follows instead. For instance, $y = x^2$ is an increasing function for $x \ge 0$, while it is decreasing for $x \le 0$. Strictly decreasing or increasing functions are called *monotonic*.

A function is *periodic* if there exists such number T' that for any x (within the function definition) $f(x + T') = f(x)$. The smallest such number T is called the *period*. For instance, $y = \sin(x)$ is a periodic function since $\sin(x + 2\pi n) = \sin(x)$ for any integer $n = 1, 2, \ldots$, i.e. the values of T' form a sequence 2π, 4π, etc.; however, only the smallest their value is the period, $T = 2\pi$.

Above we have defined a *direct* function whereby each value of x is put in the direct correspondence with a single value of y via $y = f(x)$. One can, however, solve this equation with respect to x and write it as a function of y. This will be a function different from $f(x)$ and it is called (if exists, see below) the *inverse function* $x = f^{-1}(y)$. Conventionally, however, we would like the x to be the argument and y to be the function. Hence, we swap the letters x and y and write the inverse function as $y = f^{-1}(x)$.

Example 2.1 ▶ Obtain the inverse function to $y = 5x$.

Solution. Solving $y = 5x$ for x, we get $x = \frac{1}{5}y$; interchanging x and y for the inverse function to be written in the conventional form, we get $y = \frac{1}{5}x$, i.e. the inverse function to $f(x) = 5x$ is $f^{-1}(x) = \frac{1}{5}x$. ◀

Example 2.2 ▶ Obtain the inverse function to $y = x^3$.

Solution. Solving for x and interchanging x and y, we obtain $y = f^{-1}(x) = x^{1/3}$ which is the *cube root* function (general power functions will be considered in Sect. 2.3.3). ◀

The inverse function exists if and only if there is one-to-one correspondence between x and $y = f(x)$. This means that for any $x_1 \neq x_2$, it follows that $y_1 = f(x_1) \neq y_2 = f(x_2)$. This is because for each value of the x, there must be a unique value of $y = f(x)$ for the direct function to be well defined; at the same time, for the inverse function to be defined, it is necessary that for each value of y, there will be one and only one value of $x = f^{-1}(y)$.

It is easy to see that if the direct function is increasing monotonically, then this is sufficient to establish the existence of the inverse function since the requirement for that condition is satisfied: indeed, if for any $x_1 < x_2$ (and hence $x_1 \neq x_2$) follows $y_1 = f(x_1) < y_2 = f(x_2)$, then obviously $y_1 \neq y_2$, as required. Similarly, the inverse function $f^{-1}(x)$ exists if the direct function $f(x)$ is monotonically decreasing.

Problem 2.4 Prove that if $f(x)$ is an increasing (decreasing) function, then so is its inverse (if it exists).

Problem 2.5 Let us consider a function $f(x)$ which inverse is $f^{-1}(x)$. Prove that the inverse to $f^{-1}(x)$ is $f(x)$, i.e. the two functions are mutually inverse to each other.

It is easy to see that the graphs of the two functions, $y = f(x)$ and $y = f^{-1}(x)$, are symmetric with respect to the diagonal line drawn in the first $(x, y > 0)$ and the third $(x, y < 0)$ quadrants, see Fig. 2.2. Indeed, consider a point A on the curve of the function $f(x)$ with the coordinates $(x_1, y_1 = f(x_1))$. Consequently, $x_1 = f^{-1}(y_1)$ by the definition of the inverse function. However, conventionally, we write this as $y_2 = f^{-1}(x_2)$, i.e. $y_2 = x_1$ and $x_2 = y_1$. It is seen that the point A on $f(x)$ corresponds to point B on the curve of $f^{-1}(x)$ with the coordinates (x_2, y_2). It is

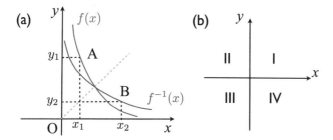

Fig. 2.2 **a** To the prove of the symmetry of $f(x)$ and $f^{-1}(x)$ with respect to the diagonal (shown with the brown dashed line) of the quadrants I and III. **b** Numeration of the quadrants of the 2D Cartesian system

Fig. 2.3 Parabola and its inverse

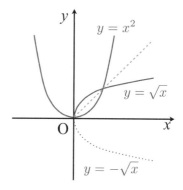

now obvious from the drawing in Fig. 2.2a that the two points A and B are located symmetrically with respect to the dashed brown diagonal line. Since the point A was chosen arbitrarily on $f(x)$, the reasoning given would correspond to any point on $f(x)$, i.e. the whole two curves, $f(x)$ and $f^{-1}(x)$, are symmetric with respect to the diagonal line of the first and third quadrants.

It follows from that simple discussion that the graph of $f^{-1}(x)$ can be obtained from the graph of $f(x)$ by reflecting the latter across the diagonal of the quadrants I and III.

As should by now become clear, not every function has an inverse. Consider, for instance, a parabola $y = x^2$, shown in Fig. 2.3 with the red line. If we reflect $y = x^2$ in the first quadrant diagonal (shown as dashed brown, as before), we obtain the blue curve containing two parts: one shown with the solid blue line which is $y = \sqrt{x} > 0$, and another shown as dotted blue line and corresponding to $y = -\sqrt{x} < 0$. In other words, in this case, the two curves (the red and blue) are rotated by 90° with respect to each other. It is easy to see, however, that to a single value of $x > 0$, there will be *two* values of $y = \pm\sqrt{x}$, which contradicts the definition of the function given above. This is of course because while solving the equation $y = x^2$ with respect to x, we obtain two values of the square root. It follows from this discussion that, strictly speaking, the function $y = x^2$ does not have an inverse. As we still would like to define an inverse to the parabola, we shall take only one value of the square root, and it is conventional to take the positive root. This way we obtain the function $y = \sqrt{x} > 0$ drawn in Fig. 2.3 with the solid blue line. This is a function since to any value of x there is only one value of $y = \sqrt{x}$ corresponding to it.

The reduction operation considered above for the parabola was necessary to define properly the inverse function to it, and the same operation is applied in some other cases as will be demonstrated in the next Section.

2.2 Infinite Numerical Sequences

2.2.1 Definitions

To continue our gradual approach to functions, we have to consider now numerical sequences and their *limit*, otherwise it would be difficult for us to define power and exponential functions which are the matter of the next Section.

Consider an infinite set of numbers $x_0, x_1, x_2, x_3, \ldots, x_n, \ldots$, which are calculated via a certain rule from the integer numbers $0, 1, 2, 3, \ldots, n, \ldots$ taken from $\mathbb{N}at$. In other words, the numbers $x_n = f(n)$ are obtained as a function depending of an integer number n. For instance, the set $1, -1, 1, -1, 1, \ldots$ can be obtained by calculating $(-1)^n$ sequentially for $n = 0, 1, 2, 3, \ldots$. Another set

$$1, \frac{1}{2}, \frac{1}{3}, \frac{1}{4}, \frac{1}{5}, \frac{1}{6}, \frac{1}{7}, \ldots \tag{2.3}$$

is obtained from $1/n$ with $n = 1, 2, 3, \ldots$. If the former set $(-1)^n$ results in an alternating sequence of numbers, the latter sequence $1/n$ obviously tends to smaller and smaller numbers and one may expect that the numbers in the sequence approach zero for large values of n, being every time slightly above it. One may say that if the former sequence does not have a definite limit when n becomes indefinitely large (we say "n tends to infinity", which is denoted as ∞, and we write it as $n \to \infty$), the latter sequence does: $1/n \to 0$ as $n \to \infty$. We use a special symbol "lim" to say that:

$$\lim_{n \to \infty} \frac{1}{n} = 0 .$$

Although intuitively our argumentation may be sound and clear, proper analysis requires a rigorous definition for the limit of an infinite numerical sequence to exist. To formulate the required definition, let us assume that a sequence $x_0, x_1, x_2, x_3, \ldots, x_n, \ldots$ approaches a. Intuitively that means that as n gets bigger and bigger, the values x_n of the sequence get closer and closer to a, i.e. the absolute value of the difference $|x_n - a|$ gets smaller and smaller (the absolute value is needed since the numbers in the sequence may appear on both sides of a, consider, e.g. the sequence $(-1)^n/n$). So, a rigorous formulation sounds like this: for any $\Delta > 0$ there always exists such positive integer N, that for all members of the sequence staying *after* it (i.e. with numbers greater than it, $n > N$) we shall have $|x_n - a| < \Delta$. This is written as follows:

$$\lim_{n \to \infty} x_n = a \ . \tag{2.4}$$

For example, consider our sequence $x_n = 1/n$ for which we suspect that it approaches $a = 0$. Let us take a small positive number $\triangle = 0.01$. Then, we have to show that starting from some number N all members of the sequence are smaller than $\triangle$, i.e. they satisfy $|x_n| < \triangle = 0.01$. Indeed, from the inequality $1/n < \triangle$ we get $n > 1/\triangle = 100$; in other words there exist such number $N = 100$ so that any member of the sequence $x_{101} = \frac{1}{101} \simeq 0.0099$, $x_{102} = \frac{1}{102} \simeq 0.0098$, $x_{103} = \frac{1}{103} \simeq 0.0097$, etc. satisfies the condition $|x_n| < 0.01$, as required. If we take even smaller value of $\triangle = 0.001$, then solving for n in $1/n < \triangle$ will result in $N = 1/\triangle = 1000$, i.e. for all $n > 1000$ all members of the sequence x_n are smaller than this smaller value of $\triangle$. It is clear that whatever value of $\triangle$ we take, however small, one can always find such number N that all members x_n of the sequence with $n > N$ will be smaller than $\triangle$. This means that indeed, according to the given above definition, the sequence $x_n = 1/n$ tends to zero.

This makes perfect sense: the smaller the value of the *deviation* from a (the value of $\triangle$), i.e. the closer we want the members of the sequence to be to that value, the larger the value of N needs to be taken, i.e. we have to consider more and more distant members of the sequence (with larger n); x_n with small values of n may deviate from the limit a significantly, but as n increases, the elements of the sequence x_n become closer and closer to it.

Problem 2.6 Prove that the infinite numerical sequence $x_n = 1/n^2$ tends as well to zero, i.e. its limit is equal to zero.

Problem 2.7 Prove, using the definition, that the infinite numerical sequence $x_n = (1 + 2n)/n$ tends to 2.

Problem 2.8 Prove that the sequence $x_n = q^n$ does not have a limit if $|q| > 1$.

Using the definition of the limit of a sequence is very tedious in practice, and it is much easier to perform algebraic manipulations on the terms of the sequence and this way find the limit of it. Consider again Problem 2.7. Each element of the sequence can alternatively be written as

$$x_n = \frac{1}{n} + 2 \ .$$

Then, considering the original sequence $x_n = y_n + z_n$ as a sum of two sequences, $y_n = 1/n$ and $z_n = 2$, we may apply the limit to each sequence separately to get $y_n = 1/n \to 0$, $z_n \to 2$, and hence $x_n \to 2$ as $n \to \infty$.

Example 2.3 ▶Consider the limit of the sequence

$$x_n = \frac{2n^2 + 3n + (-1)^n}{3n^2 - 4n + 5} .$$

Solution. Divide both the numerator and denominator by n^2:

$$x_n = \frac{2 + 3/n + (-1)^n /n^2}{3 - 4/n + 5/n^2} \to \frac{2}{3} ,$$

as all the terms containing n tend to zero as $n \to \infty$. ◀

2.2.2   Main Theorems

The validity of the operations we have performed in the examples above requires a rigorous justification. This is accomplished by the following simple theorems which we shall now discuss.

Theorem 2.1 (Addition Theorem) *If two sequences x_n and y_n have certain limits X and Y, then their sum sequence $z_n = x_n + y_n$ has a limit and it is $X + Y$.*

Proof: Since the first sequence x_n has the limit, then for any $\Delta_x = \Delta/2$ there exists such number N_x that starting from it (i.e. for any $n > N_x$) all members of the sequence would not differ from the limit by more than Δ_x, i.e. we would have $|x_n - X| < \Delta_x$. Similarly, for the second sequence: for any $\Delta_y = \Delta/2$ there would exist such N_y, so that for any $n > N_y$ we have $|y_n - Y| < \Delta_y$. Then, we can write:

$$\begin{aligned} |z_n - (X + Y)| &= |(x_n - X) + (y_n - Y)| \\ &\leq |x_n - X| + |y_n - Y| \qquad (2.5) \\ &< \Delta_x + \Delta_y = \Delta . \end{aligned}$$

Here, we used the inequality (1.195) stating that an absolute value of a sum of numbers is always less than or equal to the sum of their absolute values. Therefore, it appears that whatever value of $\Delta > 0$ we choose, one can always find an N such that the members of the sequence $z_n = x_n + y_n$ will be closer than Δ to the limiting value $X + Y$ as stated by inequality (2.5), e.g. $N = \max \{N_x, N_y\}$. This means that the limit of the sequence z_n indeed exists and the sequence converges to the sum of the individual limits, $X + Y$. **Q.E.D.**

Theorem 2.2 (Product Theorem) *If two sequences x_n and y_n have certain limits X and Y, then their product sequence $z_n = x_n y_n$ has a limit and it is XY.*

Proof: The logic is similar to that of the previous theorem; we only need to estimate the absolute value of the difference to derive the necessary value of Δ in this case:

$$
\begin{aligned}
|x_n y_n - XY| &= |(x_n - X)(y_n - Y) + X(y_n - Y) + Y(x_n - X)| \\
&\le |x_n - X||y_n - Y| + |X||y_n - Y| + |Y||x_n - X| \\
&< \Delta_x \Delta_y + |X|\Delta_y + |Y|\Delta_x = \Delta \,,
\end{aligned}
$$

i.e. by taking this value of Δ and $N = \max(N_x, N_y)$ any member of the product sequence z_n with $n > N$ will be different from the value of XY by no more than Δ, i.e. it is the limit of the product sequence. **Q.E.D.**

Theorem 2.3 (Division Theorem) *If two sequences x_n and $y_n \neq 0$ have certain limits X and $Y \neq 0$, then their ratio sequence $z_n = x_n/y_n$ has a limit and it is X/Y.*

Proof: Again, this is proven by the following manipulation:

$$
\begin{aligned}
\left| \frac{x_n}{y_n} - \frac{X}{Y} \right| &= \left| \frac{x_n Y - y_n X}{Y y_n} \right| = \left| \frac{(x_n - X)Y - (y_n - Y)X}{Y y_n} \right| \\
&\le \frac{|x_n - X||Y| + |y_n - Y||X|}{|Y||y_n|} < \frac{\Delta_x |Y| + \Delta_y |X|}{|Y||y_n|} \,.
\end{aligned}
$$

Since $y_n \neq 0$ for any n, then it should be possible to find such $\delta > 0$ that $|y_n| \geq \delta$ for any n. Hence, $1/|y_n| \leq 1/\delta$, and we can then continue:

$$\left| \frac{x_n}{y_n} - \frac{X}{Y} \right| < \frac{\Delta_x |Y| + \Delta_y |X|}{|Y||y_n|} \leq \frac{\Delta_x |Y| + \Delta_y |X|}{|Y|\delta} .$$

Choosing the expression in the right-hand side of the inequality above as Δ finally proves the theorem. **Q.E.D.**

The three theorems enable us to manipulate the limits using algebraic means, and this is exactly what we did in the previous Example:

$$\lim_{n \to \infty} x_n = \lim_{n \to \infty} \frac{2n^2 + 3n + (-1)^n}{3n^2 - 4n + 5} = \lim_{n \to \infty} \frac{2 + 3/n + (-1)^n /n^2}{3 - 4/n + 5/n^2}$$
$$= \frac{2 + \lim_{n \to \infty} (3/n) + \lim_{n \to \infty} \left[(-1)^n /n^2 \right]}{3 - \lim_{n \to \infty} (4/n) + \lim_{n \to \infty} (5/n^2)} = \frac{2 + 0 + 0}{3 - 0 + 0} \to \frac{2}{3} .$$

If we take a *subsequence* of the members of some original sequence (e.g. only terms with even n) then we shall get another infinite numerical sequence, e.g.

$$1, \frac{1}{2}, \frac{1}{4}, \frac{1}{6}, \cdots$$

is a subsequence of the sequence (2.3). It might appear to be obvious that if the original sequence has a certain limit, then its arbitrary subsequence would also converge to the same limit. This is stated by the following

---

Theorem 2.4 (Subsequence Theorem) *If a sequence x_n (with indices n taking all integer values $\mathbb{N}at$) tends to a limit X, then its arbitrary subsequence y_k (where indices k take on only a subset of values from the original set $\mathbb{N}at$ and for these indices $y_k = x_k$) tends to the same limit.*

Proof: This follows directly from the definition of the limit. Since the original sequence x_n has a limit, then for any Δ, there is such N that for any $n > N$, we have $|x_n - X| < \Delta$. Take now a subset of indices, let us call them k. On the subset using the same value of Δ, we have the same inequality, $|y_k - X| < \Delta$, is satisfied for all values of $k > N$ as these indices form a subset of the original indices $n > N$ for which it was satisfied. **Q.E.D.**

Therefore, if a sequence $x_n = 1/n$ (equal to all inverse integers) has the limit 0, then the same limit would have its subsequence $y_n = 1/(2n)$ containing only inverse even integers. However, if one can find at least two subsequences of the given sequence which converge to different limits individually, then we have to state that the original sequence does *not* have a limit.

Problem 2.9 Prove, that the sequence $x_n = (-1)^n$ does not have a limit.

If a sequence has a limit, it must be unique, it is impossible for a sequence to have two different limits. This is stated by the following

Theorem 2.5 (Uniqueness) *If a sequence x_n has a limit, it is unique.*

Proof: We shall prove this theorem by contradiction. Assume that the sequence x_n has two different limits, X_1 and $X_2 > X_1$. To prove this assumption wrong, we take a specific value of $\Delta = (X_2 - X_1)/2 > 0$. Since X_1 is the limit, than starting form some N_1 (i.e. for all $n > N_1$) we have $|x_n - X_1| < \Delta$. Similarly, since X_2 is also a limit, there must be such N_2 so that for any $n > N_2$ we have $|x_n - X_2| < \Delta$ with the same Δ. Then we write:

$$X_2 - X_1 = |X_2 - X_1| = |(X_2 - x_n) + (x_n - X_1)|$$
$$\leq |X_2 - x_n| + |X_1 - x_n| < 2\Delta = X_2 - X_1 ,$$

which is obviously incorrect. Note that we have the strict "lesser than" sign here! Hence, our assumption is wrong. The theorem is proven. **Q.E.D.**

There are several theorems related to inequalities, which are sometimes used to find particular limits. We shall consider them next.

Theorem 2.6 *If all elements of the sequence, starting from some particular element x_{n_0} (i.e. for all $n \geq n_0$), satisfy $x_n \geq A$ and the sequence has the limit X, then $X \geq A$.*

Proof: We prove this theorem by contradiction. Assume that $X < A$. Then, since the sequence has the limit, there is such $N > n_0$ that for any $n > N$ we have $|x_n - X| < \Delta$. This means that, $-\Delta < x_n - X < \Delta$. In particular, $x_n - X < \Delta$ or $x_n < \Delta + X$. Take $\Delta = A - X > 0$, then $x_n < \Delta + X = A$, which contradicts the fact that $x_n \geq A$. Our initial assumption is therefore wrong. **Q.E.D.**

Theorem 2.7 *If all elements of the sequence, starting from some particular number n_0, satisfy $x_n \leq A$ and the sequence has the limit X, then $X \leq A$.*

Problem 2.10 Prove this theorem.

Theorem 2.8 *If elements of two sequences x_n and y_n, starting from some number n_0, satisfy $x_n \leq y_n$ and their limits exist and equal to X and Y, respectively, then $X \leq Y$.*

Proof: Consider a sequence $z_n = y_n - x_n$. All elements of the new sequence, starting from some number n_0, satisfy $z_n \geq 0$ and its limit is $Z = Y - X$. Therefore, according to Theorem 2.6, $Z = Y - X \geq 0$, or $Y \geq X$, as required. **Q.E.D.**

This theorem has a simple corollary that if a sequence, y_n, is bracketed between two other sequences, $x_n \leq y_n \leq z_n$, then its limit is also bracketed between the two limits, i.e. $X \leq Y \leq Z$. In particular, if $Z = X$, then $Y = X = Z$.

2.2.3 Sum of an Infinite Numerical Series

Having defined a numerical sequence, x_n, one can consider a finite sum (cf. Sect. 1.12)

$$S_N = \sum_{n=0}^{N} x_n$$

of its first N members. These so-called *partial sums* S_N for different values of N form a numerical sequence in their own right (with respect to N), and the limit S of that sequence, if exists, should give us the sum of an infinite series built from the original sequence x_n. It is instructive to explicitly state here the definition of the convergence of the infinite series which is nothing but a rephrased definition (for S_N in this case) of the limit given above in Sect. 2.2.1 for the numerical sequence: the series $\sum_{n=0}^{\infty} x_n$ converges to the value S, if for any positive ϵ one can always find such number $\mathcal{N}$ that for any $N > \mathcal{N}$ the partial sum S_N would be different from S by no more than ϵ, i.e. $|S_N - S| < \epsilon$.

As an example of an infinite numerical series, consider an infinite geometrical progression (cf. Eq. (1.142)):

$$S_{geom} = a_0 q^0 + a_0 q^1 + a_0 q^2 + a_0 q^3 + \cdots = \lim_{N \to \infty} S_N = \lim_{N \to \infty} \frac{a_0 \left(1 - q^{N+1}\right)}{1 - q},$$
$$(2.6)$$

where we used the formula (1.143) for the sum S_N of the first N terms of the geometric series. Now, if $|q| > 1$, then $q^{N+1} \to \infty$ as $N \to \infty$ (obviously, q^N increases with N), and hence the series does not have a finite limit. If $q = 1$, then the series converges to infinity as well since it consists of identical terms a_0. If $q = -1$, then the series contains alternating a_0 and $-a_0$ terms, and hence (similarly to the case of $x_n = (-1)^n$ discussed above) does not have a limit at all. Therefore, it only makes sense to consider the case of $|q| < 1$. In this case $q^{N+1} \to 0$ as $N \to \infty$ since q^N decreases when N increases, and we obtain after taking the limit in (2.6) an important result:

$$S_{geom} = \lim_{N \to \infty} \frac{a_0 \left(1 - q^{N+1}\right)}{1 - q} = \frac{a_0}{1 - q}. \qquad (2.7)$$

Example 2.4 ▶ Determine the fractional form $r = p/q$ of the rational number, $r = 0.9\,(123)$, which contains a periodic sequence of digits in its decimal representation.

Solution. We write:

$$r = 0.9 + 0.0123 + 0.0000123 + \cdots = 0.9 + 0.0123 \left(1 + 10^{-3} + 10^{-6} + \cdots\right).$$

The expression in the brackets is nothing but the infinite geometric progression. Using (2.7) for its sum, we obtain:

$$r = 0.9 + \frac{0.0123}{1 - 10^{-3}} = \frac{9}{10} + \frac{123}{10000} \frac{1}{0.999} = \frac{9}{10} + \frac{123}{9990} = \frac{9114}{9990} \quad \blacktriangleleft.$$

Problem 2.11 Prove, that any number $r < 1$ specified via a periodic sequence of digits in the decimal representation is a rational number, i.e. it can always be written as a ratio of two integers.

2.3   Elementary Functions

Here, we shall introduce all elementary functions and their main properties. As we are sure the reader is familiar with all of them, our consideration may be rather brief; at the same time, it is worthwhile to define all functions systematically and gradually making emphasis on some specific points.

2.3.1 Polynomials

A constant is the simplest function: $y = C$ with C being some real number. Its graph is represented by a horizontal line crossing the y-axis exactly at the value of C. It is also the simplest polynomial: a polynomial of degree zero.

A linear combination of x in integer powers, from zero to n, with real coefficients, as was already discussed in detail in Sect. 1.4, gives rise to a function called a *polynomial* of degree n:

$$f(x) = a_n x^n + a_{n-1} x^{n-1} + \cdots + a_1 x + a_0 = \sum_{i=0}^{n} a_i x^i . \tag{2.8}$$

It is of special utility to know the roots of the polynomial, i.e. the values of x at which $f(x) = 0$. We have already considered finding the roots of the polynomials of the degree 1 (linear), 2 (quadratic), and 3 (cubic), see Sect. 1.11.1. We shall consider cubic and quartic equations in more detail in Sects. 3.4.5 and 3.4.6. What is important for us here is to remind the reader that equation $f(x) = 0$ for a polynomial always has at least one solution, and the maximum possible number of different roots is n. Some of the roots may be complex. However, for polynomials with real coefficients they will always come in complex conjugate pairs, $a + ib$ and $a - ib$. More precisely, and saying this a bit differently as well, one may say that this equation always has n roots some of which (or all) could be the same (repeated). This fact can be simply put by writing the function $f(x)$ in a factorised form (Sect. 1.4.1):

$$f(x) = a_n (x - x_1)^{k_1} (x - x_2)^{k_2} \cdots (x - x_m)^{k_m} , \tag{2.9}$$

which states explicitly that $f(x)$ has m distinct roots x_1 (repeated k_1 times), x_2 (repeated k_2 times), and so on ($n = k_1 + k_2 + \cdots + k_m$).

Since the complex roots always come in complex conjugate pairs, the factorisation of a polynomial (2.9) can also be written entirely via real factors:

$$f(x) = a_n (x - x_1)^{k_1} \cdots (x - x_m)^{k_m} \left(x^2 + p_1 x + q_1\right)^{l_1} \cdots \left(x^2 + p_r x + q_r\right)^{l_r} . \tag{2.10}$$

Here, we first have m simple factors $x - x_i$ corresponding to real roots of $f(x)$, which are followed by square polynomials $x^2 + p_i x + q_i$ for each ith pair of the complex conjugate roots.

Graphs of several integer power functions $y = x^n$ are shown in Fig. 2.4a. Functions with even n are even and hence symmetrical with respect to the y-axis, while functions with odd values of n are odd, i.e. for them $(-x)^n = -x^n$.

2.3.2 Rational Functions

A rational function is defined as a ratio of two polynomials. We shall assume that we deal here with real polynomials that contain only real coefficients. In many

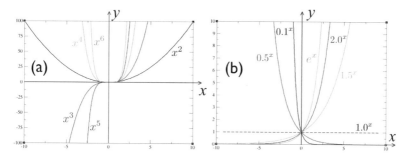

Fig. 2.4   a Power functions $y = x^n$ for selected values of n; **b** exponential functions $y = a^x$ for several values of the base a

applications, it is necessary to decompose such a function into simpler functions. It is done by writing down a desired decomposition with *undetermined coefficients* and then finding the coefficients by comparing both sides (as the equality should hold for any value of x). As an example, consider

$$f(x) = \frac{1}{x^2 - 1} \, .$$

Since $x^2 - 1$ can be factored as $(x - 1)(x + 1)$, we attempt to decompose $f(x)$ as follows:

$$\frac{1}{x^2 - 1} = \frac{A}{x - 1} + \frac{B}{x + 1} \, ,$$

with A and B being yet to be determined constants. Summing up the two terms above in the right-hand side, we obtain a fraction with the denominator identical to the original one, $x^2 - 1$, standing on the left. Let us then compare the numerators in the two sides:

$$1 = A(x + 1) + B(x - 1) \, . \tag{2.11}$$

Since this equation is to hold for any x, we may for instance put two values of x to obtain two independent equations for the two constants. For instance, take $x = 1$ and $x = -1$, yielding

$$\begin{cases} 1 = 2A \\ 1 = -2B \end{cases} \implies \begin{cases} A = 1/2 \\ B = -1/2 \end{cases} \, ,$$

so that

$$\frac{1}{x^2 - 1} = \frac{1}{2}\left(\frac{1}{x - 1} - \frac{1}{x + 1}\right) \, . \tag{2.12}$$

Another way of approaching the problem of solving Eq. (2.11) is to rearrange the right-hand side first in a form of a polynomial:

$$1 = (A - B) + (A + B)x \, .$$

In the left-hand side, the polynomial is equal to just a constant, i.e. when we compare the coefficients in both sides to the same powers of x, we obtain two equations

$$\begin{cases} 1 = A - B \\ 0 = A + B \end{cases} \implies \begin{cases} A = 1/2 \\ B = -1/2 \end{cases},$$

resulting in the same solution. This method might be preferable to the first one as it clearly uses the fact that the identity (2.11) must hold for any x.

Problem 2.12 Decompose the following fraction:

$$\frac{1}{x\,(x-1)\,(x-2)} \quad \left[\text{Answer:}\ \frac{1}{2x} - \frac{1}{x-1} + \frac{1}{2\,(x-2)} \right].$$

Generally, if the function $f(x) = P_n(x)/Q_m(x)$ is given as a ratio of two polynomials of degrees n and m, two cases must be distinguished: when $n < m$ and $n \geq m$. The latter case can always be manipulated into the first one by singling out the denominator in the numerator at its highest power term(s), as e.g. done in the following example:

$$\begin{aligned} \frac{3x^3 + 2x + 1}{x^2 + 1} &= \frac{3x\left[(x^2+1) - 1\right] + 2x + 1}{x^2 + 1} \\ &= \frac{3x\,(x^2+1) - x + 1}{x^2 + 1} = 3x - \frac{x-1}{x^2+1}. \end{aligned} \tag{2.13}$$

This way one can always represent P_n/Q_m with $n \geq m$ as (see Sect. 1.4.1)

$$\frac{P_n}{Q_m} = R_r + \frac{G_l}{Q_m}, \tag{2.14}$$

where R_r and G_l are polynomials and $l < m$.

Problem 2.13 Represent the following polynomials in the simpler form using manipulations similar to the ones applied in Eq. (2.13):

$$\frac{x^5 + 3x^2 + 3x + 1}{x^2 + 5x + 1} \quad \left[\text{Answer:}\ (x^3 - 5x^2 + 24x) - \frac{112x^2 + 21x - 1}{x^2 + 5x + 1} \right];$$
$$\tag{2.15}$$

$$\frac{x^3 + 12x^2 - x + 1}{x + 1} \quad \left[\text{Asnwer:}\ (x^2 + 11x) - \frac{12x - 1}{x + 1} \right]. \tag{2.16}$$

The rational function P_n/Q_m with $n < m$ can then be decomposed even further, as we have seen above, into a sum of simpler fractions. This process is called *partial fraction decomposition*. Without loss of generality, we shall assume that both polynomials, $P_n(x)$ and $Q_m(x)$, are monic, i.e. their coefficient to the largest power of x is one. To understand what must be the form of such decomposition, suppose that we know one root x_1 of the polynomial $Q_m(x)$ in the denominator that is repeated k_1 times. This means that (Sect. 1.4.1) the polynomial $Q_m(x)$ accepts factorisation:

$$Q_m(x) = (x - x_1)^{k_1} Q_{m-k_1}(x),$$

and hence

$$\frac{P_n(x)}{Q_m(x)} = \frac{P_n(x)}{(x - x_1)^{k_1} Q_{m-k_1}(x)},$$

where $Q_{m-k_1}(x)$ is a polynomial of the order $m - k_1$. Obviously, $Q_{m-k_1}(x_1) \neq 0$. Next, we shall perform the following identical manipulation on the numerator:

$$P_n(x) = \underbrace{\left[P_n(x) - A_{k_1} Q_{m-k_1}(x) \right]}_{D_n(x)} + A_{k_1} Q_{m-k_1}(x),$$

where $D_n(x)$ is a polynomial and $A_{k_1} = P_n(x_1)/Q_{m-k_1}(x_1)$ is a constant. This gives:

$$\frac{P_n(x)}{Q_m(x)} = \frac{D_n(x)}{(x - x_1)^{k_1} Q_{m-k_1}(x)} + \frac{A_{k_1}}{(x - x_1)^{k_1}}.$$

Note, however, that, by construction, $D_n(x_1) = 0$, i.e. x_1 is a root of this polynomial. If we assume that this root is only repeated once, then $D_n(x)$ can be factorised, $D_n(x) = (x - x_1) D_{n-1}(x)$, and one can write:

$$\frac{P_n(x)}{Q_m(x)} = \frac{A_{k_1}}{(x - x_1)^{k_1}} + \frac{D_{n-1}(x)}{(x - x_1)^{k_1-1} Q_{m-k_1}(x)}.$$

It is seen that the power of $(x - x_1)$ in the denominator of the second term is reduced by one.

Applying the same argument to the second fraction, we break up the numerator polynomial into two parts:

$$D_{n-1}(x) = \underbrace{\left[D_{n-1}(x) - A_{k_1-1} Q_{m-k_1}(x) \right]}_{T_{n-1}(x)} + A_{k_1-1} Q_{m-k_1}(x),$$

where $A_{k_1-1} = D_{n-1}(x_1)/Q_{m-k_1}(x_1)$ is a constant, and the polynomial $T_{n-1}(x)$ equals to zero at x_1 by construction. Hence, it can be factorised, $T_{n-1}(x) = (x -$

$x_1)D_{n-2}(x)$, where we again assumed that the root x_1 is repeated only once. This yields:

$$\frac{P_n(x)}{Q_m(x)} = \frac{A_{k_1}}{(x-x_1)^{k_1}} + \frac{A_{k_1-1}}{(x-x_1)^{k_1-1}} + \frac{D_{n-2}(x)}{(x-x_1)^{k_1-2}\,Q_{m-k_1}(x)},$$

where the power of $(x - x_1)$ in the denominator of the last term is reduced again. This process can be continued in a similar fashion, so that we shall eventually get:

$$\frac{P_n(x)}{Q_m(x)} = \frac{A_{k_1}}{(x-x_1)^{k_1}} + \frac{A_{k_1-1}}{(x-x_1)^{k_1-1}} + \cdots + \frac{A_1}{x-x_1} + \frac{D_{n-k_1}(x)}{Q_{m-k_1}(x)}. \qquad (2.17)$$

For simplicity, we have so far assumed that, at each step, the numerator polynomials $D_k(x)$ contain the root x_1 of the denominator polynomial $Q_m(x)$ only one time. It is easy to see that if this assumption is not made and the root is repeated, then the power of the difference $(x - x_1)$ in the fraction in the right-hand side will be reduced more than by one by means of the above procedure, and hence in the decomposition like the one in Eq. (2.17) some terms with the inverse powers of $(x - x_1)$ will be missing. In order words, we can say that the decomposition (2.17) is still valid in this more general case; however, some of the A_k coefficients might be zero.

Next we shall repeat the whole procedure for the ratio of two polynomials in the last term above, $D_{n-k_1}(x)/Q_{m-k_1}(x)$, using the second root x_2 of the polynomial $Q_m(x)$. Importantly, by construction, that is the root of the polynomial $Q_{m-k_1}(x)$ as well. If the root is repeated k_2 times, we shall then similarly obtain:

$$\frac{D_{n-k_1}(x)}{Q_{m-k_1}(x)} = \frac{B_{k_2}}{(x-x_2)^{k_2}} + \frac{B_{k_2-1}}{(x-x_2)^{k_2-1}} + \cdots + \frac{B_1}{x-x_2} + \frac{D_{n-k_1-k_2}(x)}{Q_{m-k_1-k_2}(x)}.$$

Then we shall consider the next root of $Q_m(x)$ and then the next, until we exhaust all of them. Hence, if the polynomial

$$Q_m(x) = (x-x_1)^{k_1}(x-x_2)^{k_2}\cdots(x-x_l)^{k_l} \qquad (2.18)$$

has l distinct roots x_1, x_2, etc. with their repetitions being k_1, k_2, etc., we shall eventually get:

$$\frac{P_n(x)}{Q_m(x)} = \frac{A_{k_1}}{(x-x_1)^{k_1}} + \frac{A_{k_1-1}}{(x-x_1)^{k_1-1}} + \cdots + \frac{A_1}{x-x_1}$$
$$+ \frac{B_{k_2}}{(x-x_2)^{k_2}} + \frac{B_{k_2-1}}{(x-x_2)^{k_2-1}} + \cdots + \frac{B_1}{x-x_2} + \cdots$$
$$\cdots + \frac{L_{k_l}}{(x-x_l)^{k_l}} + \frac{L_{k_l-1}}{(x-x_l)^{k_l-1}} + \cdots + \frac{L_1}{x-x_l}.$$

The obtained decomposition is unique as it uniquely follows from the procedure developed above.

In the above treatment, we have not been specific whether the roots x_1, x_2, etc. of the real polynomial $Q_m(x)$ are real or complex. Of course, the procedure we have developed is general and is valid for the polynomials even with complex coefficients in the rational function. If the two polynomials are, however, composed of exclusively real coefficients, then the decomposition can be constructed such that the obtained partial fractions would only contain real coefficients. Indeed, we know that if a root is complex, then there is definitely another root that is its complex conjugate, i.e. complex roots always happen in pairs. It is this fact that enables one to represent the decomposition entirely via real polynomials. Let z and z^* be a particular pair of two complex conjugate roots that are repeated l times. These two roots will contribute the following terms in the decomposition:

$$\left[\frac{A_l}{(x-z)^l} + \cdots + \frac{A_1}{x-z}\right] + \left[\frac{A_l^*}{(x-z^*)^l} + \cdots + \frac{A_1^*}{x-z^*}\right].$$

To ensure that the original fraction is real, we have made sure that the coefficients in the terms with z^* are being complex conjugate of the appropriate coefficients in the terms with z.

We shall rearrange this expression by combining terms with coefficients A_k and A_k^*. To see what to expect, let us start with the pair of terms containing the first inverse power:

$$\begin{aligned}\frac{A_1}{x-z} + \frac{A_1^*}{x-z^*} &= \frac{A_1(x-z^*) + A_1^*(x-z)}{(x-z)(x-z^*)} \\ &= \frac{(A_1 + A_1^*)x - (A_1 z^* + A_1^* z)}{x^2 + px + q} \\ &= \frac{Bx + C}{x^2 + px + q},\end{aligned}$$

where

$$(x-z)(x-z^*) = x^2 + px + q$$

is a real polynomial and the coefficients B and C are also real. This means that the two complex conjugate terms containing the first inverse power of $x - z$ and $x - z^*$ can be rewritten as a single fraction with the square polynomial in the denominator and a linear one in the numerator, with all coefficients being real.

What will happen if we combine the terms $A_k/(x-z)^k$ and $A_k^*/(x-z^*)^k$ with $k > 1$? Consider the case of $k = 2$:

$$\frac{A_2}{(x-z)^2} + \frac{A_2^*}{(x-z^*)^2} = \frac{A_2(x-z^*)^2 + A_2^*(x-z)^2}{\left(x^2 + px + q\right)^2}.$$

The expression in the numerator is a square polynomial. Hence, it can be written as a sum of $\alpha(x^2 + px + q)$ with a real coefficient α and a linear polynomial $C_1(x)$.

Therefore,

$$\frac{A_2}{(x-z)^2} + \frac{A_2^*}{(x-z^*)^2} = \frac{\alpha}{x^2+px+q} + \frac{C_1(x)}{\left(x^2+px+q\right)^2}.$$

Problem 2.14 Similarly show that

$$\frac{A_3}{(x-z)^3} + \frac{A_3^*}{(x-z^*)^3} = \frac{R_1(x)}{\left(x^2+px+q\right)^2} + \frac{C_1(x)}{\left(x^2+px+q\right)^3}.$$

Generally, for any $k > 2$

$$\frac{A_k}{(x-z)^k} + \frac{A_k^*}{(x-z^*)^k} = \frac{R_k(x)}{\left(x^2+px+q\right)^k}.$$

The polynomial $R_k(x)$ can be divided by the second order polynomial x^2+px+q,

$$R_k(x) = B_{k-2}(x)(x^2+px+q) + C_1(x),$$

with the residue $C_1(x)$ being generally a first-order polynomial. Hence,

$$\frac{R_k(x)}{\left(x^2+px+q\right)^k} = \frac{B_{k-2}(x)}{\left(x^2+px+q\right)^{k-1}} + \frac{C_1(x)}{\left(x^2+px+q\right)^k}.$$

Next, we divide the polynomial $B_{k-2}(x)$ by x^2+px+q again. Continuing this process, we shall eventually obtain a sum of terms each containing a first-order polynomial in the numerator and a power of (x^2+px+q) in the denominator.

Our discussion can now be formulated as the following theorem.

Theorem 2.9 *If the denominator $Q_m(x)$ with real coefficients is decomposed into a product of simple factors as in Eq. (2.18) with real and pairs of complex conjugate roots, then the fraction $P_n(x)/Q_m(x)$ (with $n < m$) can be uniquely written as a sum of real simpler fractions using the following rules:*

- *each real root $x = a$ which is repeated r times, gives rise to the following sum of terms:*

$$\frac{A_1}{x-a} + \frac{A_2}{(x-a)^2} + \cdots + \frac{A_r}{(x-a)^r}$$

with r real constants $A_1, \ldots, A_r$;

- *each pair of complex roots $z = a + ib$ and $z^* = a - ib$ corresponding to the real square polynomial $x^2 + px + q$ results in the sum of terms*

$$\frac{B_1 x + C_1}{x^2 + px + q} + \frac{B_2 x + C_2}{\left(x^2 + px + q\right)^2} + \cdots + \frac{B_l x + C_l}{\left(x^2 + px + q\right)^l}$$

with $2l$ real constants $B_1, C_1, \ldots, B_l, C_l$, where l is the repetition number for the given pair of roots.

Example 2.5 ▶ Decompose

$$R = \frac{3x^2 + x - 1}{(x - 1)^2 \left(x^2 + 1\right)} .$$

Solution. Here, the root $x = 1$ is repeated twice, and we have a pair $x = \pm i$ of complex conjugate roots as well that correspond to the real square polynomial $x^2 + 1$. Using the rules, we write:

$$R = \frac{A_1}{x - 1} + \frac{A_2}{(x - 1)^2} + \frac{Bx + C}{x^2 + 1}$$

$$= \frac{A_1 (x - 1) \left(x^2 + 1\right) + A_2 \left(x^2 + 1\right) + (Bx + C) (x - 1)^2}{(x - 1)^2 \left(x^2 + 1\right)} .$$

Opening up the brackets in the numerator above, grouping terms with the same powers of x and comparing them with the corresponding coefficients in the numerator of the original fraction $R(x)$, we obtain the following four equations for the four unknown coefficients, we list them below giving the corresponding power of the x on the left for convenience:

$$\begin{cases} x^3 : & A_1 + B = 0 \\ x^2 : -A_1 + A_2 + C - 2B = 3 \\ x^1 : & A_1 - 2C + B = 1 \\ x^0 : & -A_1 + A_2 + C = -1 \end{cases} \implies \begin{cases} A_1 = 2 \\ A_2 = 3/2 \\ B = -2 \\ C = -1/2 \end{cases} ,$$

i.e. we finally obtain the decomposition sought for:

$$\frac{3x^2 + x - 1}{(x - 1)^2 \left(x^2 + 1\right)} = \frac{2}{x - 1} + \frac{3}{2 (x - 1)^2} - \frac{4x + 1}{2 \left(x^2 + 1\right)} . \blacktriangleleft \qquad (2.19)$$

Problem 2.15 Prove the following identities corresponding to the decomposition into fractions:

$$\frac{x^2 - x + 5}{(2x - 1)\left(x^2 + 4x + 4\right)} = \frac{1}{25}\left[\frac{19}{2x - 1} + \frac{3}{x + 2} - \frac{55}{(x + 2)^2}\right] ;$$

$$\frac{x^3 + x^2 + 1}{(3x - 1)^2\left(2x^2 + 3\right)} = \frac{1}{2523}\left[\frac{137}{3x - 1} + \frac{899}{(3x - 1)^2} + \frac{189x + 237}{2x^2 + 3}\right] ;$$

$$\frac{2x - 3}{\left(x^2 + 1\right)^2\left(x^2 + 3\right)} = \frac{1}{4}\left[\frac{2x - 3}{x^2 + 3} + \frac{-2x + 3}{x^2 + 1} + \frac{4x - 6}{\left(x^2 + 1\right)^2}\right] .$$

[Hint: *note that in the first fraction* $x^2 + 4x + 4$ *should be factored first.*]

Problem 2.16 Demonstrate the following decompositions into fractions:

$$\frac{1}{x^4 + 1} = \frac{1}{2\sqrt{2}}\left(\frac{x + \sqrt{2}}{x^2 + \sqrt{2}x + 1} + \frac{-x + \sqrt{2}}{x^2 - \sqrt{2}x + 1}\right) ;$$

$$\frac{x^2}{x^4 + 1} = \frac{1}{2\sqrt{2}}\left(\frac{-x}{x^2 + \sqrt{2}x + 1} + \frac{x}{x^2 - \sqrt{2}x + 1}\right) .$$

[Hint: *note that*

$$x^4 + 1 = \left(x^4 + 2x^2 + 1\right) - 2x^2$$

$$= \left(x^2 + 1\right)^2 - \left(\sqrt{2}x\right)^2$$

$$= \left(x^2 + \sqrt{2}x + 1\right)\left(x^2 - \sqrt{2}x + 1\right) ,$$

so that the denominator in each case can be factorised. Of course, the four roots of $x^4 + 1 = 0$ *can also be formally found and then this expression be fully factored using complex roots.*]

2.3.3 General Power Function

The general power function is $y = x^\alpha$ with parameter α being from $\mathbb{R}$ (a real number). Definition of the power function to any real power is done in stages. First of all, we introduce an integer power $\alpha = n$ as

$$y = x^n = \underbrace{x \cdot x \cdot \ldots \cdot x}_{n},$$

i.e. via a repeated multiplication. In addition, we define $x^0 = 1$ for any x. It is easy to see that the power function is monotonically increasing. Indeed, consider first $n = 2$. If $x_2 > x_1$, then

$$y_2 = x_2^2 = x_2 x_2 > x_1 x_2 > x_1 x_1 = x_1^2 = y_1,$$

i.e. $y_2 > y_1$. Similarly, that property can be demonstrated for any integer n.

Then, we define the power being an inverse integer, $\alpha = 1/n$, as the solution for y of the equation $y^n = x$, which we write $y = x^{1/n}$ or $y = \sqrt[n]{x}$. In other words, $y = x^{1/n}$ is defined as an inverse function to $y = x^n$. Note, that in the cases of even values of n we only accept positive solutions for y and $x \geq 0$; this method would prevent us from arriving at multi-valued functions, as was discussed at the end of Sect. 2.1 using parabola as an example. Since the direct function, $y = x^n$, is monotonically increasing, so is the inverse integer power function, $y = x^{1/n}$, as was proven in the same section.

At the next step, we define the power being a general rational number $\alpha = n/m$ via:

$$y = x^{n/m} = \left(x^{1/m}\right)^n$$

This is an example of the composition of two functions mentioned at the beginning of Sect. 2.1: indeed, we combined two functions, $y = w^n$ and $w = x^{1/m}$, into one: $y = w^n = \left(x^{1/m}\right)^n$. As both functions involved here are monotonically increasing, so is their composition. Hence, the function $y = x^\alpha$ for any rational α increases monotonically.

Now we can extend the definition of the power function to all *positive* real numbers α from $\mathbb{R}$. Since we already covered general rational numbers (which are represented by either finite or periodic decimal fractions, Sect. 1.2), we only need to consider general irrational values of α which are represented by infinite non-periodic decimal fractions. The main idea here is to set a lower and upper bounds, r_{lower} and r_{upper}, to the irrational number α by two close rational ones: $r_{lower} < \alpha < r_{upper}$. For instance, consider α being $\sqrt{2} = 1.414213562\ldots$. This irrational number can be bound e.g. by $r_{lower} = 1.414$ and $r_{upper} = 1.415$. Then, since the function $y = x^r$ with rational r is increasing monotonically, then $x^{r_{lower}} < x^{r_{upper}}$. One then *defines* that x^α should lie between these two values to obey the monotonic behaviour: $x^{r_{lower}} < x^\alpha < x^{r_{upper}}$, or in the case of the square root of two, we write: $x^{1.414} < x^{\sqrt{2}} < x^{1.415}$. Of course, these two bounds are only approximations to the actual value of $\sqrt{2}$, so we could do much better by enclosing it in tighter bounds, e.g. $x^{1.4142} < x^{\sqrt{2}} < x^{1.4143}$, then $x^{1.41421} < x^{\sqrt{2}} < x^{1.41422}$, and so on, i.e. this process can be continued by moving the boundaries on the left and on the right of $\sqrt{2}$ closer to it simultaneously from both directions. This way we create two *numerical sequences* of numbers approaching

$\sqrt{2}$ from the left (1.414, 1.4142, 1.41421, etc.) and from the right (1.415, 1.4143, 1.41422, etc.). The sequence on the left is constructed by adding every time the next extra digit of $\sqrt{2}$ and hence is increasing gradually approaching $\sqrt{2}$ from below. The sequence on the right is composed also by adding an extra digit at the same position as in the first sequence and increasing it by 1. Obviously, this sequence is decreasing gradually approaching $\sqrt{2}$ from above. These two sequences which bound $x^{\sqrt{2}}$ on both sides converge to the same limit which, by *definition*, is taken as the exact value of $x^{\sqrt{2}}$. From the practical point of view, bounding $x^{\sqrt{2}}$ on both sides using rational approximations to $\sqrt{2}$ gives a method for calculating $x^{\sqrt{2}}$ with any required precision.

Finally, a real *negative* power is defined as follows: $y = x^{-\alpha} = 1/x^{\alpha}$ with $\alpha > 0$. This way we have defined the general power function $y = x^{\alpha}$ for *any* real α.

Let us now consider properties of the power function. For integer n and m, we obviously have $x^n x^m = x^{n+m}$ and $(x^n)^m = x^{nm}$, due to the definition of the integer power. For the inverse integer powers, we also have $\left(x^{1/n}\right)^{1/m} = x^{1/(nm)}$ since this is the inverse function of the integer power. Indeed, on the one hand, from $(y^n)^m = y^{nm} = x$ follows that $y = x^{1/(nm)}$. On the other hand, from $(y^n)^m = x$ we can first write $y^n = x^{1/m}$ and then $y = \left(x^{1/n}\right)^{1/m}$. This means that $\left(x^{1/n}\right)^{1/m} = x^{1/(nm)}$, as required. Now we can consider the case of both powers being arbitrary rational numbers (n, m, p and q are positive integers):

$$\left(x^{n/m}\right)^{p/q} = \left\{\left[\left(x^{1/m}\right)^n\right]^p\right\}^{1/q} = \left[\left(x^{1/m}\right)^{np}\right]^{1/q}$$
$$= \left[\left(x^{np}\right)^{1/m}\right]^{1/q} = \left(x^{np}\right)^{1/(mq)} = x^{np/(mq)} ,$$

as required. The other property (when the powers sum up) is proven by the following manipulation:

$$x^{n/m} x^{p/q} = x^{nq/(mq)} x^{pm/(mq)} = \left(x^{1/(mq)}\right)^{nq} \left(x^{1/(mq)}\right)^{pm}$$
$$= \left(x^{1/(mq)}\right)^{nq+pm} = x^{(nq+pm)/(mq)} = x^{(n/m)+(p/q)} .$$

Extension of these properties to arbitrary positive real numbers is achieved using the method of bounding rational sequences as already has been demonstrated above.

Problem 2.17 Extend the above properties of the powers (multiplication and summation of powers) to all real powers including negative ones.

Hence, we conclude that for any real α and β:

$$x^{\alpha} x^{\beta} = x^{\alpha\beta} \quad \text{and} \quad (x^{\alpha})^{\beta} = x^{\alpha\beta} , \tag{2.20}$$

where generally we must consider $x \geq 0$.

2.3.4   Number e

Let us now consider a very famous sequence $x_n = \left(1 + \frac{1}{n}\right)^n$. We shall show that it tends to a limit $e = 2.71828182845\ldots$ (the base of the natural logarithm).[1] To show that this sequence tends to a certain limit, let us first apply to it the binomial formula (1.153):

$$
x_n = \left(1 + \frac{1}{n}\right)^n = \sum_{k=0}^{n} \binom{n}{k} 1^{n-k} \left(\frac{1}{n}\right)^k = \sum_{k=0}^{n} \binom{n}{k} \frac{1}{n^k} =
$$

$$
= 1 + n\frac{1}{n} + \frac{n(n-1)}{2} \frac{1}{n^2} + \frac{n(n-1)(n-2)}{3!} \frac{1}{n^3} + \cdots
$$

$$
+ \frac{n(n-1)(n-2)\cdots(n-k+1)}{k!} \frac{1}{n^k} + \cdots + \frac{n!}{n!} \frac{1}{n^n},
$$

(2.21)

where we used the fact that the binomial coefficients (1.154) can be written as

$$
\binom{n}{k} = \frac{n!}{k!(n-k)!} = \frac{n(n-1)(n-2)\cdots(n-k+1)}{k!}
$$

for any $k \geq 1$. Consider now the kth binomial coefficient above: it has exactly k multipliers in the numerator, so that the kth term in the expansion (2.21) may be written as follows for any $k \geq 1$:

$$
\binom{n}{k} \frac{1}{n^k} = \frac{1}{k!} \left(\frac{n}{n}\right) \left(\frac{n-1}{n}\right) \left(\frac{n-2}{n}\right) \cdots \left(\frac{n-k+1}{n}\right)
$$

$$
= \frac{1}{k!} \left(1 - \frac{1}{n}\right) \left(1 - \frac{2}{n}\right) \cdots \left(1 - \frac{k-1}{n}\right),
$$

so that

$$
x_n = 1 + 1 + \frac{1}{2!} \left(1 - \frac{1}{n}\right) + \frac{1}{3!} \left(1 - \frac{1}{n}\right) \left(1 - \frac{2}{n}\right)
$$

$$
+ \frac{1}{4!} \left(1 - \frac{1}{n}\right) \left(1 - \frac{2}{n}\right) \left(1 - \frac{3}{n}\right) + \cdots + \frac{1}{k!} \left(1 - \frac{1}{n}\right) \left(1 - \frac{2}{n}\right) \cdots \left(1 - \frac{k-1}{n}\right)
$$

$$
+ \cdots + \frac{1}{n!} \left(1 - \frac{1}{n}\right) \left(1 - \frac{2}{n}\right) \cdots \left(1 - \frac{n-1}{n}\right),
$$

(2.22)

where above we have presented the last term of (2.21) also in the same form as the others. Note that there are $n + 1$ terms in the expression (2.22) of x_n. Now let us write in the same way x_{n+1} using (2.22):

---

[1] This is actually an irrational number although we won't be able to demonstrate this here.

$$x_{n+1} = 1 + 1 + \frac{1}{2!}\left(1 - \frac{1}{n+1}\right) + \frac{1}{3!}\left(1 - \frac{1}{n+1}\right)\left(1 - \frac{2}{n+1}\right)$$
$$+ \frac{1}{4!}\left(1 - \frac{1}{n+1}\right)\left(1 - \frac{2}{n+1}\right)\left(1 - \frac{3}{n+1}\right) + \cdots$$
$$+ \frac{1}{k!}\left(1 - \frac{1}{n+1}\right)\left(1 - \frac{2}{n+1}\right)\cdots\left(1 - \frac{k-1}{n+1}\right) + \cdots \tag{2.23}$$
$$+ \frac{1}{(n+1)!}\left(1 - \frac{1}{n+1}\right)\left(1 - \frac{2}{n+1}\right)\cdots\left(1 - \frac{n}{n+1}\right).$$

This expansion of x_{n+1} has $n + 2$ terms. Let us compare the two expansions. It is clear that for any $k \geq 1$

$$1 - \frac{k}{n+1} > 1 - \frac{k}{n}.$$

Therefore, the third term of x_{n+1} is greater than the third term of x_n, the fourth term of x_{n+1} is greater than the fourth term of x_n, and so on. In addition, the last term in x_{n+1} is absent in the expansion of x_n, but this extra term is obviously positive. Therefore, we strictly have $x_{n+1} > x_n$ for any value of n. This means that our sequence x_n is increasing with n; moreover, we see that $x_n > 2$ (notice $1 + 1$ as the first two terms in the expansions of x_n or x_{n+1}).

At the same time, the sequence is limited from above. Indeed, since $1 - k/n < 1$ for any $k \geq 1$, then

$$x_n < 1 + 1 + \frac{1}{2!} + \frac{1}{3!} + \cdots + \frac{1}{n!}.$$

To estimate the sum of the inverse factorials, we notice that for any $k \geq 1$

$$\frac{1}{k!} = \frac{1}{1 \cdot 2 \cdot 3 \cdots \cdot k} < \frac{1}{1 \cdot 2 \cdot 2 \cdots \cdot 2} = \frac{1}{2^{k-1}},$$

so that

$$x_n < 1 + 1 + \frac{1}{2} + \frac{1}{2^2} + \cdots + \frac{1}{2^{n-1}} < 1 + \sum_{k=0}^{\infty} 2^{-k},$$

where in the last passage, we replaced the final sum of the inverse powers of 2 with their infinite sum, which made the inequality even stronger (as all terms are positive). However, the infinite sum represents an infinite geometric progression, which can be calculated as we discussed above by virtue of Eq. (2.7):

$$\sum_{k=0}^{\infty} 2^{-k} = 1 + \frac{1}{2} + \frac{1}{2^2} + \cdots + \frac{1}{2^{n-1}} + \cdots = \frac{1}{1 - \frac{1}{2}} = 2,$$

so that

$$x_n < 1 + \sum_{k=1}^{\infty} 2^{-k} = 1 + 2 = 3 .$$

Hence, we showed that $2 < x_n < 3$ and x_n increases with n. Therefore, it approaches a certain limit[2] which is the number e. Hence, we can state that

$$\lim_{n \to \infty} \left(1 + \frac{1}{n} \right)^n = e . \tag{2.24}$$

From this it is easy to see that, for any positive integer m,

$$\lim_{n \to \infty} \left(1 + \frac{m}{n} \right)^n = \lim_{n \to \infty} \left(1 + \frac{1}{n/m} \right)^n = \lim_{n \to \infty} \left[\left(1 + \frac{1}{n/m} \right)^{n/m} \right]^m .$$

Here n/m may be either integer or generally rational, and we have made use of the fact that $(x^\alpha)^\beta = x^{\alpha\beta}$. However, when we take the limit $n \to \infty$, we can run n only over a subset of integer numbers, which are divisible by m. Then, $n/m = k$ will be an integer, and the limit $n \to \infty$ will be replaced with $k \to \infty$ on the subset (however, the limit should be the same, Theorem 2.4). This way we have another useful result:

$$\lim_{n \to \infty} \left(1 + \frac{m}{n} \right)^n = \lim_{k \to \infty} \left[\left(1 + \frac{1}{k} \right)^k \right]^m = \left[\lim_{k \to \infty} \left(1 + \frac{1}{k} \right)^k \right]^m = e^m . \tag{2.25}$$

Next, let us consider negative $n = -m < 0$ (and thus $m > 0$):

$$\left(1 + \frac{1}{n} \right)^n = \left(1 + \frac{1}{-m} \right)^{-m} = \left(\frac{-m+1}{-m} \right)^{-m} = \left(\frac{-m}{-m+1} \right)^m$$
$$= \left(\frac{-m+1-1}{-m+1} \right)^m = \left(1 + \frac{1}{m-1} \right)^m$$
$$= \left(1 + \frac{1}{m-1} \right)^{m-1} \left(1 + \frac{1}{m-1} \right) .$$

Consider now the limit of $m \to \infty$, which corresponds to the limit of $n \to -\infty$. We immediately see that

$$\lim_{m \to \infty} \left(1 + \frac{1}{m-1} \right)^{m-1} \left(1 + \frac{1}{m-1} \right) = \lim_{m \to \infty} \left(1 + \frac{1}{m} \right)^m \left(1 + \frac{1}{m-1} \right) = e ,$$

---

[2] A numerical sequence that is bounded from below and above by two numbers always has a limit lying between those numbers, see the note at the end of Sect. 2.2.2.

since $\left(1 + \frac{1}{m-1}\right) \to 1$. Therefore, we obtain that

$$\lim_{n \to -\infty} \left(1 + \frac{1}{n}\right)^n = e \ . \tag{2.26}$$

Problem 2.18 Prove that

$$\lim_{n \to \infty} \left(1 - \frac{m}{n}\right)^n = e^{-m} \ . \tag{2.27}$$

2.3.5 Exponential Function

Consider a general definition of $y = a^x$ for any real positive a (and avoiding the trivial case of $a = 1$) and arbitrary x from $\mathbb{R}$. The power of a for arbitrary x is defined as above by first considering rational x and then using the method of bounding numerical sequences to extend these definitions to all real numbers $-\infty < x < \infty$. It is important in this respect to establish first that the exponential function is monotonic.

Indeed, consider the case of $a > 1$. First we prove that $a^\lambda > 1$ for any positive λ. This follows immediately for $\lambda = n$ being an integer. We also have that $a^{1/m} > 1$ for any positive integer m. Indeed, assume the opposite, i.e. $a^{1/m} < 1$, i.e. $a^{1/m} = 1 - \alpha$ with some $0 < \alpha < 1$; this means that $a = (1 - \alpha)^m$. But the power function is increasing monotonically, i.e. $a = (1 - \alpha)^m < 1^m = 1$ as $1 - \alpha < 1$; hence, we arrive at $a < 1$, which contradicts our assumption of $a > 1$, and that proves the statement made above. Now it is easy to see that $a^\lambda > 1$ for any rational $\lambda = n/m$. Indeed, $a^{n/m} = (a^n)^{1/m} > 1$ since $a^n > 1$. The proof for any positive real λ is extended by applying the bounding numerical sequences.

Now returning back to the exponential function, we write for $x_2 > x_1$:

$$a^{x_2} - a^{x_1} = a^{x_1} \left(a^{x_2 - x_1} - 1\right) > 0 \ ,$$

since $a^\lambda > 1$ for any positive λ as was shown above. This means that indeed the exponential function $y = a^x$ is monotonically increasing for $a > 1$.

Problem 2.19 Prove that the exponential function for $0 < a < 1$ is monotonically decreasing.

If, as the base of the power, the number e introduced in Sect. 2.3.4 is used, then we have the function $y = e^x$.

Due to the properties of powers, Eq. (2.20), we also have:

$$a^{x_1}a^{x_2} = a^{x_1+x_2} \quad \text{and} \quad \left(a^{x_1}\right)^{x_2} = a^{x_1 x_2} .\tag{2.28}$$

Exponential functions $y = a^x$ for selected values of the base a are shown in Fig. 2.4b. We see that for $0 < a < 1$ the exponential function monotonically decreases and becomes more and more horizontal as a approaches the value of 1.0 when it becomes indeed horizontal (and hence a constant); then for $a > 1$ the function monotonically increases; the larger the value of a, the steeper the increase.

2.3.6 Hyperbolic Functions

These are combinations of exponential functions:

$$\sinh(x) = \frac{1}{2}\left(e^x - e^{-x}\right) ;\tag{2.29}$$

$$\cosh(x) = \frac{1}{2}\left(e^x + e^{-x}\right) ;\tag{2.30}$$

$$\tanh(x) = \frac{\sinh(x)}{\cosh(x)} = \frac{e^x - e^{-x}}{e^x + e^{-x}} ;\tag{2.31}$$

$$\coth(x) = \frac{\cosh(x)}{\sinh(x)} = \frac{e^x + e^{-x}}{e^x - e^{-x}} .\tag{2.32}$$

It will be shown in Vol. II that these functions are closely related to trigonometric functions $\sin(x)$, $\cos(x)$, $\tan(x)$ and $\cot(x)$, and are called, respectively, hyperbolic sine, cosine, tangent and cotangent functions.

Problem 2.20 Using the definitions of the hyperbolic functions given above, prove the following identities:

$$\cosh^2(x) - \sinh^2(x) = 1 ,\tag{2.33}$$

$$\sinh(2x) = 2\sinh(x)\cosh(x) ,\tag{2.34}$$

$$\cosh(2x) = \cosh^2(x) + \sinh^2(x) ,\tag{2.35}$$

$$\sinh(x+y) = \sinh(x)\cosh(y) + \cosh(x)\sinh(y) .\tag{2.36}$$

These identities have close analogues in trigonometric functions; for instance, the first one, Eq. (2.33), is analogous to (1.65), while the others are similar to the corresponding identities of the sine and cosine (see below).

One can also define inverse hyperbolic functions. To define those, we need to consider the logarithm first.

2.3.7 Logarithmic Function

The logarithmic function is defined as an inverse to the exponential function: from $y = a^x$ we write $x = \log_a y$, or using conventional notations, $y = \log_a x$. This function is called the logarithm of x to base a. From the properties of the exponential function, we can easily deduce the properties of the logarithmic function. First, if $y_1 = a^{x_1}$ and $y_2 = a^{x_2}$, then

$$
\begin{aligned}
y_1 y_2 = a^{x_1} a^{x_2} &= a^{x_1 + x_2} = y \\
\Rightarrow \quad \log_a y = \log_a (y_1 y_2) &= x_1 + x_2 \\
\Rightarrow \quad \log_a (y_1 y_2) &= \log_a y_1 + \log_a y_2 \; .
\end{aligned}
\tag{2.37}
$$

Then, from the definition of the logarithmic function, it follows:

$$
a^{\log_a x} = x \; .
\tag{2.38}
$$

Raise both sides of this equality into power z:

$$
\begin{aligned}
\left(a^{\log_a x} \right)^z = a^{z \log_a x} &= x^z \\
\Rightarrow \quad z \log_a x &= \log_a \left(x^z \right) \; .
\end{aligned}
\tag{2.39}
$$

Finally, we write

$$
x = a^{\log_a x} = \left(b^{\log_b a} \right)^{\log_a x} = b^{\log_a x \cdot \log_b a} \; ,
$$

from which it follows that

$$
\log_b x = \log_a x \cdot \log_b a \; ,
\tag{2.40}
$$

which is the formula that connects different bases of the logarithmic function.

If $a = e$ is used as the base, then the logarithmic function is denoted simply $\ln x = \log_e x$ and is called *natural logarithm*. This one is most frequently used. The natural logarithm $y = \ln x$ and the exponential function $y = e^x$, which is the inverse to it, are compared in Fig. 2.5. As expected, the graphs are symmetric with respect to the dashed line which bisects the first quadrant.

Now we are in a position to introduce inverse hyperbolic functions.

Fig. 2.5 Mutually inverse $y = e^x$ and $y = \ln x$ functions

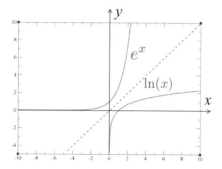

Problem 2.21 Using the definition of the hyperbolic cosine, solve equation $x = \cosh y$ with respect to $y = \cosh^{-1} x$ and hence deduce that the inverse hyperbolic cosine function accepts two branches given by:

$$\cosh^{-1}(x) = \ln\left(x \pm \sqrt{x^2 - 1}\right), \quad x \geq 1. \qquad (2.41)$$

Normally, the plus sign is used leading to positive values of the function for any $x \geq 1$. Explain the reason behind these two possibilities. Sketch both branches of the function. Show that other inverse hyperbolic functions, to be obtained in a similar manner, are single-valued and given by the following relations:

$$\sinh^{-1}(x) = \ln\left(x + \sqrt{x^2 + 1}\right), \quad -\infty < x < \infty, \qquad (2.42)$$

$$\tanh^{-1}(x) = \frac{1}{2} \ln \frac{1 + x}{1 - x}, \quad |x| < 1. \qquad (2.43)$$

$$\coth^{-1}(x) = \frac{1}{2} \ln \frac{x + 1}{x - 1}, \quad |x| > 1. \qquad (2.44)$$

Using the fact that the logarithmic function is only defined for its positive argument, explain the chosen x intervals in each case.

2.3.8 Trigonometric Functions

We have already defined sine and cosine functions in Sect. 1.6 and considered some of their properties. In particular, we found that these two functions are periodic with the period of 2π. The two functions are plotted in Fig. 2.6a. Basically, the sine function is the cosine function shifted by $\pi/2$ to the right. Both functions are periodic with the period of 2π. The tangent, $y = \tan x$, and cotangent, $y = \cot x$, functions are shown in Fig. 2.6b. They both have the period of π and are singular at

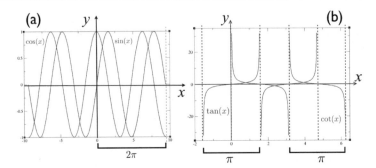

Fig. 2.6 **a** Functions $y = \sin(x)$ and $y = \cos(x)$ with the period of 2π; **b** functions $y = \tan x$ and $y = \cot x$ with the period of π. The period in each case is indicated at the bottom of the graphs

Fig. 2.7 Illustration for the proof of sum-of-angles trigonometric identities. The blue line corresponds to a unit circle centred at the origin

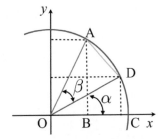

some periodically repeated points: $\tan(x)$ is singular when $\cos x = 0$ ($x = \frac{\pi}{2} + \pi n$ with $n = 0, \pm 1, \pm 2, \ldots$), while $\cot x$ is singular when $\sin x = 0$ (which happens at $x = \pi n$ with $n = 0, \pm 1, \pm 2, \ldots$).

We shall now consider other important properties of these functions.

First of all, we recall the double angle identities (1.75) and (1.76) we derived before. The sine identity can be compared with an analogous identity (2.34) we discussed above for the hyperbolic functions, while formula (2.35) for the hyperbolic functions is an analogue of the identity for the cosine function (1.76).

Next, we consider a more general formula for a sum of two angles. Referring this time to Fig. 2.7b, we write a square of AD in two different ways. The coordinates of points A and D obviously are: $A\left(\cos\left(\alpha + \beta\right), \sin\left(\alpha + \beta\right)\right)$ and $D\left(\cos\alpha, \sin\alpha\right)$, and hence

$$AD^2 = \left(\cos\left(\alpha + \beta\right) - \cos\alpha\right)^2 + \left(\sin\left(\alpha + \beta\right) - \sin\alpha\right)^2$$
$$= 2 - 2\left[\cos\left(\alpha + \beta\right)\cos\alpha + \sin\left(\alpha + \beta\right)\sin\alpha\right]$$

after opening the squares and using Eq. (1.65). At the same time, similarly to the previous case, from the isosceles triangle $\triangle OAD$ we can write $AD = 2\sin\frac{\beta}{2}$, and hence

$$2 - 2\left[\cos\left(\alpha + \beta\right)\cos\alpha + \sin\left(\alpha + \beta\right)\sin\alpha\right] = \left(2\sin\frac{\beta}{2}\right)^2 = 4\sin^2\frac{\beta}{2}$$
$$= 2\left(1 - \cos\beta\right),$$

after transforming the right-hand side using the double-angle formula (1.76). Opening the brackets and replacing $\gamma = \alpha + \beta$, we obtain:

$$\cos(\gamma - \alpha) = \cos\gamma\cos\alpha + \sin\gamma\sin\alpha \ . \tag{2.45}$$

Consequently, replacing α with $-\alpha$ in the above formula and recalling that sine and cosine functions are odd and even functions, respectively, we obtain:

$$\cos(\gamma + \alpha) = \cos\gamma\cos\alpha - \sin\gamma\sin\alpha \ . \tag{2.46}$$

Problem 2.22 Using one of the equations above, prove similar identities for the sine function as well:

$$\sin(\gamma \pm \alpha) = \sin\gamma\cos\alpha \pm \cos\gamma\sin\alpha \ . \tag{2.47}$$

The obtained equations are useful in establishing simple rules of calculating sine and cosine functions of angles $\alpha \pm \pi$ and $\alpha \pm \pi/2$ with the angle α being arbitrary. These rules can be established using definitions of the sine and cosine functions as respectively vertical and horizontal projections of the unit vector within the unit circle (Eq. 3.56) as we did in Sect. 1.6, Eq. (1.67); however, it is much easier doing that directly from the above formulae:

$$\cos(\alpha \pm \pi) = \cos\alpha \underbrace{\cos\pi}_{-1} + \sin\alpha \underbrace{\sin\pi}_{0} = -\cos\alpha \ ,$$

$$\sin(\alpha \pm \pi) = \sin\alpha \underbrace{\cos\pi}_{-1} \pm \cos\alpha \underbrace{\sin\pi}_{0} = -\sin\alpha \ ,$$

$$\cos\left(\alpha \pm \frac{\pi}{2}\right) = \cos\alpha \underbrace{\cos\frac{\pi}{2}}_{0} + \sin\alpha \underbrace{\sin\frac{\pi}{2}}_{1} = \sin\alpha \equiv \cos\left(\frac{\pi}{2} \pm \alpha\right) \ ,$$

$$\sin\left(\alpha \pm \frac{\pi}{2}\right) = \sin\alpha \underbrace{\cos\frac{\pi}{2}}_{0} \pm \cos\alpha \underbrace{\sin\frac{\pi}{2}}_{1} = \pm\cos\alpha = \pm\sin\left(\frac{\pi}{2} \pm \alpha\right) \ .$$

These expressions can be easily reproduced without repeating these calculations. In fact, the rules that we are about to formulate correspond to the sine and cosine functions to be calculated for angles $\alpha + \pi n$ and $\alpha + \pi n/2$ with any integer $n = 0, \pm 1, \pm 2, \ldots$. Assume that the angle α is between 0 and $\pi/2$. Then, consider in which quadrant of the unit circle, Fig. 2.2, the composed angle $\alpha \pm \pi n$ or $\alpha \pm \pi n/2$ lies and note whether the sine (cosine) there is positive or negative. Then, if sine (cosine) at the angle $\alpha \pm \pi n$ is considered, this is equal to the sine (cosine) of α

with the sign corresponding to that of the selected quadrant. If sine (cosine) at the angle $\alpha \pm \pi n/2$ is considered, this is equal to the cosine (sine) of α with the sign again corresponding to that of the selected quadrant. In other words, you can replace these expressions either with the same function (if $\pm \pi n$) or the opposite one (if $\pm \pi n/2$) calculated at just α and attaching the correct sign. For instance, consider $\sin(\pi - \alpha)$. Assuming α to be between 0 and $\pi/2$, the angle $\pi - \alpha$ corresponds to quadrant II where the sine is positive. Because we have the π case, we keep the same function. Hence $\sin(\pi - \alpha) = \sin \alpha$. Consider now $\cos(3\pi/2 + \alpha)$. The angle corresponds to quadrant IV where the cosine function is positive; the angle contains an addition of a multiple of $\pi/2$, hence, we have to change the cosine into sine. Therefore, $\cos(3\pi/2 + \alpha) = \sin \alpha$.

Problem 2.23 By adding (subtracting) Eqs. (2.45)–(2.47) and changing notations for the angles, derive the inverse relations:

$$\sin \alpha + \sin \beta = 2 \sin \left(\frac{\alpha + \beta}{2} \right) \cos \left(\frac{\alpha - \beta}{2} \right) ; \qquad (2.48)$$

$$\sin \alpha - \sin \beta = 2 \cos \left(\frac{\alpha + \beta}{2} \right) \sin \left(\frac{\alpha - \beta}{2} \right) ; \qquad (2.49)$$

$$\cos \alpha + \cos \beta = 2 \cos \left(\frac{\alpha + \beta}{2} \right) \cos \left(\frac{\alpha - \beta}{2} \right) ; \qquad (2.50)$$

$$\cos \alpha - \cos \beta = -2 \sin \left(\frac{\alpha + \beta}{2} \right) \sin \left(\frac{\alpha - \beta}{2} \right) . \qquad (2.51)$$

Problem 2.24 Rearrange the above identities to get the following expansions of a product of the sine and cosine functions:

$$\sin \alpha \sin \beta = \frac{1}{2} \left[\cos (\alpha - \beta) - \cos (\alpha + \beta) \right] ; \qquad (2.52)$$

$$\sin \alpha \cos \beta = \frac{1}{2} \left[\sin (\alpha + \beta) + \sin (\alpha - \beta) \right] ; \qquad (2.53)$$

$$\cos \alpha \sin \beta = \frac{1}{2} \left[\sin (\alpha + \beta) - \sin (\alpha - \beta) \right] ; \qquad (2.54)$$

$$\cos \alpha \cos \beta = \frac{1}{2} \left[\cos (\alpha - \beta) + \cos (\alpha + \beta) \right] . \qquad (2.55)$$

Using these identities, one can express trigonometric functions of multiple angles via those of only a single one. This may be useful in calculating integrals as demonstrated in later chapters. For instance, using Eqs. (2.47) and (2.46), as well as the particular cases of these for equal angles $\gamma = \alpha$, we obtain:

$$
\begin{aligned}
\sin (3\alpha) &= \sin (2\alpha + \alpha) = \sin (2\alpha) \cos \alpha + \cos (2\alpha) \sin \alpha \\
&= 2 \sin \alpha \cos^2 \alpha + \cos^2 \alpha \sin \alpha - \sin^3 \alpha \qquad (2.56) \\
&= 3 \sin \alpha \cos^2 \alpha - \sin^3 \alpha = 3 \sin \alpha - 4 \sin^3 \alpha \ ,
\end{aligned}
$$

$$
\begin{aligned}
\cos (3\alpha) &= \cos (2\alpha + \alpha) = \cos (2\alpha) \cos \alpha - \sin (2\alpha) \sin \alpha \\
&= \cos^3 \alpha - \sin^2 \alpha \cos \alpha - 2 \sin^2 \alpha \cos \alpha \qquad (2.57) \\
&= \cos^3 \alpha - 3 \sin^2 \alpha \cos \alpha = 4 \cos^3 \alpha - 3 \cos \alpha \ .
\end{aligned}
$$

Problem 2.25 Prove the following identities using the above method:

$$
\sin (4\alpha) = 4 \sin \alpha \cos^3 \alpha - 4 \sin^3 \alpha \cos \alpha \ ; \qquad (2.58)
$$

$$
\cos (4\alpha) = \cos^4 \alpha - 6 \sin^2 \alpha \cos^2 \alpha + \sin^4 \alpha \ ; \qquad (2.59)
$$

[There is also a simpler way for both above.]

$$
\sin (5\alpha) = \sin^5 \alpha - 10 \sin^3 \alpha \cos^2 \alpha + 5 \sin \alpha \cos^4 \alpha \ ; \qquad (2.60)
$$

$$
\cos (5\alpha) = \cos^5 \alpha - 10 \sin^2 \alpha \cos^3 \alpha + 5 \sin^4 \alpha \cos \alpha \ . \qquad (2.61)
$$

Problem 2.26 Using the following identity,

$$
\sin (k\Delta) = \frac{1}{2 \sin \frac{\Delta}{2}} \left\{ \cos \left[\left(k - \frac{1}{2} \right) \Delta \right] - \cos \left[\left(k + \frac{1}{2} \right) \Delta \right] \right\} , \qquad (2.62)
$$

which follows from Eq. (2.51), derive the formula below for the finite sum of sine functions:

$$
\sum_{k=1}^{n} \sin (\Delta k) = \frac{1}{\sin \frac{\Delta}{2}} \sin \frac{(n+1)\Delta}{2} \sin \frac{n\Delta}{2} . \qquad (2.63)
$$

[Hint: *use* (2.62) *for the sine functions in the sum and then observe that only the first and the last terms remain, all other terms cancel out exactly.*]

Problem 2.27 Prove the same type of identity for the cosine functions:

$$\sum_{k=1}^{n} \cos(\Delta k) = \frac{1}{2\sin\frac{\Delta}{2}}\left[\sin\left(\Delta\left(n+\frac{1}{2}\right)\right) - \sin\frac{\Delta}{2}\right] = \frac{1}{\sin\frac{\Delta}{2}}\cos\frac{(n+1)\Delta}{2}\sin\frac{n\Delta}{2}.$$

(2.64)

Problem 2.28 When studying integrals containing a rational function of sine and cosine, a special substitution replacing both of them with a tangent function is frequently found useful. Prove, using relationships between the sine and cosine functions of angles α and $\alpha/2$, the following identities:

$$\sin\alpha = \frac{2\tan\frac{\alpha}{2}}{1+\tan^2\frac{\alpha}{2}} \quad \text{and} \quad \cos\alpha = \frac{1-\tan^2\frac{\alpha}{2}}{1+\tan^2\frac{\alpha}{2}}.$$

(2.65)

Problem 2.29 Prove the following identity:

$$\cot(x+y) - \cot(x-y) = \frac{2\sin 2y}{\cos 2x - \cos 2y}.$$

(2.66)

2.3.9 Inverse Trigonometric Functions

For each of the four trigonometric functions, $\sin x$, $\cos x$, $\tan x$ and $\cot x$, one can define their corresponding inverse functions, denoted, respectively, as $\arcsin x$, $\arccos x$, $\arctan x$ and $\text{arccot} x$. These functions have the meaning inverse to that of the direct functions. For instance, $y = \arcsin x$ corresponds to $\sin y = x$, i.e. $y = \arcsin x$ is the angle sine of which is equal to x.

Because of the periodicity of the direct functions, there are different values of y (the angles) possible that give the same value of the functions x. Indeed, consider, e.g. the sine function and its inverse. Then, for all angles given by $y = y_0 + 2\pi n$ with n from all integer numbers $\mathbb{Z}$ the same value of the sine function $x = \sin y \equiv \sin y_0$ is obtained. This means that the inverse function is not well defined as there exist different values of the function $y = \arcsin x$ for the same value of its argument, x. This difficulty is similar to the one we came across before in Sect. 2.1 when considering the inverse of the $y = x^2$ function, and we realised there that by restricting the values of the argument of the direct function, y, the one-to-one correspondence can be established. The same method will be used here.

Since the sine function varies between -1 and 1 while the angles run between $-\pi/2$ and $\pi/2$, the arcsine is defined as follows: it gives an angle $-\pi/2 \le y \le \pi/2$ such that sine of it is equal to x lying in the interval $-1 \le x \le 1$. The graphs of both functions are shown in Fig. 2.8a. One can see that they are indeed symmetrical with

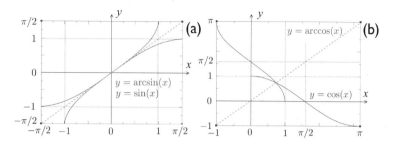

Fig. 2.8 Functions **a** $y = \sin x$, $y = \arcsin x$ and **b** $y = \cos x$, $y = \arccos x$

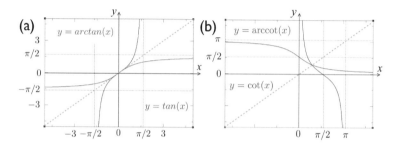

Fig. 2.9 Functions **a** $y = \tan x$, $y = \arctan x$ and **b** $y = \cot x$, $y = \mathrm{arccot} x$

respect to the diagonal of the I and III quadrants as we would expect (see the general discussion in Sect. 2.1).

Similarly, since the cosine function varies between -1 and 1 when the angle changes between 0 and π, the arccosine function and its inverse are well defined within these intervals. Namely, $y = \arccos x$ is defined for $-1 \leq x \leq 1$ and $0 \leq y \leq \pi$. Both functions are drawn in Fig. 2.8b.

Finally, functions $y = \arctan x$ and $y = \mathrm{arccot} x$ are both defined for $-\infty < x < \infty$, and the values of the arctangent lie within the interval $-\pi/2 < y < \pi/2$, while the values of the arccotangent—in the interval $0 < y < \pi$. The graphs of these two functions together with the corresponding direct functions are shown in Fig. 2.9.

The inverse trigonometric functions are useful in writing general solutions of trigonometric equations. For instance, equation $\sin x = a$ (with $-1 \leq a \leq 1$) has the general solution $x = \arcsin a + 2\pi n$, where $n = 0, \pm 1, \pm 2, \ldots$ is from $\mathbb{Z}$.

Let us now establish main properties of the inverse trigonometric functions. Obviously, it follows from their definitions that

$$\sin(\arcsin x) = x \ , \quad \cos(\arccos x) = x \text{ and } \tan(\arctan x) = x \ . \qquad (2.67)$$

Other properties are all derived from their definitions, i.e. from the properties of the corresponding direct functions. Consider, for instance, the sine function: $y = \sin(-x) = -\sin x$, from where it follows that, on the one hand, $-x = \arcsin y$, but on the other, $\arcsin(-y) = x$. This means that the arcsine function is antisymmetric (odd):

$$\arcsin(-x) = -\arcsin x \ . \tag{2.68}$$

Problem 2.30 Using the properties of the trigonometric functions, prove the following properties of the inverse trigonometric functions:

$$\arctan(-x) = -\arctan x \ , \tag{2.69}$$

$$\text{arccot}(-x) = \pi - \text{arccot}\, x \ , \tag{2.70}$$

$$\arccos(-x) = \pi - \arccos x \ , \tag{2.71}$$

$$\arcsin x + \arccos x = \frac{\pi}{2} \ , \tag{2.72}$$

$$\arctan x + \text{arccot}\, x = \frac{\pi}{2} \ , \tag{2.73}$$

$$\sin(\arccos x) = \cos(\arcsin x) = \sqrt{1 - x^2} \ , \tag{2.74}$$

$$\arctan \frac{1}{x} = \text{arccot}\, x = \arcsin \frac{1}{\sqrt{1 + x^2}} \ , \quad x > 0 \ , \tag{2.75}$$

$$\cos(\arctan x) = \frac{1}{\sqrt{1 + x^2}} \ . \tag{2.76}$$

2.4   Limit of a Function

2.4.1   Definitions

Limit of a function is one of the fundamental notions in mathematics. Several different cases need to be considered.

2.4.1.1 Limit of a Function at a Finite Point

We shall start by considering two definitions of a finite limit of a function at a point $-\infty < x < \infty$. Both definitions are completely equivalent although they sound somewhat different.

Let $f(x)$ be a function of a real variable x defined on some part of the real axis.

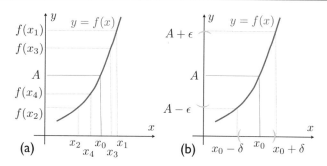

Fig. 2.10 Illustration of two equivalent definitions of the limit of a function $y = f(x)$ at the point x_0: **a** via an infinite numerical sequence $x_1, x_2, x_3, \ldots$ converging at x_0 and **b** using ϵ and δ

Definition 2.1 (*due to Heinrich Eduard Heine*) of the limit of a function $f(x)$ at x tending to x_0 (written: $x \to x_0$). If we take a numerical sequence of numbers x_n which is different from x_0 and converges to x_0, then it generates another sequence, a sequence of the values of the function,

$$f(x_1), \ f(x_2), \ f(x_3), \ \ldots, f(x_n), \ \ldots \ , \tag{2.77}$$

as illustrated in Fig. 2.10a. Then A is the limit of the function $f(x)$, i.e.

$$\lim_{x \to x_0} f(x) = A \ ,$$

if an *arbitrary* sequence of numbers converging at x_0 generates the functional sequence (2.77) converging *to the same* limit A.

Note that each sequence, if it has a limit, will converge to a unique limit (Sect. 2.2); however, what we are saying here is that if *any* sequence, converging at x_0, results in the same limit A of the generated functional sequence, only then the function $f(x)$ does have a limit at x_0. It follows from here that the function $f(x)$, if it has a limit at x_0, may only have *one* limit.

As an example, consider a function $f(x) = x$. If we take an arbitrary sequence x_n of its argument x which converges at x_0, then the corresponding functional sequence (2.77) would coincide with the sequence x_n since $f(x) = x$, and hence the functional sequence for any sequence x_n will converge to the same limit x_0.

Take now the parabolic function $f(x) = x^2$. The corresponding functional sequence becomes x_1^2, x_2^2, x_3^2, etc. This sequence can be considered as a product of two identical sequences x_n, and it tends to the limit x_0^2 (Theorem 2.2 in Sect. 2.2), where x_0 is the limit of the sequence x_n. So, if we consider an arbitrary sequence x_n converging to x_0, the functional sequence would have a limit which is x_0^2. Hence, the function $f(x) = x^2$ has a limit at x_0, equal to x_0^2.

For our next example, we consider the function $f(x) = \sin(1/x)$ that is defined everywhere except $x = 0$. Consider a specific numerical sequence $x_n = 1/(n\pi)$, where $n = 1, 2, 3, \ldots$. Obviously, the sequence tends to 0 as $n \to \infty$, and hence the functional sequence $f(x_n) = \sin(\pi n) = 0$ results in the constant sequence of zeros, converging to zero. One may think then that our function $f(x)$ tends to zero when $x \to 0$. However, this function in fact does not have a limit, and to prove that it is sufficient to build at least one example of a sequence x_n, also converging to zero, which generates the corresponding functional sequence converging to a *different* limit. And indeed, in this case, this is easy to do. Consider the second sequence, $x_n = 1/(\pi/2 + 2\pi n)$, which obviously converges to zero as well. However, generated by it functional sequence $f(x_n) = \sin(\pi/2 + 2\pi n)$ converges to 1 instead. Hence, the function $f(x) = \sin(1/x)$ does not have a limit at $x = 0$ (however, it does have a well-defined limit at any other point).

Now we shall give another, equivalent, definition of the limit of a function.

Definition 2.2 (*due to Augustin-Louis Cauchy*) of the limit of a function $f(x)$ at x_0. The limit of the function $f(x)$ is A when $x \to x_0$, if for any $\epsilon > 0$ one can indicate a vicinity of x_0 such that for any x from that vicinity $f(x)$ is different from A by no more than ϵ. Rephrased in a more formal way: for any $\epsilon > 0$ one can find such $\delta > 0$, so that for any x satisfying $|x - x_0| < \delta$ it follows $|f(x) - A| < \epsilon$. This is still exactly the same formulation: $|x - x_0| < \delta$ means that x belongs to some δ-vicinity of x_0, i.e. $x_0 - \delta < x < x_0 + \delta$, and $|f(x) - A| < \epsilon$ similarly means that $f(x)$ is in the vicinity of A by no more than ϵ as illustrated in Fig. 2.10b. In other words, no matter how close we would like to have $f(x)$ to A, we shall always find the corresponding values of x close to x_0 for which this condition will be satisfied. This second definition is sometimes called '$\epsilon - \delta$' definition of the limit of a function or the definition in terms of intervals. Note that $\delta = \delta(\epsilon)$ is a function of ϵ, i.e. it depends on the choice of ϵ.

As we said before, the two definitions are equivalent.[3] Let us now illustrate the second definition on some examples.

Consider first the limit of the function $f(x) = x$ at x_0. Choose some positive ϵ; then, by taking $\delta = \epsilon$, we immediately see that for any x satisfying $|x - x_0| < \delta$ we also have $|f(x) - x_0| = |x - x_0| < \delta = \epsilon$, i.e. $f(x)$ has the limit of x_0 at this point.

Consider now the limit of $f(x) = x^2$ at x_0. Let us prove that the limit exists and it is equal to x_0^2. Indeed, consider some vicinity $|x - x_0| < \delta$ of x_0. Then, for all x in that vicinity, we can estimate:

$$\left| f(x) - x_0^2 \right| = \left| x^2 - x_0^2 \right| = |x - x_0| \, |(x - x_0) + 2x_0|$$
$$\leq \delta \left(|x - x_0| + 2 |x_0| \right) \leq \delta \left(\delta + 2 |x_0| \right) .$$

If we now take δ as a solution of the equation $\epsilon = \delta(\delta + 2|x_0|)$ and choose its positive root, $\delta = -|x_0| + \sqrt{|x_0|^2 + \epsilon}$, then we can say that whatever value of ϵ we take, we can always find δ, e.g. the one given above, that the required condition $\left| x^2 - x_0^2 \right| < \epsilon$ is indeed satisfied for any x from the δ-vicinity of x_0. So, $f(x) = x^2$ has the limit x_0^2 when $x \to x_0$.

---

[3] We shall not prove their equivalence here.

Next, let us consider $f(x) = \sin x$. We would like to prove that it tends to $\sin x_0$ when $x \to x_0$. First of all, let us make an estimate, assuming that $|x - x_0| < \delta$. We can write:

$$\sin x - \sin x_0 = 2 \sin \frac{x - x_0}{2} \cos \frac{x + x_0}{2} \, ,$$

so that

$$|\sin x - \sin x_0| = 2 \left| \sin \frac{x - x_0}{2} \right| \cdot \left| \cos \frac{x + x_0}{2} \right|$$

$$< 2 \left| \frac{x - x_0}{2} \right| \cdot \left| \cos \frac{x + x_0}{2} \right| < |x - x_0| < \delta \, ,$$

where we used the first part of the inequality (1.196) and the fact that the absolute value of the cosine is not larger than 1. The result we have just obtained clearly shows that for any ϵ there exists such $\delta = \epsilon$, so that from $|x - x_0| < \delta$ immediately follows $|\sin x - \sin x_0| < \epsilon$, which is exactly what was required.

Finally, let us prove that

$$\lim_{x \to 0} a^x = 1 \text{ for } a > 1 \, . \tag{2.78}$$

Here $x_0 = 0$, i.e. we need to prove that from $|x| < \delta$ follows $|a^x - 1| < \epsilon$, or that

$$-\epsilon + 1 < a^x < \epsilon + 1 \, . \tag{2.79}$$

Since $a^x > 0$ for all values of x, it is sufficient to consider $0 < \epsilon < 1$. Then, as the logarithm is a monotonically increasing function, the inequality (2.79) can also be written as $\log_a (1 - \epsilon) < x < \log_a (1 + \epsilon)$ (note that $\log_a a^x = x$), where the left boundary, $\log_a (1 - \epsilon)$, is negative (since $0 < 1 - \epsilon < 1$), while the right one, $\log_a (1 + \epsilon)$, is positive ($1 + \epsilon > 1$). We see that if we take δ as the smallest absolute value of the two boundaries,

$$\delta = \min \left\{ \left| \log_a (1 - \epsilon) \right|, \left| \log_a (1 + \epsilon) \right| \right\} \, ,$$

then we obtain $-\delta < x < \delta$, i.e. $\delta = \delta(\epsilon)$ can be found from the equation above. This proves the limit (2.78).

Problem 2.31 Prove, using Eq. (2.78), that $\lim_{x \to x_0} a^x = a^{x_0}$.

Problem 2.32 Prove using the definition by intervals that $\lim_{x \to x_0} \sqrt{x} = \sqrt{x_0}$ for any $x_0 > 0$.

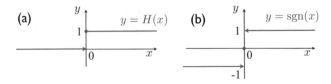

Fig. 2.11 Heaviside (**a**) and sign (**b**) functions experience a jump at $x = 0$. They do not have a certain limit at this point, although both left and right limits at $x = 0$ exist in both cases. The arrows on some lines indicate that the lines just approach the point $x = 0$, but never reach it. The fat dot indicates the value of the function at $x = 0$ according to Eqs. (2.1) and (2.2)

Fig. 2.12 Right (**a**) and left (**b**) limits of a function $y = f(x)$ at $x = x_0$ using the language of intervals

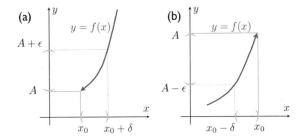

2.4.1.2 One-Side Limits

Some functions may experience a jump at point x_0. For instance, the Heaviside function $H(x)$, defined in Eq. (2.1), makes a jump at $x = 0$ from zero to one, see Fig. 2.11a. Similarly, the sign function, Eq. (2.2), experiences a jump as well, see Fig. 2.11b. For these and other cases it is useful to introduce *one-side limits*, when x tends to x_0 either from the left, being always smaller than x_0 (written: $x \to x_0 - 0$ or $x \to x_0^-$ or $x \to x_0-$), or from the right, when $x > x_0$ (written: $x \to x_0 + 0$ or $x \to x_0^+$ or $x \to x_0+$). Both definitions of the limit are easily extended to this case. For instance, using the language of intervals of the "$\epsilon - \delta$" definition, the fact that the left limit of $f(x)$ at x_0 is equal to A (written: $\lim_{x \to x_0 - 0} f(x) = A$), can be formulated as follows: for each $\epsilon > 0$ there exists $\delta > 0$ such that $x_0 - \delta < x < x_0$ implies $|f(x) - A| < \epsilon$.

Problem 2.33 Give the "$\epsilon - \delta$" definition of the right limit, $\lim_{x \to x_0^+} f(x) = A$.

Both the left and right limits are illustrated in Fig. 2.12. It is clear that if the limit of $y = f(x)$ exists at x_0 and is equal to A, then both the left and right limits exist at this point as well and both are equal to the same value A. Inversely, if both the left and right limits exist and both are equal to A, then there exists the limit of $f(x)$ at x_0 also equal to A. If, however, at point x the left and right limits exist, but are different, then at this point the limit of the function $f(x)$ does not exist. The functions $H(x)$ and sgn (x) shown in Fig. 2.11 do not have a limit at $x = 0$, although both their left and right limits exist.

As an example, let us prove that $\lim_{x \to 0+} \sqrt{x} = 0$. This function is only defined for $x \geq 0$ and hence we can only consider the right limit. We need to prove that for any $\epsilon > 0$ one can always find such $\delta > 0$ that $0 < x < \delta$ implies $\left| \sqrt{x} - 0 \right| = \sqrt{x} < \epsilon$. It is clear that this inequality is indeed fulfilled if $\delta = \epsilon^2$.

Problem 2.34 Prove that the left and right limits of the Heaviside function around $x = 0$ are 0 and 1, respectively.

Problem 2.35 Prove that the left and right limits of the sign function around $x = 0$ are -1 and 1, respectively.

Problem 2.36 Prove that the left and right limits of the function $y = |x|$ are the same and equal to 0.

2.4.2 Main Theorems

Similarly to the numerical sequences, one can operate almost algebraically with the limits of functions. This follows immediately from the first definition of the limit of a function based on numerical sequences and the fact that one can operate with sequences algebraically, as we have seen in Sect. 2.2. Therefore, similar to sequences, analogous theorems exist for the limits of the functions.

Let us formulate these theorems here for convenience; we shall do it for the limit at a point, one-side limit theorems are formulated almost identically.

Theorem 2.10 *If the function $y = f(x)$ has a limit at x_0, it is unique.*

It is proven by contradiction.

Theorem 2.11 *If two functions, $f(x)$ and $g(x)$, have limits F and G, respectively, at point x_0, then the functions $f(x) + g(x)$, $f(x)g(x)$ and $f(x)/g(x)$ have definite limits at this point, equal, respectively, to $F + G$, $F \cdot G$ and F/G (it is assumed in the last case that $G \neq 0$).*

These are easily proven using the language of intervals.

Theorem 2.12 *If* $\lim_{x \to x_0} f(x) = A$ *and* $f(x) \geq F$ *(or* $f(x) \leq F$*) in some vicinity of* x_0*, then* $A \geq F$ *(or* $A \leq F$*).*

Proven by contradiction.

Theorem 2.13 *If in some vicinity of* x_0 *the two functions* $f(x) \leq g(x)$*, and both have definite limits at* x_0*, i.e.* $\lim_{x \to x_0} f(x) = F$ *and* $\lim_{x \to x_0} g(x) = G$*, then* $F \leq G$*.*

This is a consequence of the previous theorem since $h(x) = f(x) - g(x) \leq 0$ and $\lim_{x \to x_0} h(x) = F - G \leq 0$.

It then follows from the last two theorems that if $f(x)$ is bounded by two functions, $\phi(x) \leq f(x) \leq h(x)$, and both functions $\phi(x)$ and $h(x)$ tend to the same limit A at $x \to x_0$, so is the function $f(x)$.

All these theorems can be proven similarly to those for numerical sequences. As an example, let us prove the multiplication theorem. Since the two functions have definite limits F and G, then for any positive ϵ_1 and ϵ_2 one can always find such positive δ_1 and δ_2, so that from $|x - x_0| < \delta_1$ follows $|f(x) - F| < \epsilon_1$ and from $|x - x_0| < \delta_2$ follows $|g(x) - G| < \epsilon_2$. Consider now $\delta = \min\{\delta_1, \delta_2\}$ and any value of the x within the interval $|x - x_0| < \delta$. We can then write:

$$|f(x)g(x) - FG| = |(f(x) - F)(g(x) - G) + F(g(x) - G) + G(f(x) - F)|$$
$$< \epsilon_1 \epsilon_2 + |F|\epsilon_2 + |G|\epsilon_1 ,$$

which gives an expression for the ϵ in the right-hand side above, as required.

Problem 2.37 Prove all the above theorems using the "$\epsilon - \delta$" language.

We shall need some other theorems as well.

Theorem 2.14 *Consider a function $f(x)$ that is* limited *within some vicinity of x_0, i.e. $|f(x)| \leq M$ for $|x - x_0| < \delta$. Then, if $\lim_{x \to x_0} g(x) = 0$, then $\lim_{x \to x_0} [f(x)g(x)] = 0$ as well. Note that $f(x)$ may not even have a definite limit at x_0.*

Proof: Since $g(x)$ has the zero limit, then for any x within $|x - x_0| < \delta$ we have $|g(x) - 0| = |g(x)| < \epsilon'$. Therefore, for the same values of x,

$$|f(x)g(x) - 0| = |f(x)g(x)| \leq M\,|g(x)| < M\epsilon',$$

i.e. $\epsilon = M\epsilon'$. **Q.E.D.**

As an example, consider $\lim_{x \to 0} [x \sin(1/x)]$. The sine function is not well-defined at $x = 0$ (in fact, it does not have a definite limit there, recall a discussion on sequences in Sect. 2.4.1); however, it is bounded from below and above, i.e. it is limited: $|\sin(1/x)| \leq 1$. Therefore, since the limit of x at zero is zero, so is the limit of its product with the sine function: $\lim_{x \to 0} [x \sin(1/x)] = 0$.

At variance with the sequences, two more operations exist with functions: the composition and inverse; hence, two more limit theorems follow.

Theorem 2.15 (Limit of a Composition) *Consider two functions $y = f(t)$ and $t = g(x)$; their composition is the function $h(x) = f(g(x))$. If $\lim_{x \to x_0} g(x) = t_0$ and $\lim_{t \to t_0} f(t) = F$, then $\lim_{x \to x_0} h(x) = F$.*

Proof: Since $g(x)$ has a definite limit, then there exists δ such that $|x - x_0| < \delta$ implies $|g(x) - t_0| < \epsilon_1$. On the other hand, $f(t)$ also has a definite limit at t_0, i.e. $|t - t_0| < \delta_1$ implies $|f(t) - F| < \epsilon$. However, since $t = g(x)$, the inequality $|t - t_0| < \delta_1$ is equivalent to $|g(x) - t_0| < \delta_1$. Therefore, by taking $\delta_1 = \epsilon_1$, we can write:

$$|x - x_0| < \delta \;\Rightarrow\; |g(x) - t_0| = |t - t_0| < \delta_1$$
$$\Rightarrow\; |f(t) - F| = |f(g(x)) - F| < \epsilon,$$

proving the theorem, **Q.E.D.**

This result can be used to find limits of complex functions by considering them as compositions of simpler ones, for instance,

$$\lim_{x \to x_0} \sin^2 x = \left(\lim_{x \to x_0} \sin x \right)^2 = \sin^2 x_0 \ .$$

Theorem 2.16 (Limit of the Inverse Function) *Let $f^{-1}(x)$ be an inverse function to $f(x)$; the latter has a limit at x_0 equal to F, i.e. $\lim_{x \to x_0} f(x) = F$. Then, $\lim_{x \to F} f^{-1}(x) = x_0$.*

Proof: If $y = f(x)$ is the direct function, then $x = f^{-1}(y)$. Since the direct function has the limit, then for each ϵ we can find δ such that $|x - x_0| < \delta$ implies $|f(x) - F| = |y - F| < \epsilon$. We need to prove that for any ϵ_1 there exists δ_1 such that for any y satisfying $|y - F| < \delta_1$ we may write $|x - x_0| < \epsilon_1$. By taking $\epsilon_1 = \delta$, we can see from the existence of the limit for the direct function that we can always find $\delta_1 = \epsilon$, as required, **Q.E.D.**

2.4.3 Continuous Functions

Intuitively, it seems that if one can draw a function $y = f(x)$ with a continuous line between $x = a$ and $x = b$, then the function is thought to be *continuous*. A more rigorous consideration, however, requires some care.

One of the definitions of continuity of the function $f(x)$ within an interval $a < x < b$ is that its limit at any point x_0 within this interval coincides with the value of the function there:

$$\lim_{x \to x_0} f(x) = f(x_0) \ . \tag{2.80}$$

It appears, however, that this definition follows from a less obvious statement that it is sufficient for the function $f(x)$ to have a definite limit for any x within some interval to be continuous there. Indeed, let us prove that Eq. (2.80) indeed follows from this statement. We shall prove this by contradiction: assume the opposite, that the limit of $f(x)$ at x_0 is $A \neq f(x_0)$. We start by assuming that $\lim_{x \to x_0} f(x) = A > f(x_0)$. This means that for any $\epsilon > 0$ there exists $\delta > 0$ such that $|x - x_0| < \delta$ implies $|f(x) - A| < \epsilon$. Let us choose the positive ϵ as follows: $\epsilon = \frac{1}{2}(A - f(x_0))$. Then, we have:

$$|f(x) - A| < \epsilon = \frac{1}{2}(A - f(x_0)) \ ,$$

which, if calculated at the point $x = x_0$, gives:

$$|f(x_0) - A| = A - f(x_0) < \epsilon = \frac{1}{2}(A - f(x_0)) \ ,$$

Fig. 2.13 Change of a function Δy tends to zero as change Δx of x tends to zero

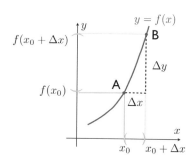

which is obviously wrong, i.e. our assumption of A being larger than $f(x_0)$ is incorrect. Similarly, one can consider the case of $A < f(x_0)$ which also is easily shown to be wrong. Hence, $A = f(x_0)$, **Q.E.D.**

This means that the function $f(x)$ is continuous within some interval of x if it has well-defined limits at any point x belonging to this interval. Moreover, the limit of the continuous function coincides with the value of it at that point.

Yet another feeling for the continuity of a function, one can get if we rewrite Eq. (2.80) in a slightly different form. Let us write $x = x_0 + \Delta x$, then $f(x) = f(x_0 + \Delta x)$ and hence instead of Eq. (2.80) we arrive at:

$$\lim_{\Delta x \to 0} [f(x_0 + \Delta x) - f(x_0)] = \lim_{\Delta x \to 0} \Delta f(x_0) = 0 . \tag{2.81}$$

Here $\Delta x = x - x_0$ is the change of the argument x_0 of the function, while $\Delta f(x_0) = f(x_0 + \Delta x) - f(x_0)$ is the corresponding change of the function, see Fig. 2.13. As Δx becomes smaller, the point B on the graph of $f(x)$ which corresponds to the coordinates $x_0 + \Delta x$ and $f(x_0 + \Delta x)$, moves continuously towards the point A corresponding to x_0 and $f(x_0)$; as Δx becomes smaller, so is $\Delta y = f(x_0 + \Delta x) - f(x_0)$. Clearly, as the limit (2.81) exists for any point x (within some interval), this kind of continuity exists at each such point; as we change x continuously between two values x_1 and $x_2 > x_1$, the function $y = f(x)$ runs continuously over all values between $f(x_1)$ and $f(x_2)$.

Since a continuous function has well-defined limits for all values of x within some interval, the function cannot jump there, i.e. it cannot be discontinuous. This means that the limits from the left and right at each such point x should exist and be equal to the value of the function at that point. If this condition is not satisfied at some point x_0, the function $f(x)$ is not continuous at this point. There are two types of *discontinuities*. If the function $f(x)$ has well defined left and right limits at point x_0, which are not equal to each other,

$$\lim_{x \to x_0^+} f(x) = f(x_0 + 0) \neq \lim_{x \to x_0^-} f(x) = f(x_0 - 0) , \tag{2.82}$$

then it is said that it has the *discontinuity of the first kind*. All other discontinuities are of the *second kind*.

Functions $H(x)$ and $\text{sgn}(x)$ have at $x = 0$ discontinuity of the first kind as both functions have well-defined limits from the left and right of this point. On the other

Fig. 2.14 The hyperbolic
function $y = 1/x$

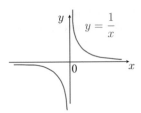

hand, the function $y = 1/x$ has infinite limits on the left (equal to $-\infty$) and on the
right (equal to $+\infty$) of the point $x = 0$, and hence has the discontinuity of the second
kind there, see Fig. 2.14.

Now we are going to establish continuity of all elementary functions. Since this
property is based on the functions having a well defined limit at each point, the limit
theorems of Sect. 2.4.2 would allow us to formulate a set of very simple statements
about the continuity of functions that are invaluable in establishing that property
in complex cases: if two functions $f(x)$ and $g(x)$ are continuous in some interval
of the values of their argument x, then their sum, difference, product and division
($f(x) \pm g(x)$, $f(x)g(x)$ and $f(x)/g(x)$) are also continuous (except at the points of
the roots of $g(x) = 0$ in the latter case). The same can be said about the composition
of the two functions, $y = f(g(x))$. Moreover, if the function $f(x)$ is monotonically
increasing or decreasing and is continuous, then its inverse (which will also be
monotonic) is also continuous.

Let us now look at the elementary functions and establish their continuity.

Polynomials. Consider first a constant function $y = C$. Obviously, this is continuous
as its limit at any $-\infty < x < \infty$ is equal to C which is the value of the function at
any of these points. Then, let us consider an integer power function $y = x^n$. Using
the binomial formula, we obtain for the change of the function at x:

$$\Delta y = (x + \Delta x)^n - x^n = \sum_{i=0}^{n} \binom{n}{i} (\Delta x)^i x^{n-i} - x^n = \sum_{i=1}^{n} \binom{n}{i} (\Delta x)^i x^{n-i} ,$$

$$(2.83)$$

where in the second passage above, we noticed that the very first term in the sum
corresponding to $i = 0$ is exactly equal to x^n since the binomial coefficient $\binom{n}{0} = 1$
for any n (see Sect. 1.13); this means that x^n cancels out exactly. Now it is seen that
each term in the sum above contains powers of Δx, and hence $\Delta y \to 0$ as $\Delta x \to 0$.
This means that the integer power function is continuous.

A polynomial is a linear combination of such functions, i.e. a sum of products
of integer powers with constant functions. Since arithmetic operations involving
continuous functions result in a continuous function, any polynomial is a continuous
function.

General power functions. Consider now $y = x^\alpha$, a general power function, where
α is an arbitrary real number. The proof of this function being continuous follows the
same logic, we used to define this function in the first place in Sect. 2.3.3. We first

consider $\alpha = 1/n$ with n being a positive integer. The function $y = x^{1/n}$ is an inverse to $y = x^n$ and hence is continuous. The function $y = x^{n/m}$ (with $\alpha = n/m$ being a positive rational number) is a composition of two functions just considered, $y = \left(x^{1/m}\right)^n$, and hence is also a continuous function. Then, consider α being an irrational number. Following the same method as in Sect. 2.3, we consider two numerical sequences α_i' and α_i'' $(i = 1, 2, 3, \ldots)$ of rational numbers bracketing the irrational α for each i (i.e. $\alpha_i' < \alpha < \alpha_i''$) and converging to α from both sides. Correspondingly, the two numerical sequences $x^{\alpha_i'}$ and $x^{\alpha_i''}$ bracket the power x^α for each i and any value of $x \geq 0$, i.e. $x^{\alpha_i'} < x^\alpha < x^{\alpha_i''}$. Consequently, due to statements proven in Sect. 2.4.2, x^α has a certain limit which is the same as the limit of both sequences, i.e. we arrive at the conclusion that $\lim_{x \to x_0} x^\alpha = x_0^\alpha$, i.e. any positive real power function is indeed continuous.

Next, the function $y = 1/x$ is obviously continuous for any $x \neq 0$:

$$\Delta y = \frac{1}{x + \Delta x} - \frac{1}{x} = -\frac{\Delta x}{x(x + \Delta x)}$$

$$\implies \lim_{\Delta x \to 0} \Delta y = - \lim_{\Delta x \to 0} \frac{\Delta x}{x(x + \Delta x)} = - \lim_{\Delta x \to 0} \frac{\Delta x}{x^2} = 0 .$$

Since for negative power functions ($\alpha < 0$) one can always write $y = x^\alpha = 1/x^{-\alpha}$, which is a composition of two continuous functions (note that $-\alpha > 0$), negative power functions are also continuous. We finally conclude that the power function with any real power α is continuous.

Trigonometric functions. We have already considered in Sect. 2.4.1.1, the limit of the sine function, $y = \sin x$, and proved that its limit at $x \to x_0$ is equal to $\sin x_0$, i.e. to the value of the sine function at this point. This proves that the sine function is continuous. The cosine function $y = \cos x$ can be composed out of the sine function, e.g. $y = \sin(\pi/2 - x)$, and hence is also continuous. The tangent and cotangent are obtained via division of two continuous functions, sine and cosine, and hence are also continuous.

Inverse trigonometric functions. These are also continuous since they are inverse of the corresponding trigonometric functions which are continuous.

Exponential functions. For any positive $a > 0$ we can write

$$\lim_{\Delta x \to 0} \left(a^{x + \Delta x} - a^x\right) = \lim_{\Delta x \to 0} a^x \left(a^{\Delta x} - 1\right)$$

$$= a^x \lim_{\Delta x \to 0} \left(a^{\Delta x} - 1\right) = a^x \left(\lim_{\Delta x \to 0} a^{\Delta x} - 1\right) = 0 ,$$

since, as we have seen above in Eq. (2.78), $\lim_{\Delta x \to 0} a^{\Delta x} = 1$. Therefore, the exponential function $y = a^x$ is continuous.

Logarithmic function. This is continuous as an inverse of the continuous exponential function.

Hyperbolic functions. These functions are derived from the exponential functions via a finite number of arithmetic operations and hence are all continuous.

Problem 2.38 Prove that the sine and cosine functions are continuous also by showing directly that their change $\Delta y \to 0$ as the change of the variable $\Delta x \to 0$.

Problem 2.39 Prove that the logarithmic function is continuous by demonstrating directly that its change $\Delta y \to 0$ as the change of the variable $\Delta x \to 0$.

Problem 2.40 Prove by direct calculation of Δy that the hyperbolic sine, cosine and tangent functions are continuous.

Concluding this section, we see that any function constructed either as a composition and/or via any number of algebraic operations from the elementary functions considered above is continuous for all values of x where all functions are defined.

2.4.4 Several Famous Theorems Related to Continuous Functions

There is a number of (actually quite famous) theorems that clarify the meaning of continuity of functions. We shall formulate them and give an idea of the proof without going into subtle mathematical details in some cases.

Theorem 2.17 (Conservation of Sign) *If a function $f(x)$ is continuous around point x_0, then there always exists such vicinity of x_0 where $f(x)$ has the same sign as $f(x_0)$.*

Proof: If the function is continuous, then for any x from a vicinity of x_0 we have $\lim_{x \to x_0} f(x) = f(x_0)$. This means that for any x satisfying $|x - x_0| < \delta$ with $\delta > 0$ we have $|f(x) - f(x_0)| < \epsilon$, which in turn means that $-\epsilon + f(x_0) < f(x) < \epsilon + f(x_0)$. Consider first the case of $f(x_0) > 0$. Then by choosing $\epsilon = f(x_0)$ one obtains $0 < f(x) < 2f(x_0)$, i.e. within (possibly) small vicinity $|x - x_0| < \delta$ of x_0 we have $f(x)$ positive. If, however, $f(x_0) < 0$, then we choose $\epsilon = -f(x_0) > 0$ yielding $-2f(x_0) < f(x) < 0$, i.e. within some vicinity of x_0 the function $f(x)$ is negative. **Q.E.D.**

Theorem 2.18 (The First Theorem of Bolzano–Cauchy—Existence of a Root)
If a continuous function $f(x)$ has different signs at the boundaries c and d of the interval $c \leq x \leq d$, then there exists a point x_0 belonging to the same interval where $f(x_0) = 0$.

Proof: Suppose for definiteness that $f(c) < 0$ and $f(d) > 0$, see Fig. 2.15. The idea of this "constructive" proof is to build a sequence of smaller and smaller intervals around x_0 such that it converges to the point x_0 itself after an infinite number of steps (if not earlier).[4] Choose the point x_1 in the middle of the interval, $x_1 = (c + d)/2$. If $f(x_1) = 0$, then $x_0 \equiv x_1$ and we stop here. Otherwise, we have to analyse the value of the function at the new point: if $f(x_1) < 0$, then we accept x_1 as the new left boundary c_1 of the interval; if $f(x_1) > 0$, we replace the right boundary d_1 with x_1. This way we arrive at a new interval $c_1 \leq x \leq d_1$, which is two times shorter, i.e. its length is $l_1 = (d - c)/2$. Then, we repeat the previous step: choose the middle point $x_2 = (c_1 + d_1)/2$ and by analysing the value of the function at point x_2 we either stop (and the theorem is proven) or construct a new interval $c_2 \leq x \leq d_2$ by replacing one of the boundary points, so that the function would still have different signs at them. The length $l_2 = l_1/2 = (d - c)/2^2$ of the new interval is again two times shorter. Repeating this process, we construct a (possibly) infinite sequence of intervals $c_n \leq x \leq d_n$ of reducing lengths $l_n = (d - c)/2^n$. It is essential that,

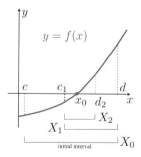

Fig. 2.15 To the "constructive" proof of the first theorem of Bolzano–Cauchy. We start with the original interval X_0 between points c and d. The first middle point between c and d is c_1, which replaces the left boundary, i.e. after the first step we have the new interval $c_1 \leq x \leq d_1$ (denoted X_1) with $c_1 = (c + d)/2$ and $d_1 = d$. At the second step, we divide the new interval into two equal parts again and choose the middle point d_2 to replace the right boundary thereby arriving at the next interval $c_2 \leq x \leq d_2$ (denoted X_2) with $c_2 = c_1$ and $d_2 = (c_1 + d_1)/2$. Continuing this process leads to narrower and narrower intervals sequence $X_0, X_1, X_2, \ldots$, which all contain the point x_0

---

[4] In fact, this so-called 'division by two' method may actually be used in practical numerical calculations for finding roots of functions $f(x)$.

by construction, all of them have their boundaries of different signs and their length becomes smaller and smaller tending to zero as $1/2^n$. Hence, in the limit, the intervals would shrink to a point, let us call it x_0, at which the function is equal to zero. The fact that such a point exists follows from the previous theorem of conservation of sign of a function: if we assume that $f(x_0) \neq 0$, it is either positive or negative and hence there will exist an interval around x_0 where the functions keeps its sign; this, however, contradicts our construction upon which the function changes its sign at the boundaries; we have to accept that $f(x_0) = 0$. **Q.E.D.**

Theorem 2.19 (The Second Theorem of Bolzano–Cauchy—Existence of Any Intermediate Value) *Suppose a continuous function $f(x)$ at the boundaries of the interval $c \leq x \leq d$ has values $f(c) = C$ and $f(d) = D$. If A is any number between C and D, then there exists a point a within the interval, $c \leq a \leq d$, where $f(a) = A$.*

Problem 2.41 Prove this theorem by defining a new function $g(x) = f(x) - A$ and then use the "existence of root" theorem.

Basically, this theorem states that the function $f(x)$ takes on all intermediate values between C and D within the interval. In other words, one can draw the graph of the function between c and d without lifting the pen off the paper.

Theorem 2.20 (Existence of a Limit at a Region Boundary for Monotonically Increasing Functions) *Consider a function $f(x)$ which is monotonically increasing within some interval $a \leq x < b$, i.e. for any $x_1 < x_2$ inside the interval $f(x_1) < f(x_2)$. If the function is limited from above within the interval, $f(x) \leq M$, then when $x \to b$ the function has a limit which is equal to M.*

Proof: (See Fig. 2.16) Let us choose some positive ϵ. Since the function $f(x)$ increases monotonically, one can always find such value x_1 that $f(x_1) > M - \epsilon$. This value of x_1 will be by some δ on the left from the boundary value b, i.e. $x_1 = b - \delta$. Then, for any $x > x_1$, i.e. for any x satisfying $|x - b| < \delta$ (and $x < b$), we would have $f(x) > f(x_1) > M - \epsilon$. However, on the other hand, $f(x) \leq M < M + \epsilon$, so that

$$M - \epsilon < f(x) < M + \epsilon \implies |f(x) - M| < \epsilon ,$$

Fig. 2.16 To the proof of
Theorem 2.20: the
monotonically increasing
function $f(x)$ is limited
within the interval
$a \le x < b$ by the number M

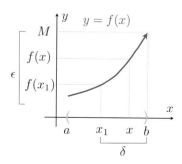

as required, i.e. the limit of the function exists and is equal to the value of M. **Q.E.D.**

Problem 2.42 Prove the analogous theorem for a monotonically increasing function at its left boundary (i.e. for $f(x) > M > 0$), as well as for a monotonically decreasing function at either of its boundaries.

2.4.5 Infinite Limits and Limits at Infinities

The limit of a function may not necessarily be finite. If for any $\epsilon > 0$, there exists $\delta > 0$ such that for any x within $|x - x_0| < \delta$ one has[5] $f(x) > \epsilon$, then it is said that the function $f(x)$ tends to $+\infty$ as $x \to x_0$. Indeed, if for any vicinity of x_0 (small δ) one can always find values of the $f(x)$ larger than any given ϵ (however large), then the function $f(x)$ grows indefinitely to infinity.

Problem 2.43 Formulate the definition for the function to tend to $-\infty$ when $x \to x_0$.

We can also consider behaviour of a function at both infinities, i.e. when x_0 is either $+\infty$ or $-\infty$. These are basically one-side limits, which are formulated as follows. The function $f(x)$ tends to F when $x \to +\infty$, if for any ϵ there exists $\delta > 0$ such that for any $x > \delta$ we have[6] $|f(x) - F| < \epsilon$.

---

[5] Note that here ϵ is not longer assumed "small", it is "large".
[6] δ is assumed to be indefinitely "large" here.

Problem 2.44 Formulate definitions for the function $f(x)$ to have: (i) a finite limit F when $x \to -\infty$, (ii) an infinite limit $F = +\infty$ (or $-\infty$) when $x \to +\infty$ (or $-\infty$).

Problem 2.45 Prove the analogue of Theorem 2.20 of the previous subsection for the limits at $+\infty$ and $-\infty$ for both monotonically increasing and decreasing functions.

Consider, for instance, the function $f(x) = 1/x^2$ at $x \to 0$. In this case $x_0 = 0$; we see that for any x within $|x - 0| = |x| < \delta$ we shall have $f(x) = 1/x^2 > 1/\delta^2 \equiv \epsilon$, i.e. one should take $\delta = 1/\sqrt{\epsilon}$ in this case. And indeed, the function grows indefinitely as we approach $x = 0$ either from the left or from the right.

Now, if $f(x) = x^2$, then at $x \to +\infty$ it tends to $+\infty$ as well. Indeed, for any $x > \delta$ we have $f(x) = x^2 > \delta^2 \equiv \epsilon$, i.e. for any ϵ we can indeed find $\delta = \sqrt{\epsilon}$.

Similarly, one can formulate infinite limits for a function at $x \to \pm\infty$.

Problem 2.46 Prove that the left and right limits of the function $y = 3^{1/x}$ at $x = 0$ are 0 and ∞, respectively, and hence this function does not have a definite limit at $x = 0$.

Problem 2.47 Prove using the "$\epsilon - \delta$" language that $\lim_{x \to +\infty} e^{-x^2}/x = 0$.

Let us prove that

$$\lim_{x \to +\infty} \left(1 + \frac{1}{x}\right)^x = e \qquad (2.84)$$

This is a generalisation to any real x of the special cases, we considered above in Sect. 2.3.4 for the base of the natural logarithm, the number e. To prove this result, we note that any positive x can be placed between two *consecutive* integers: $n \leq x \leq n + 1$. Then,

$$\frac{1}{n+1} \leq \frac{1}{x} \leq \frac{1}{n} \implies 1 + \frac{1}{n+1} \leq 1 + \frac{1}{x} \leq 1 + \frac{1}{n}.$$

Since the power function is a monotonically increasing function, we can also write:

$$\left(1 + \frac{1}{n+1}\right)^n \leq \left(1 + \frac{1}{x}\right)^x \leq \left(1 + \frac{1}{n}\right)^{n+1}$$

or

$$\frac{\left(1 + \frac{1}{n+1}\right)^{n+1}}{1 + \frac{1}{n+1}} \le \left(1 + \frac{1}{x}\right)^x \le \left(1 + \frac{1}{n}\right)^n \left(1 + \frac{1}{n}\right) .$$

As we now tend x to $+\infty$, n will also tend to the same limit. However, from Eq. (2.24) it follows that both limits on the left and right of the last inequality tend to the same limit e:

$$\frac{\left(1 + \frac{1}{n+1}\right)^{n+1}}{1 + \frac{1}{n+1}} \rightarrow \left(1 + \frac{1}{n+1}\right)^{n+1} \rightarrow e$$

and

$$\left(1 + \frac{1}{n}\right)^n \left(1 + \frac{1}{n}\right) \rightarrow \left(1 + \frac{1}{n}\right)^n \rightarrow e ,$$

so that, the function $\left(1 + \frac{1}{x}\right)^x$ will tend to the same limit. **Q.E.D.**

Problem 2.48 Prove that

$$\lim_{x \to +\infty} \left(1 + \frac{y}{x}\right)^x = e^y , \tag{2.85}$$

where y is any real number (including $y < 0$).

2.4.6   Dealing with Uncertainties

The case for the number e considered above is an example of an uncertainty. Indeed, $1 + 1/x$ tends to 1 as $x \to +\infty$, however, when raised in an infinite power $x \to +\infty$ the result is not 1, but the number $e = 2.718\ldots$. Several other types of uncertainties exist and often happen in applications, so we shall consider them here.

Firstly, let us consider yet another famous limit:

$$\lim_{x \to 0} \frac{\sin x}{x} = 1 . \tag{2.86}$$

There is an uncertainty: indeed, both $\sin x$ and x tend to zero as x tends to zero, so that we encounter the case of $0/0$ here. The result of the limit, as we shall see imminently is finite and equal to one. The starting point is the inequality (1.196), $\sin x < x < \tan x$. Dividing both sides of this inequality by $\sin x$ (and assuming $0 < x < \pi/2$), we get:

$$1 < \frac{x}{\sin x} < \frac{\tan x}{\sin x} = \frac{1}{\cos x} \quad \text{or} \quad \cos x < \frac{\sin x}{x} < 1 .$$

As $x \to 0$, the cosine tends to 1, i.e. our function of interest becomes bracketed between two equal limits of one, and hence tends to the same limit itself. **Q.E.D.**

The following examples deal with other cases of uncertainties.

Uncertainty ∞/∞:

$$\lim_{x \to +\infty} \frac{2x^2 - 1}{x^2 + 2x + 1} = \lim_{x \to +\infty} \frac{2 - 1/x^2}{1 + 2/x + 1/x^2} = \frac{2 - 0}{1 + 0 + 0} = 2 .$$

Uncertainty $\infty - \infty$:

$$\lim_{x \to +\infty} \left(\sqrt{x^2 - 1} - \sqrt{x + 1} \right) = \lim_{x \to +\infty} \left[\left(\sqrt{x^2 - 1} - \sqrt{x + 1} \right) \frac{\sqrt{x^2 - 1} + \sqrt{x + 1}}{\sqrt{x^2 - 1} + \sqrt{x + 1}} \right]$$

$$= \lim_{x \to +\infty} \frac{(x^2 - 1) - (x + 1)}{\sqrt{x^2 - 1} + \sqrt{x + 1}}$$

$$= \lim_{x \to +\infty} \frac{x^2 - x - 2}{\sqrt{x^2 - 1} + \sqrt{x + 1}}$$

$$= \lim_{x \to +\infty} \frac{x - 1 - 2/x}{\sqrt{1 - 1/x^2} + \sqrt{1/x + 1/x^2}}$$

$$= \lim_{x \to +\infty} \frac{x - 1}{1 + 0} = +\infty .$$

Uncertainty $0/0$:

$$\lim_{x \to 0} \frac{\ln(1 + x)}{x} = \lim_{x \to 0} \left[\frac{1}{x} \ln(1 + x) \right] = \lim_{x \to 0} \ln(1 + x)^{1/x} = \left| \begin{array}{c} \text{change} \\ y = 1/x \end{array} \right|$$

$$= \lim_{y \to \infty} \ln \left(1 + \frac{1}{y} \right)^y = \ln \left[\lim_{y \to \infty} \left(1 + \frac{1}{y} \right)^y \right] = \ln e = 1 .$$

Note that here we were able to take the limit inside the logarithm since the latter is a continuous function.

Uncertainty $0 \cdot \infty$:

$$\lim_{x \to 0} (x \cot x) = \lim_{x \to 0} \left(x \frac{\cos x}{\sin x} \right) = \left(\lim_{x \to 0} \frac{x}{\sin x} \right) \left(\lim_{x \to 0} \cos x \right)$$

$$= \left(\lim_{x \to 0} \frac{\sin x}{x} \right)^{-1} \cdot 1 = 1 .$$

Problem 2.49 Prove the following limits:

$$\lim_{x \to +\infty} \frac{5^x - 5^{-x}}{5^x + 5^{-x}} = 1 ; \quad \lim_{x \to 1 + 0} \left(\frac{3}{x - 1} - \frac{x + 3}{x^2 - 1} \right) = +\infty ;$$

$$\lim_{x \to 1-0} \left(\frac{3}{x-1} - \frac{x+3}{x^2-1} \right) = -\infty \; ; \quad \lim_{x \to +\infty} \left(\sqrt{x} - \sqrt{x+2} \right) = 0 \; ;$$

$$\lim_{x \to 0} \frac{1 - \cos x}{x^2} = \frac{1}{2} \; .$$

2.4.7 Partial Fraction Decomposition Revisited

Using limits may simplify finding coefficients in the decomposition of fractions. It is probably best discussing this method using examples. Let us consider decomposition of the fraction

$$\phi(x) = \frac{x^2 + 1}{x^3 + 2x^2 + x} \; .$$

First of all, we need to determine the roots of the denominator. Obviously, these are 0 and 1 (repeated twice), i.e. the denominator is $x(x+1)^2$. According to Theorem 2.9, the decomposition should be as follows:

$$\phi(x) = \frac{A}{x} + \frac{B}{x+1} + \frac{C}{(x+1)^2} \; . \tag{2.87}$$

Let us multiply both sides of this equation by x,

$$x\phi(x) = A + \frac{Bx}{x+1} + \frac{Cx}{(x+1)^2} \; ,$$

which gives the constant A in the right-hand side on its own, and take the limit $x \to 0$ on both sides. Obviously, in the right-hand side, we get simply A, while in the left-hand side the limit is

$$\lim_{x \to 0} \frac{(x^2+1)x}{x^3 + 2x^2 + x} = \lim_{x \to 0} \frac{x^2+1}{x^2 + 2x^2 + 1} = 1 \implies A = 1 \; .$$

Next, we shall eliminate B. For that, let us multiply both sides of Eq. (2.87) by $(x+1)$,

$$(x+1)\phi(x) = A\frac{x+1}{x} + B + \frac{C}{x+1} \; ,$$

and take the limit $x \to \infty$. In the right-hand side we get $A + B$, while in the left:

$$\lim_{x \to \infty} (x+1)\phi(x) = \lim_{x \to \infty} \frac{x^2+1}{x(x+1)} = \lim_{x \to \infty} \frac{1 + 1/x^2}{1 + 1/x} = 1 \implies A + B = 1 \; ,$$

and hence $B = 0$. Finally, to find C, both sides of Eq. (2.87) are multiplied by $(x + 1)^2$,

$$(x + 1)^2 \phi(x) = A \frac{(x + 1)^2}{x} + B(x + 1) + C,$$

and then we take the limit $x \to -1$. In the right-hand side we get C, while in the left:

$$\lim_{x \to -1} (x + 1)^2 \phi(x) = \lim_{x \to -1} \frac{x^2 + 1}{x} = -2 \implies C = -2.$$

Therefore,

$$\frac{x^2 + 1}{x^3 + 2x^2 + x} = \frac{1}{x} - \frac{2}{(x + 1)^2}.$$

In another example, we shall decompose a fraction with complex pair of roots in the denominator:

$$\frac{1}{x^3 + 1} = \frac{1}{(x + 1)(x^2 - x + 1)} = \frac{A}{x + 1} + \frac{Bx + C}{x^2 - x + 1}.$$

Multiplying both sides by $(x + 1)$ and taking the limit $x \to -1$, we obtain $1/3$ in the left-hand side and A in the right, so that $A = 1/3$. Multiplying both sides by $(x + 1)$ and taking this time the limit $x \to \infty$, we get 0 in the left and $A + B$ in the right-hand side, yielding $B = -1/3$. Finally, taking the limit of $x \to 0$ in the above expression (simply considering $x = 0$ in both sides), enables one to calculate $C = 1 - A = 2/3$. Hence, we obtain:

$$\frac{1}{x^3 + 1} = \frac{1/3}{x + 1} + \frac{-x/3 + 2/3}{x^2 - x + 1}.$$

In many cases, this method is the simplest one as allows avoiding cumbersome calculations.

Problem 2.50 Show using this method that

$$\frac{1}{x^3 - 1} = \frac{1/3}{x - 1} + \frac{-x/3 - 2/3}{x^2 + x + 1}.$$

Problem 2.51 Prove the decomposition:

$$\frac{2x + 3}{(x^2 - 1)(x + 3)} = -\frac{1/4}{x + 1} + \frac{5/8}{x - 1} - \frac{3/8}{x + 3}.$$

Part II
Basics

Derivatives

3

3.1   Definition of the Derivative

Very often we would like to know an *instantaneous* rate of change of something. For instance, consider a one-dimensional motion of a particle along the x-axis: at each time t we have the position of the particle given by a function $x(t)$. If at the initial moment $t = 0$ the particle was at the position $x_0 = x(0)$, then at time t it arrives at the point with the coordinate $x(t)$. If the particle moves with a constant velocity v_c, then its coordinate changes according to $x(t) = x_0 + v_c t$. This means that in this case the rate of change of $x(t)$, i.e. the velocity, can be calculated via $v_c = (x - x_0)/t$. If, however, the particle travels with different velocities along its trajectory, i.e. at some instances travels faster and at some slower, then this expression would only give an average velocity $v_{av} = (x - x_0)/t$. Although we do very often use this notion of slower or faster motion (smaller or bigger velocity) when describing a single continuous motion and seem to understand intuitively its meaning, we probably subconsciously mean average velocities over some rather short periods of time during the trajectory. For instance, we can split the whole journey time t into N equal slices of length Δt, i.e. $t_0 \equiv 0, t_1 = \Delta t, t_2 = 2\Delta t$, etc., $t_N = N\Delta t \equiv t$ (i.e. $t_i = i\Delta t$ with $\Delta t = t/N$ and $i = 0, 1, 2, \ldots, N$), when positions of the particle are correspondingly given by its coordinates $x_0, x_1, x_2, \ldots, x_i, \ldots, x_N \equiv x_t$, and then define average velocities

$$v_{av,i} = \frac{x_{i+1} - x_i}{\Delta t} = \frac{x(t_{i+1}) - x(t_i)}{\Delta t} = \frac{x(t_i + \Delta t) - x(t_i)}{\Delta t} \tag{3.1}$$

for each time interval $\Delta t = t_{i+1} - t_i$. Of course, this calculation would be able to tell us if the velocity changed along the journey; however, the actual quantitative answer would obviously depend on the number of time slices N we used: the more slices we use, the more correct this calculation would be. This simple minded exercise brings us to a method which would be able to determine the instantaneous velocity of the particle at any time along the journey: one has to tend the number of slices N to infinity ($N \to \infty$), which is equivalent to taking the time length of each slice $\Delta t \to 0$. Therefore, we arrive at the following conclusion: if we would like to calculate the velocity $v(t)$ at some time t along the journey, we need to consider the positions of the particle $x(t)$ and $x(t + \Delta t)$ at the times t and $t + \Delta t$, and then calculate

$$v(t) \simeq \frac{x(t + \Delta t) - x(t)}{\Delta t}$$

(compare Eq. (3.1)). This expression becomes more and more precise as Δt gets smaller; it becomes exact in the limit of $\Delta t \to 0$.

This kind of reasoning can be applied to many phenomena around us. Consider electrons flowing in a thin wire with cross section S. Again, let us split the time into small and equal slices Δt and count the number of electrons ΔN_i passing through S during each slice $\Delta t = t_{i+1} - t_i$. This way we can approximately calculate the electric current passing through the wire at time t_i as $J_i \simeq e \Delta N_i / \Delta t$ (e is the electron charge). This will be an average current; of course, the smaller the time slices, the better our calculation would be and, if the current changes with time, we should be able to reproduce this dependence in detail. If $N(t)$ is the total number of electrons passed through S up to time t, then their total number passed over time Δt is $\Delta N = N(t + \Delta t) - N(t)$, and we arrive again at a formula for the current "around time t" as

$$J(t) \simeq \frac{N(t + \Delta t) - N(t)}{\Delta t} ,$$

which becomes more and more exact as the time slices Δt get smaller and smaller; in the limit of $\Delta t \to 0$ this formula becomes exact.

Similarly, one can define a linear density of a metal rod by dividing the rod of length L into N slices each of length Δx (so that $L = N \Delta x$), and calculating the mass Δm_i of each i-th slice; then the average density of the slice located near x_i would be

$$\rho_i \simeq \frac{\Delta m_i}{\Delta x} = \frac{m(x_i + \Delta x) - m(x_i)}{\Delta x} ,$$

where $m(x)$ is the total mass of the rod between points $x = 0$ and $x > 0$. Again, the formula becomes exact if we tend $\Delta x \to 0$ which corresponds to the infinite number of length slices. Exactly, in the same way, one can define the linear charge density of a charged rod via the limit of its total charge within a thin slice divided by its length.

Although many more examples can be given (and the reader will find them in this book, as well as in any physics textbook), all of them can be summarised as follows: if we have a function $y = y(x)$ and we are interested in its *rate of change* at the point x, we calculate the limit

$$y'(x) = \lim_{\Delta x \to 0} \frac{\Delta y}{\Delta x} = \lim_{\Delta x \to 0} \frac{y(x + \Delta x) - y(x)}{\Delta x} , \tag{3.2}$$

called the *derivative* of the function $y(x)$ at point x. Here $\Delta y = y(x + \Delta x) - y(x)$ is the change of the function $y(x)$ over the change Δx of its argument from x to $x + \Delta x$. Sometimes, the derivative is also denoted as y' or y'_x; in the latter case, the index x emphasises specifically that the derivative is taken with respect to the variable x.

Example 3.1 ▶ Calculate the derivative of the following functions: $y = c$ (a constant), $y = x$ and $y = x^2$.

Solution. In the first case $\Delta y = c - c = 0$ and hence the derivative is zero, i.e. $(c)' = 0$. In the second case

$$\Delta y = y(x + \Delta x) - y(x) = (x + \Delta x) - x = \Delta x,$$

so that $\Delta y / \Delta x = \Delta x / \Delta x = 1$ and hence $(x)' = 1$ for any value of the x. In the last case, we need to calculate first the change of the function:

$$\Delta y = y(x + \Delta x) - y(x) = (x + \Delta x)^2 - x^2 = 2x\Delta x + (\Delta x)^2.$$

Therefore, the derivative of $y = x^2$ is

$$\left(x^2\right)' = \lim_{\Delta x \to 0} \frac{\Delta y}{\Delta x} = \lim_{\Delta x \to 0} \frac{2x\Delta x + (\Delta x)^2}{\Delta x} = \lim_{\Delta x \to 0} (2x + \Delta x) = 2x. \blacktriangleleft \quad (3.3)$$

Note that the change of the function due to the change Δx of its argument in this case,

$$\Delta y = 2x\Delta x + (\Delta x)^2 = y'(x)\Delta x + \alpha(\Delta x)\Delta x, \quad (3.4)$$

is composed of a linear part containing the derivative and an additional term, which tends to zero faster than linearly (in fact, as the square of Δx in this case) since $\alpha(\Delta x) = \Delta x$ is some function, which tends to zero as $\Delta x \to 0$.

Problem 3.1 Show using the binomial expansion (see Sect. 1.13) that

$$\left(x^n\right)' = nx^{n-1} \quad (3.5)$$

for any integer $n \geq 1$.

The derivative of a function has also a very simple geometric interpretation. Consider a function $y(x)$ plotted in Fig. 3.1a. Let us try to find an equation $y = a + kx$ for the tangent line (shown in dashed green) at the point A with coordinates x and $y(x)$. Our interest here is in obtaining the *slope* of the curve $y = y(x)$ at point A which is given by the constant k (it will, in general, depend on the x). The constant a can then be easily determined using the known coordinates of the point A. We shall

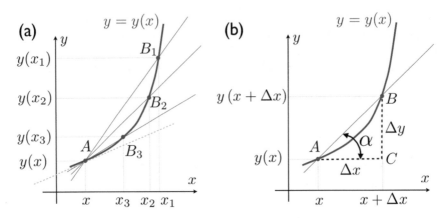

Fig. 3.1 For the determination of the tangent line to a curve of the function $y = y(x)$ at point A

be using the following procedure: let us first pick an arbitrary point B_1 on the curve of the function and draw a line connecting this point with the point A. This line may have a very different direction to the one we are seeking; however, its slope given by the tangent of AB_1 to the x-axis may already serve as the first approximation to the constant k. Let us now take a point B_2 on the graph of the function which is somewhere between the points B_1 and A; it is now closer to the point of interest. This time the line AB_2 may serve as a better approximation to the tangent line: although it still crosses the curve of the function in two close points A and B_2 (in the vicinity of A the tangent line will only *touch* the curve at the point A; of course, it may cross the curve at some other more distant point(s) if the curve sufficiently changes its direction, e.g. oscillates); still this straight line AB_2 is a much better approximation to the tangent line than the line AB_1. This process can be continued: we take a point B_3 on the curve lying between A and B_2, then a point B_4 between A and B_3, etc., each time taking the point B_N closer and closer to the point A and thereby obtaining better and better approximation to the tangent line. In the limit when the point B_N tends to the point A, our procedure would give the correct tangent line at the point A. It is easy to see that this process is equivalent to taking a point B on the curve of the function $y(x)$, see Fig. 3.1b, and moving it towards A. Therefore, to obtain the slope k at point A (or at x), we can calculate the tangent of the line AB,

$$\tan \alpha = \frac{y(x + \Delta x) - y(x)}{\Delta x} = \frac{\Delta y}{\Delta x},$$

and then take the limit as $\Delta x \to 0$ (which would exactly correspond to the point B moving towards the point A). So, in the limit, we shall get the exact value of the slope k of the tangent line at point A. Again, as one can see, we relate the slope to the derivative of the function $y(x)$ at point x, i.e. $k = y'(x)$.

Similarly to the right and left limits defined in the previous Chapter, we can also define left and right derivatives, denoted $y'_-(x) = y'(x - 0)$ and $y'_+(x) = y'(x + 0)$, respectively. In fact, the derivative defined above is exactly the right derivative as it contains only the values of the function on the right of x, the point of interest. The left derivative is defined only via points on the left of x as follows:

$$y'_-(x) = \lim_{\Delta x \to 0} \frac{y(x) - y(x - \Delta x)}{\Delta x} , \tag{3.6}$$

where $\Delta x > 0$. This expression can be written in a form very similar to that of the right derivative in Eq. (3.2) by assuming that $\Delta x < 0$ and tends to zero always being negative (this is denoted as $\Delta x \to 0 - 0$ or simply $\Delta x \to -0$):

$$y'_-(x) = \lim_{|\Delta x| \to 0} \frac{y(x) - y(x - |\Delta x|)}{|\Delta x|} \equiv \lim_{\Delta x \to -0} \frac{y(x + \Delta x) - y(x)}{\Delta x} \tag{3.7}$$

If the derivative of $y(x)$ exists at point x, then both the right and left derivatives coincide there; otherwise, the derivative does not exist.

Problem 3.2 Consider a line tangent to a curve of the function $y = y(x)$ at point x_0 which is given by the equation $y = k_\tau (x - x_0) + y_0$, with $k_\tau = y'(x_0)$ and $y_0 = y(x_0)$. Prove that the line perpendicular to the tangent line and passing through the same point (x_0, y_0) is given by the equation $y = k_\perp (x - x_0) + y_0$, with $k_\perp = -1/k_\tau = -1/y'(x_0)$.

3.2 Main Theorems

The notion of the derivative of a function is closely related to its continuity. To establish this relationship, let us first consider the change Δy of the function $y(x)$ at point x due to change Δx of its argument in a bit more detail. If the derivative $y'(x)$ exists (in other words, if the limit (3.2) exists), then one can write:

$$\Delta y = y'(x)\Delta x + \alpha \Delta x , \tag{3.8}$$

where α is an explicit function of Δx (we have seen an example of this relation above, Eq. (3.4), when considering the square function). It is essential that α tends to zero as $\Delta x \to 0$. Indeed, only in this case does the change Δy of y above yields the derivative:

$$\lim_{\Delta x \to 0} \frac{\Delta y}{\Delta x} = y'(x) + \lim_{\Delta x \to 0} \alpha = y'(x) .$$

The formula (3.8) states, quite importantly, that for any Δx the change of the function is in the first approximation given by the product of its derivative and Δx; the correction term, equal to $\alpha \Delta x$, is of a higher order with respect to Δx, and for small Δx may be neglected. Therefore, the main part of the change of the function, Δy, due to the change Δx of its argument x can be written as $\Delta y \simeq y'(x)\Delta x$. This change is called the *differential* of the function $y = y(x)$ and is denoted dy. Using dx

for the differential of the argument (this definition is consistent with the differential of the linear function $y = x$, as for it $dy = (x)' \Delta x = \Delta x$ and, since $y = x$, we have $dx = \Delta x$ as well), we can write:

$$dy = y'(x)dx \ . \tag{3.9}$$

For instance, if $y = x^2$, then $dy = \left(x^2\right)' dx = 2xdx$ or simply $d\left(x^2\right) = 2xdx$. Using the notion of differentials of the function and its argument, we can also write the derivative of the function as their ratio:

$$y'(x) \equiv \frac{dy}{dx} \ . \tag{3.10}$$

This formula has a clear meaning of the slope of the tangent line to the curve of the function $y(x)$ as using differentials already implies performing the limit $\Delta x \to 0$. In addition, the ratio of differentials also serves as a very convenient notation for the derivative which we will frequently use.

Example 3.2 ▶ Derive an explicit expression for the function $\alpha\left(\Delta x\right)$ for $y = x^n$ with positive integer $n \geq 1$.

Solution. using the binomial expansion, we write (see Eq. (2.83))

$$\Delta y = (x + \Delta x)^n - x^n = \sum_{k=0}^{n} \binom{n}{k} x^{n-k} (\Delta x)^k - x^n = \sum_{k=1}^{n} \binom{n}{k} x^{n-k} (\Delta x)^k$$

$$= \binom{n}{1} x^{n-1} \Delta x + \sum_{k=2}^{n} \binom{n}{k} x^{n-k} (\Delta x)^k$$

$$= nx^{n-1} \Delta x + \left[\sum_{k=2}^{n} \binom{n}{k} x^{n-k} (\Delta x)^{k-1} \right] \Delta x \ ,$$

which is of the required form (3.8) since $(x^n)' = nx^{n-1}$, see Eq. (3.5). Therefore, the expression in the square brackets must be the function $\alpha\left(\Delta x\right)$ we seek for; it is clearly seen that it tends to zero as $\Delta x \to 0$ (the sum starts from $k = 2$).◀

Now we are ready to formulate the relationship between the continuity of a function and the existence of its derivative:

Theorem 3.1 *If a function $y = y(x)$ has a derivative at point x, it is continuous there.*

Proof: Indeed, if $y'(x)$ exists, then one can write expression (3.8) for Δy. Taking the limit as $\Delta x \to 0$ we immediately obtain that $\Delta y \to 0$ as well at this point, which is one of the definitions of the continuity of a function, as explained in Sect. 2.4.3. **Q.E.D.**

Note that the inverse statement is not true. Indeed, consider the function $y = |x|$. This function is continuous, since the limits from left and right at $x = 0$ give one and the same number 0, and this coincides with the value of the function at $x = 0$. At the same time, the derivative at $x = 0$ does not exist. Indeed, for $x > 0$ the function is $y = x$ and hence $y' = (x)' = 1$. On the other hand, for $x < 0$ we have $y = -x$ and hence

$$y' = (-x)' = \frac{[-(x + \Delta x)] - (-x)}{\Delta x} = -1 .$$

We see that the derivatives from left and right around $x = 0$ are different, and hence the derivative at $x = 0$ does not exist.

Theorem 3.2 (Sum rule)*: if $y(x) = u(x) \pm v(x)$, then $y'(x) = u'(x) \pm v'(x)$,* i.e.

$$[u(x) \pm v(x)]' = u'(x) \pm v'(x) . \tag{3.11}$$

Proof: Consider the function $y(x)$ which can be written as a sum of two functions $u(x)$ and $v(x)$. Then, according to the definition of the derivative (3.2), we can write:

$$\frac{\Delta y}{\Delta x} = \frac{1}{\Delta x} [u(x + \Delta x) + v(x + \Delta x) - u(x) - v(x)]$$

$$= \frac{u(x + \Delta x) - u(x)}{\Delta x} + \frac{u(x + \Delta x) - u(x)}{\Delta x} = \frac{\Delta u}{\Delta x} + \frac{\Delta v}{\Delta x} .$$

After taking the limit as $\Delta x \to 0$, we obtain the required result. Similarly, one proves $(u - v)' = u' - v'$. **Q.E.D.**

Theorem 3.3 (Product rule) *If $y(x) = u(x)v(x)$, then*

$$y'(x) = [u(x)v(x)]' = u'(x)v(x) + u(x)v'(x) . \tag{3.12}$$

Proof: We need to calculate the change of the function $y(x)$ due to the change Δx of its argument:

$$\Delta y = u\,(x + \Delta x)\,v\,(x + \Delta x) - u\,(x)\,v\,(x)$$
$$= [u\,(x + \Delta x) - u\,(x)]\,v\,(x + \Delta x) + u\,(x)\,[v\,(x + \Delta x) - v\,(x)]$$
$$= (\Delta u)\,v\,(x + \Delta x) + u\,(x)\,\Delta v\ .$$

Dividing Δy by Δx and taking the limit, we obtain the required result:

$$\lim_{\Delta x \to 0} \frac{\Delta y}{\Delta x} = \lim_{\Delta x \to 0}\left[\frac{\Delta u}{\Delta x}v\,(x + \Delta x)\right] + \lim_{\Delta x \to 0}\left[u\,(x)\frac{\Delta v}{\Delta x}\right]$$
$$= \lim_{\Delta x \to 0}\frac{\Delta u}{\Delta x}\lim_{\Delta x \to 0}v\,(x + \Delta x) + u\,(x)\lim_{\Delta x \to 0}\frac{\Delta v}{\Delta x}$$
$$= u'(x)v\,(x) + u\,(x)\,v'\,(x)\ .$$

Q.E.D.

Problem 3.3 Prove that derivative of a linear combination of two functions is equal to the linear combination of their derivatives, i.e.

$$[\alpha u\,(x) + \beta v\,(x)]' = \alpha u'\,(x) + \beta v'\,(x)\ . \qquad (3.13)$$

Problem 3.4 Generalise the product rule for more than two functions:

$$[u_1(x)u_2\,(x)\cdots u_n\,(x)]' = u_1'u_2\cdots u_n + u_1u_2'\cdots u_n + \cdots + u_1u_2\cdots u_n'$$
$$\equiv \sum_{i=1}^{n}\left[u_i'\prod_{j=1,j\neq i}^{n}u_j\right]\ , \qquad (3.14)$$

where for each term in the sum over i the product, indicated by the symbol $\prod$, is taken over all values of j between 1 and n except for $j = i$.

Theorem 3.4 (Quotient rule) *If $y(x) = u(x)/v(x)$ and $v(x) \neq 0$, then*

$$y'(x) = \left[\frac{u\,(x)}{v\,(x)}\right]' = \frac{1}{v^2}\left[u'v - uv'\right]\ . \qquad (3.15)$$

Proof: Here we write

$$\Delta y = \frac{u\,(x+\Delta x)}{v\,(x+\Delta x)} - \frac{u\,(x)}{v\,(x)} = \frac{u\,(x+\Delta x)\,v\,(x) - u\,(x)\,v\,(x+\Delta x)}{v\,(x+\Delta x)\,v\,(x)}$$

$$= \frac{v(x)\,[u\,(x+\Delta x) - u\,(x)] - u\,(x)\,[v\,(x+\Delta x) - v\,(x)]}{v\,(x+\Delta x)\,v\,(x)}$$

$$= \frac{v\,(x)\,\Delta u - u\,(x)\,\Delta v}{v\,(x+\Delta x)\,v\,(x)}\ ,$$

which, after dividing by Δx and taking the limit, gives:

$$\lim_{\Delta x \to 0} \frac{\Delta y}{\Delta x} = \frac{v(x)\left(\lim_{\Delta x \to 0} \frac{\Delta u}{\Delta x}\right) - u\,(x)\left(\lim_{\Delta x \to 0} \frac{\Delta v}{\Delta x}\right)}{\lim_{\Delta x \to 0}\,[v\,(x+\Delta x)\,v\,(x)]} = \frac{vu' - uv'}{v^2}\ .$$

Q.E.D.

Problem 3.5 Show that the formula (3.5) for the derivative of positive powers can be generalised to negative integer powers as well, i.e. that $(x^n)' = nx^{n-1}$ for any n from $\mathbb{Z}$ (i.e. $n = 0, \pm 1, \pm 2, \pm 3, \ldots$).

Theorem 3.5 (Chain rule) *Consider a composite function $y = u\,(v)$ and $v = v\,(x)$, i.e. $y = u\,(v\,(x))$. Then, its derivative is obtained as follows:*

$$[u\,(v\,(x))]' = u'_v v'\ , \tag{3.16}$$

where u'_v corresponds to the derivative of the function $u\,(v)$ with respect to v (not x).

Proof: Again, we need to calculate the change Δy of the function $y(x)$:

$$\Delta y = u\,(v\,(x+\Delta x)) - u\,(v\,(x))\ .$$

Here $v\,(x+\Delta x) = \Delta v + v\,(x)$ and we can similarly write $u\,(v + \Delta v) = \Delta_v u + u\,(v)$, where $\Delta_v u$ is the corresponding change of $u\,(v)$ when v changes from v to $v + \Delta v$. Therefore,

$$\frac{\Delta y}{\Delta x} = \frac{u\,(v\,(x+\Delta x)) - u\,(v\,(x))}{\Delta x} = \frac{u\,(v + \Delta v) - u\,(v)}{\Delta x}$$

$$= \frac{u\,(v + \Delta v) - u\,(v)}{\Delta v}\,\frac{\Delta v}{\Delta x} = \frac{\Delta_v u}{\Delta v}\,\frac{\Delta v}{\Delta x}\ ,$$

so that, after taking the limit, we arrive at Eq. (3.16). **Q.E.D.**

Problem 3.6 Using the chain rule, show that $\left[u\,(x)^{-1}\right]' = -u^{-2}u'$. Does this formula agree with the quotient rule (3.15)? Then, using this result and the product rule, prove the quotient rule in a general case.

Problem 3.7 A one-dimensional movement of a particle of mass m along the x-axis is governed by the Newtonian equation of motion, $\dot{p} = -dU(x)/dx$, where $x(t)$ is the particle coordinate, $U(x)$ is its potential energy, $p = m\dot{x}$ its momentum, and dots above x and p denote the time derivative. Show that the total energy $H = p^2/2m + U(x)$ is conserved in time. (The function H is also called Hamiltonian of the particle.) [Hint: *assume that both p and x are some functions of time which would make H a composite function, so that you may use the chain rule.*]

Theorem 3.6 (Inverse function rule). *If $y = y(x)$ is the inverse of $x = x(y)$, then the derivatives of these two functions are related simply by*

$$[y(x)]' = \frac{1}{x'(y)} \,. \tag{3.17}$$

Proof: Here

$$y' = \lim_{\Delta x \to 0} \frac{\Delta y}{\Delta x} = \lim_{\Delta x \to 0} \frac{1}{\Delta x/\Delta y} = \frac{1}{\lim_{\Delta x \to 0} \frac{\Delta x}{\Delta y}} \,.$$

Since the direct function $x = x(y)$ is continuous, $\Delta x \to 0$ corresponds to $\Delta y \to 0$ as well, so that the limit in the denominator of the above formula can be replaced with $\Delta y \to 0$, which would give the derivative $x'(y) \equiv x'_y$ of $x(y)$ with respect to y, which proves the theorem. **Q.E.D.**

As an example, let us derive a formula for the derivative of the square root function $y = \sqrt{x}$. Indeed, in this case, $y(x)$ is the inverse of the function $x(y) = y^2$. Therefore, $x'_y = 2y$. Here, we have used (3.5) for the derivative of the integer power function ($n = 2$ in our case). Hence, $y' = 1/x'_y = 1/(2y) = 1/(2\sqrt{x})$. Note that this result can also be written as $\left(x^{1/2}\right)' = \frac{1}{2}x^{-1/2}$, which formally agrees with Eq. (3.5) used for $n = 1/2$.

Problem 3.8 Show, using a combination of various rules for differentiation discussed above, that formula (3.5) proved for positive powers can in fact be generalised to any rational powers as well, i.e. that $(x^\alpha)' = \alpha x^{\alpha-1}$, where $\alpha = p/q$ is a rational number (p and q are positive and/or negative integers, $q \neq 0$). [Hint: *the rational power $x^{p/q}$ can always be considered as a composite function $(x^{1/q})^p$ with positive q. Prove first, using the inverse function rule, that $(x^{1/q})' = \frac{1}{q}x^{1/q-1}$. Then, combine this result with the chain rule to extend Eq. (3.5) to all rational powers.*]

3.3 Derivatives of Elementary Functions

The rules for differentiation considered above provide us with enough muscle to tackle any function written in analytical form. But first, we need to derive formulae for the derivatives of all elementary functions.

General power function. $y = x^\alpha$ (α is a real number).

We have already considered positive and negative values of α and proved that in these cases $(x^\alpha)' = \alpha x^{\alpha-1}$. Using the definition of the inverse function and the inverse function rule, we also discussed that this formula is valid for root powers $1/q$ ($q \neq 0$) as well, and then, employing the chain rule, we extended this result for any rational power $\alpha = p/q$. It is clear then that it must be valid for any real powers α since, as we already discussed in Sects. 2.3 and 2.4.3, any irrational number can be bracketed by two infinite sequences of rational numbers converging to it, and so this can be done for the derivative as well. So, for any real α,

$$\left(x^\alpha\right)' = \alpha x^{\alpha-1} . \tag{3.18}$$

Problem 3.9 Prove that

$$n2^{n-1} = \sum_{k=1}^{n} k \binom{n}{k} .$$

[Hint: *differentiate the binomial expansion of $(1+x)^n$ with respect to x.*]

Logarithmic functions. $y = \log_a x$ and $y = \ln x$.

We write:

$$\left(\log_a x\right)' = \lim_{\Delta x \to 0} \frac{1}{\Delta x} \left[\log_a (x + \Delta x) - \log_a x\right] = \lim_{\Delta x \to 0} \frac{1}{\Delta x} \log_a \frac{x + \Delta x}{x}$$

$$= \lim_{\Delta x \to 0} \frac{1}{\Delta x} \log_a \left(1 + \frac{\Delta x}{x}\right) = \lim_{\Delta x \to 0} \log_a \left(1 + \frac{\Delta x}{x}\right)^{1/\Delta x} .$$

Denoting $\lambda = 1/\Delta x$, which tends to infinity as $\Delta x \to 0$, we can rewrite the limit as

$$\left(\log_a x\right)' = \lim_{\lambda \to \infty} \log_a \left(1 + \frac{1/x}{\lambda}\right)^{\lambda} = \log_a \left[\lim_{\lambda \to \infty} \left(1 + \frac{1/x}{\lambda}\right)^{\lambda}\right] .$$

We have already calculated the limit inside the square brackets before in Eq. (2.85), and it is equal to $e^{1/x}$. Therefore,

$$\left(\log_a x\right)' = \log_a e^{1/x} = \frac{1}{x} \log_a e = \frac{1}{x \ln a} . \tag{3.19}$$

In the last step, we made use of Eq. (2.40) to express $\log_a e$ via the logarithm with respect to the base e, i.e. $\ln e = \log_a e \ln a$, which immediately gives the required relationship due to the fact that $\ln e = 1$. In a very important case of the base $a \equiv e$, we then obtain:

$$(\ln x)' = \frac{1}{x} . \tag{3.20}$$

Exponential functions. $y = a^x$ and $y = e^x$.

To work out the derivative of $y = a^x$, it is convenient to recall that it can be considered as an inverse of the logarithmic function $x = \log_a y$. Therefore, using Eq. (3.17), we immediately have:

$$\left(a^x\right)' = \frac{1}{\left(\log_a y\right)'_y} = \frac{1}{1/(y \ln a)} = y \ln a = a^x \ln a . \tag{3.21}$$

In particular, if $a \equiv e$, we obtain a very important result:

$$\left(e^x\right)' = e^x , \tag{3.22}$$

i.e. the derivative of the e^x function is equal to itself. It is the only function that has this peculiar property.

Problem 3.10 Prove formula (3.18) by representing $y = x^\alpha$ as

$$y = e^{\ln(x^\alpha)} = e^{\alpha \ln x} .$$

This is another way of proving the formula for the derivative of the power function with arbitrary real power α.

Trigonometric functions. Let us first consider the sine function:

$$\Delta y = \sin(x + \Delta x) - \sin x = 2 \cos\left(x + \frac{\Delta x}{2}\right) \sin \frac{\Delta x}{2} ,$$

where we have used Eq. (2.49) for the difference of two sine functions. Therefore,

$$
\begin{aligned}
(\sin x)' &= \lim_{\Delta x \to 0} \frac{\Delta y}{\Delta x} = \lim_{\Delta x \to 0} \left[\cos\left(x + \frac{\Delta x}{2}\right) \frac{\sin(\Delta x/2)}{\Delta x/2} \right] \\
&= \left[\lim_{\Delta x \to 0} \cos\left(x + \frac{\Delta x}{2}\right) \right] \left[\lim_{\Delta x \to 0} \frac{\sin(\Delta x/2)}{\Delta x/2} \right] \\
&= \cos x \left[\lim_{\lambda \to 0} \frac{\sin \lambda}{\lambda} \right] ,
\end{aligned}
$$

where in the last passage we replaced the $\Delta x \to 0$ limit with the equivalent $\lambda = \Delta x/2 \to 0$ limit. The last limit we have already seen before in Eq. (2.86), and it is equal to 1. Therefore, we finally obtain:

$$(\sin x)' = \cos x . \tag{3.23}$$

The derivative of the cosine function can be obtained using the chain rule. Indeed, $\cos x = \sin(\pi/2 - x)$, i.e. $y = \cos x$ can be written as $y = \sin v$ and $v = \pi/2 - x$, so that

$$
\begin{aligned}
(\cos x)' &= (\sin v)'_v \left(\frac{\pi}{2} - x\right)' = (\cos v)(-1) = -\cos v \\
&= -\cos\left(\frac{\pi}{2} - x\right) = -\sin x .
\end{aligned}
\tag{3.24}
$$

Problem 3.11 Prove the same formula using directly the definition of the derivative and formula (2.51).

Problem 3.12 Prove that

$$(\tan x)' = \frac{1}{\cos^2 x} \tag{3.25}$$

and

$$(\cot x)' = -\frac{1}{\sin^2 x} \qquad (3.26)$$

using the definitions of the tangent and cotangent functions and the basic rules for differentiation.

Inverse trigonometric functions. We shall first consider $y = \arcsin x$. It is the inverse function to $x = \sin y$. Therefore, using Eq. (3.17), we obtain:

$$(\arcsin x)' = \frac{1}{x'_y} = \frac{1}{(\sin y)'_y} = \frac{1}{\cos y} = \frac{1}{\cos (\arcsin x)} = \frac{1}{\sqrt{1 - x^2}} , \qquad (3.27)$$

where we have used Eq. (2.74).

Problem 3.13 Prove that

$$(\arccos x)' = -\frac{1}{\sqrt{1 - x^2}} , \qquad (3.28)$$

$$(\arctan x)' = \frac{1}{1 + x^2} . \qquad (3.29)$$

You may need identities (2.74) and (2.76) here.

Hyperbolic functions. These functions are composed of exponential functions, and one can use the rules of differentiation to calculate the required derivatives. As an example, let us consider the hyperbolic sine, $y = \sinh(x)$:

$$[\sinh(x)]' = \left[\frac{1}{2} \left(e^x - e^{-x} \right) \right]' = \frac{1}{2} \left[\left(e^x \right)' - \left(e^{-x} \right)' \right]$$
$$= \frac{1}{2} \left(e^x + e^{-x} \right) = \cosh(x) . \qquad (3.30)$$

Problem 3.14 Prove the following formulae:

$$[\cosh(x)]' = \sinh(x) , \qquad (3.31)$$

$$[\tanh(x)]' = \frac{1}{\cosh^2(x)} , \qquad (3.32)$$

$$[\coth(x)]' = -\frac{1}{\sinh^2(x)} \, . \tag{3.33}$$

Problem 3.15 Prove the following formulae for the inverse hyperbolic functions:

$$\left[\sinh^{-1}(x)\right]' = \frac{1}{\sqrt{x^2 + 1}} \, ,$$

$$\left[\cosh^{-1}(x)\right]' = \frac{1}{\sqrt{x^2 - 1}} \, ,$$

$$\left[\tanh^{-1}(x)\right]' = \left[\coth^{-1}(x)\right]' = \frac{1}{1 - x^2} \, .$$

By comparing these equations with the corresponding formulae for the trigonometric functions, one can appreciate their similarity. The reason for this is quite profound and lies in the fact, proven in theory of complex functions (Vol. II), that the exponential and trigonometric functions are closely related.

Problem 3.16 A gas is stable under an isothermal compression (the temperature T is constant) if the derivative of its pressure P with respect to the volume V is negative. Show that for the van der Waals gas of n moles satisfying the equation

$$P = \frac{nRT}{V - nb} - \frac{n^2}{V^2}a,$$

where R is the universal gas constant, this condition reads: $T > 2an(V - nb)^2/RV^3$.

3.4   Complex Numbers Revisited

3.4.1   Multiplication and Division of Complex Numbers

Here, we shall continue our discussion of complex numbers started in Sect. 1.11. First of all, let us use the trigonometric form of the complex number, Eq. (1.126), to demonstrate that upon multiplication of two complex numbers,

$$z_1 = r_1 (\cos \phi_1 + i \sin \phi_1) \quad \text{and} \quad z_2 = r_2 (\cos \phi_2 + i \sin \phi_2) \, , \tag{3.34}$$

their phases add up and absolute values multiply, while upon division —their phases are subtracted and absolute values are divided.

Indeed, consider first multiplication:

$$z_1 z_2 = r_1 r_2 (a + ib) ,$$

where

$$a = \cos \phi_1 \cos \phi_2 - \sin \phi_1 \sin \phi_2 \quad \text{and} \quad b = \sin \phi_1 \cos \phi_2 + \cos \phi_1 \sin \phi_2 .$$

Using Eqs. (2.46) and (2.47) for the trigonometric functions we derived earlier, we immediately see that $a = \cos (\phi_1 + \phi_2)$ and $b = \sin (\phi_1 + \phi_2)$, i.e.

$$z_1 z_2 = r_1 r_2 [\cos (\phi_1 + \phi_2) + i \sin (\phi_1 + \phi_2)] , \qquad (3.35)$$

which proves the made statement. Division of two complex numbers is worked out using a similar argument:

$$\frac{z_1}{z_2} = \frac{r_1}{r_2} \frac{\cos \phi_1 + i \sin \phi_1}{\cos \phi_2 + i \sin \phi_2} .$$

Multiplying the denominator and numerator by $\cos \phi_2 - i \sin \phi_2$ and realising that in the denominator we then simply get

$$(\cos \phi_2 + i \sin \phi_2) (\cos \phi_2 - i \sin \phi_2) = \cos^2 \phi_2 + \sin^2 \phi_2 = 1 ,$$

we obtain:

$$\frac{z_1}{z_2} = \frac{r_1}{r_2} (\cos \phi_1 + i \sin \phi_1) (\cos \phi_2 - i \sin \phi_2) = \frac{r_1}{r_2} (a + ib) ,$$

with (see Eqs. (2.45) and (2.47))

$$a = \cos \phi_1 \cos \phi_2 + \sin \phi_1 \sin \phi_2 = \cos (\phi_1 - \phi_2) ,$$

$$b = \sin \phi_1 \cos \phi_2 - \cos \phi_1 \sin \phi_2 = \sin (\phi_1 - \phi_2) ,$$

so that

$$\frac{z_1}{z_2} = \frac{r_1}{r_2} [\cos (\phi_1 - \phi_2) + i \sin (\phi_1 - \phi_2)] . \qquad (3.36)$$

So, indeed, the phases are subtracted and the absolute values are divided.

For instance, consider $z_1 = 1 + i$ and $z_2 = -1 - i$. In this case $r_1 = r_2 = \sqrt{2}$, $\cos \phi_1 = \sin \phi_1 = 1/\sqrt{2}$, so that $\phi_1 = \frac{\pi}{4}$, and $\cos \phi_2 = \sin \phi_2 = -1/\sqrt{2}$ and hence $\phi_2 = \frac{5\pi}{4}$; we see that $z_1 z_2$ has the absolute value $\sqrt{2}\sqrt{2} = 2$ and the phase $\phi_1 + \phi_2 = \frac{3\pi}{2}$. This results in the complex number

$$z_1 z_2 = 2 \left(\cos \frac{3\pi}{2} + i \sin \frac{3\pi}{2} \right) = 2 (0 - i) = -2i ,$$

as it supposed to be, as a direct calculation shows: $(1 + i)(-1 - i) = -2i$. Similarly, z_1/z_2 has the absolute value $r_1/r_2 = 1$ and the phase $\phi_1 - \phi_2 = -\pi$, so that

$$\frac{z_1}{z_2} = \cos(-\pi) + i \sin(-\pi) = -1 .$$

Again, a direct calculation confirms this result: $(1 + i)/(-1 - i) = -(1 + i)/(1 + i) = -1$.

Problem 3.17 Using the trigonometric form, calculate the absolute values r and the phases ϕ of the complex numbers $z_1 = \left(\frac{1}{2} - i\frac{1}{2}\right)\left(\frac{1}{2} + i\frac{\sqrt{3}}{2}\right)$ and $z_2 = \left(\frac{1}{2} - i\frac{1}{2}\right)/\left(\frac{1}{2} + i\frac{\sqrt{3}}{2}\right)$. [Answer: $r_1 = 1/\sqrt{2}$, $\phi_1 = \pi/12$ and $r_2 = 1/\sqrt{2}$, $\phi_2 = -7\pi/12$.]

3.4.2 Moivre Formula

Let us now consider a square of a complex number with unit absolute value $r = 1$:

$$z^2 = (\cos\phi + i \sin\phi)^2 = \cos(2\phi) + i \sin(2\phi) ,$$

where we made use of the multiplication rule (3.35). Multiplying the square with z one more time would obviously lead to

$$z^3 = (\cos\phi + i \sin\phi)^3 = \cos(3\phi) + i \sin(3\phi) .$$

It looks plausible now that for any positive integer n the following identity is valid:

$$(\cos\phi + i \sin\phi)^n = \cos(n\phi) + i \sin(n\phi) . \tag{3.37}$$

This famous result known as the de Moivre's formula can be proven by induction.

Problem 3.18 Use induction to prove Eq. (3.37).

Hence, an integer power of a complex number can be calculated via

$$z^n = r^n [\cos(n\phi) + i \sin(n\phi)] . \tag{3.38}$$

For instance, let us calculate $\left(\sqrt{3} + i\right)^{10}$. This calculation can of course be accomplished using the binomial formula, but Eq. (3.38) is much simpler; we should only transform the complex number $z = \sqrt{3} + i$ into its trigonometric form first. We have: $r = \sqrt{3+1} = 2$, $\cos\phi = \sqrt{3}/2$ and $\sin\phi = 1/2$, so that the phase $\phi = \pi/6$. Hence,

$$\left(\sqrt{3}+i\right)^{10} = 2^{10}\left[\cos\left(10\frac{\pi}{6}\right) + i\sin\left(10\frac{\pi}{6}\right)\right] = 2^{10}\left(\cos\frac{5\pi}{3} + i\sin\frac{5\pi}{3}\right)$$

$$= 2^{10}\left(\cos\frac{\pi}{3} - i\sin\frac{\pi}{3}\right) = 2^{10}\left(\frac{1}{2} - i\frac{\sqrt{3}}{2}\right).$$

Problem 3.19 Prove the following identities:

$$\sum_{k=0}^{n}\binom{n}{k}\cos(k\phi) = 2^n\cos^n\frac{\phi}{2}\cos\frac{n\phi}{2},$$

$$\sum_{k=0}^{n}\binom{n}{k}\sin(k\phi) = 2^n\cos^n\frac{\phi}{2}\sin\frac{n\phi}{2}.$$

[Hint: *consider* $(1 + \cos\phi + i\sin\phi)^n$.]

3.4.3 Root of a Complex Number

Using the trigonometric form of a complex number and formula (3.38), it is easy to extend our discussion of the square root of Sect. 1.11.3 to an arbitrary integer root. We shall limit ourselves only to this here; a general function z^a for any real or even complex a will be considered in the second volume of this course.

The n-th root $w = \sqrt[n]{z}$ of a complex number z is defined via $w^n = z$. Hence, if z has the absolute value r and the phase ϕ, the absolute value R and the phase ψ of the complex number w are determined from $R^n = r$ and $n\psi = \phi$. Note, however, that the phase ϕ is only determined up to 2π, i.e. in the equation for ψ we can add $2\pi k$, where $k = 0, \pm1, \pm2, \ldots$:

$$n\psi = \phi + 2\pi k \implies \psi = \frac{\phi}{n} + \frac{2\pi k}{n}.$$

It is easy to see that one can use only n possible values of k, e.g. $k = 0, 1, \ldots, n-1$, to get distinct values of the phase ψ; any other value of k would give the phase ψ that has already been chosen. For instance, $k = n$ yields $2\pi k/n \to 2\pi n/n = 2\pi$, which does not change the phase ψ and is equivalent to choosing $k = 0$. Hence, there are exactly n possible values of the phase and hence n possible values of the n-th root of z:

$$\sqrt[n]{z} = \sqrt[n]{r}\left[\cos\frac{\phi + 2\pi k}{n} + i\sin\frac{\phi + 2\pi k}{n}\right], \quad k = 0, 1, 2, \ldots, n-1. \quad (3.39)$$

In the particular case of the square root, $n = 2$, we recover our formulae (1.138) and (1.139). Consider also the third, $n = 3$,

$$\left(\sqrt[3]{z}\right)_1 = \sqrt[3]{r}\left[\cos\frac{\phi}{3} + i\sin\frac{\phi}{3}\right],$$

$$\left(\sqrt[3]{z}\right)_2 = \sqrt[3]{r}\left[\cos\frac{\phi + 2\pi}{3} + i\sin\frac{\phi + 2\pi}{3}\right],$$

$$\left(\sqrt[3]{z}\right)_3 = \sqrt[3]{r}\left[\cos\frac{\phi + 4\pi}{3} + i\sin\frac{\phi + 4\pi}{3}\right] = \sqrt[3]{r}\left[\cos\frac{\phi - 2\pi}{3} + i\sin\frac{\phi - 2\pi}{3}\right],$$

and the fourth, $n = 4$, cases:

$$\left(\sqrt[4]{z}\right)_1 = \sqrt[4]{r}\left[\cos\frac{\phi}{4} + i\sin\frac{\phi}{4}\right],$$

$$\left(\sqrt[4]{z}\right)_2 = \sqrt[4]{r}\left[\cos\left(\frac{\phi}{4} + \frac{\pi}{2}\right) + i\sin\left(\frac{\phi}{4} + \frac{\pi}{2}\right)\right]$$
$$= \sqrt[4]{r}\left[-\sin\frac{\phi}{4} + i\cos\frac{\phi}{4}\right],$$

$$\left(\sqrt[4]{z}\right)_3 = \sqrt[4]{r}\left[\cos\left(\frac{\phi}{4} + \pi\right) + i\sin\left(\frac{\phi}{4} + \pi\right)\right]$$
$$= \sqrt[4]{r}\left[-\cos\frac{\phi}{4} - i\sin\frac{\phi}{4}\right] = -\left(\sqrt[4]{z}\right)_1,$$

$$\left(\sqrt[4]{z}\right)_4 = \sqrt[4]{r}\left[\cos\left(\frac{\phi}{4} + \frac{3\pi}{2}\right) + i\sin\left(\frac{\phi}{4} + \frac{3\pi}{2}\right)\right]$$
$$= \sqrt[4]{r}\left[\cos\left(\frac{\phi}{4} - \frac{\pi}{2}\right) + i\sin\left(\frac{\phi}{4} - \frac{\pi}{2}\right)\right]$$
$$= \sqrt[4]{r}\left[\sin\frac{\phi}{4} - i\cos\frac{\phi}{4}\right] = -\left(\sqrt[4]{z}\right)_2.$$

This means that the fourth root of a complex number z can generally be written as $\pm\left(\sqrt[4]{z}\right)_1, \pm\left(\sqrt[4]{z}\right)_2$.

For instance, let us calculate the cubic root of $z = -i$. In this case $r = 1$ and the phase $\phi = -\pi/2$. Hence, we obtain:

$$\left(\sqrt[3]{-i}\right)_1 = \cos\left(-\frac{\pi}{6}\right) + i\sin\left(-\frac{\pi}{6}\right) = \frac{\sqrt{3}}{2} - i\frac{1}{2},$$

$$\left(\sqrt[3]{-i}\right)_2 = \cos\left(-\frac{\pi}{6} + \frac{4\pi}{6}\right) + i\sin\left(-\frac{\pi}{6} + \frac{4\pi}{6}\right) = \cos\frac{\pi}{2} + i\sin\frac{\pi}{2} = i,$$

$$\left(\sqrt[3]{-i}\right)_3 = \cos\left(-\frac{\pi}{6} + \frac{8\pi}{6}\right) + i\sin\left(-\frac{\pi}{6} + \frac{8\pi}{6}\right)$$

$$= \cos\left(\pi + \frac{\pi}{6}\right) + i\sin\left(\pi + \frac{\pi}{6}\right) = -\frac{\sqrt{3}}{2} - i\frac{1}{2}.$$

Interestingly, even the cubic root of 1 has three values in the manifold of complex numbers.

Problem 3.20 Calculate $\sqrt[3]{1}$. [Answer: $1, -1/2 + i\sqrt{3}/2, and -1/2 - i\sqrt{3}/2$.]

Problem 3.21 Calculate $\sqrt[3]{i}$. [Answer: $-i, \sqrt{3}/2 + i/2, a -\sqrt{3}/2 + i/2$.]

Problem 3.22 Calculate $\sqrt[3]{-1}$ and hence show that $z^3 + 1$ can be factorised as follows:

$$z^3 + 1 = (z + 1)\left(z - \frac{1}{2} - i\frac{\sqrt{3}}{2}\right)\left(z - \frac{1}{2} + i\frac{\sqrt{3}}{2}\right).$$

3.4.4 Exponential Form. Euler's Formula

There is also another, very useful, form of a complex number, an exponential form, which we shall now discuss. Our consideration will be mostly intuitive; a more rigorous approach is given in Vol. II.

Let us have a look at the multiplication of two complex numbers of unit length, Eq. (3.35):

$$(\cos\phi_1 + i\sin\phi_1)(\cos\phi_2 + i\sin\phi_2) = \cos(\phi_1 + \phi_2) + i\sin(\phi_1 + \phi_2).$$

If we define a function

$$f(x) = \cos x + i\sin x, \tag{3.40}$$

then the above equation shows an interesting property of it:

$$f(x + y) = f(x) f(y) , \qquad (3.41)$$

that is valid for any real x and y. There is only one function that has this kind of property—an exponential function $f(x) = e^{kx}$, with a constant k. Indeed, let us differentiate both sides of Eq. (3.41) with respect to x. When differentiating the left-hand side, we use the chain rule:

$$\frac{d}{dx} f(x + y) = \left. \left(\frac{df}{du} \frac{du}{dx} \right) \right|_{u=x+y} = \left. \frac{df}{du} \right|_{u=x+y} ,$$

while differentiating the right-hand side, we simply get $f'(x) f(y)$. Similarly, differentiating both sides of Eq. (3.41) with respect to y, the same expression in the left-hand side is obtained, while in the right-hand side, we get $f(x) f'(y)$. Therefore, both right-hand sides must be equal:

$$f'(x) f(y) = f(x) f'(y) \quad \Longrightarrow \quad \frac{f'(x)}{f(x)} = \frac{f'(y)}{f(y)} .$$

Both sides of this equation depend on their own variables and this equality remains true for any x and y; this can only happen if $f'(x)/f(x)$ and $f'(y)/f(y)$ are equal to the same constant; let us call it k. This gives:

$$\frac{1}{f(x)} \frac{df(x)}{dx} = k \quad \Longrightarrow \quad \frac{d}{dx} (\ln f(x)) = k .$$

Only a linear function $kx + \ln C$ can have its derivative equal to a constant, with C (and hence $\ln C$) being another constant. Therefore, one can write $\ln f(x) = kx + \ln C$ leading to $f(x) = Ce^{kx}$. Substituting this result into Eq. (3.41), we find that $C = C^2$, i.e. $C = 1$. (One may also determine C by using $x = 0$ in the definition (3.40) of the function $f(x)$.) Therefore, $f(x) = e^{kx}$, an exponential function indeed.

What is left to determine is the constant k. For that, we shall differentiate $f(x)$ with respect to x and compare this result with what would have been obtained if we differentiated $f(x)$ from its definition, Eq. (3.40):

$$\frac{df}{dx} = ke^{kx} \quad \text{and} \quad \frac{df}{dx} = -\sin x + i \cos x = i (\cos x + i \sin x) = if(x) = ie^{kx} .$$

Comparing the two, we get $k = i$. So, we have just derived a famous result:

$$e^{ix} = \cos x + i \sin x , \qquad (3.42)$$

which is called Euler's formula. Using this formula, one can also express sine and cosine functions of a real variable via complex exponentials:

$$\cos x = \frac{1}{2} \left(e^{ix} + e^{-ix} \right) \quad \text{and} \quad \sin x = \frac{1}{2i} \left(e^{ix} - e^{-ix} \right) , \qquad (3.43)$$

which also have Euler's name associated with them. It is easy to see that $e^{i2\pi k} = 1$ for any integer k. Another proof of Euler's formula will be given in Sect. 7.3.6.

We have also established that

$$\frac{d}{dx}e^{ix} = ie^{ix}, \tag{3.44}$$

i.e. the complex exponential can be differentiated in the usual way by treating i as a parameter.

Hence, any complex number z can also be written in the exponential form as

$$z = re^{i\phi} \tag{3.45}$$

via its absolute value r and the phase ϕ. The $n-$th root of the complex number can then be written as

$$\sqrt[n]{z} = \sqrt[n]{r} \exp\left(i\frac{\phi + 2\pi k}{n}\right)$$
$$= \sqrt[n]{r}e^{i\phi/n}\left(\cos\frac{2\pi k}{n} + i\sin\frac{2\pi k}{n}\right), \quad k = 0, 1, \ldots, n-1. \tag{3.46}$$

Euler's formulae can be used for expressing sine or cosine of multiple angles $n\alpha$ via angles of smaller multiplicity or even of a single angle as was already discussed in Sect. 2.3.8. As an example, let us express $\sin(5\alpha)$ and $\cos(5\alpha)$ via sine and cosine functions of 3α and 2α. We have:

$$e^{i5\alpha} = \cos(5\alpha) + i\sin(5\alpha) \ .$$

At the same time,

$$e^{i5\alpha} = e^{i3\alpha}e^{i2\alpha} = (\cos(3\alpha) + i\sin(3\alpha))(\cos(2\alpha) + i\sin(2\alpha))$$

$$= [\cos(3\alpha)\cos(2\alpha) - \sin(3\alpha)\sin(2\alpha)] + i[\sin(3\alpha)\cos(2\alpha) + \cos(3\alpha)\sin(2\alpha)] \ .$$

Since the real and imaginary parts of the two expressions must be the same, we obtain:

$$\cos(5\alpha) = \cos(3\alpha)\cos(2\alpha) - \sin(3\alpha)\sin(2\alpha) \ ,$$

$$\sin(5\alpha) = \sin(3\alpha)\cos(2\alpha) + \cos(3\alpha)\sin(2\alpha) \ .$$

Of course, it should not come as a surprise that these formulae follow also directly from Eqs. (2.46) and (2.47) as the Euler's formulae are the consequence of these.

If we would like to have sine and cosine of 5α expressed via trigonometric functions of α, we should similarly split the angle 3α as the sum $2\alpha + \alpha$ and work out sine and cosine decomposition via these angles, and then similarly repeat this for 2α.

Problem 3.23 Re-derive formulae (2.58) and (2.61) using this method.

Problem 3.24 There are general formulae developed by François Viète in sixteenth century for the multiple angle sine and cosine:

$$\cos(n\alpha) = \sum_{k=0\,(\text{even})}^{n} (-1)^{k/2} \binom{n}{k} \cos^{n-k}\alpha \, \sin^k\alpha ,$$

$$\sin(n\alpha) = \sum_{k=1\,(\text{odd})}^{n} (-1)^{(k-1)/2} \binom{n}{k} \cos^{n-k}\alpha \, \sin^k\alpha ,$$

where in the first summation, we use only even values of the summation index k, while in the second—only odd. Derive these using the following method: consider

$$e^{in\alpha} = \cos(n\alpha) + i \sin(n\alpha) .$$

On the other hand, write the same complex exponential as

$$\left(e^{i\alpha}\right)^n = (\cos\alpha + i \sin\alpha)^n$$

and use the binomial formula (1.153). Basically, this method is based on using de Moivre's formula (3.37).

Problem 3.25 Prove again Eqs. (2.63) and (2.64), but this time using Euler formulae and performing summation of the geometric progressions.

3.4.5   Solving Cubic Equation

We have already given a solution to a general cubic equation $c + bx + ax^2 + x^3 = 0$ in Sect. 1.11.1. The solution consists of two steps: first, using the transformation $x = y - a/3$ the equation is transformed into the so-called depressed form $y^3 + py + q = 0$ that does not contain the square term, and second, the solution of the depressed equation is given by Eq. (1.125):

$$y = \sqrt[3]{-\frac{q}{2} + \sqrt{\frac{q^2}{4} + \frac{p^3}{27}}} + \sqrt[3]{-\frac{q}{2} - \sqrt{\frac{q^2}{4} + \frac{p^3}{27}}} . \tag{3.47}$$

If we consider the expressions under the cubic root,

$$u = -\frac{q}{2} + \sqrt{\frac{q^2}{4} + \frac{p^3}{27}} = r_u e^{i\phi_u} \quad \text{and} \quad v = -\frac{q}{2} - \sqrt{\frac{q^2}{4} + \frac{p^3}{27}} = r_v e^{i\phi_v} \, ,$$

then $y = \sqrt[3]{u} + \sqrt[3]{v}$. Each of these cube roots can take three possible values ($k = 0, 1, 2$):

$$\left(\sqrt[3]{u}\right)_k = \sqrt[3]{r_u} e^{i(\phi_u + 2\pi k)/3} = \sqrt[3]{r_u} e^{i\phi_u/3} z_k \, ,$$

where we defined three numbers $z_k = e^{i2\pi k/3}$ that are $z_0 = 1$, $z_1 = -\frac{1}{2} + i\frac{\sqrt{3}}{2}$ and $z_2 = -\frac{1}{2} - i\frac{\sqrt{3}}{2}$. Similarly for $\sqrt[3]{v}$:

$$\left(\sqrt[3]{v}\right)_k = \sqrt[3]{r_v} e^{i\phi_v/3} z_k \, , \quad k = 0, 1, 2 \, .$$

As the $y = \sqrt[3]{u} + \sqrt[3]{v}$, it seems that we have nine combinations to consider for the sum. However, it is easy to see that only three combinations need to be retained. Indeed, not all nine combinations are acceptable since (see Sect. 1.11.1) the following condition must also be satisfied: $\sqrt[3]{u}\sqrt[3]{v} = \sqrt[3]{uv} = -p/3$. Nine values of the quantity $\sqrt[3]{uv}$ are possible:

$$\left(\sqrt[3]{uv}\right)_{kl} = \sqrt[3]{r_u r_v} e^{i(\phi_u + \phi_v)/3} z_k z_l \, , \quad \text{where} \quad k, l = 0, 1, 2 \, .$$

If p and q are both real, two cases need to be discussed, corresponding to two possible signs of the expression $q^2/4 + p^3/27$ under the square root in both u and v. If $q^2/4 + p^3/27 \geq 0$, then both u and v are real numbers, hence their phases $\phi_u = \phi_v = 0$. Therefore,

$$r_u r_v = \left(-\frac{q}{2} + \sqrt{\frac{q^2}{4} + \frac{p^3}{27}}\right)\left(-\frac{q}{2} - \sqrt{\frac{q^2}{4} + \frac{p^3}{27}}\right)$$

$$= \left(-\frac{q}{2}\right)^2 - \left(\frac{q^2}{4} + \frac{p^3}{27}\right) = -\frac{p^3}{27}$$

and hence

$$\left(\sqrt[3]{uv}\right)_{kl} = \sqrt[3]{r_u r_v} z_k z_l = \sqrt[3]{-\frac{p^3}{27}} z_k z_l = -\frac{p}{3} z_k z_l \, ,$$

where $k, l = 0, 1, 2$. We see that only such values of k and l need to be chosen that give $z_k z_l = 1$. It is easy to see that there are three pairs (k, l) that give this: $(0, 0)$,

$(1, 2)$ and $(2, 1)$. Hence, we only have three possible roots that can be written as follows:

$$
y_{1,2,3} = \sqrt[3]{-\frac{q}{2} + \sqrt{\frac{q^2}{4} + \frac{p^3}{27}}} \begin{pmatrix} 1 \\ -1/2 + i\sqrt{3}/2 \\ -1/2 - i\sqrt{3}/2 \end{pmatrix}
$$

$$
+ \sqrt[3]{-\frac{q}{2} - \sqrt{\frac{q^2}{4} + \frac{p^3}{27}}} \begin{pmatrix} 1 \\ -1/2 - i\sqrt{3}/2 \\ -1/2 + i\sqrt{3}/2 \end{pmatrix}. \qquad (3.48)
$$

We see that, in this case, one root is real and the other two represent a pair of two complex conjugate numbers. This agrees with the general statement about the complex roots of a polynomial with real coefficients stated in Sect. 1.11.5.

Consider now the second case, when $q^2/4 + p^3/27 < 0$. In this case $p^3/27 < -q^2/4 < 0$,

$$
u = -\frac{q}{2} + i\sqrt{-\frac{q^2}{4} - \frac{p^3}{27}} = r e^{-i\phi}
$$

and

$$
v = -\frac{q}{2} - i\sqrt{-\frac{q^2}{4} - \frac{p^3}{27}} = r e^{i\phi},
$$

where

$$
r = \sqrt{\left(-\frac{q}{2}\right)^2 + \left(-\frac{q^2}{4} - \frac{p^3}{27}\right)} = \sqrt{-\frac{p^3}{27}}
$$

and

$$
\tan \phi = \frac{-\sqrt{-\frac{q^2}{4} - \frac{p^3}{27}}}{-\frac{q}{2}} = \sqrt{-1 - \frac{4p^3}{27q^2}}
$$

(note that expressions in all square roots here are non-negative). Next, we construct the product

$$
\left(\sqrt[3]{uv}\right)_{kl} = \sqrt[3]{r^2} e^{-i\phi/3} e^{i\phi/3} z_k z_l = \sqrt[3]{-\frac{p^3}{27}} z_k z_l = -\frac{p}{3} z_k z_l .
$$

We see that again only three pairs (k, l) of the indices k and l are possible, the same as in the previous case, yielding only three solutions:

$$
y_{1,2,3} = \sqrt{-\frac{p^3}{27}} \left[e^{-i\phi} \begin{pmatrix} 1 \\ -1/2 + i\sqrt{3}/2 \\ -1/2 - i\sqrt{3}/2 \end{pmatrix} + e^{i\phi} \begin{pmatrix} 1 \\ -1/2 - i\sqrt{3}/2 \\ -1/2 + i\sqrt{3}/2 \end{pmatrix} \right] . \quad (3.49)
$$

In this case, all three roots are real as the expression inside the square brackets contains a sum of two complex conjugate numbers.

Finally, consider the most general case of p and q being both complex. In this case, u and v are complex as well, and their product

$$uv = \left(-\frac{q}{2} + \sqrt{\frac{q^2}{4} + \frac{p^3}{27}}\right)\left(-\frac{q}{2} - \sqrt{\frac{q^2}{4} + \frac{p^3}{27}}\right) = -\frac{p^3}{27}.$$

Hence,

$$\left(\sqrt[3]{uv}\right)_{kl} = \sqrt[3]{-\frac{p^3}{27}} z_k z_l = -\frac{p}{3} z_k z_l$$

and we yet again can only choose three combinations of (k, l), exactly the same as in the two previous cases. Hence, in this most general case, the roots are:

$$y_{1,2,3} = \sqrt[3]{r_u} e^{i\phi_u/3} \begin{pmatrix} 1 \\ -1/2 + i\sqrt{3}/2 \\ -1/2 - i\sqrt{3}/2 \end{pmatrix} + \sqrt[3]{r_v} e^{i\phi_v/3} \begin{pmatrix} 1 \\ -1/2 - i\sqrt{3}/2 \\ -1/2 + i\sqrt{3}/2 \end{pmatrix}.$$

$$(3.50)$$

All three roots in this case will be complex.

Concluding, we see that a general cubic equation always has three roots (some of which may coincide). This agrees with the statement derived in Sect. 1.4.4 from the fundamental theorem of algebra, that the polynomial of order n will always have exactly n roots (some of which may coincide).

As an example, consider the depressed equation $x^3 + 3x + 2 = 0$. In this case $p = 3$ and $q = 2$, so that $q^2/4 + p^3/27 = 2 > 0$ and we can use Eq. (3.48) giving

$$y_{1,2,3} = \sqrt[3]{-1 + \sqrt{2}} \begin{pmatrix} 1 \\ -1/2 + i\sqrt{3}/2 \\ -1/2 - i\sqrt{3}/2 \end{pmatrix} + \sqrt[3]{-1 - \sqrt{2}} \begin{pmatrix} 1 \\ -1/2 - i\sqrt{3}/2 \\ -1/2 + i\sqrt{3}/2 \end{pmatrix}.$$

3.4.6   Solving Quartic Equation

As was mentioned in Sect. 1.11.1, the quartic equation is the highest order algebraic equation for which the solutions can still be written explicitly via radicals. This solution was obtained by Italian mathematician Lodovico Ferrari in 1540.[1] Here, we shall present his method in some detail.

Consider a problem of finding roots of a fourth-order polynomial

$$x^4 + a_3 x^3 + a_2 x^2 + a_1 x + a_0 = 0. \tag{3.51}$$

---

[1] Later Rene Descartes and Leonhard Euler proposed other methods for solving the quartic.

Similarly to the case of the cubic equation, it is also possible to transform the quartic equation above into its depressed form

$$x^4 + px^2 + qx + r = 0 , \tag{3.52}$$

that does not have the cubic term.

Problem 3.26 Applying the change of variable $x \to x + \gamma$ to the quartic equation, show that it is transformed into the depressed form (3.52) by choosing $\gamma = -a_3/4$, and obtain explicit expressions for the coefficients:

$$p = a_2 - \frac{3}{8}a_3^2 ,$$

$$q = a_1 - \frac{1}{2}a_2a_3 + \frac{1}{8}a_3^3$$

and

$$r = a_0 - \frac{1}{4}a_1a_3 + \frac{1}{16}a_2a_3^2 - \frac{3}{256}a_3^4 .$$

In order to solve the depressed form, we shall first rewrite our equation as follows,

$$x^4 + px^2 = -qx - r , \tag{3.53}$$

and then introduce a new variable, $m \neq 0$, which at this point is arbitrary. A certain expression containing m is then added to both sides of Eq. (3.53). To choose this expression, expand the complete square (you may use Eq. (1.11)):

$$\left(x^2 + \frac{p}{2} + m\right)^2 = x^4 + px^2 + \left(\frac{p^2}{4} + m^2 + pm + 2mx^2\right) .$$

This identity suggests that if we add to both sides of Eq. (3.53) an expression contained in the round brackets in the right-hand side of the above, we shall indeed have the complete square in the left-hand side of it. Hence, Eq. (3.53) is equivalently manipulated into the form:

$$\left(x^2 + \frac{p}{2} + m\right)^2 = -qx - r + \frac{p^2}{4} + m^2 + pm + 2mx^2 . \tag{3.54}$$

The next step is based on an observation that the right-hand side represents a quadratic polynomial with respect to x:

$$P_2(x) = 2mx^2 - qx + \left(\frac{p^2}{4} - r + pm + m^2\right) .$$

Generally, this polynomial has two distinct roots

$$x_\pm = \frac{1}{4m}\left(q \pm \sqrt{\mathcal{D}}\right),$$

where

$$\mathcal{D} = q^2 - 8m\left(\frac{p^2}{4} - r + pm + m^2\right)$$

is the corresponding discriminant. The idea is to choose the value of m (which up to this pint is still arbitrary) such that the polynomial becomes a complete square.

Problem 3.27 Show that the polynomial $P_2(x)$ has the form of a complete square,

$$P_2(x) = \left(\sqrt{2m}x - \frac{q^2}{2\sqrt{2m}}\right)^2 \qquad (3.55)$$

if m satisfies the following so-called *resolvent cubic equation*:

$$8m^3 + 8pm^2 - \left(8r - 2p^2\right)m - q^2 = 0. \qquad (3.56)$$

We know from Sects. 1.11.1 and 3.4.5 that this cubic equation always has at least one distinct root. We then choose one such root as the value of m (if p, r and q are all real, at least one real root m is expected in Eq. (3.56); one may choose this root as the value of m). Once the value of m is known, the right-hand side of Eq. (3.54) can be replaced with that of Eq. (3.55):

$$\left(x^2 + \frac{p}{2} + m\right)^2 = \left(\sqrt{2m}x - \frac{q^2}{2\sqrt{2m}}\right)^2$$

$$\implies x^2 + \frac{p}{2} + m = \pm\left(\sqrt{2m}x - \frac{q^2}{2\sqrt{2m}}\right).$$

Therefore, we obtain two independent quadratic equations for x:

$$x^2 \pm \sqrt{2m}x + \left(\frac{p}{2} + m \mp \frac{q}{2\sqrt{2m}}\right) = 0,$$

which give four solutions:

$$x_{1,2} = \frac{1}{2}\left(\sqrt{2m} \pm \sqrt{-2(p+m) - \sqrt{\frac{2}{m}}q}\right)$$

and

$$x_{3,4} = \frac{1}{2}\left(-\sqrt{2m} \pm \sqrt{-2\,(p+m) + \sqrt{\frac{2}{m}}\,q}\right).$$

The obtained formulae formally solve the problem.

Depending on the coefficients of the quartic equations, the roots could be either real, complex or both. It is possible to establish under which conditions they are real and/or complex, but we shall not dwell into this here.

3.5 Approximate Representations of Functions

Using formula (3.9), we can derive a number of very useful approximate representations of functions. These can be used for their numerical calculation, but also for all sorts of derivations.

As an example, let us show that for small x

$$\sin x \simeq x .\tag{3.57}$$

Indeed, consider the change of the function $y = \sin x$ when x changes by Δx:

$$\Delta y = \sin(x + \Delta x) - \sin x \simeq \frac{dy}{dx}\Delta x = (\cos x)\,\Delta x$$
$$\Rightarrow \quad \sin(x + \Delta x) \simeq \sin x + (\cos x)\,\Delta x .$$

Taking $x = 0$ in the last formula, we obtain $\sin \Delta x \simeq \Delta x$ as required.

Problem 3.28 Prove the following approximate formulae valid for $|x| \ll 1$:

$$\sqrt{1+x} \simeq 1 + \frac{x}{2} ,\tag{3.58}$$

$$\frac{1}{1 \pm x} \simeq 1 \mp x ,\tag{3.59}$$

$$\tan x \simeq x ,\tag{3.60}$$

$$e^x \simeq 1 + x ,\tag{3.61}$$

$$\ln(1 + x) \simeq x .\tag{3.62}$$

Problem 3.29 In a double slit experiment, Fig. 3.2, the distances the light passes from the slits to the observation point P that is a distance x from the centre of the screen (point O exactly opposite the middle distance between the

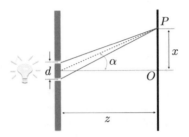

Fig. 3.2 Double slit experiment. Two slits on the left are separated by the distance d, while the screen on the right is at a distance $z \gg d$ from the slit plate. The light goes through the slits and is observed on the screen. The paths the light takes between the slits and the observation point P on the screen are different

> slits) are different. The distance between the slits is d and the distance to the screen z is much larger than d. Show by explicitly calculating the difference in the two paths Δs that for $z \gg d$ the difference is $\Delta s \approx d \sin \alpha$.

We shall learn later on how one can build up systematic corrections to the approximate expressions (3.57)–(3.62) given above; these contain additional terms with higher powers of x and hence are valid not only for very small values of x but also for x deviating significantly from zero.

3.6   Differentiation in More Difficult Cases

Direct specification of a function via $y = y(x)$ is not the only one possible. A function $y(x)$ can also be specified parametrically, e.g.

$$\begin{cases} x = x(t) \\ y = y(t) \end{cases}. \tag{3.63}$$

These types of equations may correspond to the x and y coordinates of a particle moving on a plane, in which case t (the parameter) is time. If we are interested in the derivative $y'(x)$, then it can be calculated as follows:

$$\begin{aligned} y'(x) &= \lim_{\Delta x \to 0} \frac{\Delta y}{\Delta x} = \lim_{\Delta x \to 0} \frac{y(t + \Delta t) - y(t)}{x(t + \Delta t) - x(t)} \\ &= \lim_{\Delta t \to 0} \frac{[y(t + \Delta t) - y(t)]/\Delta t}{[x(t + \Delta t) - x(t)]/\Delta t} = \frac{\lim_{\Delta t \to 0}[y(t + \Delta t) - y(t)]/\Delta t}{\lim_{\Delta t \to 0}[x(t + \Delta t) - x(t)]/\Delta t} = \frac{y'(t)}{x'(t)}. \end{aligned} \tag{3.64}$$

Here, at the first step, we explicitly indicated that the change of both y and x is due to the change of t by Δt; that is why Δy was replaced by $y(t + \Delta t) - y(t)$ and Δx by $x(t + \Delta t) - x(t)$. At the second step we changed the limit from $\Delta x \to 0$ to $\Delta t \to 0$ which is possible to do because we assume that the functions $y(t)$ and $x(t)$ are continuous (recall one of the definitions of the continuity of a function!).

At the last step, we applied the limit separately to the numerator and denominator of the fraction; this is allowed because of the properties of the limit discussed in the previous Chapter.

Problem 3.30 Prove the same result (3.64) by assuming that there exists an inverse $t = t(x)$ of the function $x = x(t)$, i.e. that the function $y = y(t) = y(t(x))$ is a composite function. [Hint: *use the chain rule (3.16) and Eq. (3.17) for the derivative of an inverse function.*]

Example 3.3 ▶ An inter-galactic rocket moves along a planar spiral trajectory away from the Sun according to the equations $x(t) = e^t \cos t$ and $y(t) = e^t \sin t$. Determine the slope of the tangent line to the trajectory as a function of time.

Solution. The slope to the trajectory will be given by the derivative $y'(x)$. Therefore, we obtain:

$$y'(x) = \frac{y'(t)}{x'(t)} = \frac{e^t \sin t + e^t \cos t}{e^t \cos t - e^t \sin t} = -\frac{1 + \cot t}{1 - \cot t} . ◀$$

Problem 3.31 Find the derivative $y'(x)$ of the function specified parametrically as $y = t^2 e^{-2t}$ and $x = (1 + t^2) e^{-2t}$. [Answer: $dy/dx = t(1 - t)/(-1 + t - t^2)$.]

A function $y = y(x)$ can also be specified implicitly via an equation $F(x, y) = 0$ that cannot be solved with respect to y (and hence the direct calculation of the derivative is impossible). Still, calculation of the derivative $y'(x)$ can be performed even in this case (although it is not guaranteed that the derivative can be obtained in a closed form).

Example 3.4 ▶ Let the function $y = y(x)$ be specified via the equation $x^2 + y^2 = R^2$ (a circle of radius R and centred at the origin). Calculate the derivative $y'(x)$.

Solution. We differentiate both sides of the equation with respect to x, treating the $y^2 = (y(x))^2$ as a composite function and hence using the chain rule. Then in the right-hand side of $x^2 + y^2 = R^2$ we have zero, while in the left-hand side $(x^2 + y^2)' = 2x + 2yy'(x)$, i.e. we obtain an equation

$$2x + 2yy'(x) = 0 ,$$

solving which with respect to the derivative finally yields: $y'(x) = -x/y(x)$. It is easy to see that this is the correct result. Indeed, in this particular case, we can directly solve the equation of the circle and obtain $y(x)$ explicitly as $y(x) = \pm\sqrt{R^2 - x^2}$, where the plus sign corresponds to the upper part of the circle ($y \geq 0$), while the minus sign corresponds to the lower part of it ($y < 0$). This expression can be differentiated easily using the chain rule:

$$y' = \pm\frac{1}{2}\left(R^2 - x^2\right)^{-1/2}(-2x) = \mp\frac{x}{\sqrt{R^2 - x^2}} = -\frac{x}{\pm\sqrt{R^2 - x^2}} = -\frac{x}{y} ,$$

i.e. the same result. ◄

Problem 3.32 Find the derivative $y'(x)$ of the function specified via equation $y^4 + 2xy^2 + e^{-x}y = 0$. [Answer: $y' = y\left(e^{-x} - 2y\right) / \left(4y^3 + 4xy + e^{-x}\right)$.]

Finally, there is a class of functions for which a special method is used to perform differentiation: these are the exponential functions containing the base which is a function of x itself, i.e. $y(x) = u(x)^{v(x)}$. In these cases, one first takes a logarithm of both sides turning the equation into an implicit with respect to y, and then differentiates both sides.

For instance, let us calculate the derivative of $y = x^x$. After taking the natural logarithm of both sides, we obtain: $\ln y = x \ln x$. Differentiating both sides (remember to use the chain rule when differentiating the left-hand side as $y = y(x)$), we obtain:

$$\frac{1}{y}y' = \ln x + x\frac{1}{x} = \ln x + 1 \quad \Rightarrow \quad y' \equiv \left(x^x\right)' = y\left(1 + \ln x\right) = x^x\left(1 + \ln x\right) .$$

Problem 3.33 Derive the general formula

$$\left(u(x)^{v(x)}\right)' = v'u^v \ln u + u'u^{v-1}v$$

for the derivative of the function $y = u(x)^{v(x)}$.

Problem 3.34 Obtain the same result by representing the function $y(x) = u(x)^{v(x)}$ as an exponential.

Problem 3.35 Find the derivative $y'(x)$ of the functions specified in the following way: (a) $y = \ln t$, $x = e^{-2t} \sinh t$; (b) $y^{-3} + 2e^{-2xy} = 1$; (c) $y = \left(1 + 2x^2\right)^{3-3x^2}$; (d) $y = \sqrt{x + \sqrt{x + \sqrt{x + y}}}$ (in the latter case, consider two methods: first, differentiate directly using the chain rule applied several times;

second, rearrange the equation in such a way as to get rid of the radicals completely and then differentiate). [Answers: *(a)* $e^{2t}/[t\cosh(t) - 2t\sinh(t)]$; *(b)* $-y/[x + (3/4) e^{2xy} y^{-4}]$; *(c)* $6xy[-\ln(1 + 2x^2) + 2(1 - x^2)/(1 + 2x^2)]$; *(d)* A/B, where $B = 1 - 8yg\sqrt{x + y}$, $A = -[1 + 2\sqrt{x + y}(2g + 1)]$, and $g = y^2 - x$.]

3.7   Higher Order Derivatives

3.7.1   Definition and Simple Examples

The derivative $y'(x)$ is a function of x as well and hence it can also be differentiated, which is denoted $y'' = (y')'$ ("double prime"). The following terminology is used: y' is called the first-order derivative, while y'' is the second order. Obviously, this process can be continued and one can talk about derivatives of the n-th order, denoted $y^{(n)}(x)$ (n is enclosed in brackets to distinguish it from the n-th power). Obviously, $y^{(n+1)} = (y^{(n)})'$. Similar notations exist for the higher order derivatives based on the symbols of differentials:

$$y'' = \frac{d^2y}{dx^2}, \quad y^{(3)} = y''' = \frac{d^3y}{dx^3}, \quad \ldots, \quad y^{(n)} = \frac{d^ny}{dx^n}.$$

Higher order derivatives are frequently used in applications. Many processes are described by equations (the so-called differential equations) involving derivatives of different orders. In mechanics acceleration, $a(t) = v'(t)$ is the rate of change of the velocity; at the same time, since the velocity $v(t) = x'(t)$ relates to the rate of change of the position, the acceleration can also be considered as the second-order derivative of the position: $a(t) = v'(t) = (x'(t))' \equiv x''(t)$.

Another important application of first and second-order derivatives is in analysing functions. We shall consider this particular point in detail later on in Sect. 3.11.

Example 3.5 ► Calculate the fifth order derivative of $y = \cos x$. Then, derive the formula for the n-th order derivative of the cosine function.

Solution. We have: $(\cos x)' = -\sin x$; $(\cos x)'' = (-\sin x)' = -\cos x$; $(\cos x)''' = (-\cos x)' = \sin x$, then $(\cos x)^{(4)} = (\sin x)' = \cos x$, and, finally, $(\cos x)^{(5)} = (\cos x)' = -\sin x$. We see that the fifth derivative is the same as the first. A general formula can be derived if we notice that $(\cos x)' = -\sin x \equiv \cos(x + \pi/2)$, $(\cos x)'' = -\cos x \equiv \cos(x + 2\pi/2)$, $(\cos x)^{(3)} = \sin x \equiv \cos(x + 3\pi/2)$, etc.

We observe that each time a phase of $\pi/2$ is added to the cosine function. Using the method of induction, we can now prove the general formula:

$$(\cos x)^{(n)} = \cos\left(x + \frac{\pi}{2}n\right) . \tag{3.65}$$

Indeed, the formula works for the cases of $n = 1, 2, 3$. We assume that it also works for some arbitrary n. Let us check if it would work for the next value $n + 1$. We have:

$$(\cos x)^{(n+1)} = \left((\cos x)^{(n)}\right)' = \left(\cos\left(x + \frac{\pi}{2}n\right)\right)' = -\sin\left(x + \frac{\pi}{2}n\right)$$
$$\equiv \cos\left(x + \frac{\pi}{2}n + \frac{\pi}{2}\right) = \cos\left(x + \frac{\pi}{2}(n+1)\right) ,$$

which is exactly the desired result, i.e. Equation (3.65) for $n \to n + 1$. The formula is proven. ◀

Problem 3.36 Prove the following formulae:

$$(\sin x)^{(n)} = \sin\left(x + \frac{\pi}{2}n\right) , \tag{3.66}$$

$$\left(e^{\alpha x}\right)^{(n)} = \alpha^n e^{\alpha x} , \tag{3.67}$$

and, assuming a general real α,

$$\left(x^\alpha\right)^{(n)} = \alpha\,(\alpha - 1)\,(\alpha - n)\cdots(\alpha - n + 1)\,x^{\alpha - n} . \tag{3.68}$$

If $\alpha \equiv m$ is a positive integer, then, in particular,

$$\left(x^m\right)^{(n)} = \frac{m!}{(m-n)!}x^{m-n} \text{ if } m \geq n \text{ and } \left(x^m\right)^{(n)} = 0 \text{ if } m < n .$$

Problem 3.37 Prove the formula:

$$\lim_{z \to 0} \frac{d^{2n}}{dz^{2n}} \left(z^2 - 1\right)^{2n} = (-1)^n \left(\frac{(2n)!}{n!}\right)^2 . \tag{3.69}$$

3.7.2 Higher Order Derivatives of Inverse Functions

We have proven above an important result in Theorem 3.2 that enabled us to calculate derivatives of some elementary functions in Sect. 3.3 such as exponential, inverse trigonometric and hyperbolic functions. Let us discuss if higher order derivatives of an inverse function can also be conveniently expressed via the derivatives of the corresponding direct functions.

Let $y = y(x)$ is a function inverse to $x = x(y)$. Its first derivative $y'_x = 1/x'_y$. Then its second derivative,

$$\frac{d^2 y}{dx^2} = \frac{d}{dx} \left(y'_x \right) = \frac{d}{dx} \frac{1}{x'_y} = -\frac{1}{\left(x'_y \right)^2} \frac{d}{dx} \left(x'_y \right).$$

The derivative over x in the last passage can be written using the chain rule as x'_y is surely some function of y:

$$\frac{d}{dx} \left(x'_y \right) = \frac{d}{dy} \left(x'_y \right) \frac{dy}{dx} = x''_y y'_x = x''_y \frac{1}{x'_y},$$

where it was recognised that the derivative of x'_y with respect to y is just the second derivative of the direct function, and y'_x was replaced with $1/x'_y$. Combining these results, yields:

$$\frac{d^2 y}{dx^2} = -\frac{x''_y}{\left(x'_y \right)^3}. \tag{3.70}$$

Problem 3.38 Derive using the method developed above the following formulae for the third and fourth-order derivatives of the inverse function:

$$\frac{d^3 y}{dx^3} = \frac{1}{\left(x'_y \right)^5} \left[3 \left(x''_y \right)^2 - x'_y x'''_y \right],$$

$$\frac{d^4 y}{dx^4} = \frac{1}{\left(x'_y \right)^7} \left[-15 \left(x''_y \right)^3 + 10 x'_y x''_y x'''_y - \left(x'_y \right)^2 x''''_y \right].$$

Problem 3.39 Consider the inverse function $y = \ln x$ with the direct function being $x = e^y$. Convince yourself using the higher order derivatives formulae obtained above that derivatives of the logarithm are the ones we expect: $y'_x = 1/x$, $y''_x = -1/x^2$, $y'''_x = 2/x^3$, and $y''''_x = -6/x^4$.

3.7.3   Leibniz Formula

Very often it is necessary to calculate the $n-$th derivative of a product of two functions. We shall now consider a formula due to Leibniz, which allows one to perform such a calculation. This result is especially useful if one of the functions in the product is a finite order polynomial as will become apparent later on.

Let $f(x) = u(x)v(x)$. Then we shall show that

$$
\frac{d^n(uv)}{dx^n} = (uv)^{(n)} = uv^{(n)} + \binom{n}{1} u'v^{(n-1)} + \binom{n}{2} u''v^{(n-2)}
$$
$$
+ \cdots + \binom{n}{k} u^{(k)}v^{(n-k)} + \cdots + u^{(n)}v \equiv \sum_{k=0}^{n} \binom{n}{k} u^{(k)}v^{(n-k)} ,
\tag{3.71}
$$

where $\binom{n}{k} = \frac{n!}{k!(n-k)!}$ are *the binomial coefficients* which we met in Sect. 1.13, and we implied that $u^{(0)} \equiv u$ and $v^{(0)} \equiv v$ are the functions themselves. In fact, we see that the Leibniz formula looks strikingly similar to the *binomial theorem* (1.153) and therefore is sometimes written in the following symbolic form:

$$
\frac{d^n(uv)}{dx^n} \equiv (uv)^{(n)} = (u + v)^{(n)} .
\tag{3.72}
$$

Let us first prove this result using a method very similar to the one we used in Sect. 1.13 when proving the binomial theorem. It is based on induction. Firstly, we check that the formula works for some initial values of $n = 1, 2, 3$ (in fact, only one value, e.g. $n = 1$, would suffice):

$$
(uv)' = u^{(1)}v + uv^{(1)} = u^{(1)}v^{(0)} + u^{(0)}v^{(1)} \quad \text{(i.e. the familiar } u'v + uv'\text{),}
$$
$$
(uv)'' = u^{(2)}v^{(0)} + 2u^{(1)}v^{(1)} + u^{(0)}v^{(2)} ,
$$
$$
(uv)''' = \left(u^{(3)}v^{(0)} + u^{(2)}v^{(1)}\right) + 2\left(u^{(2)}v^{(1)} + u^{(1)}v^{(2)}\right) + \left(u^{(1)}v^{(2)} + u^{(0)}v^{(3)}\right)
$$
$$
= u^{(3)}v^{(0)} + 3u^{(2)}v^{(1)} + 3u^{(1)}v^{(2)} + u^{(0)}v^{(3)} ,
$$

which are all consistent with Eqs. (3.71) or (3.72), since

$$
\binom{3}{0} = \frac{3!}{0!3!} = 1, \quad \binom{3}{1} = \frac{3!}{1!2!} = 3 ,
$$

$$
\binom{3}{2} = \frac{3!}{2!1!} = 3 \quad \text{and} \quad \binom{3}{3} = \frac{3!}{0!3!} = 1 .
$$

Recall that $0! = 1$ by definition. Next we assume that formula (3.71) is valid for some value n. We should now prove from it that Eq. (3.71) holds also for the next value of n, i.e. for $n \to n + 1$. Differentiating both sides of Eq. (3.71), we get:

$$(uv)^{(n+1)} = \left[\sum_{k=0}^{n} \binom{n}{k} u^{(k)} v^{(n-k)} \right]' = \sum_{k=0}^{n} \binom{n}{k} \left[u^{(k+1)} v^{(n-k)} + u^{(k)} v^{(n-k+1)} \right]$$

$$= \sum_{k=0}^{n} \binom{n}{k} u^{(k+1)} v^{(n-k)} + \sum_{k=0}^{n} \binom{n}{k} u^{(k)} v^{(n-k+1)} .$$

In the first term, we make a substitution of the summation index, $l = k + 1$, which gives (after replacing l with k again):

$$(uv)^{(n+1)} = \sum_{k=1}^{n+1} \binom{n}{k-1} u^{(k)} v^{(n-k+1)} + \sum_{k=0}^{n} \binom{n}{k} u^{(k)} v^{(n-k+1)} .$$

Then, separate out the last (with $k = n + 1$) term in the first sum and the first (the one with $k = 0$) term in the second; combine the others together:

$$(uv)^{(n+1)} = \binom{n}{n} u^{(n+1)} v^{(0)} + \sum_{k=1}^{n} \left[\binom{n}{k-1} + \binom{n}{k} \right] u^{(k)} v^{(n-k+1)}$$
$$+ \binom{n}{0} u^{(0)} v^{(n+1)} . \tag{3.73}$$

Because of Eqs. (1.156) and (1.157), we can finally rewrite Eq. (3.73) as follows:

$$(uv)^{(n+1)} = \binom{n+1}{0} u^{(0)} v^{(n+1)} + \sum_{k=1}^{n} \binom{n+1}{k} u^{(k)} v^{(n+1-k)} + \binom{n+1}{n+1} u^{(n+1)} v^{(0)}$$

$$\equiv \sum_{k=0}^{n+1} \binom{n+1}{k} u^{(k)} v^{(n+1-k)} ,$$

which is the desired result as it looks exactly as Eq. (3.71), but written for $n \to n + 1$.

Example 3.6 ► Using Leibniz formula, obtain a formula for the n-th derivative of the function $f(x) = (x^2 + 1) v(x)$.

Solution. Here $u(x) = x^2 + 1$ and hence can only be differentiated twice: $u^{(1)} = 2x$ and $u^{(2)} = 2$, while all higher derivatives give zero. Therefore, the Leibniz formula contains only the first three terms corresponding to the summation index $k = 0, 1, 2$:

$$f^{(n)}(x) = \binom{n}{0} u^{(0)} v^{(n)} + \binom{n}{1} u^{(1)} v^{(n-1)} + \binom{n}{2} u^{(2)} v^{(n-2)}$$
$$= uv^{(n)} + 2nx v^{(n-1)} + n(n-1) v^{(n-2)} ,$$

where we have used explicit expressions for the first three binomial coefficients:
$\binom{n}{0} = 1$, $\binom{n}{1} = n$ and $\binom{n}{2} = n(n-1)/2$. ◀

Problem 3.40 Consider a function $\varphi(x) = (1 - x^3)f(x)$. The n−th derivative of $\varphi(x)$ can be written in the following form:

$$\varphi^{(n)}(x) = a_0(1 - x^3)f^{(n)} + a_1 x^2 f^{(n-1)} + a_2 x f^{(n-2)} + a_3 f^{(n-3)} .$$

Show, using the Leibniz formula (3.71), that the numerical values of the coefficients are: $a_0 = 1, a_1 = -3n, a_2 = -3n(n-1)$ and $a_3 = -n(n-1)(n-2)$.

Problem 3.41 Show that the n−th derivative of the function $f(x) = x^3 \sin(2x)$, assuming n is even, is:

$$f^{(n)}(x) = (-1)^{n/2} 2^{n-3} \left\{ 2x \left[4x^2 - 3n(n-1) \right] \sin 2x + n \left[-12x^2 + (n-1)(n-2) \right] \cos 2x \right\} .$$

In some cases, the calculation of the n−th derivative could involve some tricks. We shall illustrate this by considering the n−th derivative of the $y = \arcsin(x)$ function. The first derivative $y' = (1 - x^2)^{-1/2}$. The calculation of the second derivative would result in two terms as one has to use the chain rule; more terms arise as we attempt to calculate more derivatives and there seems to be no general rule as to how to do the n−th derivative (try this!). A direct application of the Leibniz formula here does not seem to help either. The solution to this problem appears to be possible if we notice that

$$y'(x) = (1 - x)^{-1/2} (1 + x)^{-1/2} = u(x)v(x) ,$$

so that, applying the Leibniz formula now, we can write:

$$y^{(n+1)}(x) = \left(y' \right)^{(n)} = (uv)^{(n)} = \sum_{k=0}^{n} \binom{n}{k} u^{(k)} v^{(n-k)} ,$$

where (e.g. by induction)

$$
\begin{aligned}
u^{(k)} &= \left[(1-x)^{-1/2} \right]^{(k)} = \frac{1}{2} \cdot \frac{3}{2} \cdot \ldots \cdot \frac{2k-1}{2} (1-x)^{-(2k+1)/2} \\
&= \frac{(2k-1)!!}{2^k} (1-x)^{-(2k+1)/2} .
\end{aligned}
$$

Here, the double factorial means a product of all odd integers, i.e. $(2k-1)!! = 1 \cdot 3 \cdot 5 \cdot \ldots \cdot (2k-1)$. We also adopt that $(-1)!! = 1$, which makes the formula for $u^{(k)}$ also formally valid for $k = 0$, i.e. $u^{(0)} = (1-x)^{-1/2} \equiv u(x)$. The double factorial can always be expressed via single factorials and powers of 2 by inserting between odd integers the corresponding even ones:

$$(2k - 1)!! = 1 \cdot 3 \cdot 5 \cdot \ldots \cdot (2k - 1)$$

$$= \frac{1 \cdot 2 \cdot 3 \cdot 4 \cdot 5 \cdot 6 \cdot \ldots \cdot (2k - 1) \cdot (2k)}{2 \cdot 4 \cdot 6 \cdot \ldots \cdot (2k)} = \frac{(2k)!}{2^k k!} , \qquad (3.74)$$

which yields

$$u^{(k)} = \frac{(2k)!}{4^k k!} (1 - x)^{-(2k+1)/2} = \frac{1}{\sqrt{1 - x}} \frac{(2k)!}{4^k k!} (1 - x)^{-k} .$$

Correspondingly,

$$v^{(k)} = \left[(1 + x)^{-1/2} \right]^{(k)} = \left(-\frac{1}{2} \right) \cdot \left(-\frac{3}{2} \right) \cdot \ldots \cdot \left(-\frac{2k - 1}{2} \right) (1 + x)^{-(2k+1)/2}$$

$$= (-1)^k \frac{(2k - 1)!!}{2^k} (1 + x)^{-(2k+1)/2} = \frac{(-1)^k}{\sqrt{1 + x}} \frac{(2k)!}{4^k k!} (1 + x)^{-k} ,$$

which is also formally valid for $k = 0$, and therefore

$$v^{(n-k)} = \frac{(-1)^{n-k}}{\sqrt{1 + x}} \frac{(2n - 2k)!}{4^{n-k} (n - k)!} (1 + x)^{-n+k} .$$

Combining the obtained derivatives of $u(x)$ and $v(x)$ in the Leibniz formula, we finally obtain:

$$[\arcsin(x)]^{(n+1)} = \frac{1}{\sqrt{1 - x^2}} \sum_{k=0}^{n} \binom{n}{k} \frac{(2k)!}{4^k k!} \frac{(2n - 2k)!}{4^{n-k} (n - k)!} (-1)^{n-k} (1 - x)^{-k} (1 + x)^{-n+k}$$

$$= \frac{(1 + x)^{-n}}{\sqrt{1 - x^2}} \frac{(2n)!}{4^n n!} \sum_{k=0}^{n} (-1)^{n-k} \binom{n}{k}^2 \binom{2n}{2k}^{-1} \left(\frac{1 + x}{1 - x} \right)^k ,$$

$$(3.75)$$

where we expressed all factorials via the binomial coefficients.

3.7.4 Differentiation Operator

There is another proof of the Leibniz formula, which is worth giving here as it is based on rather different (and often useful!) ideas. Differentiation of a function, $f'(x)$, can be formally considered as an action of an *operator*, $D = \frac{d}{dx}$, on the function $f(x)$, i.e.

$$f'(x) = \frac{df}{dx} = \frac{d}{dx} f(x) \equiv D f(x) .$$

The n-th derivative, $f^{(n)}(x)$, can then be regarded as an action of the same operator n times:

$$f^{(n)}(x) = \left(\frac{d}{dx}\right)^n f(x) \equiv D^n f(x) \, .$$

Problem 3.42 Prove the following formulae:

$$D e^{2x} = 2e^{2x} \, ; \quad (D+1)\cos x = -\sin x + \cos x \, ;$$

$$(D+1)^2 \cos x = \left(D^2 + 2D + 1\right)\cos x = -2\sin x \, ;$$

$$\left(xD^2 + 2Dx - x\right)e^x = 2(1+x)e^x \, ,$$

but

$$\left(xD^2 + 2xD - x\right)e^x = 2xe^x \, .$$

Let us now use these types of ideas to prove the Leibniz formula. Let us introduce two operators, $D_u = \left(\frac{d}{dx}\right)_u$ and $D_v = \left(\frac{d}{dx}\right)_v$, which differentiate only the functions $u(x)$ and $v(x)$, respectively. Then, the first derivative of the product uv can be written as

$$(uv)' = (D_u + D_v)\,uv = (D_u u)\,v + u\,(D_v v) = u^{(1)}v^{(0)} + u^{(0)}v^{(1)} \, ;$$

the second derivative is obtained by acting with the same operator twice:

$$(uv)'' = (D_u + D_v)\,(D_u + D_v)\,uv = (D_u + D_v)^2\,uv$$
$$= \left(D_u^2 + D_u D_v + D_v D_u + D_v^2\right)uv = \left(D_u^2 + 2D_u D_v + D_v^2\right)uv \, ,$$

where in the last step, we recognised that the order in which operators D_u and D_v act on the product of functions u and v does not matter (since they act on different functions anyway), and hence we combined them together: $D_u D_v + D_v D_u = 2D_u D_v$. Thus, we obtain:

$$(uv)'' = \left(D_u^2 + 2D_u D_v + D_v^2\right)uv = \left(D_u^2 u\right)v + 2\,(D_u u)\,(D_v v) + u\left(D_v^2 v\right)$$
$$= u^{(2)}v^{(0)} + 2u^{(1)}v^{(1)} + u^{(0)}v^{(2)} \, .$$

This process can be continued; when we want to calculate the n-th derivative, we will have to act with the operator $(D_u + D_v)^n$ on the product uv. It must be clear now what we ought to do in the general case: since the order of operators is not important, we can treat them like numbers and expand the operator $(D_u + D_v)^n$ formally using the binomial formula. Then, associating $D_u^k u$ with $u^{(k)}$ and similarly $D_v^k v$ with $v^{(k)}$, we can write:

$$(uv)^{(n)} = (D_u + D_v)^n \, uv = \sum_{k=0}^{n} \binom{n}{k} D_u^k D_v^{n-k} uv$$

$$= \sum_{k=0}^{n} \binom{n}{k} \left(D_u^k u\right) \left(D_v^{n-k} v\right) \equiv \sum_{k=0}^{n} \binom{n}{k} u^{(k)} v^{(n-k)} \, ,$$

which is exactly the correct expression (3.71). This particular proof shows quite clearly why the Leibniz formula is so similar to the binomial formula. At the same time, we should not forget that this resemblance might be misleading: if in the binomial formula for $(a + b)^n$ we have powers and hence a^0 and b^0 are both equal to one and therefore do not need to be written explicitly, in the Leibniz formula for $(uv)^{(n)}$ the "powers" in $u^{(k)}$ and $v^{(n-k)}$ correspond to the order of the derivatives and thus $u^{(0)}$ and $v^{(0)}$ coincide with the functions $u(x)$ and $v(x)$ themselves and are *not* equal to one; these should be written explicitly whenever they appear.

3.8   Taylor's Formula

Taylor's formula[2] allows representing a function as a finite degree polynomial and a remainder term to be considered as "small". It manifests itself as one of the main results of mathematical analysis. It is used in numerous analytical proofs and estimates. It is also useful for approximate calculations of functions.

Let us first do some preliminary work which will help later on. Consider a function $f(x)$ which is a polynomial of degree n:

$$f(x) = \alpha_0 + \alpha_1 x + \alpha_2 x^2 + \cdots + \alpha_n x^n \, . \tag{3.76}$$

This polynomial contains powers of x, but can alternatively be written via powers of $(x - a)$ for any a as follows:

$$f(x) = A_0 + A_1 (x - a) + A_2 (x - a)^2 + \cdots + A_n (x - a)^n \, . \tag{3.77}$$

Problem 3.43 Using the binomial formula for each of the powers $(x - a)^k$, $k = 1, \ldots, n$, it must be possible to have the $f(x)$ written via powers of x again. By comparing terms of the same powers of x with Eq. (3.76), derive explicit expressions relating the coefficients α_i and A_k. Use this method to show that

---

[2] Named after Brook Taylor.

$$\alpha_k = \sum_{i=k}^{n} A_i \binom{i}{k} (-a)^{i-k} .$$

Problem 3.44 Using Eq. (3.77), it is seen that $f(a) = A_0$, then

$$f'(x) = A_1 + 2A_2 (x - a) + 3A_3 (x - a)^2 + \cdots \quad \Longrightarrow \quad f'(a) = A_1,$$
$$f''(x) = 2A_2 + 3 \cdot 2A_3 (x - a) + 4 \cdot 3A_4 (x - a)^2 + \cdots \quad \Longrightarrow \quad f''(a) = 2A_2,$$
$$f'''(x) = 3 \cdot 2A_3 + 4 \cdot 3 \cdot 2A_4 (x - a) + 5 \cdot 4 \cdot 3A_5 (x - a)^2 + \cdots$$
$$\Longrightarrow \quad f'''(a) = 3 \cdot 2A_3 = 6A_3,$$

etc. Show using induction that for any $k = 1, 2, \ldots, n$ one can write:

$$f^{(k)}(x) = k!A_k + \frac{(k+1)!}{1!} A_{k+1} (x - a) + \frac{(k+2)!}{2!} A_{k+2} (x - a)^2 + \cdots$$
$$\Longrightarrow \quad f^{(k)}(a) = k!A_k ,$$

so that

$$A_1 = \frac{f'(a)}{1!} , \quad A_2 = \frac{f''(a)}{2!} , \quad \ldots , \quad A_k = \frac{f^{(k)}(a)}{k!} , \quad \ldots , \quad A_n = \frac{f^{(n)}(a)}{n!} . \tag{3.78}$$

Problem 3.45 Differentiate $f(x)$ in Eq. (3.76) to express derivatives of $f(x)$ at $x = a$ via the a_k coefficients. Hence, show that the coefficients A_k can be expressed via α_i as follows:

$$A_k = \sum_{i=k}^{n} \alpha_i \binom{i}{k} a^{i-k} .$$

It follows from Eq. (3.78) that any polynomial of degree n can actually be written as a sum of powers of $(x - a)$ for any a:

$$f(x) = f(a) + \frac{f'(a)}{1!} (x - a) + \frac{f''(a)}{2!} (x - a)^2 + \cdots + \frac{f^{(n)}(a)}{n!} (x - a)^n$$
$$\equiv \sum_{k=0}^{n} \frac{f^{(k)}(a)}{k!} (x - a)^k .$$

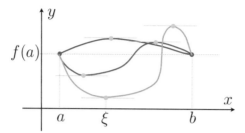

Fig. 3.3 A schematic that illustrates the point that a non-constant function $f(x)$, which has equal values at both ends of the interval $a \leq x \leq b$ will experience at least one (may be more) extremum within that interval. Here, three functions are sketched, shown by different colours, all having the same values at the edge points a and b. The extrema (minima and/or maxima) of the functions are indicated by green dots; the corresponding tangent lines at these points are also shown by horizontal green lines

Of course, if $f(x)$ is not a polynomial, it can never be written via a finite sum of powers of $(x - a)$. Let us define the following n–th degree polynomial (Taylor's polynomial)

$$F_n(x, a) = f(a) + \frac{f'(a)}{1!}(x - a) + \frac{f''(a)}{2!}(x - a)^2 + \cdots + \frac{f^{(n)}(a)}{n!}(x - a)^n$$

$$\equiv \sum_{k=0}^{n} \frac{f^{(k)}(a)}{k!}(x - a)^k \, ,$$

(3.79)

which is a function of two variables, x and a, and satisfies an obvious property $F_n(a, a) = f(a)$. Then it is clear that formally *any* function $f(x)$ can be written as

$$f(x) = F_n(x, a) + R_{n+1} \, ,$$

(3.80)

where R_{n+1}, called the *remainder term*, accounts for the full error we are making by replacing $f(x)$ with the Taylor's polynomial of degree n given by Eq. (3.79); of course, R_{n+1} depends on x as well. Taylor's theorem to be discussed below enables one to write an explicit formula for the remainder term for any $f(x)$.

But before formulating and proving Taylor's theorem, we need one more statement to be made, which, we hope, is intuitively obvious.[3] Consider some *continuous* function $f(x)$ within the interval $a \leq x \leq b$, and let the function have *identical* values at both ends: $f(a) = f(b)$. Assume first that the function $f(x)$ is not a constant (equal to its edge value). Then it is obvious that it will behave within the interval between a and b in such a way that it would have at least one minimum or maximum as illustrated in Fig. 3.3. Indeed, for instance, if the function goes down from the point $x = a$, it must eventually go up towards the point $x = b$ as its values at the edges are the same. If the function has a minimum or maximum at a point

---

[3] In mathematics literature it is called Rolle's theorem after Michel Rolle.

$x = \xi$, then its derivative is zero at that point, $f'(\xi) = 0$, hence the slope of the tangent line at that point is zero (it is horizontal). This is also illustrated in the Figure (more discussion on the extrema will be given in Sect. 3.11). To prove this statement, let us suppose that the function has a maximum at point $x = \xi$. This means that $f(\xi) \geq f(\xi - \epsilon)$ and $f(\xi) \geq f(\xi + \epsilon)$ for a sufficiently small $\epsilon > 0$, i.e. at the neighbouring points on the right and on the left of the point $x = \xi$ the function $f(x)$ has smaller or (possibly) equal values (definitely it cannot be larger at the neighbouring points). Therefore, we have:

$$\frac{f(\xi + \epsilon) - f(\xi)}{\epsilon} \leq 0 \quad \text{and} \quad \frac{f(\xi) - f(\xi - \epsilon)}{\epsilon} \geq 0 .$$

Note that the first expression is non-positive, while the second is non-negative. In the limit of $\epsilon \to 0$ both of these expressions should give the derivatives $f'(\xi)$ from the right and left. Since the function $f(x)$ is continuous, these two derivatives should be equal to each other, and that means that they should be equal to zero as otherwise it is impossible to satisfy the two inequalities written above. A similar proof is given in the case of a minimum. Note that if the function $f(x)$ is constant on an interval, all points between its edges have zero derivative and this does not contradict the above-made statement.

What is essential for us right now is that if a continuous function $f(x)$ does have equal values at two points, then there must be at least one point ξ between them where the first derivative of the function is zero: $f'(\xi) = 0$.

Taylor's theorem is based on properties of the following function of an auxiliary variable t:

$$\rho(t) = f(x) - F_n(x, t) - \frac{(x - t)^{n+1}}{(x - a)^{n+1}} [f(x) - F_n(x, a)] . \tag{3.81}$$

Let us first calculate this function at the points $t = a$ and $t = x$:

$$\rho(a) = [f(x) - F_n(x, a)] - \frac{(x - a)^{n+1}}{(x - a)^{n+1}} [f(x) - F_n(x, a)]$$

$$= [f(x) - F_n(x, a)] - [f(x) - F_n(x, a)] = 0$$

and

$$\rho(x) = [f(x) - F_n(x, x)] - \frac{(x - x)^{n+1}}{(x - a)^{n+1}} [f(x) - F_n(x, a)]$$

$$= f(x) - F_n(x, x) = 0 ,$$

as $F_n(x, x) = f(x)$ from its definition (3.79), as was already mentioned. We see that $\rho(a) = \rho(x) = 0$. Next, we calculate the derivative of $\rho(t)$:

$$\rho'(t) = -\frac{dF_n(x, t)}{dt} + (n + 1) \frac{(x - t)^n}{(x - a)^{n+1}} [f(x) - F_n(x, a)] , \qquad (3.82)$$

where

$$\frac{dF_n(x, t)}{dt} = \frac{d}{dt} \left[f(t) + \sum_{k=1}^{n} \frac{f^{(k)}(t)}{k!} (x - t)^k \right] = f'(t) + \sum_{k=1}^{n} \frac{d}{dt} \left[\frac{f^{(k)}(t)}{k!} (x - t)^k \right]$$

$$= f'(t) + \sum_{k=1}^{n} \left[\frac{f^{(k+1)}(t)}{k!} (x - t)^k + \frac{f^{(k)}(t)}{k!} (-k) (x - t)^{k-1} \right]$$

$$= f'(t) + \sum_{k=1}^{n} \left[\frac{f^{(k+1)}(t)}{k!} (x - t)^k - \frac{f^{(k)}(t)}{(k - 1)!} (x - t)^{k-1} \right]$$

$$= f'(t) + \left[f''(t) (x - t) - f'(t) \right] + \left[\frac{1}{2} f'''(t) (x - t)^2 - f''(t) (x - t) \right]$$

$$+ \left[\frac{1}{3!} f^{(4)}(t) (x - t)^3 - \frac{1}{2} f'''(t) (x - t)^2 \right]$$

$$+ \cdots + \left[\frac{f^{(n+1)}(t)}{n!} (x - t)^n - \frac{f^{(n)}(t)}{(n - 1)!} (x - t)^{n-1} \right] .$$

It is seen that the first term here is cancelled out with the third (the second term in the first square bracket), the second with the fifth, the fourth with the seventh, and so on; all terms cancel out this way apart from the first term in the last bracket, i.e.

$$\frac{dF_n(x, t)}{dt} = \frac{f^{(n+1)}(t)}{n!} (x - t)^n ,$$

and hence from (3.82) it follows that

$$\rho'(t) = -\frac{f^{(n+1)}(t)}{n!} (x - t)^n + (n + 1) \frac{(x - t)^n}{(x - a)^{n+1}} [f(x) - F_n(x, a)]$$

$$= -\frac{f^{(n+1)}(t)}{n!} (x - t)^n + (n + 1) \frac{(x - t)^n}{(x - a)^{n+1}} R_{n+1} , \qquad (3.83)$$

since R_{n+1} is the remainder term between the function $f(x)$ and the polynomial $F_n(x, a)$ by definition of Eq. (3.80).

Now we are ready to formulate and then prove the famous Taylor's theorem (formula).

Theorem 3.7 (Taylor theorem) *Let* $y = f(x)$ *have at least* $m + 1$ *derivatives* *(i.e.* $f^{(k)}(x)$ *for* $k = 1, \ldots, m + 1$*) in some vicinity of point* a*. Then this function at some point* x *within this neighbourhood can be written as the following* $(n + 1)$ *−degree polynomial (assuming* $n \leq m$*):*

$$f(x) = \left[f(a) + \frac{f'(a)}{1!}(x - a) + \frac{f''(a)}{2!}(x - a)^2 + \cdots + \frac{f^{(n)}(a)}{n!}(x - a)^n \right]$$
$$+ R_{n+1}(x) = F_n(x, a) + R_{n+1}(x),$$

(3.84)

where

$$R_{n+1}(x) = \frac{f^{(n+1)}(\xi)}{(n + 1)!}(x - a)^{n+1}$$

(3.85)

is the remainder term *with* ξ *being some point between* x *and* a*; of course, the point* ξ *generally depends on the choice of* n*.*

Proof: Let us choose a point $x > a$ in the vicinity of a (the case of $x < a$ is considered similarly) and let $F_n(x, a)$ be the first part of Taylor's formula before the remainder term. Then, consider a function $\rho(t)$ defined in Eq. (3.81). Since the function $\rho(t)$ has equal values (equal to zero) at the boundaries of the interval $a \leq t \leq x$, it has to have at least one extremum there, either a minimum or a maximum, i.e. there should exist a point ξ (where $a < \xi < x$) such that $\rho'(\xi) = 0$.

Setting the derivative from expression (3.83) at $t = \xi$ to zero, we obtain:

$$\rho'(\xi) = -\frac{f^{(n+1)}(\xi)}{n!}(x - \xi)^n + (n + 1)\frac{(x - \xi)^n}{(x - a)^{n+1}}R_{n+1} = 0$$

$$\implies \quad R_{n+1} = \frac{f^{(n+1)}(\xi)}{(n + 1)!}(x - a)^{n+1},$$

which proves the theorem. **Q.E.D.**

Note that it is often useful to write the intermediate point ξ explicitly as follows:

$$\xi = a + \vartheta(x - a), \quad \text{where} \quad 0 < \vartheta < 1.$$

This form guarantees that ξ lies between a and x. Note that the number ϑ depends on the choice of points x and a.

There are several important particular cases of Taylor's formula worth mentioning. In the special case of $a = 0$ Taylor's formula takes a simplified form:

$$f(x) = f(0) + \frac{f'(0)}{1!}x + \frac{f''(0)}{2!}x^2 + \frac{f''(0)}{3!}x^2 + \cdots + \frac{f^{(n)}(0)}{n!}x^n + R_{n+1}(x),$$

(3.86)

which is called Maclaurin's formula. Here the remainder term is given by

$$R_{n+1}(x) = \frac{f^{(n+1)}(\xi)}{(n+1)!}x^{n+1} = \frac{f^{(n+1)}(\vartheta x)}{(n+1)!}x^{n+1},$$

(3.87)

where $0 < \vartheta < 1$.

The other important formula (due to Lagrange) is obtained from the Taylor's formula in the case of $n = 0$. Indeed, in this case, the remainder term $R_1(x) = f'(\xi)(x-a)$, so that we can write:

$$
\begin{aligned}
f(x) &= f(a) + f'(\xi)(x-a) \\
&\equiv f(a) + f'(a + \vartheta(x-a))(x-a), \quad a < \xi < x, \ 0 < \vartheta < 1.
\end{aligned}
$$

(3.88)

This formula is found to be very useful when proving various results and we shall be using it often starting from the next Section.

Let us now apply the Maclaurin formula to some elementary functions.

Example 3.7 ▶ Apply the Maclaurin formula to the exponential function $y = e^x$.

Solution. Since $(e^x)' = e^x$, it is clear that $(e^x)^{(i)} = e^x$ for any $i = 1, 2, \ldots, n$. Therefore, applying formula (3.86) and noting that $e^0 = 1$, we obtain:

$$e^x = 1 + x + \frac{x^2}{2!} + \frac{x^3}{3!} + \frac{x^4}{4!} + \cdots + \frac{x^n}{n!} + \frac{e^{\vartheta x} x^{n+1}}{(n+1)!} = \sum_{k=0}^{n} \frac{x^k}{k!} + \frac{e^{\vartheta x} x^{n+1}}{(n+1)!}. \quad ◀$$

(3.89)

Problem 3.46 Prove the following Maclaurin formulae:

$$
\begin{aligned}
\ln(1+x) &= x - \frac{x^2}{2} + \frac{x^3}{3} - \cdots + (-1)^{n+1}\frac{x^n}{n} + R_{n+1} \\
&= \sum_{k=1}^{n} (-1)^{k+1}\frac{x^k}{k} + R_{n+1}
\end{aligned}
$$

(3.90)

with

$$R_{n+1} = \frac{(-1)^n}{n+1}\frac{x^{n+1}}{(1+\vartheta x)^{n+1}};$$

$$\sin x = x - \frac{x^3}{3!} + \frac{x^5}{5!} - \frac{x^7}{7!} + \cdots + \frac{(-1)^{n+1} x^{2n-1}}{(2n-1)!} + R_{2n+1}$$

$$= \sum_{k=1}^{n} \frac{(-1)^{k+1} x^{2k-1}}{(2k-1)!} + R_{2n+1} \tag{3.91}$$

with

$$R_{2n+1} = \frac{(-1)^n \cos(\vartheta x)}{(2n+1)!} x^{2n+1} \; ;$$

$$\cos x = 1 - \frac{x^2}{2!} + \frac{x^4}{4!} - \cdots + \frac{(-1)^n x^{2n}}{(2n)!} + R_{2n+2}$$

$$= \sum_{k=0}^{n} \frac{(-1)^k x^{2k}}{(2k)!} + R_{2n+2} \tag{3.92}$$

with

$$R_{2n+2} = \frac{(-1)^{n+1} \cos(\vartheta x)}{(2n+2)!} x^{2n+2} \; ;$$

$$(1+x)^{\alpha} = 1 + \alpha x + \frac{\alpha(\alpha-1)}{2} x^2 + \cdots + \frac{\alpha(\alpha-1)(\alpha-2) \cdots (\alpha-n+1)}{n!} x^n + R_{n+1}$$

$$= \sum_{k=0}^{n} \binom{\alpha}{k} x^k + R_{n+1} , \tag{3.93}$$

where

$$\binom{\alpha}{k} = \frac{\alpha(\alpha-1)(\alpha-2) \cdots (\alpha-k+1)}{k!} \tag{3.94}$$

are called *generalised binomial coefficients* and

$$R_{n+1} = \binom{\alpha}{n+1} (1 + \vartheta x)^{\alpha-n-1} x^{n+1} \; .$$

The notation $\binom{\alpha}{k}$ for the coefficients is not accidental. Indeed, show that for positive integer $\alpha \equiv m$, the formula (3.93) becomes the binomial formula. [Hint: *first, check that the coefficients* $\binom{m}{k}$ *coincide with the binomial coefficients (1.154), and second, verify that the remainder term is zero in this case, i.e. the expansion contains a finite number of terms.*]

In the following example, we shall consider a much trickier case of $y = \arcsin(x)$. It is easy to see that a direct calculation of the derivatives of the arcsine function at

$x = 0$ from Eq. (3.75) contains complicated finite sums with binomial coefficients and hence this method seems to lead us nowhere. The trick is in noticing that

$$y' = \frac{1}{\sqrt{1 - x^2}} \quad \text{and} \quad y'' = \frac{x}{(1 - x^2)^{3/2}} \equiv y' \frac{x}{1 - x^2} \quad \Longrightarrow \quad (1 - x^2) y'' = xy' .$$

Problem 3.47 Apply the Leibniz formula to both sides of the last identity to derive the following relationship:

$$(1 - x^2) y^{(n+2)} = (2n + 1) xy^{(n+1)} + n^2 y^{(n)} .$$

This is a *recurrence relation* as it allows calculating derivatives of higher orders recursively (i.e. one after another) from the previous ones.

Now apply $x = 0$ in the above recurrence relation to get $y^{(n+2)}(0) = n^2 y^{(n)}(0)$, which can be used to derive recursively all the necessary derivatives at $x = 0$ needed to write the Maclaurin formula for the arcsine function. We see that $y^{(2)}(0) = 0$ (corresponds to $n = 0$), $y^{(4)}(0) = 2^2 y^{(2)}(0) = 0$, etc., i.e. all even order derivatives are equal to zero. Then, we only need to consider odd order derivatives. We know that $y^{(1)}(0) = 1$ from $y' = (1 - x^2)^{-1/2}$; next, $y^{(3)}(0) = 1^2 y^{(1)}(0) = 1^2$, $y^{(5)}(0) = 3^2 y^{(3)}(0) = 3^2 1^2$, $y^{(7)}(0) = 5^2 y^{(5)}(0) = 5^2 3^2 1^2$. This process can be continued, but it should be clear by now (you can prove this rigorously by induction), that

$$y^{(2k+1)}(0) = [(2k - 1)!!]^2 = \left[\frac{(2k)!}{2^k k!} \right]^2 ,$$

where we have used our previous result (3.74) for the double factorial. Now we are ready to write the Maclaurin formula for the arcsine function (recall, that we only need to keep the odd order derivatives and hence only odd powers of x are to be retained):

$$
\begin{aligned}
\arcsin(x) &= \sum_{k=0}^{n} y^{(2k+1)}(0) \frac{x^{2k+1}}{(2k + 1)!} + R_{2k+2} \\
&= \sum_{k=0}^{n} \frac{(2k)!}{4^k (2k + 1) (k!)^2} x^{2k+1} + R_{2k+2} .
\end{aligned}
\tag{3.95}
$$

The constant term is missing here since $\arcsin(0) = 0$. The expression for the remainder term is rather cumbersome and is not given here. It can be worked out using Eq. (3.75) if desired.

Problem 3.48 Apply the same method for the arccosine function $y = \arccos(x)$ and prove the Maclaurin formula for it:

$$\arccos(x) = \frac{\pi}{2} - \sum_{k=0}^{n} \frac{(2k)!}{4^k (2k+1)(k!)^2} x^{2k+1} + R_{2k+2} . \qquad (3.96)$$

Problem 3.49 Apply a similar method for obtaining the Maclaurin series for the arctangent function:

$$\arctan(x) = \sum_{k=0}^{n} \frac{(-1)^k}{2k+1} x^{2k+1} + R_{2k+2} . \qquad (3.97)$$

[Hint: *since* $y' = [\arctan(x)]' = \left(1 + x^2\right)^{-1}$ *does not contain radicals, it is sufficient to consider the identity* $\left(1 + x^2\right) y' = 1$ *and then differentiate both sides of it n times by means of the Leibniz formula (3.71).*]

Compare the two Maclaurin formulae for the arcsine and arccosine with the identity (2.72): they would be identical if not for the different remainder terms. The apparent contradiction will be resolved in Sect. 7.3.3 when we shall consider infinite Taylor series and it will become clear that identity (2.72) is satisfied exactly.

3.9   Approximate Calculations of Functions

Taylor's formula is frequently used for approximate calculations of various functions. The idea is based on taking an n−term Taylor formula (the Taylor polynomial (3.79)) and omitting the remainder term. If a better approximation for a function is needed, more terms in the Taylor's formula are considered, i.e. the function is replaced by a higher order Taylor polynomial.

As an illustration of this method, we shall consider a numerical calculation of the numbers π and e. To calculate e, we shall use Taylor's formula for the exponential function (3.89) at $x = 1$:

$$e \simeq 1 + 1 + \frac{1}{2!} + \frac{1}{3!} + \frac{1}{4!} + \cdots + \frac{1}{n!} = \sum_{k=0}^{n} \frac{1}{k!}$$

$$= 2 + \sum_{k=2}^{n} e_k \quad \text{with} \quad e_{k+1} = \frac{1}{(k+1)!} = \frac{e_k}{k+1} . \qquad (3.98)$$

The recurrent formula for the terms in the sum given above allows calculating every next term, e_{k+1}, from the previous one, e_k, and is much more convenient for the practical calculation on a computer then the direct calculation from their definition

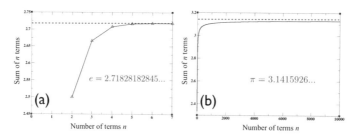

Fig. 3.4 Calculation of the numbers e (**a**) and π (**b**) using the corresponding Maclaurin finite sums (3.98) and (3.99), respectively. If the convergence of the sum for the number e is very fast, this is not so in the case of the arcsine Maclaurin series for the number π. In both cases, the exact values of the numbers e and π are shown by the horizontal red dashed lines

via the factorial. The dependence of the sum on the number of terms n is shown in Fig. 3.4a. It is really surprising that only seven terms in the sum are already sufficient to get the number $e \simeq 2.718281828459$ with the precision corresponding to 12 digits.

To calculate the number π, we consider the Maclaurin formula (3.95) for the arcsine function at $x = 1$ since $\arcsin(1) = \pi/2$:

$$\pi = 2\arcsin(1) \simeq 2\sum_{k=0}^{n} \frac{(2k)!}{4^k \, (2k+1) \, (k!)^2}$$

$$= \sum_{k=0}^{n} a_k \quad \text{with} \quad a_{k+1} = a_k \frac{(2k+1)^2}{2\,(k+1)\,(2k+3)} \, . \tag{3.99}$$

Summing this series up for bigger and bigger values of n allows obtaining better and better approximations to the number π. However, as it follows from the numerical calculation shown in Fig. 3.4b, this time the series converges much slower; even with the 10^5 terms kept in the sum the obtained approximation 3.1303099 is still with the considerable error of the order of 10^{-2}. Much better methods exist for the numerical calculation of the number π based on different principles which we cannot go into here.

Taylor's formula is frequently used for obtaining approximate representations for the functions in the vicinity of a point of interest. In other words, if one needs a simple analytical representation of a function $f(x)$ around point a, then the Taylor polynomial could be a good starting point. In Fig. 3.5, we show how this works for the exponential function $y = e^x$. Its Taylor polynomial around $x = 0$ is given by (3.89) if we drop the remainder term. It is clearly seen from the plotted graphs in (a) that a reasonable representation of this function can indeed be achieved using a small number of terms (a low order polynomial). However, the quality of the representation depends very much on the required precision (acceptable error). This is illustrated with the zoom-in image in (b) where the errors are better seen. The error gets bigger when the values of x which are further away from $x = 0$ are considered.

Generally, the higher the order of the polynomial, the better representation of the function is obtained. Also, it is a good idea to expand the function around the point of

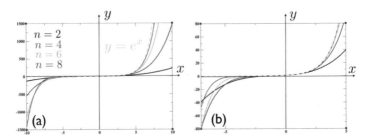

Fig. 3.5 Approximate representations of the exponential function $y = e^x$ using n−order Taylor's polynomials (3.89) around $x = 0$ (i.e. the Maclaurin formula with the remainder term dropped) for $n = 2, 4, 6, 8$ are shown together with the exact function (cyan). The central area of the graph in (**a**) is shown in (**b**) in more detail

interest, not just around $x = 0$ as we have done above for the exponential function, as this would require a smaller number of terms to achieve the same precision. However, the wider the region of x for which the analytical representation is sought, the more terms will be required.

> **Problem 3.50** Consider a mass m in the gravitational field of a very large object (e.g. the Earth) of mass M and radius R. The potential energy of the mass m at a distance d from the centre of the large object is $U(d) = -GmM/d$, where G is the gravitational constant. Show that, up to a constant, the potential energy of the mass m at the distance z above the big object surface (up to an unimportant constant) is given by $U(z) \simeq mgz$, if $z \ll R$, and hence derive an expression for the constant g. [Answer: $g = GM/R^2$.]

3.10   Calculating Limits of Functions in Difficult Cases

In Sect. 2.4.6, we looked at some simple cases of uncertainties when taking a limit. Here, we shall develop a general approach to all such cases. Taylor's formula is especially useful in calculating limits of functions in difficult cases of uncertainties $0/0$, ∞/∞ or $0 \cdot \infty$.

As an example, consider the limit (2.86) by expanding the sine function in the Maclaurin series:

$$\lim_{x \to 0} \frac{\sin x}{x} = \lim_{x \to 0} \frac{1}{x} \left[x - \frac{x^3}{3!} + \alpha(x) \right] = \lim_{x \to 0} \left[1 - \frac{x^2}{3!} + \frac{\alpha(x)}{x} \right] = 1 \,,$$

where $\alpha(x)$ corresponds to all other terms in the expansion of the sine function which do not need here to be written explicitly; what is essential for us here is that function $\alpha(x)$ tends to zero at least as x^5 when $x \to 0$. We have obtained the same result as in Sect. 2.4.6.

Problem 3.51 Use this method to prove the following limits:

$$\lim_{x \to 0} \frac{\cos x - 1}{x^2} = -\frac{1}{2} \quad ; \quad \lim_{x \to 0} \frac{\sqrt{1+x} - 1}{x} = \frac{1}{2} .$$

In fact, Taylor's theorem offers an extremely powerful general way of calculating limits with uncertainties (some special methods have already been considered in Sect. 2.4.6). Indeed, consider the limit of $f(x)/g(x)$ as $x \to x_0$ in the special case of both functions being equal to zero at the point x_0 (this is the case of $0/0$). Expanding both functions using Taylor's formula up to the $n = 1$ term and using the fact that $f(x_0) = g(x_0) = 0$, we have:

$$
\begin{aligned}
\lim_{x \to x_0} \frac{f(x)}{g(x)} &= \lim_{x \to x_0} \frac{f(x_0) + f'(x_0)(x - x_0) + \lambda(x)(x - x_0)^2}{g(x_0) + g'(x_0)(x - x_0) + \mu(x)(x - x_0)^2} \\
&= \lim_{x \to x_0} \frac{f'(x_0)(x - x_0) + \lambda(x)(x - x_0)^2}{g'(x_0)(x - x_0) + \mu(x)(x - x_0)^2} \qquad (3.100) \\
&= \lim_{x \to x_0} \frac{f'(x_0) + \lambda(x)(x - x_0)}{g'(x_0) + \mu(x)(x - x_0)} = \frac{f'(x_0)}{g'(x_0)} ,
\end{aligned}
$$

where the functions $\lambda(x)$ and $\mu(x)$ which arise in the remainder terms R_2 of Taylor's formula applied to the two functions $f(x)$ and $g(x)$, respectively, are assumed to be non-zero at x_0. The obtained result is known as L'Hôpital's rule.

This rule may need to be applied several times. This happens if the functions $\lambda(x)$ and $\mu(x)$ used above for $n = 1$ are zeroes at x_0. In other words, Taylor's formula needs to be applied to $f(x)$ and $g(x)$ keeping up to the first non-zero term in the expansion.

Problem 3.52 Let the functions $f(x)$ and $g(x)$ together with their first k derivatives be equal to zero at $x = x_0$. Prove that in this case

$$\lim_{x \to x_0} \frac{f(x)}{g(x)} = \frac{f^{(k+1)}(x_0)}{g^{(k+1)}(x_0)} ,$$

which can be considered a generalisation of L'Hôpital's rule. In practice, however, it is more convenient to apply the rule (3.100) several times until the uncertainty is fully removed.

Example 3.8 ► Calculate the limit

$$L = \lim_{x \to 0} \frac{\sin^3 x}{x^2} .$$

Solution. Apply L'Hôpital's rule

$$L = \lim_{x \to 0} \frac{(\sin^3 x)'}{(x^2)'} = \lim_{x \to 0} \frac{3 \sin^2 x \cos x}{2x} = \frac{3}{2} \lim_{x \to 0} \frac{\sin^2 x}{x} \; .$$

Now apply the rule again as we still have here 0/0:

$$L = \frac{3}{2} \lim_{x \to 0} \frac{(\sin^2 x)'}{(x)'} = \frac{3}{2} \lim_{x \to 0} \frac{2 \sin x \cos x}{1} = \frac{3}{2} \cdot 2 \cdot \lim_{x \to 0} \sin x = 0 \; . \blacktriangleleft$$

Above we considered the 0/0 case in applying L'Hôpital's rule. However, the other two cases, ∞/∞ and $0 \cdot \infty$, Can, in fact, be related to the considered 0/0 uncertainty and hence similar rules be developed. For instance, in the ∞/∞ case, when $f(x_0) = g(x_0) = \infty$, we write:

$$\lim_{x \to x_0} \frac{f(x)}{g(x)} = \lim_{x \to x_0} \frac{1/g(x)}{1/f(x)} \; .$$

The functions $1/f(x)$ and $1/g(x)$ tend to zero values when $x \to x_0$, and hence L'Hôpital's rule can be applied directly to them:

$$\lim_{x \to x_0} \frac{f(x)}{g(x)} = \lim_{x \to x_0} \frac{[1/g(x)]'}{[1/f(x)]'} = \left. \frac{[1/g(x)]'}{[1/f(x)]'} \right|_{x=x_0} \; .$$

In the same way the $0 \cdot \infty$ uncertainty is considered: if $f(x_0) = 0$ and $g(x_0) = \infty$, then in this case

$$\lim_{x \to x_0} f(x) g(x) = \lim_{x \to x_0} \frac{f(x)}{1/g(x)} = \left. \frac{f'(x)}{[1/g(x)]'} \right|_{x=x_0} \; .$$

L'Hôpital's rule is also useful in finding the limits at $\pm\infty$ in any of the cases of uncertainties. Indeed, as an example let $f(x)$ and $g(x)$ be two functions which both tend to zero as $x \to +\infty$; we would like to calculate the limit of $f(x)/g(x)$. This case can be transformed into the familiar case of the limit at zero if we introduce a new variable $t = 1/x$. In that case the limit of $x \to +\infty$ becomes equivalent to $t \to 0$, and we can write:

$$\lim_{x \to +\infty} \frac{f(x)}{g(x)} = \lim_{t \to 0} \frac{f(x(t))}{g(x(t))} = \lim_{t \to 0} \frac{F(t)}{G(t)} = \lim_{t \to 0} \frac{[F(t)]'}{[G(t)]'}$$

$$= \lim_{t \to 0} \frac{f'(x) x'(t)}{g'(x) x'(t)} = \lim_{t \to 0} \frac{f'(x)}{g'(x)} \equiv \lim_{x \to +\infty} \frac{f'(x)}{g'(x)} \; ,$$

$$(3.101)$$

where $F(t) = f(x(t))$ and $G(t) = g(x(t))$ are functions of t, and we used the chain rule when differentiating them with respect to t. As this simple derivation shows, the L'Hôpital's rule can be used even when working out the limits at infinities as well.

Problem 3.53 Prove that formula (1.143) for a finite geometrical progression is formally valid for $q = 1$. That would mean that it is actually valid for any q.

Problem 3.54 Using the L'Hôpital's rule, prove the following limits:

$$(a)\ \lim_{x \to 0} \frac{\sqrt{1 + x^2} - \sqrt{1 - x}}{x} = \frac{1}{2}\ , (b)\ \lim_{x \to 0} \frac{\ln{(1 + x)}}{x} = 1\ ,$$

$$(c)\ \lim_{x \to +\infty} \frac{\ln{(1 + x)}}{x} = 0\ , (d)\ \lim_{x \to 0} \frac{e^{\alpha x} - 1}{\sin x} = \alpha\ ,$$

$$(e)\ \lim_{x \to +\infty} \sqrt[x]{x} = 1\ ,\ (f)\ \lim_{x \to 0} \frac{\cos x - 1}{x^2} = -\frac{1}{2}\ ,$$

$$(g)\ \lim_{x \to \infty} x\left[e^{-\alpha/x} - 1\right] = -\alpha\ , (h)\ \lim_{x \to 0} \frac{xe^x - \sin x}{2x^2 \cos x} = \frac{1}{2}\ ,$$

$$(i)\ \lim_{x \to \infty} x\left[1 - x\ln\left(1 + \frac{1}{x}\right)\right] = \frac{1}{2}\ ,\ (j)\ \lim_{x \to \infty}\left(x \arctan \frac{y}{x + a}\right) = y\ .$$

[Hint: *in the case of the $x \to \infty$ limit it is convenient to introduce $t = 1/x$ so that the limit $t \to 0$ can be considered instead.*]

Problem 3.55 Prove that for any $\alpha > 0$,

$$\lim_{x \to 0} \frac{1 - \cos x^\alpha}{(1 - \cos x)^\alpha} = 2^{\alpha - 1}\quad and\quad \lim_{x \to 0} \frac{1 - \cos^2 x^\alpha}{(1 - \cos x)^\alpha} = 2^\alpha\ .$$

3.11   Analysing Behaviour of Functions

The first and second derivatives are very useful in analysing graphs of functions $y = y(x)$: one can quickly see in which intervals the function increases or decreases, or where it has extrema (minima or maxima) and saddle points.

Increase/decrease of functions It seems obvious that if a continuous function has a positive derivative in some interval $a \le x \le b$, then it increases there. The rigorous proof is provided in the following problem.

Problem 3.56 Show using the Lagrange equation (3.88) that if $f'(x) > 0$ for all x within the interval $a \le x \le b$, then $f(x)$ monotonically increases there. [Hint: *prove that* $f(x + \Delta x) > f(x)$ *for any* $\Delta x > 0$]. Similarly show that if the first derivative is negative within some interval, the function $f(x)$ monotonically decreases there.

These statements have a very simple geometrical interpretation: if the function increases, then the tangent line at any point x will make a positive angle with the x-axis; if it decreases, then this angle will be negative.

Critical points: Minima, maxima and saddle points At the beginning, let us consider an example of a function sketched in Fig. 3.6. It has two maxima (points A and C), one minimum (at B) and one saddle point (at D). At all these points, the slope of the tangent lines is zero, i.e. the tangent lines are horizontal. The saddle point, which we shall discuss below in more detail, is specifically unusual: the tangent line there is horizontal, however, it is neither a minimum nor a maximum. How can we define and distinguish different types of such points?

Let us first analyse the behaviour of a function around point A (which is at $x = x_a$) where it has a maximum. Although we already proved in Sect. 3.8 that the function has zero derivative at this point, it is worthwhile to revisit this proof again at a slightly different angle. When moving from some $x > x_a$ to the left towards x_a, our function increases and its slope is non-positive, i.e. $y'(x) = [y(x + dx) - y(x)]/dx \le 0$ (we imply the limit here as we use dx instead of Δx). If we now move towards the point x_a from the left (i.e. from some $x < x_a$), then $y(x)$ increases again; however, in this case, the slope is non-negative: $y'(x) = [y(x + dx) - y(x)]/dx \ge 0$. In other words, the slope (the first derivative of $y(x)$) *changes sign* when crossing the point x_a of the maximum being exactly equal to zero at that point; at the point x_a where the first derivative is equal to zero the tangent line (with the zero slope) is parallel

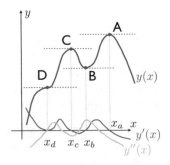

Fig. 3.6 Cartoon of a function (blue) $y = y(x)$, which has two maxima at points A and C, one minimum at point B and a saddle point at D. Its first and second derivatives, $y'(x)$ (red) and $y''(x)$ (green), are schematically drawn as well. Note that the tangent lines (dashed purple) at minima, maxima and saddle points are horizontal, i.e. have zero slope

to the x-axis. It will be clear in a moment that this is a necessary condition for a maximum (but not yet sufficient).

Importantly, if you look at the graph of $y'(x)$ around the maximum point x_a (the red curve in Fig. 3.6)), the first derivative decreases all the way starting somewhere between points x_a and x_b, passing through zero at x_a and then going down still beyond that point. In other words, the derivative of $y'(x)$, i.e. the second derivative $y''(x)$ of the original function $y(x)$, is negative in the vicinity of the maximum of a function, i.e. $y''(x_a) < 0$. We shall immediately see this as a very convenient condition of distinguishing between various critical points.

Note that the same result is valid for the other maximum at point x_c: the first derivative $y'(x_c) = 0$ there, while the second derivative $y''(x_c) < 0$ is strictly negative.

The same type of analysis can be performed around the minimum point x_b as well.

Problem 3.57 Perform similarly a general analysis of the behaviour of a function $y(x)$ around a minimum point x_b and show that: (i) $y'(x_b) = 0$ and (ii) $y''(x_b) > 0$.

We see that minima and maxima of a function (i.e. its *extrema*) can be easily found by simply solving the equation $y'(x) = 0$ with respect to x; then in many cases, one can determine whether the obtained solutions correspond to the minimum or maximum by looking at the sign of the second derivative at that point: it is positive at the minimum point(s) and negative at the maximum point(s). This would all be good, but unfortunately for some functions this method does not work as the second derivative at the extremum point of a function is zero. The method must be slightly modified as is illustrated by the following example.

Consider the function $y = x^2$. It has at the point $x = 0$ (where $y'(x) = 2x = 0$) the second derivative $y''(x) = 2 > 0$, which then definitely corresponds to a minimum. Then, consider the function $y = x^4$ shown in Fig. 3.7b together with its first two derivatives. The equation $y'(x) = 4x^3 = 0$ gives $x = 0$ as the point of the extremum, however, the second derivative $y''(x) = 12x^2$ at this point is zero and hence it is not possible to say by just looking at $y''(0)$ if $x = 0$ corresponds to the minimum or maximum of the function. In this case, one can look at the sign of the derivative on both sides of the point $x = 0$ or at the sign of the second derivative in the vicinity of this point. Indeed, consider two points very close to the point $x = 0$ of the extremum lying on both sides of it: $x = \pm\epsilon$ (with $\epsilon \to +0$). The first derivative at these points, $y'(-\epsilon) = -4\epsilon^3 < 0$ and $y'(\epsilon) = 4\epsilon^3 > 0$, changes its sign from negative to positive when moving from the left ($x = -\epsilon$) to the right ($x = \epsilon$) points, which is characteristic of the minimum. At the same time, the second derivatives at these points $y''(\pm\epsilon) = 12\epsilon^2 > 0$ are both positive, and this again is characteristic of the minimum.

Care is needed in applying the condition $y'(x) = 0$ for finding extrema of a function. This is because, firstly, the derivative may not exist. This is, e.g., the case for

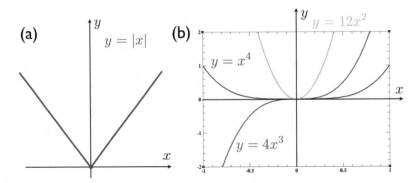

Fig. 3.7 Functions **a** $y = |x|$ and **b** $y = x^4$. In the latter case, the first $(y' = 4x^3)$ and second $(y'' = 12x^2)$ derivatives are also shown

$y = |x|$, which has a minimum at $x = 0$ where the first derivative is not defined, Fig. 3.7a. Secondly, there could also be another type of critical points, which also have a horizontal tangent line (of zero slope). These are called *saddle points* and one such point (point D) is contained on the graph of our schematic function in Fig. 3.6. As the slope is horizontal, the first derivative of the function at the saddle point $y'(x_d)$ is also equal to zero. However, the first derivative does *not* change its sign when passing this point; in addition, the second derivative at this point is zero as well: $y(x)$ increases from the left towards x_d, and then continues increasing on the right of it, i.e. the first derivative is always non-negative (and equal to zero at x_d). Hence, the second derivative is negative on the left of x_d and positive on the right of it and passes through zero at the saddle point. An important conclusion from this consideration is that the roots of the first derivative equation $y'(x) = 0$ may give any type of the critical points (minima, maxima and the saddle), and additional consideration is necessary to characterise them explicitly.

Example 3.9 ▶ As an example, consider $y(x) = xe^{-x}$.

Solution.
    Its first derivative,

$$y'(x) = e^{-x}(1 - x)$$

is equal to zero at a single point $x = 1$, hence, this is the only special point that is either a minimum, maximum or a saddle. To analyse this, we have to calculate the second derivative of $y(x)$ at this point:

$$y''(x) = e^{-x}(x - 2), \quad y''(1) = -e^{-1} < 0,$$

i.e. this point must be a maximum. ◀

Problem 3.58 Find extrema of $y(x) = \ln x / x$. [Answer: there is a maximum at $x = e$.]

Problem 3.59 Find all extrema of $y(x) = x^3 + 2x^2 - x - 2$. [Answer: *there is a minimum at* $x = (-2 + \sqrt{7})/3$ *and a maximum at* $x = (-2 - \sqrt{7})/3$.]

Problem 3.60 Consider a 2D curve specified by the function $y = f(x)$ and a point $P(x_0, y_0)$. By minimising the distance between the point P and a general point on the curve $(x, f(x))$, show that any point $(x_1, f(x_1))$ on the curve that is locally closest to P satisfies

$$(f(x_1) - y_0) f'(x_1) = x_0 - x_1. \tag{3.102}$$

This problem has a very simple geometrical interpretation, see Fig. 3.8. The point $A(x_1, f(x_1))$ is closest to the point P if the tangential line drawn at this point $y_{tan}(x)$ is perpendicular to the line PA. It is easy to see that PA is the smallest distance between the point P and the curve. This is proven by drawing a sphere around the point P such that it touches the curve exactly at A: obviously, the distance to any other point away from A will be longer than PA. Hence, the required equation for the point A can be established by relating the tangential line $y_{tan}(x) = k_\perp(x - x_1) + f(x_1)$ with $k_\perp = f'(x_1)$ and the line PA given by $y_{PA}(x) = kx + b$ with $k = (y_0 - f(x_1))/(x_0 - x_1)$. Since for two perpendicular lines $k_\perp = -1/k$ (Problem 1.217), we obtain the same Eq. (3.102) as in the problem above. The smallest distance between a point and a line is called the distance between them. There could be several such distances if Eq. (3.102) has several solutions. Each of these solutions will be *locally* the distance in a sense that the distance to any point slightly away from the solution will be longer.

In some cases, more than one solution is possible. Consider, for instance, a distance between the centre $P(x_0, y_0)$ of a circle $(x - x_0)^2 + (y - y_0)^2 = R^2$ of radius R and the circle itself. In this case $f(x) = y_0 \pm \sqrt{R^2 - (x - x_0)^2}$ with $x_0 - R \le x \le x_0 + R$. Clearly, $f'(x) = \mp(x - x_0)/\sqrt{R^2 - (x - x_0)^2}$ and Eq. (3.102) simply give an identity $-x_1 + x_0 = x_0 - x_1$, i.e. any point x_1 for which $|x_1 - x_0| \le R$ satisfies this equation, as expected.

Fig. 3.8 To the calculation of the distance to the curve from point $P(x_0, y_0)$

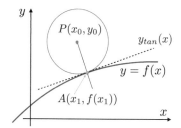

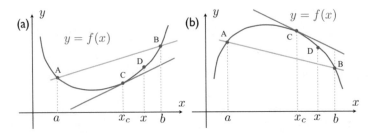

Fig. 3.9 To the two definitions of a convex (**a**) and concave (**b**) function. The tangent lines at point C are shown with magenta, while lines AB crossing the function $f(x)$ with orange in both cases

Problem 3.61 Calculate the distance between $y = \sqrt{x}$ and the point $P(2, 0)$. [Answer: *the closest point on the curve is* $A(3/2, \sqrt{3/2})$ *and the distance is* $\sqrt{7}/2$.]

Problem 3.62 What is the closest point on the parabola $y = x^2/2$ to the point $P(1, 1)$. [Answer: $A(2^{1/3}, 2^{-1/3})$.]

Problem 3.63 A projectile is fired on a flat surface with velocity v_0 and at some angle with the horizontal direction. Determine the angle that gives the maximum distance D travelled by the projectile and calculate that distance. [Answer: *the angle is* $\pi/4$ *and* $D = v_0^2/g$, *where g is the acceleration due to gravity.*]

Convex/concave behaviour of a function We know intuitively that around a minimum any function is convex, and around a maximum—concave. There exists a very simple criterion to determine if the function is concave or convex and within which interval (which may include neither minimum nor maximum, the example being the exponential function e^x which everywhere convex, but does not possess extrema, see below). Investigating the convexity of functions appears to be quite useful when analysing their general behaviour prior to plotting/sketching their graph (see below).

Note that if the function $f(x)$ is convex, its negative, $g(x) = -f(x)$, is concave. It is sufficient therefore only to consider the first (convex) case to work out the corresponding criterion.

We start by giving definitions. Let us look at Fig. 3.9a. If we plot a line AB crossing the convex function at two points, $(a, f(a))$ and $(b, f(b))$, then all points of the function between a and b would lie not higher than the corresponding values on the crossing line: $f(x) \le f_l(x)$ for any $a \le x \le b$. Let us write down the corresponding condition the function should satisfy to be convex (concave) according to this definition. The crossing line passes through two points A and B and hence has the following analytical form:

$$f_l(x) = f(a) + \frac{f(b) - f(a)}{b - a}(x - a) = \left[1 - \frac{x - a}{b - a}\right] f(a) + \frac{x - a}{b - a} f(b)$$

$$\equiv (1 - \lambda) f(a) + \lambda f(b),$$

$$(3.103)$$

where we introduced a quantity $\lambda = (x - a) / (b - a)$ lying between zero and one: $0 < \lambda < 1$. Since $x = (1 - \lambda) a + \lambda b$, the condition $f(x) \leq f_l(x)$ can equivalently be written as follows:

$$f((1 - \lambda) a + \lambda b) \leq (1 - \lambda) f(a) + \lambda f(b), \quad \text{for any } 0 < \lambda < 1. \quad (3.104)$$

In particular, for $\lambda = 1/2$:

$$f\left(\frac{a + b}{2}\right) \leq \frac{1}{2} [f(a) + f(b)]. \quad (3.105)$$

This condition shows that the convex function at the mean point between a and b is always not larger than the average of the function at both points.

The sign in these inequalities is to be reversed for the concave functions in the given interval $a \leq x \leq b$.

In fact, a simpler condition exists which can tell immediately if the function is either convex or concave. Let us look at the convex function sketched in Fig. 3.9a one more time: if we plot a tangent line at any point within the interval $a \leq x \leq b$, then it is clear that the tangent line cannot be above the graph of the function itself. In the Figure, we plotted the tangent line at point C with coordinates $(x_c, f(x_c))$, but it is readily seen that it will lie below the graph of the function $f(x)$ at any point. Hence, for any point x within the convex interval, we can simply write $f(x) \geq f_t(x)$, where

$$f_t(x) = f(x_c) + f'(x_c)(x - x_c) \quad (3.106)$$

is the tangent line at the point x_c (it passes through the point C and has the slope equal to $f'(x_c)$). Similarly, the function is concave within some interval if for any point x we have $f(x) \leq f_t(x)$, and this situation is shown schematically in Fig. 3.9b.

To work out the required criterion for the function to be convex, we shall use the $n = 1$ Taylor's formula (3.84):

$$f(x) = f(x_c) + f'(x_c)(x - x_c) + R_2 \equiv f_t(x) + R_2(x)$$

with

$$R_2(x) = \frac{1}{2} f''(\xi)(x - x_c)^2,$$

where we have used (3.85) for the remainder term in this case and ξ is somewhere between x and x_c. Therefore, from the second definition of the convex curve given above, $f(x) - f_t(x) \geq 0$, we obtain:

$$f(x) - f_t(x) = R_2(x) = \frac{1}{2} f''(\xi) (x - x_c)^2 \geq 0 \quad \Longrightarrow \quad f''(\xi) \geq 0, \quad (3.107)$$

i.e. the second derivative should not be negative. Since we used arbitrary points x_c and x, it is clear that at any point within the convex interval the second derivative of the function is not negative. Correspondingly, the second derivative is not positive within the concave interval. These are the required criteria we have been looking for.

The exponential function $y = e^x$ is convex everywhere since $y'' = e^x > 0$. The logarithmic function $y = \ln x$ is everywhere[4] concave since $y'' = (\ln x)'' = (1/x)' = -1/x^2 < 0$.

The point at which the second derivative is zero could be an *inflection point* where concave (convex) behaviour goes over into being convex (concave). Recall that $y = x^4$ has the second derivative equal to zero at $x = 0$, but has the minimum there, not the inflection point. The sufficient condition for the point to be the inflection point is that the second derivative changes sign when passing it (which is not the case for $y = x^4$).

As an example, consider the function

$$y = 4x^3 + 3x^2 - 6x + 1 . \tag{3.108}$$

Its first derivative

$$y' = 12x^2 + 6x - 6 = 6(2x - 1)(x + 1) \tag{3.109}$$

suggests that there are critical points at $x_1 = 1/2$ and $x_2 = -1$. To investigate the nature of these points, we calculate the second derivative: $y'' = 24x + 6$. At the point x_1 we have a minimum since $y''(1/2) = 18$ is positive, while at the point x_2 there is a maximum as $y''(-1) = -18$ is negative. The convex interval is obtained by solving the inequality $y'' = 24x + 6 > 0$ yielding $x > -1/4$, while the concave interval is given by $y'' = 24x + 6 < 0$ yielding $x < -1/4$. The point $x_3 = -1/4$ is therefore the inflection point. At $x \to -\infty$ the function tends to $-\infty$, while at $x \to +\infty$ it tends to the $+\infty$. The function increases where $y' > 0$ which gives two possibilities, see Eq. (3.109): either

$$\begin{cases} 2x - 1 > 0 \\ x + 1 > 0 \end{cases} \quad \Longrightarrow \quad \begin{cases} x > 1/2 \\ x > -1 \end{cases} \quad \Longrightarrow \quad x > \frac{1}{2}$$

or

$$\begin{cases} 2x - 1 < 0 \\ x + 1 < 0 \end{cases} \quad \Longrightarrow \quad \begin{cases} x < 1/2 \\ x < -1 \end{cases} \quad \Longrightarrow \quad x < -1 ,$$

---

[4] To be more precise, where it is defined, i.e. for $x > 0$.

Fig. 3.10 Graph of the
function (3.108). It has a
maximum at x_2, a minimum
at x_1 and an inflection point
at x_3

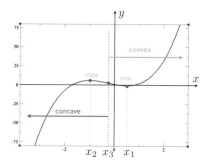

and the function decreases when $y' < 0$, which also gives two possibilities: either

$$\begin{cases} 2x - 1 > 0 \\ x + 1 < 0 \end{cases} \implies \begin{cases} x > 1/2 \\ x < -1 \end{cases} \implies \text{impossible},$$

or

$$\begin{cases} 2x - 1 < 0 \\ x + 1 > 0 \end{cases} \implies \begin{cases} x < 1/2 \\ x > -1 \end{cases} \implies -1 < x < \frac{1}{2}\,.$$

Consequently, now we should be able to sketch the function. Moving from $-\infty$ to the right towards $+\infty$, it first increases up to the point $x = -1$, where it experiences a maximum; after that point it decreases up to the point $x = 1/2$ where it has a minimum; moving further, it increases again to $+\infty$. It is convex at $x > -1/4$ and concave for $x < -1/4$. The function is plotted in Fig. 3.10.

Asymptotes of a function In order to analyse a function $y = y(x)$ at $x = \pm\infty$ and also at points x where $y(x)$ is not defined (or equal to $\pm\infty$), it is sometime possible to study its *asymptotes* (or *asymptotic curves*); they do not always exist, but if they do it helps to sketch the function and better understand its behaviour. Before giving the corresponding definition, let us first consider two functions:

$$y_1(x) = \frac{x + 2}{x + 1} = 1 + \frac{1}{x + 1} \quad \text{and} \quad y_2(x) = \frac{2x^2 + 1}{x} = 2x + \frac{1}{x}$$

to illustrate the point. Consider first $y_1(x)$; its graphs is shown in Fig. 3.11a. This function is not defined at $x = -1$, and its graph approaches the (green) vertical line $x = -1$ as x tends to -1 from both sides, i.e. when $x \to -1 - 0$ (from the left) and $x \to -1 + 0$ (from the right). As x gets closer to the value of -1, the graph of the function gets closer as well to the green vertical line. This vertical line is called the *vertical asymptote* of the function $y_1(x)$. Let us now analyse the behaviour of this function at $x \to \pm\infty$ using its alternative representation as a sum of one and the function $1/(x + 1)$ which will guide us in this discussion. We can see that on the right (for $x \to +\infty$) the function approaches the value of 1 being all the time higher than it (we say: it approaches it from above); similarly, on the left (when $x \to -\infty$) the same limiting value is being approached, this time from below, i.e. the function

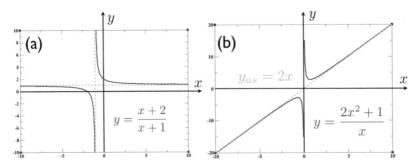

Fig. 3.11 Functions **a** $y = (x + 2) / (x + 1)$ and **b** $y = (2x^2 + 1) /x$ (blue), and their asymptotes (green)

Fig. 3.12 To the derivation
of the oblique asymptote for
the $x \rightarrow +\infty$ case

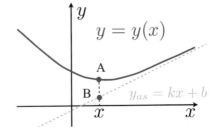

y_1 is always lower than this value. The green horizontal line $y = 1$ serves as a useful guide for this asymptotic behaviour at $\pm\infty$ and is called the *horizontal asymptote*.

Asymptotes do not need to be necessarily either vertical or horizontal, they may be of arbitrary direction as well. This is illustrated by our second function $y_2(x)$ shown in Fig. 3.11b. This function is not defined at $x = 0$ and hence the vertical line $x = 0$ serves as its vertical asymptote. However, from the way the function $y_2(x)$ is written, it is clear that as $x \rightarrow \pm\infty$, the term $1/x \rightarrow 0$ and the behaviour of the function is entirely determined by the first part $2x$, which is a straight line shown in green in the Figure. Hence, the line $y_{as} = 2x$ serves as a convenient guide to describe the behaviour of this function at $\pm\infty$ as $y_2(x)$ approaches this straight line either from above (when $x \rightarrow +\infty$) or below (when $x \rightarrow -\infty$). The line y_{as} is called an *oblique asymptote*. The horizontal asymptote can be considered as a particular case of the oblique one with the zero slope, while the vertical one—with the 90° slope.

Now we are ready to give the definitions of the asymptotes. The vertical asymptote is a vertical line $x = x_0$ with x_0 being a point at which the function $y(x)$ is not defined (tends to $\pm\infty$). The function $y(x)$ approaches the vertical asymptote as $x \rightarrow x_0$ from the left and right, i.e. the distance between the curve of the function and the asymptote tends to zero as $x \rightarrow x_0$ from either side. A horizontal or oblique asymptote $y_{as} = kx + b$ is a straight line, which approaches the curve of the function $y(x)$ at either $x \rightarrow +\infty$ or $-\infty$ (the asymptotes at both limits could be different). We shall now derive a formula for the coefficients k and b of the asymptote (obviously, $k = 0$ for the horizontal asymptote) in the case of $x \rightarrow +\infty$. To this end, let us choose an arbitrary point x as shown in Fig. 3.12. We need to calculate the distance

AB between the function $y(x)$ and the asymptote $y_{as}(x) = kx + b$ at this point. We have already solved this problem in Sect. 1.19.6 in Eq. (1.227), and the distance is given by

$$d = \frac{|kx - y(x) + b|}{\sqrt{1 + k^2}}.$$

Since d should tend to zero as $x \to +\infty$, then $kx - y(x) + b$ must be equal to some function which tends to zero at this limit:

$$\lim_{x \to +\infty} [kx - y(x) + b] = 0 \implies k = \lim_{x \to +\infty} \frac{y(x) - b}{x} = \lim_{x \to +\infty} \frac{y(x)}{x},$$
(3.110)

which is the required formula for k. The constant b is obtained similarly:

$$\lim_{x \to +\infty} [kx - y(x) + b] = 0 \implies b = \lim_{x \to +\infty} [y(x) - kx] .$$
(3.111)

In exactly the same way the limit $x \to -\infty$ is considered and exactly the same formulae are obtained in which $x \to -\infty$ is assumed.

Example 3.10 ▶ Work out the asymptotes of the function

$$y = \frac{2x^2 - 3x + 6}{x + 1}.$$
(3.112)

Solution. First we note that $x = -1$ is the vertical asymptote of the function (note that the numerator $2x^2 - 3x + 6$ is not equal to zero at $x = -1$). To find the oblique or horizontal asymptotes, we calculate the limits given by Eqs. (3.110) and (3.111) at $+\infty$ and $-\infty$:

$$k_{+\infty} = \lim_{x \to +\infty} \frac{2x^2 - 3x + 6}{x\,(x + 1)} = \lim_{x \to +\infty} \frac{2 - 3/x + 6/x^2}{1 + 1/x} = 2 ,$$

$$b_{+\infty} = \lim_{x \to +\infty} \left[\frac{2x^2 - 3x + 6}{x + 1} - 2x \right] = \lim_{x \to +\infty} \frac{2x^2 - 3x + 6 - 2x\,(x + 1)}{x + 1}$$

$$= \lim_{x \to +\infty} \frac{-5x + 6}{x + 1} = \lim_{x \to +\infty} \frac{-5 + 6/x}{1 + 1/x} = -5 ,$$

and the same limits are obviously obtained at $x \to -\infty$, i.e. one and the same asymptote $y_{as} = 2x - 5$ serves at both infinities. There are no horizontal asymptotes here as we did not find any asymptote with $k = 0$. The function together with its asymptotes is plotted in Fig. 3.13. ◀

Using critical points of a function (where extrema are) and its asymptotes, together with some additional information (e.g. crossing points with x and y axes), it is often possible to analyse the general behaviour of the function in some detail and sketch the function.

A usual procedure consists of the following steps:

Fig. 3.13 Function $y(x)$ of
Eq. (3.112) (blue) and its
asymptotes (green)

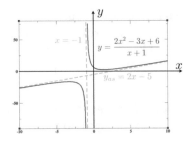

- find the domain of values of x where the function $y = y(x)$ is defined;
- check if there is any symmetry, e.g. the function is odd or even, or is periodic;
- find limits of the function at the boundaries of the domain (e.g. at $x \to \pm\infty$);
- calculate $y'(x)$ and solve the equation $y'(x) = 0$ to determine all critical points;
- determine whether the function at each critical point has a minimum, maximum or it is a saddle point;
- determine the intervals where the function increases and decreases by solving inequalities $y' > 0$ and $y' < 0$, respectively;
- calculate $y''(x)$ and determine the inflection point; then determine the intervals where the function is convex ($y'' \geq 0$) and concave ($y'' \leq 0$);
- calculate the value of the function at some selected points, especially at all critical points, the inflection points; determine the points x where the function crosses the x-axis (roots of the function, $y(x) = 0$);
- make a sketch of the function.

Problem 3.64 Analyse the function $y(x) = 3(x - 3)(x + 1)(2x - 1)$ by finding all critical and inflection points, and then making a sketch. Check your findings by plotting the function using any available software. Does this function have asymptotes at $x \to \pm\infty$? [Answer: *maximum at $x = -1/3$, minimum at $x = 2$, inflection point $x = 5/6$, no asymptotes.*]

Problem 3.65 The same for $y(x) = \left(x^2 + x + 1\right)/(x + 1)$. Does this function has an inflection point? [Answer: *maximum at $x = -2$, minimum at $x = 0$, no inflection point, asymptote $y = x$.*]

Problem 3.66 The same for $y(x) = (x - 2)^3/(x + 2)^2$. [Answer: *maximum at $x = -10$, inflection point at $x = 2$, asymptotes $x = -2$ and $y = x - 10$.*]

Problem 3.67 The same for $y = e^{-x}\sin x$. [Answer: *critical points at $x = \frac{\pi}{4} \pm \pi n$, where minima are for $n = 1, 3, \ldots$, while maxima for $n = 0, 2, 4, \ldots$, inflection points are at $x = \frac{\pi}{2} \pm \pi n$, where $n = 0, 1, 2, \ldots$; no asymptotes.*]

Problem 3.68 Consider the van der Waals equation of state of a gas

$$\left(P + \frac{a}{v^2} \right) (v - b) = RT \; ,$$

where P is pressure, T temperature, R gas constant, $v = V/N$ the volume per particle (N is the number of particles, V total volume), and a and b are constants: a indicates the strength of attraction between particles, while b corresponds to the volume a particle excludes from V (note that $v > b$). We shall consider the case of constant T to investigate the dependence $P(v)$, i.e. the isotherms of the gas.

(a) Show that the derivative

$$\frac{dP}{dv} = \frac{2a}{(v-b)^2} f(v) \; , \quad \text{where} \quad f(v) = \frac{(v-b)^2}{v^3} - \frac{RT}{2a} \; .$$

(b) Show that the derivative $f'(v)$ is positive for $b < v < 3b$, negative for $v > 3b$ and zero at $v = 3b$; hence, argue that $f(v)$ experiences a maximum at $v = 3b$, at which point it is equal to

$$f(3v) = -\frac{RT}{2a} + \frac{4}{27b} \; .$$

(c) Show that at the starting and ending points of the whole range of v the function $f(b)$ is negative, $f(b) = f(\infty) = -RT/2a$. Hence, depending on the sign of $f(3v)$, two cases need to be considered.

(d) Argue that at sufficiently large temperatures, when $RT > 8a/27b$, the function $f(v) < 0$ for any $v > b$ and hence the pressure P always decreases with the increase of the volume v.

(e) In the case of not very high temperatures, when $RT < 8a/27b$, $f(v)$ should change sign crossing the v axis twice, at volumes v_1 and v_2, where $b < v_1 < 3b$ and $3b < v_2 < \infty$ (a cubic equation needs to be solved to find these roots, but the good thing is that our general analysis does not require knowing these two critical volumes explicitly).

(f) By analysing the sign of $f(v)$ (and hence of dP/dv) in each of the intervals, $b < v < v_1$, $v_1 < v < v_2$ and $v_2 < v < \infty$, demonstrate that $P(v)$ decreases from infinity in the first interval, passing a minimum at v_1 and then starts to increase until it reaches a maximum at v_2, after which volume it decreases again until it reaches zero at $v \to \infty$.

(g) When $8a/27b = RT$, the two roots v_1 and v_2 coincide, this is called the critical point of the van der Waals equation. Show that in this case the curve $P(v)$ is decreasing with v and it has an inflection point at $v = 3b$ with the pressure $P = a/27b^2$.

(h) Sketch the graphs of the van der Waals equation using e.g. $a = 1$ and trying various values of the other two parameters (T and b).

Problem 3.69 A second-order phase transition in a ferromagnetic material between paramagnetic and ferromagnetic phases can be described, within the so-called Landau-Ginzburg theory, using the following phenomenological expression for the free energy density of the material:

$$f(m, T) = \frac{1}{2}\alpha (T - T_c) m^2 + f_4 m^4 , \qquad (3.113)$$

where m is magnetisation, T temperature, T_c phase transition temperature, f_4 and α positive constants.

(a) Show that for $T > T_c$ the free energy has a single stable state (i.e. a single minimum) with zero magnetisation only.

(b) Show that at $T < T_c$, the zero magnetisation state $m = 0$ becomes unstable (a maximum), while two states with non-zero magnetisation $m_{\pm} = \pm\sqrt{\alpha (T_c - T) / 4 f_4}$ become stable (f is minimum there). Note that the magnetisation changes continuously between the two solutions when the temperature passes the transition temperature T_c. This is characteristic for the second-order phase transition.

(c) Show that in the latter case of $T < T_c$ there are two inflection points at $m = \pm\sqrt{\alpha (T_c - T) / 12 f_4} = \pm |m_{\pm}| / \sqrt{3}$.

Problem 3.70 The same theory can be applied to the second-order phase transition between paraelectric (zero polarisation, $P = 0$) and ferroelectric ($P \neq 0$) phases of a ferroelectric material. The free energy is identical to Eq. (3.113) of a ferromagnetic if one replaces m with P. Hence, as follows from the solution of the previous Problem, the paraelectric phase exists at $T > T_c$, while ferroelectric solution becomes stable at $T < T_c$. Here, we shall consider a more general case of the material in an external electric field E:

$$f(P, T) = \frac{1}{2}\alpha (T - T_c) P^2 + f_4 P^4 - EP . \qquad (3.114)$$

Show, that the susceptibility $\chi = \partial P / \partial E$ of the paraelectric phase ($T > T_c$) at $E = 0$ is $\chi = [\alpha (T - T_c)]^{-1}$, while in the ferroelectric phase ($T < T_c$) we have $\chi = [2\alpha (T_c - T)]^{-1}$, i.e. the susceptibility jumps at the transition temperature T_c.

Problem 3.71 Consider a problem of crystal growth from the viewpoint of classical nucleation theory. In a supersaturated solution of particles (e.g. atoms), a nucleus of a crystalline phase may appear which then may grow into the bulk phase. Whether or not a spherical nucleus of radius R will continue growing or will decompose into individual particles depends on the balance of its free energy ΔG^*, which is built from two contributions, one proportional to the volume and another to the surface of the spherical nucleus:

$$\Delta G^* = \frac{4}{3}\pi R^3 \rho_B \Delta\mu + 4\pi R^2 \gamma \ .$$

The first term is negative; it accounts for the decrease of the system free energy due to growth of a more energetically favourable solid phase with $\Delta\mu < 0$ being the corresponding gain in the free energy per particle, and ρ_B the concentration of particles. This term favours formation of the crystalline (solid) phase. The second term is positive; it corresponds to the energy penalty due to formation of the surface of the nucleus with the surface energy $\gamma > 0$ per unit area. It works against the formation of the nucleus.

(a) Show that $\Delta G^*(R)$ has two extrema: at $R = 0$ and $R^* = 2\gamma/\rho_B |\Delta\mu|$.

(b) Show that the first one corresponds to a minimum while the second to a maximum.

(c) Show that at $R > 3R^*/2$ the free energy becomes negative.

(d) Show that the energy barrier for nucleation

$$\Delta G^* (R^*) = \frac{16\pi}{3} \frac{\gamma^3}{\rho^2 |\Delta\mu|^2} \ .$$

(e) Sketch the $\Delta G^*(R)$.

For nuclei sizes smaller than R^*, the growth is unfavourable; however, as R becomes larger than this critical nucleus size the free energy starts decreasing and becomes negative after $3R^*/2$, which corresponds to a very high growth rate.

Integral

4

A notion of the integral is one of the main ideas of mathematical analysis (or simply analysis). In some literature integrals are introduced in two steps: initially, the so-called *indefinite integral* is defined as an inverse operation to differentiation; then *a definite integral* is introduced via some natural applications such as an area under a curve. In this chapter, we shall use a much clearer and more rigorous route based on defining the definite integral first as a limit of an integral sum. Then the indefinite integral is introduced in a natural way simply as a convenience tool.

4.1 Definite Integral: Introduction

Consider initially a non-negative function $f(x) \geq 0$ specified in the interval $a \leq x \leq b$ as sketched in Fig. 4.1. What we would like to do is to find a way of calculating the area under the curve of the function between the boundary points a and b. If the function was a constant, $f(x) = C$, then the area is easily be calculated as $A = C(b - a)$. The area is also easily calculated in the case of a linear function,

$$f(x) = f(a) + \frac{f(b) - f(a)}{b - a}(x - a),$$

passing through the points $(a, f(a))$ and $(b, f(b))$, in which case we have a trapezoid, and hence its area $A = [f(a) + f(b)](b - a)/2$, see Eq. (1.60).

How can one calculate the area under a general curve $f(x)$ which is not a straight line? We can try to calculate the area approximately using the following steps. First, we divide the interval $a \leq x \leq b$ into n subintervals by points $x_1, x_2, \ldots, x_{n-1}$ such that

$$a \equiv x_0 < x_1 < x_2 < \cdots < x_{n-2} < x_{n-1} < x_n \equiv b,$$

where for convenience we included the boundary points a and b into the set $\{x_k\}$ as x_0 and x_n, respectively. Next, we choose an *arbitrary* point ζ_1 within the first

© The Author(s), under exclusive license to Springer Nature Switzerland AG 2022
L. Kantorovich, *Mathematics for Natural Scientists*, Undergraduate Lecture Notes in Physics, https://doi.org/10.1007/978-3-030-91222-2_4

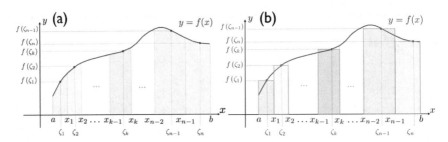

Fig. 4.1 Definition of a definite integral between a and b. The interval is divided up by n subintervals $\Delta x_k = x_k - x_{k-1}$ ($k = 1, 2, \ldots, n$), and the area of each stripe in (**a**) is replaced by the area of a rectangular of height $f(\zeta_k)$ and the width Δx_k, as shown in (**b**)

subinterval, $x_0 \leq \zeta_1 < x_1$, then another arbitrary point ζ_2 between x_1 and x_2, and so on. Altogether n points ζ_k are chosen such that $x_{k-1} \leq \zeta_k < x_k$ for all $k = 1, 2, \ldots, n$ as shown in Fig. 4.1a. This way we have divided up the area under the curve $y = f(x)$ into n strips (shown in different colours in the Figure), and hence we may calculate the whole area by summing up the areas of every such strip. The calculation of the area A_k of every strip $x_{k-1} \leq x < x_k$ can be done approximately by simply choosing the value of the function $f(\zeta_k)$ at the chosen point ζ_k within the strip as the strip height, see Fig. 4.1b, i.e. by writing

$$A_k \simeq f(\zeta_k)(x_k - x_{k-1}) = f(\zeta_k)\,\Delta x_k\,,$$

where Δx_k is the width of the strip k. This way we replace the actual strip k having a (generally) curved line at the top with a rectangular of height $f(\zeta_k)$. The error we are making may be not so large if the strip is narrow enough, i.e. its width Δx_k is relatively small (compared to the width of the whole interval $\Delta = b - a$). Summing up all contributions, we obtain an approximate value of the whole area under the curve as a sum

$$A \simeq \sum_{k=1}^{n} f(\zeta_k)\,\Delta x_k\,. \tag{4.1}$$

This is called *the integral* or *Riemann's sum* (due to Riemann). It is clear that if we make more strips, they would become narrower, the function $f(x)$ within each such interval Δx_k would change less and the approximation we are making by replacing the function within each strip with a constant $f(\zeta_k)$, less severe. So, by making more and narrower strips, one can calculate the area with better and better precision. An important point is that the *exact* result can be obtained by making an infinite number of strips (more precisely, by taking *all* widths of the strips Δx_k to zero at the same time which would require the number of strips $n \to \infty$), or ensuring that the widest of the intervals, i.e. $\max\{|\Delta x_k|\}$, goes to zero:

$$A = \lim_{\max\{|\Delta x_k|\} \to 0} \left(\sum_{k=1}^{n} f(\zeta_k)\Delta x_k \right)\,. \tag{4.2}$$

This limit (if exists) is called a *definite (Riemann) integral* and is written as follows:

$$A = \int_a^b f(x)dx , \tag{4.3}$$

where the sign $\int$ means literally "a sum" (it indeed looks like the letter "S"), and the numbers a and b are called the bottom and top limits of the integral, while the function under the integral sign the *integrand*. Note that $b > a$, i.e. the top limit is larger than the bottom one. If the limit of the integral sum exists, the function $f(x)$ is said to be *integrable*. Note that there are also more general Riemann–Stieltjes and Lebesgue integrals; we will not consider them here.

The definition for the definite integral (4.2) given above is immediately generalized for functions that are not necessarily positive. In this case the same formula (4.2) is implied, although, of course, the relation to the "area under the curve" is lost.

Moreover, the integral we considered above had a "direction" from a to b (where $b > a$), i.e. from left to right. It is also possible to consider the integral in the opposite direction, from right to left. In this case the points x_k are numbered from right to left instead and hence the differences $\Delta x_k = x_k - x_{k-1}$ become all negative. Still, the same integral sum as above can be written, which is denoted

$$A' = \int_b^a f(x)dx ,$$

i.e. it looks the same apart from the fact that the top and the bottom limits changed places: now the bottom limit b is actually larger than the top limit a since $b > a$. Assume now that $f(x)$ is positive. Then, it is clear that the above integral consists of the same "areas under the curve" as in the previous case considered above (when we integrated from a to b), but calculated with negative $\Delta x_k = -|\Delta x_k|$. Therefore, by summing up all the contributions in the integral sum we should arrive at the same numerical value of the integral as when calculating it from a to b, only with the negative sign. It is clear now that A' must be equal to $-A$, i.e. we can write the first property of the definite integral (*direction*):

$$\int_a^b f(x)dx = -\int_b^a f(x)dx . \tag{4.4}$$

Consider now a point c somewhere between a and b, and let us choose one of the division points x_k exactly at c. Then, when we add more and more points to make the subintervals smaller and smaller, we can always keep the point c as a division point. It is clear that the integral sum will split into two, one from a and c, and another from c and b; at the same time, this must be still the integral from a to b. This is because the limit of the integral sum must exist no matter how the interval $a < x < b$ is divided. Hence, we obtain the second property of the definite integral (*additivity*):

$$\int_a^b f(x)dx = \int_a^c f(x)dx + \int_c^b f(x)dx . \tag{4.5}$$

Problem 4.1 Prove that this identity still holds even if $c > b$ or $c < a$.

Next, let the function $f(x)$ be a linear combination of two other functions, $u(x)$ and $v(x)$, i.e. $f(x) = \alpha u(x) + \beta v(x)$ with some real constants α and β. Then,

$$
\lim_{\max\{|\Delta x_k|\}\to 0} \left[\sum_{k=1}^{n} f(\zeta_k)\Delta x_k \right] = \lim_{\max\{|\Delta x_k|\}\to 0} \left\{ \sum_{k=1}^{n} [\alpha u(\zeta_k) + \beta v(\zeta_k)]\,\Delta x_k \right\}
$$

$$
= \lim_{\max\{|\Delta x_k|\}\to 0} \left[\sum_{k=1}^{n} \alpha u(\zeta_k)\Delta x_k \right] + \lim_{\max\{|\Delta x_k|\}\to 0} \left[\sum_{k=1}^{n} \beta v(\zeta_k)\Delta x_k \right]
$$

$$
= \alpha \lim_{\max\{|\Delta x_k|\}\to 0} \left[\sum_{k=1}^{n} u(\zeta_k)\Delta x_k \right] + \beta \lim_{\max\{|\Delta x_k|\}\to 0} \left[\sum_{k=1}^{n} v(\zeta_k)\Delta x_k \right] ,
$$

i.e. using notations for integrals, we have:

$$
\int_a^b [\alpha u(x) + \beta v(x)]\,dx = \alpha \int_a^b u(x)dx + \beta \int_a^b v(x)dx . \qquad (4.6)
$$

Of course, the proven *linearity* property holds for a linear combination of any number of functions. In particular, it follows from this property that a constant pre-factor in the integrand can always be taken outside the integral sign.

Finally, the integral between two identical points is obviously equal to zero:

$$
\int_a^a f(x)dx = 0 .
$$

Problem 4.2 Consider two functions: one is even, $f(-x) = f(x)$, and one odd, $g(-x) = -g(x)$. Then demonstrate, using the definition of the integral, that the following identities are valid when these functions are integrated over a symmetric interval $-a \le x \le a$:

$$
\int_{-a}^a f(x)dx = 2 \int_0^a f(x)dx , \quad if \quad f(-x) = f(x) , \qquad (4.7)
$$

$$
\int_{-a}^a g(x)dx = 0 , \quad if \quad g(-x) = -g(x) . \qquad (4.8)
$$

It is sometimes useful to remember that any function $f(x)$ can always be written as a sum of even and odd functions:

$$f(x) = f_{even}(x) + f_{odd}(x) ,$$

$$f_{even}(x) = \frac{1}{2}[f(x) + f(-x)] , \quad f_{odd}(x) = \frac{1}{2}[f(x) - f(-x)] . \tag{4.9}$$

The integral defined above depends on the function $f(x)$ and the interval boundaries a and b. We did mention that the integral (if exists) should not depend on the way division of the interval is made. It might also be important how the points ζ_k (at which the function is calculated) are chosen within each subinterval. Therefore, it is clear that the limit written above is not a usual limit we considered in the previous chapters, and care is needed in taking it. However, before we start looking into these subtle complications, it is instructive to consider some examples first.

Example 4.1 ▶ Evaluate the definite integrals

$$J = \int_0^1 dx = 1 \quad \text{and} \quad I = \int_0^1 x\, dx ,$$

using subintervals of the same length and points ζ_k chosen at the end of each subinterval.

Solution. In the first integral the function $f(x) = 1$ is constant, so for any division of the interval $0 \le x \le 1$ into subintervals we have

$$J = \lim_{\max\{|\Delta x_k|\} \to 0} \left[\sum_{k=1}^n \Delta x_k \right] = \lim_{\max\{|\Delta x_k|\} \to 0} \Delta x = \Delta x = 1 ,$$

where $\Delta x = 1$ is the whole integration interval. (Note that the calculation does not depend on the choice of the points ζ_k within the subintervals because the integrand is a constant.)

In the second case $f(x) = x$. We divide the interval $0 \le x \le 1$ into n subintervals of the same length $\Delta = 1/n$ with points $x_k = k\Delta$ ($k = 0, 1, 2, \ldots, n$), and within each subinterval we choose the point $\zeta_k = x_k$ right at the end of it. Then, consider the integral sum:

$$I_n = \sum_{k=1}^n f(\zeta_k)\Delta = \sum_{k=1}^n x_k \Delta = \sum_{k=1}^n k\Delta^2 = \Delta^2 \sum_{k=1}^n k = \Delta^2 \frac{1+n}{2} n ,$$

where in the last passage we have used the formula for a sum of an arithmetic progression (Sect. 1.12, Eq. (1.147)). Since $\Delta = 1/n$ by construction, we get

$$I_n = \frac{1+n}{2n} = \frac{1}{2n} + \frac{1}{2} .$$

The limit of the widths of all subintervals going to zero corresponds to the limit of $n \to \infty$. Taking this limit in the calculated integral sum, I_n, we finally obtain $I = 1/2$. This result is to be expected as what we have just done was the calculation of the area of a right triangle of the unit bottom side and the unit height. ◄

Problem 4.3 Perform the calculation of the same integral I of the previous example by still using subintervals of the same length $\Delta = 1/n$; however, this time choose the points ζ_k either at the beginning, $\zeta_k = (k - 1) \Delta$, or in the middle, $\zeta_k = (k + 1/2) \Delta$, of each subinterval ($k = 1, 2, \ldots, n$). Make sure that you get the same result after taking the $n \to \infty$ limit, in spite of the fact that the integral sums for the given value of n are different in each case.

Problem 4.4 Generalize the above results for the intervals $a \leq x \leq b$:

$$\int_a^b dx = b - a \quad \text{and} \quad \int_a^b x dx = \frac{1}{2} \left(b^2 - a^2 \right) . \tag{4.10}$$

Problem 4.5 Using the same method (subintervals of the same length and a particular, but simple choice of the points ζ_k within them), show that the integral

$$\int_0^1 x^2 dx = \frac{1}{3} .$$

[Hint: *you will need expression* (1.148) *for the sum of the squares of integer numbers.*] Try several different choices of the points ζ_k within the subintervals, and make sure that the result is always the same in the $\Delta \to 0$ limit!

Problem 4.6 Using the subintervals of the same length, prove the following identity:

$$\int_{-1}^1 e^{\alpha x} dx = \frac{1}{\alpha} \left(e^\alpha - e^{-\alpha} \right) .$$

[Hint: *you will have to sum a geometric progression here, see* (1.143), *and calculate a limit at $n \to \infty$, which contains $0/0$ uncertainty.*]

Problem 4.7 Using the subintervals of the same length, prove the following identity:

$$\int_0^\pi \sin x\, dx = 2 \ .$$

[Hint: *use Eq.* (2.63).]

4.2   Main Theorems

Of course, the above direct method of calculating integrals seems to be extremely difficult. In many cases, analytical calculation of the required finite integral sums may be not so straightforward or even impossible to do. Fortunately, there is a much easier way whose essence is in relating the integral to the derivative which manifests itself in the main theorem of calculus to be discussed later on. However, in order to discuss it, we first have to understand in more detail the meaning of the definite integral itself.

We start from a more rigorous definition. A subtle point about existence of the limit of the integral sum (and hence of the integral) is that the same result must be obtained no matter how the division of each interval is made (e.g. equal subintervals for each particular n or of different widths) and how the points ζ_k are chosen within the subintervals. By adopting a particular way of dividing the original interval $a \le x \le b$ into n subintervals and choosing the points ζ_k within each of them we shall obtain a particular value of the integral sum

$$A(n) = \sum_{k=1}^{n} f\left(\zeta_k\right) \Delta_k \ ,$$

with $\Delta(n) = \max\{\Delta_k\}$ being the maximum width of all subintervals for the given n (assuming for definiteness that any $\Delta_k > 0$). The obtained numerical sequence, $A(1), A(2), \ldots, A(n), \ldots$, of the values of the integral sums may tend to some limit A as $n \to \infty$ (and $\Delta(n) \to 0$).

The first definition of the integral then would sound similar to the limit of the numerical sequences in Sect. 2.2: the definite integral of a function $f(x)$ on the interval $a \le x \le b$ exists (or the function $f(x)$ is *integrable* there) if any numerical sequence $\{A(n)\}$ of the integral sums obtained by choosing various ways of dividing the original interval into n subintervals, such that $\lim_{n\to\infty} \Delta(n) = 0$, and choosing points ζ_k inside the subintervals in different ways, tends to the same limit A. This definition allows us to adopt immediately all main theorems related to the limit of numerical sequences of Sect. 2.2. For instance, the limit of a sum is equal to the sum of the limits, and hence the integral of a sum of two functions is equal to the sum of individual integrals calculated for each of them.

It is possible to give an appropriate definition of the integral using the "$\epsilon - \delta$" language as well: the integral on the interval $a \le x \le b$ exists and equal A (or the function $f(x)$ is integrable there with the value of the integral being A) if for any $\epsilon > 0$ there exists such $\delta > 0$ so that for any division into subintervals satisfying $\Delta(n) < \delta$ (i.e. any of them is shorter than δ) we have

$$\left| \sum_{k=1}^{n} f(\zeta_k) \Delta_k - A \right| < \epsilon \tag{4.11}$$

for arbitrary selection of points ζ_k within the subintervals. This definition is basically saying that if we choose any precision ϵ with which we would like to estimate the value of the integral via the appropriate integral sum, then it is always possible to select an appropriate division into subintervals which are small enough to obtain the integral sum with the given precision, i.e. satisfying the condition

$$A - \epsilon < \sum_{k=1}^{n} f(\zeta_k) \Delta_k < A + \epsilon . \tag{4.12}$$

Both definitions of the limit of the integral sum are equivalent and clearly state that the integral depends only on the function $f(x)$ and the numbers a and b.

Once the rigorous definitions are given, we have to understand what conditions the functions must satisfy to be integrable.

Theorem 4.1 *To be integrable, the function $f(x)$ must be limited on the interval $a \le x \le b$, i.e. two real numbers m and $M > m$ must exist such that $m \le f(x) \le M$. This condition is the* necessary *one.*

Proof: (by contradiction): assume that the function $f(x)$ is unlimited from above (i.e. it can be as *large* as desired) within the given interval. Let us assume that it is infinite at least at one point c within the interval. Then, when we divide the interval into subinterval. One such subinterval (lets its number is k) would contain that point c and hence we can always choose the point ζ_k within that subinterval very close to c, with the value of the function $f(\zeta_k)$ there as large as we like. It is clear then that this way we can make the contribution $f(\zeta_k) \Delta_k$ of that subinterval into the integral sum as large as necessary to disobey the inequality (4.11). **Q.E.D.**

Next, we introduce new definitions due to French mathematician Darboux, called Darboux sums. Consider a specific division of the interval $a \le x \le b$ into n subintervals, and let m_k and M_k be the smallest and largest values of the function $f(x)$

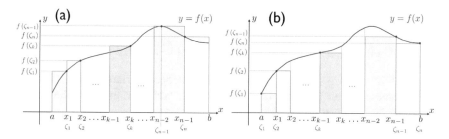

Fig. 4.2 To the definition of the upper (**a**) and lower (**b**) Darboux sums

within the $k-$th subinterval $x_{k-1} \leq x < x_k$, i.e. $m_k \leq f(x) \leq M_k$. The corresponding integrals sums ($\Delta_k = x_k - x_{k-1}$),

$$A^{low}(n) = \sum_{k=1}^{n} m_k \Delta_k \ , \tag{4.13}$$

$$A^{up}(n) = \sum_{k=1}^{n} M_k \Delta_k \ , \tag{4.14}$$

are called lower and upper Darboux sums, respectively. These sums correspond to a specific choice of the points ζ_k within each subinterval as illustrated in Fig. 4.2: for the lower sum each point ζ_k is chosen where the function $f(x)$ takes on the lowest value within each subinterval, and similarly for the upper Darboux sum. Since any other choice would give the values of the function not lower than m_k and not larger than M_k, it is clear that for an arbitrary choice of the points ζ_k and for the given division of the original interval the corresponding integral sum $A(n)$ will be not lower than $A^{low}(n)$ and not larger than $A^{up}(n)$:

$$A^{low}(n) \leq A(n) \leq A^{up}(n) \ . \tag{4.15}$$

Note that the Darboux sums only depend on the function $f(x)$ and on the division points x_k. Now let us add one more division point x'_k between the points x_{k-1} and x_k of this particular interval of length Δ_k. This way we should have $n+1$ divisions. The values of m_i and M_i for all i except for $i = k$ will be the same, however, the $k-$th subinterval now consists of two, $x_{k-1} \leq x < x'_k$ of length Δ'_k and $x'_k \leq x < x_k$ of length Δ''_k (of course, $\Delta'_k + \Delta''_k = \Delta_k$), with the lowest values of the function m'_k and m''_k and its largest values M'_k and M''_k, respectively. It is clear, however, that at least one of the lowest values will be strictly larger than the lowest value m_k for the combined subinterval $x_{k-1} \leq x < x_k$ and one of the largest values will be strictly smaller than M_k. For instance, if say m_k is taken by the function $f(x)$ at some point $x = \zeta'$ between x_{k-1} and x'_k, then in the other subinterval, $x'_k \leq x < x_k$, the function will be greater than $m_k = f(\zeta')$. Hence, the contribution of this combined interval into the lower Darboux sum can only get larger:

$$m'_k \Delta'_k + m''_k \Delta''_k > m_k \Delta'_k + m_k \Delta''_k = m_k \left(\Delta'_k + \Delta''_k \right) = m_k \Delta_k \ ,$$

and its contribution into the upper sum smaller:

$$M'_k \Delta'_k + M''_k \Delta''_k < M_k \Delta'_k + M_k \Delta''_k = M_k \left(\Delta'_k + \Delta''_k \right) = M_k \Delta_k .$$

Therefore, the corresponding lower Darboux sum becomes larger, while the upper sum becomes smaller. This should remain true if we add an arbitrary number of new divisions. We have just proved this

Theorem 4.2 *If we add more division points to the given ones, the lower Darboux sum can only increase, while the upper sum only decrease.*

We also need another theorem.

Theorem 4.3 *Any lower Darboux sum (for an arbitrary division of the interval) can never be larger than any upper sum corresponding to the same or a different division.*

Proof: Consider a particular division with the lower and upper Darboux sums being A_1^{low} and A_1^{up} (we omitted the number of divisions n here for convenience). Consider next a completely different division in which some or all division points are different from those in the first division. For the second division the Darboux sums are, respectively, A_2^{low} and A_2^{up}. Let us construct the third division by adding all new points from the second division to the first. The Darboux sums become A_3^{low} and A_3^{up}. From Theorem 4.2 it follows that $A_3^{low} \geq A_1^{low}$ and $A_3^{up} \leq A_1^{up}$. However, exactly the same combined division can be constructed by adding all new division points of the first division to the second one. Again using Theorem 4.2, we can also write: $A_3^{low} \geq A_2^{low}$ and $A_3^{up} \leq A_2^{up}$. On the other hand, the low Darboux sum is always not larger than the upper one for the same division. Therefore, we have:

$$A_1^{low} \leq A_3^{low} \quad \text{and} \quad A_3^{low} \leq A_3^{up} \quad \text{and} \quad A_3^{up} \leq A_2^{up} \quad \Longrightarrow \quad A_1^{low} \leq A_2^{up} .$$
$$(4.16)$$

This shows that the lower sum is always not larger than the upper one, even if the latter corresponds to a completely different set of divisions (of course, for the same divisions, by definition, $A^{low} \leq A^{up}$). Similarly,

$$A_1^{up} \geq A_3^{up} \quad \text{and} \quad A_3^{up} \geq A_3^{low} \quad \text{and} \quad A_3^{low} \geq A_2^{low} \quad \Longrightarrow \quad A_1^{up} \geq A_2^{low} ,$$
$$(4.17)$$

i.e. the upper sum is always not smaller than any lower sum from the same or any other division. **Q.E.D.**

The proven theorem shows that whatever divisions we make, the lower Darboux sums are always smaller than (or equal to) any of the upper ones, and, *vice versa*, the upper ones are always larger than (or equal to) any of the lower ones. This means that all possible (i.e. obtained using various divisions) lower sums are limited from above, and all possible upper sums are limited from below. Hence we denote by A_* the upper value for all the lower sums and by A^* the lowest value for all the upper sums. These are called lower and upper Darboux integrals, respectively. Obviously, $A_* \leq A^*$, and for any lower and upper sums:

$$A^{low} \leq A_* \leq A^* \leq A^{up} . \tag{4.18}$$

Note that A^{low} and A^{up} may correspond to two completely different divisions of the interval.

Theorem 4.4 (Existence of the definite integral): *For the definite integral to exist, it is necessary and sufficient that*

$$\lim_{\Delta \to 0} \left(A^{low} - A^{up} \right) = 0 , \tag{4.19}$$

i.e. for any positive ϵ there exists such positive δ that for any divisions satisfying $\Delta = \max \{\Delta_k\} < \delta$ we have $0 < A^{up} - A^{low} < \epsilon$.

Proof: We have to prove both sides of the theorem, i.e. that this condition is both necessary and sufficient. To prove the necessary condition, we should assume that the integral exists and hence the condition (4.19) follows as a consequence. If the integral exists and equals A than for *any* division satisfying $\Delta < \delta$ follows

$$A - \frac{\epsilon}{4} < \alpha < A + \frac{\epsilon}{4} \implies \alpha - A < \frac{\epsilon}{4} \text{ and } A - \alpha < \frac{\epsilon}{4} , \tag{4.20}$$

where α is the corresponding integral sum for the chosen division and we introduced $\epsilon/4$ instead of ϵ for convenience. Note that this inequality holds for any choice of the internal points ζ_k inside the divisions (subintervals). On the other hand, $\alpha > A^{low}$ and any other choice of the internal points ζ_k would result in the integral sum exceeding A^{low}. Still, for the given ϵ, we can select the internal points in such a way that the corresponding integral sum α' would satisfy

$$\alpha' < A^{low} + \frac{\epsilon}{4} \implies \alpha' - A^{low} < \frac{\epsilon}{4} . \tag{4.21}$$

Similarly, with any choice of the internal points the integral sum would not exceed A^{up}, however, one can always make a selection of the internal points such that the corresponding integral sum α'' would satisfy

$$\alpha'' > A^{up} - \frac{\epsilon}{4} \quad \Longrightarrow \quad A^{up} - \alpha'' < \frac{\epsilon}{4} . \tag{4.22}$$

We mentioned that inequalities (4.20) are valid for any choice of the internal points. In particular, they are valid for the two internal point selections we have made above, hence one can also write for them:

$$\alpha'' - A < \frac{\epsilon}{4} \quad \text{and} \quad A - \alpha' < \frac{\epsilon}{4} . \tag{4.23}$$

Therefore, we have:

$$A^{up} - A^{low} = \left(A^{up} - \alpha'' \right) + \left(\alpha'' - A \right) + \left(A - \alpha' \right) + \left(\alpha' - A^{low} \right)$$
$$< \frac{\epsilon}{4} + \frac{\epsilon}{4} + \frac{\epsilon}{4} + \frac{\epsilon}{4} < \epsilon ,$$

which means that (4.19) is indeed satisfied.

Now we prove the sufficiency condition, which means that by assuming that (4.19) is satisfied we have to show that the integral exists. For any division the lower and upper sums satisfy $A^{low} \le A_*$ and $A^* \le A^{up}$, so that

$$A^{low} + A^* \le A_* + A^{up} \quad \Longrightarrow \quad A^* - A_* \le A^{up} - A^{low} .$$

Also, $A^* \ge A_*$, from which it follows that

$$A^* - A_* \ge 0 \quad \text{and} \quad A^* - A_* \le A^{up} - A^{low} \quad \Longrightarrow \quad 0 \le A^* - A_* \le A^{up} - A^{low} .$$

On the other hand, since the condition (4.19) is satisfied, we have $A^{up} - A^{low} < \epsilon$, so that $0 \le A^* - A_* < \epsilon$ for any ϵ, meaning that $A^* = A_*$ which we shall denote simply as A. Then, we can also write

$$A^{low} \le A_* = A = A^* \le A^{up} . \tag{4.24}$$

On the other hand, for any choice of the internal points ζ_k we have the integral sum α satisfying $A^{low} \le \alpha \le A^{up}$, which, when combined with inequality (4.24), gives:

$$\alpha \le A^{up} \quad \text{and} \quad A \ge A^{low} \quad \Longrightarrow \quad \alpha - A \le A^{up} - A^{low}$$

and similarly

$$\alpha \ge A^{low} \quad \text{and} \quad A \le A^{up} \quad \Longrightarrow \quad \alpha - A \ge A^{low} - A^{up} = - \left(A^{up} - A^{low} \right) ,$$

which simply means that $|\alpha - A| \le A^{up} - A^{low}$. However, $A^{up} - A^{low} < \epsilon$ for any division obeying $\Delta < \delta$ due to condition (4.19), which yields $|\alpha - A| < \epsilon$ for any such division. But, by virtue of condition (4.11), this is exactly what we mean by saying that the integral exists and its value is A. **Q.E.D.**

At this point we are ready to formulate a sufficient condition for a function $f(x)$ to be integrable.

Theorem 4.5 *Functions which are continuous within an interval are integrable there, i.e. the definite integral (4.3) exists if $f(x)$ is continuous between a and b.*

Proof: Consider the difference of the upper and lower sums for a particular division:

$$A^{up} - A^{low} = \sum_{k=1}^{n} M_k \Delta x_k - \sum_{k=1}^{n} m_k \Delta x_k = \sum_{k=1}^{n} \Delta f_k \Delta x_k , \qquad (4.25)$$

where m_k and M_k are the minimum and maximum values of the function $f(x)$ within the subinterval $\Delta x_k = x_k - x_{k-1}$, and $\Delta f_k = M_k - m_k$ characterizes the change of the function within the same subinterval. Because the function $f(x)$ is continuous, for any $\epsilon' > 0$ one can find $\delta > 0$ such that $|x' - x''| < \delta$ implies $|f(x') - f(x'')| < \epsilon'$. Let x' be a point where $f(x)$ takes on its minimum value m_k within the interval Δx_k and x'' its maximum value M_k. Then, we can say that

$$\left| f\left(x'\right) - f\left(x''\right) \right| = |m_k - M_k| = M_k - m_k = \Delta f_k < \epsilon' \quad \text{as long as} \quad \Delta x_k < \delta .$$

Choosing $\epsilon' = \epsilon / (b - a)$, we can rewrite (4.25) as follows:

$$A^{up} - A^{low} = \sum_{k=1}^{n} \Delta f_k \Delta x_k < \sum_{k=1}^{n} \frac{\epsilon}{b-a} \Delta x_k = \frac{\epsilon}{b-a} \sum_{k=1}^{n} \Delta x_k = \frac{\epsilon}{b-a} (b-a) = \epsilon ,$$

which corresponds to the sufficient condition for the integral to exist (from the previous Theorem 4.4) to be obeyed. Therefore, continuous functions do fully satisfy that Theorem. **Q.E.D.**

In fact it appears that piecewise continuous functions are also integrable. This is because the points of discontinuity can always be chosen as some of the division points. Moreover, the value of the integral does not change if the value of the function $f(x)$ is changed in any number of *isolated points* within the integration region (see Sect. 4.5.3).

Theorem 4.6 (Average value theorem). *If the function $f(x)$ is continuous between a and b, then*

$$\int_{a}^{b} f(x)dx = f(\zeta) (b - a) , \qquad (4.26)$$

where ζ is some point within the interval.

Proof: Indeed, since the function is continuous, it should reach its minimum, m, and maximum, M, values somewhere within that interval; these could also be the boundaries of the interval. Let us now estimate the integral sum from below,

$$\sum_{k=1}^{n} f\left(\zeta_k\right) \Delta_k \geq \sum_{k=1}^{n} m \Delta_k = m \sum_{k=1}^{n} \Delta_k = m\left(b - a\right) = m\mu \;,$$

and from above,

$$\sum_{k=1}^{n} f\left(\zeta_k\right) \Delta_k \leq \sum_{k=1}^{n} M \Delta_k = M \sum_{k=1}^{n} \Delta_k = M\left(b - a\right) = M\mu \;,$$

where $\mu = b - a$ is the length of the interval. Taking the limit $\Delta = \max\{\Delta_k\} \to 0$, we then obtain

$$m\mu \leq \int_a^b f(x)dx \leq M\mu \quad \Longrightarrow \quad m \leq \frac{\int_a^b f(x)dx}{\mu} \leq M \;, \tag{4.27}$$

i.e. the value of the integral divided by the lengths of the interval μ, let us call this ratio λ, is bracketed by the minimum and maximum values of the function on the interval: $m \leq \lambda \leq M$. According to the second Bolzano–Cauchy theorem (Theorem 2.19), there should exist a point ζ within the interval at which the function $f(x)$ would take on exactly that value λ, i.e. $\lambda = f\left(\zeta\right)$, since continuous functions take on all values continuously within the interval. Recalling what the number λ is, we recover the desired result (4.26):

$$\lambda = f\left(\zeta\right) = \frac{\int_a^b f(x)dx}{\mu} \quad \Longrightarrow \quad \int_a^b f(x)dx = f\left(\zeta\right)\mu = f\left(\zeta\right)\left(b - a\right) \;.$$

Q.E.D.

This theorem is called the average value theorem because it shows that $f\left(\zeta\right)$ is equal to the integral between a and b divided by the length of the interval:

$$f\left(\zeta\right) = \frac{\int_a^b f(x)dx}{b - a} \;. \tag{4.28}$$

Theorem 4.7 (Absolute value estimate): *It is usually necessary to be able to put an upper and lower boundaries to an integral. For this purpose the following inequality is useful:*

$$\left| \int_a^b f(x)dx \right| \leq \int_a^b |f(x)|\, dx \quad \text{or} \quad -\int_a^b |f(x)|\, dx \leq \int_a^b f(x)dx \leq \int_a^b |f(x)|\, dx. \tag{4.29}$$

Proof: This follows immediately from the integral being an integral sum and in-equality (1.195):

$$\left| \sum_{k=1}^{n} f\left(\zeta_k\right) \Delta x_k \right| \leq \sum_{k=1}^{n} \left| f\left(\zeta_k\right) \right| \Delta x_k .$$

Performing the limit $\Delta \to 0$ on both sides, we obtain the result sought for. **Q.E.D.**

4.3   Main Theorem of Integration. Indefinite Integrals

We mentioned above that it is possible to relate the definite integral to a derivative, and this relationship forms a very powerful tool for calculating integrals which bypasses the calculation of the integral sums. To this end, we first have to define an integral as a function. Indeed, consider a continuous function $f(x)$ and let us integrate it between the limits a and x treating the upper limit $x > a$ as a *variable*:

$$F(x) = \int_{a}^{x} f(x')dx' . \tag{4.30}$$

We used x' inside the integral instead of x in order to distinguish it from the upper limit x; note that the integration variable (x' in our case) is a dummy variable similar to the summation index, and hence any letter can be used for it without affecting the value of the integral. It is always a good idea to follow this simple rule to avoid mistakes.

Now, the integral defined above has its upper limit as a variable and hence it is a function of that variable; we denoted this function as $F(x)$. What we would like to do is to calculate its derivative. According to the definition of the derivative,

$$F'(x) = \lim_{\Delta x \to 0} \frac{\Delta F}{\Delta x} ,$$

where

$$\Delta F = F(x + \Delta x) - F(x) = \int_{a}^{x+\Delta x} f(x')dx' - \int_{a}^{x} f(x')dx' . \tag{4.31}$$

Because of the additivity property (4.5) of the integrals, this difference is simply an integral between x and $x + \Delta x$:

$$\Delta F = \int_{x}^{x+\Delta x} f(x')dx' . \tag{4.32}$$

This result is easy to understand in terms of the integral being an "area under the curve of the function" concept and is helped by Fig. 4.3: the integral from a to $x + \Delta x$ gives an area between these two points; when we subtract from it the integral between a and x, we obtain only the area between points x and $x + \Delta x$, i.e. exactly the integral

Fig. 4.3 An illustration to
formula (4.32)

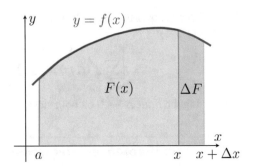

(4.32). Next, we shall use the fact that the function $f(x)$ is continuous, and hence we can benefit from Theorem 4.6 of the previous subsection to write:

$$\Delta F = \int_{x}^{x+\Delta x} f(x')dx' = f(\zeta)\,[(x+\Delta x) - x] = f(\zeta)\,\Delta x\,, \qquad (4.33)$$

where ζ is somewhere between x and $x + \Delta x$, i.e. $\zeta = x + \vartheta\,\Delta x$ with ϑ being some number between 0 and 1. Hence,

$$\begin{aligned}
F'(x) &= \lim_{\Delta x \to 0} \frac{\Delta F}{\Delta x} = \lim_{\Delta x \to 0} \frac{f(\zeta)\,\Delta x}{\Delta x} = \lim_{\Delta x \to 0} f(\zeta) \\
&= \lim_{\Delta x \to 0} f(x + \vartheta\,\Delta x) = f(x)\,.
\end{aligned} \qquad (4.34)$$

Thus, the derivative of the integral $F(x)$ considered as a function of its upper limit gives the integrand $f(x)$. In this sense, the integration is *inverse* to differentiation.

This very important result is the key for calculating integrals. Indeed, what it tells us is that in order to calculate an integral of $f(x)$ with the upper limit being x we simply need to find a function $F(x)$ which derivative is exactly equal to $f(x)$. Of course, if such a function is found, then any function $F(x) + C$, where C is a constant, would also satisfy the required condition $F'(x) = f(x)$. This general relationship between the integrand $f(x)$ and the function $F(x)$ whose derivative is equal to $f(x)$ is usually expressed via the so-called *indefinite integral*:

$$\int f(x)dx = F(x) + C \quad \Longleftrightarrow \quad F'(x) = f(x). \qquad (4.35)$$

Here $F(x)$ is any function whose derivative is equal to $f(x)$ and C is an arbitrary number (a constant). As we can calculate derivatives of arbitrary continuous functions, we can, by means of the inverse calculation, find the corresponding indefinite integrals of many functions.

Problem 4.8 Let $F(x)$ be a function whose derivative is equal to $f(x)$. Prove that other functions whose derivative is equal to $f(x)$ can only differ by a constant. [Hint: *prove by contradiction. Also note that the only function whose derivative is equal to zero is a constant function. The latter follows from the Lagrange formula* (3.88).]

Consider several examples of solving indefinite integrals. Since

$$\left(\frac{x^{\alpha+1}}{\alpha+1}\right)' = x^\alpha \quad \text{for any} \quad \alpha \neq -1 ,$$

we can immediately write this also as

$$\int x^\alpha dx = \frac{x^{\alpha+1}}{\alpha+1} + C , \quad \alpha \neq -1. \tag{4.36}$$

The case of $\alpha = -1$ is a special one, since $(\ln x)' = x^{-1}$. To make our integral valid also for negative values of x, this formula is generalized: $(\ln |x|)' = x^{-1}$. Indeed, if $x > 0$, then we get the familiar result. If, however, $x < 0$, then

$$(\ln |x|)' = (\ln(-x))' = \frac{1}{-x}(-1) = \frac{1}{x} ,$$

i.e. the same result. Therefore, we arrive at the following indefinite integral:

$$\int \frac{dx}{x} = \ln |x| + C . \tag{4.37}$$

Problem 4.9 Prove the following frequently encountered indefinite integrals:

$$\int \cos x dx = \sin x + C , \quad \int \sin x dx = -\cos x + C , \tag{4.38}$$

$$\int e^{\alpha x} dx = \frac{1}{\alpha}e^{\alpha x} + C , \tag{4.39}$$

$$\int \frac{dx}{1+x^2} = \arctan x + C , \quad \int \frac{dx}{\sqrt{1-x^2}} = \arcsin x + C , \tag{4.40}$$

$$\int \frac{dx}{\cos^2 x} = \tan x + C , \quad \int \frac{dx}{\sin^2 x} = -\cot x + C . \tag{4.41}$$

$$\int \frac{dx}{\sqrt{1+x^2}} = \ln\left|x + \sqrt{1+x^2}\right| + C . \tag{4.42}$$

We have established in Eq. (3.44) that when the complex exponential e^{ix} is differentiated with respect to x, we have ie^{ix}, i.e. i can be treated as a parameter. One can also formally[1] define the integral of the complex exponential using its definition Eq. (3.42) and the additivity property:

$$\int e^{ix}dx = \int (\cos x + i \sin x)\, dx = \sin x - i \cos x = -i\,(\cos x + i \sin x) = -ie^{ix}, \tag{4.43}$$

which can also be formally written as e^{ix}/i, i.e. in this calculation as well i can be treated as a parameter, *cf.* Eq. (4.39). We shall see later on that this fact may shortcut the calculation of certain integrals.

Now we are finally at the position to derive the formula for calculating definite integrals. For this, we consider again the definite integral with the variable upper limit. We found above that its derivative, $F'(x)$ is equal to the integrand, $f(x)$. It is therefore clear that the actual value of the integral should be

$$\int_a^x f(x')dx' = F(x) + C .$$

Here the constant C must be somehow related to the bottom limit a. To find C, we set $x = a$ in the formula above and use the fact that the integral between identical limits is equal to zero. We obtain then that $C = -F(a)$. This brings us to

$$\int_a^x f(x')dx' = F(x) - F(a) .$$

It is now convenient to replace x as the upper limit with b, and we arrive at the *main formula of the integral calculus* (the so-called Newton–Leibniz formula):

$$\int_a^b f(x)dx = F(b) - F(a) = F(x)|_a^b . \tag{4.44}$$

In the last passage above a convenient notation has been introduced for the difference of the function $F(x)$ between the end points of the integration interval, a and b, a vertical line with the two limits explicitly indicated; we shall be using it frequently.

This is our central result: calculation of a definite integral requires finding the function $F(x)$ whose derivative is equal to the integrand $f(x)$; after that the difference of the two values of $F(x)$ at b and a provides us with the final result. This means that we need to solve first for the indefinite integral, take its result as $F(x)$ (ignoring any constant C) and calculate the difference of it at b and a.

---

[1] A rigorous definition of an integral of a complex function will be give in Vol. II.

As an example, let us calculate the integral between 0 and 1 of $f_1(x) = x^2$ and $f_2(x) = 2x^2 - 1$. In the case of $f_1(x)$, we have from (4.36): $F_1(x) = x^3/3 + C$ (this is obtained with $\alpha = 2$). We can ignore the constant to obtain:

$$\int_0^1 x^2 dx = \left.\frac{x^3}{3}\right|_0^1 = \left(\frac{x^3}{3}\right)_{x=1} - \left(\frac{x^3}{3}\right)_{x=0} = \frac{1^3}{3} - \frac{0^3}{3} = \frac{1}{3},$$

which is the same result as we got in Sect. 4.1 using the method of the integral sum (Problem 4.5). The function $f_2(x)$ can be considered as a linear combination of $f_1(x)$ and $f_3(x) = 1$ (a constant function). Since the integral of a linear combination of functions is a linear combination of integrals, we can write:

$$\int_0^1 \left(2x^2 - 1\right) dx = 2\int_0^1 x^2 dx - \int_0^1 dx = 2 \cdot \frac{1}{3} - x|_0^1 = \frac{2}{3} - (1-0) = \frac{2}{3} - 1 = -\frac{1}{3}.$$

Problem 4.10 Verify the following results:

$$\int_{-1}^1 \left(-2e^x + x^3\right) dx = -2e + 2e^{-1} \; ; \; \int_0^a \left(x^2 - b\right)^3 dx = \frac{a^7}{7} - \frac{3a^5 b}{5} + a^3 b^2 - ab^3 .$$

Formula (4.44) was proven above only for continuous functions. It may fail for functions which are piece-wise continuous. As the first example to illustrate this point, consider an integral of the sign function (2.2) which is discontinuous at $x = 0$, as shown in Fig. 2.1b. For negative x the sign function is equal to -1 and hence the indefinite integral is equal to $-x + C$; at positive x the sign function is $+1$ and the indefinite integral is $x + C$. Using formally Eq. (4.44), we then would have an integral between -1 and $+1$ equal to zero:

$$\int_{-1}^1 \mathrm{sgn}(x) dx = F(x)|_{-1}^1 = x|_{x=1} - (-x)|_{x=-1} = 1 - 1 = 0,$$

where we chose to have identical arbitrary constants for both indefinite integral functions, at positive and negative x. The obtained result makes perfect sense as the sign function is an odd function, and integration of an odd function between symmetric limits should give zero, see (4.8). Moreover, we can also split the integral into two and get the same result:

$$\int_{-1}^1 \mathrm{sgn}(x) dx = \int_{-1}^0 (-1) dx + \int_0^1 (+1) dx = (-x)|_{-1}^0 + x|_0^1 = -1 + 1 = 0.$$

One may think that this example indicates the applicability of formula (4.44) even for discontinuous functions. However, remember that the correct result in our example was only obtained if the arbitrary constants were assumed to be the same when considering the indefinite integrals for negative and positive x, and there is no reason to believe that this must be the case generally, and hence one would be posed with the problem of choosing the constants in each case. As our second example, let us consider a function

$$f(x) = \begin{cases} f_1(x), & \text{when } x \geq 0 \\ f_2(x), & \text{when } x < 0 \end{cases} = \begin{cases} e^x, & \text{when } x \geq 0 \\ 2e^x, & \text{when } x < 0 \end{cases},$$

which makes a jump at $x = 0$. The indefinite integral of $f_1(x)$ is $F_1(x) = e^x + C$, while it is $F_2(x) = 2e^x + C$ for $f_2(x)$. Taking again identical constants for the two functions, we find for the definite integral between -1 and $+1$ the following value using directly formula (4.44):

$$\int_{-1}^{1} f(x)dx = e^x\big|_{x=1} - 2e^x\big|_{x=-1} = e - 2e^{-1}.$$

On the other hand, if we split the integral into two, we would get a different answer:

$$\int_{-1}^{1} f(x)dx = \int_{-1}^{0} 2e^x dx + \int_{0}^{1} e^x dx = 2e^x\big|_{-1}^{0} + e^x\big|_{0}^{1}$$
$$= 2\left(1 - e^{-1}\right) + (e - 1) = 1 - 2e^{-1} + e,$$

which differs by a constant from the result found initially with the Newton–Leibniz formula. So, a general conclusion from these examples must be that if a function makes a finite jump (has a discontinuity) at some point c which is within the integration region, $a < c < b$, then one has to split the interval into regions where the function is continuous ($a \leq x < c$ and $c \leq x \leq b$), and then integrate separately within each such region as we did above; applying the Newton–Leibniz formula directly in those cases using the boundaries of the whole region may give incorrect results.

There could only be one arbitrary constant in a single indefinite integral, even though it may consist of several functions as illustrated by this example:

$$I = \int \left(2x^2 + \frac{3}{x} - 2\sin x\right) dx.$$

We split the integral into three and take each of them individually, so that three different constants emerge:

$$I = \int 2x^2 dx + \int \frac{3}{x} dx - \int 2\sin x dx = 2\int x^2 dx + 3\int \frac{dx}{x} - 2\int \sin x dx$$
$$= \left(2\frac{x^3}{3} + C_1\right) + (3\ln|x| + C_2) - (-2\cos x + C_3)$$
$$= \frac{2}{3}x^3 + 3\ln|x| + 2\cos x + C,$$

where, as we can see, all the constants combine into one: $C = C_1 + C_2 + C_3$.

In many cases already simple algebraic manipulations of the integrand may result in a combination of elementary functions which can be integrated:

$$\int \frac{x+2}{x} dx = \int \left(1 + \frac{2}{x}\right) dx = \int dx + 2 \int \frac{dx}{x} = x + 2\ln|x| + C,$$

$$\int \frac{e^{-x}+1}{e^{-x}} dx = \int \left(1 + \frac{1}{e^{-x}}\right) dx = \int dx + \int e^x dx = x + e^x + C.$$

Problem 4.11 Prove the following identities:

$$\int \cot^2 x\, dx = -x - \cot x + C, \quad \int \tan^2 x\, dx = -x + \tan x + C.$$

[Hint: *express cotangent and tangent functions via sine and cosine functions.*]

$$\int (2x+1)^3\, dx = 2x^4 + 4x^3 + 3x^2 + x + C, \quad \int \sin(x+a)\, dx = -\cos(x+a) + C.$$

[Hint: *use the binomial formula in the first and Eqs. (2.46) and (2.47) in the second case.*]

$$\int \sin^2 \frac{x}{2} dx = \frac{1}{2}(x - \sin x) + C, \quad \int \cos^2 \frac{x}{2} dx = \frac{1}{2}(x + \sin x) + C.$$

[Hint: *use connections established in Eq. (1.76) between the sine and cosine of an angle and the cosine of the double angle.*]

However, in many cases more involved methods must be used. Some of the most frequent ones will be considered in the next section.

4.4   Indefinite Integrals: Main Techniques

As we learned in the previous section, calculation of a definite integral requires, first of all, solving the corresponding indefinite integral. This requires finding a function which, when differentiated, gives the integrand itself. Of course, one may select all sorts of functions and then differentiate them to get a "big" table of derivatives of various functions. That may be useful since by looking at this table "in reverse" it is possible to find solutions of many indefinite integrals (and it is a good idea for the reader to make his/her own Table of various indefinite integrals while reading the rest of this Chapter). The problem is that there might be no solution for a particular integral of the actual interest in the table as it is impossible to choose "all possible functions" for differentiation due to an infinite number of possibilities. We need to develop various techniques for solving integrals. This is the main theme of this section. It is of course sufficient to discuss indefinite integrals.

4.4.1   Change of Variables

Sometimes by changing the integration variable it is possible to transform the integrand into a function, which can be recognized as a linear combination of functions for which indefinite integrals are known (and hopefully listed in our big table of integrals).

Consider an integral:

$$\int f(x)dx = F(x) + C \ ,$$

where $F'(x) = f(x)$. Let $x = x(t)$ be a function of a variable t, so that $F(x) = F(x(t))$. Using the chain rule, we can write:

$$\frac{d}{dt}F(x(t)) = \frac{dF}{dx}\frac{dx}{dt} = F'(x)x'(t) = f(x)x'(t) = f(x(t))x'(t) \ .$$

This expression simply means that if we want to integrate the function $\phi(t) = f(x(t))x'(t)$ *with respect to* t, the result must be $F(x(t)) + C$, and this is true for any choice of the function $x = x(t)$. We conclude:

$$\int f(x(t))x'(t)dt = F(x(t)) + C \equiv F(x) + C = \int f(x)dx \ ,$$

i.e. it is exactly the same integral we are seeking to solve. The formal procedure we have performed can simply be considered as a replacement of x with $x(t)$ and of dx with $x'(t)dt$. We note that the differential for $x(t)$ is exactly the latter expression $x'(t)dt$ and it is indeed denoted dx. This simple argument proves the consistency of the notations for the integral we have been using.[2]

Before we consider various cases in a more systematic way, let us first look at some simple examples.

Let us calculate the integral

$$I = \int xe^{x^2}dx \ .$$

In this integral the function x^2 in the exponent is a cumbersome one, so a wise choice here is to make a substitution. The simplest choice is to introduce a new variable

---

[2] Which is due to Gottfried Leibniz and Jean Baptiste Joseph Fourier.

$t = x^2$ which turns the rather complex exponential function e^{x^2} into a simpler one, e^t. We have $dt = d\left(x^2\right) = \left(x^2\right)' dx = 2xdx$, from where it follows that $xdx = dt/2$. Alternatively, we can solve $t = x^2$ for $x(t) = \pm t^{1/2}$, yielding $dx = \left(\pm t^{1/2}\right)' dt = \pm\frac{1}{2}t^{-1/2}dt$ and hence

$$xdx = \left(\pm t^{1/2}\right)\left(\pm\frac{1}{2}t^{-1/2}\right)dt = \frac{1}{2}dt \,,$$

the same result. Here we explicitly made use of the fact that dx is the differential of x. With this substitution, we can proceed to the integral itself:

$$I = \int e^{x^2}(xdx) = \int e^t\left(\frac{1}{2}dt\right) = \frac{1}{2}\int e^t dt = \frac{1}{2}e^t + C = \frac{1}{2}e^{x^2} + C \,.$$

It is easy to check that, indeed, $\frac{1}{2}\left(e^{x^2}\right)' = \frac{1}{2}e^{x^2}\left(x^2\right)' = \frac{1}{2}e^{x^2}2x = e^{x^2}x$, exactly the integrand of I.

Next, we calculate the integral

$$I = \int \frac{x^2 - 1}{2x - 5}dx \,.$$

In this case we use $t = 2x - 5$, so that $x = (t + 5)/2$ and hence $x^2 - 1 = t^2/4 + 5t/2 + 21/4$. This substitution allows us to manipulate the numerator and calculate the integral:

$$I = \begin{vmatrix} t = 2x - 5 \\ dt = 2dx \end{vmatrix} = \int \frac{\frac{1}{4}t^2 + \frac{5}{2}t + \frac{21}{4}}{t}\frac{dt}{2}$$

$$= \int\left(\frac{1}{8}t + \frac{5}{4} + \frac{21}{8}\frac{1}{t}\right)dt = \frac{t^2}{16} + \frac{5t}{4} + \frac{21}{8}\ln|t| + C$$

$$= \frac{(2x - 5)^2}{16} + \frac{5(2x - 5)}{4} + \frac{21}{8}\ln|2x - 5| + C \,.$$

In this example a very convenient form of working out the integral by a change of the variable has been proposed, and we urge the reader to make full use of it: all the working out needed to change the integration variable is placed between two vertical lines; we only wrote there the actual change of the variable ($t = t(x)$) and the connection of the differentials dt and dx. However, the inverse transformation $x = x(t)$ (of course, only if needed and possible) together with other relevant algebraic manipulations (if not too cumbersome) may also be written there. This way of writing is convenient as it allows constructing the whole derivation without interrupting the lines of algebra and in a short and clear way. We shall be using this method frequently.

Problem 4.12 Prove the following identities using the appropriate change of variables:

$$\int a^x dx = \frac{a^x}{\ln a} + C \quad (\text{with } a > 0), \quad \int \frac{dx}{(x+3)^4} = -\frac{(x+3)^{-3}}{3} + C,$$

$$\int \sin(ax+b)\,dx = -\frac{1}{a}\cos(ax+b) + C, \quad \int \cos(ax+b)\,dx = \frac{1}{a}\sin(ax+b) + C,$$

$$\int (ax+b)^n\,dx = \frac{(ax+b)^{n+1}}{a(n+1)} + C, \quad \int \frac{dx}{ax+b} = \frac{1}{a}\ln|ax+b| + C,$$

$$\int \frac{dx}{\sqrt{a^2-x^2}} = \arcsin\frac{x}{a} + C, \quad \int \frac{dx}{x^2+a^2} = \frac{1}{a}\arctan\frac{x}{a} + C,$$

$$\int \tan x\,dx = -\ln|\cos x| + C, \quad \int \cot x\,dx = \ln|\sin x| + C,$$

$$\int x^3\sqrt{1+x^2}\,dx = \frac{1}{15}\left(1+x^2\right)^{3/2}\left(3x^2-2\right)$$

$$\int \frac{1}{x^2}\sin\frac{1}{x}\,dx = \cos\frac{1}{x} + C, \quad \int \frac{1}{x^2}\cos\frac{1}{x}\,dx = -\sin\frac{1}{x} + C.$$

Problem 4.13 Show that

$$\int \frac{dx}{e^{\alpha x}+1} = -\frac{1}{\alpha}\ln\left(1+e^{-\alpha x}\right) + C.$$

Problem 4.14 Let the function $f(x)$ be periodic with the period of X, i.e. for any x we have $f(x+X) = f(x)$. Prove that for any a

$$\int_a^{a+X} f(x)dx = \int_0^X f(x)dx. \qquad (4.45)$$

Consider specifically $f(x) = \sin(x)$ for which $X = 2\pi$, sketch the function and reconcile with this rule.

4.4.2   Integration by Parts

This method is based on the following observation: if $u(x)$ and $v(x)$ are two functions, then, as we know well, $(uv)' = uv' + u'v$. Therefore, the function $F(x) = u(x)v(x) + C$ is an indefinite integral of the function $f(x) = u(x)v'(x) + u'(x)v(x)$:

$$\int_a^b \left[uv' + u'v\right] dx = uv\big|_a^b,$$

or, since the integral of a sum is a sum of integrals:

$$\int_a^b uv' dx = uv\big|_a^b - \int_a^b vu' dx . \qquad (4.46)$$

This result is usually written in a somewhat shorter form:

$$\int_a^b u\, dv = uv\big|_a^b - \int_a^b v\, du , \qquad (4.47)$$

where we introduced the differentials $du = u'dx$ and $dv = v'dx$ of the functions $u(x)$ and $v(x)$, respectively. These were formulae for the definite integral. For indefinite integrals the analogue expressions are the following:

$$\int u\, dv = uv - \int v\, du \quad \text{or} \quad \int uv' dx = uv - \int vu' dx . \qquad (4.48)$$

There is no need here to introduce the constant as it can be done at the last stage when taking the integral $\int v\, du$ in the right-hand side.

The usefulness of this trick "by parts" is in that on the right-hand side the function $u(x)$ is differentiated and, at the same time, instead of the function v', we have $v(x)$. This interchange may help in calculating the original integral in the left hand side. Of course, in the original integral we have just a single function $f(x)$, and various "divisions" or "factorizations" of it into $u(x)$ and $v(x)$ may be possible; it is not known which one would succeed, so sometimes different divisions are needed to try and then see which one would work. In many cases neither of "reasonable" choices works and one has to think of some other "trick". In some cases more than one integration by parts is required, one after another. In some other cases algebraic equations for the integral in question may be obtained by using this method which, when solved, gives the required answer.

As an example, let us calculate the indefinite integral $I = \int xe^x dx$. The idea is to have on the right-hand side just the exponential function on its own since we know perfectly well how to integrate it. If first we try taking $u = e^x$ and $dv = xdx$, then $u' = e^x$ and $v = x^2/2$ (remember, the constant C is introduced at the last step and hence is now ignored), so that we get

$$\int xe^x dx = \int \underbrace{e^x}_{u} \underbrace{xdx}_{dv} = \underbrace{e^x}_{u} \underbrace{\frac{x^2}{2}}_{v} - \int \underbrace{\frac{x^2}{2}}_{v} \underbrace{e^x dx}_{du} = \frac{1}{2}x^2 e^x - \frac{1}{2}\int x^2 e^x dx .$$

We see that, instead of getting rid of the x in the integrand, we raised its power to two and got x^2 instead. This result shows that our choice of u and dv was not successful, and we should try another one to see if it would work. However, our unsuccessful result may give us a hint as to how to proceed: it shows (if we look from right to left) that the integral $\int x^2 e^x dx$ can in fact be made related to the integral $\int x e^x dx$ with the lower power of x:

$$\int x^2 e^x dx = x^2 e^x - 2 \int x e^x dx , \qquad (4.49)$$

if x^2 is made the $u(x)$, so that upon differentiation in the right-hand side we would get $u' = 2x$, i.e. one power less. So, coming back to our original integral, this time we make the following decision:

$$\int x e^x dx = \begin{vmatrix} u = x , & du = 1 \cdot dx = dx \\ dv = e^x dx , & v = e^x \end{vmatrix} = x e^x - \int e^x dx = x e^x - e^x + C ,$$

which is the final result. It is easy to see that

$$\left(x e^x - e^x \right)' = \left(x e^x \right)' - \left(e^x \right)' = e^x + x e^x - e^x = x e^x ,$$

i.e. the integrand of the original integral; our second choice succeeded. If we now need to calculate the integral with $x^2 e^x$, then two integrations by parts would be needed: one to reduce the power of the x from two to one as in (4.49), and then the calculation of $\int x e^x dx$ as we did above:

$$\int x^2 e^x dx = \begin{vmatrix} u = x^2 , & du = 2x dx \\ dv = e^x dx , & v = e^x \end{vmatrix} = x^2 e^x - 2 \int x e^x dx$$

$$= \begin{vmatrix} u = x , & du = dx \\ dv = e^x dx , & v = e^x \end{vmatrix} = x^2 e^x - 2 \left[x e^x - \int e^x dx \right]$$

$$= x^2 e^x - 2x e^x + 2 e^x + C = e^x \left(x^2 - 2x + 2 \right) + C .$$

Let us now calculate this integral between 0 and 10:

$$\int_0^{10} x^2 e^x dx = e^x \left(x^2 - 2x + 2 \right) \Big|_0^{10} = e^{10} \left(10^2 - 20 + 2 \right) - e^0 2 = 82 e^{10} - 2 .$$

Problem 4.15 Calculate the following integrals by parts:

$$\int (2x + 5)^3 e^{-2x} dx \quad \left[\text{Answer:} \quad - \left(4x^3 + 36x^2 + 111x + 118 \right) e^{-2x} + C \right] ;$$

$$\int x^3 e^x dx \quad \left[\text{Answer:} \quad \left(x^3 - 3x^2 + 6x - 6 \right) e^x + C \right] ;$$

$$\int x^4 e^x dx \quad \left[\text{Answer:} \quad \left(x^4 - 4x^3 + 12x^2 - 24x + 24\right) e^x + C\right].$$

Problem 4.16 Using integration by parts, derive a recurrence relation $I_n = x^n e^x - n I_{n-1}$ for the integral $I_n = \int x^n e^x dx$ (where $n = 1, 2, \ldots$). Then by repeated application of the recurrence formula, establish the following general result:

$$\int x^n e^x dx = e^x \sum_{k=0}^{n} (-1)^k \frac{n!}{(n-k)!} x^{n-k} + C,$$

valid for $n = 0, 1, 2, \ldots$. Using induction, prove that this formula is indeed correct.

Problem 4.17 Calculate

$$\int x \sin x dx \quad [\text{Answer:} \quad \sin x - x \cos x + C] \ ;$$

$$\int x \cos x dx \quad [\text{Answer:} \quad \cos x + x \sin x + C] \ ;$$

$$\int \frac{1}{x^3} \sin \frac{1}{x} dx \quad \left[\text{Answer:} \quad -\sin \frac{1}{x} + \frac{1}{x} \cos \frac{1}{x} + C\right] \ ;$$

$$\int \frac{1}{x^4} \sin \frac{1}{x} dx \quad \left[\text{Answer:} \quad -\frac{1}{x^2}\left[2x \sin \frac{1}{x} + \left(2x^2 - 1\right) \cos \frac{1}{x}\right] + C\right] \ ;$$

$$\int x^2 \cos x dx \quad [\text{Answer:} \quad 2x \cos x + \left(x^2 - 2\right) \sin x + C] \ ;$$

$$\int x^3 \cos x dx \quad [\text{Answer:} \quad 3\left(x^2 - 2\right) \cos x + x \left(x^2 - 6\right) \sin x + C] \ .$$

These examples should convince us that the integration by parts can be used for calculating integrals $\int P_n(x) e^{\alpha x} dx$, $\int P_n(x) \sin(ax + b) dx$ and $\int P_n(x) \cos(ax + b) dx$, where $P_n(x)$ is the n-th degree polynomial. Note that more general linear arguments to the exponential, sine and cosine functions can indeed be used as these arguments can always be replaced by t using the appropriate change of variables; the polynomial will remain a polynomial of the same degree (but with different coefficients).

Fortunately, integration by parts has a much wider scope. It can also be used for integrating functions which are products of a polynomial and $\ln x$ or any of the inverse trigonometric functions as illustrated below:

$$\int \arctan x dx = \begin{vmatrix} u = \arctan x , & du = \frac{1}{1+x^2} dx \\ dv = dx , & v = x \end{vmatrix} = x \arctan x - \int \frac{x dx}{1 + x^2}$$

$$= x \arctan x - \frac{1}{2} \ln \left(x^2 + 1 \right) + C ,$$

where in the last integral we have made a change of variable $t = 1 + x^2, dt = 2x dx$.

Problem 4.18 Calculate

$$\int x \arcsin x dx \quad \left[\text{Answer:} \quad \frac{x}{4}\sqrt{1 - x^2} + \frac{2x^2 - 1}{4} \arcsin x + C \right] ;$$

$$\int x^2 \arcsin x dx \quad \left[\text{Answer:} \quad \frac{x^2 + 2}{9}\sqrt{1 - x^2} + \frac{x^3}{3} \arcsin x + C \right] ;$$

$$\int \arccos x dx \quad \left[\text{Answer:} \quad -\sqrt{1 - x^2} + x \arccos x + C \right] ;$$

$$\int x \arccos x dx \quad \left[\text{Answer:} \quad -\frac{x}{4}\sqrt{1 - x^2} + \frac{x^2}{2} \arccos x + \frac{1}{4} \arcsin x + C \right] ;$$

$$\int x^3 \arccos x dx \quad \left[\text{Answer:} \quad -\frac{(2x^2 + 3) x}{32}\sqrt{1 - x^2} + \frac{x^4}{4} \arccos x + \frac{3}{32} \arcsin x + C \right] ;$$

$$\int \ln x dx \quad [\text{Answer:} \quad x \ln x - x + C] ; \quad \int x \ln x dx \quad \left[\text{Answer:} \quad -\frac{x^2}{4} \left(1 - 2 \ln x \right) + C \right] .$$

The next example which we shall consider is of the following two integrals:

$$I_S = \int e^x \sin x dx = \begin{vmatrix} u = e^x , & du = e^x dx \\ dv = \sin x dx , & v = -\cos x \end{vmatrix}$$

$$= -e^x \cos x + \int e^x \cos x dx \equiv -e^x \cos x + I_C$$

and

$$I_C = \int e^x \cos x dx = \begin{vmatrix} u = e^x , & du = e^x dx \\ dv = \cos x dx , & v = \sin x \end{vmatrix}$$

$$= e^x \sin x - \int e^x \sin x dx \equiv e^x \sin x - I_S .$$

It is seen that a single integration by parts relates one integral to the other. Therefore, double integration by parts results in an algebraic equation for one of the integrals:

$$I_S = -e^x \cos x + I_C = -e^x \cos x + \left[e^x \sin x - I_S\right] = e^x \left(-\cos x + \sin x\right) - I_S ,$$

which, when solved with respect to I_S, finally yields:

$$I_S = \frac{1}{2} e^x \left(-\cos x + \sin x\right) . \tag{4.50}$$

As a by-product, we also have

$$I_C = e^x \sin x - I_S = \frac{1}{2} e^x \left(\cos x + \sin x\right) .$$

A product of a polynomial $P_n(x)$ with several exponential or trigonometric functions is dealt with in more or less the same way:

$$\int x e^x \sin x \, dx = \begin{vmatrix} u = x , & du = dx \\ dv = e^x \sin x \, dx , & v = I_S \end{vmatrix} = x I_S - \frac{1}{2} \int e^x \left(\sin x - \cos x\right) dx ,$$

where we again denoted the intermediate integral $\int e^x \sin x \, dx$ as I_S for convenience. The integral we have arrived at is a difference of the familiar I_S and I_C integrals, which allows us to finally write:

$$\int x e^x \sin x \, dx = x I_S - \frac{1}{2} \left(I_S - I_C\right) = \left(x - \frac{1}{2}\right) I_S + \frac{1}{2} I_C$$

$$= \frac{1}{2} e^x \left[(1 - x) \cos x + x \sin x\right] + C .$$

Calculation of integrals containing a product of exponential and trigonometric functions can sometimes be greatly simplified if we resort to the appropriate complex exponential. Consider again I_S. Using Euler's formulae (3.43), we can write instead:

$$I_S = \int e^x \frac{1}{2i} \left(e^{ix} - e^{-ix}\right) dx = \frac{1}{2i} \left[\int e^{(1+i)x} dx - \int e^{(1-i)x} dx\right]$$

$$= \frac{1}{2i} \left[\frac{e^{(1+i)x}}{1+i} - \frac{e^{(1-i)x}}{1-i}\right] = \frac{e^x}{2i} \left[\frac{e^{ix}}{1+i} - \frac{e^{-ix}}{1-i}\right] .$$

Replacing back the exponentials $e^{\pm ix}$ via sine and cosine of x using Eq. (3.42), we, after some simple algebra, arrive back to our previous result (4.50). It seems that this calculation is easier as it requires only a single simple integration to be performed.

Problem 4.19 Using the method of complex exponential, calculate again the integral I_C.

Problem 4.20 Calculate

$$\int x e^x \cos x\, dx \quad \left[\text{Answer: } \frac{1}{2} e^x \left\{ (x-1)\sin x + x\cos x \right\} + C \right] ;$$

$$\int x^2 e^x \sin x\, dx \quad \left[\text{Answer: } \frac{1}{2} e^x (x-1) \left[-(x-1)\cos x + (x+1)\sin x \right] + C \right] ;$$

$$\int x^2 e^x \cos x\, dx \quad \left[\text{Answer: } \frac{1}{2} e^x (x-1) \left\{ (x-1)\sin x + (x+1)\cos x \right\} + C \right] .$$

You may use either of the methods considered above.

Problem 4.21 Show that for any positive integers n and m,

$$I_{n,m} = \int_0^1 x^n \ln^m x\, dx = \frac{(-1)^m\, m!}{(n+1)^{m+1}} . \tag{4.51}$$

[Hint: *first of all, establish the recurrence relation $I_{n,m} = -\left[m/(n+1) \right] I_{n,m-1}$.*]

Problem 4.22 Similarly, show that for any positive integers n and m,

$$J_{n,m} = \int_0^1 (1-x)^n x^m\, dx = \frac{n!\, m!}{(n+m+1)!} .$$

[Hint: *first of all, establish the recurrence relation $J_{n,m} = \left[n/(n+m+1) \right] J_{n-1,m}$ by writing $(1-x) = (1/x - 1)\, x$.*]

Problem 4.23 Show, using integration by parts, that $\int f(x) f'(x) dx = f(x)^2 + C$. This simple result can also be considered as an application of the differentials, since $df = (df/dx)\, dx = f'(x) dx$. Indeed,

$$\int f(x) f'(x) dx = \int f\, df = \frac{f^2}{2} + C = \frac{f(x)^2}{2} + C .$$

Convince yourself that this is true also by differentiating the right-hand side.

Problem 4.24 Prove, using induction and an integration by parts, that an integral of an n-th order polynomial and an exponential function is an n-th order polynomial times the same exponential function:

$$\int P^{(n)}(x) e^{ax} dx = Q^{(n)}(x) e^{ax} + C .$$

4.4.3   Integration of Rational Functions

In the case of rational functions $f(x) = P_n(x)/Q_m(x)$ with polynomials $P_n(x)$ and $Q_m(x)$ of degrees n and m, respectively, the corresponding indefinite integrals can always be calculated. As we saw in Sect. 2.3.2, a rational function can always be decomposed into a sum of simpler functions: a polynomial (which is only present if $n \geq m$) and the following functions:

$$f_k(x) = \frac{1}{(x-a)^k} \quad \text{and/or} \quad g_k(x) = \frac{Ax+B}{\left(x^2 + px + q\right)^k}$$

with $k = 1, 2, \ldots$. Note that, by construction, the polynomial $x^2 + px + q$ does not have real roots. Before we consider an example of more complex rational functions, let us first make sure that we know how to integrate the above two functions.

Integration of $f_1(x)$ (for $k = 1$) is trivial and leads to a logarithm. Integration of f_k for $k > 1$ is done using the substitution $t = x - a$ and is also simple; both of these integrals we have seen before. The integration of $g_1(x)$ can be done by a substitution which follows from making the complete square of the denominator:

$$x^2 + px + q = \left[x^2 + 2\left(\frac{p}{2}\right) x + \left(\frac{p}{2}\right)^2 \right] - \left(\frac{p}{2}\right)^2 + q = \left(x + \frac{p}{2}\right)^2 - D ,$$

where $D = p^2/4 - q$ is definitely negative because only in this case the square polynomial $x^2 + px + q$ (with D being its discriminant) corresponds to two complex conjugate roots.[3] Now, the substitution $t = x + p/2$ is made and the function $g_1(x)$ is easily integrated.

Consider, for instance, this integral:

$$\int \frac{2x+1}{x^2-2x+2}dx = \int \frac{2x+1}{(x-1)^2+1}dx = \left| \begin{array}{l} t = x-1 \\ dt = dx \end{array} \right|$$

$$= \int \frac{2t+3}{t^2+1}dt = 2\int \frac{t\,dt}{t^2+1} + 3\int \frac{dt}{t^2+1}.$$

The first integral is calculated using the substitution $y = t^2 + 1$, while the second integral leads to $\arctan t$:

$$\int \frac{2x+1}{x^2-2x+2}dx = \ln\left(1+t^2\right) + 3\arctan t + C$$

$$= \ln\left(x^2-2x+2\right) + 3\arctan(x-1) + C.$$

Problem 4.25 Calculate

$$\int \frac{x-1}{x^2+4x+6}dx \quad \left[\text{Answer:} \quad -\frac{3}{\sqrt{2}}\arctan\frac{x+2}{\sqrt{2}} + \frac{1}{2}\ln\left(x^2+4x+6\right) + C\right];$$

$$\int \frac{-3x+2}{x^2-6x+10}dx \quad \left[\text{Answer:} \quad 7\arctan(3-x) - \frac{3}{2}\ln\left(x^2-6x+10\right) + C\right].$$

Calculation of the integral $g_k(x)$ with $k \geq 2$ is a bit trickier. Firstly, we build the complete square as in the previous case of $k = 1$ and make the corresponding change of variables. Two types of integrals are then encountered which require our attention:

$$h_k = \int \frac{t\,dt}{\left(t^2+a^2\right)^k} \quad \text{and} \quad r_k = \int \frac{dt}{\left(t^2+a^2\right)^k}.$$

The first of these, h_k, is trivially calculated with the substitution $y = t^2 + a^2$:

$$h_k = \int \frac{t\,dt}{\left(t^2+a^2\right)^k} = \frac{1}{2(1-k)}\left(t^2+a^2\right)^{1-k}.$$

---

[3] Of course, when $D < 0$ the parabola appears completely above the horizontal x-axis and hence does not have real roots.

The second integral, r_k, is calculated with the following trick leading to a recurrence relation whereby r_k is expressed via the same type of integrals with lower values of k:

$$
\begin{aligned}
r_k &= \int \frac{dt}{\left(t^2 + a^2\right)^k} = \frac{1}{a^2} \int \frac{\left[(t^2 + a^2) - t^2\right] dt}{\left(t^2 + a^2\right)^k} \\
&= \frac{1}{a^2} \left[\int \frac{dt}{\left(t^2 + a^2\right)^{k-1}} - \int \frac{t^2 dt}{\left(t^2 + a^2\right)^k} \right] \\
&= \frac{r_{k-1}}{a^2} - \frac{1}{a^2} J_k ,
\end{aligned}
$$

where the integral J_k we take by parts:

$$
\begin{aligned}
J_k &= \int \frac{t^2 dt}{\left(t^2 + a^2\right)^k} = \left| \begin{array}{ll} u = t , & du = dt \\ dv = t\left(t^2 + a^2\right)^{-k} dt , & v = h_k \end{array} \right| = th_k - \int h_k dt \\
&= \frac{t}{2(1-k)} \left(t^2 + a^2\right)^{1-k} - \frac{1}{2(1-k)} \int \left(t^2 + a^2\right)^{1-k} dt \\
&= \frac{t}{2(1-k)} \left(t^2 + a^2\right)^{1-k} - \frac{r_{k-1}}{2(1-k)} ,
\end{aligned}
$$

which result finally in the following recurrence relation for r_k:

$$
r_k = \frac{r_{k-1}}{a^2} - \frac{1}{a^2} J_k = \frac{r_{k-1}}{a^2} \left[1 + \frac{1}{2(1-k)} \right] - \frac{t}{2(1-k)a^2} \left(t^2 + a^2\right)^{1-k} .
$$

It is seen that r_k is directly related to r_{k-1}. This method allows expressing recursively the given r_k via r_{k-1}, then the latter via r_{k-2}, etc., until one arrives at r_1 which we have already encountered above: $r_1 = (1/a) \arctan(x/a)$. This way it is possible to express any of the g_k integrals for any integer k via elementary functions.

Let us consider an example:

$$
\int \frac{2x + 1}{\left(x^2 - 2x + 2\right)^2} dx = \int \frac{2x + 1}{\left[(x - 1)^2 + 1\right]^2} dx = \left| \begin{array}{l} t = x - 1 \\ dt = dx \end{array} \right|
$$
$$
= \int \frac{2t + 3}{\left(t^2 + 1\right)^2} dt = 2h_2 + 3r_2 .
$$

Here

$$
h_2 = \int \frac{t dt}{\left(t^2 + 1\right)^2} = \left| \begin{array}{l} y = t^2 + 1 \\ dy = 2t dt \end{array} \right| = \frac{1}{2} \int \frac{dy}{y^2} = -\frac{1}{2y} = -\frac{1}{2\left(t^2 + 1\right)}
$$

and

$$
\begin{aligned}
r_2 &= \int \frac{dt}{\left(t^2 + 1\right)^2} = \int \frac{\left(t^2 + 1\right) - t^2}{\left(t^2 + 1\right)^2} dt = \int \frac{dt}{t^2 + 1} - \int \frac{t^2 dt}{\left(t^2 + 1\right)^2} \\
&= \arctan t - J_2 ,
\end{aligned}
$$

where the integral J_2 we calculate by parts:

$$J_2 = \int \frac{t^2 dt}{\left(t^2 + 1\right)^2} = \left| \begin{array}{ll} u = t, & du = dt \\ dv = t\left(t^2 + 1\right)^{-2} dt, & v = h_2(t) \end{array} \right|$$

$$= -\frac{t}{2\left(t^2 + 1\right)} + \frac{1}{2} \int \frac{dt}{t^2 + 1} = -\frac{t}{2\left(t^2 + 1\right)} + \frac{1}{2} J_1$$

$$= -\frac{t}{2\left(t^2 + 1\right)} + \frac{1}{2} \arctan t.$$

Combining all contributions together, we can finally state the result:

$$\int \frac{2x + 1}{\left(x^2 - 2x + 2\right)^2} dx = \frac{3x - 5}{2\left(x^2 - 2x + 2\right)} + \frac{3}{2} \arctan(x - 1) + C.$$

Problem 4.26 Calculate

$$\int \frac{x - 1}{\left(x^2 + 4x + 6\right)^3} dx \quad \left[\text{Answer:} \quad -\frac{9\sqrt{2}}{64} \arctan \frac{x + 2}{\sqrt{2}} \right.$$

$$\left. -\frac{9x^3 + 54x^2 + 138x + 140}{32\left(x^2 + 4x + 6\right)^2} + C \right];$$

$$\int \frac{-3x + 2}{\left(x^2 - 6x + 10\right)^2} dx \quad \left[\text{Answer:} \quad \frac{7}{2} \arctan(3 - x) + \frac{24 - 7x}{2\left(x^2 - 6x + 10\right)} + C \right].$$

Once we have discussed all the necessary elementary integrals, we are now ready to consider integrals of arbitrary rational functions. The calculation consists of first decomposing the rational function into elementary fractions as discussed at the beginning of this Section and especially in Sect. 2.3.2, and then, second, integrating each elementary fraction as we have just demonstrated.

As an example, we calculate the integral

$$J = \int \frac{3x^2 + x - 1}{(x - 1)^2 \left(x^2 + 1\right)} dx.$$

We start by decomposing the fraction. This was done in Eq. (2.19); therefore, we can immediately write:

$$J = 2 \int \frac{1}{x-1} dx + \frac{3}{2} \int \frac{dx}{(x-1)^2} - \frac{1}{2} \int \frac{4x+1}{x^2+1} dx$$

$$= 2 \ln |x-1| - \frac{3}{2(x-1)} - 2 \int \frac{x \, dx}{x^2+1} - \frac{1}{2} \int \frac{dx}{x^2+1}$$

$$= 2 \ln |x-1| - \frac{3}{2(x-1)} - \ln(x^2+1) - \frac{1}{2} \arctan x + C.$$

Problem 4.27 Calculate the following integrals of rational functions:

$$\int \frac{x^2-x+5}{(2x-1)(x^2+4x+4)} dx \quad \left[\text{Answer:} \quad \frac{11}{5(x+2)} + \frac{3}{25} \ln|2+x| \right.$$

$$\left. + \frac{19}{50} \ln|2x-1| + C \right];$$

$$\int \frac{x^3+x^2+1}{(x-1)^2(x^2+4)} dx \quad \left[\text{Answer:} \quad -\frac{3}{5(x-1)} + \frac{41}{50} \arctan \frac{x}{2} \right.$$

$$\left. + \frac{3}{25} \ln(x^2+4) + \frac{19}{25} \ln|x-1| + C \right];$$

$$\int \frac{dx}{1-x^3} \quad \left[\text{Answer:} \quad \frac{1}{\sqrt{3}} \arctan \frac{2x+1}{\sqrt{3}} + \frac{1}{6} \ln \frac{1+x+x^2}{(x-1)^2} + C \right];$$

$$\int \frac{dx}{1+x^4} \quad \left[\text{Answer:} \quad -\frac{1}{2\sqrt{2}} \left[\arctan \left(1 - \sqrt{2}x \right) - \arctan \left(1 + \sqrt{2}x \right) \right] \right.$$

$$\left. - \frac{1}{4\sqrt{2}} \ln \frac{1-\sqrt{2}x+x^2}{1+\sqrt{2}x+x^2} + C \right];$$

$$\int \frac{x^2 dx}{1+x^4} \quad \left[\text{Answer:} \quad -\frac{1}{2\sqrt{2}} \left[\arctan \left(1 - \sqrt{2}x \right) - \arctan \left(1 + \sqrt{2}x \right) \right] \right.$$

$$\left. + \frac{1}{4\sqrt{2}} \ln \frac{1-\sqrt{2}x+x^2}{1+\sqrt{2}x+x^2} + C \right];$$

$$\int \frac{x^5 + 2x^3 - 1}{x^2 + 4x + 4} dx \quad \left[\text{Answer:} \quad \frac{x^4}{4} - \frac{4x^3}{3} + 7x^2 - 40x \right.$$

$$\left. + \frac{49}{x+2} + 104 \ln |x + 2| + C \right].$$

4.4.4 Integration of Trigonometric Functions

Integrals containing a rational function $R(\sin x, \cos x)$ of the sine and cosine of the same angle x can always be turned into an integral of a rational function, which, as we know from the previous subsection, can always be calculated. The trick is based on replacing the sine and cosine functions by rational expressions using the substitution

$$t = \tan \frac{x}{2}, \quad \text{so that} \quad dt = \frac{dx}{2 \cos^2 \frac{x}{2}} = \frac{\sin^2 \frac{x}{2} + \cos^2 \frac{x}{2}}{2 \cos^2 \frac{x}{2}} dx$$

$$= \frac{1}{2} \left(t^2 + 1 \right) dx \quad \Longrightarrow \quad dx = \frac{2dt}{1 + t^2}, \tag{4.52}$$

so that both sine and cosine (see Eqs. (2.65)) are expressed via t:

$$\sin x = \frac{2 \tan \frac{x}{2}}{1 + \tan^2 \frac{x}{2}} = \frac{2t}{1 + t^2} \quad \text{and} \quad \cos x = \frac{1 - \tan^2 \frac{x}{2}}{1 + \tan^2 \frac{x}{2}} = \frac{1 - t^2}{1 + t^2}. \tag{4.53}$$

It is clear that any rational function of the sine and cosine functions will be transformed by means of this substitution into some rational function of t. This substitution is sometimes unnecessarily cumbersome since other methods (if known) may give the answer in a somewhat simpler way; still, it is good to know that it always leads to a solution.

As an example we shall calculate the following integral:

$$\int \frac{1 + \sin x}{1 + \cos x} dx = \left| \begin{array}{c} t = \tan \frac{x}{2} \\ dx = 2dt / \left(1 + t^2 \right) \end{array} \right| = \int \frac{1 + 2t / \left(1 + t^2 \right)}{1 + \left(1 - t^2 \right) / \left(1 + t^2 \right)} \frac{2dt}{1 + t^2}$$

$$= \int \frac{1 + t^2 + 2t}{1 + t^2} dt = \int \left(1 + \frac{2t}{1 + t^2} \right) dt = t + \int \frac{2t dt}{1 + t^2}$$

$$= \left| \begin{array}{c} y = 1 + t^2 \\ dy = 2t dt \end{array} \right| = t + \int \frac{dy}{y} = t + \ln |y| = t + \ln \left| 1 + t^2 \right|$$

$$= \tan \frac{x}{2} + \ln \left| 1 + \tan^2 \frac{x}{2} \right| = \tan \frac{x}{2} + \ln \left| \frac{1}{\cos^2 \frac{x}{2}} \right|$$

$$= \tan \frac{x}{2} - 2 \ln \left| \cos \frac{x}{2} \right| + C.$$

We have introduced the constant C only at the very last step.

Problem 4.28 Calculate the following integrals:

$$\int \frac{1+\sin^2 x}{2+\sin x + \cos x}dx \quad \left[\text{Answer:}\quad \frac{3}{\sqrt{2}}\arctan\left(\frac{1+\tan\frac{x}{2}}{\sqrt{2}}\right)\right.$$

$$\left.-\frac{1}{2}(\sin x + \cos x) + \ln(2+\sin x + \cos x) + C\right] ;$$

$$\int \frac{dx}{1+\sin x} \quad \left[\text{Answer:}\quad \frac{2}{1+\cot\frac{x}{2}} + C\right] ;$$

$$\int \frac{dx}{(1+\sin x)^2} \quad \left[\text{Answer:}\quad \frac{3\sin\frac{x}{2} - \cos\frac{3x}{2}}{3\left(\cos\frac{x}{2} + \sin\frac{x}{2}\right)^3} + C\right] ;$$

$$\int \frac{dx}{1+\cos x} \quad \left[\text{Answer:}\quad \tan\frac{x}{2} + C\right] ;$$

$$\int \frac{dx}{(1+\cos x)^2} \quad \left[\text{Answer:}\quad \frac{(2+\cos x)\sin x}{3(1+\cos x)^2} + C\right] ;$$

$$\int \frac{1+\cos^2 x}{1+\sin^2 x}dx \quad \left[\text{Answer:}\quad -x + \frac{3}{\sqrt{2}}\arctan\left(\sqrt{2}\tan x\right) + C\right] ;$$

$$\int \frac{1+\cos^2 x}{\left(1+\sin^2 x\right)^2}dx \quad \left[\text{Answer:}\quad \frac{5}{4\sqrt{2}}\arctan\left(\sqrt{2}\tan x\right)\right.$$

$$\left.-\frac{3}{4}\frac{\sin(2x)}{\cos(2x) - 3} + C\right] .$$

In many cases, as was mentioned, other substitutions may be more convenient. For instance, if the rational function $R(\sin x, \cos x)$ is odd with respect to the cosine function, i.e. $R(\sin x, -\cos x) = -R(\sin x, \cos x)$, then the substitution $t = \sin x$ can be used, for instance:

$$\int \frac{\cos x}{2+\sin x}dx = \left|\begin{matrix} t = \sin x \\ dt = \cos x dx \end{matrix}\right| = \int \frac{dt}{2+t} = \ln|2+t| + C$$

$$= \ln(2+\sin x) + C .$$

Here $R(\sin x, \cos x) = \cos x/(2+\sin x)$ and is obviously odd with respect to the cosine function. If, instead, the rational function is odd with respect to the sine function, $R\left(-\sin x, \cos x\right) = -R\left(\sin x, \cos x\right)$, then the substitution $t = \cos x$ is found to be useful:

$$\int \frac{\sin^3 x}{\cos^4 x}dx = \int \frac{\left(1 - \cos^2 x\right)\sin x}{\cos^4 x}dx = \begin{vmatrix} t = \cos x \\ dt = -\sin x dx \end{vmatrix} = \int \frac{1 - t^2}{t^4}(-dt)$$

$$= -\int t^{-4}dt + \int t^{-2}dt = \frac{1}{3}t^{-3} - t^{-1} + C = \frac{1}{3\cos^3 x} - \frac{1}{\cos x} + C .$$

Problem 4.29 Calculate the following integrals:

$$\int \sin^2 x \cos^3 x dx \quad \left[\text{Answer:} \quad \frac{1}{3}\sin^3 x - \frac{1}{5}\sin^5 x + C\right] ;$$

$$\int \sin^3 x \cos^2 x dx \quad \left[\text{Answer:} \quad -\frac{1}{3}\cos^3 x + \frac{1}{5}\cos^5 x + C\right] ;$$

$$\int \frac{\cos^2 x}{\sin x}dx \quad \left[\text{Answer:} \quad \cos x + \ln\left|\tan\frac{x}{2}\right| + C\right] .$$

In many other cases usage of trigonometric identities such as those presented in Sect. 2.3.8 may be found useful, especially if the trigonometric functions of different arguments are present. For instance, using the identity (2.52), we can easily take the following integral:

$$\int \sin(11x)\sin(9x)\,dx = \int \frac{1}{2}[\cos(2x) - \cos(20x)]\,dx$$

$$= \frac{1}{4}\sin(2x) - \frac{1}{40}\sin(20x) + C .$$

Sometimes, by reducing the power of the sine or cosine functions via sine and/or cosine functions of a different argument, integrals containing powers of the trigonometric functions can be calculated:

$$\int \cos^2 x dx = \int \frac{1}{2}[1 + \cos(2x)]\,dx = \frac{x}{2} + \frac{1}{4}\sin(2x) + C ,$$

$$\int \sin^2 x dx = \int \frac{1}{2}[1 - \cos(2x)]\,dx = \frac{x}{2} - \frac{1}{4}\sin(2x) + C ,$$

where we made use of Eq. (1.76). If we use Eq. (2.56), the following integral is easily calculated:

$$\int \frac{\sin (3x)}{\sin x} dx = \int \frac{3 \sin x \cos^2 x - \sin^3 x}{\sin x} dx = \int \left(3 \cos^2 x - \sin^2 x\right) dx$$

$$= \int \left[\frac{3}{2} (1 + \cos (2x)) - \frac{1}{2} (1 - \cos (2x))\right] dx$$

$$= \int (1 + 2 \cos(2x)) \, dx = x + \sin (2x) + C \, .$$

As a rule, if there are trigonometric functions of different arguments in the integrand, one has to first convert them into trigonometric functions of the same argument. This can always be done if the arguments are of the form nx with n being an integer.

Problem 4.30 Calculate the following integrals:

$$\int \sin (6x) \cos (5x) \, dx \quad \left[\text{Answer:} \quad -\frac{1}{2} \cos x - \frac{1}{22} \cos (11x) + C\right] ;$$

$$\int \frac{\sin (5x)}{\sin x} dx \quad \left[\text{Answer:} \quad x + \sin (2x) + \frac{1}{2} \sin (4x) + C\right] .$$

4.4.5 Integration of a Rational Function of the Exponential Function

A rational function $R (e^x)$ of the exponential function can also be transformed, using the new variable $t = e^x$, into an integral of the rational function $R_1(t) = R(t)/t$, and hence can always be integrated. For instance, consider:

$$I = \int \frac{1 + 2e^x}{1 - 2e^x} dx = \left| \begin{matrix} t = e^x \\ dt = e^x dx = t dx \end{matrix} \right| = \int \frac{1 + 2t}{t (1 - 2t)} dt \, .$$

The last integral is calculated, e.g. by decomposing the rational function into elementary functions A/t and $B/ (1 - 2t)$ (with coefficients $A = 1$ and $B = 4$),

$$\frac{1 + 2t}{t (1 - 2t)} = \frac{1}{t} - \frac{4}{2t - 1} \, ,$$

and then integrating:

$$I = \int \left(\frac{1}{t} + \frac{4}{1 - 2t}\right) dt = \ln |t| - 2 \ln |1 - 2t| = \ln e^x - 2 \ln |1 - 2e^x|$$

$$= x - \ln \left(1 - 2e^x\right)^2 + C \, .$$

Problem 4.31 Calculate the following integrals:

$$\int \sinh(x)dx \quad [\text{Answer:} \quad \cosh(x) + C] \ ;$$

$$\int \cosh(x)dx \quad [\text{Answer:} \quad \sinh(x) + C] \ ;$$

$$\int \tanh(x)dx \quad \left[\text{Answer:} \quad \ln(\cosh x) + C = -x + \ln\left(1 + e^{2x}\right) + C\right] \ ;$$

$$\int \coth(x)dx \quad \left[\text{Answer:} \quad \ln|\sinh x| + C = -x + \ln\left|e^{2x} - 1\right| + C\right] \ ;$$

$$\int \tanh^2(x)dx \quad [\text{Answer:} \quad x - \tanh x + C] \ ;$$

$$\int \tanh^3(x)dx \quad \left[\text{Answer:} \quad \ln(\cosh x) + \frac{1}{2\cosh^2 x} + C\right] \ .$$

4.4.6   Integration of Irrational Functions

Irrational functions, i.e. those which contain radicals, are generally more difficult to integrate. However, there exists a class of irrational functions which, upon an appropriate substitution, can be transformed into an integral with respect to a rational function, and hence are integrable. Consider integrals of a function $R\left(x, y(x)^{1/m}\right) = R\left(x, \sqrt[m]{y(x)}\right)$, which is rational with respect to its arguments x and $\sqrt[m]{y(x)}$, but the latter argument makes the whole function irrational. Here m is an integer and $y(x) = (ax + b)/(gx + f)$ is a simple rational function of x with some constant coefficients a, b, g and f. Note that in the function R there could be an integer power of $y^{1/m}$, i.e. $y(x)$ could be in any rational power n/m (with integer n).

These types of integrals are calculated using the following substitution:

$$t^m = \frac{ax + b}{gx + f}, \quad \text{leading to} \quad x = \omega(t) = \frac{ft^m - b}{-gt^m + a}, \tag{4.54}$$

which turns the original integral over x into a t−integral containing a rational function of t. Indeed, in this case,

$$\int R\left(x, \sqrt[m]{\frac{ax + b}{gx + f}}\right) dx = \int R\left(\omega(t), t\right) \omega'(t)dt, \tag{4.55}$$

proving the above made statement as, obviously, the derivative $\omega'(t)$ is a rational function of t and R is rational with respect to both of its arguments.

As an example, consider the following integral, where $R(x, z) = (z - 1) / (z + 1)$, $z = y^{1/2}$ (i.e. $m = 2$) and $y = x + 1$:

$$\int \frac{\sqrt{x+1}-1}{\sqrt{x+1}+1} dx = \begin{vmatrix} t^2 = x+1 \\ 2t\,dt = dx \end{vmatrix} = \int \frac{t-1}{t+1} 2t\,dt$$

$$= \begin{vmatrix} y = t+1 \\ dy = dt \end{vmatrix} = \int \frac{2}{y} (y-1)(y-2)\,dy$$

$$= 2 \int \left(y - 3 + \frac{2}{y} \right) dy = y^2 - 6y + 4\ln|y|$$

$$= x - 4\left(1 + \sqrt{x+1} \right) + 4\ln\left(1 + \sqrt{x+1} \right) + C,$$

since $y = 1 + t = 1 + \sqrt{x+1}$.

Sometimes, the irrational function in the integrand looks different to the one given above; however, it can still be manipulated into this form as demonstrated by the following example:

$$\int \frac{dx}{\sqrt{(x-1)(x+2)}} = \int \sqrt{\frac{x+2}{x-1}} \frac{dx}{x+2}$$

$$= \begin{vmatrix} t^2 = \dfrac{x+2}{x-1}, & x = \dfrac{t^2+2}{t^2-1}, & dx = \dfrac{-6t}{(t^2-1)^2}dt, & x+2 = \dfrac{3t^2}{t^2-1} \end{vmatrix}$$

$$= \int t \left(\frac{t^2-1}{3t^2} \right) \left(\frac{-6t}{(t^2-1)^2} dt \right) = -2 \int \frac{dt}{t^2-1} = \ln\left| \frac{1+t}{1-t} \right|$$

$$= \ln \left| \frac{\sqrt{x-1} + \sqrt{x+2}}{\sqrt{x-1} - \sqrt{x+2}} \right|$$

$$= \ln \left| \frac{\sqrt{x-1} + \sqrt{x+2}}{\sqrt{x-1} - \sqrt{x+2}} \cdot \frac{\sqrt{x-1} + \sqrt{x+2}}{\sqrt{x-1} + \sqrt{x+2}} \right|$$

$$= \ln \left| \frac{\left(\sqrt{x-1} + \sqrt{x+2} \right)^2}{(x-1) - (x+2)} \right| = \ln \left| \frac{1 + 2x + 2\sqrt{(x-1)(x+2)}}{-3} \right|$$

$$= \ln \left| 1 + 2x + 2\sqrt{(x-1)(x+2)} \right| - \ln 3$$

$$= \ln \left| 1 + 2x + 2\sqrt{(x-1)(x+2)} \right| + C,$$

where the constant $\ln 3$ has been dropped as we have to add an arbitrary constant C (introduced at the last step) anyway.

Problem 4.32 Calculate the following integrals:

$$\int \frac{dx}{\sqrt[3]{(x+1)(x-1)^2}} \quad \left[\text{Answer:}\ \sqrt{3}\arctan\frac{2t+1}{\sqrt{3}} + \frac{1}{2}\ln\frac{1+t+t^2}{(t-1)^2} + C\,,\right.$$

$$\left.\text{where}\ t^3 = \frac{x-1}{x+1}\right];$$

$$\int \left(\frac{x-1}{x+1}\right)^{2/3} dx \quad \left[\text{Answer:}\ -\frac{2t^2}{t^3-1} + \frac{4}{\sqrt{3}}\arctan\frac{2t+1}{\sqrt{3}}\right.$$

$$\left.-\frac{2}{3}\ln\frac{1+t+t^2}{(t-1)^2} + C\,,\ \text{where}\ t^3 = \frac{x-1}{x+1}\right].$$

Another class of integrals which can also be transformed into integrals of rational functions are the ones containing a rational function of x and a square root of a quadratic polynomial, i.e.

$$I = \int R\left(x, \sqrt{ax^2 + bx + c}\right) dx\ . \tag{4.56}$$

Euler studied these types of integrals and proposed general methods for their calculation based on special substitutions which bear Euler's name. The particular substitution to be used depends on the form of the quadratic polynomial under the square root. Three possibilities can be considered, although in practice only two would be needed.

If $a > 0$, then the first substitution can be used:

$$\sqrt{ax^2 + bx + c} = t \pm \sqrt{a}x \tag{4.57}$$

can be used (with either sign).

Problem 4.33 When applying this substitution, one needs expressions of x via t, dx via dt, and of the square root of the polynomial also via t; moreover, all these expressions should be *rational* functions in t. Demonstrate by the direct calculation that the following identities are valid:

$$x = \frac{t^2 - c}{b \mp 2\sqrt{a}t}\,,\quad dx = 2\frac{bt \mp \sqrt{a}t^2 \mp \sqrt{a}c}{(b \mp 2\sqrt{a}t)^2}dt\ \text{and}$$

$$\sqrt{ax^2 + bx + c} = \frac{bt \mp \sqrt{a}t^2 \mp \sqrt{a}c}{b \mp 2\sqrt{a}t}\,. \tag{4.58}$$

After the t integral is calculated, one has to replace it with an appropriate expression via x, which follows immediately from Eq. (4.57) as $t = \sqrt{ax^2 + bx + c} \mp \sqrt{a}x$.

Consider the following integral, for which we apply the substitution (4.57) using the minus sign for definiteness:

$$\int \sqrt{x^2 + 2x + 2}\, dx = \left| \sqrt{x^2 + 2x + 2} = \frac{2t + t^2 + 2}{2(1+t)} \, , \; dx = 2\frac{2t + t^2 + 2}{4(1+t)^2}dt \right|$$

$$= \int \left(\frac{2t + t^2 + 2}{2(1+t)} \right) \left(\frac{2t + t^2 + 2}{2(1+t)^2}dt \right) = \frac{1}{4} \int \frac{\left(2t + t^2 + 2\right)^2}{(1+t)^3}dt$$

$$= \left| \begin{array}{c} y = 1 + t \\ dy = dt \end{array} \right| = \frac{1}{4} \int \left(y^2 + 1\right)^2 \frac{dy}{y^3}$$

$$= \frac{1}{4} \int \left(y + \frac{2}{y} + \frac{1}{y^3}\right) dy = \frac{y^2}{8} - \frac{1}{8y^2} + \frac{1}{2} \ln |y| + C \, ,$$

where $y = 1 + t = \sqrt{x^2 + 2x + 2} + 1 + x$.

Problem 4.34 Using this method, calculate the following integrals:

$$\int \frac{dx}{\sqrt{x^2 + 2x + 2}} \quad \left[\text{Answer:} \quad \ln \left| 1 + x + \sqrt{x^2 + 2x + 2} \right| + C \right] ;$$

$$\int \frac{x\,dx}{\sqrt{x^2 + 2x + 2}} \left[\text{Answer:} \quad \sqrt{x^2 + 2x + 2} \right.$$

$$\left. - \ln \left| 1 + x + \sqrt{x^2 + 2x + 2} \right| + C \right] .$$

If the constant c in the quadratic polynomial is positive, Euler proposed the second substitution which also leads to a rational function integral:

$$\sqrt{ax^2 + bx + c} = xt \pm \sqrt{c} \tag{4.59}$$

Problem 4.35 Prove that in this case the required transformations are given by:

$$x = \frac{b \mp 2\sqrt{c}t}{t^2 - a} , \quad dx = 2\frac{-bt \pm \sqrt{c}t^2 \pm \sqrt{ca}}{\left(t^2 - a\right)^2}dt ,$$

$$\sqrt{ax^2 + bx + c} = \frac{bt \mp \sqrt{c}t^2 \mp \sqrt{ca}}{t^2 - a} , \quad (4.60)$$

and $t = \left(\sqrt{ax^2 + bx + c} \mp \sqrt{c}\right)/x.$

Finally, the third Euler's substitution can be applied using the roots, x_1 and x_2, of the quadratic polynomial (we only consider here the case when these roots are real):

$$ax^2 + bx + c = a\left(x - x_1\right)\left(x - x_2\right) .$$

In this case the following substitution is proposed:

$$\sqrt{ax^2 + bx + c} = t\left(x - x_1\right) \quad (4.61)$$

(the other root may be used as well).

Problem 4.36 Prove that in this case the required transformations are as follows:

$$x = \frac{t^2 x_1 - ax_2}{t^2 - a} , \quad dx = \frac{2a\left(x_2 - x_1\right)t}{\left(t^2 - a\right)^2}dt ,$$

$$\sqrt{ax^2 + bx + c} = \frac{a\left(x_1 - x_2\right)t}{t^2 - a} , \quad (4.62)$$

and $t = \sqrt{ax^2 + bx + c}/\left(x - x_1\right).$

Note that if $x_1 = x_2$ (repeated roots), then $\sqrt{ax^2 + bx + c} = \sqrt{a}\,|x - x_1|$ (a positive value of the square root is to be accepted) and the square root disappears, i.e. the integrand is a rational function without any additional transformations. Also note that the third Euler's substitution is equivalent to the one we used at the beginning of this subsection since one can always write (as we did in one of the examples above):

$$\sqrt{ax^2 + bx + c} = \sqrt{a\left(x - x_1\right)\left(x - x_2\right)}$$

$$= \sqrt{a\left(x - x_1\right)\left(x - x_2\right)}\sqrt{\frac{x - x_2}{x - x_2}} = \sqrt{a}\left(x - x_2\right)\sqrt{\frac{x - x_1}{x - x_2}} ,$$

i.e. the substitution of the type (4.54) with $m = 2$ can be used which will bring the integrand into a rational form with respect to the variable $t = \sqrt{(x - x_1) / (x - x_2)}$. Finally, it is easy to see that Euler's second substitution is in fact redundant. Indeed, if $a > 0$ then one can always use the first substitution, even when $c > 0$ when the second substitution would also be applicable. If, however, $a < 0$, then the parabola corresponding to the square polynomial is either located entirely below the x-axis (or just touches it at a single point) or its top is above it. In the former case the polynomial is everywhere negative (except may be at a single point), and this case can be disregarded within the manifold of real numbers we are interested in so far. In the latter case there must be two real roots x_1 and x_2, in which case the third substitution becomes legitimate. In the case of the two roots x_1 and x_2 being complex, the third Euler's substitution can always be used.

The consideration above leads us to a conclusion that all integrals with the function $R\left(x, \sqrt{ax^2 + bx + c}\right)$ can be transformed into a rational form and hence integrated in elementary functions.

In some rather simple cases Euler's method is not required, as much simpler substitutions exist. For instance, transforming to the complete square and then performing a straightforward change of the variable may suffice as illustrated in the example we have already considered above using a different method:

$$
\int \frac{dx}{\sqrt{x^2 + 2x + 2}} = \int \frac{dx}{\sqrt{(x + 1)^2 + 1}} = \left| \begin{matrix} t = x + 1 \\ dt = dx \end{matrix} \right| = \int \frac{dt}{\sqrt{1 + t^2}}
$$
$$
= \ln \left| t + \sqrt{1 + t^2} \right| = \ln \left| 1 + x + \sqrt{x^2 + 2x + 2} \right| + C,
$$

which is the same result as in Problem 4.34.

Problem 4.37 Using this method, calculate the following integral:

$$
\int \frac{(2x - 1)\, dx}{\sqrt{x^2 + 2x + 2}} \quad \left[\text{Answer:}\ 2\sqrt{x^2 + 2x + 2} \right.
$$
$$
\left. -3 \ln \left| 1 + x + \sqrt{x^2 + 2x + 2} \right| + C \right].
$$

Finally, integrals of the functions $R\left(x, \sqrt{a^2 - x^2}\right)$, $R\left(x, \sqrt{x^2 - a^2}\right)$ and $R\left(x, \sqrt{x^2 + a^2}\right)$ can be transformed into integrals containing rational functions of the trigonometric functions using the so-called trigonometric substitutions. Of course, these particular integrals can always be taken using the Euler's substitutions, but simpler trigonometric substitutions may result in less cumbersome algebra. Indeed, in the case of $\sqrt{a^2 - x^2}$ one uses $x = a \sin t$ (or $x = a \cos t$), e.g.

$$\int \sqrt{a^2 - x^2}dx = \begin{vmatrix} x = a\sin t\,, & \cos t = \sqrt{1 - (x/a)^2} \\ dx = a\cos t dt \end{vmatrix} = a^2 \int \sqrt{1 - \sin^2 t}\cos t dt$$

$$= a^2 \int \cos^2 t dt = a^2 \left[\frac{t}{2} + \frac{1}{4}\sin(2t) \right] = \frac{a^2}{2}[t + \sin t \cos t]$$

$$= \frac{a^2}{2}\left[\arcsin\frac{x}{a} + \frac{x}{a}\sqrt{1 - \left(\frac{x}{a}\right)^2} \right] = \frac{a^2}{2}\arcsin\frac{x}{a} + \frac{x}{2}\sqrt{a^2 - x^2} + C\,.$$

$$\tag{4.63}$$

Note that it is legitimate here to choose the positive value of the cosine function during intermediate manipulations since $-a \le x \le a$ and hence one can choose $-\pi/2 \le t \le \pi/2$ when $\cos t \ge 0$.

In the case of $\sqrt{x^2 - a^2}$, one similarly uses $x = a/\sin t$, while in the case of $\sqrt{x^2 + a^2}$ the tangent substitution, $x = a \tan t$, can be used instead since $1 + \tan^2 t = 1/\cos^2 t$ and hence the square root disappears.

Problem 4.38 Calculate the following integral:

$$\int \sqrt{x^2 - a^2}dx \quad \left[\text{Answer: } \frac{x}{2}\sqrt{x^2 - a^2} - \frac{a^2}{2}\ln\left|x + \sqrt{x^2 - a^2}\right| + C \right].$$

Problem 4.39 Calculate the integral:

$$\int \sqrt{x^2 + a^2}dx \quad \left[\text{Answer: } \frac{x}{2}\sqrt{x^2 + a^2} + \frac{a^2}{2}\ln\left|x + \sqrt{x^2 + a^2}\right| + C \right].$$

$$\tag{4.64}$$

Problem 4.40 Show that

$$\int \frac{dx}{\sqrt{x^2 - 1}} = \ln\left|x + \sqrt{x^2 - 1}\right| + C = \cosh^{-1}(x) + C\,. \tag{4.65}$$

4.5 More on Calculation of Definite Integrals

As we have already discussed in Sect. 4.3, it is easy to calculate a definite integral if the corresponding indefinite integral is known, and many methods to help with this task have been reviewed in the previous Sect. 4.4. Still, more needs to be said about some subtle points and extensions, which we are going to do in this section.

4.5.1 Change of Variables and Integration by Parts in Definite Integrals

We learned two important techniques of calculating indefinite integrals in Sect. 4.4, a change in the integration variable and integration by parts, and of course these can be used for calculating definite integrals as well. However, there are some points we must discuss, otherwise wrong results may be obtained.

Change of variables. Consider a definite integral $\int_a^b f(x)dx$. As we know from studying indefinite integrals, if a substitution $x = x(t)$ is used, then an integral with respect to the variable t is obtained, and, if we are lucky, the $t-$integral is calculated, the t is replaced back (by solving the equation $x = x(t)$ with respect to t), and that is it. Two comments are in order here as far as definite integrals are concerned.

Firstly, when a definite integral is calculated using a substitution, it is easier to work out the limits of the new variable t alongside usual steps of expressing dx via dt and x via t everywhere in the integrand. The benefit of this is that one can avoid calculating t via x and hence replacing t back with the x; instead, the boundary values of t can be immediately used after the $t-$integral is calculated to get the required result (which is a number). For instance,

$$
\int_0^a xe^{-x^2}dx = \left| \begin{array}{l} t = x^2 \,,\ dt = 2xdx \,,\ xdx = dt/2 \\ t = 0 \text{ when } x = 0 \,;\ t = a^2 \text{ when } x = a \end{array} \right|
$$
$$
= \frac{1}{2} \int_0^{a^2} e^{-t}dt = \frac{1}{2} \left(-e^{-t} \right)\big|_0^{a^2} = \frac{1}{2} \left(-e^{-a^2} + 1 \right) \,.
$$

Note that we replaced the limits in the $t-$integral when changing the variables and then used those limits directly without going back to the original x variable.

Secondly, one has to make sure that the substitution $x = x(t)$ is single-valued, continuous and its derivative $x'(t)$ is also continuous. If these conditions are not satisfied within the whole integration interval $a \leq x \leq b$, the result of the substitution could be wrong. The following examples illustrate this subtle point.

Consider the integral ($a > 0$):

$$
I_1 = \int_{-a}^a xdx = \frac{x^2}{2}\bigg|_{-a}^a = 0 \,.
$$

Of course, this integral is zero because the integrand is an odd function and the limits of integration form a symmetric interval around $x = 0$, see also Eq. (4.8), i.e.

the above obtained result makes a lot of sense. On the other hand, let us make the substitution defined by the equation $x^2 = t$:

$$I_1 = \begin{vmatrix} x^2 = t \,, \; x = \sqrt{t} \\ dx = dt/2\sqrt{t} \end{vmatrix} = \int_{a^2}^{a^2} \sqrt{t} \, \frac{dt}{2\sqrt{t}} = \frac{1}{2} \int_{a^2}^{a^2} dt = 0 \,,$$

since the two t-limits are now the same and equal to a^2. The same result as before has been obtained; good! However, let us now consider the integral of x^2 instead:

$$I_2 = \int_{-a}^{a} x^2 dx = \frac{x^3}{3} \Big|_{-a}^{a} = \frac{2}{3} a^3 \,,$$

while if we used the substitution $t = x^2$, the result would have been zero as the two limits will be the same again! This paradoxical result originates from the fact that the substitution we used defines a multivalued function $t = \pm\sqrt{x}$, and this contradicts one of the necessary conditions needed for the substitution to work. In fact, when integrating between $-a$ and 0, when x is negative, we must have used $x = -\sqrt{t}$, while $x = \sqrt{t}$ must be used for positive x. The same is true for I_1; the correct result we obtained for I_1 was accidental. That means that the integration interval must be split into two and the correct single-valued substitution is to be used in each interval individually.

Another caution needs to be given concerning the requirement that both the function $x(t)$ and its derivative $x'(t)$ are to be continuous within the integration interval.

The above points are illustrated by calculating the following integral:

$$I_3 = \int_{-\pi}^{\pi} dx = x\big|_{-\pi}^{\pi} = 2\pi \,.$$

Alternatively, let us make a substitution $t = \tan x$. When $x = -\pi$ we have $t = 0$; however, when $x = \pi$, the same result $t = 0$ is obtained. Therefore, inevitably, the integral becomes zero after the substitution is made:

$$I_3 = \begin{vmatrix} t = \tan x \\ dt = dx/\cos^2 x = \left(1 + t^2\right) dx \end{vmatrix} = \int_{0}^{0} \frac{dt}{1 + t^2} = 0 \,.$$

This result is obviously wrong, and the reason is related to the fact that the function $t = \tan x$ is not continuous over the whole range of the x values between $-\pi$ and π, and experiences two discontinuities at $x = \pm\pi/2$. Therefore, formula (4.44) is no longer valid: one has to split the x−integral into three with intervals: $-\pi \leq x < -\pi/2$, $-\pi/2 < x < \pi/2$ and $\pi/2 < x \leq \pi$ (the points $\pm\pi/2$ have to be excluded from integration, see Sect. 4.5.3 for more details). Then the correct result is obtained:

$$\int_{-\pi}^{-\pi/2} dx = \int_{0}^{\infty} \frac{dt}{1 + t^2} = \arctan x\big|_{0}^{\infty} = \frac{\pi}{2} \,,$$

$$\int_{-\pi/2}^{\pi/2} dx = \int_{-\infty}^{\infty} \frac{dt}{1+t^2} = \arctan x\big|_{-\infty}^{\infty} = \pi ,$$

$$\int_{\pi/2}^{\pi} dx = \int_{-\infty}^{0} \frac{dt}{1+t^2} = \arctan x\big|_{-\infty}^{0} = \frac{\pi}{2} ,$$

yielding[4] the correct result of 2π. The subtle points here are that in the first integral the upper limit corresponds to $x = -\frac{\pi}{2} - 0$ (as $x < -\frac{\pi}{2}$) which results in $t = \tan(-\pi/2 - 0) = +\infty$ for the upper limit of the t-integral; similarly, in the second integral the bottom limit is $-\frac{\pi}{2} + 0$ and the upper is $\frac{\pi}{2} - 0$, while in the third integral the bottom limit is $\frac{\pi}{2} + 0$.

The above examples indicate that care is needed when applying a substitution; the latter should satisfy a number of conditions and must satisfy them along with the whole integration interval as otherwise the result may be wrong!

Integration by parts. This is another important method we found for the indefinite integrals in Sect. 4.4.2. Of course, it can also be used directly when calculating definite integrals, e.g.

$$\int_{-1}^{1} x \cos x dx = \begin{vmatrix} u = x , & du = dx \\ dv = \cos x dx , & v = \sin x \end{vmatrix} = x \sin x\big|_{-1}^{1} - \int_{-1}^{1} \sin x dx$$

$$= [\sin 1 - (-1) \sin(-1)] - (-\cos x)\big|_{-1}^{1} = 0 ,$$

which is to be expected as the function $x \cos x$ is odd and the interval is symmetric.

The method of integration by parts may be found especially useful in working out general expressions for various definite integrals containing integer n as a parameter. Consider, for instance, the following integral:

$$I_n = \int_{0}^{\pi/2} \cos^n x dx , \tag{4.66}$$

where $n = 0, 1, 2, \ldots$ Integrating it by parts, we can obtain a recurrence relation connecting I_n with the integral associated with a smaller value of n:

$$I_n = \int_{0}^{\pi/2} \cos^{n-1} x \cos x dx = \begin{vmatrix} u = \cos^{n-1} x , & du = (n-1) \cos^{n-2} x (-\sin x) dx \\ dv = \cos x dx , & v = \sin x \end{vmatrix}$$

$$= \underbrace{\cos^{n-1} x \sin x\big|_{0}^{\pi/2}}_{=0} + (n-1) \int_{0}^{\pi/2} \cos^{n-2} x \sin^2 x dx$$

$$= (n-1) \int_{0}^{\pi/2} \cos^{n-2} x \left(1 - \cos^2 x\right) dx = (n-1) I_{n-2} - (n-1) I_n ,$$

---

[4] When we changed to the variable t, one or two integration limits became $\pm\infty$ leading to so-called improper integrals. We shall considered them in detail in Sect. 4.5.3 where we shall show that the same formula (4.44) can be used for their calculation as in the case of finite limits, and this is what have been done here.

which gives an equation for I_n, namely:

$$I_n = \frac{n-1}{n} I_{n-2} \ .$$

Since $I_0 = \int_0^{\pi/2} dx = \pi/2$, we obtain, for instance: $I_2 = (2-1) I_0/2 = \pi/4$, $I_4 = (4-1) I_2/4 = 3\pi/16$, etc.

For odd values of n, our recurrence relation is valid as well; however, the first (the smallest) value of n must be $n = 1$ instead, for which

$$I_1 = \int_0^{\pi/2} \cos x\, dx = \sin x \big|_0^{\pi/2} = 1 \ ,$$

and hence $I_3 = (3-1) I_1/3 = 2/3$, $I_5 = (5-1) I_3/5 = 8/15$, and so on.

Problem 4.41 Prove that for even $n = 2p$,

$$I_{2p} = \frac{\pi}{4^p} \frac{(2p-1)!}{p!\,(p-1)!} = \frac{\pi}{4^p} \binom{2p-1}{p} \ ,$$

while for odd values of $n = 2p + 1$,

$$I_{2p+1} = \frac{4^p \, (p!)^2}{(2p+1)!} \ .$$

Problem 4.42 Show that the same recurrence relation is obtained for the integral

$$J_n = \int_0^{\pi/2} \sin^n x\, dx \ ,$$

and that for both even and odd values of n the same results are obtained as for I_n.

4.5.2 Integrals Depending on a Parameter

In many applications either the limits of a definite integral $a(\lambda)$ and $b(\lambda)$ and/or the integrand $f(x, \lambda)$ itself may depend on some parameter, let us call it λ,

$$I(\lambda) = \int_{a(\lambda)}^{b(\lambda)} f(x; \lambda)dx \ , \tag{4.67}$$

and it is required to calculate the derivative of the value of the integral, which is thus a function of λ, with respect to it. We know from Sect. 4.3 that if only the upper limit depends on λ as $b \equiv \lambda$, one obtains $I'(\lambda) = f(\lambda)$, i.e. the value of the integrand at $x = \lambda$. Here we shall derive the most general formula for the λ−derivative of the integral given by Eq. (4.67) where both limits and the integrand itself may depend on λ in a general way.

We start by writing the corresponding definition of the derivative:

$$I'(\lambda) = \lim_{\Delta\lambda\to 0} \frac{I(\lambda + \Delta\lambda) - I(\lambda)}{\Delta\lambda} = \lim_{\Delta\lambda\to 0} \frac{\Delta I}{\Delta\lambda} ,$$

where

$$\begin{aligned}
\Delta I &= \int_{a+\Delta a}^{b+\Delta b} f(x; \lambda + \Delta\lambda)dx - \int_a^b f(x; \lambda)dx \\
&= \int_{a+\Delta a}^{a} f(x; \lambda + \Delta\lambda)dx + \int_a^b f(x; \lambda + \Delta\lambda)dx \\
&\quad + \int_b^{b+\Delta b} f(x; \lambda + \Delta\lambda)dx - \int_a^b f(x; \lambda)dx \\
&= \int_a^b \left[f(x; \lambda + \Delta\lambda) - f(x; \lambda) \right] dx - \int_a^{a+\Delta a} f(x; \lambda + \Delta\lambda)dx \\
&\quad + \int_b^{b+\Delta b} f(x; \lambda + \Delta\lambda)dx ,
\end{aligned}$$

with $\Delta a = a'(\lambda)\Delta\lambda$ and $\Delta b = b'(\lambda)\Delta\lambda$. According to our discussion in Sect. 4.3 (see specifically Eq. (4.33)), we can write for the second and the third integrals:

$$\begin{aligned}
\int_a^{a+\Delta a} f(x; \lambda + \Delta\lambda)dx &= f(a_1; \lambda + \Delta\lambda)\Delta a \\
&= \left[f(a_1; \lambda) + f'_\lambda(a_1; \xi)\Delta\lambda \right] a'(\lambda)\Delta\lambda ,
\end{aligned}$$

$$\begin{aligned}
\int_b^{b+\Delta b} f(x; \lambda + \Delta\lambda)dx &= f(b_1; \lambda + \Delta\lambda)\Delta b \\
&= \left[f(b_1; \lambda) + f'_\lambda(b_1; \zeta)\Delta\lambda \right] b'(\lambda)\Delta\lambda ,
\end{aligned}$$

where $a < a_1 < a + \Delta a, b < b_1 < b + \Delta b$ and in the second passage in both lines we have made use of the Lagrange formula (3.88) with respect to the variable λ:

$$f(a_1; \lambda + \Delta\lambda) = f(a_1; \lambda) + f'_\lambda(a_1; \xi)\Delta\lambda \quad \text{and} \quad f(b_1; \lambda + \Delta\lambda) = f(b_1; \lambda) + f'_\lambda(b_1; \zeta)\Delta\lambda$$

with $\lambda < \xi < \lambda + \Delta\lambda$ and $\lambda < \zeta < \lambda + \Delta\lambda$. Here f'_λ denotes the derivative of the function $f(x; \lambda)$ with respect to its second variable, λ.[5] Combining all expressions for the integrals into ΔI and dividing it by $\Delta\lambda$, we obtain:

$$
\frac{\Delta I}{\Delta\lambda} = \int_a^b \frac{f(x; \lambda + \Delta\lambda) - f(x; \lambda)}{\Delta\lambda} dx - \left[f(a_1; \lambda) + f'_\lambda(a_1; \xi)\Delta\lambda \right] a'(\lambda)
$$
$$
+ \left[f(b_1; \lambda) + f'_\lambda(b_1; \zeta)\Delta\lambda \right] b'(\lambda) .
$$

If we now take the limit $\Delta\lambda \to 0$, the integrand in the first term in the right-hand side would tend to the derivative of $f(x, \lambda)$ with respect to λ, and the points $a_1 \to a$ and $b_1 \to b$, and hence this finally yields:

$$
\frac{d}{d\lambda} \int_{a(\lambda)}^{b(\lambda)} f(x; \lambda)dx = \int_a^b \frac{df(x; \lambda)}{d\lambda}dx + f(b; \lambda) b'(\lambda) - f(a; \lambda) a'(\lambda) .
$$
$$(4.68)$$

It is seen that this result indeed generalizes the one we obtained earlier in Sect. 4.3: if a and f do not depend on λ and $b = \lambda$, formula (4.34) is recovered.

Note that above we silently assumed that we can take the limit of $\Delta\lambda \to 0$ inside the integral and hence replace

$$
\lim_{\Delta\lambda \to 0} \int_a^b \frac{f(x; \lambda + \Delta\lambda) - f(x; \lambda)}{\Delta\lambda} dx
$$

with

$$
\int_a^b \lim_{\Delta\lambda \to 0} \frac{f(x; \lambda + \Delta\lambda) - f(x; \lambda)}{\Delta\lambda} dx = \int_a^b \frac{df(x; \lambda)}{d\lambda} dx .
$$

This step, however, requires some justification. Assuming that the function $f(x, \lambda)$ has the first derivative with respect to λ, we can use the Lagrange formula (3.88) and write:

$$
f(x; \lambda + \Delta\lambda) = f(x; \lambda) + \Delta\lambda \frac{df(x; \lambda + \vartheta_1\Delta\lambda)}{d\lambda}, \quad \text{where } 0 < \vartheta_1 < 1 ,
$$

so that

$$
\frac{f(x; \lambda + \Delta\lambda) - f(x; \lambda)}{\Delta\lambda} = \frac{df(x; \lambda + \vartheta_1\Delta\lambda)}{d\lambda} .
$$

If the function $f(x, \lambda)$ has the second derivative with respect to λ as well, then one can use the Lagrange formula again[6] and write

$$
\frac{df(x; \lambda + \vartheta_1\Delta\lambda)}{d\lambda} = \frac{df(x; \lambda)}{d\lambda} + \vartheta_1\Delta\lambda \frac{d^2 f(x; \lambda + \vartheta_2\vartheta_1\Delta\lambda)}{d\lambda^2}, \quad \text{where } 0 < \vartheta_2 < 1 ,
$$

---

[5] We shall learn rather soon in Sect. 5.3 when considering functions of many variables that this is called a partial derivative with respect to λ of the function $f(x, \lambda)$ of two variables. The partial derivative is denoted $\partial f / \partial \lambda$ using the symbol ∂ rather than d.

[6] These two steps of using the Lagrange formula twice are equivalent to using the first-order ($n = 1$) Taylor expansion, Eqs. (3.84) and (3.85), for $f(x, \lambda + \Delta\lambda)$.

in which case

$$
\lim_{\Delta\lambda\to 0}\int_a^b \frac{f(x;\lambda+\Delta\lambda)-f(x;\lambda)}{\Delta\lambda}dx = \lim_{\Delta\lambda\to 0}\int_a^b \frac{df(x;\lambda+\vartheta_1\Delta\lambda)}{d\lambda}dx
$$
$$
= \int_a^b \frac{df(x;\lambda)}{d\lambda}dx + \vartheta_1 \lim_{\Delta\lambda\to 0}\left(\Delta\lambda\int_a^b \frac{d^2 f(x;\lambda+\vartheta_2\vartheta_1\Delta\lambda)}{d\lambda^2}dx\right) .
$$

The integral with the second derivative is bounded if we require that the second derivative is continuous for all $a \le x \le b$ and the values of λ between λ and $\lambda + \Delta\lambda$. Hence, it remains finite in the $\Delta\lambda \to 0$ limit, which guarantees that the last term in the expression above tends to zero in this limit. This justifies the replacement

$$
\lim_{\Delta\lambda\to 0}\int_a^b \frac{df(x;\lambda+\vartheta_1\Delta\lambda)}{d\lambda}dx \quad \text{with} \quad \int_a^b \frac{df(x;\lambda)}{d\lambda}dx ,
$$

as required.

Sometimes, by introducing a parameter artificially, the integral of interest may be related to another; if the expression for the latter integral is known, this trick allows us to calculate the former. Suppose we need to calculate the following integral

$$
I = \int_0^{\pi/2} x^3 \sin(\omega x)\, dx .
$$

Of course, this can be calculated by doing three integrations by parts; instead, we shall consider a function

$$
J(\omega) = \int_0^{\pi/2} \cos(\omega x)\, dx = \frac{1}{\omega}\sin\left(\omega\frac{\pi}{2}\right) ,
$$

and notice that upon differentiation of the cosine function $\cos(\omega x)$ in $J(\omega)$ three times with respect to ω, we get $x^3 \sin(\omega x)$, which is exactly the required integrand of I. Therefore, we can write:

$$
I = \frac{d^3}{d\omega^3} J(\omega) = \frac{d^3}{d\omega^3}\left[\frac{1}{\omega}\sin\left(\omega\frac{\pi}{2}\right)\right] = \left(\frac{1}{\omega}\right)^{(3)}\sin\left(\omega\frac{\pi}{2}\right)
$$
$$
+ 3\left(\frac{1}{\omega}\right)^{(2)}\left[\sin\left(\omega\frac{\pi}{2}\right)\right]^{(1)} + 3\left(\frac{1}{\omega}\right)^{(1)}\left[\sin\left(\omega\frac{\pi}{2}\right)\right]^{(2)} + \frac{1}{\omega}\left[\sin\left(\omega\frac{\pi}{2}\right)\right]^{(3)}
$$
$$
= \frac{-6}{\omega^4}\sin\left(\omega\frac{\pi}{2}\right) + \frac{6}{\omega^3}\frac{\pi}{2}\cos\left(\omega\frac{\pi}{2}\right) + \frac{3}{\omega^2}\left(\frac{\pi}{2}\right)^2\sin\left(\omega\frac{\pi}{2}\right)
$$
$$
- \frac{1}{\omega}\left(\frac{\pi}{2}\right)^3\cos\left(\omega\frac{\pi}{2}\right) ,
$$

which is the required expression (it can still be simplified further if desired). When performing differentiation, we have conveniently used here the Leibniz formula (3.71).

Problem 4.43 Newton's equation of motion of a Brownian particle of mass m moving in a liquid can be written as the following Langevin equation:

$$m\frac{dv}{dt} = f - \int_0^t K\left(t - \tau\right)v(\tau)d\tau \ ,$$

where f is the force due to liquid molecules hitting the particle at random, $v(t)$ particle velocity, and $K(t) = A\exp\left(-\alpha t\right)$, the so-called friction kernel which we assume decays exponentially with time. Note that the integral term stands up for its name of the friction force since it is proportional to the particle velocity and comes with the minus sign. The integral here means that the particle motion depends on all its previous velocities, i.e. the memory of its motion is essential to correctly describe its dynamics. This may be the case e.g. if the environment relaxation time is of the same order or longer than the characteristic time of the particle. Now, denote the integral term as a function $y(t)$ and show that the single Langevin equation above is equivalent to the following two equations:

$$m\frac{dv}{dt} = f - y \quad \text{and} \quad \frac{dy}{dt} = Av - \alpha y \ .$$

4.5.3 Improper Integrals

So far our definite integrals were taken between finite limits a and b; moreover, we assumed that the integrated function $f(x)$ does not have singularities anywhere within the integration interval $a \leq x \leq b$, including the boundary points themselves. We can now define more general integrals in which these conditions are relaxed.

We shall start by considering the upper limit being $+\infty$. To this end, let us consider an integral

$$I(b) = \int_a^b f(x)dx \ ,$$

which we assume exists for any finite b. This integral can be considered as a function, $I(b)$, with respect to its upper limit b. If we now consider larger and larger values of the upper limit b, then in the limit of $b \to +\infty$ we shall arrive at the so-called *improper integral* with the upper limit being $+\infty$:

$$\int_a^\infty f(x)dx = \lim_{b\to\infty}\int_a^b f(x)dx \ . \tag{4.69}$$

Let us state the definition of its convergence. The above improper integral $\int_a^\infty f(x)dx$ is said to converge to a number F, if for any $\epsilon > 0$ there always exists such $A > a$, that

$$\left| F - \int_a^A f(x)dx \right| < \epsilon .$$

This means that the smaller the value of ϵ we take, the larger the value of the upper limit A need to be taken to have the integral $\int_a^A f(x)dx$ be close (within ϵ) to F. This condition can also be formally rewritten as

$$\left| \int_A^\infty f(x)dx \right| < \epsilon ,$$

stating that the integral $\int_A^\infty f(x)dx$ tends to zero as $A \to \infty$.

Similarly, one defines the improper integrals with the bottom limit being $-\infty$,

$$\int_{-\infty}^b f(x)dx = \lim_{a \to -\infty} \int_a^b f(x)dx , \qquad (4.70)$$

or when both limits are at $\pm\infty$:

$$\int_{-\infty}^\infty f(x)dx = \lim_{b \to \infty} \left(\lim_{a \to -\infty} \int_a^b f(x)dx \right) . \qquad (4.71)$$

Problem 4.44 Give the definition of the improper integral with the bottom limit being $-\infty$.

In the case of Eq. (4.71) both limits should exist (taken in any order and independently of each other). In each case the integrals may either converge (the limits exist) or diverge (are infinite). The geometrical interpretation of the improper integrals is exactly the same as for proper ones: if $f(x)$ is a non-negative function, then any of the above improper integrals corresponds to the area under the curve of the function within the corresponding intervals $a \le x < +\infty$, $-\infty < x \le b$ or $-\infty < x < +\infty$, respectively. If any of the integrals converge, then the area under the curve is finite.

When calculating the improper integrals one can still use the Newton–Leibniz formula (4.44). Indeed, if $F'(x) = f(x)$, then, for instance,

$$\int_a^\infty f(x)dx = \lim_{b \to \infty} \int_a^b f(x)dx = \lim_{b \to \infty} [F(b) - F(a)]$$
$$= \lim_{b \to \infty} F(b) - F(a) = F(+\infty) - F(a) , \qquad (4.72)$$

and

$$
\int_{-\infty}^{\infty} f(x)dx = \lim_{b \to \infty} \left(\lim_{a \to -\infty} \int_a^b f(x)dx \right) = \lim_{b \to \infty} \left[\lim_{a \to -\infty} (F(b) - F(a)) \right]
$$
$$
= \lim_{b \to \infty} F(b) - \lim_{a \to -\infty} F(a) = F(+\infty) - F(-\infty) .
$$

(4.73)

Let us consider some examples of these types of improper integrals. As our first example let us calculate the integral which has a finite upper limit:

$$
I(b) = \int_0^b e^{-x}dx = -e^{-x}\big|_0^b = 1 - e^{-b} .
$$

Integration of the same function between 0 and $+\infty$, according to the definition given above, requires us to take the $b \to \infty$ limit. Therefore,

$$
\int_0^{\infty} e^{-x}dx = \lim_{b \to \infty} \left(1 - e^{-b} \right) = 1 .
$$

As the limit exists (i.e. it is finite), it is to be taken as the value of the integral.

Example 4.2 ▶ As another example, consider the integral

$$
I_{\lambda} = \int_1^{\infty} \frac{dx}{x^{\lambda}} ,
$$

(4.74)

where λ is positive. What are the values of λ at which this integral converges?

Solution. Let us first investigate the case of $\lambda > 1$ (strictly larger than one!). In this case

$$
I_{\lambda>1} = \lim_{b \to \infty} \int_1^b \frac{dx}{x^{\lambda}} = \lim_{b \to \infty} \frac{x^{-\lambda+1}}{-\lambda+1}\bigg|_1^b = \frac{1}{1-\lambda} \lim_{b \to \infty} \left(\frac{1}{b^{\lambda-1}} - 1 \right)
$$
$$
= -\frac{1}{1-\lambda} = \frac{1}{\lambda-1} ,
$$

since the power $\lambda - 1$ of b in $1/b^{\lambda-1}$ is positive and hence this term tends to zero in the limit. In the case of $\lambda = 1$, we similarly write:

$$
I_{\lambda=1} = \lim_{b \to \infty} \int_1^b \frac{dx}{x} = \lim_{b \to \infty} \ln x\big|_1^b = \lim_{b \to \infty} (\ln b - \ln 1) = \lim_{b \to \infty} \ln b = \infty ,
$$

i.e. in this case the integral diverges. Finally, if $0 < \lambda < 1$, then

$$
I_{\lambda<1} = \lim_{b \to \infty} \int_1^b \frac{dx}{x^{\lambda}} = \lim_{b \to \infty} \frac{x^{-\lambda+1}}{-\lambda+1}\bigg|_1^b = \frac{1}{1-\lambda} \lim_{b \to \infty} \left(b^{1-\lambda} - 1 \right) = \infty ,
$$

and the integral diverges again. We conclude that the integral (4.74) only converges for strictly $\lambda > 1$.

Let us now investigate whether the following integral exists:

$$\int_0^\infty \sin x \, dx = \lim_{b \to \infty} (-\cos x)|_0^b = \lim_{b \to \infty} (1 - \cos b) = 1 - \lim_{b \to \infty} \cos b .$$

Since the cosine function oscillates and hence does not have a definite limit as $b \to \infty$ (although its value is bounded from above and below), the integral in question does not converge (does not exist).

The above examples tell us that an improper integral can be calculated in exactly the same way as the corresponding proper ones, and the value of the integral is still given as the difference of the indefinite integral, the function $F(x)$, between the upper and lower limits. The only difference is in working out the value(s) of $F(\pm\infty)$, which requires taking the appropriate limit(s).

Problem 4.45 Check the values of the following improper integrals using their appropriate definitions:

$$\int_0^\infty e^{-x} \sin x \, dx = \frac{1}{2} ; \quad \int_0^\infty \frac{dx}{1 + x^2} = \frac{\pi}{2} ;$$

$$\int_{-\infty}^0 e^x \, dx = 1 ; \quad \int_{1/\pi}^\infty \frac{1}{x^2} \sin \frac{1}{x} \, dx = 2 .$$

Problem 4.46 Using integration by parts, derive a recurrence relation for the following integral[7]:

$$\Gamma(n) = \int_0^\infty x^{n-1} e^{-x} \, dx .$$

Show by direct calculation for several values of $n = 1, 2, 3$ and then using induction that $\Gamma(n + 1) = n!$.

---

[7] This is a particular case of one of the special functions, the so-called Gamma function.

Problem 4.47 Show by differentiating the integral below with respect to β that[8]

$$\int_0^\infty e^{-\alpha x} \frac{\sin(\beta x)}{x} dx = \arctan \frac{\beta}{\alpha} .$$

It is also possible to define an integral of a function which has a singularity[9] either at the bottom and/or top limits, and/or between the limits. If a function $f(x)$ is singular at the top limit b, i.e. $f(b) = \pm\infty$, but is finite just before it, we define a positive ϵ and define the integral via the limit as:

$$\int_a^b f(x)dx = \lim_{\epsilon \to 0} \int_a^{b-\epsilon} f(x)dx . \tag{4.75}$$

Here the function is finite for $a \le x \le b - \epsilon$ for any $\epsilon > 0$, and hence the integral before the limit is well defined. If the limit exists, then the integral converges. Similarly, if the singularity of $f(x)$ coincides with the bottom limit, $f(a) = \pm\infty$, then we consider

$$\int_a^b f(x)dx = \lim_{\epsilon \to 0} \int_{a+\epsilon}^b f(x)dx . \tag{4.76}$$

Finally, if the singularity occurs at a point c somewhere between a and b, one breaks the whole integration interval into two subintervals $a \le x < c$ and $c < x \le b$, so that the initial integral is split into two, and then the above definitions are applied to each of the integrals separately:

$$\int_a^b f(x)dx = \lim_{\epsilon_1 \to 0} \int_a^{c-\epsilon_1} f(x)dx + \lim_{\epsilon_2 \to 0} \int_{c+\epsilon_2}^b f(x)dx . \tag{4.77}$$

If both limits exist, the integral on the left-hand side is said to converge (or to exist). It is essential here that both limits are taken independently; this is indicated by using $\epsilon_1 \ne \epsilon_2$ in the two integrals.[10] This type of reasoning also explains why changing the value of the function $f(x)$ at a single isolated point $x = c$ will not change the value of the integral. Indeed, in this case the integral can be defined by virtue of Eq. (4.77); clearly, the value of the function $f(x)$ at $x = c$ does not appear in this case at all; it could in fact be different from either of the two limits $\lim_{x \to c \pm 0} f(x)$.

---

[8] Note that this integral depends on β as a parameter, and one has to justify differentiation with respect to this parameter inside the integral sign. In this case this can be justified because the integral converges *uniformly*. We shall consider uniform convergence of integrals in Sect. 7.2.4, where we shall also discuss this particular case in detail.

[9] I.e. the function is equal to $\pm\infty$ at an isolated point between $-\infty$ and $+\infty$.

[10] Compare with Sect. 4.5.4.

Example 4.3 ► As an example of a singularity inside the integration interval, consider the function $f(x) = 1/x^\lambda$ with positive λ. This function is singular at $x = 0$, so that whether it is integratable around this point needs to be specifically investigated.

Solution. Firstly, let $\lambda > 1$ and consider the integral

$$I_{\lambda>1} = \int_0^1 \frac{dx}{x^\lambda} = \lim_{\epsilon \to 0} \int_{0+\epsilon}^1 \frac{dx}{x^\lambda} = \lim_{\epsilon \to 0} \frac{1}{1-\lambda} \left[1 - \frac{1}{\epsilon^{\lambda-1}} \right] .$$

Since $\lambda - 1 > 0$, the limit of $1/\epsilon^{\lambda-1}$ is equal to infinity, i.e. the integral does not converge for any $\lambda > 1$. In the case of $\lambda = 1$, we have

$$I_{\lambda=1} = \int_0^1 \frac{dx}{x} = \lim_{\epsilon \to 0} \int_\epsilon^1 \frac{dx}{x} = \lim_{\epsilon \to 0} (\ln 1 - \ln \epsilon) = - \lim_{\epsilon \to 0} \ln \epsilon = \infty ,$$

i.e. the limit is not finite again, and hence the integral diverges[11] for $\lambda = 1$. Finally, consider the case of $0 < \lambda < 1$. In this case

$$I_{\lambda<1} = \int_0^1 \frac{dx}{x^\lambda} = \lim_{\epsilon \to 0} \int_\epsilon^1 \frac{dx}{x^\lambda} = \lim_{\epsilon \to 0} \frac{1}{1-\lambda} \left[1 - \epsilon^{1-\lambda} \right] = \frac{1}{1-\lambda} ,$$

since $1 - \lambda > 0$ and hence the limit of $\epsilon^{1-\lambda}$ is well defined and equal to zero. So, the above integral only converges for $0 < \lambda < 1$.

Problem 4.48 Check whether the following integrals converge and then calculate them if they do:

$$\int_0^1 \frac{dx}{\sqrt{1-x^2}} ; \quad \int_{-1}^1 \frac{dx}{\sqrt{1-x^2}} ; \quad \int_0^2 \ln x\, dx ; \quad \int_1^2 \frac{1}{x \ln x} dx .$$

[Hint: *use the substitution* $t = \ln x$ *in the last two cases. Answers:* $\pi/2$, π, $2(\ln 2 - 1)$, *diverges at* $x = 1$.].

---

[11] It is said that it diverges logarithmically.

Problem 4.49 Consider a function $f(x)$ on the interval $a \leq x \leq b$ which has singularities at a finite number of points $c_1, c_2, \ldots, c_n$ between a and b. If the integrals of $f(x)$ between these boundaries exist and the indefinite integral $F(x)$ is continuous in any of the singular points $(F'(x) = f(x))$, show that formally the Newton–Leibniz formula (4.44) is still valid, i.e. that

$$\int_a^b f(x)dx = F(b) - F(a) . \tag{4.78}$$

[Hint: *first consider a single singularity ($n = 1$), and then generalize your result for their arbitrary finite n.*]

An essential condition for the above formula to be valid is that the indefinite integral $F(x)$ is continuous at all singularity points as otherwise the limits from the left and right at the singularity points would be different and not cancel out. This point is illustrated by the following problem.

Problem 4.50 Show using the definition of the improper integral of a function with a singularity at $x = 0$ that, after splitting the integral into two around the $x = 0$ point,

$$\int_{-1}^1 x^{-1/3} dx = 0 .$$

On the other hand, show using the Newton–Leibniz formula (4.78), that the same result is obtained.

This is because the indefinite integral $F(x) = x^{2/3}$ is continuous at $x = 0$.

The integral of $f(x)$ may not always be calculated using the Newton–Leibniz formula as this requires knowing the indefinite integral, i.e. the function $F(x)$ such that $F'(x) = f(x)$. Still, it is frequently useful to know if the given improper integral converges even though its analytical calculation is difficult or even impossible. Therefore, it is useful to be able to assess the convergence of an integral without actually calculating it. Below we shall consider several useful criteria which can be used to test the convergence of improper integrals.

Theorem 4.8 (Comparison test) *Consider two continuous functions $f(x)$ and $g(x)$ such that $0 \leq f(x) \leq g(x)$ for all x within the interval $a < x < b$, where a could be either finite or $-\infty$ and b is also either finite or equal to $+\infty$. Then, if the integral $\int_a^b g(x)dx$ converges, then the integral $\int_a^b f(x)dx$ also converges; conversely, if the latter integral diverges, the former one diverges as well.*

Proof: For definiteness consider both integrals with a finite bottom limit a and the infinite upper limit $b = +\infty$. We define two functions of the upper limit:

$$F(X) = \int_a^X f(x)dx \quad \text{and} \quad G(X) = \int_a^X g(x)dx \;.$$

Since $f(x)$ and $g(x)$ are positive, both functions $F(X)$ and $G(X)$ are increasing monotonically. Indeed, for instance by taking any positive $\Delta X > 0$ we can write:

$$F(X + \Delta X) = \int_a^{X+\Delta X} f(x)dx = \int_a^X f(x)dx + \int_X^{X+\Delta X} f(x)dx$$
$$= F(X) + \int_X^{X+\Delta X} f(x)dx = F(X) + \Delta F \geq F(X) \;,$$

where the integral ΔF is obviously positive since $f(x) > 0$ within the interval $X \leq x \leq X + \Delta X$.

Now since $f(x) \leq g(x)$, then $F(X) \leq G(X)$ as integrals are defined as the integral sums. Next, assume that the integral of $g(x)$ converges. This means that the limit of $G(X)$ when $X \to +\infty$ exists; moreover, since $G(X)$ increases with the increase of X, it must be limited from above, i.e. $G(+\infty) < M$ with some finite M. The function $F(X)$ is never larger than $G(X)$, and hence is also limited from above, e.g. by the same value M. Hence, the function $F(X)$ must have a definite limit when $x \to +\infty$ (see Theorem 2.20 in Sect. 2.4.4), i.e. $F(+\infty)$ can only be finite, i.e. the integral of $f(x)$ also converges. This proves the first part of the theorem that from convergence of the integral of $g(x)$ follows convergence of the integral with $f(x)$.

The second part of the theorem is proven by contradiction: let the integral of $f(x)$ diverges. Then, if we assume that the integral of $g(x)$ converges, the integral of $f(x)$ will converge as well by virtue of the first part of the theorem. However, this contradicts with the fact that the integral of $f(x)$ actually diverges. Therefore, our assumption concerning the convergence of the integral of $g(x)$ is incorrect; this integral diverges. **Q.E.D.**

Other cases (the bottom limit is $-\infty$ and/or the functions have singularities between $-\infty$ and $+\infty$) are considered similarly. Note that the theorem cannot establish convergence of the integral of $f(x)$ if the integral of $g(x)$ diverges; the former integral may either diverge or converge and this requires further consideration.

As an example consider the convergence of the integral of the function $f(x) = e^{-x} \sin^2 x$ containing the infinite upper limit. Since $\sin^2 x \leq 1$, we can introduce a positive function $g(x) = e^{-x}$ such that for any x we have $f(x) \leq g(x)$ since $\sin^2 x \leq 1$. Then, because the integral

$$\int_a^\infty e^{-x}dx = -e^{-x}\Big|_a^\infty = e^a$$

converges (here a is any finite number), so must be the case for the integral $\int_a^\infty e^{-x} \sin^2 x dx$.

Problem 4.51 Investigate the convergence of the integral

$$\int_1^\infty \frac{x^{-\alpha}}{1+x^2} dx$$

for various values of a positive parameter α. [Hint: *find the upper limit of the function* $1/(1+x^2)$ *for* $x \geq 1$. Answer: $\alpha > -1$.]

Problem 4.52 Investigate the convergence of the integral

$$\int_{-1}^1 \frac{\sin^2 x}{\sqrt{x}} dx .$$

[Answer: *converges.*]

Problem 4.53 Show using multiple integrations by parts that for any integer n,

$$\int_0^\infty x^n e^{-\lambda x} dx = \frac{n!}{\lambda^{n+1}} .$$

When investigating convergence, it is useful to approximate the integrand around the singularity point to see if the integral with the limits just around that singularity converges. For instance, in the above two problems, we have the integrands given by $f_1(x) = x^{-\alpha}/(1+x^2)$ and $f_2(x) = x^{-1/2} \sin^2 x$. In the first case we have to investigate the convergence at $x \to \infty$. For large x one may approximate $1+x^2$ simply with x^2, leading to the integral of $x^{-\alpha-2}$ which, when integrated, results in a function proportional to $x^{-\alpha-1}$. This function is well defined at $x \to \infty$ (and tends to zero there) only if $\alpha > -1$. In the second case the singularity occurs at $x = 0$. For small x we can replace $\sin x$ with x, which results in the integrand proportional to $x^{-1/2+2} = x^{3/2}$, which is integrable around $x = 0$.

The criterion stated by Theorem 4.8 requires the two functions to be necessarily non-negative. If the given function $f(x)$ changes its sign, then this test cannot be applied.

Theorem 4.9 *Consider an improper integral $\int_a^b f(x)dx$ (where either the function $f(x)$ has singularities within the interval and/or one or both limits a and b are infinite). This integral converges if the integral $\int_a^b |f(x)|\, dx$ of the absolute value of $f(x)$ converges.*

Proof: Indeed, since the integral of $|f(x)|$ converges, then $\int_a^b |f(x)|\, dx < M$ with some positive M. On the other hand, the integral is a limit of the integral sum, and hence one can write (see Eq. (4.29) in Sect. 4.2)

$$\left| \int_a^b f(x)dx \right| \le \int_a^b |f(x)|\, dx < M \, ,$$

i.e. the improper integral under study, $\int_a^b f(x)dx$, is limited and hence converges. **Q.E.D.**

The proven theorem tells us that from the convergence of the integral $\int_a^b |f(x)|\, dx$ follows convergence of $\int_a^b f(x)dx$; in this case it is said the latter integral converges *absolutely*. This condition of convergence is stronger than an ordinary convergence of the integral $\int_a^b f(x)dx$, since the integral $\int_a^b |f(x)|\, dx$ may not converge even though the integral $\int_a^b f(x)dx$ does, i.e. the statement of the theorem does not work in the opposite direction.

Problem 4.54 Consider an improper integral $\int_a^b f(x)dx$ that converges absolutely and a function $g(x)$ bounded from below and above, $|g(x)| < M$. Prove that the improper integral $\int_a^b f(x)g(x)dx$ also converges absolutely.

Problem 4.55 In particular, investigate the convergence of the integral $\int_1^\infty g(x)\, x^{-\lambda}dx$, where $\lambda > 0$ and the function $g(x)$ is limited from below and above. [Answer: *converges for $\lambda > 1$.*]

Problem 4.56 Similarly, investigate the convergence of the integral $\int_0^\infty g(x)e^{-x}\, dx$. [Answer: *converges.*]

Problem 4.57 *(Cauchy's test)* Consider an integral $F = \int_a^b f(x)dx$ between two finite limits, and let the function $f(x)$ be singular at the upper limit. Prove that if in the vicinity of the point $x = b$ (i.e. for points satisfying $b - \epsilon < x < b$ with some—may be even small—positive ϵ) the function $f(x)$ satisfies an estimate $|f(x)| < M/(b-x)^\lambda$ (M is positive), then the integral F converges absolutely for $\lambda < 1$.

For instance, if we consider an integral $\int_0^1 \sin(x)/\sqrt{1-x}dx$, then according to the above Cauchy's test it converges, and converges absolutely, since

$$\left| \frac{\sin(x)}{\sqrt{1-x}} \right| < \frac{1}{(1-x)^{1/2}}$$

with $\lambda = 1/2 < 1$.

One can also consider the case of the function $f(x)$ being not bound in its absolute value in the vicinity of $x = b$, i.e. when there $|f(x)| > M/(b-x)^\lambda$ with some $\lambda \geq 1$. Cauchy's test in that case can be extended in saying that the integral $F = \int_a^b f(x)dx$ would diverge. Indeed, in this case we may consider an interval $b - \Delta \leq x \leq b - \epsilon$ (where $\Delta > \epsilon > 0$ and we are supposed to consider the limit $\epsilon \to 0$) and make an estimate of the part of the integral taken over this interval. Since the function in its absolute value is greater than some positive number within the interval $b - \Delta \leq x \leq b - \epsilon$, it cannot change its sign and must remain either positive or negative. Consider for definiteness the function being positive. Then, if $\lambda > 1$, we can write:

$$\int_{b-\Delta}^{b-\epsilon} f(x)dx > M \int_{b-\Delta}^{b-\epsilon} \frac{dx}{(b-x)^\lambda} = M \int_\epsilon^\Delta \frac{dt}{t^\lambda} = -\frac{M}{\lambda-1}\left(\frac{1}{\Delta^{\lambda-1}} - \frac{1}{\epsilon^{\lambda-1}}\right).$$

The expression in the square brackets in the $\epsilon \to 0$ limit tends to infinity if $\lambda > 1$, i.e. the integral can take a value that is arbitrarily large, i.e. it diverges. It is also easy to see that the integral diverges (and diverges logarithmically) if $\lambda = 1$, since

$$\int_{b-\Delta}^{b-\epsilon} f(x)dx > M \int_\epsilon^\Delta \frac{dt}{t} = M \ln \frac{\Delta}{\epsilon}.$$

Hence, concluding, if $|f(x)| > M/(b-x)^\lambda$, then the integral $F = \int_a^b f(x)dx$ diverges for $\lambda \geq 1$. The considered convergence tests are illustrated by our discussion in Example 4.3.

Similarly one can consider an integration of the function $f(x)$ that is singular at the lower limit, $x = a$. The same criteria apply in this case. If the function $f(x)$ is singular at some point $x = c$ within the interval $a < x < b$, then the whole integral can be split into two, one between a and c and another between c and b, and then it is necessary to apply Cauchy's test for the upper limit of the first integral and the lower limit of the second; if both integrals converge, the initial integral from a to b also converges.

Finally, Cauchy's tests can be extended to the upper (lower) limits being plus (minus) infinity. For definiteness, we shall formulate this test for the integral $f = \int_a^\infty f(x)dx$ with the infinite upper limit: if $|f(x)| < M/x^\lambda$ with $\lambda > 1$, then the integral converges. If, however, $|f(x)| > M/x^\lambda$ and $\lambda \le 1$, then the integral diverges (compare with Example 4.2). The proofs are practically the same as given above for the final interval case and are offered for the reader to try. The same tests are valid for the bottom limit to be $-\infty$. If both limits are infinite, then the convergence must be investigated at each of them separately.

In practice, one can either consider the behaviour of the function $f(x)$ at the singularity point and/or infinite upper/lower limit by extracting leading terms, or by considering instead if there exists such $\lambda > 1$ that the expression $f(x)x^\lambda$ has a finite value there. Indeed, considering the infinite upper limit for definiteness, from the existence of the limit $\lim_{x\to\infty} x^\lambda f(x) = A$ (with A being finite) follows that $x^\lambda f(x)$ must be bounded at this limit, i.e. the Cauchy test $|f(x)| < M/x^\lambda$ is satisfied.

As an example, consider the integral

$$\int_0^\infty \frac{3x^2 + 4x + 1}{x^3 + 1}dx \,.$$

At $x = 0$ the integrand is finite, $x^3 + 1 > 0$ for any $x \ge 0$, so one only needs to consider the convergence of the integral at the upper limit. For large x the denominator behaves like x^3, while the numerator—as $3x^2$. Hence, the integrand at large values of the x behaves like $3x^2/x^3 = 3/x$, and hence diverges logarithmically at the infinite upper limit. One can come to the same conclusion by multiplying the integrand with x (using $\lambda = 1$) and considering the limit. In this case the function

$$x\frac{3x^2 + 4x + 1}{x^3 + 1} = \frac{3x^3 + 4x^2 + x}{x^3 + 1} = \frac{3 + 4/x + 1/x^2}{1 + 1/x^3}$$

has a finite limit at $x \to \infty$. Since in this case $\lambda = 1$, the integral diverges.

Problem 4.58 Consider an integral

$$\int_{-\infty}^\infty \frac{Q_n(x)}{P_m(x)}dx \,,$$

where $Q_n(x)$ and $P_m(x)$ are polynomials of the orders n and m, respectively. Show that the integral converges if $m - n \ge 2$ and diverges if $m - n \le 1$.

Problem 4.59 Investigate the convergence of the following improper integrals:

$$\int_0^\infty \frac{x^3 + x + 1}{x^5 + x + 2} dx \; ; \quad \int_{-\infty}^0 \frac{2x^2 + 1}{x^3 + x - 2} dx \; ; \quad \int_0^\infty \frac{x^3 + x + 1}{x^2 + x + 2} \frac{dx}{\sqrt{x + 1}} .$$

[Answers: *converges; diverges; diverges*]

Problem 4.60 Investigate the convergence of the following integrals:

$$\int_0^\infty \frac{\sin x}{1 + x^2} dx \quad \text{and} \quad \int_1^\infty \frac{\cos x}{x^\lambda} dx \quad \text{with} \quad \lambda > 0 .$$

[Answers: *converges; converges for* $\lambda > 1$.]

The Cauchy tests considered above correspond to checking the absolute convergence. However, this is only a sufficient condition: even if the Cauchy test fails, the integral $\int f(x)dx$ may still converge. There exists another test that could be useful if the absolute convergence test fails (the integral $\int |f(x)|\,dx$ diverges).

Consider the infinite upper limit for definiteness, i.e. the integral $\int_a^\infty f(x)dx$ with $a > 0$. Let us introduce a function $F(x) = \int_a^x f(t)dt$. Then, if the function $F(x)$ remains bounded when $x \to \infty$, then the integral $\int_a^\infty t^{-\lambda} f(t)dt$ converges for any $\lambda > 0$. Indeed, for any finite $x > a$,

$$\int_a^x \frac{f(t)}{t^\lambda} dt = \left[\frac{1}{t^\lambda} F(t) \right]_a^x - \int_a^x \frac{-\lambda}{t^{\lambda+1}} F(t)dt = \frac{F(x)}{x^\lambda} + \lambda \int_a^x \frac{F(t)}{t^{\lambda+1}} dt \; ,$$

where we have used the integration by parts and the fact that $F(a) = 0$. Since $F(x)$ is bounded in the $x \to \infty$ limit, the first term on the right-hand side for $\lambda > 0$ goes to zero; the second term in the limit is the integral $\int_a^\infty F(t)/t^{\lambda+1}dt$ that converges by the Cauchy test as $\lambda + 1 > 1$. This concludes the proof.

The simplest illustration of this test would be the integral $\int_0^\infty \frac{\sin x}{x}dx$. At $x = 0$ the integrand is finite (equals to one), so we have to consider the convergence at the upper limit. Here $F(x) = \int_0^x \sin(t)dt = 1 - \cos(x)$ and remains bounded for any x. Since for our integral $\lambda = 1$, it converges.[12]

---

[12] We shall learn in Chap. 2 of Vol. II that this integral is equal to $\pi/2$.

4.5.4 Cauchy Principal Value

Consider a function $f(x)$ which lies within the interval $a < x < b$ and has there a singularity at an internal point c (here we assume that a and b are either finite or infinite). The corresponding improper integral is defined by Eq. (4.77) with the two limits taken independently of each other, i.e. with $\epsilon_1 \neq \epsilon_2$. If this improper integrals diverges, it might in some cases be useful to constrain the limiting procedure to having $\epsilon_1 = \epsilon_2 \equiv \epsilon$ and hence perform a *single* limit $\epsilon \to +0$. If this limit exists, the integral is said to converge in the sense of the *Cauchy principal value*:

$$\mathcal{P} \int_a^b f(x)dx \quad \text{or} \quad \fint_a^b f(x)dx = \lim_{\epsilon \to 0} \left(\int_a^{c-\epsilon} f(x)dx + \int_{c+\epsilon}^b f(x)dx \right).$$
(4.79)

This value may be assigned to some of the improper integrals which otherwise diverge. It is a special form of *regularization* of the otherwise ill-defined integrals.

Consider for instance the following improper integral ($a < c < b$):

$$I = \int_a^b \frac{dx}{x - c}.$$

In the sense of the formal definition of Sect. 4.5.3 this integral is at least ill-defined:

$$I = \lim_{\epsilon_1 \to 0} \int_a^{c-\epsilon_1} \frac{dx}{x - c} + \lim_{\epsilon_2 \to 0} \int_{c+\epsilon_2}^b \frac{dx}{x - c} = \lim_{\epsilon_1 \to 0} \ln \frac{\epsilon_1}{c - a} + \lim_{\epsilon_2 \to 0} \ln \frac{b - c}{\epsilon_2}$$

$$= \ln \frac{b - c}{c - a} + \left[\lim_{\epsilon_1 \to 0} \ln \epsilon_1 - \lim_{\epsilon_2 \to 0} \ln \epsilon_2 \right]$$

$$= \ln \frac{b - c}{c - a} + [(-\infty) - (-\infty)] = \ln \frac{b - c}{c - a} + (-\infty + \infty).$$

At the same time, its principal value is well defined:

$$I = \lim_{\epsilon \to 0} \left[\int_a^{c-\epsilon} \frac{dx}{x - c} + \int_{c+\epsilon}^b \frac{dx}{x - c} \right] = \lim_{\epsilon \to 0} \left[\ln \frac{\epsilon}{c - a} + \ln \frac{b - c}{\epsilon} \right] = \ln \frac{b - c}{c - a},$$

as the two terms with $\ln \epsilon$ cancel out exactly.

Problem 4.61 Calculate the principal value of $\int_a^b (x - c)^{-n} dx$ with $n > 1$ being a positive integer and c lying between a and b. Consider the cases of n odd and even. When does this integral exist? [Answer n *must be odd;* $\left[(b - c)^{1-n} - (a - c)^{1-n} \right] / (1 - n).$]

It can also be shown that a more general integral

$$I = \int_a^b \frac{f(x)}{x - c} dx \; ,$$

in which the function $f(x)$ is continuous everywhere between a and b, is well defined in the principal value sense. Indeed, let us rewrite it as follows:

$$I = \int_a^b \frac{f(x) - f(c)}{x - c} dx + f(c) \int_a^b \frac{dx}{x - c} \; .$$

Here the first integral is not improper anymore since the integrand is a continuous function at $x = c$ with the limiting value of $f'(c)$ (use L'Hôpital's rule, Sect. 3.10, to check this!); the second integral is well defined in the principal sense.

The regularization we performed above, when discussing the improper integrals with the integrand having a singularity, is not the only possible. Indeed, consider the integral (4.71) with both infinite boundaries. Its usual definition contains two limits and in some cases these may yield divergence if applied independently. However, the following regularization also called the principal value, may result in a well-defined limit:

$$\mathcal{P} \int_{-\infty}^{\infty} f(x) dx = \lim_{a \to \infty} \int_{-a}^{a} f(x) dx \; . \tag{4.80}$$

As an example, consider the following integral:

$$\begin{aligned}
I &= \int_{-\infty}^{\infty} \frac{x dx}{x^2 + 1} = \lim_{a \to -\infty} \left\{ \lim_{b \to +\infty} \int_a^b \frac{x dx}{x^2 + 1} \right\} \\
&= \lim_{a \to -\infty} \left\{ \lim_{b \to +\infty} \frac{1}{2} \left[\ln \left(b^2 + 1 \right) - \ln \left(a^2 + 1 \right) \right] \right\} \\
&= \frac{1}{2} \left[\lim_{b \to \infty} \ln \left(b^2 + 1 \right) - \lim_{a \to -\infty} \ln \left(a^2 + 1 \right) \right] \; .
\end{aligned}$$

This expression is ill-defined as it contains a difference of two infinities. At the same time, the principal value of this integral is well defined:

$$I = \lim_{a \to +\infty} \int_{-a}^{a} \frac{x dx}{x^2 + 1} = \lim_{a \to +\infty} \frac{1}{2} \ln \frac{1 + a^2}{1 + a^2} = 0 \; ,$$

which in fact is what one would assign to this integral as the integrand is an odd function.

Problem 4.62 Investigate whether improper integrals $\int_{-\infty}^{\infty} \sin x dx$ and $\int_{-\infty}^{\infty} \cos x dx$ exist in the principal value sense. [Answer: 0; *diverges*.]

Problem 4.63 Show that

$$\mathcal{P} \int_{-E}^{E} \frac{dx}{(x-a)(x-b)} = \frac{1}{a-b} \ln \left| \frac{(E-a)(E+b)}{(E+a)(E-b)} \right|$$

irrespective of whether the points a and/or b are inside or outside the integration interval.

Problem 4.64 Show that $(0 < a < A)$

$$\mathcal{P} \int_{0}^{A} \frac{x^2 f(x)}{x^2 - a^2} dx = \int_{0}^{A} \frac{x^2 (f(x) - f(a))}{x^2 - a^2} dx + f(a) \left[A + \frac{a}{2} \ln \left| \frac{A-a}{A+a} \right| \right].$$

4.6   Convolution and Correlation Functions

Using integration, new functions can be defined. We shall consider here two such functions that are defined in a similar way and are widely used in physical applications.

We shall start from a function called *convolution*. that is a result of an operation on two functions. Given two functions $f(x)$ and $g(x)$, both defined on the whole x-axis, $-\infty < x < \infty$, their convolution, usually denoted as $f(x) * g(x)$ or $(f * g)(x)$, is defined via the integral:

$$f(x) * g(x) = \int_{-\infty}^{\infty} f(x')g(x - x')dx'. \tag{4.81}$$

It is a function of x that is defined on the whole x-axis as well. If both functions are defined for the interval $0 \le x < \infty$, then the following definition is used:

$$f(x) * g(x) = \int_{0}^{t} f(x')g(x - x')dx'. \tag{4.82}$$

It is easy to see that in this case the convolution is also defined for all positive values of x including zero.

The convolution represents a modification of a signal $f(x)$ by some "weighting function" $g(x)$ (or *vice versa*). A few examples given below should illustrate some of its applications. But before that, let us list the obvious properties of the convolution:

$$f * g = g * f \quad \text{(symmetric)} \tag{4.83}$$

$$f * (g * h) = (f * g) * h \quad \text{(associative)} \tag{4.84}$$

$$f * (g + h) = f * g + f * h \quad \text{(distributive)}. \tag{4.85}$$

These properties are valid for both definitions of the convolution given above. The distributivity is trivial; the symmetry property is proven by an obvious change of the integration variable $x' \to x'' = x - x'$. The proof of associativity is given in the problem below.

Problem 4.65 Prove associativity of the convolution in both definitions. [Hint: *you need to write one of the expressions explicitly using the definition of convolution, either the one in the left- or right-hand sides, this would look like an integral inside another integral (this is what we call a double integral[13]); then make an appropriate change of the variable in one of the integrals to get the other form.*]

The convolution is different from multiplication even though it satisfies the above properties that are characteristic for the multiplication operation. For instance, obviously, $(1 * f)(x) \neq f(x)$. Also, if we adopt the definition (4.82), then

$$(1 * 1)(x) = \int_0^x dx' = x$$

that is different from 1. Also,

$$(x * x)(x) = \int_0^x t(x - t)dt = x \int_0^x t \, dt - \int_0^x t^2 dt = \frac{x^3}{6}.$$

Obviously, these examples are not appropriate for the first definition (4.81) of the convolution as in this case the integrals would not exist.

To have some idea of what the convolution might be, let us consider the weighting function $g(x)$ being a narrow rectangular impulse of unit area centred at $x = a$:

$$g(x) = \Delta(x - a), \quad \text{where} \quad \Delta(x) = \frac{1}{2\delta} \begin{cases} 1, & |x| \leq \delta \\ 0, & |x| > \delta \end{cases}.$$

---

[13] More on these in Sect. 6.1.

Using this function,[14] we obtain for the convolution of $f(x)$ with it (using the definition (4.81) with infinite boundaries):

$$(f * g)(x) = \int_{-\infty}^{\infty} f(x')\Delta(x - x' - a)dx' = \int_{-\infty}^{\infty} f(x - a - t)\Delta(t)dt$$

$$= \frac{1}{2\delta} \int_{-\delta}^{\delta} f(x - a - t)dt .$$

If δ is very small, then for $f(x)$ being sufficiently smooth and changing little within the narrow interval of $|t| \leq \delta$, we can replace it by its value $f(x - a)$ at $t = 0$, which gives $(f * g)(x) \approx f(x - a)$. So, for $g(x)$ being an impulse at $x = a$, the convolution of $f(x)$ with it simply shifts by a to the right along the x-axis. Any general function $g(x)$ can be thought of as composed of such impulses and hence the convolution with it would transform the function $f(x)$ in a particular way.

Problem 4.66 Consider a function called truncated exponential,

$$E(x) = \begin{cases} e^{-x}, & x \geq 0 \\ 0, & x < 0 \end{cases} . \tag{4.86}$$

Show that its convolution with itself (using both definitions) is $(E * E)(x) = xE(x)$.

Problem 4.67 Using the definition of the truncated exponential (4.86), show that

$$E(\alpha x) * E(\beta x) = \frac{1}{\beta - \alpha} [E(\alpha x) - E(\beta x)] .$$

Check explicitly by taking an appropriate limit that in the case of $\alpha = \beta$ this result coincide with the one in Problem 4.66.

---

[14] This function in the limit of $\delta \to 0$ is called Dirac delta function and will be considered in more detail in Chapter 4 of Vol. II.

Fig. 4.4 The hat function of
Problem 4.68

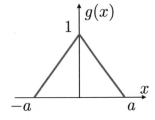

Problem 4.68 Consider a hat function $g(x)$ shown in Fig. 4.4. Show that for any integrable $f(x)$

$$(f * g)(x) = \int_{-a}^{a} f(x - x')dx' - \frac{1}{a}\int_{0}^{a} x'\left[f(x + x') + f(x - x')\right]dx'.$$

Finally, let us consider differentiation of the convolution. This operation is trivial for the convolution (4.81) defined with infinite limits, provided that one can use formula (4.68). The situation is however a bit more complicated for the convolution defined via Eq. (4.82) since the upper limit in this case also depends on the variable.

Problem 4.69 Prove that

$$\frac{d}{dx}(f * g)(x) = f(x)g(0) + (f * g')(x),$$

$$\frac{d^2}{dx^2}(f * g)(x) = f'(x)g(0) + f(x)g'(0) + (f * g'')(x).$$

Then prove, using induction, that generally, for any integer $n \geq 1$

$$\frac{d^n}{dx^n}(f * g)(x) = \sum_{i=0}^{n-1} f^{(i)}(x)g^{(n-1-i)}(0) + \left(f * g^{(n)}\right)(x),$$

where $f^{(n)}(x)$ is the n-th derivative of the function $f(x)$; it is assumed that $f^{(0)}(x) = f(x)$.

Another function that is frequently used in physical applications is the correlation function (autocorrelation function if $f(t) = g(t)$)

$$C_{fg}(t) = \int_{-\infty}^{\infty} f(t')g(t + t')dt'. \tag{4.87}$$

Here we have used t for it instead of x as in most applications this function is defined with respect to time. This function characterizes the degree at which two functions are correlated with each other in time. For instance, consider a velocity autocorrelation function of particle in a liquid, $C_{vv}(t)$. If the liquid is at equilibrium, the correlation function at long times will tend to zero as there will be no correlation between the velocity $v(t')$ of the particle at any time t' and its value at a very long time t after that. At small times the integral (4.87) for $C_{vv}(t)$ will not be equal to zero but is expected to decay to zero as a function of time.

Problem 4.70 Show that the autocorrelation function of the truncated exponent $E(\alpha t)$, see Eq. (4.86), is

$$C_{EE}(t) = \frac{1}{2\alpha} e^{-\alpha|t|}.$$

[Hint: *consider separately the cases of positive and negative times t.*]

We see from this problem that at long times this correlation function decays to zero exponentially and symmetrically in both directions ($t \to \pm\infty$).

It is possible to show that any autocorrelation function has its values at $|t| > 0$ smaller that at $t = 0$, i.e. the autocorrelation function reaches its maximum at $t = 0$.

Theorem 4.10 *Any autocorrelation function $C_{ff}(t)$ satisfies the following inequality:*

$$\left| C_{ff}(t) \right| \le C_{ff}(0), \tag{4.88}$$

with the equality reached only at $t = 0$.

Proof: The method is very similar to the one used in Sect. 1.18 to prove the famous Cauchy–Bunyakovsky–Schwarz inequality. Consider the non-negative integral:

$$\int_{-\infty}^{\infty} \left[f(t') + \lambda f(t + t') \right]^2 dt' = \int_{-\infty}^{\infty} \left[f(t') \right]^2 dt' + \lambda^2 \int_{-\infty}^{\infty} \left[f(t + t') \right]^2 dt' + 2\lambda C_{ff}(t) \ge 0.$$

The second integral in the right-hand side is identical to the first, hence, we obtain:

$$a\lambda^2 + 2\lambda C_{ff} + a \ge 0, \text{ where } a = \int_{-\infty}^{\infty} \left[f(t') \right]^2 dt' = C_{ff}(0).$$

We have obtained a quadratic polynomial which is non-negative. For this to happen, its discriminant $\mathcal{D} = \left[2C_{ff}(t) \right]^2 - 4a^2$ must be non-positive, i.e. we should have

$|C_{ff}(t)| \le a$, which is the required result. The zero value of the discriminant is reached when $C_{ff}(t) = a = C_{ff}(0)$, which is only possible when $t = 0$. **Q.E.D.**

If the function $f(t)$ decays monotonously (i.e. for any $t_2 > t_1$ we have $f(t_2) < f(t_1)$), then its autocorrelation function also monotonously decays. Indeed, considering the function $f(t)$ being non-negative, we can write for $t_2 > t_1$ that

$$C_{ff}(t_2) = \int_{-\infty}^{\infty} f(t')f(t_2 + t')dt' < \int_{-\infty}^{\infty} f(t')f(t_1 + t')dt' = C_{ff}(t_1).$$

Hence, it follows that for monotonously decreasing functions the correlation function decays monotonously from its maximum value at $t = 0$.

Problem 4.71 Consider an autocorrelation of a function $f(t)$ that is zero for $t < 0$ and for $t > 0$ decays to zero as $t \to \infty$ at least as quickly as the decaying exponential function $e^{-\alpha t}$ with $\alpha > 0$, i.e. $|f(t)| \le e^{-\alpha t}$. Show that

$$\left| C_{ff}(t) \right| < \frac{1}{2\alpha} e^{-\alpha t}$$

and hence the correlation function decays to zero at least as fast.

4.7   Applications of Definite Integrals

Here we shall consider some applications of definite integrals in geometry and physics. The concept of integration is so general and widespread in engineering and physics (as well as, e.g. in chemistry and biology where physical principles are applied), that it is absolutely impossible to give here a complete account of all possible uses of it; only some most obvious examples will be given.

The underlying idea in all these applications is the same; therefore, it is essential to get the general principle, and then it would be straightforward to apply it in any other situation: we have to sum up small quantities of something to get the whole of it, e.g. small lengths along a curved line to calculate the length of the whole line, small volumes within some big volume bounded by some curved surface or the work done over a small distance summed over the whole distance. There is always a parameter in the problem at hand, let us call it the "driving coordinate" and denote by q, such that if its entire change Δq_{all} is multiplied with the other quantity of interest, say F, it would give the required result: $A = F \Delta q_{all}$. The problem here is that the formula $A = F \Delta q_{all}$ would only make sense if F was constant; if this is not the case and $F(q)$ is a function of q, the simple product $A = F \Delta q_{all}$ would not make any sense. So, the idea is to discretize the values of q and construct the Riemann sum similar to the one in Eq. (4.1) which we discussed when doing the calculation of an area under the curve. So, we choose n points $q_0 \equiv q_{ini}, q_1, q_2, \ldots, q_n \equiv q_{fin}$

between the initial, q_{ini}, and final, q_{fin}, values of the parameter q. Most easily, these points can be chosen equidistant, i.e. $q_i = q_{ini} + i\Delta q$, where $i = 0, 1, 2, \ldots, n$ and $\Delta q = (q_{fin} - q_{ini})/n$. Then, the quantity of interest corresponding to a small change $\Delta q = q_{i+1} - q_i$ of q can approximately be calculated as $\Delta A_i \simeq F(q_i)\Delta q$ since for sufficiently small values of Δq, i.e. between sufficiently close values q_i and q_{i+1}, the function $F(q)$ can be approximated as a constant $F(q_i)$. The final quantity is then obtained by summing up all the contributions,

$$A \simeq \sum_{i=1}^{n} F(q_i)\Delta q \ ,$$

and then taking the limit of $n \to \infty$ (or $\Delta q \to 0$) which would give the exact result in the case of the variable function $F(q)$.

4.7.1 Length of a Curved Line

Consider a curved line in the $x - y$ plane specified by the function $y = f(x)$ within the interval $a \le x \le b$ as shown in Fig. 4.5a. We would like to calculate the length of the line. To do this, we first divide the line into n small segments by points A_i, where $i = 0, 1, 2, \ldots, n$, and then connect the points by straight lines (shown as dashed). The length of each segment between two neighbouring points A_i and A_{i+1} is then approximated by $\Delta l_i = \sqrt{(\Delta x_i)^2 + (\Delta y_i)^2}$, where $\Delta x_i = x_{i+1} - x_i$ and $\Delta y_i = y_{i+1} - y_i$ are the corresponding increments of the x and y coordinates of the two points, as shown in Fig. 4.5b. Since the change Δy_i can be related to the derivative of the function $y = f(x)$ at point x_i, i.e. $\Delta y_i \simeq f'(x_i)\Delta x_i$ (recall our discussion in Sect. 3.2, Eq. (3.9), and Sect. 3.9), we can write

$$\Delta l_i \simeq \sqrt{(\Delta x_i)^2 + (f'(x_i)\Delta x_i)^2} = \sqrt{1 + (f'(x_i))^2}\,\Delta x_i \ .$$

Summing up these lengths along the whole curve and taking the limit of $n \to \infty$ (or $\max_i \{\Delta x_i\} \to 0$), we arrive at the useful exact formula for calculating the length of a curved line:

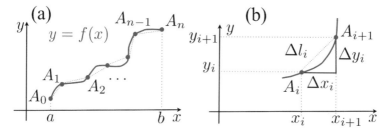

Fig. 4.5 a A curved line between points $(a, f(a))$ and $(b, f(b))$ is divided by points A_i ($i = 0, \ldots, n$) into n curved segments that are then replaced by straight lines (shown by the dashed lines). **b** The i-th such segment is shown. Its length is approximated by the length Δl_i of the straight line connecting the points A_i and A_{i+1}

$$l = \int_a^b \sqrt{1 + \left(\frac{dy}{dx}\right)^2} \, dx = \int_a^b \sqrt{1 + (f'(x))^2} \, dx \ . \tag{4.89}$$

The role of the *driving* coordinate is here served by x. To make sure that this calculation always yields a positive result, one has to integrate from the smallest to the largest values of the x.

As an example, let us calculate the length of a circle of radius R. The equation of a circle centred at the origin is $x^2 + y^2 = R^2$. Its length is most easily calculated by considering a quarter of the circle in the first quadrant, where both $x > 0$ and $y > 0$ and $y = \sqrt{R^2 - x^2}$. This will have to be multiplied by 4 to give the whole circle circumference. We have $y' = -x/\sqrt{R^2 - x^2}$ and hence

$$\frac{l}{4} = \int_0^R \sqrt{1 + (y')^2} \, dx = R \int_0^R \frac{dx}{\sqrt{R^2 - x^2}} = R \int_0^1 \frac{dt}{\sqrt{1 - t^2}}$$
$$= R \arcsin t |_0^1 = R \frac{\pi}{2} \ ,$$

which yields for the whole length a well-known expression of $l = 2\pi R$.

Note that the same integral over x can also be calculated using the substitution $x = R \cos \phi$:

$$\frac{l}{4} = R \int_0^R \frac{dx}{\sqrt{R^2 - x^2}} = \left| \begin{array}{c} x = R \cos \phi \\ dx = -R \sin \phi d\phi \end{array} \right| = R \int_{\pi/2}^0 \frac{-R \sin \phi d\phi}{R \sin \phi}$$
$$= R \int_0^{\pi/2} d\phi = R \frac{\pi}{2} \ ,$$

which of course is the same result. Although what we have just done can formally be considered as yet another substitution, we have actually used the polar coordinates (see Sect. 1.19.2) with the ϕ being the angle, since in these coordinates $x = R \cos \phi$ and $y = R \sin \phi$.

This consideration brings us to the following point: a line can also be considered by means of the polar coordinates in which case the angle ϕ is used as the driving coordinate along the line. In polar coordinates the line is specified via the equation (Sect. 1.19.3) $r = r(\phi)$. Since $x = r(\phi) \cos \phi$ and $y = r(\phi) \sin \phi$, we can write:

$$\begin{aligned} dx &= x'(\phi)d\phi = \left[r'(\phi) \cos \phi - r \sin \phi \right] d\phi \ , \\ dy &= y'(\phi)d\phi = \left[r'(\phi) \sin \phi + r \cos \phi \right] d\phi \ , \end{aligned} \tag{4.90}$$

and therefore

$$l = \int_{\phi_{min}}^{\phi_{max}} dl = \int_{\phi_{min}}^{\phi_{max}} \sqrt{dx^2 + dy^2} = \int_{\phi_{min}}^{\phi_{max}} \sqrt{r'(\phi)^2 + r^2} d\phi \ . \tag{4.91}$$

Check this simple result in the integrand using Eqs. (4.90).

Let us apply this rather powerful formula to the same problem of calculating the circumference of the circle of radius R. The equation of the circle is simply $r = R$ (i.e. r does not depend on ϕ), so that $r' = 0$ and

$$l = \int_0^{2\pi} \sqrt{0 + R^2} d\phi = 2\pi R .$$

Problem 4.72 Show that the length of the spiral in Fig. 1.38d given in the polar coordinates as $r(\phi) = e^{-\phi}$ (with ϕ changing from 0 to ∞) is equal to $\sqrt{2}$.

Problem 4.73 Show that the length of the single revolution of the Archimedean spiral, Fig. 1.38c, given in polar coordinates as $r = a\phi$ (with ϕ changing from 0 to 2π), is equal to

$$l = \frac{a}{2}\left[2\pi\sqrt{1 + 4\pi^2} + \ln\left(2\pi + \sqrt{1 + 4\pi^2}\right)\right] .$$

Problem 4.74 Show that the length of the snail curve specified via $r = a(1 + \cos\phi)$, see Fig. 1.38b, is equal to $8a$.

More generally, if a curve on the $x - y$ plane is specified parametrically as $x = x(t)$ and $y = y(t)$, then

$$l = \int_{t_{min}}^{t_{max}} \sqrt{x'(t)^2 + y'(t)^2} dt , \tag{4.92}$$

since $dx = x'(t)dt$ and $dy = y'(t)dt$. This result can easily be generalized to a curve in three dimensions which is specified parametrically as $x = x(t)$, $y = y(t)$ and $z = z(t)$. In this case the length dl of a small section of the curve obtained by incrementing the three coordinates by dx, dy and dz, is obtained as

$$dl = \sqrt{dl_1^2 + dz^2} = \sqrt{(dx^2 + dy^2) + dz^2} = \sqrt{x'(t)^2 + y'(t)^2 + z'(t)^2} dt \tag{4.93}$$

(see Fig. 4.6) and therefore the length of the line

$$l = \int_{t_{min}}^{t_{max}} \sqrt{x'(t)^2 + y'(t)^2 + z'(t)^2} dt . \tag{4.94}$$

Fig. 4.6 For the derivation of the diagonal dl of a three-dimensional cuboid formed by three sides (increments along the three Cartesian axes) dx, dy and dz

Fig. 4.7 **a** Three revolutions of the spiral. **b** Three coils of wire of radius r wound around a cylinder of radius R

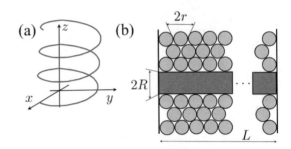

Problem 4.75 Show that the length of the three revolutions of the spiral, specified via $x(\phi) = R\cos\phi$, $y(\phi) = R\sin\phi$ and $z(\phi) = b\phi$ and shown in Fig. 4.7a, is given by $l = 6\pi\sqrt{R^2 + b^2}$.

Problem 4.76 Derive a formula for the maximum length of a wire of radius r that can be coiled around a cylinder of radius R and length L in a single layer. Then, consider the second, third, and so on layers wound one after another, see Fig. 4.7b.

4.7.2 Area of a Plane Figure

We know that if $y = y(x)$ is a positive function, then the area under it (i.e. the area bounded by the curve above and the x-axis below) and between the vertical lines $x = a$ and $x = b$ is given by the definite integral $\int_a^b y(x)dx$. This result is easily generalized for an area bounded by two curves $y = y_1(x)$ and $y = y_2(x)$ as (see Fig. 4.8):

$$A = \int_a^b [y_1(x) - y_2(x)]\, dx .$$

(4.95)

Fig. 4.8 Area between two curves $y = y_1(x)$ and $y = y_2(x)$ intercepting at $x = a$ and $x = b$

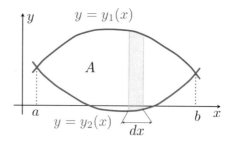

Problem 4.77 Write a similar formula for a figure enclosed between two curves $x = x_1(y)$ and $x = x_2(y)$.

These formulae, however, are not very convenient if the figure has a more complex shape. For instance, this may happen if the figure (i.e. the closed curve which bounds it) is specified parametrically as $x = x(t)$ and $y = y(t)$, see schematics in Fig. 4.9a. To derive the corresponding formula in this case, we have to carefully consider regions where for the same value of x there are several possible values of y, i.e. the shaded regions in Fig. 4.9a. To this end, let us analyse several possible cases noticing that with the chosen direction of the traverse of the bounding curve (which corresponds to the increase of the parameter t) the body of the figure is always on the right; regions which are on the left of the traverse direction must not be included in the calculation. The green shaded region in case A in Fig. 4.9b is to be fully included, and its area is $dA = y(t)[x(t + dt) - x(t)] = y(t)x'(t)dt$ and is positive. The to-be-excluded area in case B can formally be written using the same expression, but dA in this case will be negative (since $x(t + dt) < x(t)$ and will then be subtracted). Similar analysis shows that the area is negative in case C and positive in case D. It is clear now that if we sum up all these areas, the correct expression for the whole

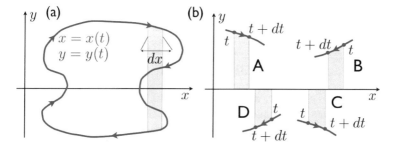

Fig. 4.9 For the calculation of the area of a figure whose bounding curve is specified by parametric equations $x = x(t)$ and $y = y(t)$ with the parameter t. **a** A sketch of a figure with a complex shape; green shaded areas should be included into the calculated area, while the orange shaded areas are to be excluded. **b** Several particular cases (see text). The direction in which the bounding curve is traversed when the parameter t is increased is also indicated

Fig. 4.10 For the derivation of formula (4.97) for the area of a plane figure specified via equation $r = r(\phi)$ in polar coordinates

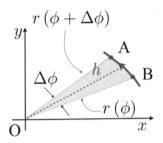

area enclosed by the bounding curve is obtained:

$$A = \left| \int_{t_a}^{t_b} y(t) x'(t) dt \right| , \tag{4.96}$$

where t_a and t_b are the initial (usually, zero) and final values of the t required to draw the complete figure. The regions that need to be excluded are excluded in this case automatically (explain, why). We use the absolute value in the above equation since, depending on the direction of the traverse, the result may be negative.

As an example, let us calculate the area enclosed by a circle of radius R centred at the origin. In this case the parametric equations for the circle are $x(\phi) = R \cos \phi$ and $y(\phi) = R \sin \phi$, so that we obtain:

$$A_{circle} = \left| \int_0^{2\pi} (R \sin \phi)(-R \sin \phi) \, d\phi \right| = R^2 \int_0^{2\pi} \sin^2 \phi d\phi$$

$$= R^2 \int_0^{2\pi} \frac{1}{2} [1 - \cos(2\phi)] \, d\phi = R^2 \frac{1}{2} (2\pi - 0) = \pi R^2 .$$

Problem 4.78 Show that the area of an ellipse, given parametrically via $x(\phi) = a \cos \phi$ and $y(\phi) = b \sin \phi$, is equal to $A_{ellipse} = \pi ab$. Note that when $a = b$, the ellipse turns into a circle of radius a, and the correct expression for its area, πa^2, is indeed recovered.

Finally, we shall consider yet another useful result for the area of a figure specified in polar coordinates via an equation $r = r(\phi)$ (see Sect. 1.19.3). In this case one can use Eq. (4.96) with the polar angle ϕ as the parameter (see Problem 4.80). However, a simpler method exists. Consider the green shaded sector in Fig. 4.10 corresponding to a small change $\Delta\phi$ of the polar angle ϕ from ϕ to $\phi + d\phi$. The area of the sector is then calculated assuming that the figure OAB can be treated approximately as a triangle with the height $h \simeq r(\phi)$ which is perpendicular to the side $AB \simeq 2r(\phi) \tan(\Delta\phi/2) \simeq r(\phi)\Delta\phi$ of the triangle (recall that for small angles, $\tan \alpha \simeq \alpha$, see Eq. (3.60)). Therefore, the area of the triangle $\Delta A \simeq h \cdot AB/2 = r^2(\phi)\Delta\phi/2$.

Summing up these areas and tending $\Delta\phi \to 0$, i.e. integrating over the angle ϕ, an exact expression is obtained:

$$A = \frac{1}{2} \int_{\phi_1}^{\phi_b} r^2(\phi)d\phi \ . \tag{4.97}$$

It is easy to see that for a circle $r(\phi) = R$ is a constant, the angle ϕ changes between 0 and 2π, so that the same expression for the circle area is obtained as above, $A = \frac{1}{2}R^2 2\pi = \pi R^2$.

Problem 4.79 Here we shall perform a more careful derivation of Eq. (4.97). Let (x, y) be the two Cartesian coordinates of the point B in Fig. 4.10, where $x = r(\phi)\cos\phi$ and $y = r(\phi)\sin\phi$, while the point A similarly has coordinates corresponding to the angle $\phi + \Delta\phi$. Calculate the area of the triangle OAB by considering the absolute value of the vector product of the vectors $\overrightarrow{OA}$ and $\overrightarrow{OB}$, and then expand the obtained expression in terms of $\Delta\phi$ recovering $\Delta A = r^2(\phi)\Delta\phi/2$ to the first order in $\Delta\phi$.

Problem 4.80 Derive Eq. (4.97) from Eq. (4.96). [Hint: *use $x(\phi) = r(\phi)\cos\phi$ and $y(\phi) = r(\phi)\sin\phi$ and integration by parts (only in the term containing $r'(\phi)$); also make use of the fact that the angle ϕ changes by an integer amount of π when traversing around the whole figure.*]

Problem 4.81 Show that the area enclosed by the three-leaf rose shown in Fig. 1.38a is $A = \pi a^2/4$.

4.7.3   Volume of Three-Dimensional Bodies

Later on in Chap. 6 after developing concepts of a function of several variables and of double and triple integrals, we shall learn how to calculate volumes of general three-dimensional bodies. However, in simple cases it is possible to calculate the volume of some figures using a single definite integral. Here we shall consider some of the most common cases.

We start with the simplest case when one can select a direction through the figure (let us draw the z-axis along it) such that the area $A(z)$ of the cross section at each value of z is known, see Fig. 4.11a. The volume inside a small cylinder of height dz enclosed between cross sections at z and $z + dz$, is obviously equal to $A(z)dz$, so

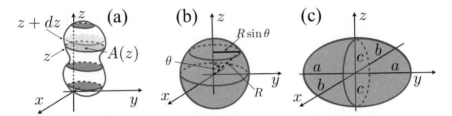

Fig. 4.11 Three-dimensional figures for which the area of its cross section at each value of z is known to be $A(z)$. **a** A general figure. **b** A sphere of radius R. **c** An ellipsoid with parameters a, b and c

that the whole volume is given by:

$$V = \int_{z_1}^{z_2} A(z)dz \,, \tag{4.98}$$

where z_1 and z_2 and the smallest and the largest values of the z which are necessary to specify the object under discussion.

As a simple application of this result, let us first consider a cylinder of radius R and length (height) h. If we choose the z-axis along the symmetry axis of the cylinder, then the cross section at any point z along the cylinder height will be a circle of the same radius R with the area $A(z) = \pi R^2$. It does not depend on the z, leading to the volume

$$V_{cylinder} = \int_0^h \pi R^2 dz = \pi R^2 h \,. \tag{4.99}$$

As a less trivial example, consider now a sphere of radius R centred at the origin, Fig. 4.11b. A cross section of the sphere perpendicular to the z-axis is a circle of radius $r = R \sin \theta$, where the angle θ is related to z via $z = R \cos \theta$, i.e.

$$r = R\sqrt{1 - (z/R)^2} = \sqrt{R^2 - z^2}$$

(which is basically the familiar Pythagoras theorem), so that the area of the cross section $A(z) = \pi r^2 = \pi \left(R^2 - z^2 \right)$, and we obtain for the volume:

$$V_{sphere} = \int_{-R}^R \pi \left(R^2 - z^2 \right) dz = \pi \left(2R^3 - 2\frac{R^3}{3} \right) = \frac{4}{3}\pi R^3 \,. \tag{4.100}$$

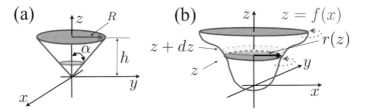

Fig. 4.12 **a** A cone of radius R and height h. **b** A solid obtained by a full revolution around the z-axis of the curve $z = f(x)$

Problem 4.82 Show that the volume of an ellipsoid shown in Fig. 4.11c is

$$V_{ellipsoid} = \frac{4}{3} \pi abc . \tag{4.101}$$

Note that the equation of the ellipsoid is given by

$$\left(\frac{x}{a}\right)^2 + \left(\frac{y}{b}\right)^2 + \left(\frac{z}{c}\right)^2 = 1 . \tag{4.102}$$

The simple result (4.101) makes perfect sense: when the ellipsoid is fully symmetric, i.e. when $a = b = c$, it becomes a sphere, and the volume in this case goes directly into the expression (4.100) for the sphere.

Problem 4.83 Show that the volume of a cone shown in Fig. 4.12a is

$$V_{cone} = \frac{1}{3} \pi R^2 h = \frac{1}{3} \pi h^3 \tan^2 \alpha . \tag{4.103}$$

Problem 4.84 Show that the volume of a tetrahedron $A_1 B_1 C_1 D$ (also called a triangular pyramid), shown in Fig. 4.13, is given by the following formula:

$$V_{tetrahedron} = \frac{1}{3} S_\triangle c \sin \alpha , \tag{4.104}$$

where $S_\triangle$ is the area of its base (i.e. of the triangle $\triangle A_1 B_1 C_1$), $c = A_1 D$ is one of its sides starting from the base and going towards the pyramid summit, and α is the angle between the side and the base. [Hint: *Let the base $A_1 B_1 C_1$ of the pyramid be in the $x - y$ plane. Consider a point A_2 along the side $A_1 D = c$, and let $A_1 A_2 = t$, where t changes from 0 to c. Demonstrate, using the similarity of the triangles $\triangle A_1 B_1 C_1$ and $\triangle A_2 B_2 C_2$, that the area of the intermediate triangle $\triangle A_2 B_2 C_2$, expressed as a function of the parameter t, is given by $S_\triangle (t) = \left(ah/2c^2\right) (c - t)^2$, where $h = A_1 F_1$ is the height of the*

Fig. 4.13 For the calculation of the volume of a triangular pyramid (tetrahedron) $A_1 B_1 C_1 D$

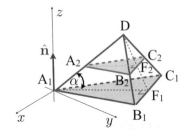

base drawn towards the side $a = B_1 C_1$. Then relate dz and dt, and obtain the volume by integrating over dt.]

Problem 4.85 Next consider the triangular pyramid of the previous problem as created by three vectors: $\overrightarrow{A_1 B_1}$, $\overrightarrow{A_1 C_1}$ and $\overrightarrow{A_1 D}$. Show that the volume of the pyramid (4.104) can also be written as

$$V_{tetrahedron} = \frac{1}{6}\left|\left(\overrightarrow{A_1 D} \cdot \left[\overrightarrow{A_1 B_1} \times \overrightarrow{A_1 C_1}\right]\right)\right| = \frac{1}{6}\left|\left[\overrightarrow{A_1 B_1}, \overrightarrow{A_1 C_1}, \overrightarrow{A_1 D}\right]\right| , \tag{4.105}$$

i.e. it can be related to the mixed product of the three vectors. Since the triple product corresponds to the volume of the parallelepiped formed by the three vectors, see Eq. (1.111), we see that the volume of the tetrahedron is one-sixth of the volume of the corresponding parallelepiped.

The volume can also be easily expressed via a definite integral for another class of symmetric (in fact, cylinder-like) three-dimensional objects which are obtained by rotating a two-dimensional curve $z = f(x)$ around the z-axis, the so-called *solids of revolution*, see Fig. 4.12b. In this case one can directly use Eq. (4.98) in which the radius of the circle in the cross section, $r(z)$, is obtained by solving the equation $z = f(x)$ for the x, i.e. $r \equiv x = f^{-1}(z)$. Therefore,

$$V = \int_{z_1}^{z_2} \pi x^2 dz , \quad x = f^{-1}(z). \tag{4.106}$$

The cone in Fig. 4.12a is the simplest example of such a figure as it can be obtained by rotating the straight line $z = kx$ with $k = \cot \alpha$ around the z-axis. Therefore, in this case $x(z) = z/k = (\tan \alpha) z$ and the integration above yields exactly the same result as in Eq. (4.103).

Problem 4.86 Show that the volume of a solid of revolution obtained by rotating around the z-axis the parabola $z = ax^2$ with $0 \le x \le R$ is $V = \frac{1}{2}\pi a R^4$.

Problem 4.87 Show that the volume of a solid of revolution obtained by rotating around the z-axis the curve $z = ax^4$ with $0 \leq x \leq R$ is $V = (2\pi/3)\, R^6 a$.

4.7.4 A Surface of Revolution

A general consideration of the surface area of three-dimensional bodies requires developing a number of special mathematical instruments, and we shall do it in Chap. 6. These will allow us to calculate the area of the outer surface of arbitrary bodies. Here we only show that in some simple cases this problem can be solved with the tools which are at our disposal already.

Consider an object created by rotating around the z-axis a curve $z = f(x)$, see Fig. 4.12b. Above we have considered its volume. Now we turn to the calculation of its outer surface. Looking at the surface layer created between the cross sections at z and $z + dz$, we can calculate its area as a product of $2\pi r(z)$ (which is the length of the circumference of the cross section at z with $r(z) = x(z) = f^{-1}(z)$, as above) and the layer width, dl. The latter is not just dz as the curve $z = f(x)$ may have a curvature. The calculation of dl has already been done in Sect. 4.7.1 (and, specifically, see Fig. 4.5b) when considering the length of a curve. Hence, we can write down the result immediately:

$$dl = \sqrt{(dx)^2 + (dz)^2} = \sqrt{1 + x'(z)^2}\, dz \ .$$

Therefore, the surface area of the solid of revolution enclosed between z_1 and z_2 is obtained by summing up all these contributions, i.e. by integrating:

$$A = \int_{z_1}^{z_2} 2\pi x(z)\sqrt{1 + x'(z)^2}\, dz \ . \tag{4.107}$$

As an example, consider the surface of a sphere. The equation of a sphere of radius R and centred at the origin, see Fig. 4.11b, is expressed by $x^2 + y^2 + z^2 = R^2$. It is convenient to consider the upper part of the sphere ($z > 0$) and then multiply the result by 2. The upper hemisphere of the sphere can be considered as a revolution of the curve $z = \sqrt{R^2 - x^2}$ around the z-axis. The latter equation is easily solved with respect to the radius of the cross section: $x(z) = \sqrt{R^2 - z^2}$. Therefore, $x'(z) = -z/\sqrt{R^2 - z^2}$ and we can write:

$$\begin{aligned}
A_{sphere} &= 2 \int_0^R 2\pi \sqrt{R^2 - z^2}\sqrt{1 + \frac{z^2}{R^2 - z^2}}\, dz \\
&= 4\pi \int_0^R \sqrt{R^2 - z^2}\sqrt{\frac{R^2}{R^2 - z^2}}\, dz = 4\pi R \int_0^R dz = 4\pi R^2 \ .
\end{aligned} \tag{4.108}$$

Fig. 4.14 A torus of radius c with the tube of radius R can be obtained by rotating a coloured circle of radius R displaced from the origin by the distance c along the x-axis

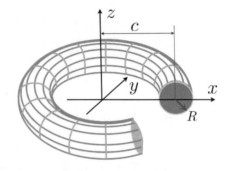

Problem 4.88 Show that the area of the side surface of a cone in Fig. 4.12a, obtained by rotating the line $z = kx$ with $k = \cot \alpha$, is

$$A_{cone} = \frac{\pi h^2}{k^2} \sqrt{1 + k^2} = \pi h^2 \frac{\sin \alpha}{\cos^2 \alpha} = \pi R \sqrt{h^2 + R^2} \,. \qquad (4.109)$$

Problem 4.89 Show that the area of the side surface of a solid of revolution obtained by rotating around the z-axis the parabola $z = ax^2$ with $0 \leq x \leq R$ is

$$A = \frac{\pi}{6a^2} \left[(1 + 4ah)^{3/2} - 1 \right] \,,$$

where $h = aR^2$ is the height of the object.

Problem 4.90 A torus of radius c with the tube of radius R, see Fig. 4.14, can be formed by rotating the circle of radius R (shown in the Figure as the cross section of the torus tube in the $x - z$ plane) around the z-axis. The centre of the circle is displaced from the origin by c along the x-axis. The equation of the circle $(x - c)^2 + z^2 = R^2$, which gives $r \equiv x = c \pm \sqrt{R^2 - z^2}$ for the radius of the cross-sectional perpendicular to the z-axis. The plus sign corresponds to the outer part of the surface obtained by rotating the right half of the circle with $c \leq x \leq c + R$, while the minus sign gives the inner surface of the torus due to revolution of the left side of the circle, $c - R \leq x \leq c$, around the z-axis. Because the function $x(z)$ is multivalued, the two surfaces have to be considered separately. Show that their areas are given by $A_{\pm} = 2\pi R (\pi c \pm 2R)$, where the plus sign corresponds to the outer and the minus to the inner, surfaces. Hence, the total surface of the torus appears to be

$$A_{torus} = 4\pi^2 Rc \,. \qquad (4.110)$$

Problem 4.91 Consider a solid of revolution created by the line $z = f(x)$. Let the line be given parametrically as $z = z(t)$ and $x = x(t)$. Show that the area of the surface of revolution can be calculated via the integral:

$$A = \int_{t_1}^{t_2} 2\pi x(t) \sqrt{x'(t)^2 + z'(t)^2}\, dt \,, \qquad (4.111)$$

where t_1 and t_2 are the corresponding initial and final values of the parameter t. Note that this formula is more general than (4.107) since it can also be used when a rotation of the curve is performed around an arbitrary axis. It could be any line in space, not necessarily the Cartesian axes, if the appropriate parametric equations are provided.

Problem 4.92 Consider the torus of Fig. 4.14 again, but this time let us specify the equation of the circle parametrically: $x = c + R\cos\phi$ and $z = R\sin\phi$, where $0 \le \phi \le 2\pi$. The convenience of this representation becomes apparent when using Eq. (4.111) instead of (4.107); indeed, the calculation of the surface area can be performed in a single step, there is no need to consider the inner and outer torus surfaces separately as we did in the Problem above. Show that the same result (4.110) for the surface area is obtained.

4.7.5 Probability Distributions

We have introduced the concept of probability in Sect. 1.17. What we considered there were discrete events.

Suppose there is a continuous random variable x, and one can define the probability $dP(x)$ for this variable to be within an interval between x and $x + dx$. In this case it is convenient to introduce the *probability distribution function* $f(x)$ such that $dP(x) = f(x)dx$ and hence $f(x) = dP/dx$. Then, the probability for the variable x to be within the interval $a \le x \le b$ is given, by virtue of probability addition Theorem 1.10, by (all events are independent)

$$P\,(a \le x \le b) = \int_a^b f(x)dx \,.$$

If the complete interval for x is, say, the whole real axis, then $P\,(-\infty \le x \le \infty) = 1$, and hence the distribution function must be normalized:

$$\int_{-\infty}^{\infty} f(x)dx = 1 \,.$$

Problem 4.93 It is known (see Problem 6.24) that the probability distribution $f(v_x)$ of the molecules to have their velocities in the x direction between v_x and $v_x + dv_x$ is given by

$$f(v_x) = A \exp\left(-\alpha v_x^2\right) .$$

Show that the normalization constant $A = (\alpha/\pi)^{1/2}$. You will need the value of the Gaussian integral Eq. (6.16).

Problem 4.94 In theory of a free electron gas in metals, the total number of electrons n (per unit volume) can be written via an energy integral

$$\int_0^\infty f(\epsilon) D(\epsilon) d\epsilon = n ,$$

where $D(\epsilon)$ is the density of states ($D(\epsilon)d\epsilon$ gives the number of states electrons can occupy between energies ϵ and $\epsilon + d\epsilon$) and $f(\epsilon) = [\exp((\epsilon - \mu)/k_B T) + 1]^{-1}$ is the probability to find an electron with energy ϵ (the Fermi-Dirac distribution), μ is the chemical potential, k_B Boltzmann constant and T temperature. In the case of a two-dimensional electron gas the density of states $D(\epsilon) = D$ is a constant, not depending on the energy. Using an appropriate substitution, calculate the integral and hence obtain the following expression for the chemical potential of the gas:

$$\mu = k_B T \ln\left(e^{n/(Dk_B T)} - 1\right) .$$

4.7.6   Simple Applications in Physics

Of course, the language of physics is mathematics and hence integrals (comprising an essential part of the language) play an extremely important role in it. Here we shall give some obvious examples to illustrate this point.

Mechanics. Consider a particle performing a one-dimensional movement along the x-axis. Suppose we know the velocity of the particle v, then the distance the particle would travel between times t_1 and t_2 is obviously $s = v(t_1 - t_2)$. The problem with this simple formula is that it can only be used if the velocity of the particle was constant over the whole distance. To generalize this result for the case of a variable velocity, we can use the concept of integration: (i) divide up the time into small intervals Δt_i, (ii) assuming that the velocity v_i within each interval is constant, calculate the distance $\Delta x_i \simeq v_i \Delta t_i$ the particle passed over each such small time interval (Δx_i calculated that way is approximate for finite Δt_i), and (iii) sum these distances up to get the whole distance passed over the whole time. In the limit of all $\Delta t_i \to 0$ (or infinite number of divisions of the whole time) the calculation becomes exact with the distance expressed via the integral:

$$s \simeq \sum_{i=1}^{n} v_i \Delta t_i \implies s = \int_{t_1}^{t_2} v(t) dt . \tag{4.112}$$

For instance, consider a particle moving with the speed $v(t) = v_0 + at$, which linearly increases $(a > 0)$ or decreases $(a < 0)$ with time from the value of v_0 (at $t = 0$). In this case the distance s is

$$s = \int_0^t (v_0 + a\tau) d\tau = v_0 t + \frac{1}{2}at^2 .$$

Here the acceleration $a = F/m$ is assumed to be constant, which happens when the force F acting on the particle of mass m is constant; e.g. during a vertical motion under the Earth's gravity or because of the friction during the sliding motion on a surface.

Another example is related to calculating work on a particle done by a variable force that moves it along the x-axis: if the force F changes during a process, then, again, the work ΔW after a small distance Δx can be considered approximately constant yielding $\Delta W \simeq F(x)\Delta x$. To calculate the total work done by the force on the particle in displacing it from the initial position x_1 to its final position x_2, we integrate:

$$W = \int_{x_1}^{x_2} F dx . \tag{4.113}$$

Problem 4.95 A body of mass m is taken from the surface of the Earth to infinity. The Earth's mass is M and radius R. The gravitational force acting between two bodies which are a distance x apart is $F = -GmM/x^2$. Show that the work needed to be done against the Earth's gravity is $W = -GmM/R$. What is the minimum possible velocity the rocket needs to develop in order to overcome the Earth's gravity?

If two functions are equal for all values of x within some interval, $f_1(x) = f_2(x)$, then obviously their integrals are equal as well, $\int_a^b f_1(x)dx = \int_a^b f_2(x)dx$, where the points a and b lie within the interval. The passage from the former equality to the latter is usually described as "integrating the equality" or "integrating both sides of" $f_1(x) = f_2(x)$ with respect to x from a to b.

Problem 4.96 Consider a particle of mass m moving along the x-axis under the influence of the force $F(x)$. The Newtonian equation of motion is $\dot{p} = F(x)$, where $\dot{p}$ is the time derivative of the particle's momentum, p. Integrate both sides of this equation with respect to x between the initial, x_0, and final, x_1, points along particle trajectory to show that

$$\frac{p_1^2}{2m} - \frac{p_0^2}{2m} = \int_{x_0}^{x_1} F(x)dx .$$

Interpret your result. [Hint: *express $\dot{p}$ via the derivative of p over x first.*]

Another class of problems is related to the calculation of the mass of an object from its mass density. For linear objects such as a rod of length L oriented along the x-axis and positioned between coordinates $x = 0$ and $x = L$, the linear density is defined as the derivative $\rho(x) = dm/dx$. Here $m(x)$ is the mass of a part of the rod between 0 and x. Correspondingly, $\Delta m \simeq \rho(x)\Delta x$ will be the mass of a small piece of the rod of length Δx. The total mass is calculated by summing up these contributions along the whole length of the rod and taking the limit $\Delta x \to 0$, i.e. by the integral

$$m = \int_{x_1}^{x_2} \rho(x)dx . \tag{4.114}$$

If we are interested in calculating the centre of mass of the rod, this can also be related to the appropriate definite integral. For a system of n particles of mass m_i arranged along the x-axis at the positions x_i, the centre of mass is determined as

$$x_{cm} = \frac{1}{m} \sum_{i=1}^{n} x_i m_i$$

with $m = \sum_{i=1}^{n} m_i$ being the total mass of all particles. For a continuous rod, we split it into small sections of length Δx and mass $\Delta m \simeq \rho(x)\Delta x$ that are positioned between x and $x + \Delta x$. Each of these sections can be treated as a point mass for sufficiently short sections. Hence, using the above formula with substitutions $m_i \to \Delta m$ and $x_i \to x$ and taking the limit $\Delta x \to 0$, we arrive at the corresponding expression for the continuous rod:

$$x_{cm} = \frac{1}{m} \int_{x_1}^{x_2} x\rho(x)dx \quad \text{with} \quad m = \int_{x_1}^{x_2} \rho(x)dx . \tag{4.115}$$

Let us consider calculating the centre of mass of a thin curved line specified on the plane by equation $y = f(x)$, see Fig. 4.15a. As the line is in two dimensions, there

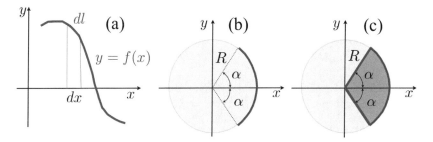

Fig. 4.15 To the calculation of the centre of mass of: **a** a thin curved line specified by equation $y = f(x)$; **b** an arch (blue) of the angle 2α and radius R, as well as **c** a corresponding sector of the circle (also blue)

will be two coordinates of the centre of mass, (x_{xm}, y_{cm}), which are determined by considering small lengths

$$dl = \sqrt{(dx)^2 + (dy)^2} = \sqrt{1 + f'(x)^2}\, dx$$

along the line and then summing up all the contributions:

$$x_{cm} = \frac{1}{m} \int_{x_1}^{x_2} x\rho(x)\sqrt{1 + f'(x)^2}\, dx ,$$

$$y_{cm} = \frac{1}{m} \int_{x_1}^{x_2} f(x)\rho(x)\sqrt{1 + f'(x)^2}\, dx ,$$

(4.116)

where m is the mass of the line and $\rho(x)$ is the line density of the rod at the point with coordinates $(x, f(x))$.

Problem 4.97 Show that if the line is specified parametrically in the three-dimensional space as $x = x(t)$, $y = y(t)$ and $z = z(t)$ (or, simply, in the vector form, as $\mathbf{r} = \mathbf{r}(t)$), then the position $\mathbf{r}^{(cm)} = \left(r_\alpha^{(cm)} \right) = (x_{cm}, y_{cm}, z_{cm})$ of the centre of mass is given by

$$r_\alpha^{(cm)} = \frac{1}{m} \int_{t_1}^{t_2} r_\alpha(t)\rho(t)\sqrt{x'(t)^2 + y'(t)^2 + z'(t)^2}\, dt ,$$

where $\alpha = x, y, z$ designates the three Cartesian components of the vectors and $\rho(t)$ is the corresponding line density.

Problem 4.98 Show that the coordinates of the centre of mass of an arch of a unit linear density, radius R and angle 2α, shown in Fig. 4.15b, is given by $(x_{cm}, 0)$, where $x_{cm} = R \sin\alpha / \alpha$.

Problem 4.99 Show that the coordinates of the centre of mass of a uniform sector of radius R and angle 2α, shown in Fig. 4.15c, is given by $(x_{cm}, 0)$, where

$$x_{cm} = \frac{2R\sin\alpha}{3\alpha}.$$

Problem 4.100 Consider a cylinder of the uniform density ρ, radius R and height h, rotating around its symmetry axis with the angular velocity ω. Show that the kinetic energy of the cylinder is $E_{KE} = \pi\rho h R^4\omega^2/4$. [Hint: *particles of the cylinder positioned distance r from its symmetry axis has the velocity* $v(r) = \omega r$.]

Problem 4.101 Consider a circular disk of radius R placed in water vertically such that its centre is at the depth of $h > R$. If ρ is the water density, show that the total force exerted on any side surface of the disk is $F = \pi h\rho g R^2$, where g is the Earth's gravitational constant.

Problem 4.102 Generalize the result of the previous problem for the disk which is only partially vertically immersed into water ($h < R$). Show that in this case

$$F = \rho g \left\{ hR^2\left(\frac{\pi}{2} + \arcsin\frac{h}{R}\right) + h^2\sqrt{R^2 - h^2} + \frac{2}{3}\left(R^2 - h^2\right)^{3/2}\right\}.$$

However, the mass and the centre of mass of more complex two- and even three-dimensional objects are defined via their corresponding surface and volume densities. As such, these are much easier to consider later on after we learn a bit more of necessary mathematics.

Gas theory. An equation of the ideal gas reads $PV = nRT$, where P is pressure, V volume, T temperature, n the number of moles and R is the gas constant. In a process in which the volume of the gas changes from V_1 to V_2, the work done by the gas is $W = P(V_2 - V_1)$. Provided, of course, that the pressure is the same during the process in question. This happens in the isobaric process when the V changes due to a change in temperature, but the pressure remains the same. However, if the process is not isobaric, this formula is not longer applicable and one has to take into consideration explicitly the change of the pressure during the gas expansion (or contraction). Again, the solution to this problem can easily be found using integration: divide the whole process into small volume changes ΔV during which the pressure can approximately be considered constant, calculate the work $\Delta W \simeq P\Delta V$ done during this small volume change, and then sum up all this changes and take the limit

$\Delta V \to 0$, which results in the integral:

$$W_{gas} = \int_{V_1}^{V_2} P dV \,, \tag{4.117}$$

where the pressure P is a function of the volume, $P = P(V)$.

Problem 4.103 Show that the work done by the gas in an adiabatic process, $PV^{\gamma} = Const$, where the adiabatic index $\gamma = C_P/C_V$ is given as the ratio of the respective specific heats, is

$$W_{adiab} = \frac{P_1 V_1}{\gamma - 1}\left[1 - \left(\frac{V_1}{V_2}\right)^{\gamma-1}\right].$$

Problem 4.104 Consider a gas of molecules of mass m under influence of the Earth's gravity. The number density of the gas would be greatest near the Earth's surface and then it will get smaller and smaller as the distance z from the surface is increased:

$$\rho(z) = \rho_0 \exp\left(-\frac{mgz}{k_B T}\right) = \rho_0 \exp\left(-\frac{Mgz}{RT}\right) \,,$$

where k_B is Boltzmann's constant, $R = k_B N_A$ the universal gas constant, N_A number of atoms in a mole of gas (Avogadro's number), $M = m N_A$ the mole mass, T absolute temperature and g the standard acceleration near the Earth's surface (assumed constant). Show that the weight $\mathcal{P} = m_{gas} g$ of the air in a vertical tube above the Earth's surface of cross-sectional area A and a small height H, where m_{gas} is the total mass of air in the tube, is $\mathcal{P} = A(P_0 - P_H)$, where P_0 and P_H are the gas pressures at the bottom and top of the tube. Rewrite this equation as $\mathcal{P}/A + P_H = P_0$ and then interpret this result.

Electrostatics and magnetism. Consider a continuous distribution of charge along a line, specified parametrically as $\mathbf{r} = \mathbf{r}(t)$ in a three-dimensional space. The linear charge density $\rho(t) = dq/dl$ may vary along the line. Here dq is an amount of charge contained along a small length

$$dl = \sqrt{x'(t)^2 + y'(t)^2 + z'(t)^2}\, dt$$

along the line, see Fig. 4.16a. The electrostatic field $d\mathbf{E}$ at point M, whose position is given by the vector $\mathbf{r}_M$, is equal to $d\mathbf{E} = dq\, \mathbf{g}/g^3$, where $\mathbf{g} = \mathbf{r}_M - \mathbf{r}$ is the vector connecting dl and the point M and $g = |\mathbf{g}| = |\mathbf{r}_M - \mathbf{r}|$. The total field is obtained

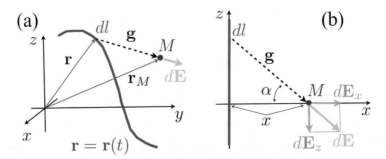

Fig. 4.16 For the calculation of the electrostatic field at point M of a charged line: **a** a general case of a line $\mathbf{r} = \mathbf{r}(t)$ specified parametrically, and **b** a vertical uniformly charged straight line with the field calculated distance x away from it

by summing all contributions coming from all little fragments dl along the whole charged line, i.e. by integrating along the length of the line:

$$
\begin{aligned}
\mathbf{E}\left(\mathbf{r}_M\right) &= \int_{line} \frac{\mathbf{g}}{g^3} dq = \int_{line} \frac{\mathbf{g}}{g^3} \rho dl \\
&= \int_{t_0}^{t_1} \frac{\mathbf{r}_M - \mathbf{r}(t)}{|\mathbf{r}_M - \mathbf{r}(t)|^3} \rho(t) \sqrt{x'(t)^2 + y'(t)^2 + z'(t)^2}\, dt\ .
\end{aligned}
\tag{4.118}
$$

We have ignored here the irrelevant pre-factor $1/4\pi\epsilon_0$.

As an example of the application of this method, consider the field due to an infinite uniformly charged straight line a distance x from it as shown in Fig. 4.16b. Let the element $dl \equiv dz$ of the line with the charge $dq = \rho dl = \rho dz$ be distance z from the point on the line which is closest to the point M. One can see that $z = g\sin\alpha = x\tan\alpha$. It is clear that the field components along the z-axis from the two-line elements dl symmetrically placed at both sides of the line at z and $-z$ would cancel each other. Hence, after integrating, there will only be the component of the field along the x-axis. Then, the vector $\mathbf{g} = (x, 0, -z)$ and the required component of the field E_x is:

$$
\begin{aligned}
E_x &= \int_{-\infty}^{\infty} \frac{\rho x}{\left(x^2 + z^2\right)^{3/2}} dz = \left| \begin{array}{c} z = x\tan\alpha \\ dz = xd\alpha/\cos^2\alpha \end{array} \right| \\
&= \rho x \int_{-\pi/2}^{\pi/2} \frac{1}{\left(x^2 + x^2\tan^2\alpha\right)^{3/2}} \frac{xd\alpha}{\cos^2\alpha} \\
&= \frac{\rho}{x} \int_{-\pi/2}^{\pi/2} \frac{\cos^3\alpha}{\cos^2\alpha} d\alpha = \frac{\rho}{x} \int_{-\pi/2}^{\pi/2} \cos\alpha d\alpha = \frac{2\rho}{x}\ ,
\end{aligned}
$$

i.e. the field decays with the distance x from the charged line as $1/x$.

Similarly one can consider the magnetic field $\mathbf{B}$ due to a curved wire given parametrically as $\mathbf{r} = \mathbf{r}(t)$ in a three-dimensional space, see Fig. 4.17a. According to the Biot–Savart law, the magnetic field $d\mathbf{B}$ at point M due to directed element $d\mathbf{l} = \mathbf{n}dl$

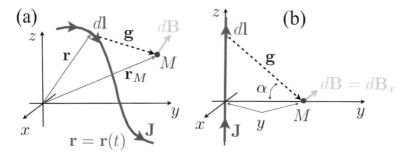

Fig. 4.17 For the calculation of the magnetic field **B** at point M due to the wire in which the current J is running: **a** a general case of a line $\mathbf{r} = \mathbf{r}(t)$ given parametrically, and **b** a vertical wire, and the observation point M which is the distance y from the wire

($\mathbf{n}$ is the unit vector directed along the current J), which is separated from the point M by the vector $\mathbf{g}$, is

$$d\mathbf{B} = \frac{J\left[d\mathbf{l} \times \mathbf{g}\right]}{g^3} = \frac{J\left[\mathbf{n} \times \mathbf{g}\right]}{g^3} dl$$

(up to an irrelevant pre-factor), so that, integrating along the whole length of the wire, we obtain for the field due to the entire wire:

$$\begin{aligned}
\mathbf{B}\left(\mathbf{r}_M\right) &= \int_{line} \frac{J\left[\mathbf{n} \times \mathbf{g}\right]}{g^3} dl \\
&= J \int_{t_0}^{t_1} \frac{\left[\mathbf{n}(t) \times \mathbf{g}(t)\right]}{g(t)^3} \sqrt{x'(t)^2 + y'(t)^2 + z'(t)^2} dt .
\end{aligned}$$

(4.119)

Problem 4.105 Show that the magnetic field at distance y from the infinitely long wire with the current J, as shown in Fig. 4.17b, is directed along the x-axis and is equal to

$$\mathbf{B} = -\frac{2J}{y}\mathbf{i} ,$$

where $\mathbf{i}$ is the unit vector along the x-axis.

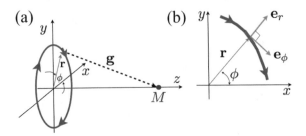

Fig. 4.18 a For the calculation of the magnetic field at a point M along the z-axis due to a single coil of wire placed within the $x - y$ plane and with its centre at the origin. **b** The close-up of a piece of the wire within the $x - y$ plane. Orthogonal unit vectors $\mathbf{e}_r$ and $\mathbf{e}_\phi$ are along radial and tangential directions, respectively, to the circle of the wire

Problem 4.106 Consider a single coil of wire of radius R placed within the $x - y$ plane as shown in Fig. 4.18a. The current in the wire is equal to J. Show that the magnetic field **B** at a point M along the symmetry axis of the circle (i.e. along the z-axis) at a distance z from it is

$$\mathbf{B} = (0, 0, B_z) \quad \text{with} \quad B_z = \frac{2\pi J R^2}{\left(R^2 + z^2\right)^{3/2}} \, .$$

[Hint: *the vector* $\mathbf{n} \equiv \mathbf{e}_\phi$ *is tangential to the circle as shown in Fig.* 4.18*b; also,* $\mathbf{g} = \mathbf{z} - \mathbf{r}$, *where* $\mathbf{r} = R\mathbf{e}_r$ ($\mathbf{e}_r$ *being a unit vector along* $\mathbf{r}$)*, and* $\mathbf{z} = z\mathbf{k}$ *is the vector along the* z *axis. The two unit vectors* $\mathbf{e}_r$ *and* $\mathbf{e}_\phi$ *are perpendicular to each other; due to symmetry, only the* z *component of the field remains; hence, only the* z *component of the vector* $[\mathbf{n} \times \mathbf{g}] = [\mathbf{e}_\phi \times (z\mathbf{k} - R\mathbf{e}_r)]$ *is needed.*]

Examples given above serve to illustrate the general concept of integration which is in summing up small contributions of some cumulative quantity such as length of a curve, surface area, electric field, etc. Physics presents numerous examples of such quantities, but in order to consider them we need to build up our mathematical muscle by considering, in the first instance, functions of many variables and multiple integration. Moreover, integration also appears when solving the so-called differential equations which connect derivatives of functions of interest and the functions themselves into a single (or a set of) equation(s). Since most physical laws are given via these kinds of equations, one may say that solutions of equations in physics involve integration in a natural way. You may appreciate now that differentiation and integration, together with infinite series and differential equations (still yet to be considered), represent the basic language of the calculus used in physics.

4.8 Summary

As it should have become clear from the previous chapter, any "well-behaved" function written in an analytical form can be differentiated, however complex. This is not the case for integration: there is no magical general recipe to integrate an arbitrary function.

There are many methods to integrate specific classes of functions, some of which we considered in this chapter:

- the change of the variable, and there are a number of different ones suited for particular types (classes) of functions;
- integration by parts, several of them may be needed;
- sometimes integration by parts establishes an equation for the integral being calculated;
- some integrals can be calculated by introducing parameters in the integrand; if the differentiation with respect to the parameter yields a known integral, then the required integral is obtained by differentiating it; not one but many differentiations are sometimes required.

More methods exist for solving integrals. For instance, indefinite integrals can be expanded into infinite series; some *definite integrals* can be calculated using more sophisticated methods such as integral transforms or complex calculus. Examples of indefinite integrals which cannot be expressed via a finite number of elementary functions (meaning: *solved*): $\int e^{-\alpha x^2} dx$, $\int \cos x^2 dx$, $\int \frac{\sin x}{x} dx$, $\int \frac{1}{x} e^{\alpha x} dx$. Some unsolvable indefinite integrals can be calculated analytically for special sets of limits, e.g.

$$\int_0^\infty e^{-\alpha x^2} dx = \frac{1}{2}\sqrt{\frac{\pi}{\alpha}}, \quad \int_0^\infty \frac{\sin(x)}{x} dx = \frac{\pi}{2}.$$

Many integrals have been studied and solved over the last centuries, and several comprehensive tables of integrals exist which list these results. For example, the following book is an absolutely excellent reference: I.S. Gradshteyn and I.M. Ryzhik, *Table of Integrals, Series and Products,* Academic Press, 1980. It is a good idea to consult this and other books when needed. There are also computer programs designed to solve integrals, both indefinite and definite, such as Mathematica. Finally, it is a good idea to remember that definite snail.

Functions of Many Variables: Differentiation

Mathematical ideas of the previous chapters, such as limits, derivatives and one-dimensional integrals, are the tools we developed for functions $y = f(x)$ of a single variable. These functions, however, present only particular cases. In practical applications, functions of more than one variable are frequently encountered. For instance, a temperature of an extended object is a function of all its coordinates (x, y, z) and may also be a function of time, t, i.e. it depends on four variables. Electric and magnetic fields, $\mathbf{E}$ and $\mathbf{B}$, in general, will also depend on all three coordinates and time.[1] Therefore, one has to generalise our definitions of the limits and derivatives for function of many variables. It is the purpose of this chapter to lie down the necessary principles of the theory of differentiation extending the foundations we developed so far to functions of many variables. In the next chapter, the notion of integration will be extended to this case as well.

5.1 Specification of Functions of Many Variables

If a *unique* number z corresponds to a pair x and y of variables, defined within their respective intervals, then it is said that a function of two variables is defined. This correspondence is written either as $z = z(x, y)$ or $z = f(x, y)$. This is a direct way in which a function can be defined, e.g. by a formula like $z = \left(x^2 + y\right)^2$. Alternatively, a function of two variables can be defined by an equation $f(x, y, z) = 0$, a solution z of which for every pair of x and y provides the required correspondence (if proven unique, i.e. that there is one and only one value of z for every pair of the variables x

---

[1] In fact, in the latter case, we have three functions, not one, for either of the fields, as each component of the field is an independent function of all these variables.

© The Author(s), under exclusive license to Springer Nature Switzerland AG 2022
L. Kantorovich, *Mathematics for Natural Scientists*, Undergraduate Lecture Notes in Physics, https://doi.org/10.1007/978-3-030-91222-2_5

and y). A function of two variables $z = z(x, y)$ can also be specified parametrically as $x = x(u, v)$, $y = y(u, v)$ using *two* parameters u and v defined within their respective intervals. Functions of more than two variables are defined in a similar way.

When a function of a single variable is defined, $y = f(x)$, it can be represented as a graph of y vs. x, i.e. the function defines a line in a two-dimensional space (on a plane). Correspondingly, functions $z = f(x, y)$ of two variables x and y can be represented in the *three-dimensional space* by points with the coordinates (x, y, z). It is clear that a surface will be obtained this way: as the coordinates x and y change, only a single z value is found each time.

Before making a general statement on functions of arbitrary number of variables, let us consider here some examples of functions of two variables, which is the simplest case. More specifically, we shall dwell on *surfaces of the second order*, which are generally given by the equation

$$a (x - x_0)^2 + b (y - y_0)^2 + c (z - z_0)^2 = 1 .$$

The equation defines not one but two functions $z = f(x, y)$ given explicitly as

$$z = z_0 \pm \sqrt{\frac{1}{c} \left[1 - a (x - x_0)^2 - b (y - y_0)^2 \right]} .$$

Recall that only a single value of the z can be put into correspondence to any pair of x and y, and hence there are two functions: one gives a single surface in the upper $z \geq z_0$ half-space, while the other in the lower $z \leq z_0$ half-space. Depending on the coefficients a, b and c, different types of surfaces are obtained. The point (x_0, y_0, z_0) defines the central point of the surfaces and can be chosen to coincide with the origin of the coordinate system, i.e. when $x_0 = y_0 = z_0 = 0$; their non-zero values simply correspond to a shift of the surfaces by the vector $\mathbf{r}_0 = (x_0, y_0, z_0)$. Hence, we shall assume in the following that $x_0 = y_0 = z_0 = 0$, and consider a simpler equation

$$ax^2 + by^2 + cz^2 = 1 . \tag{5.1}$$

Sphere

If $a = b = c = 1/R^2$, we obtain the equation of a sphere of radius R and centred at the origin: $x^2 + y^2 + z^2 = R^2$ (see Sect. 1.19.5). There are two surfaces here describing the upper ($z \geq 0$) and the lower ($z \leq 0$) hemispheres via $z = \pm\sqrt{R^2 - x^2 - y^2}$.

Ellipsoid

A more elaborate example is presented by a choice of still positive but non-equal coefficients via $a = 1/\alpha^2$, $b = 1/\beta^2$ and $c = 1/\gamma^2$, which yields the equation

$$\left(\frac{x}{\alpha}\right)^2 + \left(\frac{y}{\beta}\right)^2 + \left(\frac{z}{\gamma}\right)^2 = 1 \tag{5.2}$$

Fig. 5.1 Some of the second-order surfaces: **a** ellipsoid; **b** one-pole (one sheet) hyperboloid; **c** two-pole (two sheets) hyperboloid; **d** saddle-like elliptic paraboloid

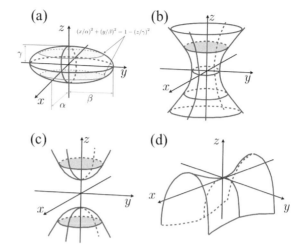

of an ellipsoid shown in Fig. 5.1a. To convince ourselves of this, let us consider cross sections of this shape with various planes perpendicular to the Cartesian axes. First of all we note that the absolute values of x/α, y/β and z/γ must be all less or equal to one, as otherwise the left-hand side in Eq. (5.2) would be larger than the right-hand side. Therefore, we conclude that $-\alpha \le x \le \alpha$, $-\beta \le y \le \beta$ and $-\gamma \le z \le \gamma$. Then, consider cross sections of the ellipsoid with planes $z = g$ (where $0 < g < \gamma$) which are parallel to the $x - y$ plane. In the cross section by either of these planes, we obtain a line with the equation $(x/\alpha)^2 + (y/\beta)^2 = d^2$, where $d^2 = 1 - (g/\gamma)^2 > 0$ is some positive number. This line is an *ellipse*

$$\left(\frac{x}{A}\right)^2 + \left(\frac{y}{B}\right)^2 = 1$$

with the semi-axes $A = \alpha d$ and $B = \beta d$. As g increases from 0 to its maximum value of γ, the parameter d decreases from 1 to 0, i.e. the semi-axes decrease from their corresponding maximum values (equal to α and β, respectively) to zeros. Therefore, at $z = \gamma$, the ellipse is simply a point, but then, as the plane $z = g$ moves down closer to the $x - y$ plane, we observe the ellipse in the cross section whose size gradually increases as we approach the $x - y$ plane. As the crossing plane moves further towards negative values of g, the size of the ellipse in the cross section gradually decreases and turns into a point at $g = -\gamma$.

One-Pole (One Sheet) Hyperboloid
If now $a = 1/\alpha^2 > 0$, $b = 1/\beta^2 > 0$, but $c = -1/\gamma^2 < 0$, then we obtain the equation

$$\left(\frac{x}{\alpha}\right)^2 + \left(\frac{y}{\beta}\right)^2 - \left(\frac{z}{\gamma}\right)^2 = 1 , \tag{5.3}$$

which corresponds to the figure shown in Fig. 5.1b called one-pole (or one sheet) hyperboloid.

Two-Pole (Two Sheet) Hyperboloid

When choosing -1 in the right-hand side of (5.3), a very different figure is obtained called two pole (or two sheet) hyperboloid which is shown in Fig. 5.1c. Its equation is

$$\left(\frac{x}{\alpha}\right)^2 + \left(\frac{y}{\beta}\right)^2 - \left(\frac{z}{\gamma}\right)^2 = -1 . \tag{5.4}$$

Problem 5.1 Using the method of sections with planes perpendicular to the coordinate axes, investigate the shape of the figure given by Eq. (5.3).

Problem 5.2 Investigate the shape of the figure given by Eq. (5.4). Are there any limitations for the values of z?

Hyperbolic Paraboloid

This figure is described by the equation

$$\left(\frac{x}{\alpha}\right)^2 - \left(\frac{y}{\beta}\right)^2 = z . \tag{5.5}$$

Let us investigate its shape. Considering first crossing planes $x = g$ (which are perpendicular to the x-axis), we obtain the lines $z = -y^2/\beta^2 + (g/\alpha)^2$, which are parabolas facing downwards; as g increases from zero in either the positive or negative direction, the top of the parabolas lifts up. Similarly, cutting with planes $y = g$ (which are perpendicular to the y-axis) results in parabolas $z = x^2/\alpha^2 - (g/\beta)^2$ which face upwards; increasing the value of g from zero results in the parabolas shifting downwards. Finally, planes $z = g > 0$ (parallel to the $x - y$ plane) result in hyperbolas $(x/A)^2 - (y/B)^2 = 1$ with $A = \alpha\sqrt{g}$ and $B = \beta\sqrt{g}$. When $z = -g < 0$, then hyperbolas $- (x/A)^2 + (y/B)^2 = 1$ are obtained which are rotated by $90°$ with respect to the first set. The obtained figure has a peculiar shape of a *saddle* shown in Fig. 5.1d: the centre of the coordinate system $x = y = z = 0$ is a minimum along the x-axis direction, but it is a maximum when one moves along the perpendicular direction. This point is usually called the saddle point or, in physics, the *transition state* (TS).

Problem 5.3 Consider two cross sections of the hyperbolic paraboloid Eq. (5.5): by $x = 0$ plane and $y = 0$ plane. Show that $z(x, 0)$ reaches a minimum at $x = 0$, while $z(0, y)$ reaches a maximum at $y = 0$.

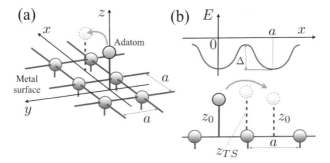

Fig. 5.2 Transition of an adatom between two hollow sites on a simple cubic lattice. The distance between the nearest atoms on the surface is a. **a** Three-dimensional view in which the initial and final (dimmed) positions of the adatom are shown. **b** Minimum energy path along the x direction with the energy barrier Δ between the two positions indicated. The x coordinate serves as the reaction coordinate here

This situation is very common in many physical processes: the total energy E of a system is a function of atomic coordinates and may contain many minima.[2] Each minimum corresponds to a stable configuration of the system. If we are interested in transitions between different states, i.e. between two neighbouring minima, then we have to consider the minimum energy path connecting the two minima. The energy barrier Δ along the path would determine the rate with which such a transition can happen: the rate is approximately proportional to $\exp\left(-\Delta/k_B T\right)$, where k_B is the Boltzmann's constant and T the absolute temperature. Sometimes, there is only a single so-called "reaction" coordinate connecting the two minima along the minimum energy path, and going along this direction requires overcoming the energy barrier, i.e. along this direction one has to move over a maximum. At the same time, since this is the minimum energy path, each point on it is a minimum with respect to the rest of the coordinates of the system. In other words, the crossing point is nothing but a saddle on the complicated potential energy surface E considered as a function of its all coordinates.

The simplest example of this situation can be found in atomic diffusion on a metal surface: the energy $E(x, y, z)$ of an adatom on the surface can be considered merely as a function of its two Cartesian coordinates, x and y, determining its lateral position, and of its height z above the surface as schematically shown in Fig. 5.2. Assuming the simplest square lattice of the metal surface with the distance between nearest atoms to be a, and the adatom occupying a hollow site with distance z_0 above the surface as shown in the Figure, the transition between two such sites $(0, 0, z_0)$ and $(a, 0, z_0)$ (which are equivalent) would require the adatom to move along the x-axis overcoming an energy barrier Δ at the transition point $(a/2, 0, z_{TS})$, which is exactly half way through; it corresponds to the bridge position of the adatom. Here

---

[2]We shall define more accurately minimum (and maximum) points of functions of many variables later on as points where the function has the lowest (the highest) values within a finite vicinity of their arguments.

z_{TS} is the height of the adatom at the transition state, and the x coordinate appears to serve as the reaction coordinate. In practice, however, a single Cartesian coordinate rarely serves as the reaction coordinate for the transition and the latter is instead specified in a more complicated way.

5.2   Limit and Continuity of a Function of Several Variables

Let us recall how we defined the limit of a function $y = f(x)$ of a single variable at a point x_0 (Sect. 2.4.1.1). Actually, we had two definitions that are equivalent. One definition was based on building a sequence of points

$$\{x_n, n = 1, 2, 3, \ldots\} \equiv \{x_1, x_2, x_3, \ldots, x_n, \ldots\}$$

which converges to the point x_0; this sequence generates the corresponding functional sequence of values

$$\{f(x_i), i = 1, 2, 3, \ldots\}$$

of $f(x)$, and the limit of the function as $x \to x_0$ is the limit of this functional sequence. Note that no matter which particular sequence is chosen, as long as it converges to the same value of x_0, the *same limit* of the functional sequence is required for the limit to exist. Another definition was based on the "$\epsilon - \delta$" language: we called A the limit of $f(x)$ when $x \to x_0$ if for any $\epsilon > 0$ one can always find a positive $\delta = \delta(\epsilon)$, such that for any x within the distance δ from x_0 (i.e. for any x satisfying $|x - x_0| < \delta$) the number A differs from $f(x)$ by no more than ϵ, i.e. $|f(x) - A| < \epsilon$ is implied. Note that one might take the values of x either below or above x_0, still the same limit should be achieved.

Both these definitions are immediately generalised to the case of a functions of more than one variable if we note that a function $f(x_1, x_2, \ldots, x_N)$ of N variables can formally be considered as a function of an N-dimensional vector $\mathbf{x} = (x_1, \ldots, x_N)$, i.e. as $f(\mathbf{x})$, and hence we are interested in the definition of the limit of $f(\mathbf{x})$ when $\mathbf{x} \to \mathbf{x}_0$, with $\mathbf{x}_0$ being some other vector. The generalisation is simply done by replacing a single variable x in the definitions corresponding to the cases of a single variable function, with the N-dimensional vector $\mathbf{x} = (x_1, \ldots, x_N)$; correspondingly, the distance $|x - x_0|$ between the points x and x_0 in the one-dimensional space is replaced with the distance (1.118) between vectors $\mathbf{x}$ and $\mathbf{x}_0$, i.e. with $d(\mathbf{x}, \mathbf{x}_0) = |\mathbf{x} - \mathbf{x}_0|$. With these substitutions, either definition of the limit of the function $f(\mathbf{x})$ when $\mathbf{x} \to \mathbf{x}_0$ sounds exactly the same.

The first definition—via a sequence—has a simple geometrical interpretation: points $\mathbf{x}_1, \mathbf{x}_2, \ldots, \mathbf{x}_n, \ldots$ in the sequence form a line in the N-dimensional space, which converges to a single point $\mathbf{x}_0$. The limit would exist if and only if any *path* towards $\mathbf{x}_0$, as depicted in Fig. 5.3, generates the corresponding functional sequence converging to the same limit A.

This condition is not always satisfied. Consider as our first example the limit of the function of two variables $f(x, y) = xy/(x + y)$ at the point $(x, y) = (0, 0)$. We

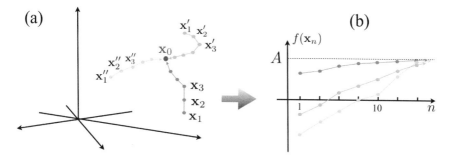

Fig. 5.3 This illustration shows that for a function $f(\mathbf{x})$, which has a well-defined limit when $\mathbf{x} \to \mathbf{x}_0$, any path to the limit point $\mathbf{x}_0$ made out of a sequence of points in the N-dimensional space, as shown in **a**, results in the same limit A for the functional sequence presented schematically in **b**

shall do this by constraining the y using $y = kx$ with some constant k when $x \to 0$. By changing the constant k different paths towards the centre of the coordinate system are taken; however, a well-defined limit exists for any value of k. Indeed,

$$\lim_{x \to 0, y \to 0} \frac{xy}{x+y} = \lim_{x \to 0} \frac{kx^2}{x(1+k)} = \frac{k}{1+k} \lim_{x \to 0} x = 0 .$$

Consider now, using the same set of paths, the function $f(x, y) = xy/\left(x^2 + y^2\right)$. For this function the limit does depend on the value of the constant k:

$$\lim_{x \to 0, y \to 0} \frac{xy}{x^2 + y^2} = \lim_{x \to 0} \frac{kx^2}{x^2 \left(1 + k^2\right)} = \frac{k}{1 + k^2} .$$

Therefore, different limiting values can be obtained using various values of k and hence, we must conclude the limit of this function at the point $(0, 0)$ does *not* exist.

Problem 5.4 Investigate whether the function $f(x, y) = \arctan(x/y)$ has well-defined limits at the following (x, y) points: (a) $(1, 1)$; (b) $(0, 1)$; (c) $(1, 0)$; (d) $(0, 0)$. Calculate the limits when they do exist. [Answers: *(a)* $\pi/4$; *(b)* 0; *(c,d) do not exist.*]

Most of the limit theorems discussed in Sect. 2.4.2 are immediately generalised to the case of functions of many variables, e.g. if the limit exists it is unique and that one can manipulate limits algebraically.

The notion of *continuity* of a function of many variables is introduced similarly to the case of a function of a single variable (see Sect. 2.4.3): namely, the function $f(\mathbf{x})$ is continuous at the point $\mathbf{x}_0$ if its limit at $\mathbf{x} \to \mathbf{x}_0$ coincides with the value of the function at the point $\mathbf{x}_0$, i.e.

$$\lim_{\mathbf{x} \to \mathbf{x}_0} f(\mathbf{x}) = f(\mathbf{x}_0) .\tag{5.6}$$

Points where the function $f(\mathbf{x})$ is not continuous are called points of discontinuity. For continuous functions, the change Δy of the function $y = f(\mathbf{x})$ by going from the point $\mathbf{x}$ to the point $\mathbf{x}' = \mathbf{x} + \Delta \mathbf{x}$ should approach zero when the point $\mathbf{x}'$ approaches $\mathbf{x}$, i.e. when $\Delta \mathbf{x} \to 0$:

$$\lim_{\Delta \mathbf{x} \to 0} \Delta y = \lim_{\Delta \mathbf{x} \to 0} [f(\mathbf{x} + \Delta \mathbf{x}) - f(\mathbf{x})] = 0 . \tag{5.7}$$

Many properties of continuous functions of several variables are also analogous to those of functions of a single variable. For instance, if a function $f(\mathbf{x})$ is continuous within some region in the N-dimensional space (where $\mathbf{x}$ is defined) and takes there two values A and B, then it takes in this region all values between A and B, i.e. for any C lying between A and B there exists a point $\mathbf{x}_C$ such that $f(\mathbf{x}_C) = C$. In particular, if $A < 0$ and $B > 0$, then there exists at least one point $\mathbf{x}_0$ where the function $f(\mathbf{x}_0) = 0$.

Problem 5.5 Consider the function $z = f(x, y) = xy^2$. Show that the change of the function Δz corresponding to the change Δx and Δy of its variables can be written in the form:

$$\Delta z = y^2 \Delta x + 2xy \Delta y + \alpha \Delta x + \beta \Delta y ,$$

where both α and β tend to zero as $(\Delta x, \Delta y) \to \mathbf{0}$. Note that the choice of α and β here is not unique.

5.3   Partial Derivatives. Differentiability

When in Sect. 3.1, we considered a function of a single variable, $y = f(x)$, we defined its derivative as the ratio $\Delta y / \Delta x$ of the change of the function Δy due to the change Δx of its variable, taken at the limit $\Delta x \to 0$. This definition cannot be directly generalised to the case of functions of many variables as $\Delta \mathbf{x}$ in this case is an N-dimensional vector.

At the same time, we can easily turn a function of many variables into a function of just one variable by "freezing" all other variables. Indeed, let us fix all variables but x_1; one can then consider the change of the function $y = f(x_1, x_2, \ldots, x_N) = f(\mathbf{x})$ due to its single variable x_1, the so-called partial change,

$$\Delta_1 y = f(x_1 + \Delta x_1, x_2, \ldots, x_N) - f(x_1, x_2, \ldots, x_N) ,$$

and then define the "usual" derivative

$$\lim_{\Delta x_1 \to 0} \frac{\Delta_1 y}{\Delta x_1} \equiv \left. \frac{dy}{dx} \right|_{x_2, \ldots, x_N} .$$

The latter notation, consisting of the variables written as a subscript to the vertical line, explicitly specifies that the variables $x_2, x_3, \ldots, x_N$ are fixed. This derivative is called *partial derivative* of the function y with respect to its first variable x_1, and is denoted $\partial y / \partial x_1$. It shows the rate of change of the function $y = f(\mathbf{x})$ due to its single variable x_1.

Similarly, one can consider the partial change $\Delta_i y$ of the function with respect to any variable x_i, where $i = 1, 2, \ldots, N$. Correspondingly, dividing the partial change of the function $\Delta_i y$ by the change of the corresponding variable, Δx_i, and taking the limit, we arrive at the partial derivative

$$\frac{\partial y}{\partial x_i} = \lim_{\Delta x_i \to 0} \frac{\Delta_i y}{\Delta x_i} \equiv \left. \frac{dy}{dx} \right|_{x_1, \ldots, x_{i-1}, x_{i+1}, \ldots, x_N}$$

with respect to the variable x_i. The partial derivative $\partial y / \partial x_i$ is sometimes also denoted as y'_{x_i} or $\partial_x y$.

Problem 5.6 Calculate all partial derivatives of the functions:

$f(x, y) = \sin x \sin y \quad \left[\text{Answer: } \dfrac{\partial f}{\partial x} = \cos x \sin y, \quad \dfrac{\partial f}{\partial y} = \sin x \cos y \right]$;

$f(x, y) = e^{x^2 + y^2} \quad \left[\text{Answer: } \dfrac{\partial f}{\partial x} = 2x e^{x^2 + y^2}, \quad \dfrac{\partial f}{\partial y} = 2y e^{x^2 + y^2} \right]$;

$f(x, y, z) = \dfrac{1}{\sqrt{x^2 + y^2 + z^2}}$

$\left[\text{Answer: } \dfrac{\partial f}{\partial x} = -\dfrac{x}{r^3}, \quad \dfrac{\partial f}{\partial y} = -\dfrac{y}{r^3}, \quad \dfrac{\partial f}{\partial z} = -\dfrac{z}{r^3}, \quad r = \sqrt{x^2 + y^2 + z^2} \right]$.

Problem 5.7 The three components F_x, F_y and F_z of the force acting on a particle in an external field (or potential) $U(\mathbf{r}) = U(x, y, z)$ are obtained via partial derivatives of the minus potential, i.e. $F_x = -\partial U / \partial x$ and similarly for the other two components. A satellite of mass M which is orbiting the Earth of mass M_0 experiences the potential field $U(\mathbf{r}) = -\alpha / |\mathbf{r} - \mathbf{r}_0|$, where $\alpha \propto M_0 M$ and $\mathbf{r}_0$ is the position vector of the Earth. Determine the force acting on the satellite.

Partial derivatives can be considered as derivatives along the directions of the Cartesian axes; it is also possible to define a derivative along an arbitrary direction; this will be done later on in Sect. 5.8.

In Sect. 3.2, we obtained a general expression for a change Δy of the function $y = f(x)$ of a single variable and related this to the differentiability of the function. There exists a generalisation of this result for a function of many variables. For simplicity of notations, we shall consider the case of a function of two variables, $z = f(x, y)$; a more general case is straightforward.

We say that the function $z = f(x, y)$ is differentiable at (x, y) if its change Δz due to the changes Δx and Δy of its variables can be written, similarly to Eq. (3.8), as:

$$\Delta z = A\Delta x + B\Delta y + \alpha\Delta x + \beta\Delta y , \tag{5.8}$$

where α and β tend to zero as both Δx and Δy tend to zero, while A and B depend only on the point (x, y), but not on Δx and Δy. This formula states that the first two terms in the right-hand side are of the first order with respect to the changes of the variables, Δx and Δy, while the last two terms are at least of the second order (cf. Problem 5.5).

Problem 5.8 Consider the function $z = x^2 y + xy^2$ and derive expressions for A, B, α and β in the above formula (5.8). Check explicitly that both α and β tend to zero as both Δx and Δy tend to zero.

It is obvious from this definition that if the function is differentiable, i.e. if its change can be recast in the form (5.8), then $\Delta z \to 0$ when Δx and Δy tend to zero, i.e. the function $z = f(x, y)$ is continuous. It is also clear that both partial derivatives exist at the point (x, y). Indeed,

$$\frac{\partial z}{\partial x} = \lim_{\Delta x \to 0} \frac{\Delta z}{\Delta x}\bigg|_{\Delta y=0} = \lim_{\Delta x \to 0} \left[A + B\frac{\Delta y}{\Delta x} + \alpha + \beta\frac{\Delta y}{\Delta x} \right]_{\Delta y=0}$$
$$= \lim_{\Delta x \to 0} (A + \alpha)_{\Delta y=0} = A + \lim_{\Delta x \to 0} \alpha|_{\Delta y=0} = A ,$$

and similarly

$$\frac{\partial z}{\partial y} = \lim_{\Delta y \to 0} \frac{\Delta z}{\Delta y}\bigg|_{\Delta x=0} = \lim_{\Delta y \to 0} \left[A\frac{\Delta x}{\Delta y} + B + \alpha\frac{\Delta x}{\Delta y} + \beta \right]_{\Delta x=0}$$
$$= B + \lim_{\Delta y \to 0} \beta|_{\Delta x=0} = B ,$$

where, while calculating the partial derivative, we kept the other variable constant, i.e. considered its change to be equal to zero *before* taking the limit. Therefore, both partial derivatives exist and hence the change of the function (5.8) can be written in a more detailed form as

$$\Delta z = \frac{\partial z}{\partial x}\Delta x + \frac{\partial z}{\partial y}\Delta y + \alpha\Delta x + \beta\Delta y . \tag{5.9}$$

The first two terms in the above formula can be used to estimate numerically the change of the function due to small changes Δx, Δy of its arguments.

In a general case of a function $y = y(\mathbf{x})$ of N variables $\mathbf{x} = (x_1, \ldots, x_N)$, the previous formula is straightforwardly generalised as follows:

$$\Delta y = y(\mathbf{x} + \Delta \mathbf{x}) - y(\mathbf{x}) = \sum_{i=1}^{N} \left(\frac{\partial y}{\partial x_i} \Delta x_i + \alpha_i \Delta x_i \right), \qquad (5.10)$$

where all α_i tend to zero when the vector $\Delta \mathbf{x} \to \mathbf{0}$.

Importantly, continuity does not yet guarantee the differentiability of the function. For instance, the function $z = \sqrt{x^2 + y^2}$ is continuous at the point $(0, 0)$, however its partial derivatives do not exist there: indeed, when, for instance, we are interested in the $\partial z / \partial x$, then at this point $y = 0$ and hence $z = \sqrt{x^2} = |x|$, which does not have the derivative at $x = 0$ (since the derivatives from the left and right are different). Since the partial derivatives do not exist at this point, the function is not differentiable there. Also, similarly to the case of functions of a single variable, differentiability is a stronger condition than continuity: continuity of a function at a point is not yet sufficient for it to be differentiable there.

Theorem 5.1 (Sufficient Condition of Differentiability). *For the function $z = f(x, y)$ to be differentiable at the point (x, y) it is sufficient that z has continuous partial derivatives there.*

Proof: Consider a total change $\Delta z = f(x + \Delta x, y + \Delta y) - f(x, y)$ of the function z due to changes Δx and Δy of its both variables, which we shall rewrite as follows:

$$\Delta z = [f(x + \Delta x, y + \Delta y) - f(x, y + \Delta y)] + [f(x, y + \Delta y) - f(x, y)]$$
$$= \Delta_x z|_{y + \Delta y} + \Delta_y z|_x .$$

$$(5.11)$$

Here, the first term is the partial change $\Delta_x z$ of the function z with its second variable kept constant at the value of $y + \Delta y$, while the second term gives the partial change $\Delta_y z$ with the first variable kept constant at x. Since, in both cases, we effectively deal with functions of a single variable (as the other variable is frozen), and since we assumed that partial derivatives exist, we can apply the Lagrange formula (3.88) for the corresponding partial changes:

$$\Delta_x z|_{y + \Delta y} = f'_x(x + \vartheta \Delta x, y + \Delta y) \Delta x , \quad \Delta_y z|_x = f'_y(x, y + \theta \Delta y) \Delta y ,$$
$$(5.12)$$

where ϑ and θ are both somewhere between zero and one. Note that derivatives above are, in fact, the partial derivatives, and we have used simplified notations for them: f'_x and f'_y. (The subscript indicates which variable is being differentiated upon.) As noted above, these kinds of simplified notations are frequently met in the literature and we shall be using them frequently as well.

Since the partial derivatives are continuous, we can write:

$$f_x' \left(x + \vartheta \Delta x, y + \Delta y \right) = f_x'(x, y) + \alpha = \frac{\partial z}{\partial x} + \alpha \quad \text{and}$$

$$f_y' \left(x, y + \theta \Delta y \right) = f_y'(x, y) + \beta = \frac{\partial z}{\partial y} + \beta \,,$$

where both functions, α and β, tend to zero when Δx or Δy tend to zero. Substituting the above results into Eqs. (5.11) and (5.12) leads exactly to the required form (5.9), proving the differentiability of the function z. **Q.E.D.**

Hence, if the partial derivatives exist and are continuous, the function of two variables is differentiable.

Problem 5.9 Prove the above theorem for a function of three variables, and then for any number of variables.

Above we have introduced first-order partial derivatives. However, a partial derivative of a function is a function itself and hence may also be differentiated. Consider $z = f(x, y)$. Taking the partial derivative with respect to the variable x of the $x-$partial derivative yields the second-order partial derivative:

$$\frac{\partial}{\partial x} \left(\frac{\partial z}{\partial x} \right) = \frac{\partial^2 z}{\partial x^2} \equiv f_{xx}'' \,.$$

However, if we take a partial derivative with respect to y of the partial derivative $\partial z / \partial x$, or *vice versa*, two mixed second-order partial derivatives are obtained:

$$\frac{\partial}{\partial y} \left(\frac{\partial z}{\partial x} \right) = \frac{\partial^2 z}{\partial y \partial x} \equiv f_{yx}'' \quad \text{and} \quad \frac{\partial}{\partial x} \left(\frac{\partial z}{\partial y} \right) = \frac{\partial^2 z}{\partial x \partial y} \equiv f_{xy}'' \,.$$

If this process is continued, various partial derivatives can be obtained.

As an example, consider $z(x, y) = \exp \left[\alpha \left(x - y \right) \right]$. We have after a simple calculation:

$$z_x' = \alpha e^{\alpha(x-y)} \,; \; z_y' = -\alpha e^{\alpha(x-y)} \,;$$

$$z_{xy}'' = \frac{\partial}{\partial x} \left(z_y' \right) = -\alpha^2 e^{\alpha(x-y)} \,; \; z_{yx}'' = \frac{\partial}{\partial y} \left(z_x' \right) = -\alpha^2 e^{\alpha(x-y)} \equiv z_{xy}'' \,;$$

$$z_{xx}'' = \frac{\partial}{\partial x} \left(z_x' \right) = \alpha^2 e^{\alpha(x-y)} \,; z_{yy}'' = \frac{\partial}{\partial y} \left(z_y' \right) = \alpha^2 e^{\alpha(x-y)} \,;$$

$$z'''_{xyy} = \frac{\partial}{\partial x} \left(z''_{yy} \right) = \alpha^3 e^{\alpha(x-y)} \; ; \quad z'''_{yxy} = \frac{\partial}{\partial y} \left(z''_{xy} \right) = \alpha^3 e^{\alpha(x-y)} \equiv z'''_{xyy} \; ;$$

$$z'''_{yyx} = \frac{\partial}{\partial y} \left(z''_{yx} \right) = \alpha^3 e^{\alpha(x-y)} \equiv z'''_{xyy} \equiv z'''_{yxy} \; , \quad \text{etc.}$$

Problem 5.10 Calculate all non-zero partial derivatives of the function $z(x, y) = x^3 y^2 + x^2 y^3$.

Problem 5.11 In a one-dimensional (along the x-axis) Brownian motion the probability to find a particle between x and $x + dx$ at time t is defined by the probability distribution function

$$w(x, t) = \frac{1}{\sqrt{4\pi Dt}} e^{-x^2/4Dt} \; , \tag{5.13}$$

where D is the diffusion constant. Check by direct differentiation that this function satisfies the following (so-called *partial differential*) equation:

$$\frac{1}{D} \frac{\partial w}{\partial t} = \frac{\partial^2 w}{\partial x^2} \; . \tag{5.14}$$

This is called the diffusion equation.

It follows from the example and the problem that the partial derivatives can be taken in any order, the result does not depend on it. The validity of this important statement follows from the following theorem which we shall formulate and prove for the case of a function of two variables and for the derivatives of the second order.

Theorem 5.2 (On Mixed Derivatives) *Consider a function* $z = f(x, y)$. *If its mixed second-order partial derivatives* f''_{xy} *and* f''_{yx} *exist and are continuous in some vicinity of a point* (x, y), *then they are equal at that point:*

$$f''_{xy} = f''_{yx} \; . \tag{5.15}$$

Proof: The first partial derivative f'_x requires calculation of the partial change of the function

$$\Delta_x f = f(x + \Delta x, y) - f(x, y) \equiv \phi(y) ,$$

which we can formally consider as a function $\phi(y)$ of the other variable y. The mixed derivative f''_{yx} will then require calculating the partial difference of the function $\phi(y)$; hence the total change needed for the mixed derivative will be

$$\Delta_{yx} f = \Delta_y \phi = \phi(y + \Delta y) - \phi(y)$$
$$= [f(x + \Delta x, y + \Delta y) - f(x, y + \Delta y)] - [f(x + \Delta x, y) - f(x, y)] .$$
$$(5.16)$$

This expression, after dividing by $\Delta x \Delta y$ and taking the limits $\Delta x \to 0$ and $\Delta y \to 0$, results in the mixed derivative f''_{yx}.

Similarly, let us construct the full change $\Delta_{xy} f$ of the function f required for the calculation of the mixed derivative f''_{xy}. We first calculate the change

$$\Delta_y f = f(x, y + \Delta y) - f(x, y) \equiv \varphi(x) ,$$

and then the additional change with respect to the other variable:

$$\Delta_{xy} f = \Delta_x \varphi = \varphi(x + \Delta x) - \varphi(x)$$
$$= [f(x + \Delta x, y + \Delta y) - f(x + \Delta x, y)] - [f(x, y + \Delta y) - f(x, y)] .$$
$$(5.17)$$

Dividing this expression by $\Delta x \Delta y$ and taking the limits would give the other mixed derivative f''_{xy}. One can see now that the two expressions (5.16) and (5.17) are exactly the same, $\Delta_{xy} f = \Delta_{yx} f$ (although with terms arranged in a different order), which already suggests that the mixed derivatives must be the same.

To proceed more rigorously, we use the Lagrange formula (3.88) for each of the changes of the function:

$$\phi(y) = \Delta_x f = f'_x(x + \theta_1 \Delta x, y)\Delta x \quad \text{and}$$
$$\Delta_{yx} f = \Delta_y \phi = \phi'_y(y + \theta_2 \Delta y)\Delta y \equiv f''_{yx}(x + \theta_1 \Delta x, y + \theta_2 \Delta y)\Delta x \Delta y$$

and, similarly,

$$\varphi(x) = \Delta_y f = f'_y(x, y + \vartheta_1 \Delta y)\Delta y \quad \text{and}$$
$$\Delta_{xy} f = \Delta_x \varphi = \varphi'_x(x + \vartheta_2 \Delta x)\Delta x \equiv f''_{xy}(x + \vartheta_2 \Delta x, y + \vartheta_1 \Delta y)\Delta x \Delta y ,$$

where numbers $\vartheta_1, \vartheta_2, \theta_1$ and θ_2 are each between 0 and 1. Note that the application of the Lagrange formula for each individual step in either of the expressions above is perfectly legitimate since each step corresponds to the change of the function with respect to a different variable.

Since both changes of the function $f(x, y)$ are the same, we can write:

$$f''_{xy}(x + \vartheta_2 \Delta x, y + \vartheta_1 \Delta y) = f''_{yx}(x + \theta_1 \Delta x, y + \theta_2 \Delta y) .$$

Both mixed derivatives are continuous functions, hence taking the limits $\Delta x \to 0$ and $\Delta y \to 0$ yields

$$\lim_{\Delta x \to 0, \Delta y \to 0} f''_{yx}(x + \theta_1 \Delta x, y + \theta_2 \Delta y) = f''_{yx}(x, y) ,$$

$$\lim_{\Delta x \to 0, \Delta y \to 0} f''_{xy}(x + \vartheta_2 \Delta x, y + \vartheta_1 \Delta y) = f''_{xy}(x, y) ,$$

which establishes the required result, $f''_{xy}(x, y) = f''_{yx}(x, y)$, **Q.E.D.**

Problem 5.12 Generalise this theorem for a function of $N > 2$ variables. [Hint: *try to make use of the already proven theorem.*]

Problem 5.13 Generalise this theorem for mixed derivatives of any order.

5.4 A Surface Normal. Tangent Plane

As an application of the developed formalism, let us derive an expression for the normal **n** to a surface $z = z(x, y)$ at point $M_0(x_0, y_0, z_0)$, see Fig. 5.4a. To this end, we shall consider lines lying within the surface. These lines are obtained by keeping either x or y at their values x_0 or y_0, respectively. The line DC is obtained by changing x and keeping $y = y_0$, its equation is $z = z(x, y_0)$. Similarly, the line AB with equation $z = z(x_0, y)$ is obtained by keeping $x = x_0$ and changing y. Both lines cross exactly at the point M_0.

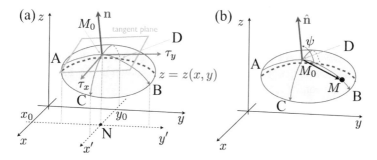

Fig. 5.4 To the derivation of the normal (red) to the surface (blue), $z = z(x, y)$, at point M_0. The plane tangential to the surface at point M_0 is shown in light green

Any point on our surface can be represented by the vector $\mathbf{r} = x\mathbf{i} + y\mathbf{j} + z(x, y)\mathbf{k}$. Consider a point M (with coordinates x_0, y, $z(x_0, y)$) along the AB curve ($x = x_0$, y is variable), see Fig. 5.4b. The vector $\overrightarrow{M_0 M}$ makes an angle ψ with the normal $\mathbf{n}$. As the point M approaches the point M_0, the angle ψ must approach the value of $\pi/2$. The cosine of ψ can be obtained by considering the dot product of vectors $\mathbf{n} = (n_x, n_y, n_z)$ and $\overrightarrow{M_0 M} = (0, \Delta y, \Delta z)$, where $\Delta y = y - y_0$ and $\Delta z = z(x_0, y) - z(x_0, y_0)$:

$$
\cos \psi = \frac{\overrightarrow{M_0 M} \cdot \mathbf{n}}{\left| \overrightarrow{M_0 M} \right| |\mathbf{n}|} = \frac{n_y \Delta y + n_z \Delta z}{|\mathbf{n}| \sqrt{\Delta y^2 + \Delta z^2}} = \frac{n_y + n_z \frac{\Delta z}{\Delta y}}{|\mathbf{n}| \sqrt{1 + \left(\frac{\Delta z}{\Delta y} \right)^2}} .
$$

In the limit when the point M approaches M_0 (or, which is equivalent, $\Delta y \to 0$), the angle $\psi \to \pi/2$, i.e. $\cos \psi \to 0$. Hence,

$$
\lim_{\Delta y \to 0} \frac{n_y + n_z \frac{\Delta z}{\Delta y}}{|\mathbf{n}| \sqrt{1 + \left(\frac{\Delta z}{\Delta y} \right)^2}} = \frac{n_y + n_z \lim_{\Delta y \to 0} \frac{\Delta z}{\Delta y}}{|\mathbf{n}| \sqrt{1 + \left(\lim_{\Delta y \to 0} \frac{\Delta z}{\Delta y} \right)^2}} = \frac{n_y + n_z \frac{\partial z}{\partial y}}{|\mathbf{n}| \sqrt{1 + \left(\frac{\partial z}{\partial y} \right)^2}} = 0 ,
$$

which means that $n_y + n_z \frac{\partial z}{\partial y} = 0$. Similarly, considering a point M somewhere on the DC line, when $y = y_0$ and x is allowed to change, we obtain an equation $n_x + n_z \frac{\partial z}{\partial x} = 0$. Assuming $n_z \neq 0$ (which corresponds to the surface $z = z(x, y)$ which is not parallel to the z-axis anywhere in the vicinity of the given point), we obtain for the normal:

$$
\begin{aligned}
\mathbf{n} = (n_x, n_y, n_z) &= \left(-n_z \frac{\partial z}{\partial x}, -n_z \frac{\partial z}{\partial y}, n_z \right) = -n_z \left(\frac{\partial z}{\partial x}, \frac{\partial z}{\partial y}, -1 \right) \\
\implies \quad \mathbf{n} &= \left(\frac{\partial z}{\partial x}, \frac{\partial z}{\partial y}, -1 \right) ,
\end{aligned}
\tag{5.18}
$$

where in the last step we removed n_z as this would only affect the (non-important) length of the normal vector, not its direction.

Once we have an expression for the normal, we can also write an equation for the tangent plane touching our surface at point M_0 as shown in Fig. 5.4a. According to Eq. (1.218), the equation for the plane has the form:

$$
n_x (x - x_0) + n_y (y - y_0) + n_z (z - z_0) = 0 ,
$$

where the components of the normal are given by Eq. (5.18). Therefore, we can rewrite the equation for the tangential plane in more detail as follows:

$$
z - z_0 = (x - x_0) \left(\frac{\partial z}{\partial x} \right)_{M_0} + (y - y_0) \left(\frac{\partial z}{\partial y} \right)_{M_0} .
\tag{5.19}
$$

Note that derivatives above are calculated at point M_0 and hence are constants.

As an example of the application of this result, consider the cone specified by the equation $z = \sqrt{x^2 + y^2}$. Using Eq. (5.18), we have: $\mathbf{n} = (x/z, y/z, -1)$.

Problem 5.14 Show that the normal vector to the sphere $x^2 + y^2 + z^2 = R^2$ is always directed radially from the sphere centre, i.e. it goes along the vector $\mathbf{r} = (x, y, z)$.

Problem 5.15 Obtain the normal vector to the one-pole hyperboloid Eq. (5.3) at the point (x_0, y_0, z_0) which is on its surface.

Problem 5.16 The same for the two-pole hyperboloid Eq. (5.4).

Problem 5.17 The same for the saddle Eq. (5.5).

Problem 5.18 Derive the result Eq. (5.18) using a slightly different method. First, consider the vectors[3] $\boldsymbol{\tau}_x = \partial \mathbf{r}/\partial x$ and $\boldsymbol{\tau}_y = \partial \mathbf{r}/\partial y$, where $\mathbf{r} = x\mathbf{i} + y\mathbf{j} + z(x, y)\mathbf{k}$. These vectors, see Fig. 5.4(a), are tangential to the curves DC and AB, respectively, at the point M_0. Therefore, the normal is obtained by calculating their vector product: $\mathbf{n} = \boldsymbol{\tau}_x \times \boldsymbol{\tau}_y$. Up to a scale factor and (possibly) a sign this is exactly the same result as in Eq. (5.18).

Here, we have considered a specific surface $z = z(x, y)$. Using the same method, one can also consider surfaces $x = x(y, z)$ and $y = y(z, x)$. An arbitrary surface requires a somewhat more general approach; this will be done later on in Sect. 6.5.1.

5.5 Exact Differentials

The linear part of the change of the function $z = f(x, y)$ due to changes of its variables dx and dy, see Eq. (5.9), is called its *exact differential* (cf. Sect. 3.2). It is written as follows:

$$dz = \frac{\partial z}{\partial x} dx + \frac{\partial z}{\partial y} dy . \qquad (5.20)$$

---

[3]Notation $\partial \mathbf{r}/\partial \lambda$ corresponds to the vector $(\partial x/\partial \lambda, \partial y/\partial \lambda, \partial z/\partial \lambda)$.

In the general case of a function $y = f(\mathbf{x})$ of N variables $\mathbf{x} = (x_1, \ldots, x_N)$, the differential is

$$dy = \sum_{i=1}^{N} \frac{\partial f}{\partial x_i} dx_i \ . \tag{5.21}$$

The independence of the mixed derivatives on the order in which differentiation is performed allows establishing a simple necessary condition for a form such as $A(x, y)dx + B(x, y)dy$ to be the exact differential of *some* function $z = z(x, y)$. This means that there exists such a function $z = z(x, y)$ whose differential (5.20) is equal exactly to $A(x, y)dx + B(x, y)dy$. Of course, it is not at all obvious that for any functions A and B this would be necessarily true; if this is not true, the form is called *inexact differential*. We shall now establish the necessary and sufficient condition for the form to be the exact differential.

Theorem 5.3 *The form $A(x, y)dx + B(x, y)dy$ is the exact differential dz of some function $z = z(x, y)$ if and only if*

$$\frac{\partial A}{\partial y} = \frac{\partial B}{\partial x} \ . \tag{5.22}$$

Proof: We first prove that this is a necessary condition. Indeed, if there was a function z such that its differential dz is given precisely by formula (5.20), then we should have that $A(x, y) \equiv \partial z/\partial x$ and $B(x, y) \equiv \partial z/\partial y$. Let us now differentiate A with respect to y and B with respect to x:

$$\frac{\partial A}{\partial y} = \frac{\partial}{\partial y}\left(\frac{\partial z}{\partial x}\right) = \frac{\partial^2 z}{\partial y \partial x} \quad \text{and} \quad \frac{\partial B}{\partial x} = \frac{\partial}{\partial x}\left(\frac{\partial z}{\partial y}\right) = \frac{\partial^2 z}{\partial x \partial y} \ .$$

We have arrived at two mixed derivatives where differentiation is performed in the opposite order. However, if the function $z = z(x, y)$ satisfies the required conditions of Theorem 5.2, these mixed derivatives must be the same, i.e. the functions A and B must then satisfy the condition (5.22).

To prove the sufficiency, we assume that the condition (5.22) is satisfied. We have to show that there exists a function $z = z(x, y)$ such that its differential dz has exactly the given form $Adx + Bdy$. Let us define a function $g(x, y)$ such that

$$A(x, y) \equiv \frac{\partial g}{\partial x} \ . \tag{5.23}$$

This can always be done. Then we can write:

$$\frac{\partial A}{\partial y} = \frac{\partial^2 g}{\partial y \partial x} \equiv \frac{\partial^2 g}{\partial x \partial y} = \frac{\partial}{\partial x}\left(\frac{\partial g}{\partial y}\right) \ .$$

On the other hand, because of our assumption (5.22), this is also equal to $\partial B/\partial x$, which means that

$$\frac{\partial B}{\partial x} = \frac{\partial}{\partial x}\left(\frac{\partial g}{\partial y}\right) \implies \frac{\partial}{\partial x}\left(B - \frac{\partial g}{\partial y}\right) = 0 \implies B = \frac{\partial g}{\partial y} + h(y), \quad (5.24)$$

where $h(y)$ must be some function of y only. The latter result is a direct consequence of our assumption (5.22). Next, we define a function $p(y)$ such that $p'(y) = h(y)$. The function $p(y)$ can be always found, at least in principle, as it is an indefinite integral of $h(y)$.

Now we are ready to provide a working guess for the function $z(x, y)$ as follows: $z(x, y) = g(x, y) + p(y)$. This function satisfies the required properties. Indeed,

$$\frac{\partial z}{\partial x} = \frac{\partial g}{\partial x} \equiv A \quad \text{and} \quad \frac{\partial z}{\partial y} = \frac{\partial g}{\partial y} + p'(y) = \frac{\partial g}{\partial y} + h(y) \equiv B \,,$$

as required. **Q.E.D.**

The second part of the Theorem suggests also a constructive method of finding the function $z = z(x, y)$, provided, of course, that the condition (5.22) is satisfied.

We shall illustrate the method by considering the differential form $\left(1 + 3x^2 y^2\right)$ $dx + 2x^3 y dy$. Here $A(x, y) = 1 + 3x^2 y^2$ and $B(x, y) = 2x^3 y$. We would like to know if this form is an exact differential, and if it is, of which function $z(x, y)$. We have:

$$\frac{\partial A}{\partial y} = 6x^2 y \quad \text{and} \quad \frac{\partial B}{\partial x} = 6x^2 y \,,$$

which are the same and hence there must be a function z such that its differential $dz = \left(1 + 3x^2 y^2\right) dx + 2x^3 y dy$. To find this function, we first consider the condition $\partial z/\partial x = A = 1 + 3x^2 y^2$. It follows from here that z is an indefinite integral of A with respect to the variable x, with y considered as a constant since we have a partial derivative of z with respect to x. Thus,

$$z(x, y) = \int A(x, y)dx = x + x^3 y^2 + C(y) \,,$$

where we have added an arbitrary function $C(y)$ since its derivative with respect to x is obviously zero. Now, we shall use the expression for B and the condition that it must be equal to $\partial z/\partial y$, i.e.

$$\frac{\partial z}{\partial y} = \frac{\partial}{\partial y}\left[x + x^3 y^2 + C(y)\right] = 2x^3 y + C'(y) \equiv B = 2x^3 y \,,$$

from which it follows that $C'(y) = 0$, so $C(y)$ does not depend on y at all and must be a constant C. This results in the final expression for the function sought for: $z = x + x^3 y^2 + C$. It is easily checked that, indeed, $\partial z/\partial x = 1 + 3x^2 y^2$ and $\partial z/\partial y = 2x^3 y$, as it should be.

Although in the above example $C'(y) = 0$, in other cases, it may not be equal to zero, so that $C(y)$ may still be some function of y.

Problem 5.19 Determine if the expression $2xy\,dx + (x^2 + y^2)\,dy$ corresponds to the exact differential dz of some function $z = z(x, y)$, and then find this function. [Answer: $z = x^2 y + y^3/3 + C$.]

Problem 5.20 The same for the form $(6xy + 5y^2)\,dx + (3x^2 + 10xy)\,dy$. [Answer: $z = 3x^2 y + 5xy^2 + C$.]

Problem 5.21 The same for $2x(1 + y^2)dx + 2x^2 y\,dy$. [Answer: $z = x^2(1 + y^2)$.]

Problem 5.22 Show that the criterion (5.22) for the case of a function of N variables is generalised as follows: for the expression

$$\sum_{i=1}^{N} A_i(\mathbf{x})dx_i$$

to be the exact differential of some function $y = f(\mathbf{x})$, the following conditions need to be satisfied:

$$\frac{\partial A_i}{\partial x_j} = \frac{\partial A_j}{\partial x_i} \quad \text{for any} \ \ i \neq j \ .$$

Argue that there will be exactly $N(N-1)/2$ such equations to be satisfied.

5.6   Derivatives of Composite Functions

Consider a differentiable function $y = f(\mathbf{x})$ of N variables $\mathbf{x} = (x_1, \ldots, x_N)$. Let each of the variables be a continuous function of a single variable t, i.e. $x_i = x_i(t)$, and each of these functions can be differentiated with respect to t. This makes $y = y(t)$ a function of a single variable t and it is reasonable to ask a question of how to find its derivative dy/dt. In order to proceed, we need to calculate the change of the function $\Delta y = y(t + \Delta t) - y(t)$ due to the change Δt of its variable t. Once t is changed by Δt, each of the functions $x_i(t)$ is changed as well by $\Delta x_i = x_i(t + \Delta t) - x_i(t)$. Because the function $y = f(\mathbf{x})$ is differentiable, we can use a generalisation of Eq.

(5.10) to relate the change in y to the changes in x_i:

$$\Delta y = \sum_{i=1}^{N} \left(\frac{\partial y}{\partial x_i} + \alpha_i \right) \Delta x_i \; ,$$

and hence

$$y'(t) = \lim_{\Delta t \to 0} \frac{\Delta y}{\Delta t} = \lim_{\Delta t \to 0} \sum_{i=1}^{N} \left(\frac{\partial y}{\partial x_i} + \alpha_i \right) \frac{\Delta x_i}{\Delta t}$$

$$= \sum_{i=1}^{N} \left(\frac{\partial y}{\partial x_i} + \lim_{\Delta t \to 0} \alpha_i \right) \lim_{\Delta t \to 0} \frac{\Delta x_i}{\Delta t} \; .$$

When $\Delta t \to 0$, all $\Delta x_i \to 0$ since the functions $x_i(t)$ are continuous. Also, all $\alpha_i \to 0$ when $\Delta \mathbf{x} \to 0$, and hence we obtain:

$$y'(t) = \frac{dy}{dt} = \sum_{i=1}^{N} \frac{\partial y}{\partial x_i} \lim_{\Delta t \to 0} \frac{\Delta x_i}{\Delta t} = \sum_{i=1}^{N} \frac{\partial y}{\partial x_i} \frac{dx_i}{dt} \; . \qquad (5.25)$$

This formula has a very simple meaning: if arguments of a function of N variables in turn depend on a single variable t, one should differentiate y with respect to each of its direct variables x_i (partial derivatives with ∂), multiply them by the derivatives $x_i'(t)$ and then sum up all the contributions. If the function y was considered as a function of a *single* variable x_i, in this case y would be a composition $y = y(x_i) = y(x_i(t))$, and its derivative is $y'(t) = y'(x_i)x'(t)$ according to the chain rule (3.16), the result for a function of a single variable. Therefore, formula (5.25) tells us that one has to use the chain rule for each argument x_i of the function $y = f(\mathbf{x})$ individually (i.e. assuming that other arguments are constants, and hence using the partial derivative symbol for the $y'(x_i) \Rightarrow \partial y / \partial x_i$), and then sum up all such contributions. It is clear from this discussion that our result (5.25) is the straightforward generalisation of Eq. (3.16) for functions of more than one variable, i.e. it is a chain rule for this case.

As an example, consider $z = x^2 y^2 + \sin x \sin y$ with $x = t^2$ and $y = t^3$. Using the formula just derived, we write:

$$\frac{dz}{dt} = \frac{\partial}{\partial x} \left(x^2 y^2 + \sin x \sin y \right) \frac{d\left(t^2\right)}{dt} + \frac{\partial}{\partial y} \left(x^2 y^2 + \sin x \sin y \right) \frac{d\left(t^3\right)}{dt}$$

$$= \left(2xy^2 + \cos x \sin y \right) 2t + \left(2x^2 y + \sin x \cos y \right) 3t^2$$

$$= 2t \left(2t^8 + \cos t^2 \sin t^3 \right) + 3t^2 \left(2t^7 + \sin t^2 \cos t^3 \right)$$

$$= 10t^9 + 2t \cos t^2 \sin t^3 + 3t^2 \sin t^2 \cos t^3 \; .$$

It is instructive to see that exactly the same result is obtained if we replace the functions x and y in $z = z(x, y)$ with their respective expressions via t at the very beginning and then differentiate with respect to t:

$$z(t) = \left(t^2\right)^2 \left(t^3\right)^2 + \sin t^2 \sin t^3 = t^{10} + \sin t^2 \sin t^3 \; ,$$

which, as can easily be checked by direct differentiation (do it!), results exactly in the same expression obtained before using the chain rule.

As a more physical example, let us consider a gas of atoms. The atoms in the gas all have different positions and momenta (velocity times atom mass, m) which change with time t. Statistical properties of such a gas can be described by the distribution function $f(\mathbf{r}, \mathbf{p}, t)$ defined in the following way. We define a six-dimensional phase space formed by the particles coordinates $\mathbf{r} = (x, y, z)$ (the coordinate subspace) and momentum $\mathbf{p} = (p_x, p_y, y_z)$ (the momentum subspace). Then $f(\mathbf{r}, \mathbf{p}, t)d\mathbf{r}d\mathbf{p}$ gives the number of atoms of the gas at time t which coordinates fall within a small box between $\mathbf{r}$ and $\mathbf{r} + d\mathbf{r}$ of the coordinate subspace (i.e. within the Cartesian coordinates ranging between x and $x + dx$, y and $y + dy$, z and $z + dz$), and their momenta are within a small box in the momentum subspace between $\mathbf{p}$ and $\mathbf{p} + d\mathbf{p}$.

The total change of the number of atoms in that six-dimensional box in time can be due to a possible intrinsic time dependence of the distribution function itself, and because the atoms move in and out of the box. The total change, per unit time, is given by the total derivative df/dt of the function $f(\mathbf{r}, \mathbf{p}, t)$ with respect to time, considering all seven its arguments $f(\mathbf{r}, \mathbf{p}, t) = f(x, y, z, p_x, p_y, p_z, t)$. Therefore, f can actually be written as $f(\mathbf{x})$, where $\mathbf{x} = (\mathbf{r}(t), \mathbf{p}(t), t)$ is a seven-dimensional vector. The time dependence of the $\mathbf{r}$ and $\mathbf{p}$ follow from the definition of the velocity and Newton's equations of motion: $\dot{\mathbf{r}}(t) = d\mathbf{r}/dt = \mathbf{p}/m$ and $\dot{\mathbf{p}}(t) = d\mathbf{p}/dt = \mathbf{F}$, where $\mathbf{F}$ is the force (due to an external field and interaction between atoms). Therefore, the total derivative of this composite function, according to Eq. (5.25), is:

$$\left(\frac{df}{dt}\right)_{tot} = \frac{\partial f}{\partial t}\frac{dt}{dt} + \sum_{\alpha=1}^{3}\frac{\partial f}{\partial r_\alpha}\frac{dr_\alpha}{dt} + \sum_{\alpha=1}^{3}\frac{\partial f}{\partial p_\alpha}\frac{dp_\alpha}{dt} = \frac{\partial f}{\partial t} + \sum_{\alpha=1}^{3}\frac{\partial f}{\partial r_\alpha}\frac{p_\alpha}{m} + \sum_{\alpha=1}^{3}\frac{\partial f}{\partial p_\alpha}F_\alpha,$$

where $\alpha = 1, 2, 3$ designates the three Cartesian components of the vectors. The expression above can be written in a more compact vector form as follows:

$$\left(\frac{df}{dt}\right)_{tot} = \frac{\partial f}{\partial t} + \frac{\partial f}{\partial \mathbf{r}}\cdot\frac{\mathbf{p}}{m} + \frac{\partial f}{\partial \mathbf{p}}\cdot\mathbf{F}. \tag{5.26}$$

Here the second and the third terms in the right-hand side are scalar products of vectors: $\partial f/\partial \mathbf{r}$ is the vector $(\partial f/\partial x, \partial f/\partial y, \partial f/\partial z)$ and $\partial f/\partial \mathbf{p}$ is similarly the vector $(\partial f/\partial p_x, \partial f/\partial p_y, \partial f/\partial p_z)$.[4] The change of f in time is due to collisions between atoms. Equating the obtained expression for the derivative of f to the so-called collision integral results in the (classical) Boltzmann's kinetic equation for the gas of atoms.

---

[4]We shall learn in Sect. 5.8 that these vectors are called gradients.

Problem 5.23 Calculate $z'(t)$ for the following function: $z = \exp\left(-x^2 - y^2\right)$, where $x = e^{-t}$ and $y = 2e^{-2t}$. [Answer: $2ze^{-2t}\left(1 + 8e^{-2t}\right)$.]

Problem 5.24 The Hamiltonian[5] of a particle in an external field $U(\mathbf{r}, t)$ is a function of both particle momentum $\mathbf{p}$ and its position $\mathbf{r}$, i.e. $H(\mathbf{r}, \mathbf{p}) = \mathbf{p}^2/2m + U(\mathbf{r}, t)$. Show that the Hamiltonian function is conserved in time only if the external potential does not depend explicitly on time. [Hint: *the Newton's equations of motion read* $\dot{\mathbf{p}} = \mathbf{F} = -\partial U/\partial \mathbf{r}$, *where* $\dot{\mathbf{p}} \equiv d\mathbf{p}/dt$.]

The derivative (5.25) is often called *total derivative* of the function y with respect to t since it includes contributions from all its variables into its overall change with the t. Sometimes this formula is written in a symbolic form via *operators*:

$$\frac{dy}{dt} = \sum_{i=1}^{N} \frac{dx_i}{dt} \frac{\partial y}{\partial x_i} = \left(\sum_{i=1}^{N} \frac{dx_i}{dt} \frac{\partial}{\partial x_i}\right) y = \widehat{D} y . \tag{5.27}$$

Here the operator

$$\widehat{D} = \sum_{i=1}^{N} \frac{dx_i}{dt} \frac{\partial}{\partial x_i} = x_1'(t) \frac{\partial}{\partial x_1} + x_2'(t) \frac{\partial}{\partial x_2} + \cdots + x_N'(t) \frac{\partial}{\partial x_N} \tag{5.28}$$

contains a sum of partial derivatives with respect to all variables x_1, x_2, etc., multiplied by t-derivatives of the variables themselves. Frequently a hat above a symbol is used to indicate that it is an operator.

Now we can consider a more general case of the composite functions. Let us first look at the simplest case of a function of only two variables, $z = f(x, y)$, when the variables are both functions not of one (as we have done before) but two variables: $x = x(u, v)$ and $y = y(u, v)$. Then, effectively, the function z becomes a function of two variables, u and v, and it is legitimate to ask ourselves what would be its partial (this time!) derivatives with respect to these two variables. Consider first the partial derivative of $z = z(x(u, v), y(u, v))$ with respect to u, i.e. the variable v is kept constant. In that case z may be thought of as effectively depending only on a single variable u, and hence our general result (5.25) can be directly applied:

$$\frac{\partial z}{\partial u} = \frac{\partial z}{\partial x} \frac{\partial x}{\partial u} + \frac{\partial z}{\partial y} \frac{\partial y}{\partial u} \quad \text{or} \quad z_u' = z_x' x_u' + z_y' y_u' , \tag{5.29}$$

where we have again used simplified notations for the partial derivatives, e.g. $y_u' = \partial y/\partial u$, etc.

---

[5]Due to Sir William Rowan Hamilton.

Similarly,

$$\frac{\partial z}{\partial v} = \frac{\partial z}{\partial x}\frac{\partial x}{\partial v} + \frac{\partial z}{\partial y}\frac{\partial y}{\partial v} \quad \text{or} \quad z'_v = z'_x x'_v + z'_y y'_v \,. \tag{5.30}$$

This result is easily generalised to a more general case of functions of more than two variables each of which is also a function of arbitrary number of variables.

Problem 5.25 Derive analogous formulae for the partial derivatives z'_u and z'_v of $z = z(x)$, where $x = x(u, v)$.

Problem 5.26 Similarly for $z = z(u, x)$, where $x = x(u, v)$.

Problem 5.27 Similarly for $z = z(u, v, x)$, where $x = x(u, v)$.

Problem 5.28 Consider $z = z(x, y)$ with $x = x(u, v)$ and $y = y(u, v)$. Prove the following expressions for the second-order derivatives of z with respect to u and v:

$$z''_{uu} = z'_x x''_{uu} + z'_y y''_{uu} + \left[z''_{xx}\left(x'_u\right)^2 + 2z''_{xy}x'_u y'_u + z''_{yy}\left(y'_u\right)^2 \right] , \tag{5.31}$$

$$z''_{vv} = z'_x x''_{vv} + z'_y y''_{vv} + \left[z''_{xx}\left(x'_v\right)^2 + 2z''_{xy}x'_v y'_v + z''_{yy}\left(y'_v\right)^2 \right] , \tag{5.32}$$

$$z''_{uv} = z'_x x''_{uv} + z'_y y''_{uv} + \left[z''_{xx}x'_u x'_v + z''_{xy}\left(x'_u y'_v + x'_v y'_u\right) + z''_{yy}y'_u y'_v \right] , \tag{5.33}$$

where the simplified notations for the second-order partial derivatives were again used, e.g. $z''_{xy} = \partial^2 z/\partial x \partial y$.

Problem 5.29 Polar coordinates on the plane (r, φ) are defined via $x = r\cos\varphi$ and $y = r\sin\varphi$. Show that differential operators with respect to the Cartesian coordinates and polar coordinates are related as follows:

$$\frac{\partial}{\partial r} = \cos\varphi\frac{\partial}{\partial x} + \sin\varphi\frac{\partial}{\partial y} \quad \text{and} \quad \frac{\partial}{\partial\varphi} = -r\sin\varphi\frac{\partial}{\partial x} + r\cos\varphi\frac{\partial}{\partial y} \,. \tag{5.34}$$

[Hint: *assume a function* $z = f(x, y)$ *with* $x = x(r, \varphi)$ *and* $y = y(r, \varphi)$ *as given by the equations for the polar coordinats, and then use Eqs. (5.29) and (5.30)*.] Then apply these relationships to $z = x^2 y^2$ and calculate its partial derivatives with respect to r and φ. Verify your result by differentiating directly the function $z = x^2 y^2 = r^4 \cos^2\varphi\sin^2\varphi$.

Finally, let us consider *implicit functions* and their derivatives. Yet again, we start from the simplest case of a function of two variables, $z = z(x, y)$, which is specified by means of a function of three variables $f(x, y, z)$ via the following algebraic equation:

$$f(x, y, z(x, y)) = 0 . \tag{5.35}$$

If we could solve this equation for the function $z = z(x, y)$, then there would have been no problem as the required partial derivatives $\partial z/\partial x$ and $\partial z/\partial y$ would be available for the calculation directly. However, this may not be the case (i.e. the equation above cannot be solved analytically) and hence a general method is required. The idea of this method is based on the fact that the function f is a constant (equal to zero), so that its derivative with respect to either x or y (not forgetting the fact that its third argument z is also a function of these two) is zero. Therefore, one can write:

$$\left. \frac{\partial f}{\partial x} \right|_{tot} = \frac{\partial f}{\partial x} + \frac{\partial f}{\partial z} \frac{\partial z}{\partial x} = 0 \quad \Longrightarrow \quad \frac{\partial z}{\partial x} = -\frac{\partial f/\partial x}{\partial f/\partial z} . \tag{5.36}$$

Note that the partial derivative $\partial f/\partial x|_{tot}$ in the left-hand side corresponds to the total derivative of f with respect to the variable x, including its dependence on x via its third variable z. In the right-hand side of the final formula for $\partial z/\partial x$, however, the partial derivative $\partial f/\partial x$ is taken only with respect to the first (explicit) argument of f, i.e. keeping not only y but also z constant. Similarly one obtains

$$\frac{\partial z}{\partial y} = -\frac{\partial f/\partial y}{\partial f/\partial z} . \tag{5.37}$$

Problem 5.30 Show that the results above are formally correct even if a function of more than two variables is similarly specified via an algebraic equation, e.g. $f(x, y, z, w) = 0$, where $w = w(x, y, z)$.

Problem 5.31 Consider a function $w = w(x, y) = h(x, y, z(x, y))$. As we already know, the first-order derivative is given by:

$$\frac{\partial w}{\partial x} = \frac{\partial h}{\partial x} + \frac{\partial h}{\partial z}\frac{\partial z}{\partial x} \,. \tag{5.38}$$

Show that for the second-order derivatives we get:

$$\frac{\partial^2 w}{\partial x^2} = \frac{\partial h}{\partial z}\frac{\partial^2 z}{\partial x^2} + \frac{\partial^2 h}{\partial x^2} + 2\frac{\partial^2 h}{\partial x \partial z}\frac{\partial z}{\partial x} + \frac{\partial^2 h}{\partial z^2}\left(\frac{\partial z}{\partial x}\right)^2 \,, \tag{5.39}$$

and similarly for the derivatives with respect to y. Also, for the mixed derivative,

$$\frac{\partial^2 w}{\partial x \partial y} = \frac{\partial h}{\partial z}\frac{\partial^2 z}{\partial x \partial y} + \frac{\partial^2 h}{\partial x \partial y} + \frac{\partial^2 h}{\partial x \partial z}\frac{\partial z}{\partial y} + \frac{\partial^2 h}{\partial y \partial z}\frac{\partial z}{\partial x} + \frac{\partial^2 h}{\partial z^2}\frac{\partial z}{\partial x}\frac{\partial z}{\partial y} \,. \tag{5.40}$$

As an example, consider the function $z = z(x, y)$ specified via the equation $x^2 + y^2 + z^2 = 1$ and the condition $z \geq 0$. In this case $f = x^2 + y^2 + z^2 - 1$, and we obtain:

$$\frac{\partial z}{\partial x} = -\frac{\partial f/\partial x}{\partial f/\partial z} = -\frac{2x}{2z} = -\frac{x}{z} \quad \text{and} \quad \frac{\partial z}{\partial y} = -\frac{y}{z} \,.$$

This result can be compared with that obtained via the direct calculation: solving the algebraic equation for z, we obtain $z = +\sqrt{1 - x^2 - y^2}$, so that the required partial derivatives can be immediately calculated. The results, as the reader no doubt can easily check, are exactly the same as above (do it!).

Problem 5.32 Consider equations $x = r \cos\varphi$ and $y = r \sin\varphi$, defining the polar coordinates. Differentiate both sides of the equations with respect to x and y, remembering that both r and φ are indirect functions of x and y. Solve the obtained algebraic equations with respect to the derivatives of r and φ, and hence show that:

$$\frac{\partial r}{\partial x} = \cos\varphi \,, \quad \frac{\partial r}{\partial y} = \sin\varphi \,, \quad \frac{\partial \varphi}{\partial x} = -\frac{\sin\varphi}{r} \quad \text{and} \quad \frac{\partial \varphi}{\partial y} = \frac{\cos\varphi}{r} \,. \tag{5.41}$$

Using the above equations (5.34), we can also derive relations relating second order *operator derivatives* with respect to r and φ with those with respect to x and y. Indeed,

$$\frac{\partial^2 f}{\partial r^2} = \frac{\partial}{\partial r}\left(\frac{\partial f}{\partial r}\right) = \frac{\partial}{\partial r}\left(\cos\varphi\frac{\partial f}{\partial x} + \sin\varphi\frac{\partial f}{\partial y}\right) \,.$$

Since we differentiate with respect to r, the other independent variable φ is to be kept constant, and hence the cosine and sine functions need to be left alone:

$$\frac{\partial^2 f}{\partial r^2} = \cos\varphi \frac{\partial}{\partial r}\left(\frac{\partial f}{\partial x}\right) + \sin\varphi \frac{\partial}{\partial r}\left(\frac{\partial f}{\partial y}\right).$$

The functions $\partial f/\partial x$ and $\partial f/\partial y$ are some functions of x and y, and hence we can again use Eq. (5.34), yielding

$$\frac{\partial^2 f}{\partial r^2} = \cos\varphi \left(\cos\varphi \frac{\partial}{\partial x} + \sin\varphi \frac{\partial}{\partial y}\right)\frac{\partial f}{\partial x} + \sin\varphi \left(\cos\varphi \frac{\partial}{\partial x} + \sin\varphi \frac{\partial}{\partial y}\right)\frac{\partial f}{\partial y}$$

$$= \cos^2\varphi \frac{\partial^2 f}{\partial x^2} + \cos\varphi \sin\varphi \frac{\partial^2 f}{\partial y \partial x} + \sin\varphi \cos\varphi \frac{\partial^2 f}{\partial x \partial y} + \sin^2\varphi \frac{\partial^2 f}{\partial y^2},$$

and, since the mixed derivatives do not depend on the order in which the partial derivatives are taken, we obtain:

$$\frac{\partial^2 f}{\partial r^2} = \cos^2\varphi \frac{\partial^2 f}{\partial x^2} + 2\cos\varphi \sin\varphi \frac{\partial^2 f}{\partial y \partial x} + \sin^2\varphi \frac{\partial^2 f}{\partial y^2},$$

which can be written symbolically with operators as

$$\frac{\partial^2}{\partial r^2} = \cos^2\varphi \frac{\partial^2}{\partial x^2} + 2\cos\varphi \sin\varphi \frac{\partial^2}{\partial y \partial x} + \sin^2\varphi \frac{\partial^2}{\partial y^2}$$

$$= \left(\cos\varphi \frac{\partial}{\partial x} + \sin\varphi \frac{\partial}{\partial y}\right)^2. \tag{5.42}$$

This is of course the same as repeating twice the operator $\partial/\partial r$, Eq. (5.34), i.e. $(\partial/\partial r)^2$, as it should be.

Problem 5.33 Show that

$$\frac{\partial^2}{\partial \varphi^2} = -r\left(\cos\varphi \frac{\partial}{\partial x} + \sin\varphi \frac{\partial}{\partial y}\right) + \left(-r\sin\varphi \frac{\partial}{\partial x} + r\cos\varphi \frac{\partial}{\partial y}\right)^2, \tag{5.43}$$

$$\frac{\partial^2}{\partial r \partial \varphi} = -\sin\varphi \frac{\partial}{\partial x} + \cos\varphi \frac{\partial}{\partial y}$$

$$+ \left(\cos\varphi \frac{\partial}{\partial x} + \sin\varphi \frac{\partial}{\partial y}\right)\left(-r\sin\varphi \frac{\partial}{\partial x} + r\cos\varphi \frac{\partial}{\partial y}\right) \equiv \frac{\partial^2}{\partial \varphi \partial r}. \tag{5.44}$$

Problem 5.34 Calculate all three second derivatives of the function $z(x, y) = x \sin y + y \sin x$ with respect to the polar coordinates using Eqs. (5.42)–(5.44). Verify your expressions by replacing x and y directly in z and performing differentiation.

Problem 5.35 Using the above formulae, demonstrate by a direct calculation that

$$\frac{\partial^2 f}{\partial r^2} + \frac{1}{r}\frac{\partial f}{\partial r} + \frac{1}{r^2}\frac{\partial^2 f}{\partial \varphi^2} = \frac{\partial^2 f}{\partial x^2} + \frac{\partial^2 f}{\partial y^2} . \tag{5.45}$$

Here, the expression in the right-hand side represents a two-dimensional Laplacian[6] of the function $f(x, y)$. This identity shows how the Laplacian written in Cartesian coordinates (in the right-hand side) can equivalently be written using polar coordinates (in the left-hand side). There are much more powerful methods to change variables in this kind of differential expressions that we shall learn in Chap. 7 of Vol. II.

Problem 5.36 Consider the spherical coordinates

$$x = r \sin\theta \cos\phi , \quad y = r \sin\theta \sin\phi \quad \text{and} \quad z = r \cos\theta .$$

Show by differentiating these equations with respect to x, y and z that

$$\frac{\partial r}{\partial x} = \sin\theta \cos\phi , \quad \frac{\partial\theta}{\partial x} = \frac{1}{r}\cos\theta\cos\phi , \quad \frac{\partial\phi}{\partial x} = -\frac{1}{r}\frac{\sin\phi}{\sin\theta} ,$$

$$\frac{\partial r}{\partial y} = \sin\theta \sin\phi , \quad \frac{\partial\theta}{\partial y} = \frac{1}{r}\cos\theta\sin\phi , \quad \frac{\partial\phi}{\partial y} = \frac{1}{r}\frac{\cos\phi}{\sin\theta} ,$$

$$\frac{\partial r}{\partial z} = \cos\theta , \quad \frac{\partial\theta}{\partial z} = -\frac{1}{r}\sin\theta , \quad \frac{\partial\phi}{\partial z} = 0 .$$

Problem 5.37 In quantum mechanics, Cartesian components of the operator of the angular momentum of an electron are defined as follows:

$$\widehat{L}_x = -i\hbar\left(y\frac{\partial}{\partial z} - z\frac{\partial}{\partial y}\right) , \quad \widehat{L}_y = -i\hbar\left(z\frac{\partial}{\partial x} - x\frac{\partial}{\partial z}\right) , \quad \widehat{L}_z = -i\hbar\left(x\frac{\partial}{\partial y} - y\frac{\partial}{\partial x}\right) .$$

---

[6]Named after Pierre-Simon, marquis de Laplace.

Show that, in the spherical coordinates, the operators are (these operators act on functions depending only on the angles so that the r dependence can be ignored):

$$\widehat{L}_x = i\hbar \left(\sin\phi \frac{\partial}{\partial\theta} + \frac{\cos\phi}{\tan\theta} \frac{\partial}{\partial\phi} \right) , \quad \widehat{L}_y = i\hbar \left(-\cos\phi \frac{\partial}{\partial\theta} + \frac{\sin\phi}{\tan\theta} \frac{\partial}{\partial\phi} \right) ,$$

$$\widehat{L}_z = -i\hbar \frac{\partial}{\partial\phi} .$$

Finally, demonstrate that the square of the operator of the total angular momentum is:

$$\widehat{\mathbf{L}}^2 = \widehat{L}_x^2 + \widehat{L}_y^2 + \widehat{L}_z^2 = -\hbar^2 \left[\frac{1}{\sin\theta} \frac{\partial}{\partial\theta} \left(\sin\theta \frac{\partial}{\partial\theta} \right) + \frac{1}{\sin^2\theta} \frac{\partial^2}{\partial\phi^2} \right] .$$

Consider now, quite generally, three quantities x, y and z are related by a single equation $f(x, y, z) = 0$. This means that any one of them may be considered as a function of the other two. Then, writing differentials for $z = z(x, y)$ and $x = x(y, z)$, we obtain:

$$dz = \left(\frac{\partial z}{\partial x} \right)_y dx + \left(\frac{\partial z}{\partial y} \right)_x dy \tag{5.46}$$

and

$$dx = \left(\frac{\partial x}{\partial y} \right)_z dy + \left(\frac{\partial x}{\partial z} \right)_y dz . \tag{5.47}$$

Here, the variable written as a subscript to the partial derivatives is considered constant while differentiating. Use now dz from (5.46) in the second term in the right-hand side of the expression for dx above:

$$dx = \left(\frac{\partial x}{\partial y} \right)_z dy + \left(\frac{\partial x}{\partial z} \right)_y \left[\left(\frac{\partial z}{\partial x} \right)_y dx + \left(\frac{\partial z}{\partial y} \right)_x dy \right]$$

$$= \left(\frac{\partial x}{\partial z} \right)_y \left(\frac{\partial z}{\partial x} \right)_y dx + \left[\left(\frac{\partial x}{\partial y} \right)_z + \left(\frac{\partial x}{\partial z} \right)_y \left(\frac{\partial z}{\partial y} \right)_x \right] dy .$$

Examine the coefficient to dx. The variable y is kept constant, and hence x can be considered as a function of only z, while z can be considered as a function of x via the inverse function; however, as we know from Eq. (3.17), $z'_x = 1/x'_z$, which proves that the coefficient to dx is in fact equal to one. To have the left-hand side equal to the right one, we then need the expression in the square brackets to be zero, which gives us an interesting identity between various partial derivatives in this case:

$$\left(\frac{\partial x}{\partial y} \right)_z + \left(\frac{\partial x}{\partial z} \right)_y \left(\frac{\partial z}{\partial y} \right)_x = 0 . \tag{5.48}$$

Problem 5.38 Prove that Eq. (5.48) can also be written in a more symmetric form:

$$\left(\frac{\partial x}{\partial z}\right)_y \left(\frac{\partial z}{\partial y}\right)_x \left(\frac{\partial y}{\partial x}\right)_z = -1 \, . \tag{5.49}$$

Next, let us examine relationships between derivatives of four functions x, y, z and t, from which only two (any two) are independent, i.e. any other two can be considered as functions of them. Then, we can write, considering t as a function of x and y and using dx from (5.47):

$$\begin{aligned}
dt &= \left(\frac{\partial t}{\partial x}\right)_y dx + \left(\frac{\partial t}{\partial y}\right)_x dy \\
&= \left(\frac{\partial t}{\partial x}\right)_y \left[\left(\frac{\partial x}{\partial y}\right)_z dy + \left(\frac{\partial x}{\partial z}\right)_y dz\right] + \left(\frac{\partial t}{\partial y}\right)_x dy \\
&= \left[\left(\frac{\partial t}{\partial x}\right)_y \left(\frac{\partial x}{\partial y}\right)_z + \left(\frac{\partial t}{\partial y}\right)_x\right] dy + \left(\frac{\partial t}{\partial x}\right)_y \left(\frac{\partial x}{\partial z}\right)_y dz \, .
\end{aligned}$$

In the last term, y is kept constant and hence t can be considered a function of $x = x(z)$; hence the coefficient to dz can be recognised to be the chain rule for $(\partial t / \partial z)_y$:

$$dt = \left[\left(\frac{\partial t}{\partial x}\right)_y \left(\frac{\partial x}{\partial y}\right)_z + \left(\frac{\partial t}{\partial y}\right)_x\right] dy + \left(\frac{\partial t}{\partial z}\right)_y dz \, .$$

Now, we compare this result with the one in which $t = t(y, z)$:

$$dt = \left(\frac{\partial t}{\partial y}\right)_z dy + \left(\frac{\partial t}{\partial z}\right)_y dz \, .$$

Comparison immediately yields a useful identity:

$$\left(\frac{\partial t}{\partial y}\right)_z = \left(\frac{\partial t}{\partial y}\right)_x + \left(\frac{\partial t}{\partial x}\right)_y \left(\frac{\partial x}{\partial y}\right)_z \, . \tag{5.50}$$

Numerous applications of the derived identities (5.48) and (5.50) can be found in thermodynamics, and several typical examples of these the reader can find in the next section.

5.7   Applications in Thermodynamics

The formalism we have developed is quite useful in thermodynamics. To illustrate these kinds of applications, let us assume that the state of a system is uniquely determined by a pair of variables selected from the following four of them: pressure P, temperature T, volume V and entropy S. Once a particular pair of variables is chosen, any other variable appears a function of these two. For each pair, the so-called *thermodynamic potential* is defined, which conveniently describes the state of the system. The thermodynamic potential $A = A(X, Y)$ is a unique function of the chosen two variables X and Y and hence its total differential has the form

$$dA = \left(\frac{\partial A}{\partial X}\right)_Y dX + \left(\frac{\partial A}{\partial Y}\right)_X dY \ . \tag{5.51}$$

Above, for convenience, as was done in the previous section and is customarily done in thermodynamics, we have written partial derivatives with a subscript indicating which particular variable is kept constant. We shall constantly use these notations throughout this section. Note that above dA is the exact differential with respect to the corresponding variables X and Y.

We shall now consider four most useful thermodynamic potentials indicating in each case the appropriate pair of variables corresponding to them: (i) the Helmholtz free energy

$$F = U - TS \quad \text{[variables } T, V] \ , \tag{5.52}$$

where U is the internal energy; (ii) the Gibbs free energy

$$G = U + PV - TS \quad \text{[variables } T, P] \ ; \tag{5.53}$$

(iii) the enthalpy

$$H = U + PV \quad \text{[variables } S, P] \ ; \tag{5.54}$$

and, finally, (iv) internal energy U for which the appropriate variables are S and V. Note that entropy can also be considered as a thermodynamic potential.

From the second law of thermodynamics, it follows that $TdS = dU + PdV$, meaning physically that the applied heat $\Delta Q = TdS$ during an infinitesimal process, in general, goes towards increasing the system free energy dU and for doing some work PdV. Note that the heat Q is not a thermodynamic potential and its change $\Delta Q = TdS$ is *not* an exact differential (that is why instead of dQ we write ΔQ), but $dS = \Delta Q/T$ is.[7]

Let us now write explicitly total differentials for each thermodynamic potential and establish simple exact relationships between various thermodynamic quantities.

---

[7]The factor $1/T$, which turns an inexact differential into an exact one, is called the integrating factor. We shall come across these when considering differential equations.

Subtracting[8] $d(TS) = TdS + SdT$ from dU, we obtain $dU - d(TS) = d(U-TS) \equiv dF$, which, from the second law above, yields

$$dF = dU - d(TS) = (TdS - PdV) - (TdS + SdT) = -SdT - PdV . \quad (5.55)$$

Since dF is an exact differential, the following identities must then be rigorously obeyed:

$$S = -\left(\frac{\partial F}{\partial T}\right)_V \quad \text{and} \quad P = -\left(\frac{\partial F}{\partial V}\right)_T . \quad (5.56)$$

Note that the last equation is in fact an *equation of state* of the system, which relates P, T and V together, making any one of them a function of the other two. Since the mixed derivatives do not depend on the order in which each individual partial derivative is taken, one more identity can be established here. Indeed,

$$\left(\frac{\partial S}{\partial V}\right)_T = -\frac{\partial^2 F}{\partial V \partial T} , \quad \left(\frac{\partial P}{\partial T}\right)_V = -\frac{\partial^2 F}{\partial T \partial V} \quad \Longrightarrow \quad \left(\frac{\partial S}{\partial V}\right)_T = \left(\frac{\partial P}{\partial T}\right)_V . \quad (5.57)$$

This last formula is called the Maxwell relation.

Problem 5.39 Establish similarly exact differentials and the corresponding identities for the internal energy:

$$dU = TdS - PdV , \quad T = \left(\frac{\partial U}{\partial S}\right)_V , \quad P = -\left(\frac{\partial U}{\partial V}\right)_S , \quad \left(\frac{\partial T}{\partial V}\right)_S = -\left(\frac{\partial P}{\partial S}\right)_V . \quad (5.58)$$

Problem 5.40 Similarly for the Gibbs free energy:

$$dG = VdP - SdT , \quad V = \left(\frac{\partial G}{\partial P}\right)_T , \quad S = -\left(\frac{\partial G}{\partial T}\right)_P , \quad \left(\frac{\partial V}{\partial T}\right)_P = -\left(\frac{\partial S}{\partial P}\right)_T . \quad (5.59)$$

Problem 5.41 Similarly for the enthalpy:

$$dH = TdS + VdP , \quad T = \left(\frac{\partial H}{\partial S}\right)_P , \quad V = \left(\frac{\partial H}{\partial P}\right)_S , \quad \left(\frac{\partial T}{\partial P}\right)_S = \left(\frac{\partial V}{\partial S}\right)_P . \quad (5.60)$$

---

[8]We are using here the product rule written directly for the differential of the function TS.

Problem 5.42 Prove the following relationships:

$$U = F - T\left(\frac{\partial F}{\partial T}\right)_V, \quad \left(\frac{\partial U}{\partial V}\right)_T = -P + T\left(\frac{\partial P}{\partial T}\right)_V, \quad \left[\frac{\partial}{\partial T}\left(\frac{F}{T}\right)\right]_V = -\frac{U}{T^2};$$

$$(5.61)$$

$$H = G - T\left(\frac{\partial G}{\partial T}\right)_P, \quad \left(\frac{\partial H}{\partial P}\right)_T = V - T\left(\frac{\partial V}{\partial T}\right)_P, \quad \left[\frac{\partial}{\partial T}\left(\frac{G}{T}\right)\right]_P = -\frac{H}{T^2}.$$

$$(5.62)$$

At the next step, we introduce various thermodynamic observable quantities, which can be measured experimentally. These are: (i) heat capacities at constant volume and pressure, c_V and c_P, defined via

$$c_V = \left(\frac{\partial Q}{\partial T}\right)_V = \left(\frac{\partial U}{\partial T}\right)_V = T\left(\frac{\partial S}{\partial T}\right)_V,$$

$$c_P = \left(\frac{\partial Q}{\partial T}\right)_P = \left(\frac{\partial H}{\partial T}\right)_P = T\left(\frac{\partial S}{\partial T}\right)_P.$$

$$(5.63)$$

Note that when V is constant, $\Delta Q = dU = TdS$, while when P is constant,

$$\Delta Q = dU + PdV = d(U + PV) = dH = TdS,$$

as follows from Eqs. (5.58) and (5.60)); (ii) thermal expansion and thermal pressure coefficients

$$\alpha_V = \frac{1}{V}\left(\frac{\partial V}{\partial T}\right)_P \quad \text{and} \quad \alpha_P = \frac{1}{P}\left(\frac{\partial P}{\partial T}\right)_V; \qquad (5.64)$$

(iii) isothermal and adiabatic compressibilites

$$\gamma_T = -\frac{1}{V}\left(\frac{\partial V}{\partial P}\right)_T \quad \text{and} \quad \gamma_S = -\frac{1}{V}\left(\frac{\partial V}{\partial P}\right)_S. \qquad (5.65)$$

Simple relationships exist between all these quantities, which can be established using the mathematical devices, we developed at the end of the previous Section, see Eqs. (5.48) and (5.50). We start by considering V as a function of P and T, i.e. $V = V(T, P)$, and applying Eq. (5.48) with the correspondence: $(V, T, P) \implies (x, z, y)$. We have:

$$\left(\frac{\partial V}{\partial T}\right)_P\left(\frac{\partial T}{\partial P}\right)_V + \left(\frac{\partial V}{\partial P}\right)_T = 0 \implies \frac{(\partial V/\partial T)_P}{(\partial P/\partial T)_V} + \left(\frac{\partial V}{\partial P}\right)_T = 0,$$

where we inverted the derivative $(\partial T/\partial P)_V$ using the inverse function relationship. Upon employing the definitions for α_V, γ_T and α_P given above, this immediately yields:

$$\frac{V\alpha_V}{P\alpha_P} - V\gamma_T = 0 \quad \Longrightarrow \quad \alpha_V = P\alpha_P\gamma_T \ . \tag{5.66}$$

Another useful relationship is obtained by first applying (5.48) with the correspondence $(S, P, V) \Longrightarrow (x, y, z)$, giving

$$\left(\frac{\partial S}{\partial V}\right)_P \left(\frac{\partial V}{\partial P}\right)_S + \left(\frac{\partial S}{\partial P}\right)_V = 0 \quad \Longrightarrow \quad \left(\frac{\partial V}{\partial P}\right)_S = -\frac{(\partial S/\partial P)_V}{(\partial S/\partial V)_P} \ ; \tag{5.67}$$

similarly, using the correspondence $(P, V, T) \Longrightarrow (y, z, x)$, we get

$$\left(\frac{\partial T}{\partial V}\right)_P \left(\frac{\partial V}{\partial P}\right)_T + \left(\frac{\partial T}{\partial P}\right)_V = 0 \quad \Longrightarrow \quad \left(\frac{\partial V}{\partial P}\right)_T = -\frac{(\partial T/\partial P)_V}{(\partial T/\partial V)_P} = -\frac{(\partial V/\partial T)_P}{(\partial P/\partial T)_V} \ . \tag{5.68}$$

In the last passage, we reversed both derivatives using the inverse function relation. Therefore,

$$\frac{\gamma_S}{\gamma_T} = \frac{(\partial V/\partial P)_S}{(\partial V/\partial P)_T} = \frac{(\partial S/\partial P)_V \, (\partial P/\partial T)_V}{(\partial S/\partial V)_P \, (\partial V/\partial T)_P} = \frac{(\partial S/\partial T)_V}{(\partial S/\partial T)_P} \equiv \frac{c_V}{c_P} \ .$$

So far we have used formula (5.48) which connects three quantities, with any two being independent. Let us now illustrate an application of Eq. (5.50) which connects four functions. We choose $(S, T, P, V) \Longrightarrow (t, y, z, x)$, which gives:

$$\left(\frac{\partial S}{\partial T}\right)_P = \left(\frac{\partial S}{\partial T}\right)_V + \left(\frac{\partial S}{\partial V}\right)_T \left(\frac{\partial V}{\partial T}\right)_P = \left(\frac{\partial S}{\partial T}\right)_V + \left(\frac{\partial P}{\partial T}\right)_V \left(\frac{\partial V}{\partial T}\right)_P \ ,$$

where we have used (5.57) in the last passage. Using now definitions of the heat capacities c_V and c_P, as well as of α_P and α_V, we finally obtain:

$$c_P = c_V + TVP\alpha_V\alpha_P = c_V + \frac{TV\alpha_V^2}{\gamma_T} \ ,$$

where we have used Eq. (5.66) in the last passage.

Problem 5.43 Prove the following relationship:

$$\gamma_T = \gamma_S + \frac{TV\alpha_V^2}{c_P} \ .$$

[Hint: *first, use formula (5.50) with the correspondence* $(V, P, S, T) \implies (t, y, z, x)$; *then employ Eq. (5.48) for* $(T, P, S) \implies (x, y, z)$ *and the last formula in Eq. (5.59).*]

Problem 5.44 Prove that

$$\left(\frac{\partial S}{\partial T}\right)_V = \left(\frac{\partial S}{\partial T}\right)_P + \left(\frac{\partial P}{\partial T}\right)_V \left(\frac{\partial S}{\partial P}\right)_T .$$

5.8 Directional Derivative and the Gradient of a Scalar Field

Here, we shall generalise further our definition of the partial derivative. Indeed, the partial derivatives introduced above for a function of many variables correspond to the rate of change of the function along the Cartesian axes. In fact, it is also possible to define the partial derivative along an arbitrary direction in space. This is what we shall accomplish here.

Suppose there is a scalar function $U(\mathbf{r}) = U(x, y, z)$ (it is frequently said a "*scalar field*") that is continuous in some region V in a three-dimensional space (generalisation to spaces of arbitrary dimensions is straightforward). Consider a change (per unit length) of U along some curved line L between points M_0 and M:

$$\frac{\Delta U}{\Delta l} = \frac{U(M) - U(M_0)}{\Delta l} ,$$

where Δl is the length $M_0 M$ along the line L. Now, suppose the line L is defined parametrically in a natural way as $x = x(l)$, $y = y(l)$ and $z = z(l)$ via the length l along the curve (measured relative to some arbitrary fixed point on the curve); the value l corresponds to the point M_0, while $l + \Delta l$ to the point M as shown in Fig. 5.5. This means that the change of U along the curve can be written, in the limit of the point M approaching M_0, as

$$\frac{\partial U}{\partial l} = \lim_{\Delta l \to 0} \frac{\Delta U}{\Delta l} = \lim_{\Delta l \to 0} \frac{U(\mathbf{r}(l + \Delta l)) - U(\mathbf{r}(l))}{\Delta l} = \frac{\partial U}{\partial x} \frac{dx}{dl} + \frac{\partial U}{\partial y} \frac{dy}{dl} + \frac{\partial U}{\partial z} \frac{dz}{dl}$$

by the rule of differentiation of a composite function $U(x(l), y(l), z(l))$. The derivatives dx/dl, dy/dl and dz/dl correspond to the cosines of the tangent vector $d\mathbf{l}$ to the curve L at the point M_0 with the x, y and z axes, see Fig. 5.5. Therefore,

$$\frac{\partial U}{\partial l} = \frac{\partial U}{\partial x} l_x + \frac{\partial U}{\partial y} l_y + \frac{\partial U}{\partial z} l_z = \frac{\partial U}{\partial x} \cos \alpha + \frac{\partial U}{\partial y} \cos \beta + \frac{\partial U}{\partial z} \cos \gamma , \quad (5.69)$$

where $\mathbf{l} = (l_x, l_y, l_z)$ is the unit vector tangent to the curve L at point M_0, and α, β and γ are the angles it forms with the three Cartesian axes.

Fig. 5.5 To the calculation
of the directional derivative
of the scalar field $U(\mathbf{r})$ along
the curve L

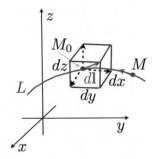

Problem 5.45 Prove that

$$\cos^2 \alpha + \cos^2 \beta + \cos^2 \gamma = 1 \; ,$$

i.e. the vector $\mathbf{l}$ is indeed a unit vector.

The derivative considered above is called *directional derivative* of the scalar field U along the curve L. It depends only on the point M_0 and the chosen direction $\mathbf{l}$. For some directions, the directional derivative at the same point M_0 can be smaller, for some larger. Let us find the direction of the maximum *increase* of U at the point M_0. If we introduce the *gradient* of the scalar field U as the vector

$$\mathbf{g} = \operatorname{grad} U(\mathbf{r}) = \frac{\partial U}{\partial x}\mathbf{i} + \frac{\partial U}{\partial y}\mathbf{j} + \frac{\partial U}{\partial z}\mathbf{k} \; , \qquad (5.70)$$

then it follows from Eq. (5.69) that the directional derivative can be written as a scalar product

$$\frac{\partial U}{\partial l} = (\operatorname{grad} U, \mathbf{l}) = \mathbf{g} \cdot \mathbf{l} = |\mathbf{g}| \cos \vartheta \; , \qquad (5.71)$$

where ϑ is the angle between $\mathbf{g}$ and $\mathbf{l}$. It is seen that the maximum value the directional derivative is obtained when the direction $\mathbf{l}$ is chosen exactly along the gradient vector $\mathbf{g}$. Thus, the gradient of a scalar field points in the direction of the most rapid increase of the field, and its magnitude gives the rate of the increase. Correspondingly, the most rapid decrease of U is provided by $-\operatorname{grad} U$, i.e. the direction which is opposite to the gradient.

The gradient of U can also be written, quite formally, as a product of the so-called *del operator*

$$\nabla = \frac{\partial}{\partial x}\mathbf{i} + \frac{\partial}{\partial y}\mathbf{j} + \frac{\partial}{\partial z}\mathbf{k} \qquad (5.72)$$

and the field U. Indeed,

$$\nabla U = \left(\frac{\partial}{\partial x} \mathbf{i} + \frac{\partial}{\partial y} \mathbf{j} + \frac{\partial}{\partial z} \mathbf{k} \right) U = \frac{\partial U}{\partial x} \mathbf{i} + \frac{\partial U}{\partial y} \mathbf{j} + \frac{\partial U}{\partial z} \mathbf{k} ,$$

which is exactly grad U. This notation is quite useful and is frequently used.

As an example, let us calculate the gradient of $U(\mathbf{r}) = f(r)$, where $f(r)$ is some function of $r = |\mathbf{r}|$. The calculation of the gradient requires calculating derivatives of the function $f(r)$ with respect to Cartesian components x, y and z; these are easily done since $r = \sqrt{x^2 + y^2 + z^2}$:

$$\frac{\partial f}{\partial x} = \frac{df}{dr} \frac{\partial r}{\partial x} = \frac{df}{dr} \frac{x}{r} , \quad \frac{\partial f}{\partial y} = \frac{df}{dr} \frac{y}{r} \quad \text{and} \quad \frac{\partial f}{\partial z} = \frac{df}{dr} \frac{z}{r} ,$$

so that

$$\nabla U = \frac{\partial f}{\partial x} \mathbf{i} + \frac{\partial f}{\partial y} \mathbf{j} + \frac{\partial f}{\partial z} \mathbf{k} = \frac{df}{dr} \frac{x\mathbf{i} + y\mathbf{j} + z\mathbf{k}}{r} = \frac{df}{dr} \frac{\mathbf{r}}{r} ,$$

i.e. the gradient of a scalar field which depends only on the distance to the centre of the coordinate system has a radial character, i.e. it is directed along the vector $\mathbf{r}$.

Problem 5.46 The gravitational potential of a galaxy decays with the distance r from it as $\varphi(r) = -\alpha/r$, where α is a positive constant. Calculate the force $\mathbf{F} = -m\nabla\varphi(r)$ with which the galaxy acts on a spaceship of mass m at distance r. What is the direction of the force? State whether the interaction between the two objects corresponds to repulsion or attraction.[Answer: $\mathbf{F} = -m\alpha\mathbf{r}/r^3$.]

Problem 5.47 Calculate the directional derivative of $\phi(x, y, z) = xyz$ at the point $(1, -1, 0)$ in the direction $\mathbf{e} = (1, 1, 1)$. [Answer: $-1/\sqrt{3}$.]

Problem 5.48 Consider the scalar field $\phi(x, y, z) = 2xy^3 + z^2$. (a) Evaluate the derivative of ϕ along the direction $\mathbf{e} = (1, 2, 0)$ at arbitrary point (x, y, z); (b) find the direction of the most rapid *increase* of ϕ at the point $(2, 2, 0)$; (c) what is the direction of the most rapid *decrease* of ϕ at the same point? [Answer: (a) $2y^2 (y + 6x)/\sqrt{5}$; (b) $(16, 48, 0)$; (c) $(-16, -48, 0)$.]

Problem 5.49 Given $\phi = x^2 - y^2 z$, calculate $\nabla\phi$ at point $(1, 1, 1)$. Hence, find the directional derivative of ϕ at that point in the direction $\mathbf{e} = (1, -2, 1)$. [Answer: $(2, -2, -1)$ *and* $5/\sqrt{6}$.]

The gradient (in fact, its generalisation to many variables) is useful when one wants to find a minimum (or maximum) of a function $f(\mathbf{r})$ in p dimensions. Indeed, if we start from some initial point $\mathbf{r}_0$, then $\mathbf{g}_0 = -\text{grad} f(\mathbf{r})$ would indicate the direction of the most rapid decrease of f from that point. Take a step $\delta\mathbf{r} = \lambda_1 \mathbf{g}_0$

along $\mathbf{g}_0$ with some λ_1 (the step length) and arrive at the point $\mathbf{r}_1 = \mathbf{r}_0 + \delta\mathbf{r}$. Calculate the gradient there, $\mathbf{g}_1 = -\text{grad} f(\mathbf{r}_1)$, and then move to the next point $\mathbf{r}_2 = \mathbf{r}_1 + \lambda_2\mathbf{g}_1$, and so on. Near the minimum, the gradient will be very small (exactly equal to zero at the minimum, see Sect. 5.10), the steps $\delta\mathbf{r}$ would become smaller and smaller, so that after some finite number of steps a point very close to the exact minimum will be reached. This method is called *steepest descent*. Note that the steps λ_1, λ_2, etc. should be taken small enough as compared to the characteristic dimensions of the well of the function $f(\mathbf{r})$ containing the minimum point.

For instance, this method may be used to calculate the mechanical equilibrium of a collection of atoms (e.g. in a molecule or a solid). This is because in physics the gradient of an external potential $U(\mathbf{r})$ is related to the force $\mathbf{F}$ acting on a particle moving in it via $\mathbf{F} = -\text{grad}\,U = -\partial U/\partial\mathbf{r} = -\nabla U$ (all these notations are often used). If we can calculate the total potential energy of a set of atoms, then the forces acting on atoms at the given atomic configuration are determined by the components of the minus gradient of the energy. Moving atoms in the direction of forces leads to a lower energy. To find mechanical equilibrium of the system of atoms at zero temperature, one has to move atoms in the direction of forces unless the forces become less than some predefined tolerance. This idea is at the heart of all modern methods of material modelling.

To finish this Section, we shall mention one more application of the gradient. Consider a surface in 3D that is specified by the equation $f(x, y, z) = C$, where C is a constant (e.g. for a sphere $x^2 + y^2 + z^2 = R^2$).

Theorem 5.4 *The normal vector to the surface $f(x, y, z) = 0$ can be obtained from the gradient of the function $f(x, y, z)$.*

Proof: Because of the equation $f(x, y, z) = 0$, only two coordinates are independent on the surface. Let us assume that the coordinates x and y are independent, i.e. $z = z(x, y)$. Then, consider the gradient of f:

$$\nabla f = \frac{\partial f}{\partial x}\mathbf{i} + \frac{\partial f}{\partial y}\mathbf{j} + \frac{\partial f}{\partial z}\mathbf{k} \ .$$

To calculate the necessary derivatives, we first differentiate both sides of the equation $f(x, y, z) = f(x, y, z(x, y)) = 0$ with respect to x and y, taking into account the $z = z(x, y)$ dependence explicitly:

$$\left(\frac{\partial f}{\partial x}\right)_{tot} = \frac{\partial f}{\partial x} + \frac{\partial f}{\partial z}\frac{\partial z}{\partial x} = 0 \quad \text{and} \quad \left(\frac{\partial f}{\partial y}\right)_{tot} = \frac{\partial f}{\partial y} + \frac{\partial f}{\partial z}\frac{\partial z}{\partial y} = 0$$

(zeros in the right-hand sides are due to the zero right-hand side of $f(x, y, z) = 0$). Hence, the partial derivatives entering the gradient above are:

$$\frac{\partial f}{\partial x} = -\frac{\partial f}{\partial z}\frac{\partial z}{\partial x} \quad \text{and} \quad \frac{\partial f}{\partial y} = -\frac{\partial f}{\partial z}\frac{\partial z}{\partial y},$$

and hence the gradient becomes:

$$\nabla f = \frac{\partial f}{\partial z}\left(-\frac{\partial z}{\partial x}\mathbf{i} - \frac{\partial z}{\partial y}\mathbf{j} + \mathbf{k}\right).$$

Comparing this result with Eq. (5.18), we immediately see that ∇f is indeed directed along the normal $\mathbf{n}$ to the surface $z = z(x, y)$ (or $f(x, y, z) = 0$). **Q. E. D.**

As an example of this application of the gradient, we calculate the normal to the surface $x^2 + y^2 = z$ at the point $A(1, 1, 2)$. Here, $f(x, y, z) = x^2 + y^2 - z$ and $\nabla f = 2x\mathbf{i} + 2y\mathbf{j} - \mathbf{k}$, so that the unit normal (of unit length)

$$\mathbf{n} = \frac{\nabla f}{|\nabla f|} = \frac{2x\mathbf{i} + 2y\mathbf{j} - \mathbf{k}}{\sqrt{4x^2 + 4y^2 + 1}},$$

which at the point A takes on the value of $\mathbf{n}(A) = 2(\mathbf{i} + \mathbf{j})/3 - \mathbf{k}/3$.

Problem 5.50 Using the definition of the gradient, prove its following properties:

$$\nabla(U + V) = \nabla U + \nabla V, \quad \nabla(UV) = U\nabla V + V\nabla U, \quad \nabla U(V) = \frac{dU}{dV}\nabla V,$$
$$(5.73)$$

where U and V are scalar fields.

5.9   Taylor's Theorem for Functions of Many Variables

Similarly to the case of functions of a single variable considered in Sect. 3.8, there is an equivalent formula for functions of many variables. For simplicity, we shall only consider the case of a function $z = f(x, y)$ of two variables; the formulae for functions of more than two variables can be obtained similarly although this may be quite cumbersome.

What we would like to do is to write an expansion for $f(x_0 + \Delta x, y_0 + \Delta y)$ around the point (x_0, y_0) in terms of $\Delta x = x - x_0$ and $\Delta y = y - y_0$ which are variations of the function arguments around that point. Instead of generalising the method we applied in Sect. 3.8, we shall use a very simple trick which makes use of the Taylor formula (3.84) for the functions of a single variable which we already know. To this end, we construct the following auxiliary function

$$F(t) = f(x_0 + t\Delta x, y_0 + t\Delta y) \tag{5.74}$$

of a single variable $0 \le t \le 1$. The required change Δz of our function $z = f(x, y)$ is obtained by calculating $F(1)$, i.e. by taking $t = 1$ in $F(t)$, and then taking away $f(x_0, y_0) = F(0)$. Sine $F(t)$ is a function of a single variable, it can be expanded according to the Maclaurin formula (3.86), i.e. the Taylor's formula around the point $t = 0$, in terms of $\Delta t = t$:

$$F(t) = F(0) + F'(0)t + \frac{F''(0)t^2}{2} + \cdots + \frac{F^{(n)}(0)t^n}{n!} + \frac{F^{(n+1)}(\xi)t^{n+1}}{(n+1)!} , \tag{5.75}$$

where the last term is the remainder $R_{n+1}(\xi)$ with $\xi = \vartheta t$ and $0 < \vartheta < 1$. By taking $t = 1$, the function $F(1)$ becomes equal to $f(x_0 + \Delta x, y_0 + \Delta y) = f(x, y)$, and the expansion above turns into the required Taylor expansion for the function $f(x, y)$ around the point (x_0, y_0).

To calculate $F'(t)$, we notice that $F(t)$ can be considered as a composite function $F(t) = f(x(t), y(t))$ with the arguments being linear functions of t, i.e. $x(t) = x_0 + t\Delta x$ and $y(t) = y_0 + t\Delta y$ with $\partial x/\partial t = \Delta x$ and $\partial y/\partial t = \Delta y$. Therefore, applying formula (5.25) for the derivative of a composite function of two variables $(N = 2)$, we have:

$$F'(t) = \frac{\partial F}{\partial x}\frac{\partial x}{\partial t} + \frac{\partial F}{\partial y}\frac{\partial y}{\partial t} = \frac{\partial F}{\partial x}\Delta x + \frac{\partial F}{\partial y}\Delta y = \left(\Delta x\frac{\partial}{\partial x} + \Delta y\frac{\partial}{\partial y}\right)F(t) . \tag{5.76}$$

Here, in the last passage, we employed a simplified notation using an operator (within the brackets) acting on the function $F(t)$. This notation, as will be seen presently, proves to be very convenient in writing higher order derivatives. Indeed, the second derivative can be calculated by differentiating the just obtained first derivative and noticing that both functions $\partial F/\partial x$ and $\partial F/\partial y$ are again composite functions of t:

$$
\begin{aligned}
F''(t) &= \frac{d}{dt}\left(\frac{\partial F}{\partial x}\Delta x + \frac{\partial F}{\partial y}\Delta y\right) = \Delta x\frac{d}{dt}\left(\frac{\partial F}{\partial x}\right) + \Delta y\frac{d}{dt}\left(\frac{\partial F}{\partial y}\right) \\
&= \Delta x\left(\frac{\partial^2 F}{\partial x^2}\Delta x + \frac{\partial^2 F}{\partial y\partial x}\Delta y\right) + \Delta y\left(\frac{\partial^2 F}{\partial x\partial y}\Delta x + \frac{\partial^2 F}{\partial y^2}\Delta y\right) \\
&= (\Delta x)^2\frac{\partial^2 F}{\partial x^2} + 2\Delta x\Delta y\frac{\partial^2 F}{\partial x\partial y} + (\Delta y)^2\frac{\partial^2 F}{\partial y^2} \\
&= \left(\Delta x\frac{\partial}{\partial x} + \Delta y\frac{\partial}{\partial y}\right)^2 F(t) ,
\end{aligned}
$$

where again we have written the result of the differentiation using an operator; in fact, this is the same operator as in our result (5.76) for the first derivative, but squared.[9]

Problem 5.51 Use induction to prove that generally

$$F^{(n)}(t) = \left(\Delta x \frac{\partial}{\partial x} + \Delta y \frac{\partial}{\partial y} \right)^n F(t) \tag{5.77}$$

for any $n = 1, 2, 3, \ldots$.

The obtained formulae allow us to write:

$$
\begin{aligned}
F^{(n)}(0) &= \left(\Delta x \frac{\partial}{\partial x} + \Delta y \frac{\partial}{\partial y} \right)^n F(t) \bigg|_{t=0} \\
&= \left(\Delta x \frac{\partial}{\partial x} + \Delta y \frac{\partial}{\partial y} \right)^n f(x, y) \bigg|_{x=x_0, y=y_0} ,
\end{aligned} \tag{5.78}
$$

while the remainder term at $t = 1$ yields

$$
\begin{aligned}
R_{n+1}(\vartheta) &= \frac{1}{(n+1)!} \left(\Delta x \frac{\partial}{\partial x} + \Delta y \frac{\partial}{\partial y} \right)^{n+1} F(\vartheta t) \bigg|_{t=1} \\
&= \frac{1}{(n+1)!} \left(\Delta x \frac{\partial}{\partial x} + \Delta y \frac{\partial}{\partial y} \right)^{n+1} f(x_0 + \vartheta \Delta x, y_0 + \vartheta \Delta y) .
\end{aligned} \tag{5.79}
$$

Correspondingly, the Taylor expansion, after taking $t = 1$ in Eq. (5.75), reads:

$$
\begin{aligned}
f(x, y) &= f(x_0, y_0) + F'(0) + \frac{F''(0)}{2} + \cdots + \frac{F^{(n)}(0)}{n!} + R_{n+1} \\
&= f(x_0, y_0) + \sum_{k=1}^{n} \frac{1}{k!} \left(\Delta x \frac{\partial}{\partial x} + \Delta y \frac{\partial}{\partial y} \right)^n f(x, y) \bigg|_{x=x_0, y=y_0} + R_{n+1} ,
\end{aligned} \tag{5.80}
$$

which is our final result.

The formulae we have obtained here can in fact be generalised to functions of any number of variables; this however requires developing more algebraic results; in particular, a generalisation of the Binomial expansion to forms $(a_1 + \cdots + a_N)^n$ containing $N > 2$ terms (this has been done in Sect. 1.16) is required and an investigation of the properties of the corresponding generalised multinomial coefficients. Then, the method developed above can be basically repeated almost without change.

---

[9]Recall that a similar trick with operators we have already used in Sect. 3.7.4 when illustrating the derivation of the Leibniz formula (3.71).

To see where this is heading, let us write down a few first terms of the above expansion for the function $f(\mathbf{r}) = f(x_1, x_2)$ of two variables $\mathbf{r} = (x_1, x_2)$ around the point $\mathbf{a} = (a_1, a_2)$ explicitly:

$$f(\mathbf{r}) = f(\mathbf{a}) + \sum_{k=1}^{2} \left(\frac{\partial f}{\partial x_k}\right)_{\mathbf{a}} \Delta x_k + \frac{1}{2!} \sum_{k,k'=1}^{2} \left(\frac{\partial^2 f}{\partial x_k \partial x_{k'}}\right)_{\mathbf{a}} \Delta x_k \Delta x_{k'}$$

$$+ \frac{1}{3!} \sum_{k,k',k''=1}^{2} \left(\frac{\partial^3 f}{\partial x_k \partial x_{k'} \partial x_{k''}}\right)_{\mathbf{a}} \Delta x_k \Delta x_{k'} \Delta x_{k''} + \cdots + R_{n+1},$$

where the derivatives are calculated at the point $\mathbf{a}$ and $\Delta x_k = x_k - a_k$. Written in this form, the Taylor formula can immediately be generalised to a function of $N > 2$ variables by simply extending all summations to N.

As an important example, let us derive an analogue of the Lagrange formula (3.88) for the case of a function of two variables. Applying Eqs. (5.77)–(5.80) for $n = 0$, we get:

$$f(x, y) = f(x_0, y_0) + \left(\Delta x \frac{\partial}{\partial x} + \Delta y \frac{\partial}{\partial y}\right) f(x_0 + \vartheta \Delta x, y_0 + \vartheta \Delta y)$$

$$= f(x_0, y_0) + f'_x(x', y') \Delta x + f'_y(x', y') \Delta y, \qquad (5.81)$$

which is the required result. Here, we differentiate $f(x, y)$ with respect to x and y and then replace the variables according to $x \to x' = x_0 + \vartheta \Delta x$ and $y \to y' = y_0 + \vartheta \Delta x$, respectively. Another important formula that is frequently needed corresponds to the $n = 1$ case:

$$f(x, y) = f(x_0, y_0) + \left[f'_x(x_0, y_0) \Delta x + f'_y(x_0, y_0) \Delta y\right]$$

$$+ \frac{1}{2}\left[f''_{xx}(x', y')(\Delta x)^2 + 2 f''_{xy}(x', y') \Delta x \Delta y + f''_{yy}(x', y')(\Delta y)^2\right], \quad (5.82)$$

where the second derivatives are calculated at $x' = x_0 + \vartheta \Delta x$ and $y' = y_0 + \vartheta \Delta y$.

Problem 5.52 Derive the $n = 2$ Taylor expansion of $f(x, y) = e^{x+y}$.

5.10   Introduction to Finding an Extremum of a Function

Now we are in a position to discuss finding maxima and minima (extrema) of functions of more than one variable. We shall only consider here in detail the case of functions of two variables, $z = f(x, y)$.

Similarly to functions of a single variable, we say that the function $z = f(x, y)$ has a maximum at point $M_0 (x_0, y_0)$, if there exists a positive radius R such that for *any* point $M(x, y)$ for which the distance to the point M_0 is smaller than R (i.e. $d(M, M_0) = \sqrt{(x - x_0)^2 + (y - y_0)^2} < R$), the value of the function $f(x, y) < f(x_0, y_0)$. The definition of the minimum is defined similarly.

5.10.1 Necessary Condition: Stationary Points

The necessary condition for the minimum or maximum can easily be established. Consider Fig. 5.6 where the function $z = f(x, y)$ has a maximum at point M_0. If we fix the first variable x of the function at x_0 and allow only the other variable to change, we would draw the line AB on the surface associated with our function; this line will correspond to the function $g(y) = f(x_0, y)$, which is a function of only a single variable y. Since for any point y within some vicinity of y_0 the value of that function is smaller than its value $g(y_0) = f(x_0, y_0)$ (since according to our assumption the point M_0 is a maximum), the function $g(y)$ has a maximum at the point y_0. Therefore, from the familiar necessary condition for the maximum of a function of one variable, the derivative $g'(y)$ is equal to zero at $y = y_0$. But this derivative is in fact the partial derivative $f'_y (x_0, y)$ calculated at the point $y = y_0$. Similarly, fixing $y = y_0$ and considering the function $h(x) = f(x, y_0)$, we arrive at the line DC in Fig. 5.6 and the necessary condition $h'(x_0) = f'_x (x_0, y_0) = 0$. Therefore, we conclude that the necessary condition for a function to have a maximum is that its partial derivatives are equal to zero at that point:

$$f'_x (x_0, y_0) = f'_y (x_0, y_0) = 0 . \tag{5.83}$$

In the case of a minimum, a similar analysis yields exactly the same condition, but in this case, each line $x = x_0$ or $y = y_0$ would correspond to a minimum of a function of a single variable. We see that the condition (5.83) generalises directly the corresponding condition for functions of a single variable of Sect. 3.11. Solving two equations (5.83), gives the *stationary point* (x_0, y_0) which might be a minimum or maximum of the function. More than one solution is possible (or none[10]). Note that the stationary point(s) obtained may be either minimum, maximum or neither of these (see below) since these conditions are only the necessary ones.

It is also easy to realise that this simple result is immediately generalised to the case of a function of any number of variables $y = f(\mathbf{x})$, where $\mathbf{x} = (x_1, \ldots, x_N)$.

---

[10]If a function is defined within a finite region, than a maximum or minimum may still exist at the region boundary.

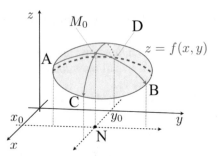

Fig. 5.6 For the proof of the necessary condition of a maximum of a function $z = f(x, y)$ of two variables. Here the function has a maximum at the point M_0 and the lines AB and DC along the surface $z = f(x, y)$, which is associated with the function, correspond to one variable to be fixed and another allowed to change

Indeed, if at the point $\mathbf{x}_0 = \left(x_1^0, \ldots, x_N^0\right)$ the function has a minimum (maximum), then by fixing all variables but one, say x_i, we arrive at a function of a single variable $g(x_i) = f\left(x_1^0, \ldots, x_{i-1}^0, x_i, x_{i+1}^0, \ldots, x_N^0\right)$, which has a minimum (maximum) at the point x_i^0, and hence the corresponding partial derivative must be zero:

$$\left(\frac{\partial g}{\partial x_i}\right)_{x_i^0} = f_{x_i}'\left(\mathbf{x}^0\right) = 0 \, . \tag{5.84}$$

By repeating this process for every $i = 1, \ldots, N$ we arrive at N such conditions, which give N equations; solving these equations yields the stationary point(s) $\mathbf{x}^0$ which might correspond to extrema of the function. Note that the total differential of the function at the stationary point is zero as all its partial derivatives vanish:

$$df = \sum_{i=1}^{N} \frac{\partial f}{\partial x_i} dx_i = 0 \, . \tag{5.85}$$

As an example, let us find the minimum distance d between two lines specified by the vector equations $\mathbf{r} = \mathbf{r}_1 + t_1\mathbf{a}_1$ and $\mathbf{r} = \mathbf{r}_2 + t_2\mathbf{a}_2$, where t_1 and t_2 are the corresponding parameters. We need to minimise a function of the square of the distance,

$$
\begin{aligned}
d^2(t_1, t_2) &= (\mathbf{r}_1 + t_1\mathbf{a}_1 - \mathbf{r}_2 - t_2\mathbf{a}_2)^2 = (x_1 + t_1 a_{1x} - x_2 - t_2 a_{2x})^2 \\
&\quad + \left(y_1 + t_1 a_{1y} - y_2 - t_2 a_{2y}\right)^2 + (z_1 + t_1 a_{1z} - z_2 - t_2 a_{2z})^2 \, ,
\end{aligned}
\tag{5.86}
$$

with respect to the two parameters t_1 and t_2. Calculating the partial derivatives and setting them to zero, two equations are obtained for the parameters:

$$\frac{\partial d^2}{\partial t_1} = 2\mathbf{a}_1 \cdot (\mathbf{r}_1 + t_1\mathbf{a}_1 - \mathbf{r}_2 - t_2\mathbf{a}_2) = 0 \qquad \Rightarrow t_1 a_1^2 - t_2\mathbf{a}_1 \cdot \mathbf{a}_2 = -\mathbf{a}_1 \cdot \mathbf{r}_{12} \, ,$$

$$\frac{\partial d^2}{\partial t_2} = -2\mathbf{a}_2 \cdot (\mathbf{r}_1 + t_1\mathbf{a}_1 - \mathbf{r}_2 - t_2\mathbf{a}_2) = 0 \qquad \Rightarrow t_1\mathbf{a}_1 \cdot \mathbf{a}_2 - t_2 a_2^2 = -\mathbf{a}_2 \cdot \mathbf{r}_{12} \, ,$$

where $\mathbf{r}_{12} = \mathbf{r}_1 - \mathbf{r}_2$, $a_1^2 = \mathbf{a}_1 \cdot \mathbf{a}_1$ and $a_2^2 = \mathbf{a}_2 \cdot \mathbf{a}_2$. Solving these equations with respect to the two parameters (e.g. solve for t_1 in the first equation and substitute into the second), we obtain

$$t_1 = \frac{-a_2^2 (\mathbf{a}_1 \cdot \mathbf{r}_{12}) + (\mathbf{a}_1 \cdot \mathbf{a}_2)(\mathbf{a}_2 \cdot \mathbf{r}_{12})}{a_1^2 a_2^2 - (\mathbf{a}_1 \cdot \mathbf{a}_2)^2}, \quad t_2 = \frac{a_1^2 (\mathbf{a}_2 \cdot \mathbf{r}_{12}) - (\mathbf{a}_1 \cdot \mathbf{a}_2)(\mathbf{a}_1 \cdot \mathbf{r}_{12})}{a_1^2 a_2^2 - (\mathbf{a}_1 \cdot \mathbf{a}_2)^2}.$$

Problem 5.53 Substituting the found parameters into the distance square in (5.86) show that the minimum distance between the two lines is indeed given correctly by Eq. (1.232).

5.10.2 Characterising Stationary Points: Sufficient Conditions

Of course, similar to the case of a single variable function, we still have to establish the sufficient conditions for the minimum and maximum. Recall that the function $y = x^3$ does have its first derivative $y' = 3x^2$ equal to zero at $x = 0$, however, this point is neither minimum nor maximum as its second derivative $y'' = 6x$ is neither positive nor negative at $x = 0$. Analogous to this situation is the saddle in Fig. 5.1(d). Indeed, the function $z = x^2 - y^2$ of the particular case of the hyperbolic paraboloid of Eq. (5.5) has both its partial derivatives $z'_x = 2x$ and $z'_y = -2y$ equal to zero at $x = y = 0$, however, this point is neither minimum nor maximum: it is minimum along some directions and a maximum along the others. If we, however, consider a similar function $z = x^2 + y^2$, then at the same point it obviously has a minimum: indeed, $z'_x = 2x$ and $z'_y = 2y$ give a single solution $x = y = 0$ at which $z(0, 0) = 0$, and at any other point the function $z(x, y) > 0$, i.e. the function at any point in the vicinity of the point $(0, 0)$ is definitely larger than zero and hence this point must be a minimum. Hence, the necessary conditions (5.84) must be supplemented with the corresponding *sufficient* conditions which should tell us if each point found by solving Eqs. (5.84) is minimum, maximum or neither of these.

Consider a function $z = f(x, y)$ which has both partial derivatives $f'_x(x_0, y_0) = f'_y(x_0, y_0) = 0$ at some stationary point (x_0, y_0). The change $\Delta z = f(x, y) - f(x_0, y_0)$ of the function at the point $M(x, y)$ somewhere in a close vicinity of the stationary point can then be written using the $n = 2$ case Taylor expansion in which the first-order terms should be dropped as the corresponding partial derivatives are zero:

$$\Delta z = \frac{1}{2} \left[F_{xx}^0 (\Delta x)^2 + 2F_{xy}^0 \Delta x \Delta y + F_{yy}^0 (\Delta y)^2 \right] + R_3(x_0, y_0, \Delta x, \Delta y), \quad (5.87)$$

where $R_3(x_0, y_0, \Delta x, \Delta y)$ is the remainder that tends to zero in the $\Delta x, \Delta y \to 0$ limit at least as their third power, and

$$F^0_{xx} = f''_{xx}(x_0, y_0) \ , \ \ F^0_{xy} = f''_{xy}(x_0, y_0) \ , \ \text{ and } F^0_{yy} = f''_{yy}(x_0, y_0)$$

are the appropriate second-order derivatives calculated at the point (x_0, y_0). The important point relevant for our discussion is that the remainder $R_3(x_0, y_0, \Delta x, \Delta y)$ is of a higher order than the second-order terms written explicitly in Eq. (5.87), and hence one can always choose such a vicinity of the point (x_0, y_0) that the contribution of these higher order terms is deemed to be less important than of the second-order terms. Hence, in most cases (see below), the terms $R_3(x_0, y_0, \Delta x, \Delta y)$ can be neglected.

Dropping the higher order terms, we can rearrange the second order terms in Eq. (5.87) to build the full square. Assuming first that $F^0_{xx} \neq 0$, we write:

$$\Delta z = \frac{F^0_{xx}}{2} \left\{ \left[(\Delta x)^2 + 2\frac{F^0_{xy}}{F^0_{xx}} \Delta x \Delta y + \left(\frac{F^0_{xy}}{F^0_{xx}} \Delta y \right)^2 \right] + \left[-\left(\frac{F^0_{xy}}{F^0_{xx}} \Delta y \right)^2 + \frac{F^0_{yy}}{F^0_{xx}} (\Delta y)^2 \right] \right\}$$

$$= \frac{F^0_{xx}}{2} \left[\left(\Delta x + \frac{F^0_{xy}}{F^0_{xx}} \Delta y \right)^2 + \left(\frac{\Delta y}{F^0_{xx}} \right)^2 \mathcal{D}^0 \right] ,$$

$$(5.88)$$

where $\mathcal{D}^0 = -\left(F^0_{xy} \right)^2 + F^0_{xx} F^0_{yy}$.

Assume first that $F^0_{xx} > 0$ and $\mathcal{D}^0 > 0$. It follows now from Eq. (5.88) that $\Delta z > 0$ for *any* deviation $(\Delta x, \Delta y)$ from the point (x_0, y_0), meaning that this point must be a minimum: indeed, by departing from that point in any direction, the function $f(x, y)$ only increases from its value $f(x_0, y_0)$. Conversely if $F^0_{xx} < 0$ (but still $\mathcal{D}^0 > 0$), then we have a maximum.

Next, consider the case of $\mathcal{D}^0 < 0$ with F^0_{xx} being still non-zero. This case corresponds to neither minimum nor maximum as we shall specifically investigate presently. Indeed, let us go along a particular direction in the $x - y$ plane, given by $\Delta y = k\Delta x$ with some fixed k (any real number). Along this direction, the change of the function (5.87) (with the higher order terms dropped) reads

$$\Delta z = \frac{1}{2} (\Delta x)^2 \left[F^0_{xx} + 2F^0_{xy} k + F^0_{yy} k^2 \right] .$$

The square polynomial in the brackets has two real roots $k_{1,2} = \left(-F^0_{xy} \pm \sqrt{D} \right) / F^0_{yy}$ since its discriminant $D = -\mathcal{D}^0 = \left(F^0_{xy} \right)^2 - F^0_{xx} F^0_{yy}$ is positive. This means that the parabola $g(k) = F^0_{xx} + 2F^0_{xy} k + F^0_{yy} k^2$ definitely crosses the k-axis at two points k_1 and k_2. Then, depending on the sign of F^0_{yy}, the parabola either has a minimum or maximum as a function of k; either way, however, it is positive for some values of k and negative for some others. Therefore, going along any direction of the k in which the parabola is positive, Δz increases (as $\Delta z \sim (\Delta x)^2$), while going along other directions in which the parabola is negative we have $\Delta z \sim -(\Delta x)^2$ and it is decreasing. Hence, in some directions, the function $f(x, y)$ goes up as we depart from the point (x_0, y_0), as in the others it is going down, i.e. the function has neither minimum nor maximum at this point.

There are a few more special cases that might arise. First of all, consider the case of $\mathcal{D}^0 = 0$ with $F_{xx}^0 \neq 0$. It is seen from Eq. (5.88), that in this case

$$\Delta z = \frac{F_{xx}^0}{2}\left(\Delta x + \frac{F_{xy}^0}{F_{xx}^0}\Delta y\right)^2 .$$

If $F_{xx}^0 > 0$ ($F_{xx}^0 < 0$), then for a sufficiently small departure from the point (x_0, y_0), and in *any* direction apart from one, the second-order terms in Δz will be positive (negative). Nevertheless, this case is special as there exist a direction, $k = -F_{xx}^0/F_{xy}^0$, such that

$$\Delta x + \frac{F_{xy}^0}{F_{xx}^0}\Delta y = \Delta x\left(1 + \frac{F_{xy}^0}{F_{xx}^0}k\right) = 0 ,$$

and hence the second-order approximation to Δz will be zero for *any* deviation Δx along this direction, at least in the close neighbourhood of the point (x_0, y_0). Hence, here higher order terms must be of importance. Therefore, no general conclusion can be made in this case and a special analysis is required in each particular case.

For all other possible cases, Δz changes sign in different directions $\Delta y = k\Delta x$ and hence the stationary point will be neither minimum nor maximum. Consider the case of $F_{xx}^0 = 0$, while $F_{yy}^0 \neq 0$. We can write keeping the second-order terms:

$$\Delta z = \frac{\Delta y}{2}\left(2F_{xy}^0\Delta x + F_{yy}^0\Delta y\right) = \frac{k}{2}\left(2F_{xy}^0 + kF_{yy}^0\right)(\Delta x)^2 .$$

It is clearly seen that, depending on the chosen direction k, the leading second-order terms in Δz may either make it positive or negative. Similarly, if $F_{xx}^0 = F_{yy}^0 = 0$, then

$$\Delta z = F_{xy}^0\Delta x\Delta y = F_{xy}^0 k(\Delta x)^2$$

and yet again, depending on the direction in which the point (x_0, y_0) is crossed, Δz might either be positive ($k > 0$), negative ($k < 0$) or zero ($k = 0$). Hence, in this case, we should not expect a minimum or a maximum.

Summarising, the question of whether the function $z = f(x, y)$ has a minimum or a maximum at the stationary point (x_0, y_0), where it has zero partial derivatives, is solved in the following way: it is a minimum if

$$F_{xx}^0 = f_{xx}'' > 0 \quad \text{and} \quad \mathcal{D}^0 = F_{xx}^0 F_{yy}^0 - \left(F_{xy}^0\right)^2 = \begin{vmatrix} f_{xx}'' & f_{xy}'' \\ f_{xy}'' & f_{yy}'' \end{vmatrix} > 0 \quad \text{(Min)} \quad (5.89)$$

at the point (x_0, y_0), while we have a maximum there if

$$F_{xx}^0 = f_{xx}'' < 0 \quad \text{and} \quad \mathcal{D}^0 = F_{xx}^0 F_{yy}^0 - \left(F_{xy}^0 \right)^2 = \begin{vmatrix} f_{xx}'' & f_{xy}'' \\ f_{xy}'' & f_{yy}'' \end{vmatrix} > 0 \quad \text{(Max)} . \quad (5.90)$$

The case of $\mathcal{D}^0 = 0$ requires special consideration. In all other cases, the function does not have an extremum.

Generalisation of this result to functions of more than two variables will be formulated in Vol. II.

As an example of the application of the developed method, let us investigate the extrema of the function $z = 5x^2 - 3xy + 6y^2$. Setting its first derivatives $z_x' = 10x - 3y$ and $z_y' = -3x + 12y$ to zero, we obtain a possible point for a minimum or a maximum as $x = y = 0$. The function $z = 0$ at this point. To characterise this point, we calculate the second derivatives, $z_{xx}'' = 10$, $z_{xy}'' = -3$ and $z_{yy}'' = 12$, and the discriminant $\mathcal{D} = 10 \cdot 12 - (-3)^2 = 111$. Since $\mathcal{D} > 0$ and $z_{xx}'' > 0$, we have a minimum at the point $(0, 0)$. It is easy to see that this result makes perfect sense. Indeed, a simple manipulation of our function yields:

$$z = 5 \left(x - \frac{3}{10} y \right)^2 + \frac{111}{20} y^2 ,$$

which clearly shows that the function is positive everywhere around the $(0, 0)$ point and hence has a minimum there.

To illustrate the uncertainty in the case of $\mathcal{D}^0 = 0$, consider two simple functions: $z = x^4 + y^4$ and $h = x^3 + y^3$. In both cases, the partial derivative method suggests the stationary point at $(0, 0)$, where the function is equal to zero and (please, check!) all second-order derivatives are equal to zero as well. Hence, $F_{xx}^0 = \mathcal{D}^0 = 0$ in both cases. However, as is easily seen, the function $z(x, y) > 0$ away from the point $(0, 0)$ and hence has a well-defined minimum, while the function $h(x, y)$ changes sign when crossing zero (e.g. take $y = 0$ and move x from negative to positive values) and hence has neither minimum nor maximum. Both functions are shown in Fig. 5.7.

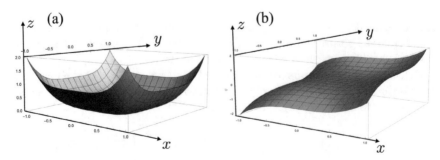

Fig. 5.7 Functions **a** $z(x, y) = x^4 + y^4$ and **b** $z = x^3 + y^3$

Fig. 5.8 A saddle function
$z =$
$27x^2 - 25y^2 + 30xy - 60x$
makes at the point
$(5/6, 1/2)$, see Problem 5.56

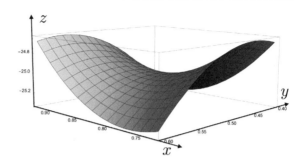

Problem 5.54 Show that the function $z = 2x^2 + 2xy + 3y^2 - 2x - y$ has a minimum at the point $(1/2, 0)$.

Problem 5.55 Show that the function $z = -9x^2 - 5y^2 + 6xy + 12x - 5$ has a maximum at the point $(5/6, 1/2)$.

Problem 5.56 Show that the function $z = 27x^2 - 25y^2 + 30xy - 60x$ does not have an extremum at the point $(5/6, 1/2)$. Plot this function and analyse the behaviour of $z(x, y)$ around this point. Convince yourself that it is a saddle, as shown in Fig. 5.8.

Problem 5.57 Show that the function $w = x^2 + y^2 + (1 - x - y)^2$ has a minimum at the point $(1/3, 1/3)$.

Problem 5.58 (*The least square method*) Suppose an experiment is performed N times by measuring a property y for different values of some parameter x. This way N pairs of data points (x_i, y_i) for $i = 1, \ldots, N$ become available. Assuming a linear dependence of y over x, i.e. $y = ax + b$, the "best" fit to the measured data is obtained by minimising the error function

$$\epsilon(a, b) = \sum_{i=1}^{N} (y_i - y(x_i))^2 = \sum_{i=1}^{N} (y_i - ax_i - b)^2 . \qquad (5.91)$$

Demonstrate by direct calculation that the "best" constants a and b are determined by solving the equations

$$S_{xx}a + S_x b = S_{xy} \quad \text{and} \quad S_x a + Nb = S_y ,$$

where

$$S_x = \sum_{i=1}^{N} x_i , \quad S_y = \sum_{i=1}^{N} y_i , \quad S_{xx} = \sum_{i=1}^{N} x_i^2 \quad \text{and} \quad S_{xy} = \sum_{i=1}^{N} x_i y_i .$$

By construction, the error function is non-negative and hence may only have a minimum at the values of a and b to be determined by the two equations given above. Still, show using the tools developed in this subsection that the error $\epsilon(a, b)$ has the minimum at the corresponding values of a and b. [Hint: *note that*

$$N S_{xx} - S_x^2 = \frac{1}{2} \sum_{i=1}^{N} \sum_{j=1}^{N} \left(x_i - x_j\right)^2$$

and is hence always positive.]

5.10.3 Finding Extrema Subject to Additional Conditions

Very often it is necessary to find a minimum (or maximum) of a function "with restrains", i.e. when the variables are constrained by some conditions, for instance, when an extremum of $h = h(x, y, z)$ is sought on a surface specified by the equation $g(x, y, z) = 0$.

Sometimes, this problem can simply be solved by relating one of the variables to the others by solving the equation of the condition, $z = z(x, y)$, and hence formulating the problem without a constraint for a function of a smaller number of variables, i.e. by considering the function $w(x, y) = h(x, y, z(x, y))$.

As an example, consider an extremum of $h = x^2 + y^2 + z^2$ on the plane specified by $x + y + z = 1$. From this condition, $z = 1 - x - y$, and hence we have to find an extremum of the function of two variables, $w = x^2 + y^2 + (1 - x - y)^2$, without any constraints. We know how to do this, see Problem 5.57: at $x = y = 1/3$ we have a minimum; hence, at this point $z = 1/3$ as well. Therefore, at the point $(1/3, 1/3, 1/3)$, the constrained function $h(x, y, z)$ experiences a minimum.

Problem 5.59 Find a conditional extremum of the function

$$h(x, y, z) = -x^2 - y^2 - z^2 + xy + yz + xz + 3(x + y + z)$$

on the plane $x + y + z = 1$. [Answer: *a maximum at* $(1/3, 1/3)$.]

Problem 5.60 Find a conditional extremum of the function

$$h(x, y, z) = x^2 + y^2 + z^2 - 2(x + y + z)$$

on the cone $z = x^2 + y^2$. [Answer: *a minimum at* $\left(x_0, x_0, 2x_0^2\right)$, *where* $x_0 \simeq$ 0.7607 *and is determined by solving the algebraic equation* $4x^3 - x - 1 = 0$.]

The method formulated above is only useful if the constraint equation(s) can be solved analytically with respect to some of the variables so that these could be eliminated from the function explicitly. In many cases, this may be either impossible or inconvenient to do, and hence another method must be developed. To this end, we consider a problem of finding extrema of a function $h = h(x, y, z)$ subject to a general condition $g(x, y, z) = 0$. The condition formally gives $z = z(x, y)$ and hence defines a function of two variables $w(x, y) = h(x, y, z(x, y))$. We have already discussed calculating partial derivatives of such a function in Sect. 5.6. Therefore, one can immediately use formula (5.38) for calculating the required first-order derivatives needed for finding the stationary point(s):

$$\frac{\partial w}{\partial x} = \frac{\partial h}{\partial x} + \frac{\partial h}{\partial z}\frac{\partial z}{\partial x} = 0 \text{ and } \frac{\partial w}{\partial y} = \frac{\partial h}{\partial y} + \frac{\partial h}{\partial z}\frac{\partial z}{\partial y} = 0. \tag{5.92}$$

Next, formulae (5.39) and (5.40) can be employed for calculating the corresponding second-order derivatives and hence characterising the stationary point(s). What is still missing is that we need derivatives of z with respect to x and y, but we may not know the explicit dependence $z = z(x, y)$. This, however, should not be a problem as we can differentiate the equation of the condition with respect to x and y and get the required derivatives. Differentiating both sides of the condition equation $g(x, y, z) = 0$ with respect to x and y, we obtain:

$$\frac{\partial g}{\partial x} + \frac{\partial g}{\partial z}\frac{\partial z}{\partial x} = 0 \implies z_x' = -g_x'/g_z', \tag{5.93}$$

$$\frac{\partial g}{\partial y} + \frac{\partial g}{\partial z}\frac{\partial z}{\partial y} = 0 \implies z_y' = -g_y'/g_z', \tag{5.94}$$

which, of course, are exactly the same as Eqs. (5.36) and (5.37) in Sect. 5.6.

We still need the second-order derivatives of $z(x, y)$. These can be calculated by differentiating the condition equation twice for which we can use Eqs. (5.39) and (5.40):

$$\frac{\partial g}{\partial z}\frac{\partial^2 z}{\partial x^2} + \frac{\partial^2 g}{\partial x^2} + 2\frac{\partial^2 g}{\partial x \partial z}\frac{\partial z}{\partial x} + \frac{\partial^2 g}{\partial z^2}\left(\frac{\partial z}{\partial x}\right)^2 = 0$$

$$\implies z_{xx}'' = -\frac{1}{g_z'}\left[g_{xx}'' + 2g_{xz}''z_x' + g_{zz}''\left(z_x'\right)^2\right], \tag{5.95}$$

$$\frac{\partial g}{\partial z}\frac{\partial^2 z}{\partial y^2} + \frac{\partial^2 g}{\partial y^2} + 2\frac{\partial^2 g}{\partial y \partial z}\frac{\partial z}{\partial y} + \frac{\partial^2 g}{\partial z^2}\left(\frac{\partial z}{\partial y}\right)^2 = 0$$

$$\implies z''_{yy} = -\frac{1}{g'_z}\left[g''_{yy} + 2g''_{yz}z'_y + g''_{zz}\left(z'_y\right)^2\right], \tag{5.96}$$

$$\frac{\partial g}{\partial z}\frac{\partial^2 z}{\partial x \partial y} + \frac{\partial^2 g}{\partial x \partial y} + \frac{\partial^2 g}{\partial x \partial z}\frac{\partial z}{\partial y} + \frac{\partial^2 g}{\partial y \partial z}\frac{\partial z}{\partial x} + \frac{\partial^2 g}{\partial z^2}\frac{\partial z}{\partial x}\frac{\partial z}{\partial y} = 0$$

$$\implies z''_{xy} = -\frac{1}{g'_z}\left[g''_{xy} + g''_{xz}z'_y + g''_{yz}z'_x + g''_{zz}z'_x z'_y\right]. \tag{5.97}$$

The obtained equations fully solve the problem. Indeed, the stationary point(s) is(are) obtained from Eqs. (5.92)–(5.94) yielding

$$h'_x - h'_z\frac{g'_x}{g'_z} = 0 \implies \frac{h'_x}{g'_x} = \frac{h'_z}{g'_z},$$

and similarly for the y derivatives. Hence, we can write the required equations for the stationary point(s) as follows:

$$\frac{h'_x}{g'_x} = \frac{h'_y}{g'_y} = \frac{h'_z}{g'_z}. \tag{5.98}$$

These two equations are to be solved together with the condition $g(x, y, z) = 0$ giving one or more stationary points (or none).

As an example, let us consider again the same problem as at the beginning of this Section: we would like to optimise the function $h = x^2 + y^2 + z^2$ on the plane specified by $x + y + z = 1$. Here $g = x + y + z - 1$, and hence $g'_x = g'_y = g'_z = 1$ with all second-order derivatives of g equal to zero. Therefore, $z'_x = z'_y = -1$ with all second-order derivatives equal to zero as well. Since $h'_x = 2x$, $h'_y = 2y$ and $h'_z = 2z$, the stationary point is obtained by solving Eq. (5.98), yielding

$$\frac{2x}{1} = \frac{2y}{1} = \frac{2z}{1},$$

i.e. $x = y = z$, which, together with the constraint equation, gives the same point as we obtained above using the direct calculation: $(1/3, 1/3, 1/3)$. Now, we need to calculate the second-order derivatives to characterise the stationary point. Since the second-order derivatives of $z(x, y)$ are equal to zero, we obtain from Eqs. (5.39) and (5.40):

$$w''_{xx} = h''_{xx} + 2h''_{xz}z'_x + h''_{zz}\left(z'_x\right)^2 = 2 + 2\cdot 0\cdot(-1) + 2(-1)^2 = 4,$$

similarly $w''_{yy} = 4$, and

$$w''_{xy} = h''_{xy} + h''_{xz}z'_y + h''_{yz}z'_x + h''_{zz}z'_x z'_y = 0 + 0\cdot z'_y + 0\cdot z'_x + 2(-1)(-1) = 2,$$

which are again exactly the same as given by the direct calculation performed above. Hence, we make the same conclusion that the stationary point found corresponds to a minimum. Indeed, $w''_{xx} = 4 > 0$ and $\mathcal{D}^0 = w''_{xx} w''_{yy} - \left(w''_{xy}\right)^2 = 4 \cdot 4 - 2^2 = 12 > 0$, see Eq. (5.89).

Problem 5.61 Find a conditional extremum of the function $h(x, y, z) = 2x^3 y^2 z$, defined in the region $\{0 < x \le 1, 0 < y \le 1\}$ on the plane $x + y + z = 1$ using the above method. [Answer: *a maximum at* $(1/2, 1/3, 1/6)$.]

Problem 5.62 Find a conditional extremum of the function $h(x, y, z) = x^2 + y^2 + z^2 - 2(x + y + z)$ on the cone $z = x^2 + y^2$ using the above method. Compare the result with that obtained in Problem 5.60.

5.10.4 Method of Lagrange Multipliers

Let us now have a closer look at the equations (5.98) (essentially, there are two of them) which are to be solved (together with the constraint equations) to give the stationary point(s). Note that sometimes it is more convenient to consider three equations instead of the two by introducing the fourth unknown λ equal to either of the fractions in (5.98):

$$h'_x = \lambda g'_x , \quad h'_y = \lambda g'_y , \quad h'_z = \lambda g'_z . \tag{5.99}$$

The three equations above together with the condition equation itself are sufficient to find the required four unknowns: x, y, z and λ. Alternatively, we can consider an auxiliary function

$$\Phi(x, y, z) = h(x, y, z) - \lambda \, g(x, y, z) . \tag{5.100}$$

Treating now all three variables x, y and z as being independent, we obtain the same three equations (5.99) for the stationary point(s), e.g. $\Phi'_x = h'_x - \lambda \, g'_x = 0$, etc. The factor λ is called a Lagrange multiplier, and this method of *Lagrange multipliers* is frequently used in practical calculations since it is extremely convenient: it treats all the variables on an equal footing and also allows the introduction of more than one condition easily. It is also applicable to functions of arbitrary number of variables. Therefore, it is essential that we consider this method for a general case.

Consider a function $w = w(\mathbf{x}) = w\left(x_1, \ldots, x_{n+p}\right)$ of $n + p = N$ variables. We would like to determine all its stationary points subject to p (where $1 \le p \le N - 1$) conditions $g_k\left(x_1, \ldots, x_{n+p}\right) = 0$, where $k = 1, \ldots, p$. If the point $\mathbf{x} = \left(x_1, \ldots, x_{n+p}\right)$ corresponds to an extremum of the function w, then the total differential of w at this point must vanish (see Eq. (5.85)):

$$dw = \sum_{i=1}^{n+p} \frac{\partial w}{\partial x_i} dx_i = 0 .$$ (5.101)

The total differential of every condition must also be zero because $g_k = 0$:

$$dg_k = \sum_{i=1}^{n+p} \frac{\partial g_k}{\partial x_i} dx_i = 0 , \quad k = 1, \ldots, p .$$ (5.102)

Therefore, an arbitrary linear combination of the differentials will also be zero:

$$dw + \sum_{k=1}^{p} \lambda_k dg_k = \sum_{i=1}^{n+p} \left[\frac{\partial w}{\partial x_i} + \sum_{k=1}^{p} \lambda_k \frac{\partial g_k}{\partial x_i} \right] dx_i = \sum_{i=1}^{n+p} \frac{\partial \Phi}{\partial x_i} dx_i = 0 , \quad (5.103)$$

where λ_k (with $k = 1, \ldots, p$) are arbitrary numbers and we introduced an auxiliary function

$$\Phi = w(x_1, \ldots, x_N) + \sum_{k=1}^{p} \lambda_k g_k(x_1, \ldots, x_N) .$$ (5.104)

We determine the multipliers λ_k by assuming that the last p variables $x_{n+1}, \ldots, x_{n+p}$ depend on the first n of them (i.e. on $x_1, \ldots, x_n$), and hence the latter can be considered as independent. Then, we request that the contribution to the differential $d\Phi$ in (5.103) due to each "dependent" variable dx_i (for $i = n + 1, n + 2, \ldots, n + p$) be individually equal to zero, i.e. that

$$\frac{\partial w}{\partial x_i} + \sum_{k=1}^{p} \lambda_k \frac{\partial g_k}{\partial x_i} = \frac{\partial \Phi}{\partial x_i} = 0 , \quad i = n + 1, \ldots, n + p .$$ (5.105)

Then, we are left with the condition

$$\sum_{i=1}^{n} \left[\frac{\partial w}{\partial x_i} + \sum_{k=1}^{p} \lambda_k \frac{\partial g_k}{\partial x_i} \right] dx_i = \sum_{i=1}^{n} \frac{\partial \Phi}{\partial x_i} dx_i = 0 .$$

Since the first n variables $x_1, \ldots, x_n$ are independent and the above expression should be zero for any dx_i, we arrive at the n equations:

$$\frac{\partial \Phi}{\partial x_i} = 0 , \quad i = 1, \ldots, n .$$ (5.106)

We see by inspecting Eqs. (5.105) and (5.106), that both conditions can be written simply as

$$\frac{\partial \Phi}{\partial x_i} = 0 , \quad i = 1, \ldots, n + p ,$$ (5.107)

i.e. in order to find the stationary point(s), we can consider the auxiliary function (5.104) instead of the original function $w(\mathbf{x})$ ignoring the fact that some variables depend on the others due to constraints (conditions). In other words, all variables can be considered as independent and all constraints (conditions) combined into the auxiliary function (5.104); then the problem of finding the stationary point(s) is solved by considering Φ. Note that there are exactly $n + 2p = N + p$ equations for N variables $\{x_i\}$ and p Lagrange multipliers: N equations (5.107) and p conditions $g_k\left(x_1, \ldots, x_{n+p}\right) = 0$ (where $k = 1, \ldots, p$).

As an example of using Lagrange multipliers, let us find the minimum distance between a point $M(x_M, y_M, z_M)$ and a plane $Ax + By + Cz = D$. If $P(x, y, z)$ is a point on the plane, then we have to find the minimum of the distance squared,

$$d\left(M, P\right)^2 = (x - x_M)^2 + (y - y_M)^2 + (z - z_M)^2 \ ,$$

subject to the additional condition that the point P is on the plane, i.e. it satisfies the equation $Ax + By + Cz = D$. As we only have a single condition, there will be one Lagrange multiplier λ, and we consider the auxiliary function

$$\Phi = (x - x_M)^2 + (y - y_M)^2 + (z - z_M)^2 - \lambda\left(Ax + By + Cz - D\right) \ ,$$

for which the stationary point is determined by three equations:

$$\frac{\partial \Phi}{\partial x} = 2\left(x - x_M\right) - \lambda A = 0 \ , \quad \frac{\partial \Phi}{\partial y} = 2\left(y - y_M\right) - \lambda B = 0 \ ,$$

$$\frac{\partial \Phi}{\partial z} = 2\left(z - z_M\right) - \lambda C = 0 \ , \tag{5.108}$$

and the equation of the plane. Multiplying the first equation above by A, the second by B and the third by C, summing them up and using the equation of the plane, we obtain an explicit expression for the Lagrange multiplier:

$$\lambda = -\frac{2\left(Ax_M + By_M + Cz_M - D\right)}{A^2 + B^2 + C^2} \ . \tag{5.109}$$

Using now Eq. (5.108), we can obtain the coordinates of the point P on the plane which is nearest to the point M:

$$x = x_M + \frac{1}{2}\lambda A \ , \quad y = y_M + \frac{1}{2}\lambda B \ , \quad z = z_M + \frac{1}{2}\lambda C \ . \tag{5.110}$$

It is easy to see that the distance d to the plane coincides with that given by Eq. (1.229).

Functions of Many Variables: Integration

6

6.1 Double Integrals

We recall that one-dimensional definite integral introduced in Sect. 4.1 is related
to functions of a single variable. Here we shall generalise the idea of integration to
functions of more than one variable. We start from the simplest case—integration of
functions of two variables.

6.1.1 Definition and Intuitive Approach

Consider a thin 2D plate lying in the $x - y$ plane, Fig. 6.1a, the area density of which
is $\rho(x, y)$. We would like to calculate the total mass of the plate. Here the density
$\rho(x, y) = \rho(M)$ is the function of a point M inside the plate. If the plate was uniform,
then its total mass could be obtained simply by multiplying its surface density ρ by
its area. However, since the density may be highly non-uniform across the plate, we
divide the plate area by horizontal and vertical lines into small area segments (regions)
as shown in Fig. 6.1b: points x_k divide the x-axis with divisions Δx_k, while the points
y_j divide the y-axis into intervals Δy_j, so each little rectangular segment has a
double index with the area $\Delta A_{kj} = \Delta x_k \Delta y_j$. Within each such segment the density
can be approximately treated as a constant yielding its mass equal to $\rho\left(M_{kj}\right) \Delta A_{kj}$,
where the point $M_{kj}\left(\xi_k, \zeta_j\right)$ is chosen somewhere inside the (kj)-segment with the
coordinates ξ_k and ζ_j, where $x_{k-1} \leq \xi_k < x_k$ and $y_{j-1} \leq \zeta_j < y_j$. The total mass
of the whole plate is then equal to the sum over all segments:

$$\sum_{k,j} \rho\left(M_{kj}\right) \Delta A_{kj} = \sum_{k,j} \rho\left(\xi_k, \zeta_j\right) \Delta x_k \Delta y_j .$$

© The Author(s), under exclusive license to Springer Nature Switzerland AG 2022
L. Kantorovich, *Mathematics for Natural Scientists*, Undergraduate Lecture Notes
in Physics, https://doi.org/10.1007/978-3-030-91222-2_6

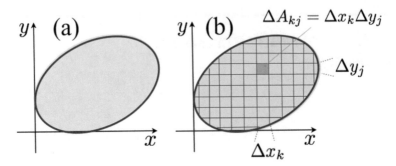

Fig. 6.1 2D plate lying in the $x - y$ plane

Fig. 6.2 The calculation of
the volume of a cylinder
oriented along the z-axis

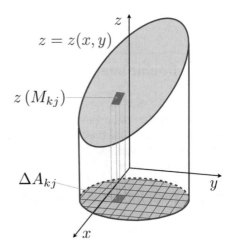

Then, we consider all segments becoming smaller and smaller (so that their number increases), and *in the limit* of infinitesimally small segments we arrive at the so-called *double integral*

$$\int\int_A \rho(x, y)dxdy = \lim_{\max \Delta A_{kj} \to 0} \sum_{k,j} \rho\left(M_{kj}\right) \Delta A_{kj} , \qquad (6.1)$$

which gives the *exact* mass of the plate. Above, the symbol A standing by the symbol of the double integral denotes the two-dimensional region in the $x - y$ plane corresponding to the plate.

Another example where a similar concept is employed is encountered when we would like to calculate the volume of a cylindrical body bounded by the $x - y$ plane at the bottom and some surface $z = z(x, y) > 0$ at the top, see Fig. 6.2. Again, we divide the bottom surface into small regions of the area ΔA_{kj}, choose a point $M_{kj}(\xi_k, \zeta_j)$ inside each such region, calculate the height of the cylinder at that point,

$z\left(M_{kj}\right)$ and then estimate the volume as a sum

$$V \simeq \sum_{k,j} z\left(M_{kj}\right) \Delta A_{kj} = \sum_{k,j} z\left(\xi_k, \zeta_j\right) \Delta x_k \Delta y_j ,$$

which in the limit of infinitesimally small regions tends to the double integral

$$V = \int\int_A z(x, y)dxdy$$

that gives the exact volume of the cylindrical body.

Since the double integral is defined as a limit of an integral sum, it has the same properties as a usual (one-dimensional) definite integral. In particular, in order for the integral sum to converge, it is *necessary* that the function which is integrated be bounded within the integration region A. Theory of the Darboux sum developed in Sect. 4.2 can be directly applied here as well almost without change.[1] In particular, a (rather mild) sufficient *condition* for a function $f(x, y)$ to be integrable in region A is its continuity, although a wider class of functions can also be integrated. Also, the average value Theorem 4.6 in Sect. 4.2 is valid for double integrals, i.e.

$$\int\int_A f(x, y)dxdy = f(\xi, \zeta) A , \tag{6.2}$$

where A is the total area of region A and the function in the right hand side is calculated at some point $M(\xi, \zeta)$ belonging to the region A, i.e. the point M lies somewhere inside A.

6.1.2 Calculation via Iterated Integral

Double integrals $\int\int_A f(x, y)dxdy$ can be calculated simply by considering one definite integral after another, and this, as we shall argue presently, may be done in any order in the cases of *proper* integrals (for improper integrals see Sect. 6.1.3).

Indeed, consider a plane region A bounded from the bottom and the top by functions $y_1(x)$ and $y_2(x)$, respectively, as depicted in Fig. 6.3a. Horizontally, the whole region stretches from $x = a$ to $x = b$. The integral (6.1) is considered as a limit of an integral sum, and obviously, the sum would only make sense if it does not depend on the order in which we add contributions from each region. This is ensured if the integral sum converges, which we assume is the case. However, as we shall show below, a certain *order* in summing up the regions will allow us to devise a simple method of calculating the double integral via usual definite integrals calculated one after another.

---

[1] After all, in Sect. 4.2 we did not actually made use of the fact that what we sum corresponds to a one-dimensional integral; we simply considered the limit of an integral sum.

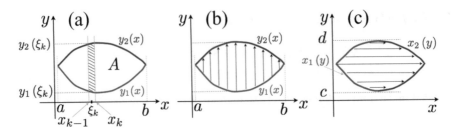

Fig. 6.3 The discussion concerning the order of integration in double integrals

Let us divide up the whole integration region A into small regions as we did above with the index k numerating divisions along the x-axis, while j along the y. Let m_{kj} and M_{kj} be the minimum and the maximum values of the function $f(x, y)$ within the region (kj), i.e. $m_{kj} \leq f(x, y) \leq M_{kj}$ for any x, y within the region (kj). Then, integrating this inequality with respect to dy over the vertical span of the small (kj) region (i.e. between y_{j-1} and y_j) we arrive at the inequality:

$$m_{kj}\Delta y_j \leq \int_{y_{j-1}}^{y_j} f(\xi_k, y)\, dy \leq M_{kj}\Delta y_j, \quad \text{where} \quad x_{k-1} \leq \xi_k < x_k,$$

i.e. the integral in the middle of the inequality is bounded from the bottom and the top.[2] Since this is the case for any j, we can sum these inequalities for all j values, giving:

$$\sum_j m_{kj}\Delta y_j \leq \int_{y_1(\xi_k)}^{y_2(\xi_k)} f(\xi_k, y)\, dy \leq \sum_j M_{kj}\Delta y_j$$

$$\implies \sum_j m_{kj}\Delta y_j \leq g(\xi_k) \leq \sum_j M_{kj}\Delta y_j, \tag{6.3}$$

where the integral in the middle is now taken between the two y-values at the lower and upper boundaries of the region, $y_1(\xi_k)$ and $y_2(\xi_k)$, corresponding to the given value of $x = \xi_k$.

The integral in the middle of the inequality is some function $g(x)$ at $x = \xi_k$. Since the inequality is valid for any choice of the value of ξ_k (as long as it lies within the interval between x_{k-1} and x_k), it can be replaced simply by x lying within the interval:

$$\sum_j m_{kj}\Delta y_j \leq g(x) \leq \sum_j M_{kj}\Delta y_j, \quad \text{where} \quad x_{k-1} \leq x < x_k.$$

---

[2] Note that on the left and right of the inequality prior to integration we had constant functions m_{kj} and M_{kj} which do not depend on x, so the integration simply corresponds to multiplying by Δy_j in either part.

Next, we integrate both sides of the above inequality with respect to dx between x_{k-1} and x_k, giving:

$$\left(\sum_j m_{kj}\Delta y_j\right)\Delta x_k \leq \int_{x_{k-1}}^{x_k} g(x)\,dx \leq \left(\sum_j M_{kj}\Delta y_j\right)\Delta x_k .$$

Since this is valid for any k, we can sum up these inequalities over all values of k, which yields:

$$\sum_k\sum_j m_{kj}\Delta x_k\Delta y_j \leq \int_a^b g(x)dx \leq \sum_k\sum_j M_{kj}\Delta x_k\Delta y_j .$$

On the left and on the right we have the lower and upper Darboux sums. Now we take the limit of all regions $\Delta A_{kj} = \Delta x_k \Delta y_j$ tending to zero; both Darboux sums converge to the value of the double integral, since according to our assumption the double integral exists. Therefore, the integral in the middle of the inequality is bounded by the same value and hence exists and is equal to the double integral:

$$\iint_A f(x,y)dxdy = \int_a^b g(x)dx = \int_a^b\left(\int_{y_1(x)}^{y_2(x)} f(x,y)dy\right)dx$$

$$= \int_a^b dx\int_{y_1(x)}^{y_2(x)} f(x,y)dy .$$

$$(6.4)$$

Here we recalled that the $g(x)$ is an integral of $f(x,y)$ over dy taken between $y_1(x)$ and $y_2(x)$.

Hence, in order to calculate the double integral, we have to perform integration over y for a fixed value of x first between the smallest, $y_1(x)$, and the largest, $y_2(x)$, values of the y for the value of x in question, and then integrate the result over all possible values of x. In this sense y serves as a "fast" and x as a "slow" variable. The integration over the fast variable is depicted by vertical arrows in Fig. 6.3b.

Two ways of writing the double integral shown on the right in Eq. (6.4) demonstrate this explicitly in two different ways (both are frequently used). The order of integration in the last passage in the above equation is from right to left: the x integration is performed *after* the y integration.

The method we just described basically suggests that in order to calculate the double integral, we may first integrate over a (hatched) vertical stripe in region A shown in Fig. 6.3a, which corresponds to the limit of the corresponding integral sum over j for the fixed ξ_k:

$$\Delta S(\xi_k) = \Delta x_k \sum_j f\left(\xi_k, \zeta_j\right)\Delta y_j \implies \Delta x_k\int_{y_1(\xi_k)}^{y_2(\xi_k)} f(\xi_k, y)dy = \Delta x_k g(\xi_k) ,$$

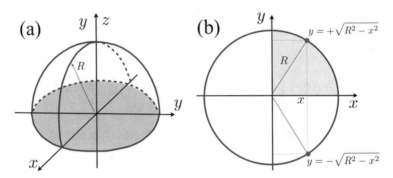

Fig. 6.4 a A hemisphere bounded by the surfaces $z = 0$ and $z = \sqrt{R^2 - x^2 - y^2}$ used in Example 6.1. **b** The base of the hemisphere showing the span of y (the vertical thin dashed line) for the given value of x. The shaded quarter of the circle is used for the actual integration

and then sum up contributions from all such vertical stripes:

$$\sum_k \Delta S(\xi_k) = \sum_k g(\xi_k) \Delta x_k \; .$$

Finally, taking the limit max $\Delta x_k \to 0$, we arrive at a definite integral with respect to x taken between a and b which gives the value of the double integral we seek. The arrows in Fig. 6.3b serve to indicate symbolically that the first integration is performed in the direction of the arrows (i.e. along y), and then the contributions from all such integrations are summed up by means of taking the x integral.

This way of calculating the double integral via a sequence of definite integrals is called an *iterated integral*. The obtained formula provides us with a practical recipe for calculating double integrals: first, we fix the value of x and calculate the y integral between $y_1(x)$ and $y_2(x)$ (these give the span of y for this particular value of x); then, we perform the x integration between a and b.

Above, we assumed that the integration region is such that it can be bounded from below and above by two functions $y_1(x)$ and $y_2(x)$ for which there is only one pair $y_1(\xi_k)$ and $y_2(\xi_k)$ of boundary values for the given division of the x-axis ($y_1(\xi_k)$ and $y_2(\xi_k)$ may coincide, e.g. at the boundaries). The consideration is easily generalised to more complex integration regions when this is not possible: in those cases the region should be divided up into smaller regions each satisfying the required condition. Since the integral is a sum, the final result is obtained by simply summing up the results of the integration over each of the regions.

Example 6.1 ▶ Calculate the volume of the hemisphere of radius R centred at zero with $z \geq 0$, see Fig. 6.4a.

Solution. For the given x, y, the height of the hemisphere is $z(x, y) = \sqrt{R^2 - x^2 - y^2}$ as the equation of the sphere is $x^2 + y^2 + z^2 = R^2$. Thus, the volume of the hemisphere is obtained from the double integral

$$V = \int\int z(x, y)dxdy = \int_{-R}^{R} dx \int_{-\sqrt{R^2-x^2}}^{\sqrt{R^2-x^2}} \sqrt{R^2 - x^2 - y^2}dy \ .$$

Here x changes between $-R$ and R in total. For the given x, the variable y spans between $-\sqrt{R^2 - x^2}$ and $+\sqrt{R^2 - x^2}$, see Fig. 6.4b, since at the base of the hemisphere we have a circle $x^2 + y^2 = R^2$. Also, due to symmetry, we may only consider a quarter of the whole volume in the actual integration (and hence a quarter of the area of the base, i.e. the integration over x will run between 0 and R, while the integration over y between 0 and $+\sqrt{R^2 - x^2}$; see the shaded region in Fig. 6.4b) and then introduce a factor of four[3] :

$$V = 4\int_0^R dx \int_0^{\sqrt{R^2-x^2}} \sqrt{R^2 - x^2 - y^2}dy$$

$$= 4\int_0^R dx \left\{ \frac{1}{2} \left(y\sqrt{R^2 - x^2 - y^2} + (R^2 - x^2)\arcsin\frac{y}{\sqrt{R^2 - x^2}} \right) \right\}_{y=0}^{y=\sqrt{R^2-x^2}}$$

$$= 4\int_0^R \frac{1}{2} \left\{ (R^2 - x^2)\frac{\pi}{2} \right\} dx = \pi \left(R^3 - \frac{R^3}{3} \right) = \frac{2\pi}{3}R^3 \ .$$

As it supposed to be, the volume of the whole sphere is two times larger, i.e. $4\pi R^3/3$.
◄

Above, we performed the y integration first (the fast variable), followed by the x integration (the slow variable). Alternatively, we can do the integrations in the reverse order. Indeed, assuming that the same region in Fig. 6.3a can also be described by the boundary lines $x = x_1(y)$ and $x = x_2(y)$ bounding the whole region from the left and right, respectively, see Fig. 6.3c, we can start by summing up contributions along the x-axis first, i.e. from all little regions that have the same value of ζ_j lying somewhere between the division points y_{j-1} and y_j, and then sum up all these contributions as shown schematically in Fig. 6.3c. Obviously, this way we arrive at an alternative formula:

$$\int\int_A f(x, y)dxdy = \int_c^d r(y)dy = \int_c^d \left(\int_{x_1(y)}^{x_2(y)} f(x, y)dx \right) dy$$

$$= \int_c^d dy \int_{x_1(y)}^{x_2(y)} f(x, y)dx \ ,$$

(6.5)

where the whole span of y is between c and d. Note that the horizontal integration (over x) gives a function $r(y)$. In this case the "fast" variable is x, and the "slow"

---

[3] We will use the following result: $\int \sqrt{a^2 - t^2}dt = \frac{1}{2}\left[t\sqrt{a^2 - t^2} + a^2 \arcsin\frac{t}{a} \right] + C$, see Eq. (4.63).

Fig. 6.5 The integration
region in Example 6.2

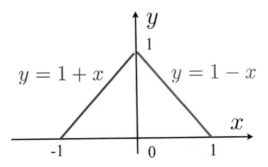

is y, and arrows in Fig. 6.3c indicate schematically the direction of integration over
the slow variable for different values of the fast variable, i.e. the way we "scan" the
integration region (sum the terms in the integral sum).

Thus, either order of integration can be used (either over x first and then over
y, or *vice versa*) and the same result will be obtained. In some cases useful general
relationships can be obtained by using this "symmetry" property. We can also use
it to our advantage to simplify the calculation of a double integral; for example, it
is easier to recognise the bounding functions $x_1(y)$ and $x_2(y)$ being on the left and
right rather than $y_1(x)$ and $y_2(x)$ as the lower and upper boundaries, respectively.

Example 6.2 ▶ Calculate the area bounded by curves shown in Fig. 6.5.

Solution. First of all, we note that it is a triangle, and hence its area should be equal
to one (the horizontal base is equal to 2 and the vertical height to 1). Let us now apply
the double integrals to do this calculation. We first try the method of Fig. 6.3b. The
integral is to be split into two regions, $-1 \le x \le 0$ and $0 \le x \le 1$, since the upper
boundary line is different in each region as indicated in the figure:

$$\int\int dx dy = \int_{-1}^{0} dx \int_{0}^{1+x} dy + \int_{0}^{1} dx \int_{0}^{1-x} dy$$

$$= \int_{-1}^{0} (1+x)\, dx + \int_{0}^{1} (1-x)\, dx = 1\,,$$

the correct result. Now, if we use the other method of Fig. 6.3c, we need to calculate
only a single double integral as the whole integration region can be dealt with at
once. Indeed, if we draw a horizontal line at any value of y between 0 and 1, then it
will cross the lines $y = 1 + x$ and $y = 1 - x$ at values of $x = y - 1$ and $x = 1 - y$,
respectively; no need to split the integration into two. Hence, we obtain:

$$\int_{0}^{1} dy \int_{y-1}^{1-y} dx = \int_{0}^{1} [1 - y - (y-1)]\, dy = 2 \int_{0}^{1} (1-y)\, dy = 1\,.$$

The result is the same, although the calculation may seem easier. ◀

Problem 6.1 Calculate the area enclosed between the curves $y = x^2$ and $y = 2 - x^2$ using the double integral. Use both methods, when either the y or the x integrations are performed first. [Answer: 8/3.]

In the following example we shall learn that in some cases the order in which the integration is performed could be the key in actually calculating the integral analytically.

Example 6.3 ▶Calculate the double integral:

$$ I = \int_0^1 dy \int_y^1 \frac{ye^x}{x} dx = \int_0^1 y\, dy \int_y^1 \frac{e^x}{x} dx \ . $$

Solution. In this double integral the integration over x is implied to be done first (with specific limits y and 1), with the integration over y to follow. Unfortunately, there is a problem: the indefinite integral $\int x^{-1}e^x dx$ cannot be written via elementary functions. Hence, we won't be able to get the function $g(y) = \int_y^1 x^{-1}e^x dx$ in the analytical form and perform the y integration, $\int_0^1 yg(y)dy$. We know, however, that one does not need necessarily to follow this order in taking the double integral; the same result is obtained if we perform the integration in the reverse order: perform the y integration first and then do the x one. It is not guaranteed that this way we would succeed in calculating this integral, but it is something worth trying!

So, what we would like to do is to rewrite this integral as

$$ I = \int_a^b \frac{e^x}{x} \left(\int_{y_1(x)}^{y_2(x)} y\, dy \right) dx \ , \tag{6.6} $$

where we kept the functions themselves intact as the two integrands of course will not change; but the limits will, and this is directly indicated. So, before we attempt to perform this integration, we need to work out the new limits for both integrals. To do this, let us try to work out the integration region in the $x - y$ plane. This region is in fact encoded in our original formula for the integral: the total span of the y variable is from 0 to 1, and for each value of y the other variable, x, changes between y and 1. This means that the boundaries of the x integration are the functions $x(y) = y$, $x(y) = 1$, $y = 0$ and $y = 1$. Drawing these boundary curves as shown in Fig. 6.6a, we find that they enclose a triangle (shaded light blue); hence, this is the actual integration region in the $x - y$ plane. The blue arrows indicate, for each value of y, the direction of the x integration which is to be performed first. This case corresponds to the one depicted in Fig. 6.3c. Once the integration region is established, it should not be difficult to work out the boundaries of the integral in Eq. (6.6) in which the fast and slow variables swapped. Indeed, we see that the overall span of x is from 0 to 1, and for each value of x the fast variable, y, changes between the dashed green

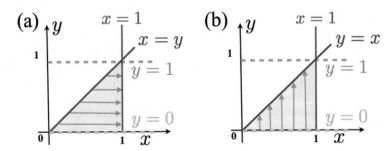

Fig. 6.6 Integration region in Example 6.3 is enclosed by four lines shown by different colours. Light blue arrows indicate the way the region is scanned during the integration with the direction of the arrows corresponding to the fast variable in each case

at the bottom $(y = 0)$ and the blue line at the top $(y = x)$, see Fig. 6.6b. Hence, we may now write:

$$I = \int_0^1 \frac{e^x}{x} \left(\int_0^x y\,dy \right) dx = \int_0^1 \frac{e^x}{x} \left(\frac{y^2}{2} \Big|_0^x \right) dx$$

$$= \int_0^1 \frac{e^x}{x} \frac{x^2}{2} dx = \frac{1}{2} \int_0^1 xe^x dx = \frac{1}{2} (x - 1) e^x \Big|_0^1 = \frac{1}{2}.$$

So, by swapping the integrals, we have been able to calculate the double integral! ◄

Problem 6.2 Prove the following formula:

$$\int_0^1 dx \int_{x^2}^{\sqrt{x}} f(x, y)dy = \int_0^1 dy \int_{y^2}^{\sqrt{y}} f(x, y)dx \,,$$

valid for any integrable $f(x, y)$ (it is obvious for a symmetric function $f(x, y) = f(y, x)$). Draw the integration region.

Note that swapping of integrals (interchanging of their order) is trivial if the limits of both integrations are constants:

$$\int_a^b dx \int_{a'}^{b'} f(x, y)dy = \int_{a'}^{b'} dy \int_a^b f(x, y)dx \,.$$

In this simple case the integration region is a rectangular with sides $a \leq x \leq b$ and $a' \leq y \leq b'$.

6.1.3 Improper Integrals

So far we have considered double integrals with finite limits. We have argued that one can do the integration in any order for these integrals: either over x first (the fast variable) and over y second (the slow), or the other way round. In practice, it is often necessary to deal with double integrals in which one (both) integration(s) is (are) performed using one or both limits to be infinities, i.e. one (both) of the integrals is (are) improper, Sect. 4.5.3. Here we shall briefly consider the questions of whether one can interchange the order of integration if one of the integrals is improper and under which conditions it is possible to do so.

For that let us consider the following improper integral with respect to t containing the function $f(x, t)$ of two variables:

$$F(x) = \int_a^\infty f(x, t)\, dt \ . \tag{6.7}$$

This integral can be treated as an improper integral depending on the parameter x (cf. Sect. 4.5.2). We shall assume that x may take any value between finite numbers x_1 and x_2, i.e. $x_1 \le x \le x_2$. We are going to investigate such improper integrals here.

First of all, let us state an important definition of the so-called *uniform convergence* of an improper integral which is central to our discussion, to follow. We have already given the definition of convergence of improper integrals in Sect. 4.5.3. Let us apply this definition to our case: the integral (6.7) is said to converge to $F(x)$ for the given value of x, if for any $\epsilon > 0$ there always exists $A > a$ such that

$$\left| F(x) - \int_a^A f(x, t) dt \right| < \epsilon \quad \text{or} \quad \left| \int_A^\infty f(x, t) dt \right| < \epsilon \ . \tag{6.8}$$

This condition states that the integral converges to $F(x)$ for the given x, and there is no reason to believe in general that A will not depend on x; in fact, in many cases it will. However, in some cases it won't. If the value of A (for the given ϵ) can be chosen the same for all values of x between x_1 and x_2, then the integral (6.7) is said to converge to $F(x)$ *uniformly* with respect to x.

There is a very simple test one can apply in order to check whether the convergence is uniform or not, due to Weierstrass.

Theorem 6.1 (Weierstrass Test) *If the function $f(x, t)$ which is continuous with respect to both of its variables can be estimated from above as $|f(x, t)| \le M(t)$ for all values of x and t, and the function $M(t)$ is also continuous, then the integral (6.7) converges uniformly with respect to x.*

Proof: Indeed, what we need to prove is that the integral $\int_A^\infty f(x, t)dt$ appearing in Eq. (6.8) can be estimated by some chosen ϵ using the bottom limit A which is independent of x. Indeed,

$$\left| \int_A^\infty f(x, t)dt \right| \leq \int_A^\infty |f(x, t|\, dt \leq \int_A^\infty M(t)dt = \epsilon \,.$$

It is seen that for the given ϵ one can always choose such value of A that the integral of $M(t)$ between A and ∞ is equal to ϵ. The value of A chosen in this way does not depend on x, and hence, the integral converges uniformly. **Q.E.D.**

This simple test can be illustrated on the integral $\int_0^\infty e^{-xt}dt$. In this case $f(x, t) = e^{-xt}$ and can be estimated from above for any $x > 1$ as $f(x, t) \leq e^{-t}$. Therefore, $M(t) = e^{-t}$ and

$$\int_A^\infty e^{-t}dt = e^{-A} \quad \Longrightarrow \quad e^{-A} = \epsilon \quad \Longrightarrow \quad A = -\ln \epsilon \,.$$

It is seen that A can be chosen for the given ϵ in the same way for any value of x for $x > 1$. The integral therefore converges uniformly within this interval of x.

Problem 6.3 Check uniform convergence of the following improper integrals:

$$(a) \ \int_\alpha^\infty \frac{dt}{x^2 + t^2} \ (\alpha > 0) \ ; \quad (b) \ \int_0^\infty \frac{\sin^2 (xt)}{a^2 + t^2}dt \ ;$$

$$(c) \ \int_\alpha^\infty \frac{\sin (xt)}{x^2 + t^2}dt \ (\alpha > 0) \ ; \quad (d) \ \int_0^\infty e^{-x^2 t^2} \cos^3 (xt)\, dt \ (x > 1);$$

$$(e) \ \int_0^\infty \frac{\sin (xt)}{t^\nu}dt \ (\nu > 1) \ ; \quad (f) \ \int_\alpha^\infty \frac{\cos (x^2 t^3)}{(x + t)^\nu}dt \ (\nu > 1) \,.$$

[Answers: *yes in all cases.*]

Problem 6.4 Show that the integral

$$E_1(x) = \int_x^\infty \frac{e^{-t}}{t}dt$$

converges uniformly with respect to $x > 1$. $E_1(x)$ is a special function called the exponential integral   [Hint: *first of all, prove that for $x > 1$ we have $e^{-xu}/u < e^{-u}$; then make the substitution $u = t/x$ in the integral, estimate the integrand and hence prove the uniform convergence.*]

Next, we shall prove a theorem which states under which conditions the function $F(x)$ defined by the improper integral (6.7) is continuous.

Theorem 6.2 *If $f(x, t)$ is continuous with respect to both of its variables, and the integral (6.7) converges uniformly, then it defines a continuous function $F(x)$.*

Proof: Indeed, the function $F(x)$ is continuous at a point x_0 if $\lim_{x \to x_0} F(x) = F(x_0)$. This means that for any $\epsilon > 0$ one can find a positive δ such that for any x in the proximity to x_0, i.e. $|x - x_0| < \delta$, one has $|F(x) - F(x_0)| < \epsilon$. Let us split the integration between a and $+\infty$ by some A which is larger than a and does not depend on x. We then have:

$$|F(x) - F(x_0)| = \left| \int_a^\infty f(x, t)dt - \int_a^\infty f(x_0, t)\, dt \right|$$

$$= \left| \int_a^A [f(x, t) - f(x_0, t)]\, dt + \int_A^\infty f(x, t)dt - \int_A^\infty f(x_0, t)\, dt \right|$$

$$\leq \int_a^A |f(x, t) - f(x_0, t)|\, dt + \left| \int_A^\infty f(x, t)dt \right| + \left| \int_A^\infty f(x_0, t)\, dt \right| .$$

Since $f(x, t)$ is continuous, for any $\epsilon_1 > 0$ one can find such $\delta > 0$ that $|f(x, t) - f(x_0, t)| < \epsilon_1$. This fact allows estimating the first integral in the right-hand side above as less than or equal to $\epsilon_1 (A - a)$. The other two integrals can be estimated using some $\epsilon_2 > 0$ due to uniform convergence of the improper integral in question. Hence,

$$|F(x) - F(x_0)| \leq \epsilon_1 (A - a) + 2\epsilon_2 = \epsilon ,$$

as required. Indeed, for the chosen ϵ one can define separately $\epsilon_2 < \epsilon$ and determine the value of A from the condition of the uniform convergence; then, ϵ_1 is determined from the above equation relating all these quantities. Once ϵ_1 is known, then the value of δ is determined from the continuity of the function $f(x, t)$. **Q.E.D.**

The other theorem we would like to prove is related to the fact that one can change the order of integration in a double integral if one of the integrals is improper with respect to its upper limit.

Theorem 6.3 *If $f(x, t)$ is continuous with respect to both of its variables and the integral (6.7) $F(x)$ converges uniformly, then*

$$\int_{x_1}^{x_2} F(x)dx = \int_a^\infty dt \left[\int_{x_1}^{x_2} dx \, f(x, t) \right] .$$

Proof: Here

$$\int_{x_1}^{x_2} F(x)dx = \int_{x_1}^{x_2} dx \int_a^\infty dt \, f(x, t) .$$

We need to prove that the difference

$$L_A = \int_{x_1}^{x_2} F(x)dx - \int_a^A dt \int_{x_1}^{x_2} dx \, f(x, t)$$

tends to zero as $A \to \infty$. Consider this difference carefully:

$$
\begin{aligned}
L_A &= \int_{x_1}^{x_2} dx \int_a^\infty dt \, f(x, t) - \int_a^A dt \int_{x_1}^{x_2} dx \, f(x, t) \\
&= \int_{x_1}^{x_2} dx \int_a^A dt \, f(x, t) + \int_{x_1}^{x_2} dx \int_A^\infty dt \, f(x, t) - \int_a^A dt \int_{x_1}^{x_2} dx \, f(x, t) .
\end{aligned}
$$

The first and the last double integrals are the proper ones, and since a proper double integral does not depend on the order of integration, they cancel out. We obtain

$$L_A = \int_{x_1}^{x_2} dx \int_A^\infty dt \, f(x, t) . \tag{6.9}$$

Let us estimate this double integral:

$$|L_A| = \left| \int_{x_1}^{x_2} dx \int_A^\infty dt \, f(x, t) \right| \leq \int_{x_1}^{x_2} dx \underbrace{\left| \int_A^\infty dt \, f(x, t) \right|}_{\leq \epsilon_1} \leq \epsilon_1 (x_2 - x_1) = \epsilon ,$$

since the integral over t of $f(x, t)$ converges uniformly. Therefore, by choosing arbitrary $\epsilon > 0$, we calculate $\epsilon_1 = \epsilon / (x_2 - x_1)$ and then determine the appropriate value of A to satisfy the inequality above. This means that for any $\epsilon > 0$ one can always find such $A > a$ that $|L_A| < \epsilon$. This simply means that L_A in the expression (6.9) converges to zero in the $A \to \infty$ limit, as required. **Q.E.D.**

Problem 6.5 Prove that (see also Problem 6.4)

$$\int_0^\infty E_1(x)dx = 1 , \quad \text{where} \quad E_1(x) = \int_x^\infty t^{-1}e^{-t}dt .$$

[Hint: *change the order of the integration.*]

The theorems considered above are of course valid for the bottom infinite limit or for both integration limits in one or both integrals being $\pm\infty$. Moreover, they are also valid for the improper integrals due to singularity of the integrand $f(x,t)$ somewhere inside the considered x and t intervals.

6.1.4 Change of Variables: Jacobian

We know from Chap. 4 that the change of variables in single integrals is an extremely powerful tool which enables their calculation in many cases. Remarkably, it is also possible to perform a change of variables in double integrals, but this time we have to change two variables at once. To explain how this can be done is the purpose of this section.

Let us assume that the calculation of a double integral can be simplified by introducing new variables u and v via some so-called *connection equations*:

$$\begin{cases} x = x(u, v) \\ y = y(u, v) \end{cases} . \tag{6.10}$$

If the new variables are chosen properly, the integration function (the integrand) may be significantly simplified; moreover, and more importantly, the integration region on the $x - y$ plane changes to a very different shape in the (u, v) coordinate system. If the latter becomes more like a rectangular, this transformation may simplify the calculation of the integral, as illustrated in Fig. 6.7.

In order to change the integration variables $(x, y) \rightarrow (u, v)$, we have to find the relationship between the areas $dxdy$ and $dudv$ in the two systems. To this end, consider a parallelepiped $ABCD$ in the (u, v) system, Fig. 6.8a. The coordinates of the points of its four vertices are: $A(u, v)$, $B(u, v + dv)$, $C(u + du, v + dv)$ and

Fig. 6.7 The shape of the integration region in the (u, v) system may be much simpler than in the (x, y) system, and thus may make the calculation of the integral possible

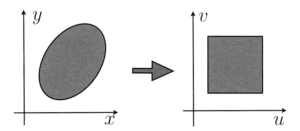

Fig. 6.8 The calculation of the relationship between the small areas in the two coordinate systems

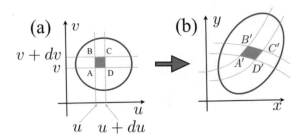

$D(u + du, v)$, and they form a parallelogram with the area $du dv$. These points will transform into points A', B', C' and D', respectively, in the (x, y) system whose coordinates are obtained explicitly via the connection relations (6.10). If the point A' is given by the vector $\mathbf{r} = \mathbf{r}(A) = (x, y, 0)$, where $x = x(u, v)$ and $y = y(u, v)$, then the other three points are:

$$
\mathbf{r}(B') = \mathbf{r} + \frac{\partial \mathbf{r}}{\partial v} dv , \quad \mathbf{r}(C') = \mathbf{r} + \frac{\partial \mathbf{r}}{\partial v} dv + \frac{\partial \mathbf{r}}{\partial u} du \quad \text{and}
$$
$$
\mathbf{r}(D') = \mathbf{r} + \frac{\partial \mathbf{r}}{\partial u} du .
$$

(6.11)

Indeed, for example, the point B' is the image of the point B in the (u, v) system and is obtained by advancing point A by dv there; therefore,

$$
x(B') = x(u, v + dv) = x(u, v) + \frac{\partial x}{\partial v} dv ,
$$

and similarly for the y coordinate. Therefore, in the vector form we obtain the first relation in (6.11). All other points are obtained accordingly, and in the vector notations their coordinates are given by Eq. (6.11). Once we know the points of the chosen small area in the (x, y) system, which can approximately be considered as a parallelogram, we can calculate the vectors corresponding to its sides:

$$
\overrightarrow{A'B'} = \mathbf{r}(B') - \mathbf{r}(A') = \frac{\partial \mathbf{r}}{\partial v} dv = \left(\frac{\partial x}{\partial v} dv, \frac{\partial y}{\partial v} dv, 0 \right) ,
$$
$$
\overrightarrow{A'D'} = \mathbf{r}(D') - \mathbf{r}(A') = \frac{\partial \mathbf{r}}{\partial u} du = \left(\frac{\partial x}{\partial u} du, \frac{\partial y}{\partial u} du, 0 \right) .
$$

Since the shaded region in the (x, y) system, which is the image of the corresponding shaded rectangular region in the (u, v) system in Fig. 6.8, is a parallelogram, its area can be calculated using the absolute value of the vector product of its two adjacent sides:

$$
\overrightarrow{A'B'} \times \overrightarrow{A'D'} = \begin{vmatrix} \mathbf{i} & \mathbf{j} & \mathbf{k} \\ (\partial x/\partial v) dv & (\partial y/\partial v) dv & 0 \\ (\partial x/\partial u) du & (\partial y/\partial u) du & 0 \end{vmatrix} = \begin{vmatrix} (\partial x/\partial v) dv & (\partial y/\partial v) dv \\ (\partial x/\partial u) du & (\partial y/\partial u) du \end{vmatrix} \mathbf{k} ,
$$

Fig. 6.9 A quarter of a circle in the (x, y) system becomes a rectangular in the (r, ϕ) system of the polar coordinates

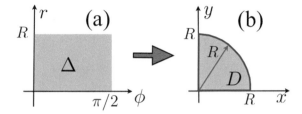

and hence the area $dA = \left| \overrightarrow{A'B'} \times \overrightarrow{A'D'} \right|$ becomes:

$$dA = \left| \begin{matrix} (\partial x/\partial v)\, dv & (\partial y/\partial v)\, dv \\ (\partial x/\partial u)\, du & (\partial y/\partial u)\, du \end{matrix} \right| = \left| \frac{\partial x}{\partial v} \frac{\partial y}{\partial u} - \frac{\partial x}{\partial u} \frac{\partial y}{\partial v} \right| dudv = |J(u, v)|\, dudv \,,$$

where

$$J(u, v) = \frac{\partial(x, y)}{\partial(u, v)} = \frac{\partial x}{\partial v} \frac{\partial y}{\partial u} - \frac{\partial x}{\partial u} \frac{\partial y}{\partial v} = \left| \begin{matrix} \partial x/\partial v & \partial y/\partial v \\ \partial x/\partial u & \partial y/\partial u \end{matrix} \right| \qquad (6.12)$$

is called *Jacobian*[4] of the transformation. The notation $\frac{\partial(x,y)}{\partial(u,v)}$ for the Jacobian is very convenient as it shows symbolically how it should be calculated. Thus, the areas in the two systems, which correspond to each other by means of the transformation (6.10), are related via the Jacobian:

$$dA = dxdy = \left| \frac{\partial(x, y)}{\partial(u, v)} \right| dudv \,, \qquad (6.13)$$

and the change of the variables in the double integral is thus performed using the following rule:

$$\int \int_{\Sigma} f(x, y)dxdy = \int \int_{\Delta} f(x(u, v), y(u, v)) \left| \frac{\partial(x, y)}{\partial(u, v)} \right| dudv \,, \qquad (6.14)$$

where the region Σ in the (x, y) system transforms into the region Δ in the (u, v) system. Note that the *absolute value* of the Jacobian must be used in the formula since the area must be positive.

As an example, let us calculate the integral $I = \int \int_D xydxdy$ over the region D shown in Fig. 6.9b. We shall transform the integral to the polar coordinates (r, ϕ) in which the integration region D (a quarter of a circle) turns into a rectangle with sides R and $\pi/2$, shown in Fig. 6.9a. Indeed, if we take any value of r between 0 and R within region D in the $x - y$ plane, we can draw a quarter of a circle by changing the angle ϕ from 0 to $\pi/2$. Hence, the span of ϕ is the same for any r, and hence in the (r, ϕ) coordinate system we shall have a rectangular. The old coordinates are related

---

[4] Named after Carl Gustav Jacob Jacobi.

Fig. 6.10 The derivation of
the surface area in polar
coordinates

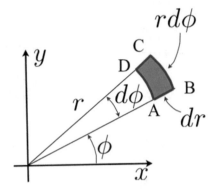

to the new ones via $x = r \cos \phi$, $y = r \sin \phi$ so that the Jacobian can be calculated
as

$$\frac{\partial(x, y)}{\partial(u, v)} = \begin{vmatrix} \partial x/\partial r & \partial y/\partial r \\ \partial x/\partial \phi & \partial y/\partial \phi \end{vmatrix} = \begin{vmatrix} \cos \phi & \sin \phi \\ -r \sin \phi & r \cos \phi \end{vmatrix} = r \cos^2 \phi + r \sin^2 \phi = r \,,$$

and the double integral becomes:

$$I = \int \int (r \cos \phi) \, (r \sin \phi) \, (r dr d\phi)$$

with r changing from 0 to R and ϕ from 0 to $\pi/2$. The integrand is a product of
two functions; the limits are constants so that it does not matter which integral is
calculated first:

$$I = \left(\int_0^R r^3 dr \right) \left(\int_0^{\pi/2} \cos \phi \sin \phi d\phi \right) = \left(\frac{R^4}{4} \right) \frac{\sin^2 \phi}{2} \Big|_0^{\pi/2} = \frac{R^4}{8} \,.$$

Let us remember, for the polar coordinate system the Jacobian is equal to r and the
area element in this system

$$dA = r dr d\phi \,. \tag{6.15}$$

It is instructive to obtain this result also using a very simple geometrical consideration
illustrated in Fig. 6.10. The hatched region $ABCD$ is obtained by advancing point
A along the two polar coordinates r and ϕ: point D is obtained by advancing A
by $d\phi$ and point B by advancing by dr. Then, the area dA of the hatched region
$ABCD$ (which can be approximately considered rectangular) is a product of the
sides AB and AD, where $AB = dr$ and $AD = 2r \sin(d\phi/2) = 2r d\phi/2 = r d\phi$, so
that $dA = r dr d\phi$, exactly as in (6.15).

Example 6.4 ► Find the volume of the hemisphere in Fig. 6.4 using the change of variables method.

Solution. It is wise to introduce polar coordinates in this problem as well, since in these coordinates $x^2 + y^2 = r^2$. The Jacobian is $J = r$, and the volume becomes:

$$V = \int\int \sqrt{R^2 - x^2 - y^2}\,dxdy = \int\int \sqrt{R^2 - r^2}\,rdrd\phi$$

$$= \int_0^R \sqrt{R^2 - r^2}\,rdr \int_0^{2\pi} d\phi = 2\pi \int_0^R \sqrt{R^2 - r^2}\,rdr \ .$$

By making the substitution $\zeta = R^2 - r^2$, the r-integral can then be calculated:

$$V = -\pi \int_{R^2}^0 \zeta^{1/2}d\zeta = \pi \left(\frac{\zeta^{3/2}}{3/2}\right)\Big|_0^{R^2} = \frac{2\pi R^3}{3},$$

i.e. the same result as we obtained in Example 6.1 in the previous section using direct integration over x and y. However, the calculation appeared to be extremely simple this time! ◄

Problem 6.6 Calculate the integral

$$\int\int \frac{e^{-\alpha\sqrt{x^2+y^2}}}{\sqrt{x^2 + y^2}}\,dxdy$$

over the entire $x - y$ plane using an appropriate change of coordinates. [Answer: $2\pi/\alpha$.]

Problem 6.7 Using polar coordinates, show that

$$\int\int_S \frac{dxdy}{x^2 + y^2} = 2\pi \ln \frac{R_2}{R_1},$$

where region S is bound by two concentric circles of radii R_1 and $R_2 > R_1$, both centred at the origin.

Problem 6.8 Show using Cartesian coordinates that the double integral $\int\int_S ydxdy = 0$, where S is the $x \geq 0$ half circle of radius R centred at the origin. Then repeat the calculation in polar coordinates.

Fig. 6.11 The calculation of
the electrostatic potential
from a disk in Problem 6.10

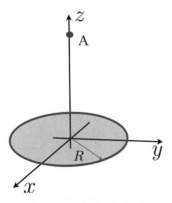

Problem 6.9 Show that the integral of the previous problem is equal to $2R^3/3$ if S is the $y \geq 0$ half circle. Do the calculation using both methods as well.

Problem 6.10 Calculate the electrostatic potential ϕ of a circular disk of radius R at the point A along its symmetry axis, separated by the distance z from the disk, as shown in Fig. 6.11. The disk is uniformly charged with the surface charge density σ. [Hint: *use polar coordinates when integrating over the disk surface.*] [Answer: $\phi(z) = 2\pi\sigma\left(\sqrt{z^2 + R^2} - z\right)$].

Problem 6.11 A thin disk of radius R placed in the $x - y$ plane with its centre at the origin has the surface mass density $\rho(x, y) = \rho_0 e^{-\alpha r}$, where $r = \sqrt{x^2 + y^2}$. Assuming that the density decays sufficiently quickly with increasing r, i.e. that $\alpha \gg 1/R$, calculate the total disk mass using the polar coordinates. [Answer: $2\pi\rho_0/\alpha^2$.]

In physics applications, the following so-called *Gaussian integral* is frequently used:

$$\int_{-\infty}^{\infty} e^{-\alpha x^2} dx = \sqrt{\frac{\pi}{\alpha}} . \tag{6.16}$$

There is a very nice method based on a double integral and polar coordinates which allows this integral to be calculated.[5] Consider a product of two Gaussian integrals, in which we shall use x as the integration variable in the first of them and y in the

---

[5] The corresponding indefinite integral cannot be represented by a combination of elementary analytical functions.

second:

$$G^2 = \int_{-\infty}^{\infty} e^{-\alpha x^2} dx \int_{-\infty}^{\infty} e^{-\alpha y^2} dy .$$

Of course, it does not matter which letters are used as integration variables, but using these makes the argument more transparent: the product of these two integrals can be also considered as a double integral across the whole $x - y$ plane, in which case we can use the polar coordinates in an attempt of calculating it:

$$G^2 = \int \int_{-\infty}^{+\infty} e^{-\alpha(x^2+y^2)} dx dy = \int_0^{2\pi} d\phi \int_0^{\infty} e^{-\alpha r^2} r dr$$

$$= 2\pi \int_0^{\infty} e^{-\alpha r^2} r dr = \left| \begin{matrix} t = r^2 \\ dt = 2r dr \end{matrix} \right| = \pi \int_0^{\infty} e^{-\alpha t} dt = \frac{\pi}{\alpha}$$

so that $G = \sqrt{\pi/\alpha}$ as required.

Problem 6.12 Prove that

$$\int_{-\infty}^{\infty} \exp\left[-a (x - z)^2 - b (x - y)^2\right] dx = \sqrt{\frac{\pi}{a+b}} \exp\left[-\frac{ab}{a+b} (y - z)^2\right] .$$

(6.17)

Problem 6.13 Prove that the diffusion probability for the Brownian motion of a particle in a solution, introduced by Eq. (5.13), satisfies the equation

$$w (x - x_0, t - t_0) = \int_{-\infty}^{\infty} w (x - x', t - t') w (x' - x_0, t' - t_0) dx' .$$

(6.18)

This equation shows that the probability to find the particle at the point x at time t, if it was at x_0 at time t_0, can be represented as a sum of all possible trajectories via an intermediate point x' at some intermediate time t', where $t_0 < t' < t$. The above equation lies at the heart of the so-called Markov stochastic processes and is called Einstein–Smoluchowski–Kolmogorov–Chapman equation.

Problem 6.14 The van der Waals interaction of an atom with another atom that is distance r away from it is given by $-C_{vdW}/r^6$. In this problem we shall calculate the interaction of an atom with an infinite plane that is placed distance h away from it. In doing so, we shall assume that the interaction energy of the atom with the surface area $dA = dx dy$ of the plane is given by $-C_{vdW}\sigma/r^6$, where σ is the plane surface density and r is the distance between the atom and that area. Show by performing an appropriate integration over the plane using the polar coordinates that the total interaction of the atom with that plane is

$$V(z) = -\frac{C_{vdW}\pi\sigma}{2h^4}.$$

Next use this result to show that the interaction of an atom that is placed above a semi-infinite crystal surface at the height h is

$$V(z) = -\frac{C_{vdW}\rho\pi}{6h^3}, \tag{6.19}$$

where ρ is the volume density of atoms in the crystal. Notice that the van der Waals interaction decays with the distance much slower now after we integrated over the volume of the crystal: if between atoms it decays as r^{-6}, its decay is reduced to the inverse third power of the distance between the atom and the surface.

6.2   Volume (Triple) Integrals

6.2.1   Definition and Calculation

Double integrals were justified by problems based on areas and two-dimensional planar objects. Similarly, triple integrals appear in problems related to volumes of three-dimensional objects. Similarly, to all other cases, these are defined as a limit of an integral sum,

$$\sum_{k,j,l} f(\xi_k, \zeta_j, \eta_l)\Delta x_k \Delta y_j \Delta z_l ,$$

where ξ_k, ζ_j and η_l are some points within the intervals Δx_k, Δy_j and Δz_l along the Cartesian axes. The integrals are calculated exactly in the same way as two-dimensional integrals by chaining one integration to another (the *iterated integral*). For *proper* integrals it does not matter which integration (over x, y or z) goes first, second and third. This is dictated entirely by convenience of the calculation (the integrand and the shape of the integration region). For instance, the volume integral of a function $f(\mathbf{r}) = f(x, y, z)$ over a volume region V,

$$I = \int\int\int_V f(x, y, z)dxdydz , \tag{6.20}$$

Fig. 6.12 The calculation of
the volume of a hemisphere

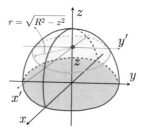

can be calculated in the following sequence of three ordinary definite integrals:

$$
I = \int_a^b \left[\int_{y_1(x)}^{y_2(x)} \left(\int_{z_1(x,y)}^{z_2(x,y)} f(x, y, z)dz \right) dy \right] dx
$$
$$
= \int_a^b dx \int_{y_1(x)}^{y_2(x)} dy \int_{z_1(x,y)}^{z_2(x,y)} f(x, y, z)dz ,
$$

(6.21)

where we first fix x and y and calculate the integral over z between the corresponding minimum and maximum z values z_1 and z_2 for the given x and y; after that the y integration is performed for the fixed x value and the x integration is performed last. When integrating in this order, the fastest variable is z and the slowest is x.

Example 6.5 ▶ Calculate the volume of the hemisphere in Fig. 6.4 using the volume integral.

Solution. The volume is given by the triple integral $\int \int \int dx dy dz$ performed over the hemisphere. However, in order to use the method of iterated integral, we have to work out the limits for each of the three integrals. First of all, we need to decide on the order in which the integrations are performed. In the case of the hemisphere it is simpler to have the z integration to be run last (i.e. to have the z variable being the slowest) because if you fix z, the obtained cross section will become a circle of radius $r = \sqrt{R^2 - z^2}$, see Fig. 6.12. Hence, the integration is to be performed as follows:

$$
V = \int dz \int dx \int dy .
$$

While changing the value of the z from 0 to R, the radius of the circle in the cross section continuously reduces from R to 0. While the region in the $x - y$ plane corresponding to the fixed value of z is known (the circle), we can discuss the integration limits for the remaining x and y integrals. First, x (the variable in the middle of the three integrals) changes between $\pm r = \pm\sqrt{R^2 - z^2}$, and we are left with the y, the slowest variable, that is to change between $\pm\sqrt{r^2 - x^2} = \pm\sqrt{R^2 - z^2 - x^2}$, compare with Fig. 6.4b. Hence, we can finally write:

$$
V = \int_0^R dz \int_{-\sqrt{R^2-z^2}}^{\sqrt{R^2-z^2}} dx \int_{-\sqrt{R^2-z^2-x^2}}^{\sqrt{R^2-z^2-x^2}} dy .
$$

By performing the calculation of each integral in order from right to left (and the dy integral is trivial), we get[6]:

$$V = \int_0^R dz \int_{-\sqrt{R^2-z^2}}^{\sqrt{R^2-z^2}} 2\sqrt{R^2 - x^2 - z^2} dx =$$

$$= 2 \int_0^R dz \frac{1}{2} \left\{ x\sqrt{R^2 - x^2 - z^2} + \left(R^2 - z^2\right) \arcsin \frac{x}{\sqrt{R^2 - z^2}} \right\} \Big|_{x=-\sqrt{R^2-z^2}}^{x=\sqrt{R^2-z^2}}$$

$$= \int_0^R dz \left\{ \left(R^2 - z^2\right) \arcsin(1) - \left(R^2 - z^2\right) \arcsin(-1) \right\}$$

$$= \int_0^R dz \left\{ \left(R^2 - z^2\right) \frac{\pi}{2} - \left(R^2 - z^2\right) \left(-\frac{\pi}{2}\right) \right\} = \pi \int_0^R \left(R^2 - z^2\right) dz = \pi \frac{2R^3}{3} ,$$

again, the same result as before. Note that this calculation performed in Cartesian coordinates was rather cumbersome. ◄

Problem 6.15 The integration limits with another choice of the order in which the iterative integration is performed are shown below:

$$V = \int_{-R}^R dx \int_{-\sqrt{R^2-x^2}}^{\sqrt{R^2-x^2}} dy \int_0^{\sqrt{R^2-y^2-x^2}} dz .$$

What is the cross section of the hemisphere corresponding to the y, z integration (fixed x)? Explain the limits in this case.

6.2.2   Change of Variables: Jacobian

Similar to the case of double integrals, we can perform a change of variables in volume integrals as well. Suppose, we would like to introduce new variables (u, v, g) instead of (x, y, z) using *connection equations*:

$$x = x(u, v, g), \quad y = y(u, v, g) \text{ and } z = z(u, v, g) . \tag{6.22}$$

The integration region V in the (x, y, z) system of coordinates will transform into a region Ω in the (u, v, g) system. To obtain the required relationship between the infinitesimal volumes $dV = dxdydz$ and $d\Omega = dudvdg$ in the two systems, let us consider a cuboid in the (u, v, g) system that is obtained by advancing by du, dv and dg along the corresponding coordinate axes u, v and g from the point $A(u, v, g)$

---

[6] Once again, we need to use the integral from Eq. (4.63).

Fig. 6.13 A small cuboid in the (u, v, g) system

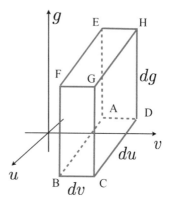

(corresponding to the point $A'(x, y, z)$ in the (x, y, z) system), see Fig. 6.13. In the (x, y, z) system the cuboid $ABCDEFGH$ transforms into the parallelepiped $A'B'C'D'E'F'G'H'$. The vectors corresponding to its three sides and running from the point $A'(x, y, z)$ are:

$$\overrightarrow{A'B'} = \frac{\partial \mathbf{r}}{\partial u} du \, , \quad \overrightarrow{A'D'} = \frac{\partial \mathbf{r}}{\partial v} dv \quad \text{and} \quad \overrightarrow{A'E'} = \frac{\partial \mathbf{r}}{\partial g} dg \, .$$

Then the volume of the parallelepiped is given by the absolute value of the mixed product of these three vectors (see Eq. (1.111)):

$$
\begin{aligned}
dV &= \left| \left(\frac{\partial \mathbf{r}}{\partial u} du \cdot \left[\frac{\partial \mathbf{r}}{\partial v} dv \times \frac{\partial \mathbf{r}}{\partial g} dg \right] \right) \right| \\
&= \begin{vmatrix} (\partial x/\partial u)\, du & (\partial y/\partial u)\, du & (\partial z/\partial u)\, du \\ (\partial x/\partial v)\, dv & (\partial y/\partial v)\, dv & (\partial z/\partial v)\, dv \\ (\partial x/\partial g)\, dg & (\partial y/\partial g)\, dg & (\partial z/\partial g)\, dg \end{vmatrix} \\
&= |J(u, v, g)|\, du\, dv\, dg \, ,
\end{aligned}
\tag{6.23}
$$

where

$$
J(u, v, g) = \frac{\partial(x, y, z)}{\partial(u, v, g)} = \begin{vmatrix} \partial x/\partial u & \partial y/\partial u & \partial z/\partial u \\ \partial x/\partial v & \partial y/\partial v & \partial z/\partial v \\ \partial x/\partial g & \partial y/\partial g & \partial z/\partial g \end{vmatrix}
\tag{6.24}
$$

is the corresponding 3×3 Jacobian. As you can see, for three dimensions it is defined in a very similar way to the two-dimensional case of Eq. (6.12). Then, the volume integral

$$V = \int \int \int_\Omega f(\mathbf{r}(u, v, g)) \left| \frac{\partial(x, y, z)}{\partial(u, v, g)} \right| du\, dv\, dg \, .
\tag{6.25}$$

As an example, let us again calculate the volume of the hemisphere in Fig. 6.4. It is advantageous to use the spherical coordinates (r, θ, ϕ) here,

$$x = r \sin \theta \cos \phi \, , \quad y = r \sin \theta \sin \phi \quad \text{and} \quad z = r \cos \theta \, ,$$

since in this system the hemisphere transforms into a parallelepiped that can be easily integrated: $0 \le r \le R$, $0 \le \theta \le \pi/2$ and $0 \le \phi < 2\pi$. The Jacobian of the transformation

$$\frac{\partial(x, y, z)}{\partial(r, \theta, \phi)} = \begin{vmatrix} \partial x/\partial r & \partial y/\partial r & \partial z/\partial r \\ \partial x/\partial \theta & \partial y/\partial \theta & \partial z/\partial \theta \\ \partial x/\partial \phi & \partial y/\partial \phi & \partial z/\partial \phi \end{vmatrix}$$

$$= \begin{vmatrix} \sin\theta\cos\phi & \sin\theta\sin\phi & \cos\theta \\ r\cos\theta\cos\phi & r\cos\theta\sin\phi & -r\sin\theta \\ -r\sin\theta\sin\phi & r\sin\theta\cos\phi & 0 \end{vmatrix} = r^2\sin\theta .$$

It is a useful exercise to use the definition of the 3×3 determinant (Sect. 1.9) and verify the above result for the Jacobian. Hence, our integral in the spherical system becomes:

$$V = \int_0^R dr \int_0^{\pi/2} d\theta \int_0^{2\pi} \left|r^2\sin\theta\right| d\phi = \int_0^R r^2 dr \int_0^{2\pi} d\phi \int_0^{\pi/2} \sin\theta d\theta$$

$$= \frac{R^3}{3} 2\pi \left(-\cos\theta\right)_0^{\pi/2} = \frac{2\pi}{3} R^3 .$$

As you see, all three integrals are completely independent yielding the same result as in Examples 6.1, 6.4 and 6.5 again! This is by far the simplest calculation of the sphere volume we have ever performed!

It is useful to remember, the Jacobian for the spherical coordinate system is $r^2\sin\theta$, and the volume element

$$dV = dxdydz = r^2\sin\theta dr d\theta d\phi . \tag{6.26}$$

This result can also be obtained using a simple geometrical argument, see Fig. 6.14. Indeed, consider a small cuboid with the bottom surface $ABCD$. It is obtained in the following way: point D has spherical coordinates (r, θ, ϕ). Point A is obtained from D by advancing the angle θ by $d\theta$, while point C by advancing the angle ϕ by $d\phi$. Hence, $CD = FD \cdot d\phi = r\sin\theta d\phi$ and $DA = rd\theta$. These lengths provide

Fig. 6.14 The geometrical derivation of the volume element in the spherical coordinate system

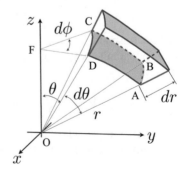

two sides of the cuboid. The height of it is obtained by simply advancing each of the corners of the bottom surface $ABCD$ by dr. Hence, the volume becomes

$$dV = CD \cdot DA \cdot dr = r^2 \sin\theta d\theta d\phi dr ,$$

i.e. the same expression as the one above in Eq. (6.26) obtained using the Jacobian.

Problem 6.16 The moment of inertia of a solid body of density $\rho(x, y, z)$ about the z-axis is given by

$$I_z = \int \int \int \left(x^2 + y^2 \right) \rho(x, y, z) dx dy dz ,$$

and

$$M = \int \int \int \rho(x, y, z) dx dy dz$$

is the body mass. Calculate the mass and the z-moment of inertia of a cylinder of uniform density $\rho = 1$ of radius R and height h whose symmetry axis is running along the z-axis, Fig. 6.16. [Hint: *use the cylindrical coordinates (r, ϕ, z) with the connection equations $x = r \cos\phi$, $y = r \sin\phi$ and $z = z$.*] [Answer: $M = \pi R^2 h$, $I_z = \frac{1}{2} M R^2$.]

Problem 6.17 Calculate the moment of inertia about the z-axis of a uniform sphere of radius R and density $\rho = 1$ centred at the origin using the cylindrical coordinates (r, ϕ, z). [Answer: $M = \frac{4}{3}\pi R^3$, $I_z = \frac{2}{5} M R^2$.]

Problem 6.18 A uniform sphere of radius R and density ρ is rotated around an axis passing through its centre with the angular velocity ω. Show that the kinetic energy of the sphere is $E_{KE} = (4/15)\pi R^5 \rho \omega^2$.

Problem 6.19 Reproduce Eq. (6.19) of Problem 6.14 using an appropriate volume integration in spherical coordinates.

Problem 6.20 Consider a van der Waals interaction of a sphere of radius R and volume density ρ_t with a semi-infinite crystal of volume density ρ_s that is distance h away from the sphere as shown in Fig. 6.15. A general point P within the sphere would interact with the crystal according to Eq. (6.19). Integrate over the sphere using spherical coordinates to show that the interaction energy

Fig. 6.15 A sphere of radius R and volume density ρ_t is placed above a semi-infinite crystal surface of the volume density ρ_s

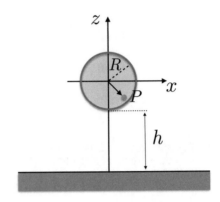

Fig. 6.16 The calculation of the moment of inertia of a cylinder in Problem 6.16

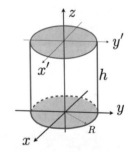

is

$$V(h) = \frac{\pi^2 C_{vdW}\rho_t\rho_s}{6}\left[\ln\left(1 + \frac{2R}{h}\right) - \frac{2R(R+h)}{h(2R+h)}\right].$$

Plot this as a function of $x = R/h$ and convince yourself that it is negative (attraction). Moreover, the potential energy gets more negative if x increases (e.g. the sphere gets closer to the surface). Next, show in the limit of $R \gg h$ that the force acting on the sphere due to the surface is

$$F(h) = -\frac{\partial V(h)}{\partial h} \approx -\frac{HR}{6h^2},$$

where $H = \pi^2 C_{vdW}\rho_t\rho_s$ is the so-called Hamaker constant that, as we have seen, appears naturally when considering van der Waals interaction between two bodies. This interaction is essential, for instance, in the theory of atomic force microscopy (AFM). As can be seen, the force is directed downwards, i.e. it attracts the sphere to the surface, as expected.

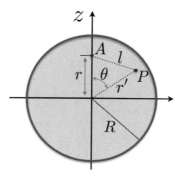

Fig. 6.17 The calculation of the gravitational potential at point A inside a uniform sphere of radius R and density ρ. The points A and a general point P inside the sphere have spherical coordinates $(r, 0, 0)$ and (r', θ, ϕ), respectively, so that the distance l to the observation point A can be expressed via r', r and θ using the cosine theorem Eq. (1.70)

Problem 6.21 Consider the gravitational potential

$$\psi(\mathbf{r}) = -G \int \int \int \frac{\rho}{|\mathbf{r} - \mathbf{r}'|} d\mathbf{r}'$$

due to a uniform sphere of radius R and constant density ρ. Here the integration is performed over the volume V of the sphere and G is the gravitational constant. To perform the volume integration, you may use the spherical coordinates and benefit from the symmetry of the problem realising that the potential will only depend on the distance from the centre and hence the observation point A at distance r from the centre can be chosen along the z-axis as shown in Fig. 6.17. Show the potential distance r from the centre of the sphere is given by the following expression:

$$\psi(r) = -2\pi G\rho \begin{cases} 2R^3/3r, & r \geq R \\ R^2 - r^2/3, & 0 \leq r \leq R \end{cases}.$$

Problem 6.22 A tunnel is dug between two opposite points on the Earth surface passing through the Earth's centre, and a small body of mass m is placed into the tunnel and released as shown in Fig. 6.18. Using results of Problem 6.21, show that the body will move inside the tunnel according to the harmonic differential equation $\ddot{r} + \omega^2 r = 0$, where $r(t)$ is the distance from the centre, $\omega = \sqrt{4\pi\rho_E G/3}$ and ρ_E is the Earth density. The differential equation has the harmonic solution $r(t) = A \sin(\omega t + \phi)$ with some amplitude A and phase ϕ. Hence, the body will oscillate in the tunnel with frequency ω. Note that the latter does not depend on the mass of the body.

Problem 6.23 Consider a one-dimensional diffusion of a particle along the x-axis. Initially at time t_0 the particle was at the position x_0, and we are interested in finding a transition probability for the particle to be found at the position x at time t. Let us divide up the interval between t_0 and t into $N + 1$ equidistant time intervals of duration $\epsilon = (t - t_0)/(N + 1)$, and consider the diffusion as a sequence of $N + 1$ short diffusions $t_0 \to t_1 \to t_2 \to \ldots \to t_N \to t$, where $t_k = t_0 + k\epsilon$. The final transition rate is then given as a product of the rates $w(x_1 - x_0, \epsilon)$, $w(x_2 - x_1, \epsilon)$, etc. for the particle to arrive at x_1 from x_0 during the first time interval $\epsilon = t_1 - t_0$, then to move from x_1 to x_2 over the same time ϵ, and so on, i.e.

$$
w(x - x_0, t - t_0) = \int_{-\infty}^{\infty} dx_1 \cdots \int_{-\infty}^{\infty} dx_N \, w(x - x_N, \epsilon) \times
$$
$$
\times w(x_N - x_{N-1}, \epsilon) \ldots w(x_2 - x_1, \epsilon) \, w(x_1 - x_0, \epsilon) . \tag{6.27}
$$

It is easy to see that this equation can be obtained by the recursive application of the Einstein–Smoluchowski–Kolmogorov–Chapman equation (6.18) which was written for a single intermediate point. The equation above is an extension of it for many such intermediate points. It also means that the diffusion from x_0 to x can be thought of as a sum over all possible trajectories as illustrated in Fig. 6.19, and is called Wiener[7] path integral.

The individual transition probability is given by Eq. (5.13):

$$
w(x_i - x_j, \epsilon) = \frac{1}{\sqrt{4\pi D\epsilon}} e^{-(x_i - x_j)^2/4D\epsilon} , \tag{6.28}
$$

where D is the diffusion coefficient. As was shown in Problem 5.11, this transition rate satisfies the diffusion equation (5.14).

Integrate explicitly the transition probability N times in Eq. (6.27) to show that $w(x - x_0, t - t_0)$ is given by Eq. (6.28) with $\epsilon \to t - t_0$ and $x_i - x_j \to x - x_0$. [Hint: *integrate first over x_1 using Eq. (6.17), then use this result to integrate over x_2, then integrate over x_3, and so on. You should quickly observe a very simple rule which should enable you to propose a formula for the result of the integration over $x_1, \ldots, x_k$. Then, use induction to show that the integration over x_{k+1} gives the same formula after the substitution $x_{k+1} \to x_{k+2}$ and $(k + 1) \to (k + 2)$. The required result then follows immediately.*]

---

[7] Named after Norbert Wiener.

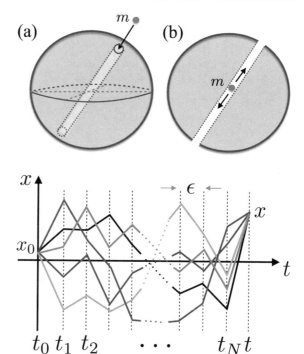

Fig. 6.18 a A thin tunnel is dug through the Earth between two opposite points on its surface and through its centre. Then a small body of mass m is placed into the tunnel and released. **b** Due to gravity, the body will oscillate in the tunnel

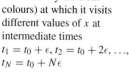

Fig. 6.19 Diffusion of a particle from x_0 to x over time $t - t_0$ can be envisaged as a sampling of different trajectories (some of which are shown by different colours) at which it visits different values of x at intermediate times $t_1 = t_0 + \epsilon$, $t_2 = t_0 + 2\epsilon$, ..., $t_N = t_0 + N\epsilon$

6.3   Applications in Physics: Kinetic Theory of Dilute Gases

In this section, we shall consider elements of the classical kinetic theory of dilute gases. Specifically, we shall derive a familiar equation for the pressure of the dilute gas by considering explicitly collisions of the gas molecules with a wall, the heat, mass and momentum transfer by the gas molecules per surface area and per unit time, when there is a distribution of temperature and molecular concentration in the gas, or of the gas molecules velocities, respectively. These applications require performing multiple integrations using spherical coordinates and may serve as an excellent illustration of the techniques we have developed above.

6.3.1   Maxwell Distribution

We shall start by discussing the Maxwell distribution given the probability to find a molecule with velocity $\mathbf{v} = (v_x, v_y, v_z)$ within the interval between $\mathbf{v}$ and $\mathbf{v} + d\mathbf{v}$. To derive the Maxwell distribution, we shall start from the following problem.

Problem 6.24 Let the probability distribution for a molecule to have its x component of the velocity v_x between v_x and $v_x + dv_x$ be $f(v_x^2)dv_x$, and similarly for the other components v_y and v_z. Due to equivalence of both directions along each of the axes, the probability distributions must not depend on the sign of the components; hence, the distributions can be made a function of the velocity components squared. At the same time, the probability distribution for an atom to have the velocity $v^2 = v_x^2 + v_y^2 + v_z^2$ should also be given by the same function $f(v^2)$ that depends on the full velocity squared. Obviously, $f(v^2)$ must be equal to the product of individual probabilities along each of the Cartesian axes:

$$f(v_x^2 + v_y^2 + v_z^2) = f(v_x^2)f(v_y^2)f(v_z^2).$$

By differentiating this equation, find that the function $f(x) = Ce^{kx^2}$, i.e. it can only depend exponentially on the velocity squared.

Hence, the probability to find a molecule with velocities within the interval between $\mathbf{v}$ and $\mathbf{v} + d\mathbf{v}$ is

$$df = f\left(v_x, v_y, v_z\right) dv_x dv_y dv_z = Ce^{-\alpha\left(v_x^2 + v_y^2 + v_z^2\right)} dv_x dv_y dv_z.$$

If we integrate over all possible velocities, the probability must be equal to one as this event is certain.

Problem 6.25 Using spherical coordinates, perform the integration over all velocities to show that the normalisation constant $C = (\alpha/\pi)^{3/2}$. Note that when going to the spherical coordinates, the velocity volume element $dv_x dv_y dv_z = v^2 dv \sin\theta d\theta d\phi$ according to Eq. (6.26). You will need the first of the Gaussian integrals in Eq. (7.28) in performing the v integration.

To obtain a physically meaningful exponent α, we should consider the average square of the velocity

$$\overline{v^2} = \int \int \int v^2 f\left(v_x, v_y, v_z\right) dv_x dv_y dv_z,$$

since, due to equipartition theorem, $m\overline{v^2}/2 = (3/2)k_BT$, where m is the mass of the molecule, k_B the Boltzmann constant and T temperature.

Problem 6.26 Perform the calculation of the average velocity square to find that $\alpha = m/2k_BT$. You will need another Gaussian integral from Eq. (7.28) to perform the v integration.

Hence, the Maxwell distribution is

$$f\left(v_x, v_y, v_z\right) = \left(\frac{m}{2\pi k_B T}\right)^{3/2} \exp\left[-\frac{m}{2k_B T}\left(v_x^2 + v_y^2 + v_z^2\right)\right]. \tag{6.29}$$

Problem 6.27 Show that the mean (average) speed of the gas molecules

$$\overline{v} = \int\int\int vf\left(v_x, v_y, v_z\right) dv_x dv_y dv_z = \sqrt{\frac{8k_B T}{\pi m}}. \tag{6.30}$$

6.3.2   Gas Equation

Consider the pressure gas molecules exert on a wall of a container. This pressure arises due to collisions of the gas molecules with the wall: the molecules change their momentum upon the collisions exerting a force on the wall which is the origin of the pressure. Let the wall be located in the $x - y$ plane as shown in Fig. 6.20a. Consider all molecules with velocities v that make the angle θ to the surface normal (the z-axis), which collide with the surface area ΔA of the wall over time Δt. All these molecules would fit into the blue tilted cylinder of the side $v\Delta t$. Its volume $\Delta V = h\Delta A = v\Delta t \Delta A \cos\theta$, where $h = v\Delta t \cos\theta$ is the cylinder height. The total number of such molecules whose velocities lie within the interval between $\mathbf{v}$ and $\mathbf{v} + d\mathbf{v}$ is (n is the molecules concentration)

$$\begin{aligned}
\Delta N &= (n\,\Delta V)\, f\left(v_x, v_y, v_z\right) dv_x dv_y dv_z \\
&= n\Delta A\Delta t\, v\cos\theta\, f\left(v_x, v_y, v_z\right) dv_x dv_y dv_z \\
&= n\Delta A\Delta t\left(\frac{m}{2\pi k_B T}\right)^{3/2} \exp\left[-\frac{mv^2}{2k_B T}\right] v^3 dv \cos\theta \sin\theta d\theta d\phi,
\end{aligned} \tag{6.31}$$

where in the last line we have introduced spherical coordinates. Each of these molecules upon collision with the surface area, as shown in Fig. 6.20b, changes its momentum by

$$m\mathbf{v}_f - m\mathbf{v}_i = 2mv\cos\theta\,\widehat{\mathbf{k}},$$

where $\mathbf{v}_i = \mathbf{v}$ is the initial velocity and $\mathbf{v}_f$ final velocity, after the elastic collision, and the change of the momentum is directed along the z-axis (along the unit vector $\widehat{\mathbf{k}}$). Hence, the total momentum transferred to the surface area is $\Delta K = \Delta N\, 2mv\cos\theta$. Integrating over all velocities that can reach the surface area, i.e. integrating over $0 \leq v < \infty, 0 \leq \theta \leq \pi/2$ and $0 \leq \phi \leq 2\pi$, we should be able to calculate the total

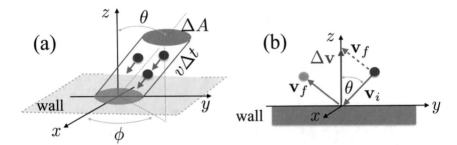

Fig. 6.20 The calculation of the gas pressure on the wall presented by the $x - y$ plane: **a** molecules inside the volume indicated by the bent cylinder and with the velocity $\mathbf{v} = (v \sin \theta \cos \phi, v \sin \theta \sin \phi, v \cos \theta)$ can only hit the area ΔA of the wall over time Δt. **b** A molecule hitting the wall with velocity $\mathbf{v} = \mathbf{v}_i$ is elastically reflected from the wall with the velocity $\mathbf{v}_f$, where $|\mathbf{v}_i| = |\mathbf{v}_f|$, so that the change of the velocity $\Delta \mathbf{v} = \mathbf{v}_f - \mathbf{v}_i$ (shown in blue) is directed exactly perpendicular to the wall surface (along z)

momentum transfer due to all molecules of the gas above the wall, per unit area and time, as

$$\frac{\Delta K}{\Delta A \Delta t} = 2mn \left(\frac{m}{2\pi k_B T} \right)^{3/2} \int_0^\infty \exp \left[-\frac{mv^2}{2k_B T} \right] v^4 dv$$

$$\times \int_0^{\pi/2} \cos^2 \theta \sin \theta d\theta \int_0^{2\pi} d\phi .$$

Performing the three integrations (do it!), we obtain:

$$\frac{\Delta K}{\Delta A \Delta t} = nk_B T .$$

This is exactly the pressure P as it is the force ΔF exerted per unit area, $P = \Delta F / \Delta A$, and the force is given by the change of the momentum per unit time, $\Delta F = \Delta K / \Delta t$. Hence, we obtain the familiar ideal gas equation $P = nk_B T$.

6.3.3   Kinetic Coefficients

Consider first the phenomenon of the heat transport in an ideal (dilute) gas. Let us assume that there is a temperature gradient along the z-axis.

We choose an imaginary horizontal $x - y$ plane at the position z_0 along the z-axis (such a plane for $z_0 = 0$ is shown in Fig. 6.20a), and then calculate the number of gas molecules coming to some area ΔA of that plane from above it. The total number of molecules coming to the plane from above is given by Eq. (6.31) of the preceding subsection. These molecules come with a certain energy ϵ they acquired after the last collision. If the mean free path of the molecules in the gas (the mean distance between two consecutive collisions) is λ, then the molecules with velocities v making the angle θ with the vertical (as in the figure) had their last collision at the height $z = z_0 + \lambda \cos \theta$ above the imaginary plane and hence their energy (assuming

Fig. 6.21 Molecules coming
to the imaginary $x - y$ plane
from below at the angle θ to
the vertical

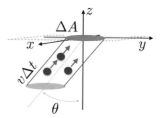

λ is sufficiently small, but still much larger than $v\Delta t$, and the length of the cylinder
in Fig. 6.20a) is

$$\epsilon(z) = \epsilon(z_0 + \lambda \cos\theta) \approx \epsilon(z_0) + \lambda \cos\theta \frac{d\epsilon}{dz} = \epsilon(z_0) + \lambda \cos\theta \frac{d\epsilon}{dT}\frac{dT}{dz}$$

with the derivative dT/dz being calculated at the height z_0 of the plane (we have
just applied the Taylor formula keeping the first two terms). The derivative of the
molecular energy with respect to the temperature is equal to mc_V, where c_V is
the specific heat capacity. Hence, the energy $\Delta Q_\downarrow$ that is carried by the molecules
through the plane from above, per unit area and time, is

$$\frac{\Delta Q_\downarrow}{\Delta A \Delta t} = \int\int\int n\left[\epsilon(z_0) + mc_V \lambda \cos\theta \frac{dT}{dz}\right] f(v)v^3 dv \cos\theta \sin\theta d\theta d\phi,$$

where we integrate over all possible velocities $0 \le v < \infty$ and all angles $0 \le \theta \le$
$\pi/2$ and $0 \le \phi \le 2\pi$. Obviously, the chosen range of the angle θ guarantees only
molecules coming from above.

Problem 6.28 Perform the triple integration to show that

$$\frac{\Delta Q_\downarrow}{\Delta A \Delta t} = \frac{1}{4}n\bar{v}\left[\epsilon(z_0) + \frac{2}{3}mc_V \lambda \frac{dT}{dz}\right],$$

where $\bar{v}$ is the mean velocity of Eq. (6.30).

The obtained expression gives the energy flux carried by the molecules through
the area ΔA per unit time from above. There will also be molecules passing this
imaginary surface from below as well, and we can perform similarly the calculation
for them. The flux of molecules from below, see Fig. 6.21, would follow the same
expression as for the molecules coming from the above, with the acute angle θ
velocities make with the vertical. Each of the molecules carries the energy

$$\epsilon(z) = \epsilon(z_0 - \lambda \cos\theta) \approx \epsilon(z_0) - mc_V \lambda \cos\theta \frac{dT}{dz}.$$

Hence, the only difference with the previous case is the minus sign before λ. Hence, the energy flux becomes

$$\frac{\Delta Q_\uparrow}{\Delta A \Delta t} = \frac{1}{4} n \bar{v} \left[\epsilon(z_0) - \frac{2}{3} m c_V \lambda \frac{dT}{dz} \right].$$

The net flux of energy, per unit area and time, through the imaginary surface at z_0 is therefore the difference of the two fluxes:

$$\frac{\Delta Q}{\Delta A \Delta t} = \frac{\Delta Q_\downarrow - \Delta Q_\uparrow}{\Delta A \Delta t} = \left(\frac{1}{3} n \bar{v} m c_V \lambda \right) \frac{dT}{dz}.$$

In the round brackets we have the heat transport coefficient given by this elementary kinetic theory. Hence, the energy flux through a surface area is proportional to the rate dT/dz in which the temperature changes.

Next, we shall consider diffusion along the z-axis. Similarly, to the previous case of heat transport, in the consideration of the mass transport we face the situation in which there is a concentration gradient along this axis, i.e. the concentration of the molecules $n(z)$ depends on z.

Problem 6.29 Calculate the difference of the molecular fluxes, one from the above and another from the below, i.e. a net flux through an imaginary $x - y$ plane at position z_0. Using a similar consideration to the heat transport, show that the net flux of the number of molecules, per unit area and time, is

$$\frac{\Delta N}{\Delta A \Delta t} = \frac{\Delta N_\downarrow - \Delta N_\uparrow}{\Delta A \Delta t} = \left(\frac{1}{3} \bar{v} \lambda \right) \frac{dn}{dz}.$$

Here $D = \bar{v}\lambda/3$ is the diffusion coefficient. The flux is found to be proportional to the rate of change, dn/dz, of the molecular concentration (the first Fick's law).

Finally, let us consider viscosity. Suppose, we have our gas confined between two parallel plates, both are perpendicular to the z-axis. If we assume that the bottom plate is fixed and the upper plate moves with a certain velocity parallel to the bottom one, gas molecules would acquire some collective (systematic) horizontal velocity $u(z)$ (with respect to the bottom plate which is at rest), which, due to viscosity, will be the largest near the upper plate, see Fig. 6.22.

Consider an imaginary $x - y$ plane at some position z_0 between the two plates. Molecules that experienced their last collision at the height $z = z_0 \pm \lambda \cos \theta$ have their horizontal momentum

$$p_\downarrow(z) = m u_\downarrow(z) = m \left[u(z_0) + \lambda \cos \theta \frac{du}{dz} \right]$$

Fig. 6.22 The top plate moves parallel to the bottom plate causing the gas molecules confined inside to acquire an additional collective horizontal velocity $u(z)$ that depends on their vertical position z between the plates

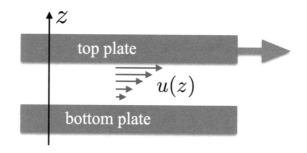

and

$$p_\uparrow(z) = mu_\uparrow(z) = m\left[u\,(z_0) - \lambda\cos\theta\frac{du}{dz}\right],$$

for the molecules coming from above and below the imaginary plane, respectively, similar to our previous examples.

Problem 6.30 Calculate the difference of the total (integrated over all velocities) momenta $\Delta P = P_\downarrow - P_\uparrow$, per unit area and time, brought by the two fluxes of the molecules, and show that the shear stress is

$$\frac{\Delta P}{\Delta A\Delta t} = \left(\frac{1}{3}mn\bar{v}\lambda\right)\frac{du}{dz},$$

where the coefficient in the round brackets represents the shear viscosity coefficient.

6.4   Line Integrals

6.4.1   Line Integrals for Scalar Fields

The consideration in this section generalises the one we made in Sect. 4.7.1 when we considered the calculation of the length of a curved line.

Consider a scalar function (a *scalar field*) $f(\mathbf{r}) = f(x, y, z)$ that is specified in the three-dimensional space. This could be a linear mass density of a wire ρ, temperature T, potential energy U, and so on. Then, consider a curved line AB that is specified parametrically as

$$x = x(u)\,, \quad y = y(u)\,, \quad z = z(u)\,, \tag{6.32}$$

where u is a parameter (e.g. time t in which case the line may be a particle trajectory in the 3D space). We assume that $a \le u \le b$. Then, let us divide the line into n sections by points $M_1, M_2, \ldots, M_{n-1}$, the beginning and the end of the line are marked by points $M_0 \equiv A$ and $M_n \equiv B$, respectively. Hence, the curve is specified by points $M_0 = A, M_1, M_2, \ldots, M_{n-1}, M_n = B$ as shown in Fig. 6.23. These division points correspond to the values $u_0, u_1, u_2, \ldots, u_{n-1}, u_n$ of the parameter u. Link the nearest points with straight lines (to form a red broken line as shown in the figure), and choose a point N_i somewhere inside each section $M_{i-1}M_i$ that corresponds to the value ξ_i of the parameter u. Finally, consider the sum

$$\sum_{i=1}^{n} f(N_i)\Delta_i = \sum_{i=1}^{n} f\left(x(\xi_i), y(\xi_i), z(\xi_i)\right) \Delta_i \,,$$

where

$$\Delta_i = |M_{i-1}M_i| = \sqrt{(x_i - x_{i-1})^2 + (y_i - y_{i-1})^2 + (z_i - z_{i-1})^2}$$

is the length of the ith section of the broken line. Next, we replace the differences of the coordinates using derivatives of the functions (6.32) with respect to the parameter,[8] e.g.,

$$x_i - x_{i-1} = x(u_i) - x(u_{i-1}) = x(u_{i-1} + \Delta u_i) - x(u_{i-1}) \approx x'(\xi_i)\,\Delta u_i \,,$$

where $\Delta u_i = u_i - u_{i-1}$ is the change of the parameter u when going from the point M_{i-1} to M_i. Repeating this for the y and z, we can write

$$\Delta_i \approx \sqrt{x'(\xi_i)^2 + y'(\xi_i)^2 + z'(\xi_i)^2}\,\Delta u_i$$

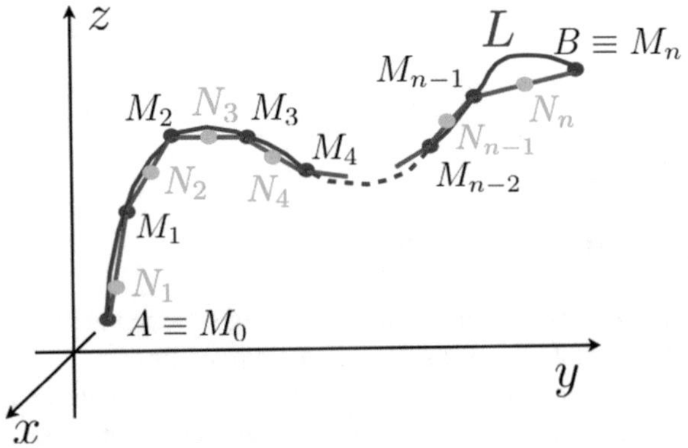

Fig. 6.23 A curved line (blue) between points A and B specified via the parameter u spanning between a and b is divided by points M_i $(i = 1, \ldots, n-1)$ into n curved segments that are then replaced by straight segments (red). Points $N_1, N_2, \ldots, N_n$ are chosen somewhere inside each such segment

---

[8] This corresponds to the terms of the order $(\Delta u_i)^2$, see Sect. 3.8.

and the integral sum becomes

$$\sum_{i=1}^{n} f(N_i)\Delta_i \approx \sum_{i=1}^{n} f(x(\xi_i), y(\xi_i), z(\xi_i)) \sqrt{x'(\xi_i)^2 + y'(\xi_i)^2 + z'(\xi_i)^2}\Delta u_i .$$

In the limit of all $\Delta u_i \to 0$ (whereby the number of segments tends to infinity), the obtained expression represents an ordinary integral sum of the definite integral over u,

$$I = \int_a^b f(x(u), y(u), z(u))\sqrt{x'(u)^2 + y'(u)^2 + z'(u)^2}du , \qquad (6.33)$$

and hence all the theorems discussed in Sect. 4.2 of the existence of a definite integral are directly applicable here. If the limit exists, then it does not depend on the way the curved line is divided up and how the points N_i are chosen within each segment, and the sum converges to what is called a *line integral* between points A and B. It is denoted as follows:

$$I = \int_{AB} f(x, y, z)dl = \int_{AB} f(l)dl , \qquad (6.34)$$

where

$$dl = \sqrt{dx^2 + dy^2 + dz^2} = \sqrt{x'(u)^2 + y'(u)^2 + z'(u)^2}du$$

is the length of the differential section of the line related to the changes dx, dy and dz of the Cartesian coordinates due to the increment du of the parameter (e.g. $dx = (dx/du)\, du = x'(u)du$ and so on), i.e. the differential length (4.93) (see also Fig. 5.5), and $f(l)$ is the value of the function $f(x, y, z)$ on the line at the point where the length l is reached (e.g. measured from the initial point A); obviously, the value of l is directly related to the value of u. This line integral is called the *line integral for scalar fields* or of *the first kind*.

If the line is *closed*, then the following notation is frequently used:

$$I = \oint_L f(l)dl .$$

Formula (6.33) can be used to calculate the integral in practice.

The above defined line integral possesses all properties of definite integrals. There is only one exception: the integral does *not* change sign upon changing the direction of integration as we deal with the scalar field and the positive length dl:

$$\int_{AB} f(x, y, z)dl = \int_{BA} f(x, y, z)dl . \qquad (6.35)$$

To make sure that this is the case in practice, you must follow the rule: if you integrate from smaller to larger values of u, then the change $du > 0$ and hence dl will automatically be positive. However, if you integrate in the opposite direction,

Fig. 6.24 The derivation of
the elementary length dl on a
circle of radius R

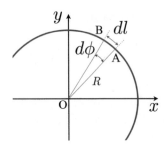

from larger to smaller values of u, then $du < 0$ and hence the minus sign should then
be assigned to it (and hence to the integral) to ensure that $dl > 0$.

Let us now consider some examples.

First, we shall consider the calculation of the length of a circle of radius R. The
circle can be specified on the $x - y$ plane as $x = R \cos \phi$ and $y = R \sin \phi$ with ϕ
changing from zero to 2π. So in this problem the role of the parameter u is played
by the angle ϕ. Note that since this is a planar problem, $z'(\phi) = 0$. Then, the length
is obtained as a line integral with $f(x, y, z) = 1$:

$$l = \int_0^{2\pi} \sqrt{(x'(\phi))^2 + (y'(\phi))^2} d\phi = \int_0^{2\pi} \sqrt{R^2 \sin^2 \phi + R^2 \cos^2 \phi} d\phi$$

$$= \int_0^{2\pi} R d\phi = 2\pi R ,$$

as expected. Note that the line element in polar coordinates appears to be simply
$dl = R d\phi$, and summing up such line elements along the circle, as is clear from the
above, gives the whole circle length. This value of the elementary length dl on a
circle of radius R becomes evident also from Fig. 6.24. Indeed, dl is the side of the
isosceles $\triangle ABO$ with the sides of length R and the base $BA = 2R \sin \frac{d\phi}{2}$. Since
$\sin x \simeq x$ for small x, we immediately get $dl = BA = R d\phi$, the result anticipated.

In our second example we shall determine the mass of a non-uniform wire which
has a shape of a spiral and is given in polar coordinates as $r = e^{-\phi}$, with the angle ϕ
changing from 0 to ∞, see Fig. 1.38d. We shall assume that the linear mass density of
the spiral is proportional to the revolution angle: $\rho(\phi) = \alpha \phi$, where α is a parameter.
The mass is given by the integral $M = \int \rho dl$. Introducing polar coordinates, we
arrive at the following definite integral:

$$l = \alpha \int_0^{\infty} \phi \sqrt{(x'(\phi))^2 + (y'(\phi))^2} d\phi ,$$

where $x(\phi) = r(\phi) \cos \phi = e^{-\phi} \cos \phi$ and $y(\phi) = e^{-\phi} \sin \phi$ so that

$$l = \alpha \int_0^{\infty} \phi \sqrt{\left(-e^{-\phi} \cos \phi - e^{-\phi} \sin \phi\right)^2 + \left(-e^{-\phi} \sin \phi + e^{-\phi} \cos \phi\right)^2} d\phi$$

$$= \alpha \int_0^{\infty} \phi e^{-\phi} \sqrt{2 \left(\cos^2 \phi + \sin^2 \phi\right)} d\phi = \alpha \sqrt{2} \int_0^{\infty} \phi e^{-\phi} d\phi = \alpha \sqrt{2} .$$

Fig. 6.25 Triangle $\triangle ABO$

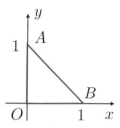

Interestingly, despite of the fact that the wire becomes heavier and heavier as it revolves indefinitely around the centre of the coordinate system, its mass remains finite.

In our third example, we will calculate the line integral

$$I = \oint_L (x + 2y)\, dl$$

along the closed path L shown in Fig. 6.25. We split the integral into three contributions:

$$\oint_L (x + 2y)\, dl = \int_{OB} (x + 2y)\, dl + \int_{BA} (x + 2y)\, dl + \int_{AO} (x + 2y)\, dl .$$

We shall use x as a parameter u for the line intervals OB and AB with the length increment

$$dl = \sqrt{dx^2 + dy^2} = \sqrt{dx^2 \left(1 + \left(\frac{dy}{dx}\right)^2\right)} = \sqrt{1 + y'(x)^2}dx ,$$

while y should be used for AO as the parameter, with the line increment being

$$dl = \sqrt{dx^2 + dy^2} = \sqrt{dy^2 \left(1 + \left(\frac{dx}{dy}\right)^2\right)} = \sqrt{1 + x'(y)^2}dy .$$

Remember that the increment of length should always be positive, and the integral does not depend on the direction along the line! Along the line OB: $y = 0$ and $y'(x) = 0$:

$$\int_{OB} (x + 2y)\, dl = \int_0^1 (x + 2 \cdot 0) \sqrt{1 + 0^2}dx = \int_0^1 x\, dx = \frac{x^2}{2}\Big|_0^1 = \frac{1}{2} .$$

Along the line BA we have: $y = -x + 1$, $y'(x) = -1$, so

$$\int_{BA} (x + 2y)\, dl = \int_{AB} (x + 2y)\, dl = \int_0^1 (x + 2(1 - x)) \sqrt{1 + (-1)^2}dx$$

$$= \sqrt{2} \int_0^1 (2 - x)\, dx = 2\sqrt{2} - \sqrt{2}\frac{1}{2} = \frac{3}{\sqrt{2}} .$$

Note that we integrated from $x = 0$ to $x = 1$ to ensure that $dl > 0$. Finally, along the line AO we have $x = 0$ and $x'(y) = 0$, so (recall that in this case y is the parameter):

$$\int_{AO} (x + 2y)\, dl = \int_{OA} (x + 2y)\, dl = \int_0^1 (0 + 2y)\, dy = \int_0^1 2y\, dy = 1 .$$

Again, we integrated from the smaller ($y = 0$) to the larger ($y = 1$) value of the y here to ensure the positiveness of dl. Adding up all the three contributions, we obtain the final result: $I = \frac{1}{2} + \frac{3}{\sqrt{2}} + 1 = \frac{3}{2}\left(1 + \sqrt{2}\right)$.

Problem 6.31 Find the length of a two-dimensional spiral specified parametrically in polar coordinates (r, ϕ) as $r^2 = ae^{-\alpha\phi}$, where $a > 0$, $\alpha > 0$ and $0 \le \phi \le \infty$. [Answer: $2\sqrt{a}\left(1 + \alpha^2/4\right)/\alpha$.]

Problem 6.32 Find the length of a line specified parametrically as $x = e^{-\varphi}\cos\varphi$, $y = e^{-\varphi}\sin\varphi$ and $z = e^{-\varphi}$, where $0 < \varphi < \infty$. [*Answer:* $\sqrt{3}$.]

Problem 6.33 Calculate the length of the parabola $y = x^2$ between points $(x, y) = (0, 0)$ and $(1, 1)$. [Hint: *the integral (4.64) may be of use.*] [Answer: $\left[2\sqrt{5} + \ln\left(2 + \sqrt{5}\right)\right]/4$.]

Problem 6.34 Calculate the mass of ten revolutions of the unfolding helix wire given parametrically by $x = e^{\phi}\cos\phi$, $y = e^{\phi}\sin\phi$ and $z = e^{\phi}$ and with the linear mass density $\rho = \alpha\phi$. [Answer: $\alpha\sqrt{3}\left[(20\pi - 1)e^{20\pi} + 1\right]$.]

6.4.2　Line Integrals for Vector Fields

Above, a scalar field was considered. However, one may also consider a vector function $\mathbf{F}(\mathbf{r}) = \mathbf{F}(x, y, z)$ specified in the 3D space, the so-called *vector field*. For instance, it could be a force acting on a particle moving along some trajectory L. If we would like to calculate the work done by the force on the particle, we may split the trajectory into little *directional straight* segments $\Delta\mathbf{l}$ that would run along the actual trajectory (see again Fig. 6.23), calculate the work $\Delta A = \mathbf{F}(l) \cdot \Delta\mathbf{l}$ done along each of them, and then sum up all the contributions:

$$A \simeq \sum_{i=1}^n \mathbf{F}(l_i) \cdot \Delta\mathbf{l}_i ,$$

where $\mathbf{F}\,(l_i)$ is the value of the vector field at some point $(x(\xi_i),\,y(\xi_i),\,z(\xi_i))$ within the segment $M_{i-1}M_i$ corresponding to the length l_i of the line measured from the initial point $M_0 = A$, see Fig. 6.23, and

$$\Delta \mathbf{l}_i = (\Delta x_i, \Delta y_i, \Delta z_i) \approx \left(x'\,(\xi_i)\,\Delta u_i, \, y'\,(\xi_i)\,\Delta u_i, \, z'\,(\xi_i)\,\Delta u_i\right)$$

is the vector of the line increment connecting the points M_{i-1} and M_i, see also Fig. 5.5. Hence,

$$A \approx \sum_{i=1}^{n} \left[F_x\,(l_i)\,x'\,(\xi_i) + F_y\,(l_i)\,y'\,(\xi_i) + F_z\,(l_i)\,z'\,(\xi_i)\right] \Delta u_i$$

is an integral sum for an appropriate definite integral over the parameter u that appears in the limit when all the segments are getting smaller and smaller. Hence, as each of the segments length tend to zero, we shall arrive at the definite integral

$$I = \int_L \left[\widetilde{F}_x(u)x'(u) + \widetilde{F}_y(u)y'(u) + \widetilde{F}_z(u)z'(u)\right] du \,, \tag{6.36}$$

where $\widetilde{F}_x(u) \equiv F_x\,(x(u), y(u), z(u))$, and similarly $\widetilde{F}_y(u)$ and $\widetilde{F}_z(u)$. These are functions of u and as such they are calculated *along* the line L. This is called a line integral and is denoted by

$$I = \int_L \mathbf{F}(l) \cdot d\mathbf{l} = \int_L F_x(l)dx + F_y(l)dy + F_z(l)dz \,. \tag{6.37}$$

In the second equality we have written the dot product of $\mathbf{F}$ and $d\mathbf{l}$ explicitly via the three components dx, dy and dz of the vector $d\mathbf{l}$. Note that the increments dx, dy and dz are not independent, but correspond to the *change along the line*, i.e. all three depend on the single parameter u. This line integral is also called a *line integral for vector fields* or of *the second kind*.

Two methods for the calculation of the line integral may be of use: using a parametric representation of the line L or in Cartesian coordinates. In the first method the line L is specified parametrically as $x = x(u)$, $y = y(u)$ and $z = z(u)$, as we have done above, and hence the integral is calculated as the du integral simply by noticing that $dx = x'(u)du$, and similarly for dy and dz; in other words, Eq. (6.36) is used. Of course, the integral can also be split into three independent integrals (since the whole integral is a limit of an integral sum, and the sum can be split into three), each of them to be calculated independently, e.g.

$$\int_L F_x(l)dx = \int_L F_x\,(x(u), y(u), z(u))\,x'(u)du = \int_L \widetilde{F}_x(u)x'(u)du \,,$$

and similarly for the other two integrals. Either of these methods gives a simple recipe to calculate this type of line integral, and these techniques are the most convenient in practice.

In the second (Cartesian) method, projections of the line L on the Cartesian planes $y - z$, $x - z$ and $x - y$ are used. Indeed, first the line integral is split into three:

$$\int_L F_x(l)dx + F_y(l)dy + F_z(l)dz = \int_L F_x(l)dx + \int_L F_y(l)dy + \int_L F_z(l)dz .$$
$$(6.38)$$

Here, each of the integrals in the right-hand side is completely independent and can be calculated separately. When calculating any of the three integrals, e.g. the first one $\int_L F_x dx$, we have to assume that the other two variables, y and z, are some functions of the x; this is possible as we integrate *alone* the line, i.e. x is simply used to traverse the line (instead of l or u). Basically, this method is similar to the first one; the main difference is that $u = x$ is used for the calculation of the dx integral, $u = y$ for the dy, and $u = z$ for the dz integrals, respectively, i.e. different parameters are employed for each part of the line integral.

It may seem that the line integral we just introduced is quite different from the one defined in the previous Sect. 6.4.1 by Eq. (6.34). Indeed, for instance, our new integral *does* change sign when the direction of the calculation along the same line is reversed:

$$\int_{AB} \mathbf{F}(l) \cdot d\mathbf{l} = - \int_{BA} \mathbf{F}(l) \cdot d\mathbf{l} ,$$
$$(6.39)$$

since the directional differential length $d\mathbf{l}$ changes its sign in this case. However, the two integrals are still closely related. By noting that $d\mathbf{l} = \mathbf{m}dl$, where $\mathbf{m}$ is the unit vector directed tangentially to the line L at each point and along the direction of integration, and dl is the differential length, we can write:

$$\int_L \mathbf{F}(l) \cdot d\mathbf{l} = \int_L [\mathbf{F}(l) \cdot \mathbf{m}] dl ,$$
$$(6.40)$$

where we have now arrived at the familiar line integral for the scalar field $f(l) = \mathbf{F}(l) \cdot \mathbf{m}$ in the right-hand side.

Consider several examples.

Example 6.6 ► Let us calculate the line integral

$$J = \int_{AB} xdx - ydy + zdz$$

along one revolution of a helical line in Fig. 4.7a that is specified parametrically as $x = R \cos \phi$, $y = R \sin \phi$ and $z = b\phi$.

Solution. One revolution corresponds to a change of ϕ from zero to 2π. We have:

$$dx = -R \sin \phi d\phi, \quad dy = R \cos \phi d\phi \quad \text{and} \quad dz = bd\phi ,$$

so

$$J = \int_0^{2\pi} [-xR \sin \phi - yR \cos \phi + zb] \, d\phi = \int_0^{2\pi} \left[-2R^2 \sin \phi \cos \phi + b^2 \phi \right] d\phi$$

$$= -2R^2 \int_0^0 \sin \phi d (\sin \phi) + b^2 \int_0^{2\pi} \phi d\phi = 0 + b^2 \left. \frac{\phi^2}{2} \right|_0^{2\pi} = 2\pi^2 b^2 . \blacktriangleleft$$

Example 6.7 ▶ In our second example we shall calculate the integral

$$J = \oint_L 2xy \, dx + x^2 \, dy$$

along the closed loop shown in Fig. 6.25 which is traced anticlockwise.

Solution. Consider each of the three parts of the path separately. Along OB $dy = 0$ and

$$J_{OB} = \int_0^1 2xy \, dx = 0 \ \text{ since } \ y = 0 .$$

Along BA $y = 1 - x$, so by considering x as the parameter u, we have $dy = y'(x) dx = -dx$, and hence

$$J_{BA} = \int_1^0 \left[2xy \, dx + x^2 \, (-dx) \right] = \int_1^0 \left[2x \, (1 - x) - x^2 \right] dx$$

$$= \left. \left(2 \frac{x^2}{2} - 3 \frac{x^3}{3} \right) \right|_1^0 = 0 .$$

Note that, opposite to the example in the previous subsection, the direction of integration here does matter,[9] and hence along the path BA the integral was calculated from $x = 1$ to $x = 0$ corresponding to the change of x when moving from point B to A.

Finally, along AO we have $dx = 0$ and, therefore,

$$J_{AO} = \int_1^0 x^2 \, dy = 0 \ \text{ since } \ x = 0 .$$

Thus, the integral $J = 0$. ◀

Note that, as we shall see later on, this result is not accidental as the integrand is the total differential of the function $F(x, y) = x^2 y$. Indeed,

$$dF = \frac{\partial F}{\partial x} dx + \frac{\partial F}{\partial y} dy = 2xy \, dx + x^2 \, dy .$$

---

[9] Although in this particular case it does not affect the final zero result!.

Fig. 6.26 A square region

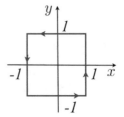

Problem 6.35 Calculate the line integral $\int_L \mathbf{F} \cdot d\mathbf{l}$ along the straight line connecting points $(0, 0, 0)$ and $(1, 1, 1)$ for the vector field $\mathbf{F} = (z, y, x)$. [Answer: $3/2$.]

Problem 6.36 Calculate the work done by the force field $F = (0, x, 0)$ in moving a particle between two points $A(1, 0, 0)$ and $B(0, 1, 0)$ along a circle of unit radius and centred at the origin. [Answer: $\pi/4$.]

Problem 6.37 Evaluate the line integral $\int -x^2 dx + 3xy dy$ along the sides of the triangle ABC in the anticlockwise direction. The triangle lies in the $x - y$ plane and its vertices are: $A(0, 0)$, $B(2, 0)$ and $C(2, 2)$. [Answer: 4.]

Problem 6.38 Evaluate $\int \mathbf{F} \cdot d\mathbf{l}$ along the parabola $y = 2x^2$ between points $A(0, 0)$ and $B(1, 2)$ if $\mathbf{F} = xy\mathbf{i} - y^2\mathbf{j}$. [Answer: $-13/6$.]

Problem 6.39 A particle moves in the anticlockwise direction along a square, see Fig. 6.26, by the force $\mathbf{F} = xy\mathbf{i} + x^2 y^2 \mathbf{j}$. Calculate the work done by the force along the closed loop. [Answer: 0.]

Problem 6.40 Calculate the line integral $\oint \mathbf{F} \cdot d\mathbf{l}$ in the *clockwise* direction along the unit circle centred at the origin, if $\mathbf{F} = (-y\mathbf{i} + x\mathbf{j}) / (x^2 + y^2)$. [Answer: -2π.]

Problem 6.41 Evaluate the integral $\oint_L \mathbf{F} \cdot d\mathbf{l}$ taken in the anticlockwise direction along the circle of radius R centred at $(a, 0)$ if $\mathbf{F} = xy\mathbf{i} + y^2\mathbf{j}$. [Answer: $-\pi a R^2$.]

Problem 6.42 Calculate the work done by the force $\mathbf{F} = (-y, x, z)$ in moving a particle from $(1,0,0)$ to $(-1, 0, \pi)$ along the helical trajectory specified via $x = \cos t,\ y = \sin t,\ z = t$. [Answer: $\pi \left(1 + \pi/2\right)$.]

6.4.3   Two-Dimensional Case: Green's Formula

A plane region is said to be *simply connected* if it does not have holes in it, see Fig. 6.27. Holes may appear if, for instance, a function has singularities at some points which must be removed. It is possible to relate a double integral over some simply connected plane region S to a line integral taken over the boundary of this region.

Theorem 6.4 (Green's Theorem) *For any simply connected region S, the following* Green's formula *is valid:*

$$\int\!\!\int_S \left(\frac{\partial Q}{\partial x} - \frac{\partial P}{\partial y} \right) dx\, dy = \oint_L P(x, y)dx + Q(x, y)dy , \qquad (6.41)$$

where the double integral in the left-hand side is taken over the region S, while the line integral in the right-hand side is taken along its boundary L in the anticlockwise *direction.*

Proof: Consider a simply connected region S bounded from the bottom and the top by two functions $y = y_1(x)$ and $y = y_2(x)$, Fig. 6.28a, and two functions $P(x, y)$ and $Q(x, y)$ defined there. Note that the region S we are considering right now is

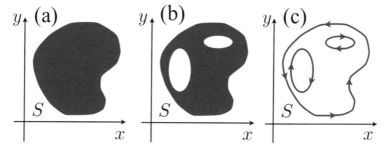

Fig. 6.27 Simply connected (**a**) and not simply connected (**b**) regions; (**c**) directions in the line integral in the Green's formula in the case of the region in (**b**) that contains two holes

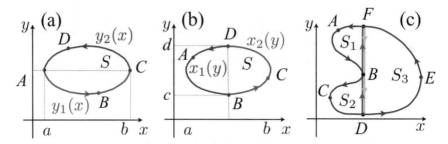

Fig. 6.28 The derivation of the Green's formula

very special: it is such that any line parallel to the coordinate axes would cross the boundary of S in *no more* than two points. Then, the double integral of $\partial P/\partial y$ over the region is

$$\iint_S \frac{\partial P}{\partial y} dxdy = \int_a^b dx \int_{y_1(x)}^{y_2(x)} \frac{\partial P(x,y)}{\partial y} dy = \int_a^b [P(x,y_2(x)) - P(x,y_1(x))] dx$$

$$= \int_a^b P(x,y_2(x)) dx - \int_a^b P(x,y_1(x)) dx$$

$$= -\int_b^a P(x,y_2(x)) dx - \int_a^b P(x,y_1(x)) dx$$

$$= -\int_{CDA} P(x,y) dx - \int_{ABC} P(x,y) dx = -\oint_{ABCDA} P(x,y) dx,$$

where $ABCDA$ is the boundary of the region S. In the left-hand side we have a double integral, while in the right a line integral taken in the *anticlockwise* direction around the region's boundary.

On the other hand, this special region can be considered as bounded by functions $x = x_1(y)$ and $x = x_2(y)$ on the left and the right of the region S, as shown in Fig. 6.28b. In this case, we can similarly write:

$$\iint_S \frac{\partial Q}{\partial x} dxdy = \int_c^d dy \int_{x_1(y)}^{x_2(y)} \frac{\partial Q(x,y)}{\partial x} dx = \int_c^d [Q(x_2(y),y) - Q(x_1(y),y)] dy$$

$$= \int_c^d Q(x_2(y),y) dy - \int_c^d Q(x_1(y),y) dy$$

$$= \int_c^d Q(x_2(y),y) dy + \int_d^c Q(x_1(y),y) dy$$

$$= \int_{BCD} Q(x,y) dy + \int_{DAB} Q(x,y) dy = \oint_{ABCDA} Q(x,y) dy,$$

where the line integral in the right-hand side is taken along the whole closed boundary $ABCDA$ of region S, traversed again in the *anticlockwise* direction. Now we subtract one expression from the other to get the required Eq. (6.41).

The formula we have just proved is rather limited as the derivation has been done for a very special region S shown in Fig. 6.28a, b. This result is, however, valid for *any* region, since it is always possible to split the arbitrarily complicated region into subregions each of the same type as considered above, i.e. that vertical and horizontal lines cross the boundary of each part at two points at most. The double integrals over all regions sum up into one double integral taken over the whole region, while the line integrals over all internal lines used to break the whole region into subregions would cancel out so that the line integral only over the outside boundary of S remains. To illustrate this idea of generalising the region shape, consider the region shown in Fig. 6.28c: the region S is split into three regions S_1, S_2 and S_3 by a vertical line. Let us consider the line integrals along each of the boundaries: $ABFA$ for S_1, $BCDB$ for S_2 and $BDEFB$ for S_3. Thus, for each of the regions the above Green's formula has been proven to be valid:

$$\int\int_{S_1} \left(\frac{\partial Q}{\partial x} - \frac{\partial P}{\partial y} \right) dxdy = \oint_{ABFA} P(x, y)dx + Q(x, y)dy \ ,$$

$$\int\int_{S_2} \left(\frac{\partial Q}{\partial x} - \frac{\partial P}{\partial y} \right) dxdy = \oint_{BCDB} P(x, y)dx + Q(x, y)dy \ ,$$

$$\int\int_{S_3} \left(\frac{\partial Q}{\partial x} - \frac{\partial P}{\partial y} \right) dxdy = \oint_{BDEFB} P(x, y)dx + Q(x, y)dy \ .$$

If we now sum up these equations, we will have the double integral over the whole region S in the left-hand side, and the line integral over all boundaries in the right-hand side. However, the line integral does change sign when the direction is changed. Therefore, the contributions from the added (vertical) line, which is run twice in the opposite directions, will exactly cancel out, and we arrive at the line integral taken only over the actual boundary of S. **Q.D.E.**

Green's formula is actually valid even for regions with holes, but in this case it is necessary to take the line integrals over the *inner boundary* of the region (the holes boundaries) as well going in the clockwise direction as shown in Fig. 6.27c. The idea of the proof is actually simple: we first consider an extension of the functions $P(x, y)$ and $Q(x, y)$ inside the holes (any will do), then apply Green's theorem to the whole region including the holes and subtract from it Green's formula for each of the holes.

As a simple example which demonstrates the validity of Green's formula, let us verify it for functions $P(x, y) = 2xy$ and $Q(x, y) = x^2$ and the triangle region shown in Fig. 6.25. We know from Example 6.7 of the previous section that the line integral is equal to zero. Consider now the left-hand side of Green's formula:

$$\int\int \left(\frac{\partial Q}{\partial x} - \frac{\partial P}{\partial y} \right) dxdy = \int\int (2x - 2x)\, dxdy = 0 \ ,$$

Fig. 6.29 Semicircle region

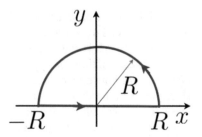

i.e. the same result as expected from Green's formula.

In our next example, we shall verify the Green's formula for $P(x, y) = x^2 + y^2$ and $Q(x, y) = 4xy$ and the region which is a semicircle of radius R lying in the $y \geq 0$ region as shown in Fig. 6.29. We will first calculate the double integral using polar coordinates $x = r \cos \phi$ and $y = r \sin \phi$ (recall that the corresponding Jacobian is r):

$$\int \int \left(\frac{\partial Q}{\partial x} - \frac{\partial P}{\partial y} \right) dxdy = \int \int (4y - 2y) \, dxdy = \int \int 2y dxdy$$

$$= \int \int (2r \sin \phi) \, rdrd\phi = 2 \int_0^R r^2 dr \int_0^\pi \sin \phi d\phi$$

$$= \frac{2R^3}{3} (-\cos \phi)_0^\pi = \frac{4R^3}{3}.$$

On the other hand, consider the line integral

$$\oint \left(x^2 + y^2 \right) dx + 4xydy$$

along the boundary of the region. The boundary consists of two parts: along the semicircle and along the x-axis from $-R$ to R. The first integral is most easily calculated using the polar coordinates $x = R \cos \phi$ and $y = R \sin \phi$ (using ϕ as a parameter representing the semicircle):

$$\int_{semicircle} \left[(R^2) x'(\phi) + \left(4R^2 \cos \phi \sin \phi \right) y'(\phi) \right] d\phi =$$

$$= \int_0^\pi \left[-R^3 \sin \phi + 4R^3 \cos^2 \phi \sin \phi \right] d\phi = R^3 \int_0^\pi \left(4 \cos^2 \phi - 1 \right) \sin \phi d\phi$$

$$= \left| \begin{array}{c} t = \cos \phi \\ dt = -\sin \phi d\phi \end{array} \right| = R^3 \int_1^{-1} \left(4t^2 - 1 \right) (-dt) = R^3 \left(\frac{4t^3}{3} - t \right)_{-1}^1 = \frac{2R^3}{3}.$$

The line integral along the bottom boundary of the semicircle is most easily calculated in the Cartesian coordinates since $y = 0$ (then $dy = 0$):

$$\int_{bottom} \left(x^2 + y^2 \right) dx + 4xydy = \int_{-R}^R x^2 dx = \frac{2R^3}{3}.$$

Summing up the two contributions, we arrive at the same result as given by the double integral above.

Problem 6.43 Consider a planar figure S with the boundary line L. Using the Green's formula, derive the following alternative expressions for the area of the figure via line integrals:

$$A = \iint_S dx\,dy \equiv \oint_L -y\,dx \equiv \oint_L x\,dy \equiv \frac{1}{2}\oint_L -y\,dx + x\,dy \ . \tag{6.42}$$

[Hint: *in each case find appropriate functions* $P(x, y)$ *and* $Q(x, y)$.]

Problem 6.44 Suppose that a planar figure S is specified in polar coordinates as $r = r(\phi)$ and contains the origin inside it. Show that the area of S is

$$A = \frac{1}{2}\int_{\phi_{min}}^{\phi_{max}} r^2(\phi)d\phi \ , \tag{6.43}$$

where the integration is performed over the appropriate range of the angle ϕ. Derive this expression using both the first and the last formulae for the area given in (6.42).

Some problems related to the application of Eq. (6.43) can be found in Sect. 4.7.2.

6.4.4 Exact Differentials

We found in Example 6.7 that the line integral over a specific closed path was zero. However, it can be shown that this particular line integral will be zero for *any* closed path. One may wonder when line integrals over a closed loop are equal to zero for any choice of the contour L. More precisely, the necessary and sufficient conditions for this are established by the following theorem.

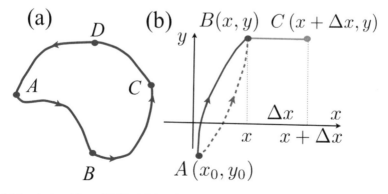

Fig. 6.30 a A closed-loop $ABCDA$. **b** For the proof of the step $2\rightarrow3$ of Theorem 6.5

Theorem 6.5 *Consider a simply connected region S in the x − y plane. Then, the following four statements are equivalent to each other:*

1. *For any closed loop inside region S, the integral*

$$\oint_L P(x, y)dx + Q(x, y)dy = 0 ;$$

(6.44)

2. *The line integral $\int_C P(x, y)dx + Q(x, y)dy$ does not depend on the integration curve C; it only depends on its starting and ending points;*
3. *$dU(x, y) = P(x, y)dx + Q(x, y)dy$ is the exact differential, and*
4. *the following condition is satisfied:*

$$\frac{\partial P}{\partial y} = \frac{\partial Q}{\partial x} .$$

(6.45)

Proof: We shall prove this theorem by using the scheme: from 1 follows 2 (i.e. 1→2), then 2→3, 3→4 and then, finally, 4→1.

1→2. We assume that Eq. (6.44) is satisfied for *any* closed loop L. We have to prove from this that the statement 2 is true. Indeed, take a closed-loop $ABCDA$ as shown in Fig. 6.30a. We have:

$$\oint_{ABCDA} P(x, y)dx + Q(x, y)dy = 0 ,$$

or

$$\int_{ABC} P(x, y)dx + Q(x, y)dy + \int_{CDA} P(x, y)dx + Q(x, y)dy = 0 ,$$

$$\int_{ABC} P(x, y)dx + Q(x, y)dy - \int_{ADC} P(x, y)dx + Q(x, y)dy = 0 ,$$

$$\int_{ABC} P(x, y)dx + Q(x, y)dy = \int_{ADC} P(x, y)dx + Q(x, y)dy ,$$

i.e. the line integral is the same along the lines ABC and ADC. Since the initial closed loop was chosen arbitrarily, then the integral will be the same along any path from A to C.

2→3. Now we are given that the line integral does not depend on the particular path, and we have to prove that $dU = P(x, y)dx + Q(x, y)dy$ is the exact differential. Let us choose a point $A(x_0, y_0)$. Then, for any point $B(x, y)$, we can define a function

$$U(x, y) = \int_{A(x_0, y_0)}^{B(x, y)} P(x_1, y_1)dx_1 + Q(x_1, y_1)dy_1 .$$

(6.46)

Importantly, we are able to define such a function that is dependent only on the coordinates of the final point[10] (x, y) since we assumed that the line integral does not depend on the particular path. Note also that we have used integration variables x_1 and y_1 here to distinguish them from the coordinates of the point B. Now consider the partial derivatives of the function $U(x, y)$. For the x-derivative we have, see Fig. 6.30b:

$$\frac{\partial U}{\partial x} = \lim_{\Delta x \to 0} \frac{\Delta U}{\Delta x} = \lim_{\Delta x \to 0} \frac{U(x + \Delta x, y) - U(x, y)}{\Delta x}$$

$$= \lim_{\Delta x \to 0} \frac{1}{\Delta x} \left(\int_{A(x_0, y_0)}^{C(x+\Delta x, y)} \cdots - \int_{A(x_0, y_0)}^{B(x, y)} \cdots \right),$$

where we omitted the integrand for simplicity of notations. The first integral above in the right-hand side is taken along the path from A to C, while the second one goes from A to B. Since, according to our assumption 2, the actual path does not matter, the path $A \to C$ in the first integral can be taken via point B as $A \to B \to C$, with the transition from B to C being passed horizontally along the x-axis. Then, the part of the path $A \to B$ cancels out, and we have

$$\frac{\Delta U}{\Delta x} = \frac{1}{\Delta x} \int_{B(x, y)}^{C(x+\Delta x, y)} P(x_1, y_1) dx_1 + Q(x_1, y_1) dy_1 = \frac{1}{\Delta x} \int_{B(x, y)}^{C(x+\Delta x, y)} P(x_1, y) dx_1$$

$$= \frac{1}{\Delta x} \int_x^{x+\Delta x} P(x_1, y) dx_1 = P(\zeta, y) ,$$

where $\zeta = x + \theta \Delta x$ is some point between x and $x + \Delta x$ with $0 < \theta < 1$. Here we have used the (average value) Theorem 4.6 for definite integrals in Sect. 4.2. Note also that above the dy_1 part of the line integral disappeared since along the chosen path $dy_1 = 0$ and y_1 should be set to y. In the limit of $\Delta x \to 0$ the point $\zeta \to x$, hence we have that $\partial U / \partial x = P(x, y)$.

Similarly, one can show that $\partial U / \partial y = Q(x, y)$ (do it!). Therefore, we find that the differential of the function $U(x, y)$ is

$$dU = \frac{\partial U}{\partial x} dx + \frac{\partial U}{\partial y} dy = P dx + Q dy .$$

We see that it is indeed exactly equal to $P dx + Q dy$, as required.

3→4. This has already been proven for exact differentials in Sect. 5.5 (see Eq. (5.22)), but can easily be repeated here. Since dU is an exact differential, then $\partial U / \partial x = P(x, y)$ and $\partial U / \partial y = Q(x, y)$, and hence

$$\frac{\partial}{\partial y} \left(\frac{\partial U}{\partial x} \right) = \frac{\partial P}{\partial y} \quad \text{must be equal to} \quad \frac{\partial}{\partial x} \left(\frac{\partial U}{\partial y} \right) = \frac{\partial Q}{\partial x} ,$$

---

[10] This function also depends on the initial point coordinates (x_0, y_0); however, this dependence is not of interest for us here.

as required (since the mixed derivatives do not depend on the order of differentiation).

4→1. Since the region S is simply connected, we can use the Green's formula (6.41) for any closed loop L that lies inside S:

$$\oint_L Q(x, y)dx + P(x, y)dy = \int\int_{S_L} \left(\frac{\partial Q}{\partial x} - \frac{\partial P}{\partial y}\right) dxdy \, ,$$

where S_L is the region for which L is the bounding curve. However, due to condition 4, Eq. (6.45), the double integral in the right-hand side above is zero for any choice of L, i.e. the closed-loop integral for any L is zero, i.e. the statement 1 is true. The theorem has been completely proven. **Q.E.D.**

The condition imposed on region S (simply connected) is very important here as otherwise we will have to use a more general Green's formula (see the comment just after the proof of the Green's formula in Sect. 6.4.3). For instance, region S cannot have holes. A classic example is the vector field

$$\mathbf{A} = -\frac{y}{x^2 + y^2}\mathbf{i} + \frac{x}{x^2 + y^2}\mathbf{j} \, . \tag{6.47}$$

It is seen that

$$\frac{\partial A_x}{\partial y} = \frac{y^2 - x^2}{\left(x^2 + y^2\right)^2} = \frac{\partial A_y}{\partial x}$$

so that $dU = A_x dx + A_y dy$ is the exact differential and thus the closed-loop integral appears to be zero according to Theorem 6.5. Consider, however, directly a closed-loop integral around a circle of a unit radius *centred at zero* and taken in the anticlockwise direction using polar coordinates $x = \cos\phi$ and $y = \sin\phi$:

$$\oint A_x dx + A_y dy = \oint -\frac{y}{x^2 + y^2}dx + \frac{x}{x^2 + y^2}dy$$
$$= \int_0^{2\pi} \left[-\frac{\sin\phi}{1}(-\sin\phi) + \frac{\cos\phi}{1}(\cos\phi)\right] d\phi = \int_0^{2\pi} d\phi = 2\pi \, .$$

It is not equal to zero in contrast to what the theorem would have suggested! The reason is that the functions $A_x(x, y)$ and $A_y(x, y)$ have a singularity at $x = y = 0$, so the region S has a hole at the origin and hence is not simply connected. The theorem is not valid since the contour (the unit radius circle) has the origin inside it. If we calculated the line integral avoiding the origin, the integral would be equal to zero as all the conditions of the theorem would be satisfied. This point is illustrated well by a discussion initiated by the following problem.

Problem 6.45 Consider a line integral for the vector field (6.47) along the contour shown in Fig. 6.31. The contour consists of two closed loops: ABA that is going anticlockwise and ACA going clockwise. Argue that the line

Fig. 6.31 A closed contour around the centre of the coordinate system based on two closed loops that are run in the opposite directions

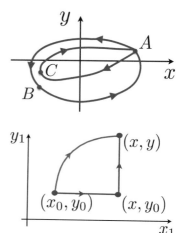

Fig. 6.32 The calculation of the exact differential

integral of the vector field **A** over the clockwise loop is equal to -2π, while the integral over the other loop going in the anticlockwise direction is equal to 2π, so the integral is zero if one contour is traversed after the other (which also constitutes a closed loop). Docs this illustrate the validity of Theorem 6.5? Explain why the internal (external) closed loop can be deformed into the shape of the external (internal) loop without changing the value of the integral over it (if traversed in the same direction)?

We have already discussed in Sect. 5.5 a method of restoring the function $U(x, y)$ from its exact (total) differential $dU = P(x, y)dx + Q(x, y)dy$. Formula (6.46) suggests a much simpler method of doing the same job. Moreover, as the integral can be taken along an arbitrary path connecting points (x_0, y_0) and (x, y), we can find the simplest route to make the calculation the easiest. For instance, one may perform the integration along the following path consisting of two straight lines parallel to the Cartesian axes: $(x_0, y_0) \to (x, y_0) \to (x, y)$ as shown in Fig. 6.32. Since along the first part $y_1 = y_0$ and is constant, while along the second part $x_1 = x$ is fixed, the integral is transformed into a sum of two ordinary definite integrals:

$$U(x, y) = \int_{x_0}^{x} P(x_1, y_0)dx_1 + \int_{y_0}^{y} Q(x, y_1)dy_1 . \tag{6.48}$$

Note that $U(x, y)$ is defined up to an arbitrary constant since the choice of the initial point (x_0, y_0) is arbitrary. The information about the initial point should combine into a single term to serve as such a constant.

Consider calculation of $U(x, y)$ from its differential $dU = 2xydx + x^2dy$ as an example. First, we check that condition (6.45) is satisfied:

$$\frac{\partial P}{\partial y} = \frac{\partial}{\partial y}(2xy) = 2x \text{ and } \frac{\partial Q}{\partial x} = \frac{\partial}{\partial x}(x^2) = 2x \text{ as well.}$$

This means that dU is indeed the exact differential. Then, using Eq. (6.48), we obtain

$$U(x, y) = \int_{x_0}^x 2x_1 y_0 dx_1 + \int_{y_0}^y x^2 dy_1$$

$$= 2y_0 \left(\frac{x^2}{2} - \frac{x_0^2}{2} \right) + x^2 (y - y_0) = x^2 y - x_0^2 y_0 .$$

Since $x_0^2 y_0$ is arbitrary, it serves as an arbitrary constant C, i.e. we conclude that any function $U(x, y) = x^2 y + C$ is a solution. This is also easily checked by the direct calculation: $\partial U / \partial x = 2xy$ and $\partial U / \partial y = x^2$, exactly as required.

Problem 6.46 Verify that $x^2 y^3 dx + x^3 y^2 dy$ is the exact differential of some function $U(x, y)$ and hence explain why the integral $\oint_L x^2 y^3 dx + x^3 y^2 dy = 0$ for any contour L.

Problem 6.47 Find the function $U(x, y)$ that gives rise to the exact differential $dU = x^2 y^3 dx + x^3 y^2 dy$.

Problem 6.48 Calculate the differential of $U(x, y) = x^3 y^3 / 3 + C$ and compare it with the one given in the previous problem.

Problem 6.49 Explain why the line integral

$$U(x, y) = \int_{A(x_0, y_0)}^{B(x,y)} - \sin x_1 \sin y_1 dx_1 + \cos x_1 \cos y_1 dy_1$$

does not depend on the particular path that connects two points $A(x_0, y_0)$ and $B(x, y)$. Show that the function $U = \cos x \sin y + C$.

Problem 6.50 Verify the Green's formula for $P(x, y) = x^2 + y^2$ and $Q(x, y) = 2x^2 y$ and the region of a square of side 2 shown in Fig. 6.26.

6.5   Surface Integrals

So far we have discussed double integrals that are calculated on planar regions. However, one may consider a more general problem of the integration over non-planar surfaces. This is the subject of this section. We shall start, however, from a general discussion of how to define a surface and how to obtain its normal vector (or simply a *normal*).

6.5.1 Surfaces

Generally, a line in space can be specified parametrically as $x = x(u)$, $y = y(u)$ and $z = z(u)$, where u is a real parameter. A surface S can also be specified in the same way, but using *two* parameters:

$$x = x(u, v), \quad y = y(u, v) \text{ and } z = z(u, v), \quad (6.49)$$

so that any point lying on the surface is represented by a vector

$$\mathbf{r}(u, v) = x(u, v)\mathbf{i} + y(u, v)\mathbf{j} + z(u, v)\mathbf{k}. \quad (6.50)$$

Note that we assume a direct correspondence $(u, v) \rightarrow (x, y, z)$ here, i.e. there exists only one unique set of (x, y, z) for each pair (u, v) (note that it is allowed that several different sets (u, v) give the same point (x, y, z)).

For instance, a spherical surface of radius R can be specified by equations:

$$x = R \sin \theta \cos \phi, \quad y = R \sin \theta \sin \phi, \quad z = R \cos \theta,$$

where $0 \leq \theta \leq \pi$ and $0 \leq \phi < 2\pi$. These are the spherical angles defined as in Fig. 1.37c. They play the role of the two parameters that specify this surface. For instance, if we restrict θ to the interval $0 \leq \theta \leq \pi/2$ only, then we shall obtain the upper hemisphere, but to specify the whole sphere θ has to be between 0 and π. Note also that for $\theta = 0$ any value of ϕ results in the same point $x = y = 0, z = R$.

We say that $\mathbf{r}(u, v)$ describes a *smooth surface* if all three functions (6.49) and their first derivatives

$$\frac{\partial \mathbf{r}}{\partial u} = \frac{\partial x}{\partial u}\mathbf{i} + \frac{\partial y}{\partial u}\mathbf{j} + \frac{\partial z}{\partial u}\mathbf{k} \text{ and } \frac{\partial \mathbf{r}}{\partial v} = \frac{\partial x}{\partial v}\mathbf{i} + \frac{\partial y}{\partial v}\mathbf{j} + \frac{\partial z}{\partial v}\mathbf{k} \quad (6.51)$$

with respect to u and v are continuous functions in the domain of u and v, and are *nonparallel* everywhere there.

In order to understand what this means, let us calculate a normal to the surface at a point $M_0(x_0, y_0, z_0)$, corresponding to parameters u_0, v_0 as shown in Fig. 6.33. The line $CM_0D = \mathbf{r}(u, v_0)$ ($v = v_0$ is fixed, u alone is allowed to change!) passes through the point M_0, as does the other line $AM_0B = \mathbf{r}(u_0, v)$ obtained by setting $u = u_0$ and allowing only v to change; they both meet at the point M_0. The vectors calculated at the point M_0,

$$\boldsymbol{\tau}_u = \frac{\partial \mathbf{r}(u, v_0)}{\partial u}\bigg|_{u=u_0} = \left(\frac{\partial x}{\partial u}, \frac{\partial y}{\partial u}, \frac{\partial z}{\partial u}\right)_{u=u_0, v=v_0} = \left(x'(u), y'(u), z'(u)\right)_{u=u_0, v=v_0},$$

$$\boldsymbol{\tau}_v = \frac{\partial \mathbf{r}(u_0, v)}{\partial v}\bigg|_{v=v_0} = \left(\frac{\partial x}{\partial v}, \frac{\partial y}{\partial v}, \frac{\partial z}{\partial v}\right)_{u=u_0, v=v_0} = \left(x'(v), y'(v), z'(v)\right)_{u=u_0, v=v_0},$$

Fig. 6.33 Surface $\mathbf{r}(u, v)$.
Lines AB and CD that run
across the surface
correspond to constant
values of $u = u_0$ and $v = v_0$,
respectively. Both lines cross
at the point M_0

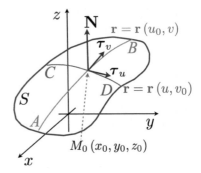

are tangent to the two lines and lie in the *tangent plane* to our surface at point M_0
(see also Sect. 5.4). The normal vector to the plane can be calculated as a vector
product of the two tangent vectors:

$$\mathbf{N}(u_0, v_0) = \boldsymbol{\tau}_u \times \boldsymbol{\tau}_v = \begin{vmatrix} \mathbf{i} & \mathbf{j} & \mathbf{k} \\ x'(u) & y'(u) & z'(u) \\ x'(v) & y'(v) & z'(v) \end{vmatrix} . \tag{6.52}$$

Expanding the determinant along the first row, we obtain:

$$\begin{aligned} \mathbf{N}(u_0, v_0) &= \begin{vmatrix} y'(u) & z'(u) \\ y'(v) & z'(v) \end{vmatrix} \mathbf{i} + \begin{vmatrix} z'(u) & x'(u) \\ z'(v) & x'(v) \end{vmatrix} \mathbf{j} + \begin{vmatrix} x'(u) & y'(u) \\ x'(v) & y'(v) \end{vmatrix} \mathbf{k} \\ &= \frac{\partial(y, z)}{\partial(u, v)} \mathbf{i} + \frac{\partial(z, x)}{\partial(u, v)} \mathbf{j} + \frac{\partial(x, y)}{\partial(u, v)} \mathbf{k} \\ &= J_x(u, v) \mathbf{i} + J_y(u, v) \mathbf{j} + J_z(u, v) \mathbf{k} . \end{aligned} \tag{6.53}$$

It is seen that the normal is expressed via the Jacobians that form components of it
in Eq. (6.53). Note that the Cartesian coordinates in the definition of the Jacobians
together with the Jacobians' subscripts form the cyclic sequence $x \to y \to z \to$
$x \to \ldots$, e.g.

$$J_x = \frac{\partial(y, z)}{\partial(u, v)} .$$

This helps to memorise this formula. Equivalently, one can use formula (6.52) for the
whole vector $\mathbf{N}$ instead, which also has a form simple enough to remember. Also note
that the normal $\mathbf{N}$ may not (and most likely will never) be of unit length, although
one can always construct the unit normal $\mathbf{n} = \mathbf{N}/|\mathbf{N}|$ from it if required.

Thus, we are now at the position to clarify the statement made above about smooth
surfaces: the conditions formulated above actually mean that the normal to the smooth
surface exists and is not zero. The latter is ensured by the condition that the derivatives
(6.51) are non-parallel at each point on the surface (otherwise the normal as the vector
product would be zero) and changes continuously as we move the point M_0 across
the surface.

Let us now see how this technique can be used to obtain the normal of the side surface of a cylinder of radius R of infinite length running along the z-axis. The surface of the cylinder can be described using the polar coordinates by means of the following equations:

$$x = R \cos \phi, \quad y = R \sin \phi \quad \text{and} \quad z = z \,,$$

i.e. the two parameters describing the surface are (ϕ, z) with $0 \le \phi < 2\pi$ and $-\infty < z < \infty$. To calculate the normal, we should calculate the three Jacobians from (6.53):

$$
\begin{aligned}
J_x &= \frac{\partial(y, z)}{\partial(\phi, z)} = \begin{vmatrix} R \cos \phi & 0 \\ 0 & 1 \end{vmatrix} = R \cos \phi \,, \\
J_y &= \frac{\partial(z, x)}{\partial(\phi, z)} = \begin{vmatrix} 0 & -R \sin \phi \\ 1 & 0 \end{vmatrix} = R \sin \phi \,, \\
J_z &= \frac{\partial(x, y)}{\partial(\phi, z)} = \begin{vmatrix} -R \sin \phi & R \cos \phi \\ 0 & 0 \end{vmatrix} = 0 \,,
\end{aligned}
\tag{6.54}
$$

leading to the normal vector

$$\mathbf{N} = (R \cos \phi)\, \mathbf{i} + (R \sin \phi)\, \mathbf{j} + 0\mathbf{k} = x\mathbf{i} + y\mathbf{j} \equiv \mathbf{r}_\perp \,.$$

It coincides with the vector perpendicular to the cylinder axis for it lies in the plane cutting the cylinder across it. For instance, for $\phi = \pi/2$ we have $\mathbf{r}_\perp = R\mathbf{j}$ and for $\phi = \pi$ we obtain $\mathbf{r}_\perp = -R\mathbf{i}$.

In some applications the surface is specified by the equation $z = z(x, y)$. To accommodate this case, we can treat x and y themselves as the parameters u and v in Eq. (6.49) and define the surface by the equations:

$$x = u, \quad y = v, \quad z = z(u, v) \,,$$

which gives for the Jacobians:

$$
\begin{aligned}
J_x &= \frac{\partial(y, z)}{\partial(u, v)} = \frac{\partial(y, z)}{\partial(x, y)} = \begin{vmatrix} y'_x & z'_x \\ y'_y & z'_y \end{vmatrix} = \begin{vmatrix} 0 & z'_x \\ 1 & z'_y \end{vmatrix} = -z'_x = -\frac{\partial z}{\partial x} \,, \\
J_y &= \frac{\partial(z, x)}{\partial(u, v)} = \frac{\partial(z, x)}{\partial(x, y)} = \begin{vmatrix} z'_x & x'_x \\ z'_y & x'_y \end{vmatrix} = \begin{vmatrix} z'_x & 1 \\ z'_y & 0 \end{vmatrix} = -z'_y = -\frac{\partial z}{\partial y} \,, \\
J_z &= \frac{\partial(x, y)}{\partial(u, v)} = \frac{\partial(x, y)}{\partial(x, y)} = \begin{vmatrix} 1 & 0 \\ 0 & 1 \end{vmatrix} = 1 \,,
\end{aligned}
\tag{6.55}
$$

where we replaced back u, v by x, y, so that

$$\mathbf{N}(x, y) = -\frac{\partial z}{\partial x}\mathbf{i} - \frac{\partial z}{\partial y}\mathbf{j} + \mathbf{k} \,,
\tag{6.56}$$

Recall that an expression for the surface normal was obtained by us earlier in Eq. (5.18), the result which was derived using a rather different consideration. The two formulae differ only by a sign and hence describe two possible (opposite) directions of the normal to the surface (see also a comment at the end of this subsection).

Problem 6.51 Consider a surface specified by the equation $z(x, y) = 3(x^4 + y^4)$. Using x and y as the coordinates u and v specifying the surface, calculate the appropriate Jacobians and obtain an expression for the surface outward normal as a function of x and y. [Answer: $\mathbf{N} = \left(-12x^3, -12y^3, 1\right)$.]

Problem 6.52 A surface is specified parametrically by coordinates (u, ϕ) via the connection relations: $x = u^2 - 1$, $y = u \sin \phi$ and $z = u \cos \phi$. Calculate the three Jacobians J_x, J_y and J_z for this surface and hence work out the expression for its normal. [Answer: $\mathbf{N} = \left(-u, 2u^2 \sin \phi, 2u^2 \cos \phi\right)$.]

Some surfaces may consist of pieces of smooth surfaces, e.g. a cube has six such surfaces. The surfaces consisting of such smooth pieces are called *piecewise smooth* surfaces.

Also, the surfaces may have *sides* or *orientations*. Indeed, if we take a membrane, i.e. a surface that caps a closed line L, see Fig. 6.34a, then two normals $\mathbf{n}_1$ and $\mathbf{n}_2 = -\mathbf{n}_1$ can be defined at each point. However, if we move along the surface, the normal on the upper side of the membrane, $\mathbf{n}_1$, will move continuously on the upper side; it will never come to the lower side. The same is true for the other (lower side) normal $\mathbf{n}_2$. Such surfaces are called *two-sided* or *orientable*. A classic example of a non-orientable (one-sided) surface is the "Möbius[11] strip" (or "leaf") shown in Fig. 6.35. In practically all physical problems the surfaces are two-sided so that we stick to those.

6.5.2   Area of a Surface

Now we would like to calculate the area of a surface S specified parametrically as above, i.e. $\mathbf{r} = \mathbf{r}(u, v)$, using two parameters u and v. The problem here is that the surface may not be planar and we have to figure out how to calculate its area using some sort of integration. The driving idea to solve this problem is as follows. We consider the (u, v) coordinate system, in which the values of u and v form a plane surface region Σ, and draw horizontal and vertical lines $u = u_i$ and $v = v_j$ there to form a grid as shown in Fig. 6.36a. The (u, v) grid lines will form coordinate lines

---

[11] The strip was discovered independently by August Ferdinand Möbius and Johann Benedict Listing.

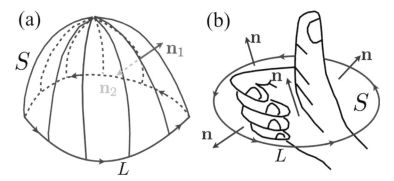

Fig. 6.34 a The surface S caps a closed-loop line L that serves as its boundary. Two oppositely directed surface normal vectors $\mathbf{n}_1$ and $\mathbf{n}_2$ can be defined at each point M of S that specify the two sides (orientations) of the surface. The traverse direction along the line L corresponding to the upper side (the normal vector $\mathbf{n}_1$) is also shown: at each point along the path the surface is on the left which corresponds to the *right-hand rule* shown in **b** (the thumb indicates a general orientation of the normal vectors of the surface, while the four fingers indicate the traverse direction along the surface boundary L)

Fig. 6.35 Möbius strip

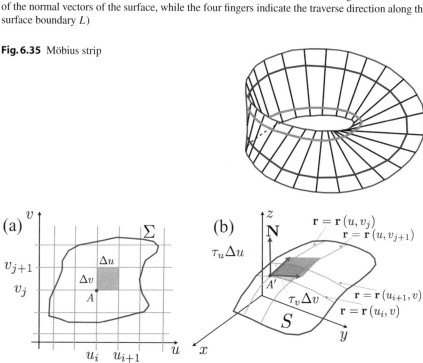

Fig. 6.36 The mapping of the (u, v) system into the (x, y, z) system: to the definition of the area of a general surface

$\mathbf{r}_i(v) = \mathbf{r}(u_i, v)$ and $\mathbf{r}_j(u) = \mathbf{r}(u, v_j)$ on the actual surface S in the (x, y, z) 3D space as schematically shown in Fig. 6.36b.

Consider now a small surface element $\Delta u \Delta v$ in the (u, v) system around the point $A(u_i, v_j)$, the hatched area in Fig. 6.36a. It will transform into a small surface element in the surface S; in particular, the point $A \to A'$, Fig. 6.36b. For small Δu and Δv the actual surface element (on the actual surface in Fig. 6.36b) can be replaced by the corresponding surface element in the *tangent surface* built by the tangent vectors (hatched)

$$
\boldsymbol{\tau}_u \Delta u = \frac{\partial \mathbf{r}(u, v_j)}{\partial u} \Delta u \quad \text{and} \quad \boldsymbol{\tau}_v \Delta v = \frac{\partial \mathbf{r}(u_i, v)}{\partial v} \Delta v ,
$$

as shown in Fig. 6.36b, and its area is

$$
\Delta A_{ij} = |\boldsymbol{\tau}_u \Delta u \times \boldsymbol{\tau}_v \Delta v| = |\boldsymbol{\tau}_u \times \boldsymbol{\tau}_v| \Delta u \Delta v = \left| \mathbf{N}(u_i, v_j) \right| \Delta u \Delta v
$$

$$
= \sqrt{J_x^2(u_i, v_j) + J_y^2(u_i, v_j) + J_z^2(u_i, v_j)} \Delta u \Delta v ,
$$

where we have used Eq. (6.53) for the normal vector to the surface. This way the whole surface S will be represented by small planar tiles smoothly going around it.

Summing areas of all these little sections (or tiles) up, we obtain an estimate for the surface area as the integral sum:

$$
A \simeq \sum_{i,j} \Delta A_{ij} = \sum_{i,j} \sqrt{J_x^2(u_i, v_j) + J_y^2(u_i, v_j) + J_z^2(u_i, v_j)} \Delta u \Delta v ,
$$

which, in the limit of little sections tending to zero, gives the required *exact* formula for the surface area as a usual double integral over all allowed values of u and v:

$$
A = \int \int_{\Sigma} \sqrt{J_x^2(u, v) + J_y^2(u, v) + J_z^2(u, v)} \, du dv = \int \int_{\Sigma} |\mathbf{N}(u, v)| \, du dv ,
$$

$$
\tag{6.57}
$$

the final result sought for.

Let us illustrate the formula we have just derived by calculating the surface area of a hemisphere shown in Fig. 6.4a. The hemisphere is specified parametrically by spherical coordinates

$$
x = R \sin \theta \cos \phi, \quad y = R \sin \theta \sin \phi \quad \text{and} \quad z = R \cos \theta ,
$$

with the parameters $0 \le \theta \le \pi/2$ and $0 \le \phi < 2\pi$. The required Jacobians in Eq. (6.53) are calculated to be:

$$
J_x = \frac{\partial(y, z)}{\partial(\theta, \phi)} = \begin{vmatrix} R \cos \theta \sin \phi & -R \sin \theta \\ R \sin \theta \cos \phi & 0 \end{vmatrix} = R^2 \sin^2 \theta \cos \phi ,
$$

$$
J_y = \frac{\partial(z, x)}{\partial(\theta, \phi)} = \begin{vmatrix} -R \sin \theta & R \cos \theta \cos \phi \\ 0 & -R \sin \theta \sin \phi \end{vmatrix} = R^2 \sin^2 \theta \sin \phi ,
$$

$$
J_z = \frac{\partial(x, y)}{\partial(\theta, \phi)} = \begin{vmatrix} R \cos \theta \cos \phi & R \cos \theta \sin \phi \\ -R \sin \theta \sin \phi & R \sin \theta \cos \phi \end{vmatrix}
$$

$$
= R^2 \sin \theta \cos \theta \left(\sin^2 \phi + \cos^2 \phi \right) = R^2 \sin \theta \cos \theta .
$$

The normal is then given by[12]

$$\mathbf{N}(\theta, \phi) = R^2 \left(\sin^2 \theta \cos \phi \, \mathbf{i} + \sin^2 \theta \sin \phi \, \mathbf{j} + \sin \theta \cos \theta \, \mathbf{k} \right)$$

$$= R \sin \theta \left(\underbrace{R \sin \theta \cos \phi}_{x} \, \mathbf{i} + \underbrace{R \sin \theta \sin \phi}_{y} \, \mathbf{j} + \underbrace{R \cos \theta}_{z} \, \mathbf{k} \right) = (R \sin \theta) \, \mathbf{r} \,,$$

(6.58)

and thus its length is $|\mathbf{N}(\theta, \phi)| = (R \sin \theta) R = R^2 \sin \theta$. Therefore, the surface of the hemisphere, according to Eq. (6.57), is

$$A = \int \int dA = \int \int \left(R^2 \sin \theta \right) d\theta d\phi = R^2 \int_0^{\pi/2} \sin \theta d\theta \int_0^{2\pi} d\phi$$

$$= 2\pi R^2 \left(-\cos \theta \right)_0^{\pi/2} = 2\pi R^2 \,,$$

as expected: the total surface area of the complete sphere must be two times larger, $4\pi R^2$.

Remember this expression for the differential surface area in the spherical coordinates,

$$dA = R^2 \sin \theta d\theta d\phi \,. \tag{6.59}$$

This result can also be obtained using a rather simple geometrical consideration identical to the one we actually used above in Sect. 6.2.2 for the derivation of the volume element. In spherical coordinates (and assuming a constant radius R) we have $x = R \sin \theta \cos \phi$, $y = R \sin \theta \sin \phi$ and $z = R \cos \theta$. Consider a spherical region $ABCD$ in Fig. 6.37. It can be approximately considered rectangular with the sides $AD = R d\theta$ and $CD = FD \cdot d\phi = R \sin \theta d\phi$ so that the area becomes $dA = AD \cdot CD = R^2 \sin \theta d\theta d\phi$, which is exactly the same expression as obtained above using Jacobians.

Problem 6.53 Show that if the surface is specified by the equation $z = z(x, y)$, then the surface area (6.57) becomes:

$$A = \int \int_\Sigma \sqrt{1 + \left(\frac{\partial z}{\partial x} \right)^2 + \left(\frac{\partial z}{\partial y} \right)^2} \, dx dy \,, \tag{6.60}$$

where Σ is the projection of the surface S on the $x - y$ plane.

---

[12] Note that $\mathbf{N} \sim \mathbf{r}$ as one would intuitively expect for the normal to a sphere!.

Fig. 6.37 The geometrical
derivation of the differential
surface area in spherical
coordinates

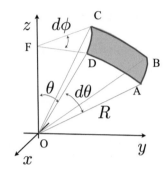

Fig. 6.38 a An elliptic cone
bound from above by the
plane $z = R$; **b** a paraboloid
of the height h^2

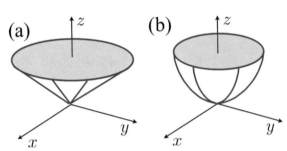

As an illustration of the application of Eq. (6.60), we shall calculate the surface
area of the elliptic cone $z = \sqrt{x^2 + y^2} > 0$ cut by the horizontal plane $z = R$, see
Fig. 6.38a. We have:

$$\frac{\partial z}{\partial x} = \frac{x}{\sqrt{x^2 + y^2}} \quad \text{and} \quad \frac{\partial z}{\partial y} = \frac{y}{\sqrt{x^2 + y^2}} \,,$$

and thus

$$\sqrt{1 + \left(\frac{\partial z}{\partial x}\right)^2 + \left(\frac{\partial z}{\partial y}\right)^2} = \sqrt{1 + \frac{x^2}{x^2 + y^2} + \frac{y^2}{x^2 + y^2}} = \sqrt{2} \,.$$

The projection of the surface S on the $x - y$ plane (serving as the $u - v$ plane) is a
circle of radius R since at the top of the cone ($z = R$) we have $x^2 + y^2 = R^2$, the
equation of the circle. Therefore, the surface area

$$A = \int\int \sqrt{2} dx dy = \sqrt{2} \int\int_{circle} dx dy = \sqrt{2} \pi R^2 \,,$$

where we have used πR^2 for the area of a circle of radius R (the "spread" of x, y).

Problem 6.54 The paraboloid in Fig. 6.38b is described by the equation $z = x^2 + y^2$. Write the equation for the surface in a parametric form $x = x(u, v)$,
$y = y(u, v)$ and $z = z(u, v)$ using x and y as the simplest choice of the pa-

rameters u and v. Obtain the expression for the *outer* normal to the surface
N. [Hint: *to choose the correct sign for the outer normal, consider a particular point on the surface (choose the simplest one; note that the surface is
two-sided).*] [Answer: $\mathbf{N} = (2x, 2y, -1)$.]

Problem 6.55 Repeat the previous problem by considering the polar coordinates (r, ϕ) as u and v. Notice that the length of **N** will change, but not the
direction.

Problem 6.56 For the paraboloid of Fig. 6.38b, calculate the surface area for
the part of the figure that is bound by the planes $z = 0$ and $z = h^2$. [Answer:
$\pi \left[\left(4h^2 + 1 \right)^{3/2} - 1 \right] /6$.]

Problem 6.57 Repeat the previous problem using the polar coordinates instead of x and y as u and v. Note that the same result is obtained although the
length of the vector **N** changed with this choice of the parameters describing
the surface.

Problem 6.58 The torus in Fig. 4.14 is specified by the parametric equations:

$$x = (c + R \cos \phi) \cos \theta , \quad y = (c + R \cos \phi) \sin \theta \quad \text{and} \quad z = R \sin \phi ,$$
$$(6.61)$$

where $0 \leq \theta < 2\pi$ and $0 \leq \phi < 2\pi$. Calculate (a) the normal to the torus
$\mathbf{N}(\theta, \phi)$ and (b) the entire surface area. [Answer: *(a)* $\mathbf{N} = R(c + R \cos \phi)$
$(\cos \phi \cos \theta, \cos \phi \sin \theta, \sin \phi)$; *(b)* $4\pi^2 cR$.]

6.5.3 Surface Integrals for Scalar Fields

If we are given a scalar field $f(\mathbf{r})$, defined on a surface S, one can define an integral
sum

$$\sum_{ij} f(M_{ij}) \Delta A_{ij}$$

with ΔA_{ij} being the surface area of the ij-th surface element. Then the expression
above in the limit of all little surface areas tending to zero yields the *surface integral*:

$$\int \int_S f(\mathbf{r}) dA . \qquad (6.62)$$

Fig. 6.39 The calculation of
the electrostatic potential
from a thin spherical layer at
point A in Problem 6.59

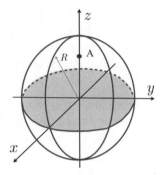

Here points M_{ij} are on the surface. The area of the surface S we considered above is an example of the surface integral with the unit scalar field $f \equiv 1$ everywhere on the surface. Therefore, the surface integral (6.62) is a natural generalisation of our previous definition. If the surface is specified parametrically via Eq. (6.50), the surface integral is calculated similarly to Eq. (6.57) as follows:

$$
\iint_S f(\mathbf{r}) dA = \iint_\Sigma f(\mathbf{r}(u, v)) \sqrt{J_x^2(u, v) + J_y^2(u, v) + J_z^2(u, v)} du dv
$$

$$
= \iint_\Sigma f(\mathbf{r}(u, v)) |\mathbf{N}(u, v)| du dv .
$$

(6.63)

In particular, if the surface is defined via equation $z = z(x, y)$, we then have instead

$$
\iint_S f(\mathbf{r}) dA = \iint_\Sigma f(x, y, z(x, y)) \sqrt{1 + \left(\frac{\partial z}{\partial x}\right)^2 + \left(\frac{\partial z}{\partial y}\right)^2} dx dy .
$$

(6.64)

As an example, let us evaluate the surface integral $J = \iint x dA$ over that part of the sphere $x^2 + y^2 + z^2 = R^2$ that lies in the region $x \geq 0$, $y \geq 0$ and $z \geq 0$ (the first octant). We use spherical coordinates to specify the surface, i.e. using the angles θ and ϕ in place of the two parameters u and v. Then the surface element in the spherical coordinates $dA = R^2 \sin\theta d\theta d\phi$, see Eq. (6.59). In the first octant $0 \leq \phi \leq \pi/2$ and $0 \leq \theta \leq \pi/2$, so the integral becomes:

$$
J = \iint (R \sin\theta \cos\phi) \left(R^2 \sin\theta d\theta d\phi\right) = R^3 \int_0^{\pi/2} \sin^2\theta d\theta \int_0^{\pi/2} \cos\phi d\phi
$$

$$
= R^3 \left(\frac{\theta}{2} - \frac{1}{4}\sin(2\theta)\right)_0^{\pi/2} (\sin\phi)_0^{\pi/2} = R^3 \frac{\pi}{4} .
$$

Problem 6.59 Consider a thin spherical surface layer of radius R which is uniformly charged with the surface charge density σ. Show by calculating the electrostatic potential

$$\phi(\mathbf{r}) = \sigma \int \int_S \frac{dA'}{|\mathbf{r} - \mathbf{r}'|}$$

produced by the layer at a point $\mathbf{r}$, that φ is constant anywhere *inside* the sphere (i.e. it does not depend on the position $\mathbf{r} = (x, y, z)$ there). [Hint: *place the point* $\mathbf{r}$ *on the z axis as shown by point A in Fig.* 6.39 *and use the spherical coordinates* (θ, ϕ) *when integrating over the surface of the sphere.*]

Problem 6.60 Show that the potential of the sphere from the previous problem at the distance z outside the sphere (i.e. for $z > R$) is given by $\phi = Q/z$, where $Q = 4\pi R^2 \sigma$ is the total charge of the sphere.

Problem 6.61 Consider now the potential

$$\phi(\mathbf{r}) = \int \int_S \frac{\sigma(\mathbf{r}')}{|\mathbf{r} - \mathbf{r}'|} dA'$$

produced at point $\mathbf{r}$ by a non-uniformly charged sphere S with the radius R and the surface charge density $\sigma(\mathbf{r}) = \alpha (x^2 + y^2)$. In the equation above for the potential points $\mathbf{r}'$ lie on the surface S of the sphere and dA' is its differential surface area. Show that outside the sphere and at the distance D (where $D > R$) from its centre

$$\phi(D) = \frac{8\pi\alpha R^4}{3D} \left(1 - \frac{R^2}{5D^2}\right).$$

Show also that the total charge on the sphere $Q = \int \int_S \sigma(\mathbf{r}) dA = 8\pi\alpha R^4/3$, and hence the potential far away from the sphere $\phi \simeq Q/D$, as it would be for a point charge Q localised in the sphere centre.

6.5.4 Surface Integrals for Vector Fields

Similar to line integrals, which were generalised to vectors fields in Sect. 6.4.2, one can also consider surface integrals for vector fields.

Let $\mathbf{F}(\mathbf{r})$ be a vector field defined on a surface S for which we have a well-defined *unit* normal $\mathbf{n}(M) = \mathbf{N}(M)/|\mathbf{N}(M)|$ at each point M belonging to S. Then, a *flux* of $\mathbf{F}$ through the surface S is defined as the following surface integral:

$$\Phi = \int \int (\mathbf{F} \cdot \mathbf{n}) \, dA = \int \int F_n dA = \int \int \mathbf{F} \cdot d\mathbf{A}. \qquad (6.65)$$

Here, as it is customarily done, we introduced a *directed* surface element $d\mathbf{A} = \mathbf{n}\,dA$ whose direction is chosen along the vector $\mathbf{n}$, and $F_n = \mathbf{F} \cdot \mathbf{n}$ is the projection of $\mathbf{F}$ on the unit normal vector $\mathbf{n}$. If the flux integral is taken over a *closed* surface S, the notation $\oiint$ for the surface integral is frequently used instead.

To see just one application where this type of integral appears (many more will be given in Sects. 6.7 and 6.9), let us consider a flow of a liquid. If we take a plane surface area ΔA and assume that the velocity distribution of the liquid particles $\mathbf{v}$ across the surface is uniform, then the volume of the liquid flown through the surface during the time dt, Fig. 6.40, is:

$$\frac{dV}{dt} = \frac{\Delta A v_n dt}{dt} = v_n \Delta A \ , \tag{6.66}$$

where $v_n = \mathbf{v} \cdot \mathbf{n}$ is the projection of the velocity vector on the normal $\mathbf{n}$ to the surface. Indeed, the amount of the liquid passed through ΔA over time dt would correspond to the volume dV of the cylinder in the figure whose base has the area of ΔA and the height $v_n dt$ (as $|\mathbf{v}|\,dt$ is the length of its side which may not necessarily be parallel to the normal $\mathbf{n}$). If, however, the surface S is *not* planar and the velocity distribution *not* uniform, then the same result would have been obtained only for a small surface area ΔA which can be approximately assumed planar, and the velocity distribution across it is uniform. Dividing the whole surface S into small surface elements ΔA and summing up contributions from each of them, $\sum v_n \Delta A$, we arrive at the flux integral $\int \int v_n dA = \int \mathbf{v} \cdot d\mathbf{A}$ in the limit. Hence, the flux integral of the velocity field through a surface gives a rate of flow of the liquid through that surface area, i.e. the volume of the liquid flown through it per unit time.

There is a very simple (and general!) formula for the calculation of the surface (flux) integrals for vector fields. If the surface S is specified parametrically as $x = x(u, v)$, $y = y(u, v)$ and $z = z(u, v)$, then, according to Eq. (6.57), $dA = |\mathbf{N}|\,dudv$.

Fig. 6.40 Flow of a liquid through the surface area ΔA during time dt

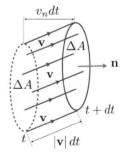

At the same time, the unit normal to the surface $\mathbf{n} = \mathbf{N}/|\mathbf{N}|$, hence we can write:

$$\int\int_S \mathbf{F} \cdot d\mathbf{A} = \int\int (\mathbf{F} \cdot \mathbf{n}) \, dA = \int\int \left(\mathbf{F} \cdot \frac{\mathbf{N}}{|\mathbf{N}|}\right) (|\mathbf{N}| \, dudv) = \int (\mathbf{F} \cdot \mathbf{N}) \, dudv$$

$$= \int\int [\mathbf{F}(x(u,v), y(u,v), z(u,v)) \cdot \mathbf{N}(u,v)] \, dudv$$

$$= \int\int \left(F_x J_x + F_y J_y + F_z J_z\right) dudv \ .$$

$$(6.67)$$

To illustrate application of this powerful result, let us find the flux of the field $\mathbf{F} = (yz, xz, xy)$ through the outer surface of the cylinder $x^2 + y^2 = R^2$, $0 \le z \le h$. We use cylindrical coordinates ($r = R$ (is fixed), ϕ and z), in which Cartesian coordinates are defined via

$$x = R\cos\phi, \quad y = R\sin\phi \ \text{ and } \ z = z \ .$$

The coordinates ϕ and z specify the surface of the cylinder. Thus, in this case $u \equiv \phi$ and $v \equiv z$, so that, see Eq. (6.54),

$$J_x = R\cos\phi, \quad J_y = R\sin\phi \ \text{ and } \ J_z = 0 \ ,$$

and we obtain

$$\text{Flux} = \int\int d\phi dz \, (yzR\cos\phi + xzR\sin\phi)$$

$$= \int_0^h dz \int_0^{2\pi} d\phi \left(R^2 z \sin\phi\cos\phi + R^2 z \cos\phi\sin\phi\right)$$

$$= 2R^2 \int_0^h z \, dz \int_0^{2\pi} \sin\phi\cos\phi \, d\phi = 2R^2 \left(\frac{h^2}{2}\right) \left(\frac{\sin^2\phi}{2}\right)_0^{2\pi} = 0 \ .$$

Problem 6.62 Calculate the surface integral $\int\int \mathbf{F} \cdot d\mathbf{A}$ over the triangular surface bound by the points $(2, 0, 0)$, $(0, 2, 0)$ and $(0, 0, 2)$ as shown in Fig. 6.41, where $\mathbf{F} = (x, y, z)$ and the outward normal to the surface is used. [Hint: *the general equation for the plane is $ax + by + cz = 1$, where the constants a, b and c can be found from the known points on the plane.*] [Answer: 4.]

There is also another way of calculating the surface integrals (6.65) for vector fields in which it is expressed via a sum of double integrals taken over Cartesian

Fig. 6.41 Triangle surface area used for the flux calculation in Problem 6.62

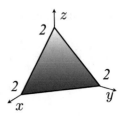

planes $x - y$, $y - z$ and $x - z$. This method is sometimes more convenient. Consider the vector field $\mathbf{F} = (F_x, F_y, F_z)$, then

$$\int\int \mathbf{F} \cdot d\mathbf{A} = \int\int \left(F_x n_x + F_y n_y + F_z n_z\right) dA$$
$$= \int\int F_x n_x dA + \int\int F_y n_y dA + \int\int F_z n_z dA,$$
(6.68)

since at the end of the day, the integral is an integral sum and the latter can always be split into the sum of the three separate integral sums leading to the expression above, with the three integrals on the right-hand side. Note that in each of them the integration is performed over the actual surface, i.e. x, y and z are related to each other by the equation of the surface. Above, as follows from Eq. (6.53), we have:

$$n_x = \cos\alpha = \cos(\widehat{\mathbf{n}, \mathbf{i}}) = \frac{N_x}{|\mathbf{N}|} = \frac{J_x(u, v)}{\sqrt{J_x^2(u, v) + J_y^2(u, v) + J_z^2(u, v)}}, \quad (6.69)$$

$$n_y = \cos\beta = \cos(\widehat{\mathbf{n}, \mathbf{j}}) = \frac{N_y}{|\mathbf{N}|} = \frac{J_y(u, v)}{\sqrt{J_x^2(u, v) + J_y^2(u, v) + J_z^2(u, v)}}, \quad (6.70)$$

$$n_z = \cos\gamma = \cos(\widehat{\mathbf{n}, \mathbf{k}}) = \frac{N_z}{|\mathbf{N}|} = \frac{J_z(u, v)}{\sqrt{J_x^2(u, v) + J_y^2(u, v) + J_z^2(u, v)}}. \quad (6.71)$$

These are components of the normal vector $\mathbf{n}$ along the x, y and z axes, and $\alpha = (\widehat{\mathbf{n}, \mathbf{i}})$, $\beta = (\widehat{\mathbf{n}, \mathbf{j}})$ and $\gamma = (\widehat{\mathbf{n}, \mathbf{k}})$ are the corresponding angles the normal makes with the Cartesian axes. Then, $n_z dA = dA \cos\gamma$ projects the surface area dA on the $x - y$ plane, Fig. 6.42, and thus can simply be replaced by $\pm dxdy$. Note the sign: depending on which side of the surface is considered (upper with n_z being positive or lower with n_z negative), a plus or minus sign should be attached to the surface element (since dA is always positive) as stressed in the figure. Similarly, $n_x dA = \pm dydz$ and $n_y dA = \pm dzdx$, so the integrals (6.68) can be rewritten as

$$\int\int_S \mathbf{F} \cdot d\mathbf{A} = \pm \int\int_{S_{yz}} F_x dydz \pm \int\int_{S_{xz}} F_y dzdx \pm \int\int_{S_{xy}} F_z dxdy, \quad (6.72)$$

Fig. 6.42 Projection of the surface element dA on the $x - y$ plane for **a** $n_z > 0$ (directed upwards) and **b** $n_z < 0$ (directed downwards)

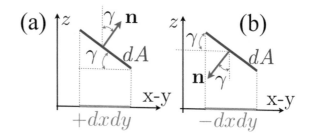

where S_{xy}, S_{xz} and S_{yz} are projections of the surface S onto the corresponding coordinate planes.

This expression gives a practical way of calculating the flux integrals by essentially breaking them down into three double integrals with respect to the three projections on coordinate surfaces, and then calculating each one of them via

$$\int\int_{S_{yz}} F_x(x, y, z)dydz = \int\int F_x\,(x(y, z), y, z)\,dydz\ ,$$

$$\int\int_{S_{xz}} F_y(x, y, z)dzdx = \int\int F_y\,(x, y(x, z), z)\,dzdx\ ,$$

$$\int\int_{S_{xy}} F_z(x, y, z)dxdy = \int\int F_z\,(x, y, z(x, y))\,dxdy\ .$$

As an example, we calculate the flux of the force field $\mathbf{F} = (x^2 z, xy^2, z)$ through the outer surface of $z = x^2 + y^2$, $0 \le z \le 1$, $x \ge 0$ and $y \ge 0$, see Fig. 6.43. The integral

$$I = \int\int_S \mathbf{F} \cdot d\mathbf{A} = I_x + I_y + I_z$$

$$= \int\int_{S_{yz}} x^2 z\,dydz + \int\int_{S_{xz}} xy^2\,dxdz - \int\int_{S_{xy}} z\,dxdy\ ,$$

(a)

$z = x^2 + y^2$

(b)

S_{xz} S_{yz}

S_{xy}

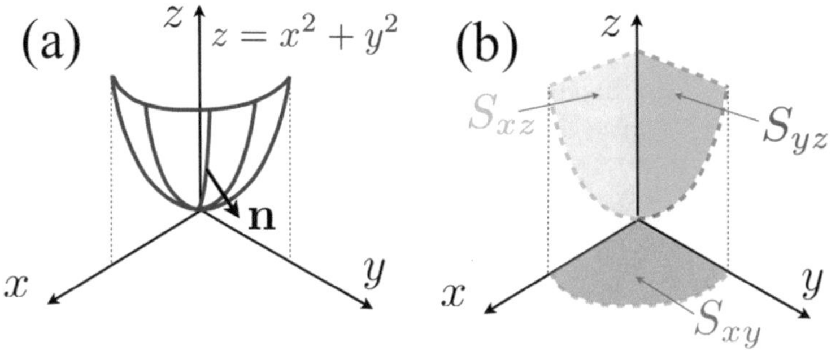

Fig. 6.43 a The surface $z = x^2 + y^2$ bounded by $0 \le z \le 1, x \ge 0$ and $y \ge 0$ (blue). Its projections on the three Cartesian planes are also shown in **b** by light brown (S_{xy}), green (S_{xz}) and red (S_{yz}) colours

where the surfaces S_{xy} (with $z = 0$), S_{xz} (with $y = 0$) and S_{yz} (with $x = 0$) are explicitly indicated in the figure. Note the minus sign before the $dxdy$ integral. This is because in the cases of I_x and I_y the normal is directed along the x and y axes, while in the case of I_z it is directed downwards (in the direction of $-z$), so we have to take the negative sign for the $dxdy$ integral.

The individual integrals are calculated by remembering that the third variable is related to the other two used in the integration via the equation $z = x^2 + y^2$ of the surface, and hence can be easily expressed. We obtain:

$$
I_x = \int\!\!\int_{S_{yz}} x^2 z\, dy\, dz = \int\!\!\int \left(z - y^2\right) z\, dy\, dz
$$
$$
= \int_0^1 dy \int_{y^2}^1 dz\, (z - y^2)\, z = \int_0^1 dy \left(\frac{1}{3} + \frac{y^6}{6} - \frac{y^2}{2}\right) = \frac{4}{21},
$$
$$
I_y = \int\!\!\int_{S_{xz}} xy^2 dx\, dz = \int\!\!\int x\left(z - x^2\right) dx\, dz
$$
$$
= \int_0^1 dx \int_{x^2}^1 dz\, (z - x^2)\, x = \int_0^1 dx \left(\frac{x}{2} - x^3 + \frac{x^5}{2}\right) = \frac{1}{12},
$$

and, finally, the last integral I_z is most easily calculated using polar coordinates (r, ϕ) (recall that the two-dimensional Jacobian is equal to r in this case):

$$
I_z = -\int\!\!\int_{S_{xy}} z\, dx\, dy = -\int\!\!\int \underbrace{\left(x^2 + y^2\right)}_{r^2} \underbrace{dx\, dy}_{r\, dr\, d\phi}
$$
$$
= -\int_0^1 r^3 dr \int_0^{\pi/2} d\phi = -\left.\frac{r^4}{4}\right|_0^1 \left(\frac{\pi}{2}\right) = -\frac{\pi}{8}.
$$

Thus, $I = 4/21 + 1/12 - \pi/8$.

Problem 6.63 Try to reproduce this result using the first method of calculating the flux integral based on Eq. (6.67).

Problem 6.64 Calculate the flux of the magnetic field $\mathbf{B} = (2xy, yz, 1)$ through the surface of the planar circle of radius R centred at the origin and located in the $x - y$ plane. [Answer: πR^2.]

Problem 6.65 Calculate the flux of the vector field $\mathbf{F} = (0, 0, z)$ through the external surface (i.e. in the outward direction) of the circular paraboloid $z = x^2 + y^2$ shown in Fig. 6.38b, bound by the planes $z = 0$ and $z = 1$. [Answer: $-\pi/2$.]

Problem 6.66 Evaluate the flux integral over the cube $0 < x < 1$, $0 < y < 1$ and $0 < z < 1$ for the field $\mathbf{F} = (x, y, 2)$ and the outward direction for each face of the cube. [Answer: 2.]

Problem 6.67 Evaluate the surface (flux) integral over the outward surface of the unit sphere centred at the origin with $\mathbf{F} = \mathbf{r}$. [Answer: 4π.]

6.5.5 Relationship Between Line and Surface Integrals. Stokes's Theorem

We recall that there is a relationship between the line integral on the plane around some surface area and the double integral over that area (Green's formula), see Sect. 6.4.3. In fact, this, essentially two-dimensional result, can be generalised into all three dimensions and is called *Stokes's formula* (theorem).

Theorem 6.6 (Stokes's Theorem) *Let S be a simply connected non-planar piecewise smooth orientable (two-sided, see Fig. 6.34a) surface with the closed boundary line L. If three functions $F_x(\mathbf{r})$, $F_y(\mathbf{r})$ and $F_z(\mathbf{r})$ are all continuous together with their first derivatives everywhere in S including its boundary, then*

$$\oint_L \mathbf{F} \cdot d\mathbf{l} = \int\int_S \mathrm{curl}\, \mathbf{F} \cdot d\mathbf{A} , \qquad (6.73)$$

where $\mathbf{F}(\mathbf{r}) = F_x(\mathbf{r})\mathbf{i} + F_y(\mathbf{r})\mathbf{j} + F_z(\mathbf{r})\mathbf{k}$ is the vector field, and we also defined another vector:

$$\mathrm{curl}\, \mathbf{F}(\mathbf{r}) = \left(\frac{\partial F_z}{\partial y} - \frac{\partial F_y}{\partial z}\right)\mathbf{i} + \left(\frac{\partial F_x}{\partial z} - \frac{\partial F_z}{\partial x}\right)\mathbf{j} + \left(\frac{\partial F_y}{\partial x} - \frac{\partial F_x}{\partial y}\right)\mathbf{k} .$$

$$(6.74)$$

This vector is called curl *of $\mathbf{F}$; it is composed of partial derivatives of the components of the vector field with respect to the Cartesian coordinates. Note that the surface boundary line L is traversed in a direction that corresponds to the chosen side of the surface to be always on the left (according to the right-hand rule, see Fig. 6.34b). Note that formula (6.73) can also be written differently as*

$$\oint_L F_x dx + F_y dy + F_z dz$$

$$= \int\int_S \left(\frac{\partial F_z}{\partial y} - \frac{\partial F_y}{\partial z}\right) dy dz + \left(\frac{\partial F_x}{\partial z} - \frac{\partial F_z}{\partial x}\right) dx dz + \left(\frac{\partial F_y}{\partial x} - \frac{\partial F_x}{\partial y}\right) dx dy ,$$

$$(6.75)$$

> where in the right-hand side we employed an alternative form for the surface
> (flux) integral, see Sect. 6.5.4.

Proof: Consider first a surface S of a rather special shape which is specified by
the equation $z = z(x, y)$ and let S_{xy} be its projection on the $x - y$ plane, Fig. 6.44.
The surface boundary L will become the boundary L_{xy} in the projection S_{xy}. We
can specify the boundary L by an additional equation relating y to x, i.e. $y = y(x)$,
which we assume does exist and which, together with the equation for the surface
$z = z(x, y)$, would give uniquely the boundary L. Then, consider the x-part of the
line integral (see the text around Eq. (6.38)):

$$J_P = \oint_L F_x(x, y, z)dx = \oint_{L_{xy}} F_x\left(x, y(x), z\left(x, y(x)\right)\right) dx . \qquad (6.76)$$

We have rewritten it as a line integral over the projection L_{xy} of the line L onto
the $x - y$ plane, see again the figure. Using Green's formula (6.41) valid for planar
regions, we obtain that the line integral J_P must be equal to the double integral over
the planar region S_{xy} (note that $P \equiv F_x$ and $Q \equiv 0$ in our case):

$$J_P = \int\int_{S_{xy}} \left(-\frac{\partial F_x}{\partial y}\right)_{tot} dxdy = -\int\int_{S_{xy}} \left(\frac{\partial F_x}{\partial y} + \frac{\partial F_x}{\partial z}\frac{\partial z}{\partial y}\right) dxdy . \qquad (6.77)$$

Note that the derivative $(\partial F_x/\partial y)_{tot}$ is the full derivative with respect to y, and,
when calculating it, we must take account of the fact that $z = z(x, y)$ also depends
on y. Next, consider the components of the normal vector $\mathbf{n}$ to the surface, Eqs.
(6.69)–(6.71). Since for that surface x and y serve as its parameters u and v, the
corresponding Jacobians are given by Eqs. (6.55); hence,

$$n_y = \cos(\widehat{\mathbf{n}, \mathbf{j}}) = \frac{-\partial z/\partial y}{\sqrt{1 + (\partial z/\partial x)^2 + (\partial z/\partial y)^2}} ,$$

$$n_z = \cos(\widehat{\mathbf{n}, \mathbf{k}}) = \frac{1}{\sqrt{1 + (\partial z/\partial x)^2 + (\partial z/\partial y)^2}} ,$$

and therefore,

$$\frac{\partial z}{\partial y} = -\frac{n_y}{n_z} = -\frac{\cos(\widehat{\mathbf{n}, \mathbf{j}})}{\cos(\widehat{\mathbf{n}, \mathbf{k}})} .$$

This result enables us to rewrite Eq. (6.77) as

$$J_P = -\int\int_{S_{xy}} \left(\frac{\partial F_x}{\partial y} - \frac{\partial F_x}{\partial z}\frac{\cos(\widehat{\mathbf{n}, \mathbf{j}})}{\cos(\widehat{\mathbf{n}, \mathbf{k}})}\right) dxdy . \qquad (6.78)$$

Fig. 6.44 The proof of the Stokes's theorem: the surface S with the boundary L is projected on the $x - y$ plane forming a planar surface S_{xy} with the planar boundary L_{xy}

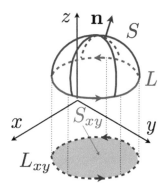

On the other hand, consider the following surface integral over the surface S:

$$\int\int_S \left(-\frac{\partial F_x}{\partial y} \cos(\widehat{\mathbf{n}, \mathbf{k}}) + \frac{\partial F_x}{\partial z} \cos(\widehat{\mathbf{n}, \mathbf{j}}) \right) dA$$

$$= -\int\int_S \left(\frac{\partial F_x}{\partial y} - \frac{\partial F_x}{\partial z} \frac{\cos(\widehat{\mathbf{n}, \mathbf{j}})}{\cos(\widehat{\mathbf{n}, \mathbf{k}})} \right) \cos(\widehat{\mathbf{n}, \mathbf{k}}) dA$$

$$= -\int\int_{S_{xy}} \left(\frac{\partial F_x}{\partial y} - \frac{\partial F_x}{\partial z} \frac{\cos(\widehat{\mathbf{n}, \mathbf{j}})}{\cos(\widehat{\mathbf{n}, \mathbf{k}})} \right) dxdy.$$

Here in the last passage we have been able to replace $\cos(\widehat{\mathbf{n}, \mathbf{k}})dA$ with $+dxdy$ since n_z is positive, Fig. 6.44. Thus, we notice that the same expression (6.78) has been obtained, i.e. we have just proven that

$$J_P = \oint_L F_x(x, y, z)dx = \int\int_S \left(\frac{\partial F_x}{\partial z} \cos(\widehat{\mathbf{n}, \mathbf{j}}) - \frac{\partial F_x}{\partial y} \cos(\widehat{\mathbf{n}, \mathbf{k}}) \right) dA$$

$$= \int\int_S \left(\frac{\partial F_x}{\partial z} n_y - \frac{\partial F_x}{\partial y} n_z \right) dA.$$

(6.79)

A similar consideration *for the same surface S* but specified via $y = y(x, z)$ and the function $F_z(\mathbf{r})$ gives

$$J_R = \oint_L F_z(x, y, z)dz = \int\int_S \left(\frac{\partial F_z}{\partial y} \cos(\widehat{\mathbf{n}, \mathbf{i}}) - \frac{\partial F_z}{\partial x} \cos(\widehat{\mathbf{n}, \mathbf{j}}) \right) dA$$

$$= \int\int_S \left(\frac{\partial F_z}{\partial y} n_x - \frac{\partial F_z}{\partial x} n_y \right) dA,$$

(6.80)

and if *the same surface S* is specified via $x = x(y, z)$ instead and we consider the function $F_y(\mathbf{r})$, one obtains:

$$J_Q = \oint_L F_y(x, y, z)dy = \int\int_S \left(\frac{\partial F_y}{\partial x} \cos(\widehat{\mathbf{n}, \mathbf{k}}) - \frac{\partial F_y}{\partial z} \cos(\widehat{\mathbf{n}, \mathbf{i}}) \right) dA$$

$$= \int\int_S \left(\frac{\partial F_y}{\partial x} n_z - \frac{\partial F_y}{\partial z} n_x \right) dA.$$

(6.81)

Fig. 6.45 Surface S contains a hole in it. The traverse directions, chosen with respect to the right-hand rule for the given direction of the normal **n** (outward), correspond to the surface being always on the left

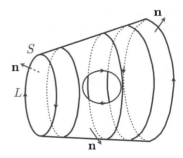

Any piecewise smooth simply connected surface can always be represented as a combination of elementary surfaces considered above. Therefore, the equations above are valid for any such S. Let us add these three identities (6.79)–(6.81) together. In the left-hand side we then get the line integral over the boundary L from all three functions (components of **F**), while in the right-hand side the sum of the surface integrals:

$$
\oint_L F_x dx + F_y dy + F_z dz \equiv \int\int_S \left(\frac{\partial F_x}{\partial z} n_y - \frac{\partial F_x}{\partial y} n_z \right) dA
$$
$$
+ \int\int_S \left(\frac{\partial F_z}{\partial y} n_x - \frac{\partial F_z}{\partial x} n_y \right) dA + \int\int_S \left(\frac{\partial F_y}{\partial x} n_z - \frac{\partial F_y}{\partial z} n_x \right) dA
$$
$$
= \int\int_S \left[\left(\frac{\partial F_z}{\partial y} - \frac{\partial F_y}{\partial z} \right) n_x + \left(\frac{\partial F_x}{\partial z} - \frac{\partial F_z}{\partial x} \right) n_y + \left(\frac{\partial F_y}{\partial x} - \frac{\partial F_x}{\partial y} \right) n_z \right] dA
$$
$$
= \int\int_S (\mathrm{curl}\, \mathbf{F} \cdot \mathbf{n})\, dA = \int\int_S \mathrm{curl}\, \mathbf{F} \cdot d\mathbf{A} .
$$

If a more complex surface is considered, it can always be divided up into finite fragments for each of which the above identity is valid separately. Summing up such identities for all such fragments, we have in the right-hand side the surface integral over the entire surface, while in the left-hand side the line integral over all boundary lines including the internal lines used to divide the surface up. However, the internal boundaries will be contributing twice, each time traversed in the opposite directions, and hence there will be no contribution from them since the line integral changes sign if taken in the opposite direction (cf. our discussion of Green's theorem in Sect. 6.4.3 and especially Fig. 6.28c). Therefore, the formula is valid for any surface, which is the required general result. **Q.E.D.**

Note that the Stokes's theorem is also valid for more complicated surfaces as well, e.g. the ones that contain holes, Fig. 6.45. In this case the line integral $\oint_L$ is considered along all boundaries, including the inner ones, and the traverse directions should be chosen in accord with the right-hand rule as illustrated in the figure.

The curl of $\mathbf{F}$ can also be formally written as a determinant, and this form is easier to remember:

$$\operatorname{curl} \mathbf{F}(\mathbf{r}) = \begin{vmatrix} \mathbf{i} & \mathbf{j} & \mathbf{k} \\ \partial/\partial x & \partial/\partial y & \partial/\partial z \\ F_x & F_y & F_z \end{vmatrix} . \tag{6.82}$$

This determinant notation is to be understood as follows: the determinant has to be opened along the first row with the operators in the second row to appear on the left from the components of the vector field $\mathbf{F}$ occupying the third row. Then, their "product" should be interpreted as an *action* of the operator on the corresponding component of $\mathbf{F}$. For instance, the x component of the curl of $\mathbf{F}$ is obtained as

$$(\operatorname{curl} \mathbf{F})_x = \begin{vmatrix} \partial/\partial y & \partial/\partial z \\ F_y & F_z \end{vmatrix} = \frac{\partial}{\partial y} F_z - \frac{\partial}{\partial z} F_y \longmapsto \frac{\partial F_z}{\partial y} - \frac{\partial F_y}{\partial z} ,$$

which is exactly the x component of the curl appearing in the original formula (6.74). It is easy to check that it works for the other two components as well.

As an example, we shall verify Stokes's theorem by considering the flux of the vector field $\mathbf{F} = -y\mathbf{i} + x\mathbf{j} + \mathbf{k}$ through the hemisphere $z = \sqrt{R^2 - x^2 - y^2} \geq 0$ centred at the origin. We shall first consider the line integral in the left-hand side of Stokes's theorem (6.73). Here the surface boundary L is a circle $x^2 + y^2 = R^2$ of radius R. Using polar coordinates, $x = R \cos \phi$ and $y = R \sin \phi$ with the angle ϕ as a parameter, we obtain (note that $z = 0$ for the boundary line):

$$\oint_L \mathbf{F} \cdot d\mathbf{l} = \oint_L F_x dx + F_y dy$$

$$= \int_0^{2\pi} [(-R \sin \phi)(-R \sin \phi) + (R \cos \phi)(R \cos \phi)] \, d\phi$$

$$= R^2 \int_0^{2\pi} d\phi = 2\pi R^2 .$$

On the other hand, let us now consider the surface integral in the right-hand side of Eq. (6.73). The curl of $\mathbf{F}$ is

$$\operatorname{curl} \mathbf{F} = \begin{vmatrix} \mathbf{i} & \mathbf{j} & \mathbf{k} \\ \partial/\partial x & \partial/\partial y & \partial/\partial z \\ -y & x & 1 \end{vmatrix}$$

$$= \left(\frac{\partial 1}{\partial y} - \frac{\partial x}{\partial z} \right) \mathbf{i} - \left(\frac{\partial 1}{\partial x} - \frac{\partial(-y)}{\partial z} \right) \mathbf{j} + \left(\frac{\partial x}{\partial x} - \frac{\partial(-y)}{\partial y} \right) \mathbf{k}$$

$$= 0\mathbf{i} - 0\mathbf{j} + (1 - (-1))\mathbf{k} = 2\mathbf{k} ,$$

and the normal to the hemisphere is given by $\mathbf{N} = (R \sin \theta) \, \mathbf{r}$, see Eq. (6.58), so that, following Eq. (6.67), and using spherical coordinates,

$$\int \int_S \operatorname{curl} \mathbf{F} \cdot d\mathbf{A} = \int \int (\operatorname{curl} \mathbf{F} \cdot \mathbf{N}) \, d\theta d\phi = \int \int 2R \sin \theta \underbrace{(\mathbf{k} \cdot \mathbf{r})}_{z} \, d\theta d\phi$$

$$= 2R \int \int (\sin \theta) \, z \, d\theta d\phi = 2R \int \int (\sin \theta) \, (R \cos \theta) \, d\theta d\phi$$

$$= 2R^2 \int_0^{\pi/2} \sin \theta \cos \theta d\theta \int_0^{2\pi} d\phi = 2\pi R^2 .$$

The same result is obtained also using Eq. (6.68):

$$\int \int_S \operatorname{curl} \mathbf{F} \cdot d\mathbf{A} = \int \int 2 \, (\mathbf{k} \cdot \mathbf{n}) \, dA = 2 \int \int n_z dA$$

$$= 2 \int \int_{circle} dx dy = 2\pi R^2 .$$

Note that, according to the Stokes's theorem, the surface (flux) integral of a vector field $\mathbf{V} = \operatorname{curl} \mathbf{F}$ does not depend on the actual surface, only on its bounding curve. Therefore, if a vector field can be written as $\mathbf{V} = \operatorname{curl} \mathbf{F}$, then the surface integral

$$\int \int_S (\mathbf{V} \cdot \mathbf{n}) \, dA = \int \int_S (\operatorname{curl} \mathbf{F} \cdot \mathbf{n}) \, dA = \oint_L \mathbf{F} \cdot d\mathbf{l}$$

will not depend on the actual surface, only on its bounding curve L. This property can be used to calculate some of the surface integrals by replacing them with simpler line integrals, as is illustrated by the surface integral $\int \int_S \operatorname{curl} \mathbf{F} \cdot d\mathbf{A}$ with $\mathbf{F} = (xyz, -x, z)$ and the surface S being a hemisphere $x^2 + y^2 + z^2 = R^2$ bounded below by the $z = 0$ plane. According to Stokes's theorem, the integral in question can be replaced by the line integral

$$\oint_{circle} F_x dx + F_y dy + F_z dz$$

over the bounding circle $x^2 + y^2 = R^2$. This is most easily calculated in polar coordinates ($x = R \cos \phi$, $y = R \sin \phi$ and $z = 0$):

$$\oint_{circle} = \int_0^{2\pi} [xyz \, (-R \sin \phi) - x \, (R \cos \phi)] \, d\phi = \int_0^{2\pi} [-R \cos \phi \, (R \cos \phi)] \, d\phi$$

$$= -R^2 \int_0^{2\pi} \cos^2 \phi d\phi = -R^2 \left(\frac{\phi}{2} + \frac{1}{4} \sin (2\phi) \right) \Big|_0^{2\pi} = -\pi R^2 .$$

Problem 6.68 Verify Stokes's theorem for $\mathbf{F} = (x, 0, z)$ and the surface S being the triangle in Fig. 6.41.

Problem 6.69 Verify Stokes's theorem for $\mathbf{F} = (xy, yz, xz)$ and the surface S being a square with corners at $(\pm 1, \pm 1, 0)$. [Answer: *either of the integrals is zero.*]

Problem 6.70 Consider a cylinder $x^2 + y^2 = R^2$ of height h. Its upper end at $z = h$ is terminating with the plane, while its other end at $z = 0$ is opened. Verify Stokes's theorem for the surface of the cylinder if $\mathbf{F} = (0, -y, z)$. [Answer: *either of the integrals is zero.*]

Problem 6.71 Prove that the *curl* $[f(r)\mathbf{r}] = 0$, where $f(r)$ is an arbitrary function of $r = |\mathbf{r}|$.

Problem 6.72 Calculate *curl* $\mathbf{F}$, where $\mathbf{F} = (yz, xz, xy)$. [Answer: **0**.]

Problem 6.73 Calculate *curl* $\mathbf{F}$ at the point $(1, 1, 1)$, where $\mathbf{F} = (2x^2z, 2xy, xz)$. [Answer: $(0, 1, 2)$.]

Problem 6.74 Use Stokes's theorem to evaluate the line integral $\oint_L \mathbf{F} \cdot d\mathbf{l}$, where $\mathbf{F} = (-y, x, 2z)$ and the contour L is the unit circle $x^2 + y^2 = 1$ in the $z = 2$ plane. [Hint: *cap the circle with the hemisphere.*] [Answer: 2π.]

Problem 6.75 Calculate, using Stokes's theorem, the surface integral $\int \int_S \mathbf{F} \cdot d\mathbf{A}$, where $\mathbf{F} = $ curl $\mathbf{B}$ with $\mathbf{B} = (3y, -2x, xy)$, and S is the hemispherical surface $x^2 + y^2 + z^2 = R^2$ with $z \geq 0$. Will the integral change if a different surface is chosen which caps *the same* bounding curve? [Answer: $-5\pi R^2$; *no.*]

Problem 6.76 Choose the planar surface S in the previous problem (i.e. the disk of radius R) that has the same bounding line $x^2 + y^2 = R^2$. Calculate $\mathbf{F}$=curl $\mathbf{B}$ and show by the direct calculation of the surface integral, $\int \int_S \mathbf{F} \cdot d\mathbf{A}$, that it is equal to the same value as in the previous problem.

Fig. 6.46 Circular ellipsoid
$z = 1 - x^2 - y^2, z \geq 0$

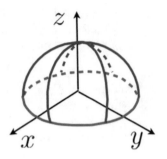

Problem 6.77 Verify Stokes's theorem if $\mathbf{F} = (y^2, z^2, x^2)$ and S is the surface of the circular ellipsoid described by $z = 1 - x^2 - y^2$ with $z \geq 0$, see Fig. 6.46. [Answer: *either of the integrals is zero.*]

Problem 6.78 Use Stokes's theorem to evaluate the line integral $\oint_L \mathbf{F} \cdot d\mathbf{l}$ over the unit circle $x^2 + y^2 = 1$ in the $z = 5$ plane, where $\mathbf{F} = (x + y, z - 2x + y, y - z)$. [Answer: -3π.]

6.5.6   Three-Dimensional Case: Exact Differentials

As in the planar case, Sect. 6.4.4, there exists a very general theorem in the three-dimensional case as well, that states:

Theorem 6.7 *Let D be a simply connected region in the three-dimensional space, and there is a vector field $\mathbf{F}(\mathbf{r}) = (F_x, F_y, F_z)$ that is continuous in D together with its first derivatives. Then, the following four statements are equivalent to each other:*

1. For any closed loop L inside D

$$\oint_L F_x dx + F_y dy + F_z dz = \oint_L \mathbf{F} \cdot d\mathbf{l} = 0 \; ; \qquad (6.83)$$

2. The line integral $\int_C \mathbf{F} \cdot d\mathbf{l}$ does not depend on the actual integration curve C, only on its starting and ending points;
3. $dU(x, y, z) = F_x(x, y, z)dx + F_y(x, y, z)dy + F_z(x, y, z)dz$ is the exact differential, and

---

> 4. *The following three conditions are satisfied:*
>
> $$\frac{\partial F_x}{\partial y} = \frac{\partial F_y}{\partial x}, \quad \frac{\partial F_z}{\partial y} = \frac{\partial F_y}{\partial z} \quad and \quad \frac{\partial F_x}{\partial z} = \frac{\partial F_z}{\partial x}, \qquad (6.84)$$
>
> *or, which is equivalent (see Eq. (6.74)), curl* $F = 0$.

Proof: This theorem is proven similarly to the planar case of Sect. 6.4.4 by following the logic: $1 \rightarrow 2$, $2 \rightarrow 3$, $3 \rightarrow 4$ and then, finally, $4 \rightarrow 1$.

- $1 \rightarrow 2$: the proof is identical to the planar case;
- $2 \rightarrow 3$: essentially the same as in the planar case, the function $U(x, y, z)$ can be defined that gives rise to the exact differential dU, and this function is verified to be the line integral

$$U(x, y, z) = \int_{A(x_0, y_0, z_0)}^{B(x, y, z)} F_x dx + F_y dy + F_z dz, \qquad (6.85)$$

where $A(x_0, y_0, z_0)$ is a fixed point;
- $3 \rightarrow 4$: the same as in the planar case;
- $4 \rightarrow 1$: this is slightly different (although the idea is the same): we take a closed contour L and "dress" on it a surface S so that L would serve as its boundary. Then we use the Stokes's formula: $\oint_L \cdots = \int \int_S \cdots$. However, the curl of $\mathbf{F}$ is equal to zero due to our assumption (property 4). Therefore, $\oint_L \cdots = 0$. Since the choice of L was arbitrary, this result is valid for any L in D. **Q.E.D.**

Note that Eq. (6.85) is indispensable for finding exact differentials, similar to the two-dimensional case. Since the integral does not depend on the path, we can choose it to simplify the calculation. In many cases this is most easily done by taking the integration path $(x_0, y_0, z_0) \rightarrow (x, y_0, z_0) \rightarrow (x, y, z_0) \rightarrow (x, y, z)$ which contains straight lines going parallel to the Cartesian axis; it would split the line integral into three ordinary definite integrals:

$$U(x, y, z) = \int_{x_0}^{x} F_x(x_1, y_0, z_0) dx_1 + \int_{y_0}^{y} F_y(x, y_1, z_0) dy_1 + \int_{z_0}^{z} F_z(x, y, z_1) dz_1. \qquad (6.86)$$

An arbitrary constant appears naturally due to an arbitrary choice of the initial point $A(x_0, y_0, z_0)$. This results in a direct generalisation of formula (6.48) of the two-dimensional case.

Note that Eq. (6.85) can be formally considered as appearing after integrating both sides of the exact differential $dU = F_x dx + F_y dy + F_z dz$ between points

$A\,(x_0,\,y_0,\,z_0)$ and $B(x,\,y,\,z)$:

$$\int_A^B dU = \int_A^B F_x dx + F_y dy + F_z dz$$

$$\implies\ U(x,\,y,\,z) = \int_A^B F_x dx + F_y dy + F_z dz\,.$$

Of course, this operation only makes sense in the case of the validity of the theorem, i.e. when the line integral that appears in the right-hand side of such integration depends only on the initial and final points. Note that the function U defined this way contains also an additive constant containing the information about the point A; this constant can be considered as an arbitrary constant C. In other words, formula (6.85) defines a family of functions that all give rise to the same exact differential.

For instance, consider the expression $yzdx + zxdy + xydz$. First of all, we check that this expression is the exact differential. Here $F_x = yz$, $F_y = zx$ and $F_z = xy$. Therefore, if $\mathbf{F} = (F_x,\,F_y,\,F_z)$, then

$$\mathrm{curl}\,\mathbf{F} = \begin{vmatrix} \mathbf{i} & \mathbf{j} & \mathbf{k} \\ \partial/\partial x & \partial/\partial y & \partial/\partial z \\ yz & zx & xy \end{vmatrix} = \mathbf{i}\,(x-x) - \mathbf{j}\,(y-y) + \mathbf{k}\,(z-z) = 0\,,$$

which means that we can apply the theorem. In particular, we can use Eq. (6.86) to find $U(x,\,y,\,z)$ which would give rise to the differential $dU = yzdx + zxdy + xydz$:

$$U = \int_{x_0}^x y_0 z_0 dx_1 + \int_{y_0}^y z_0 x dy_1 + \int_{z_0}^z xy dz_1$$

$$= y_0 z_0\,(x-x_0) + z_0 x\,(y-y_0) + xy\,(z-z_0) = xyz - x_0 y_0 z_0 = xyz + C\,.$$

So, in this case we have obtained a family of functions $U(x,\,y,\,z) = xyz + C$ that all give the same exact differential. It is readily checked that

$$dU = \frac{\partial U}{\partial x} dx + \frac{\partial U}{\partial y} dy + \frac{\partial U}{\partial z} dz = yzdx + xzdy + xydz\,,$$

which indeed is our expression.

6.5.7   Ostrogradsky–Gauss Theorem

We have seen above that there exists a simple relationship between the surface integral over some, generally non-planar, surface S and its bounding contour L (Stokes's theorem). Similarly, there is also a very general result relating the integral over some volume V and its bounding surface S which bears the names of Ostrogradsky and Gauss.

Fig. 6.47 A cylindrical space region bounded from below and above by surfaces $z = z_1(x, y)$ and $z = z_2(x, y)$, respectively. The upper cap forms a surface S_2, the lower one S_1 and the cylindrical part S_3

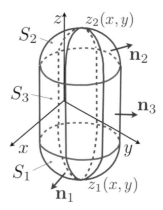

Theorem 6.8 (Divergence Theorem). *Let $\boldsymbol{F} = (F_x, F_y, F_z)$ be a vector field which is continuous (together with its first derivatives) in some three-dimensional region D. Then,*

$$\int\int\int_D \operatorname{div}\boldsymbol{F}\,dxdydz = \oiint_S \boldsymbol{F} \cdot d\boldsymbol{A} , \tag{6.87}$$

where S is the bounding surface to D, with the outward normal $\boldsymbol{n}$, and the scalar

$$\operatorname{div}\boldsymbol{F} = \nabla \cdot \boldsymbol{F} = \frac{\partial F_x}{\partial x} + \frac{\partial F_y}{\partial y} + \frac{\partial F_z}{\partial z} \tag{6.88}$$

is called divergence *of the vector field $\boldsymbol{F}$.*

Note also another formal notation for the divergence that is frequently used, $\nabla \cdot \boldsymbol{F}$, which represents a dot product of the vector of the del operator Eq. (5.72) and the vector $\boldsymbol{F}$ of the field:

$$\nabla \cdot \boldsymbol{F} = \left(\frac{\partial}{\partial x}\boldsymbol{i} + \frac{\partial}{\partial y}\boldsymbol{j} + \frac{\partial}{\partial z}\boldsymbol{k}\right) \cdot \left(F_x\boldsymbol{i} + F_y\boldsymbol{j} + F_z\boldsymbol{k}\right)$$

$$= \frac{\partial}{\partial x}F_x + \frac{\partial}{\partial y}F_y + \frac{\partial}{\partial z}F_z = \frac{\partial F_x}{\partial x} + \frac{\partial F_y}{\partial y} + \frac{\partial F_z}{\partial z} ,$$

which is the same as in Eq. (6.88).

Proof: Consider a region D that is cylindrical along the z-axis and capped from below and above by surfaces $z = z_1(x, y)$ and $z = z_2(x, y)$, respectively, Fig. 6.47.

Consider then a volume integral

$$\int\int\int_D \frac{\partial F_z}{\partial z} dxdydz = \int\int dxdy \int_{z_1(x,y)}^{z_2(x,y)} \frac{\partial F_z}{\partial z} dz$$

$$= \int\int_{S_{xy}} \left[F_z\left(x, y, z_2(x, y)\right) - F_z\left(x, y, z_1(x, y)\right) \right] dxdy ,$$

where S_{xy} is the projection of the surfaces S_1 or S_2 on the $x - y$ plane (both projections are the same due to the specific shape of D).

On the other hand, the surface integral over the whole surface of the cylinder $S = S_1 + S_2 + S_3$ (see the figure) for the vector field $\mathbf{G} = (0, 0, F_z)$ is

$$\oiint_S \mathbf{G} \cdot d\mathbf{A} = \int\int_S F_z n_z dA = \int\int_{S_1} F_z n_z dA + \int\int_{S_2} F_z n_z dA + \int\int_{S_3} F_z n_z dA .$$

The integral over S_3 is zero since $n_z = 0$ there; the other two integrals transform into:

$$\int\int_{S_1} F_z n_z dA = - \int\int_{S_{xy}} F_z\left(x, y, z_1(x, y)\right) dxdy$$

and

$$\int\int_{S_2} F_z n_z dA = + \int\int_{S_{xy}} F_z\left(x, y, z_2(x, y)\right) dxdy .$$

There is the minus sign chosen for the S_1 integral since $n_z < 0$ there, i.e. its normal leans towards the negative z direction. Therefore,

$$\oiint_S \mathbf{G} \cdot d\mathbf{A} = \int\int_{S_{xy}} \left[F_z\left(x, y, z_2(x, y)\right) - F_z\left(x, y, z_1(x, y)\right) \right] dxdy ,$$

which is the same as the volume integral above. Hence,

$$\int\int\int_D \frac{\partial F_z}{\partial z} dxdydz = \oiint_S \begin{pmatrix} 0 \\ 0 \\ F_z \end{pmatrix} \cdot d\mathbf{A} , \tag{6.89}$$

where the vector $\mathbf{G}$ has been shown explicitly via its components. This result should be valid for any region D that can be broken down into cylindrical regions as above. This is because the volume integrals for these regions will just sum up (the left-hand side), while the surface integrals will be taken over all surfaces. However, for all *internal* surfaces (where different cylinders touch) the two integrals will be taken over different sides of the same surface (the normal vectors are opposite), and thus will cancel out as the surface integral changes sign if the other side of the surface is used.

Similarly, if the region D has a cylindrical shape with the axis along the x-axis and the caps are specified by the equations $x = x_1(y, z)$ and $x = x_2(y, z)$, then we obtain

$$\int\int\int_D \frac{\partial F_x}{\partial x} dx dy dz = \oiint_S \begin{pmatrix} F_x \\ 0 \\ 0 \end{pmatrix} \cdot d\mathbf{A} . \tag{6.90}$$

Finally, for cylindrical regions along the y-axis with the two caps specified by $y = y_1(x, z)$ and $y = y_2(x, z)$, we obtain

$$\int\int\int_D \frac{\partial F_y}{\partial y} dx dy dz = \oiint_S \begin{pmatrix} 0 \\ F_y \\ 0 \end{pmatrix} \cdot d\mathbf{A} . \tag{6.91}$$

As any volume can be broken down by means of either of these three ways, i.e. using touching cylindrical volumes oriented along either of the three coordinate axes, these three results are valid at the same time for any volume. Therefore, summing up all three expressions (6.89)–(6.91), we arrive at the final formula (6.87) sought for. **Q.E.D.**

As an example of calculating the divergence, consider the vector field $\mathbf{F} = e^{-r^2}(yz\mathbf{i} + xz\mathbf{j} + xy\mathbf{k})$ at the point $A(1, 1, 1)$. Here, $r^2 = x^2 + y^2 + z^2$, so the divergence is easily calculated from its definition using the necessary partial differentiation:

$$\nabla \cdot \mathbf{F} = \frac{\partial}{\partial x}\left(e^{-r^2}yz\right) + \frac{\partial}{\partial y}\left(e^{-r^2}xz\right) + \frac{\partial}{\partial z}\left(e^{-r^2}xy\right)$$
$$= e^{-r^2}(-2xyz) + e^{-r^2}(-2xyz) + e^{-r^2}(-2xyz) = -6xyze^{-r^2} ,$$

and at the point A we have: $\nabla \cdot \mathbf{F}(A) = -6e^{-3}$.

The divergence obeys some important properties. First of all, it is obviously additive:

$$\nabla \cdot (\mathbf{F}_1 + \mathbf{F}_2) = \nabla \cdot \mathbf{F}_1 + \nabla \cdot \mathbf{F}_2 .$$

Some other properties are given in the problems below.

Problem 6.79 Show that

$$\nabla f(\phi) = \frac{df}{d\phi}\nabla\phi ,$$

and that, in particular, $\nabla e^{\lambda\phi} = \lambda e^{\lambda\phi}\nabla\phi$.

Problem 6.80 Using the definition of the divergence in Cartesian coordinates, prove that

$$\nabla \cdot (U\mathbf{F}) = U\,(\nabla \cdot \mathbf{F}) + \mathbf{F} \cdot (\nabla U) \,, \tag{6.92}$$

where in the first term in the right-hand side there is a divergence of a vector field $\mathbf{F}$, while in the second term the gradient of the scalar field U. Using this result, calculate $\nabla \cdot (xy\mathbf{F})$ with $\mathbf{F} = \left(y^2, 0, xz\right)$. Compare your result with direct calculation of the divergence of the vector field $\mathbf{G} = \left(xy^3, 0, x^2yz\right)$, [Answer: $y\left(y^2 + x^2\right)$.]

Problem 6.81 Verify the divergence theorem for $\mathbf{F} = (xy, yz, zx)$ and the volume region of a cube with vertices at the points $(\pm 1, \pm 1, \pm 1)$. [Answer: *each integral is zero.*]

Problem 6.82 Verify the divergence theorem if $\mathbf{F} = (xy^2, x^2y, y)$ and S is a circular cylinder $x^2 + y^2 = 1$ oriented with its axis along the z-axis, $-1 \le z \le 1$. [Answer: *each of the integrals is equal to π.*]

Problem 6.83 Use the divergence theorem to calculate the surface flux integral $\oiint_S \mathbf{F} \cdot d\mathbf{A}$ over the entire surface of the cylinder $x^2 + y^2 = R^2$, $-h/2 \le z \le h/2$, where $\mathbf{F} = (x^3, y^3, z^3)$. [Answer: $\pi h R^2 \left(6R^2 + h^2\right)/4$.]

Problem 6.84 Consider a closed surface formed by the side surface of the paraboloid $z = x^2 + y^2$ of Problem 6.65 and its upper circle at $z = 1$, see Fig. 6.38b. Calculate the outer flux through its surface, $\oiint_S \mathbf{F} \cdot d\mathbf{A}$ with $\mathbf{F} = (0, 0, z)$, using two methods: (a) directly, by calculating the flux through both surfaces, and (b) using the divergence theorem. [Answer: $\pi/2$.]

There are also some variants of the Gauss theorem. We shall discuss only two of them. Consider three specific vector fields $\mathbf{F}_1 = (P, 0, 0)$, $\mathbf{F}_2 = (0, P, 0)$ and $\mathbf{F}_3 = (0, 0, P)$. Applying the divergence theorem to each of them separately, we obtain three integral identities:

$$\oiint_S P n_x dA = \int\int\int_V \frac{\partial P}{\partial x} dV \,, \quad \oiint_S P n_y dA = \int\int\int_V \frac{\partial P}{\partial y} dV$$

and

$$\oiint_S P n_z dA = \int\int\int_V \frac{\partial P}{\partial z} dV \,.$$

Multiplying each of these equations by the unit base vectors $\mathbf{i}$, $\mathbf{j}$ and $\mathbf{k}$, respectively, and then summing them up, we obtain a useful integral formula:

$$\oiint_S P\mathbf{n}\, dA = \int\int\int_V \left(\frac{\partial P}{\partial x}\mathbf{i} + \frac{\partial P}{\partial y}\mathbf{j} + \frac{\partial P}{\partial z}\mathbf{k}\right) dV \equiv \int\int\int_V \mathrm{grad}\, P\, dV \ . \tag{6.93}$$

We shall make use of this form of the Gauss's theorem later on in Sect. 6.9.3. Above, we have introduced a vector field

$$\mathrm{grad}\, P = \nabla P = \frac{\partial P}{\partial x}\mathbf{i} + \frac{\partial P}{\partial y}\mathbf{j} + \frac{\partial P}{\partial z}\mathbf{k} \ ,$$

which is the gradient of the scalar field $P(\mathbf{r})$, see Sect. 5.8.

Problem 6.85 Use a similar method to prove another integral identity:

$$\oiint_S (\mathbf{a} \cdot \mathbf{n})\, \mathbf{b}\, dA = \int\int\int_V [\mathbf{b}\, \nabla \cdot \mathbf{a} + (\mathbf{a} \cdot \nabla)\, \mathbf{b}]\, dV \ , \tag{6.94}$$

where

$$(\mathbf{a} \cdot \nabla)\, \mathbf{b} = (\mathbf{a} \cdot \nabla b_x)\, \mathbf{i} + (\mathbf{a} \cdot \nabla b_y)\, \mathbf{j} + (\mathbf{a} \cdot \nabla b_z)\, \mathbf{k}$$

and

$$\mathbf{a} \cdot \nabla f = a_x \frac{\partial f}{\partial x} + a_y \frac{\partial f}{\partial y} + a_z \frac{\partial f}{\partial z}$$

is a dot product of the vector $\mathbf{a}$ and $\nabla\, f(\mathbf{r})$ [Hint: *choose vector fields* $\mathbf{G}_1 = b_x\mathbf{a}$, $\mathbf{G}_2 = b_y\mathbf{a}$ *and* $\mathbf{G}_3 = b_z\mathbf{a}$.]

Problem 6.86 Using Eq. (6.94), show that

$$\int\int\int_V P\, dV = \oiint_S \mathbf{r}\, (\mathbf{P} \cdot d\mathbf{A}) - \int\int\int_V \mathbf{r} \cdot \mathrm{div}\, \mathbf{P}\, dV \ .$$

6.6 Comparison of Line and Surface Integrals

Above, we have defined line integrals of scalar (the first kind) and vector (the second kind) fields taken over a curved line L by the equations

$$\int_L f(\mathbf{r})dl = \int_a^b f(\mathbf{r}(u))\sqrt{x'(u)^2 + y'(u)^2 + z'(u)^2}du \tag{6.95}$$

and

$$\int_L \mathbf{F}(\mathbf{r}) \cdot d\mathbf{l} = \int_L \left[F_x(\mathbf{r}(u))x'(u) + F_y(\mathbf{r}(u))y'(u) + F_z(\mathbf{r}(u))z'(u) \right] du , \quad (6.96)$$

respectively. We have also defined integrals of the scalar and vector fields (of the first and second kind) taken over a curved surface S via

$$\int\int_S f(\mathbf{r})dA = \int\int_\Sigma f(\mathbf{r}(u, v))\sqrt{J_x^2(u, v) + J_y^2(u, v) + J_z^2(u, v)}dudv \quad (6.97)$$

and

$$\int\int_S \mathbf{F}(\mathbf{r}) \cdot d\mathbf{A} = \int\int \left[F_x(\mathbf{r}(u, v))J_x(u, v) + F_y(\mathbf{r}(u, v))J_y(u, v) + F_z(\mathbf{r}(u, v))J_z(u, v) \right] dudv ,$$
$$(6.98)$$

where J_x, J_y and J_z are components of the vector $\mathbf{N}$ normal to the surface that are given by the appropriate Jacobians, e.g. $J_x = \partial(y, z)/\partial(u, v)$.

The purpose of this very short section is to direct attention of a reader to the fact that the expressions for the line and surface integrals of the same kind are rather similar. Indeed, comparing Eqs. (6.95) and (6.97) we notice that they have a very similar structure since the Jacobians in the surface integral can be considered as generalisations of the derivatives of the Cartesian coordinates in the line integral to the two-dimensional case. Similarly, one can compare Eqs. (6.96) and (6.98) for the line and surface integrals of the second kind. In both cases we have a dot product of the vector field $\mathbf{F}$ calculated along the line (or surface) with either the components of the vector $\mathbf{l}$ along the line in the line integral or the components of the vector $\mathbf{N}$ normal to the surface in the surface integral. This analogy or similarity may no doubt be useful in remembering equations (6.95)–(6.98) in practical applications.

6.7   Application of Integral Theorems in Physics. Part I

The Gauss and Stock's theorems are very powerful results that are frequently used in many areas of physics and engineering. Here we shall consider some of them. More applications are postponed until Sect. 6.9.

6.7.1   Continuity Equation

As our first example, we shall derive the so-called *continuity equation*

$$\frac{\partial \rho}{\partial t} + \text{div}\,(\rho \mathbf{v}) = 0 , \quad (6.99)$$

that relates the time change of the density ρ of a fluid with its velocity distribution $\mathbf{v}(\mathbf{r})$.

The continuity equation corresponds to the conservation of mass in this case. Consider an arbitrary volume V of the fluid with the bounding surface S. The change of mass in the volume due to the flow of liquid through the element dA of the bounding surface during time dt is

$$dM = \rho\,(v_n dt)\,dA = \rho\,(\mathbf{v}\cdot\mathbf{n})\,dA dt = \rho\mathbf{v}\cdot d\mathbf{A} dt\;. \tag{6.100}$$

We have used here that the flow of fluid through a surface area $d\mathbf{A}$ results in the change in mass per unit time equal to $\rho v_n dA = \rho\,(\mathbf{v}\cdot\mathbf{n})\,dA = \rho\mathbf{v}\cdot d\mathbf{A}$, where $\mathbf{n}$ is the *outer* normal to the surface area dA, see Eq. (6.66). The mass flow is assumed positive if the fluid flows out of the volume. The total change (loss) of mass during time dt due to the flow of fluid through the whole surface S that bounds the volume V is then the surface integral

$$dM = dt \oiint_S \rho\mathbf{v}\cdot d\mathbf{A}\;. \tag{6.101}$$

On the other hand, dM, i.e. the loss of mass in the volume, would correspond to the change of density within it during this time, i.e. the same change of mass can also be calculated by integrating the density change over time dt across the whole volume V:

$$dM = M(t) - M(t+dt) = -\int\int\int_V [\rho(t+dt) - \rho(t)]\,dx dy dz$$
$$= -dt \int\int\int_V \frac{\partial\rho}{\partial t} dx dy dz\;,$$

where we subtracted the mass at time $t+dt$ from the mass at time t to ensure that the loss is positive. The two expressions should be equal to each other due to *mass conservation*:

$$\oiint_S \rho\mathbf{v}\cdot d\mathbf{A} = -\int\int\int_V \frac{\partial\rho}{\partial t} dx dy dz\;.$$

Let us now transform the surface integral using the Gauss theorem of Eq. (6.87):

$$\int\int\int_V \operatorname{div}(\rho\mathbf{v})\,dx dy dz = -\int\int\int_V \frac{\partial\rho}{\partial t} dx dy dz$$
$$\Longrightarrow\quad \int\int\int_V \left[\operatorname{div}(\rho\mathbf{v}) + \frac{\partial\rho}{\partial t}\right] dx dy dz = 0\;.$$

Since the 3D region V is arbitrary, this integral can only be equal to zero if the integrand is zero, i.e. we arrive at the continuity equation (6.99).

Note that the divergence term in the continuity equation (6.99) can also be written as:

$$
\begin{aligned}
\text{div}\,(\rho\mathbf{v}) &= \frac{\partial\,(\rho v_x)}{\partial x} + \frac{\partial\left(\rho v_y\right)}{\partial y} + \frac{\partial\,(\rho v_z)}{\partial z} \\
&= \frac{\partial v_x}{\partial x}\rho + \frac{\partial\rho}{\partial x}v_x + \frac{\partial v_y}{\partial y}\rho + \frac{\partial\rho}{\partial y}v_y + \frac{\partial v_z}{\partial z}\rho + \frac{\partial\rho}{\partial z}v_z \\
&= \rho\left(\frac{\partial v_x}{\partial x} + \frac{\partial v_y}{\partial y} + \frac{\partial v_z}{\partial z}\right) + \left(\frac{\partial\rho}{\partial x}v_x + \frac{\partial\rho}{\partial y}v_y + \frac{\partial\rho}{\partial z}v_z\right) \\
&= \rho\,\text{div}\,\mathbf{v} + \mathbf{v}\cdot\text{grad}\,\rho\,,
\end{aligned}
$$

so the continuity equation may also be written as follows:

$$
\frac{\partial\rho}{\partial t} + \rho\,\text{div}\,\mathbf{v} + \mathbf{v}\cdot\text{grad}\,\rho = 0\,. \tag{6.102}
$$

The continuity equation is also applicable in other situations. For instance, due to conservation of charge, the flux of charge in or out of a certain volume should be reflected by the change of the charge density inside the volume, which is precisely what the continuity equation (6.99) tells us. Indeed, consider a flow of charged particles through an imaginable tube of cross-section dA (infinitesimally small). The total charge passed through dA per time dt will be then $dQ = \rho v_n dA dt = \rho\,(\mathbf{v}\cdot d\mathbf{A})\,dt$, where ρ is the volume charge density; this result is absolutely identical to the one above, Eq. (6.100), which we have derived for the mass of the fluid. Therefore, Eq. (6.99) must also be valid. However, in this particular case it is convenient to rewrite the continuity equation in a slightly different form using the current density $\mathbf{j}$. It is defined in such a way that the current through a surface S can be written via the flux integral of the current density:

$$
I = \int\!\!\int_S \mathbf{j}\cdot d\mathbf{A} = \int\!\!\int_S \mathbf{j}\cdot\mathbf{n}\,dA\,. \tag{6.103}
$$

The current corresponds to the amount of charge dQ flown per unit time, i.e.

$$
dQ = I dt = dt \int\!\!\int_S \mathbf{j}\cdot d\mathbf{A}\,.
$$

Comparing this with expression (6.101) for the mass, we see that $\mathbf{j} = \rho\mathbf{v}$, i.e. the charge passed through a small area dA per unit time is in fact $dQ = \mathbf{j}\cdot d\mathbf{A}dt$ which makes perfect sense. Hence, the continuity equation can be rewritten as follows:

$$
\frac{\partial\rho}{\partial t} + \text{div}\,\mathbf{j} = 0\,. \tag{6.104}
$$

6.7.2   Archimedes Law

It is well known that the pulling force acting on an object of volume V in a uniform liquid of density ρ equals the weight of the liquid replaced by the object. This is in essence the law established by Archimedes. We shall now prove this statement for arbitrary shape of the volume V.

Indeed, the value of the force acting on a small area dA of the object surface that is located at the depth z (see Fig. 6.48) is $dF = \rho g z \, dA$, where g is the Earth gravity constant. Since the force acts in the direction opposite to the outer normal vector $\mathbf{n}$ of the surface, we can write in the vector form that the force $d\mathbf{F} = -\rho g z \mathbf{n} \, dA$. The components of the total force acting on the object from the liquid are obtained by summing up all contributions from all elements of the object surface:

$$F_x = -\oiint_S \rho g z n_x \, dA, \quad F_y = -\oiint_S \rho g z n_y \, dA \quad \text{and} \quad F_z = -\oiint_S \rho g z n_z \, dA.$$

Now, consider the first integral. If we take $\mathbf{G} = (-\rho g z, 0, 0)$ as a vector field, then the force F_x can be written as the surface integral and hence we can use the divergence theorem (6.87):

$$F_x = \oiint_S G_x n_x \, dA = \oiint_S \mathbf{G} \cdot d\mathbf{A} = \int\int\int_V \operatorname{div} \mathbf{G} \, dx \, dy \, dz = \int\int\int_V \frac{\partial G_x}{\partial x} dx \, dy \, dz.$$

However, $\partial G_x / \partial x = 0$, since $G_x = -\rho g z$ depends only on z; hence, $F_x = 0$. Similarly, the force in the y direction is also zero, $F_y = 0$, so only the vertical force remains. If we introduce the vector field $\mathbf{G} = (0, 0, -\rho g z)$, then we can write:

$$F_z = \oiint_S G_z n_z \, dA = \oiint_S \mathbf{G} \cdot d\mathbf{A} = \int\int\int_V \operatorname{div} \mathbf{G} \, dx \, dy \, dz = \int\int\int_V \frac{\partial G_z}{\partial z} dx \, dy \, dz$$

$$= \int\int\int_V (-\rho g) \, dx \, dy \, dz = -\rho g \int\int\int_V dx \, dy \, dz = -\rho g V.$$

Thus, the force is directed upwards (the positive direction of the z-axis was chosen downwards in Fig. 6.48), and is equal exactly to the mass of the displaced liquid. This is a very powerful result since it is proven for an arbitrary shape of the object immersed in the liquid!

Fig. 6.48 An object is placed inside a uniform liquid. Note that the positive direction of the z-axis is *down*

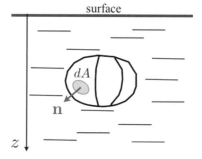

Problem 6.87 An engineer is designing an air balloon with a specific ratio $\lambda = R/d < 1$ of the radius R of its horizontal cross-section in the middle and of its half-height, d. The shape of the balloon can be described by the equation $x^2 + y^2 + \lambda^2 z^2 = R^2$. Show that the Cartesian coordinates of the points on the balloon surface can be expressed in terms of the z and the polar angle ϕ as follows:

$$x = \lambda\sqrt{d^2 - z^2}\cos\phi \ , \quad y = \lambda\sqrt{d^2 - z^2}\sin\phi \ , \quad z = z \ .$$

Then show that the corresponding Jacobians for the surface normal in terms of the coordinates (z, ϕ) are:

$$J_x = \lambda\sqrt{d^2 - z^2}\cos\phi \ , \quad J_y = \lambda\sqrt{d^2 - z^2}\sin\phi \ , \quad J_z = \lambda^2 z \ .$$

Hence, show that the surface area of the balloon is the following:

$$A = \frac{2\pi R d}{\sqrt{1 - \lambda^2}}\left(\lambda\sqrt{1 - \lambda^2} + \arcsin\sqrt{1 - \lambda^2}\right) \ ,$$

and its volume $V = 4\pi\lambda^2 d^3/3$. If m is the maximum mass of a basket intended to be used in the flights, σ is the surface density of the material to be used for the balloon, and ρ is the air density, which gives advice to the engineer concerning the necessary conditions that the parameters λ and d must satisfy in order for the balloon to be able to take off.

Problem 6.88 Consider a cuboid-shaped float in water with the horizontal area S and height l shown in Fig. 6.49a. Note that the float is only partially immersed into water. If ρ_W and ρ are densities of water and of the material of the float that is assumed uniform, show that the equation of the vertical motion of the float is $\ddot{u} + \omega^2 u = 0$, where $u(t)$ is the float vertical displacement and $\omega = \sqrt{g\rho_W/l\rho}$. Here g is the gravitational constant. This motion is called harmonic since the solution of this so-called differential equation[13] is $u(t) = A\sin(\omega t + \phi)$ with A being an amplitude and ϕ phase. Note that the motion of the float is harmonic for any of its vertical displacement. In other words, if we displace the float from its equilibrium position by any amount $|u| < l$, it will oscillate vertically with the same frequency ω. Of course, in reality, due to various energy dissipation channels the float will lose energy very quickly and will arrive back at equilibrium.

---

[13] We shall study differential equations in more detail in Chap. 8.

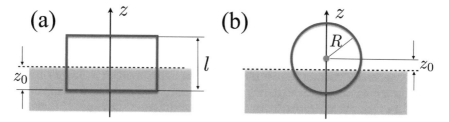

Fig. 6.49 Cuboid (**a**) and spherical (**b**) floats in water

Problem 6.89 As an example of a generally non-harmonic motion, consider a uniform float in the form of a sphere of radius R shown in Fig. 6.49b. (i) First, show that at equilibrium the position of the centre of the sphere z_0 that is measured with respect to the water level is determined by the cubic equation

$$\frac{4\rho}{\rho_W} = (1 - \lambda)(2 - \lambda - \lambda^2),$$

where $\lambda = z_0/R$. Here g is the gravitational constant, ρ and ρ_W are densities of the sphere and water, respectively. (ii) Second, show that the motion of the float is described by the following non-linear differential equation:

$$\ddot{u} + \frac{3g\rho_W}{4\rho R^3}\left[(R^2 - z_0^2) - z_0 u - \frac{1}{3}u^2\right]u .$$

Note that this equation is non-linear as it contains u^2 and u^3 terms. (iii) Third, consider small oscillations of the sphere, when $|u| \ll R$ and $|u| \ll z_0$. Show that in this case the oscillations are harmonic with the frequency $\omega = \sqrt{g\rho_W S_0/\rho V}$, where $S_0 = \pi(R^2 - z_0^2)$ is the area of cross-section of the sphere (a circle) at the water level at equilibrium and $V = 4\pi R^3/3$ is the sphere volume. (iv) Finally, argue that the same result is to be obtained for small oscillations of a float of arbitrary shape of volume V and the cross-section area S_0.

6.8   Vector Calculus

6.8.1   Divergence of a Vector Field

We have already introduced the divergence of a vector field: if there is a vector field $\mathbf{F} = (F_x, F_y, F_z)$, then its divergence is given by Eq. (6.88). However, this specific definition is only useful in Cartesian coordinates. A more general definition of the divergence is needed in order to write the divergence in an alternative coordinate system, such as cylindrical or spherical coordinates. These are called *curvilinear coordinate* systems (we shall discuss them in detail in Vol. II, Chap. 7). So, if one would like to be able to write the divergence in any of these coordinate systems, a more general definition of the divergence is required.

This general definition can be formulated via a flux through a closed surface. Consider a point M, surround it by a closed surface S and calculate the ratio of the flux of the vector field $\mathbf{F}$ through that surface to the volume of the region V enclosed:

$$\frac{\text{flux}}{\text{volume}} = \frac{\oiint_S \mathbf{F} \cdot d\mathbf{A}}{V} .$$

Using the Gauss theorem (6.87), the surface integral is expressed via a volume integral:

$$\frac{\text{flux}}{\text{volume}} = \frac{\int \int \int_V \text{div} \, \mathbf{F} dV}{V} ,$$

where $dV = dx dy dz$ is a differential volume element. The average value theorem which we formulated in Eq. (6.2) for double integrals is also valid for volume integrals. Therefore,

$$\int \int \int \text{div} \, \mathbf{F}(\mathbf{r}) dV = \text{div} \, \mathbf{F}(\mathbf{p}) V \quad \Longrightarrow \quad \frac{\text{flux}}{\text{volume}} = \text{div} \, \mathbf{F}(\mathbf{p}),$$

where the point $\mathbf{p}$ is somewhere inside the volume V. Now, take the limit of the volume V around the point M tending to zero. As the volume goes to zero, the point $\mathbf{p}$ remains inside the volume, and in the limit tends to the point M. Therefore, the ratio of the flux to the volume in the limit results exactly in div $\mathbf{F}(M)$. Thus, we have just proven that

$$\lim_{V \to 0} \frac{1}{V} \oiint_S \mathbf{F} \cdot d\mathbf{A} = \text{div} \, \mathbf{F}(M) . \tag{6.105}$$

This definition of the divergence can be used to derive a working expression for it in general curvilinear coordinates (Vol. II, Chap. 7).

As was mentioned earlier, the divergence can also be formally written via the del operator (5.72) as a dot product with the field:

$$\nabla \cdot \mathbf{F} \quad \longmapsto \quad \text{div} \, \mathbf{F} . \tag{6.106}$$

One can also apply the divergence to the gradient of some scalar field U, i.e. consider an operation div (grad U).

Problem 6.90 Show that

$$\nabla \cdot (\nabla U) = \Delta U = \left(\frac{\partial^2}{\partial x^2} + \frac{\partial^2}{\partial y^2} + \frac{\partial^2}{\partial z^2} \right) U = \frac{\partial^2 U}{\partial x^2} + \frac{\partial^2 U}{\partial y^2} + \frac{\partial^2 U}{\partial z^2} \,. \tag{6.107}$$

Problem 6.91 Prove that

$$\nabla \cdot (\phi \nabla \psi) = \phi \Delta \psi + \nabla \phi \cdot \nabla \psi \,,$$

where ϕ and ψ are scalar fields.

The construction (6.107) has a special name, *Laplacian*, and various notations for it are frequently used in the literature. Apart from the notation Δ for the *Laplace operator* which appears within the round brackets in (6.107), the notation $\nabla^2 \equiv \nabla \cdot \nabla$ is also often used since it can be easily recognised that the dot product of the del operator with itself is in fact Δ:

$$\nabla \cdot \nabla = \left(\frac{\partial}{\partial x}\mathbf{i} + \frac{\partial}{\partial y}\mathbf{j} + \frac{\partial}{\partial z}\mathbf{k} \right) \cdot \left(\frac{\partial}{\partial x}\mathbf{i} + \frac{\partial}{\partial y}\mathbf{j} + \frac{\partial}{\partial z}\mathbf{k} \right)$$

$$\longmapsto \frac{\partial}{\partial x}\frac{\partial}{\partial x} + \frac{\partial}{\partial y}\frac{\partial}{\partial y} + \frac{\partial}{\partial z}\frac{\partial}{\partial z} = \frac{\partial^2}{\partial x^2} + \frac{\partial^2}{\partial y^2} + \frac{\partial^2}{\partial z^2} \equiv \Delta \,.$$

Problem 6.92 Prove that

$$\Delta(\phi\psi) = (\Delta\phi)\,\psi + 2\nabla\phi \cdot \nabla\psi + \phi(\Delta\psi) \,,$$

and

$$\Delta e^{\lambda\phi} = \lambda\,[\Delta\phi + \lambda\,(\nabla\phi \cdot \nabla\phi)]\,e^{\lambda\phi} \,.$$

Problem 6.93 Show that the Shrödinger equation

$$\left[\frac{1}{2m} \left(\mathbf{p} - \frac{q}{c}\mathbf{A} \right)^2 + qV \right] \psi = i\frac{\partial}{\partial t}\psi$$

for the wavefunction $\psi(\mathbf{r}, t)$ of an electron of charge q and mass m in the electro-magnetic field specified by the scalar $V(\mathbf{r}, t)$ and vector $\mathbf{A}(\mathbf{r}, t)$ poten-

Fig. 6.50 A closed contour
L around the point M on the
surface Σ

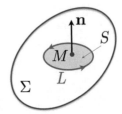

tials is invariant with respect to the so-called *gauge transformation:*

$$\mathbf{A} \to \mathbf{A}' = \mathbf{A} + \nabla U , \; V \to V' = V - \frac{1}{c}\frac{\partial U}{\partial t} \text{ and } \psi \to \psi' = \psi e^{iqU/c} ,$$

(6.108)

where $U(\mathbf{r}, t)$ is an arbitrary function, i.e. we also have an identical Shrödinger
equation in the primed quantities,

$$\left[\frac{1}{2m} \left(\mathbf{p} - \frac{q}{c}\mathbf{A}' \right)^2 + qV' \right] \psi' = i\frac{\partial}{\partial t}\psi' .$$

Here c is the speed of light and $\mathbf{p} = -i\nabla$ is the momentum operator of the
electron. The tricky point about this problem is to recognise that the momentum
operator acts on *all* functions staying on the right of it, and also that

$$\left(\mathbf{p} - \frac{q}{c}\mathbf{A}' \right)^2 = \left(\mathbf{p} - \frac{q}{c}\mathbf{A}' \right)\left(\mathbf{p} - \frac{q}{c}\mathbf{A}' \right) = \left(i\nabla + \frac{q}{c}\mathbf{A}' \right)\left(i\nabla + \frac{q}{c}\mathbf{A}' \right) .$$

6.8.2   Curl of a Vector Field

We have already formally introduced the *curl* of a vector field $\mathbf{F}(\mathbf{r})$ by formula (6.74).
Similarly, to the divergence, a more general definition to the curl can also be given,
and this is important as a tool to derive alternative expressions in other coordinate
systems.

Consider a point M on a surface Σ, Fig. 6.50. Enclose M by a closed line L lying
in the surface and consider the line integral of $\mathbf{F}$ along it divided by the area A of the
closed loop, $\oint_L \mathbf{F} \cdot d\mathbf{l}/A$. Using Stock's theorem (6.73), we can write:

$$\frac{1}{A}\oint_L \mathbf{F} \cdot d\mathbf{l} = \frac{1}{A}\int\int_S \text{curl}\,\mathbf{F} \cdot d\mathbf{A} = \frac{1}{A}\int\int_S (\text{curl}\,\mathbf{F} \cdot \mathbf{n})\,dA .$$

(6.109)

Above, S is the surface on Σ for which the line L serves as the boundary. Note the
direction of the traverse along L is related to the normal of the surface Σ at M in the

usual way (the right hand rule). According to the average value theorem,

$$\frac{1}{A} \int\!\!\int_{S} (\text{curl}\,\mathbf{F} \cdot \mathbf{n})\, dA = \frac{1}{A} (\text{curl}\,\mathbf{F}(P) \cdot \mathbf{n})\, A = \text{curl}\,\mathbf{F}(P) \cdot \mathbf{n}\,,$$

where the point P lies somewhere inside surface S. It is seen then that in the limit of the area A tending to zero (assuming the point M is always inside S), the point $P \to M$, and we arrive at the dot product of the curl $\mathbf{F}(M)$ at the point M and the normal vector there:

$$\lim_{A \to 0} \frac{1}{A} \oint_{L} \mathbf{F} \cdot d\mathbf{l} = \text{curl}\,\mathbf{F} \cdot \mathbf{n}\,. \tag{6.110}$$

This is the general expression for the curl we have been looking for, as it does not depend on the particular coordinate system. It can be used for the derivation of the curl in a general curvilinear coordinate system (Chap. 7, Vol. II).

To understand better what does the curl mean, let us consider a fluid with the velocity distribution $\mathbf{v}(\mathbf{r})$. An *average* velocity *along* some circular contour L can be defined as

$$v_{av} = \frac{1}{2\pi R} \oint_{L} \mathbf{v} \cdot d\mathbf{l}\,,$$

where R is the circle radius. The intensity of the rotation of fluid particles can be characterised by their angular velocity

$$\omega \sim \frac{v_{av}}{R} = \frac{1}{2\pi R^2} \oint_{L} \mathbf{v} \cdot d\mathbf{l} \sim \frac{1}{\pi R^2} \oint_{L} \mathbf{v} \cdot d\mathbf{l} = \frac{1}{A} \oint_{L} \mathbf{v} \cdot d\mathbf{l}\,,$$

where $A = \pi R^2$ is the area of the circle. This result can now be compared with Eqs. (6.109) and (6.110). We see that curl$\mathbf{v}(\mathbf{r})$ characterises the *intensity* of the fluid rotation around a given point $\mathbf{r}$ about the chosen direction given by the normal $\mathbf{n}$ to the closed loop L. The maximum intensity is achieved along the direction of the curl of $\mathbf{v}$.

Interestingly, curl of $\mathbf{F}$ can also be written via the del operator (5.72) as a *vector product* of ∇ and $\mathbf{F}$:

$$\text{curl}\,\mathbf{F} = \nabla \times \mathbf{F}\,, \tag{6.111}$$

which is checked directly by looking at Eq. (6.82).

As an example, consider the curl of the vector field $\mathbf{F}(\mathbf{r}) = \mathbf{r}$. We have:

$$\text{curl}\,\mathbf{F} = \begin{vmatrix} \mathbf{i} & \mathbf{j} & \mathbf{k} \\ \partial/\partial x & \partial/\partial y & \partial/\partial z \\ x & y & z \end{vmatrix}$$

$$= \left(\frac{\partial z}{\partial y} - \frac{\partial y}{\partial z} \right) \mathbf{i} - \left(\frac{\partial z}{\partial x} - \frac{\partial x}{\partial z} \right) \mathbf{j} + \left(\frac{\partial y}{\partial x} - \frac{\partial x}{\partial y} \right) \mathbf{k} = \mathbf{0}\,,$$

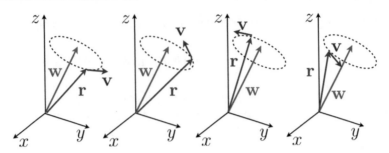

Fig. 6.51 The vector field $\mathbf{v}(\mathbf{r}) = \mathbf{w} \times \mathbf{r}$ corresponds to a vortex around the fixed vector $\mathbf{w}$ as the vector $\mathbf{r}$ samples the three-dimensional space

which is consistent with the intuitive understanding of the curl, as the field $\mathbf{r}$ is radial (it corresponds to rays coming directly out of the centre of the coordinate system) and hence does not have rotations.

Now, let us look at another vector field

$$\mathbf{v}(\mathbf{r}) = \mathbf{w} \times \mathbf{r} = \begin{vmatrix} \mathbf{i} & \mathbf{j} & \mathbf{k} \\ w_x & w_y & w_z \\ x & y & z \end{vmatrix}$$

$$= \left(w_y z - w_z y\right)\mathbf{i} + (w_z x - w_x z)\mathbf{j} + \left(w_x y - w_y x\right)\mathbf{k} ,$$

where $\mathbf{w}$ is a fixed vector. This may correspond, for example, to a velocity distribution of a fluid in a vortex as shown in Fig. 6.51. Using again the general formula (6.82) for the curl, we write:

$$\text{curl } \mathbf{v} = \begin{vmatrix} \mathbf{i} & \mathbf{j} & \mathbf{k} \\ \partial/\partial x & \partial/\partial y & \partial/\partial z \\ w_y z - w_z y & w_z x - w_x z & w_x y - w_y x \end{vmatrix}$$

$$= \left[\frac{\partial}{\partial y}\left(w_x y - w_y x\right) - \frac{\partial}{\partial z}\left(w_z x - w_x z\right)\right]\mathbf{i}$$

$$- \left[\frac{\partial}{\partial x}\left(w_x y - w_y x\right) - \frac{\partial}{\partial z}\left(w_y z - w_z y\right)\right]\mathbf{j}$$

$$+ \left[\frac{\partial}{\partial x}\left(w_z x - w_x z\right) - \frac{\partial}{\partial y}\left(w_y z - w_z y\right)\right]\mathbf{k}$$

$$= 2w_x\mathbf{i} + 2w_y\mathbf{j} + 2w_z\mathbf{k} = 2\mathbf{w} .$$

This result is in complete agreement with our intuitive understanding of the curl: it shows the direction (along the vector $\mathbf{w}$) and the strength (the velocity is proportional to the magnitude $w = |\mathbf{w}|$ of $\mathbf{w}$) of the rotation of the fluid around the direction given by the vector $\mathbf{w}$.

The curl possesses some important properties, e.g. additivity:

$$\text{curl }(\mathbf{F}_1 + \mathbf{F}_2) = \text{curl }\mathbf{F}_1 + \text{curl }\mathbf{F}_2 . \tag{6.112}$$

Another useful property is this:

$$\text{div curl } \mathbf{F} = \nabla \cdot (\nabla \times \mathbf{F}) = 0 \,. \qquad (6.113)$$

This identity can be checked directly using an explicit expression (6.74) for the curl and the definition (6.88) of the div:

$$\nabla \cdot (\nabla \times \mathbf{F}) = \frac{\partial}{\partial x}\left(\frac{\partial F_z}{\partial y} - \frac{\partial F_y}{\partial z}\right) + \frac{\partial}{\partial y}\left(\frac{\partial F_x}{\partial z} - \frac{\partial F_z}{\partial x}\right) + \frac{\partial}{\partial z}\left(\frac{\partial F_y}{\partial x} - \frac{\partial F_x}{\partial y}\right)$$

$$= \left(\frac{\partial^2 F_z}{\partial x \partial y} - \frac{\partial^2 F_y}{\partial x \partial z}\right) + \left(\frac{\partial^2 F_x}{\partial y \partial z} - \frac{\partial^2 F_z}{\partial y \partial x}\right) + \left(\frac{\partial^2 F_y}{\partial z \partial x} - \frac{\partial^2 F_x}{\partial z \partial y}\right) = 0 \,.$$

Problem 6.94 Using the definition of the curl in Cartesian coordinates, prove that

$$\nabla \times (U\mathbf{F}) = U\nabla \times \mathbf{F} - \mathbf{F} \times \nabla U \,, \qquad (6.114)$$

where in the first term in the right-hand side there is a curl of a vector field $\mathbf{F}$, while in the second term there is a gradient of a scalar field U.

Problem 6.95 Similarly, prove the identities:

$$\nabla \times \nabla U = \mathbf{0} \,; \qquad (6.115)$$

$$\nabla \times \nabla \times \mathbf{F} = \nabla (\nabla \cdot \mathbf{F}) - \Delta \mathbf{F} \,, \qquad (6.116)$$

where $\Delta \mathbf{F} = \Delta F_x \mathbf{i} + \Delta F_y \mathbf{j} + \Delta F_z \mathbf{k}$ is the vector Laplacian.

6.8.3   Vector Fields: Scalar and Vector Potentials

In applications related to vector fields, it is important to be able to characterise them. Here, we shall consider the so-called conservative and solenoidal vector fields since these possess certain essential properties. The field $\mathbf{F}(\mathbf{r})$ is called *conservative*, if one can introduce a *scalar potential* $\phi(\mathbf{r})$ such that $\mathbf{F} = \text{grad } \phi = \nabla \phi$. The field $\mathbf{F}(\mathbf{r})$ is called *solenoidal*, if there exists a *vector potential* $\mathbf{G}(\mathbf{r})$ such that $\mathbf{F} = \text{curl } \mathbf{G} = \nabla \times \mathbf{G}$. It appears that any vector field can always be represented as a sum of some conservative and solenoidal fields. This theorem, which we shall discuss in more detail below, plays an extremely important role in many applications in physics.

6.8.3.1 Conservative Fields

If the field $\mathbf{F}$ is conservative with the potential $\phi(\mathbf{r})$, i.e. $\mathbf{F} = \nabla\phi$, then the scalar function $\phi'(\mathbf{r}) = \phi(\mathbf{r}) + C$ is also a potential for any constant C. Thus, the conservative fields are determined by a single scalar function $\phi(\mathbf{r})$ which is defined up to a constant.

It is easy to see for instance that the electrostatic field $\mathbf{E}(\mathbf{r}) = -q\mathbf{r}/r^3$, where $r = |\mathbf{r}| = \sqrt{x^2 + y^2 + z^2}$, is conservative. Consider a function $\phi(\mathbf{r}) = q/r = q \left(x^2 + y^2 + z^2\right)^{-1/2}$ and let us show that it can be chosen as a potential for $\mathbf{E}$. We have $\partial\phi/\partial x = -qx/r^3$, etc., so we can write

$$\nabla\phi = \frac{\partial\phi}{\partial x}\mathbf{i} + \frac{\partial\phi}{\partial y}\mathbf{j} + \frac{\partial\phi}{\partial z}\mathbf{k} = \frac{-xq}{r^3}\mathbf{i} + \frac{-yq}{r^3}\mathbf{j} + \frac{-zq}{r^3}\mathbf{k}$$

$$= -\frac{q}{r^3}(x\mathbf{i} + y\mathbf{j} + z\mathbf{k}) = -\frac{q}{r^3}\mathbf{r} \equiv \mathbf{E}(\mathbf{r}),$$

as required. So, the electrostatic potential in electrostatics (no moving charges) is basically, from the mathematical point of view, a potential of the conservative electric field (see Sect. 6.9.1 for more details).

The conservative field has a number of important properties that are provided by the following theorem

---

Theorem 6.9 *Let the field $\mathbf{F}(\mathbf{r})$ be defined in a simply connected region D. Then the following three conditions are equivalent:*

1. *The field is conservative, i.e. there exists a scalar potential function $\phi(\mathbf{r})$ such that $\mathbf{F} = \nabla\phi$;*
2. *The field is irrotational, i.e. $\nabla \times \mathbf{F} = 0$;*
3. *The line integral $\oint_L \mathbf{F} \cdot d\mathbf{l} = 0$ for any closed contour L.*

Proof: We shall prove the theorem using the familiar scheme: $1 \to 2 \to 3 \to 1$.

- $1 \to 2$: if $\mathbf{F} = \nabla\phi = \operatorname{grad}\phi$, then $\operatorname{curl}\mathbf{F} = \operatorname{curl}\operatorname{grad}\phi = 0$ according to Eq. (6.115).
- $2 \to 3$: take a closed contour L, then, using the Stokes's theorem (6.73),

$$\oint_L \mathbf{F} \cdot d\mathbf{l} = \int\int_S \operatorname{curl}\mathbf{F} \cdot d\mathbf{A} = 0,$$

since $\operatorname{curl}\mathbf{F} = 0$.

- $3 \to 1$: since $\oint_L \mathbf{F} \cdot d\mathbf{l} = 0$ for any contour L, then the line integral $\int_A^B \mathbf{F} \cdot d\mathbf{l}$ does not depend on the particular integration path $A(x_0, y_0, z_0) \to B(x, y, z)$, but only on the initial and final points A and B (see Sect. 6.5.6). Therefore, the function

$$\phi(x, y, z) = \int_{A(x_0, y_0, z_0)}^{B(x,y,z)} \mathbf{F} \cdot d\mathbf{l} \tag{6.117}$$

is the exact differential, and thus $\partial\phi/\partial x = F_x$, $\partial\phi/\partial y = F_y$ and $\partial\phi/\partial z = F_z$, i.e. $\mathbf{F} = \nabla\phi$, as required. **Q.E.D.**

Equation (6.117) gives a practical way of calculating the potential $\phi(\mathbf{r})$ for a conservative field from the field itself. As we know from the previous sections, this integral is only defined up to a constant, and this is how it should be for the potential, as was mentioned above.

> **Problem 6.96** Demonstrate that in the 2D case when the two components $F_x(x, y)$ and $F_y(x, y)$ of the vector field $\mathbf{F}$ depend only on x and y, and the z component is zero, $F_z = 0$, we arrive at 2D results of Sect. 6.4.4 for determining the function $\Phi(x, y)$ of the exact differential $d\Phi = \frac{\partial F_x}{\partial y}dx + \frac{\partial F_y}{\partial x}dy$.

It is essential that the region D is simply connected. The classic example is a magnetic field due to an infinite wire stretched along the z-axis (*cf.* Eq. (6.47) and the text that follows):

$$\mathbf{H}(\mathbf{r}) = \frac{\lambda\left(-y\widehat{i} + x\widehat{j}\right)}{\rho^2},$$

where $\rho = \sqrt{x^2 + y^2}$ is a distance from the wire. It is easy to see that

$$\mathrm{curl}\,\mathbf{H} = \begin{vmatrix} \mathbf{i} & \mathbf{j} & \mathbf{k} \\ \partial/\partial x & \partial/\partial y & \partial/\partial z \\ -\lambda y/\rho^2 & \lambda x/\rho^2 & 0 \end{vmatrix}$$

$$= \left(-\frac{\partial}{\partial z}\frac{\lambda x}{\rho^2}\right)\mathbf{i} - \left(\frac{\partial}{\partial z}\frac{\lambda y}{\rho^2}\right)\mathbf{j} + \left(\frac{\partial}{\partial x}\frac{\lambda x}{\rho^2} + \frac{\partial}{\partial y}\frac{\lambda y}{\rho^2}\right)\mathbf{k} = 0$$

(perform differentiation and prove the result above!). However, the line integral $\oint_L \mathbf{H}\cdot d\mathbf{l}$ taken along the circle $x^2 + y^2 = 1$ (e.g. for $z = 0$) is not equal to zero. Using polar coordinates $x = \cos\phi$ and $y = \sin\phi$, we have:

$$\oint_L \mathbf{H}\cdot d\mathbf{l} = \oint_L \frac{\lambda}{x^2 + y^2}(-ydx + xdy)$$

$$= \int_0^{2\pi} \frac{\lambda}{1^2}[-\sin\phi\,(-\sin\phi) + \cos\phi\,(\cos\phi)]\,d\phi$$

$$= \lambda\int_0^{2\pi} d\phi = 2\pi\lambda \neq 0.$$

The obtained non-zero result for the line integral is due to the fact that the region is not simply connected as the field $\mathbf{H}$ is infinite at any point along the wire $x = y = 0$. Hence, the whole z-axis must be removed from D which makes the region not simply connected. Of course, any closed-loop line integral which does not have the z-axis inside the loop will be equal to zero as required by Theorem 6.9.

Problem 6.97 Consider the vector field $\mathbf{F} = (2x, 2y, 2z)$. Prove that it is conservative. Then, using the line integral, show that its scalar potential $\phi = x^2 + y^2 + z^2 + C$. Check that indeed $\mathbf{F} = \nabla\phi$.

Problem 6.98 The same for $\mathbf{F} = (yz, xz, xy)$. [Answer: $\phi = xyz + C$.]

Problem 6.99 The same for $\mathbf{F} = (y, x, -z)$. [Answer: $\phi = xy - z^2/2 + C$.]

Problem 6.100 Given the vector field $\mathbf{F} = (2xz, 2yz, x^2 + y^2)$, show that $\mathbf{F}$ is conservative. Explicitly calculate the line integral along the closed path $x^2 + z^2 = 1$ and $y = 0$ using polar coordinates. Verify that $dU = F_x dx + F_y dy + F_z dz$ is the exact differential, and next show that the function $U(x, y, z) = (x^2 + y^2) z + C$. State the relationship between the field $\mathbf{F}$ and the function $U(x, y, z)$.

Problem 6.101 Repeat all steps of the previous problem but for the vector field $\mathbf{F} = (y + z, x + z, x + y)$, showing that in this case $U = xz + yz + xy + C$.

6.8.3.2 Solenoidal Fields

As was already mentioned, if a vector field $\mathbf{A}$ can be represented as a curl of another field, $\mathbf{B}$, called its *vector potential*, i.e. $\mathbf{A} = \text{curl } \mathbf{B} = \nabla \times \mathbf{B}$, then the field $\mathbf{A}$ is called *solenoidal*. For this field, see Eq. (6.113),

$$\text{div}\mathbf{A} = \text{div curl } \mathbf{B} = 0 \text{ or } \nabla \cdot \mathbf{A} = 0 . \tag{6.118}$$

This condition serves as another definition of the solenoidal field. However, it is not only the necessary condition as proven by the above manipulation. It is also a sufficient one as the reverse statement is also valid: if $\text{div}\mathbf{A} = 0$, then there exists a vector potential $\mathbf{B}$, such that $\mathbf{A} = \text{curl } \mathbf{B}$. To prove the existence, choose two points $M_0(x_0, y_0, z_0)$ and $M(x, y, z)$, and then consider a vector function $\mathbf{B} = (B_x, B_y, B_z)$ defined in the following way:

$$
\begin{aligned}
B_x &= \int_{z_0}^{z} A_y(x, y, z_1) dz_1 , \\
B_y &= \int_{x_0}^{x} A_z(x_1, y, z_0) dx_1 - \int_{z_0}^{z} A_x(x, y, z_1) dz_1 , \\
B_z &= 0 .
\end{aligned}
\tag{6.119}
$$

Consider the curl of this vector field. Starting from the x component of the curl, we write:

$$(\text{curl } \mathbf{B})_x = \frac{\partial B_z}{\partial y} - \frac{\partial B_y}{\partial z} = -\frac{\partial B_y}{\partial z}$$

$$= -\frac{\partial}{\partial z} \left(\int_{x_0}^x A_z(x_1, y, z_0) dx_1 - \int_{z_0}^z A_x(x, y, z_1) dz_1 \right).$$

The first term does not contribute since it does not depend on z; by differentiating the second integral over the upper limit we get the integrand, see Sect. 4.3. Therefore,

$$(\text{curl } \mathbf{B})_x = \frac{\partial}{\partial z} \int_{z_0}^z A_x(x, y, z_1) dz_1 = A_x(x, y, z).$$

Similarly,

$$(\text{curl } \mathbf{B})_y = \frac{\partial B_x}{\partial z} - \frac{\partial B_z}{\partial x} = \frac{\partial B_x}{\partial z} = \frac{\partial}{\partial z} \left(\int_{z_0}^z A_y(x, y, z_1) dz_1 \right) = A_y(x, y, z),$$

and

$$(\text{curl } \mathbf{B})_z = \frac{\partial B_y}{\partial x} - \frac{\partial B_x}{\partial y}$$

$$= \frac{\partial}{\partial x} \left(\int_{x_0}^x A_z(x_1, y, z_0) dx_1 - \int_{z_0}^z A_x(x, y, z_1) dz_1 \right)$$

$$- \frac{\partial}{\partial y} \left(\int_{z_0}^z A_y(x, y, z_1) dz_1 \right),$$

where differentiation of the first integral gives $A_z(x, y, z_0)$ in the same way as for the above two components, but in the two integrals taken over z_1 the differentiation is to be performed with respect to the variable which happens to be only in the integrand. According to Sect. 4.5.2, the operators $\partial/\partial x$ and $\partial/\partial y$ can then be inserted inside the integrals since the integration limits do not depend on the variables with respect to which the differentiation is performed. Hence,

$$(\text{curl } \mathbf{B})_z = A_z(x, y, z_0) - \int_{z_0}^z \frac{\partial A_x(x, y, z_1)}{\partial x} dz_1 - \int_{z_0}^z \frac{\partial A_y(x, y, z_1)}{\partial y} dz_1$$

$$= A_z(x, y, z_0) - \int_{z_0}^z \left[\frac{\partial A_x(x, y, z_1)}{\partial x} + \frac{\partial A_y(x, y, z_1)}{\partial y} \right] dz_1.$$

Since we know that $\text{div}\mathbf{A} = \partial A_x/\partial x + \partial A_x/\partial y + \partial A_x/\partial z = 0$, the expression in the square brackets above can be replaced by $-\partial A_z(x, y, z_1)/\partial z_1$, yielding finally

$$(\text{curl } \mathbf{B})_z = A_z(x, y, z_0) + \int_{z_0}^z \frac{\partial A_z(x, y, z_1)}{\partial z_1} dz_1$$

$$= A_z(x, y, z_0) + \left[A_z(x, y, z) - A_z(x, y, z_0) \right] = A_z(x, y, z).$$

Thus, we have shown that indeed curl$\mathbf{B} = \mathbf{A}$. This proves the above made statement that there *exists* a vector field $\mathbf{B}$ that can serve as a vector potential for the solenoidal field $\mathbf{A}$. Of course, the vector potential (6.119) is not the only one which can do this job (see below).

Thus, the two conditions for $\mathbf{A}$ to be a solenoidal field, namely that $\mathbf{A}$ is curl of some $\mathbf{B}$ and that div$\mathbf{A} = 0$, are completely equivalent.

The choice of the vector potential $\mathbf{B}$ given above is not unique: any vector $\mathbf{B}' = \mathbf{B} + \text{grad}\, U(x, y, z)$ with arbitrary scalar field $U(x, y, z)$ is also a vector potential since curl grad $U = \nabla \times \nabla U = \mathbf{0}$ for any U, see Eq. (6.115), yielding curl $\mathbf{B} = $ curl $\mathbf{B}'$. Thus, the vector potential is defined only up to the gradient of an arbitrary scalar field.

Problem 6.102 Show that the vector field $\mathbf{F} = \left(-xz \cosh y, z \sinh y, x^2 y\right)$ is solenoidal.

Problem 6.103 Consider a point charge q at a position given by the vector $\mathbf{R} = (X, Y, Z)$. It creates an electric field

$$\mathbf{E}(\mathbf{r}) = q \frac{\mathbf{r} - \mathbf{R}}{|\mathbf{r} - \mathbf{R}|^3}$$

at point $\mathbf{r}$, where $|\mathbf{r} - \mathbf{R}| = \sqrt{(x - X)^2 + (y - Y)^2 + (z - Z)^2}$. Show that the field is solenoidal anywhere outside the position of the charge q, i.e. that

$$\nabla \cdot \mathbf{E}(\mathbf{r}) = 0 \quad \text{if} \quad \mathbf{r} \neq \mathbf{R} . \tag{6.120}$$

Of course, the above statement is immediately generalised for the electrostatic field of arbitrary number of point charges.

Any solenoidal field has a number of properties as stated by the following theorem.

Theorem 6.10 *If $\mathbf{F}$ is a solenoidal field, then its flux through any smooth closed surface is equal to zero.*

Proof: First, consider a simply connected region D and a closed smooth surface S anywhere inside it. Then, the flux through that surface

$$\oiint_S \mathbf{F} \cdot d\mathbf{A} = \int \int \int_V \text{div} \mathbf{F}\, dx dy dz = 0 ,$$

Fig. 6.52 Region D (blue) contains a hole inside (orange). The plane P cuts it into two regions D_1 and D_2, and crosses the surface S at the line L

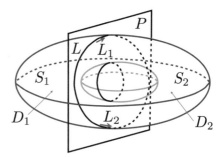

since the field is solenoidal, $\mathrm{div}\,\mathbf{F} = 0$. Here we have used the Gauss theorem, Eq. (6.87), and V is the region that is covered by S.

Next, consider a more general region D with the closed surface S, which contains a hole inside, as schematically shown in Fig. 6.52. Let us divide it into two parts D_1 and D_2 by cutting through D with a plane P. The plane will cross S at the line L, and the surface S will also be broken down into two surfaces S_1 (left) and S_2 (right) that cap the line L on both sides. Then, the surface integral over the entire outer surface of D,

$$\oiint_S \mathbf{F} \cdot d\mathbf{A} = \int\!\!\int_{S_1} \mathbf{F} \cdot d\mathbf{A} + \int\!\!\int_{S_2} \mathbf{F} \cdot d\mathbf{A} \,,$$

can be represented as a sum of two flux integrals: over the surface S_1 and over S_2. Since $\mathrm{div}\,\mathbf{F} = 0$, there exists a vector potential $\mathbf{B}$ such that $\mathbf{F} = \mathrm{curl}\,\mathbf{B}$. Therefore, using the Stokes's theorem, Eq. (6.73), we have for each of the surfaces:

$$\int\!\!\int_{S_1} \mathbf{F} \cdot d\mathbf{A} = \int\!\!\int_{S_1} \mathrm{curl}\,\mathbf{B} \cdot d\mathbf{A} = \oint_{L_1} \mathbf{B} \cdot d\mathbf{l} \,,$$

where L_1 is the boundary of S_1 along the line L in the direction, as indicated in Fig. 6.52. Similarly,

$$\int\!\!\int_{S_2} \mathbf{F} \cdot d\mathbf{A} = \int\!\!\int_{S_2} \mathrm{curl}\,\mathbf{B} \cdot d\mathbf{A} = \oint_{L_2} \mathbf{B} \cdot d\mathbf{l} \,,$$

L_2 is the boundary of the capping surface S_2. L_2 also goes along the line L but with the traverse direction opposite to that of L_1, Fig. 6.52. Since L_1 is opposite to L_2, the sum of the two line integrals is zero which proves that even in this case the flux is also zero. Above, we have only discussed the case with a single "hole"; obviously, the case of many "holes" can be considered along the same lines. **Q.E.D.**

Theorem 6.10 has a very simple consequence. Consider a flow of a fluid described by a vector field of velocity $\mathbf{v}$ that is solenoidal. Let us choose a surface S_0 around its lines of flow, Fig. 6.53, in such a way that the normal $\mathbf{n}_0$ to S_0 at every point is perpendicular to $\mathbf{v}$. It means that every line of flow does not cross S_0 and remains inside it so that S_0 forms a sort of a tube. Now, consider two surfaces S_1 and S_2 on

Fig. 6.53 Lines of flow of a liquid in a tube

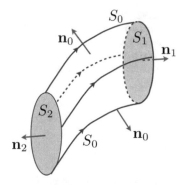

both sides of the tube with normals $\mathbf{n}_1$ and $\mathbf{n}_2$ directed outward as shown. A flux through the whole surface $S_0 + S_1 + S_2$ is zero for the solenoidal flow as was proven above. However, the flux through S_0 is also zero since at each point on the surface, $\mathbf{v}$ is perpendicular to $\mathbf{n}_0$. Therefore, the flux through $S_1 + S_2$ is zero:

$$\int\int_{S_1} \mathbf{v} \cdot \mathbf{n}_1 dA + \int\int_{S_2} \mathbf{v} \cdot \mathbf{n}_2 dA = \int\int_{S_1} \mathbf{v} \cdot \mathbf{n}_1 dA - \int\int_{S_2} \mathbf{v} \cdot (-\mathbf{n}_2) dA = 0$$

or

$$\int\int_{S_1} \mathbf{v} \cdot \mathbf{n}_1 dA = \int\int_{S_2} \mathbf{v} \cdot (-\mathbf{n}_2) dA \,,$$

where $-\mathbf{n}_2$ is the normal to S_2 directed along the line of flow similar to $\mathbf{n}_1$. Thus, the flux through any surface cutting the tube is conserved along the tube. This also means that the lines of flow cannot converge into a point since in this case the flux through the tube will become zero which is impossible (as it should be conserved).

6.8.3.3 Expansion of a General Field into Solenoidal and Conservative Components

Consider a vector field $\mathbf{A}(\mathbf{r})$ defined in some *simply connected* region D which has non-zero divergence and curl, i.e. $\mathrm{div}\mathbf{A}(\mathbf{r}) = f(\mathbf{r}) \neq 0$ and $\mathrm{curl}\,\mathbf{A}(\mathbf{r}) = \mathbf{g}(\mathbf{r}) \neq 0$, where $f(\mathbf{r})$ and $\mathbf{g}(\mathbf{r})$ are some rather general scalar and vector functions, respectively.

Theorem 6.11 *Any vector field $\mathbf{A}(\mathbf{r})$ defined in a simply connected region D can always be represented as a sum of a conservative, $\mathbf{A}_c(\mathbf{r})$, and solenoidal, $\mathbf{A}_s(\mathbf{r})$, fields, i.e.*

$$\mathbf{A}(\mathbf{r}) = \mathbf{A}_c(\mathbf{r}) + \mathbf{A}_s(\mathbf{r}) \ \text{with} \ \nabla \times \mathbf{A}_c(\mathbf{r}) = 0 \ \text{and} \ \nabla \cdot \mathbf{A}_s(\mathbf{r}) = 0 \,. \quad (6.121)$$

Proof: Let us first determine the conservative field $\mathbf{A}_c$ from the equations

$$\text{curl}\,\mathbf{A}_c = 0 \quad \text{and} \quad \text{div}\mathbf{A}_c = f(\mathbf{r})\ .$$

Since the field is conservative (due to our choice, $\text{curl}\,\mathbf{A}_c = 0$), there exists a scalar potential U such that $\mathbf{A}_c = \text{grad}\,U$. Therefore, for U we have an equation

$$\text{div}\mathbf{A}_c = \text{div}\,\text{grad}\,U = \Delta U = f(\mathbf{r})\ . \tag{6.122}$$

This is the so-called Poisson's equation, which always has a solution (see below), i.e. one can determine $\mathbf{A}_c$ by solving it.

Next, we consider the difference $\mathbf{A} - \mathbf{A}_c = \mathbf{A}_s$, and let us calculate its div and curl:

$$\text{div}\mathbf{A}_s = \text{div}\mathbf{A} - \text{div}\mathbf{A}_c = f - f = 0$$

and

$$\text{curl}\,\mathbf{A}_s = \text{curl}\,\mathbf{A} - \text{curl}\,\mathbf{A}_c = \text{curl}\,\mathbf{A} = \mathbf{g}\ .$$

We see that the difference, $\mathbf{A} - \mathbf{A}_c$ is, in fact, solenoidal. **Q.E.D.**

We have assumed above that the functions $\mathbf{g}(\mathbf{r})$ and $f(\mathbf{r})$ are not equal identically to zero in region D. If they were, then we would have $\text{curl}\mathbf{A} = \mathbf{0}$ and $\text{div}\mathbf{A} = 0$ anywhere in D, i.e. the field $\mathbf{A}(\mathbf{r})$ would be conservative and solenoidal at the same time. Since $\text{curl}\,\mathbf{A} = \mathbf{0}$, then there exists a scalar potential U, such that $\mathbf{A} = \text{grad}\,U$. However, since $\mathbf{A}$ is also solenoidal, $\text{div}\mathbf{A} = \text{div}\,\text{grad}\,U = \Delta U = 0$, which means that the scalar potential U satisfies the so-called Laplace equation $\Delta U = 0$. Functions that are solutions of the Laplace equation are called *harmonic functions*. Interestingly, the vector field satisfying $\text{curl}\,\mathbf{A} = 0$ and $\text{div}\mathbf{A} = 0$ is determined up to the gradient of a harmonic potential: $\mathbf{A}' = \mathbf{A} + \nabla U$. Indeed,

$$\text{div}\mathbf{A}' - \text{div}\mathbf{A} = \text{div}\,\text{grad}\,U = \Delta U = 0$$

for a harmonic function. The other condition,

$$\text{curl}\,\mathbf{A}' - \text{curl}\,\mathbf{A} = \text{curl}\,\text{grad}\,U = \mathbf{0}\ ,$$

is satisfied for any U since curl of grad is always equal to zero, Eq. (6.115).

It is possible to obtain the decomposition of a vector field $\mathbf{A} = \mathbf{A}_c + \mathbf{A}_s$ explicitly into a conservative and solenoidal components, given that

$$\text{div}\mathbf{A} = f(\mathbf{r}) \quad \text{and} \quad \text{curl}\,\mathbf{A} = \mathbf{g}(\mathbf{r})\ . \tag{6.123}$$

The method is based on the fact that a general solution of the Laplace equation $\Delta\phi = \rho$ can be written via a volume integral:

$$\phi(\mathbf{r}) = -\frac{1}{4\pi}\iiint_V \frac{\rho(\mathbf{r}')\,d\mathbf{r}'}{|\mathbf{r} - \mathbf{r}'|}\ . \tag{6.124}$$

Here the integration is performed over the volume of the spacial region V where $\rho(\mathbf{r})$ is defined. The notation for the volume differential, $d\mathbf{r}'$, used above for the volume integral is frequently used in physics literature; it simply means that $d\mathbf{r}' = dx'dy'dz' \equiv dV'$; we use the prime here because $\mathbf{r}$ is already used in the left-hand side. We will not prove Eq. (6.124) here.[14] However, this formula can be easily illustrated using known concepts of electrostatics discussed below in Sect. 6.9.1. It is discussed there, how the solution (6.124) of the Laplace equation $\Delta\phi = \rho$ can be justified, see specifically Eq. (6.133).[15]

Since the conservative contribution $\mathbf{A}_c$ satisfies curl $\mathbf{A}_c = 0$, one can introduce a potential $U(\mathbf{r})$ such that $\mathbf{A}_c = \text{grad } U$. Also, since $\text{div}\mathbf{A}_c = \text{div}\mathbf{A} = f(\mathbf{r})$ (note that $\text{div}\mathbf{A}_s = 0$ by construction), the potential U satisfies the Laplace equation,

$$\text{div}\mathbf{A}_c = \text{div grad } U = \Delta U = f(\mathbf{r}) \, ,$$

with the general solution

$$U(\mathbf{r}) = -\frac{1}{4\pi} \int \int \int_V \frac{f(\mathbf{r}') \, d\mathbf{r}'}{|\mathbf{r} - \mathbf{r}'|} \, , \tag{6.125}$$

This formula solves for the $\mathbf{A}_c$ part, as U totally defines $\mathbf{A}_c$.

Now we need to find a general solution for the solenoidal component. The latter satisfies the equations:

$$\text{div}\mathbf{A}_s = 0 \quad \text{and} \quad \text{curl } \mathbf{A}_s = \text{curl } \mathbf{A} = \mathbf{g}(\mathbf{r}) \, , \tag{6.126}$$

since curl $\mathbf{A}_c = 0$. Because the divergence of $\mathbf{A}_s$ is zero, there exists a vector potential $\mathbf{G}$ such that $\mathbf{A}_s = \text{curl } \mathbf{G}$. Moreover, the choice of $\mathbf{G}$ is not unique, as $\mathbf{G}' = \mathbf{G} + \text{grad } \Psi$ is also perfectly acceptable for any scalar field $\Psi(\mathbf{r})$ as we established above. On the other hand,

$$\text{div}\mathbf{G}' = \text{div}\mathbf{G} + \text{div grad } \Psi = \text{div}\mathbf{G} + \Delta\Psi \, .$$

The field Ψ is arbitrary; we shall select it in such a way that $\Delta\Psi = -\text{div}\mathbf{G}$. This is simply the Laplace equation with the right-hand side equal to $\text{div}\mathbf{G}$. This means that we can always choose the vector potential $\mathbf{G}'$ being solenoidal, i.e. with $\text{div}\mathbf{G}' = 0$. Hence, we have:

$$\text{curl } \mathbf{A}_s = \mathbf{g} \, , \quad \text{div}\mathbf{A}_s = 0 \quad \text{with} \quad \mathbf{A}_s = \text{curl } \mathbf{G}' \quad \text{and} \quad \text{div}\mathbf{G}' = 0 \, .$$

From the first equation,

$$\text{curl } \mathbf{A}_s = \text{curl curl } \mathbf{G}' = \text{grad div}\mathbf{G}' - \Delta\mathbf{G}' = -\Delta\mathbf{G}' \equiv \mathbf{g} \, ,$$

---

[14] The full proof is provided in Chap. 5 of Vol. II.
[15] Note that the Laplace equations here and in electrostatics differ by an unimportant factor of -4π.

where we used Eq. (6.116) and the fact that $\mathrm{div}\mathbf{G}' = 0$. It is seen that $\mathbf{G}'$ satisfies the vector Laplace equation, i.e. for each of its components we write a formula analogous to Eq. (6.125); in the vector form we have then:

$$\mathbf{G}'(\mathbf{r}) = \frac{1}{4\pi} \int \int \int_V \frac{\mathbf{g}\left(\mathbf{r}'\right) d\mathbf{r}'}{|\mathbf{r} - \mathbf{r}'|} . \tag{6.127}$$

Equations (6.125) and (6.127) fully solve the decomposition problem of an arbitrary vector field $\mathbf{A}$ since the potentials U and $\mathbf{G}'$ define the conservative and solenoidal components of it, and hence we can write:

$$\mathbf{A}(\mathbf{r}) = \mathrm{grad}\, U(\mathbf{r}) + \mathrm{curl}\, \mathbf{G}'(\mathbf{r}) . \tag{6.128}$$

6.9 Application of Integral Theorems in Physics. Part II

In this section more applications in physics of the calculus we have developed in this chapter will be briefly outlined.

6.9.1 Maxwell's Equations

A nice application of integral theorems is provided by Maxwell's equations of electromagnetism. We shall denote the electric and magnetic fields in vacuum as $\mathbf{E}$ and $\mathbf{H}$ here.

Consider first electrostatics which is the simplest case. A point charge q creates around itself an electric field $\mathbf{E}(\mathbf{r}) = q\mathbf{r}/r^3$. A probe charge q_0 placed in this field experiences a force $\mathbf{F} = q_0\mathbf{E}$, where $\mathbf{E}$ is the electrostatic field due to q.

Problem 6.104 Consider a sphere S_{sph} of radius R with the charge q placed at its centre. Show that the flux integral of $\mathbf{E}$ due to the charge q through the surface of the sphere is

$$\oiint_{S_{sph}} \mathbf{E} \cdot d\mathbf{A} = 4\pi q .$$

Note that the flux does not depend on the sphere radius. [Hint: *choose the coordinate system with q in its centre and note that for any point on the sphere the normal* $\mathbf{n}$ *is proportional to* $\mathbf{r}$.]

In fact, it is easy to see that the result is valid for any surface S, not necessarily the sphere. Indeed, as we have seen in Sect. 6.8.3.2, the electrostatic field is solenoidal anywhere outside the charge. Therefore, if we surround the charge with a sphere

Fig. 6.54 Point charge q is
surrounded by a sphere S_{sph}
which in turn is surrounded
by an arbitrary surface S

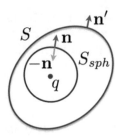

S_{sph} which lies completely inside the surface S as shown in Fig. 6.54, then the flux
through the sphere can be easily seen to be the same as the flux through S:

$$\oiint_S \mathbf{E} \cdot d\mathbf{A} = \left(\oiint_S \mathbf{E} \cdot d\mathbf{A} - \oiint_{S_{sph}} \mathbf{E} \cdot d\mathbf{A} \right) + \oiint_{S_{sph}} \mathbf{E} \cdot d\mathbf{A}$$

$$= \oiint_{S_{sph}} \mathbf{E} \cdot d\mathbf{A} = 4\pi q \ ,$$

since the expression in the brackets represents a flux through a closed surface $S +
S_{sph}$ (with the outer normal $-\mathbf{n}$ used for the S_{sph} because of the minus sign before the
flux integral) which is zero. This is because inside the region of space ΔV enclosed
by the two surfaces S_{sph} and S the field $\mathbf{E}$ is solenoidal, and hence the flux is zero
due to Theorem 6.10.

Next, if there are more than one charge inside some region, this result would
be valid for each of the charges if the surface S contains all of them inside. But the
fields due to every individual charge sum up (the superposition principle). Therefore,
generally, the flux of the total field $\mathbf{E}$ through a closed surface S containing charges
q_1, q_2, etc. must satisfy the following equation:

$$\oiint_S \mathbf{E} \cdot d\mathbf{A} = 4\pi \sum_i q_i \ . \tag{6.129}$$

For a continuous distribution of the charge, we shall have instead in the right-hand
side the total charge enclosed by the surface S, which is

$$Q = \int \int \int_V \rho dV \ ,$$

where $dV = dxdydz$ is the differential volume element, V is the volume for which
S is its boundary and ρ the charge density. Therefore,

$$\oiint_S \mathbf{E} \cdot d\mathbf{A} = 4\pi \int \int \int_V \rho dV \ .$$

The derived equation can in principle allow one to find the field $\mathbf{E}$ due to the charge
distribution ρ. However, it is inconvenient in practical calculations as it contains

integrals, it is much better to have an equation relating the field at each point to the charge density at that point. Such an equation can be obtained by applying the Gauss (divergence) theorem to the flux integral in the left-hand side:

$$\oint_S \mathbf{E} \cdot d\mathbf{A} = \int\int\int_V \operatorname{div}\mathbf{E} \, dV \implies \int\int\int (\operatorname{div}\mathbf{E} - 4\pi\rho) \, dV = 0 \, .$$

This result is valid for any volume V. Therefore, it must be that

$$\operatorname{div}\mathbf{E} = 4\pi\rho \, . \tag{6.130}$$

This fundamental equation of electrostatics is the Maxwell's first equation.

The electrostatic field is also a conservative field for which one can choose[16] a potential ϕ such that $\mathbf{E} = -\operatorname{grad} \phi$ (see Sect. 6.8.3.1; note that there only a single charge was considered; generalisation to many charges is however straightforward). Therefore, according to Theorem 6.9 of Sect. 6.8.3.1,

$$\operatorname{curl} \mathbf{E} = 0 \, . \tag{6.131}$$

This is the Maxwell's second equation. The corresponding differential equation for the potential $\phi(\mathbf{r})$ is obtained by using $\mathbf{E} = -\operatorname{grad} \phi$ in Eq. (6.130):

$$\operatorname{div} \operatorname{grad} \phi = \Delta\phi = -4\pi\rho \, , \tag{6.132}$$

which is the familiar *Poisson's equation*. It is straightforward to check explicitly that for a single point charge q located at $\mathbf{r}_q$, the potential at point $\mathbf{r}$ is $\phi(\mathbf{r}) = q/|\mathbf{r} - \mathbf{r}_q|$. Indeed, the electric field is easily calculated to be:

$$\begin{aligned}
\mathbf{E} &= -\operatorname{grad} \frac{q}{|\mathbf{r} - \mathbf{r}_q|} \\
&= -\left(\mathbf{i}\frac{\partial}{\partial x} + \mathbf{j}\frac{\partial}{\partial y} + \mathbf{k}\frac{\partial}{\partial z}\right) \frac{q}{\sqrt{(x - x_q)^2 + (y - y_q)^2 + (z - z_q)^2}} \\
&= \frac{q(\mathbf{r} - \mathbf{r}_q)}{|\mathbf{r} - \mathbf{r}_q|^3} \, ,
\end{aligned}$$

as required. Then, we know that the result for many charges is obtained simply by summing up contributions from all charges. If the charge is distributed continuously, then we divide up the space into small volumes $dV' \equiv d\mathbf{r}'$ with the charge $dQ = \rho(\mathbf{r}') \, dV'$ in each, where the volume is positioned near the point $\mathbf{r}'$ with the charge

---

[16] Note that it is customary in electrostatics to choose the minus sign when defining the potential.

Fig. 6.55 a The line integral
of the magnetic field **H** is
taken around a wire with the
current density **j**. **b** A surface
S is chosen so that the line L
is its boundary

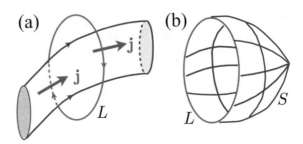

density being $\rho\left(\mathbf{r}'\right)$. Then the potential at point $\mathbf{r}$ due to this infinitesimal charge will
be

$$d\phi = \frac{dQ}{|\mathbf{r} - \mathbf{r}'|} = \frac{\rho\left(\mathbf{r}'\right)}{|\mathbf{r} - \mathbf{r}'|} dV' .$$

The total potential due to the whole charge distribution is then given by the integral

$$\phi\left(\mathbf{r}\right) = \int\int\int_V \frac{\rho\left(\mathbf{r}'\right) dV'}{|\mathbf{r} - \mathbf{r}'|} . \tag{6.133}$$

This formula can be considered as a general solution of the Laplace equation (6.132).

Moving charges create the magnetic field **H**, and the force acting on a probe charge
q_0 moving with the velocity **v** in the magnetic field is given by the Lorentz formula:
$\mathbf{F} = (q_0/c) [\mathbf{v} \times \mathbf{H}]$, where c is the speed of light. It was established experimentally
by Ampère that a closed-loop line integral of **H** around a wire with the current I is
proportional to the current, see Fig. 6.55a:

$$\oint_L \mathbf{H} \cdot d\mathbf{l} = \frac{4\pi}{c} I . \tag{6.134}$$

This is Ampère's circuital law which allows, e.g., to calculate directly the magnetic
field around an infinite wire.

> **Problem 6.105** Show using Eq. (6.134) that the magnitude of the field **H**
> around an infinite wire of the current I at a distance r from the wire is $H = 2I/cr$.

Maxwell generalised the result (6.134) for any flow of the current and postulated
that the formula is valid for any closed-loop contour L surrounding the current I.
Since the current can be expressed via the current density **j** using Eq. (6.103) with
some surface S which has L as its boundary, Fig. 6.55b, then, upon using Stock's
theorem (6.73) for the line integral, we obtain:

$$\oint_L \mathbf{H} \cdot d\mathbf{l} = \int\int_S \operatorname{curl} \mathbf{H} \cdot d\mathbf{A} \implies \int\int_S \operatorname{curl} \mathbf{H} \cdot d\mathbf{A} = \frac{4\pi}{c} \int\int_S \mathbf{j} \cdot d\mathbf{A} ,$$

and hence

$$\int\int_S \left(\text{curl }\mathbf{H} - \frac{4\pi}{c}\mathbf{j}\right) \cdot d\mathbf{A} = 0 \quad \Longrightarrow \quad \text{curl }\mathbf{H} = \frac{4\pi}{c}\mathbf{j} , \tag{6.135}$$

since the surface S (as well as the contour L) can be chosen arbitrarily. This is the next Maxwell's equation corresponding to stationary conditions (no time dependence).

Another experimental result states that the flux of the magnetic field through any closed surface S is zero.

Problem 6.106 Prove from this that the magnetic field is solenoidal, i.e.

$$\nabla \cdot \mathbf{H} = 0 . \tag{6.136}$$

[Hint: *use the divergence theorem.*] This is yet another Maxwell's equation.

Problem 6.107 Prove from Eq. (6.135) that $\nabla \cdot \mathbf{j} = 0$. This corresponds to the continuity equation (6.104) under stationary conditions. [Hint: *take the divergence of both sides.*]

Equations (6.130), (6.131), (6.135) and (6.136) form the complete set of equations for the electric and magnetic fields in a vacuum under stationary conditions. As we have seen in the last problem, they do not contradict the continuity equation.

Consider now the general case of the charge $\rho(\mathbf{r}, t)$ and current $\mathbf{j}(\mathbf{r}, t)$ densities depending not only on the spatial position $\mathbf{r}$ but also on time, t. This results in the fields $\mathbf{H}(\mathbf{r}, t)$ and $\mathbf{E}(\mathbf{r}, t)$ depending both on $\mathbf{r}$ and t. Faraday discovered the law of induction stating that the current will flow in a closed loop of wire L if the flux of the magnetic field through a surface S having L as its boundary, Fig. 6.55b, changes in time. This observation can be formulated in the following way: the rate of change of the magnetic flux through S is equal to the electromotive force, i.e.

$$-\frac{1}{c}\frac{\partial}{\partial t}\int\int_S \mathbf{H} \cdot d\mathbf{A} = \oint_L \mathbf{E} \cdot d\mathbf{l} . \tag{6.137}$$

Problem 6.108 Derive from this integral formulation of the induction law its differential analog:

$$-\frac{1}{c}\frac{\partial \mathbf{H}}{\partial t} = \nabla \times \mathbf{E} . \tag{6.138}$$

This equation replaces (6.131) for the case of non-stationary sources (charge and/or current densities).

Finally, Maxwell conceived the equation which was to replace Eq. (6.135) of the stationary case. He realised that a displacement current needs to be added related to the rate of change of the electric flux through the surface S:

$$\oint_L \mathbf{H} \cdot d\mathbf{l} = \frac{4\pi}{c} I + \frac{1}{c} \frac{\partial}{\partial t} \int \int_S \mathbf{E} \cdot d\mathbf{A} \ . \tag{6.139}$$

Problem 6.109 Derive from this integral equation its differential analog:

$$\nabla \times \mathbf{H} = \frac{4\pi}{c} \mathbf{j} + \frac{1}{c} \frac{\partial \mathbf{E}}{\partial t} \ . \tag{6.140}$$

Problem 6.110 Show that Eq. (6.138) does not contradict Eq. (6.136). [Hint: *take divergence of both sides.*]

Problem 6.111 Show that Maxwell equations do not contradict the full continuity equation (6.104). [Hint: *apply divergence to both sides of Eq. (6.140).*]

Equations (6.130), (6.136), (6.138) and (6.140) form the complete set of equations of classical electrodynamics. Introduction of the displacement current (the last term on the right-hand side of Eq. (6.140)) was necessary at least for two reasons. First, without that term the continuity equation would not be satisfied. Second, this term is necessary for establishing electro-magnetic waves. Discussing now this latter point, let us consider Maxwell's equations in a vacuum where there are no charges and currents, i.e. $\rho = 0$ and $\mathbf{j} = \mathbf{0}$. Then, take a partial time derivative of both sides of Eq. (6.140),

$$\text{curl}\ \frac{\partial \mathbf{H}}{\partial t} = \frac{1}{c} \frac{\partial^2 \mathbf{E}}{\partial t^2} \ .$$

Then, we replace $\partial \mathbf{H}/\partial t$ using the other Maxwell's equation (6.138),

$$-c\,(\text{curl curl}\,\mathbf{E}) = \frac{1}{c} \frac{\partial^2 \mathbf{E}}{\partial t^2} \ ,$$

calculate the curl of the curl using Eq. (6.116),

$$\text{curl curl}\,\mathbf{E} = \text{grad div}\,\mathbf{E} - \Delta \mathbf{E} \ ,$$

in which only the second term remains in the right-hand side due to Eq. (6.130) (recall that there are no charges). This finally gives:

$$\frac{1}{c^2} \frac{\partial^2 \mathbf{E}}{\partial t^2} = \Delta \mathbf{E} \ . \tag{6.141}$$

This is a *wave equation* for the three components of the electric field **E**.

Problem 6.112 Starting from Eq. (6.138), derive similarly the wave equation for the magnetic field:

$$\frac{1}{c^2}\frac{\partial^2 \mathbf{H}}{\partial t^2} = \Delta \mathbf{H} . \qquad (6.142)$$

Equations (6.141) and (6.142) describe oscillations of **E** and **H** which propagate in space with the speed c.

Another interesting point worth mentioning is the symmetry of the equations with respect to the two fields, **E** and **H** (compare Eq. (6.130) with (6.136), and also (6.138) with (6.140)); the only difference (apart from the sign in some places) comes from the charge density ρ and the current density **j**. The absence of their magnetic analogues in Eqs. (6.136) and (6.138) corresponds to the experimental fact that magnetic charges do not exist, and hence there is also no magnetic current.

In practice, instead of the six quantities (three components of **E** and three of **H**) it is more convenient to work with only four fields: one scalar and one vector. These are introduced in the following way. Since **H** is solenoidal, Eq. (6.136), there exists a vector potential $\mathbf{A}(\mathbf{r}, t)$ such that

$$\mathbf{H} = \operatorname{curl} \mathbf{A} = \nabla \times \mathbf{A} . \qquad (6.143)$$

Replacing **H** in Eq. (6.138), we obtain:

$$\operatorname{curl}\left(\frac{1}{c}\frac{\partial \mathbf{A}}{\partial t} + \mathbf{E}\right) = 0 ,$$

i.e. the field inside the brackets is conservative. This in turn means that there exists a scalar field (potential) $-\phi(\mathbf{r}, t)$ (the minus sign is chosen due to historical reasons) such that

$$\frac{1}{c}\frac{\partial \mathbf{A}}{\partial t} + \mathbf{E} = -\operatorname{grad} \phi \quad \Longrightarrow \quad \mathbf{E} = -\operatorname{grad} \phi - \frac{1}{c}\frac{\partial \mathbf{A}}{\partial t} . \qquad (6.144)$$

Problem 6.113 Show now that replacing **A** and ϕ with (the already familiar gauge transformation, see Eq. (6.108))

$$\mathbf{A}' = \mathbf{A} + \nabla U \quad \text{and} \quad \phi' = \phi - \frac{1}{c}\frac{\partial U}{\partial t} , \qquad (6.145)$$

does not change the fields $\mathbf{E}$ and $\mathbf{H}$. This means that Maxwell equations are invariant with respect to the gauge transformation.

This shows that the choice of the scalar and vector potentials is not unique.

Problem 6.114 In this problem we shall derive the so-called Jefimenko equations[17] of classical electromagnetism. We shall start from the retarded solutions of the Maxwell equations for the scalar, $\phi(\mathbf{r}, t)$, and vector, $\mathbf{A}(\mathbf{r}, t)$, potentials (see Vol. II, Chap. 5):

$$\phi(\mathbf{r}, t) = \int \frac{d\mathbf{r}'}{|\mathbf{r} - \mathbf{r}'|} \rho(\mathbf{r}', t_r), \qquad (6.146)$$

$$\mathbf{A}(\mathbf{r}, t) = \frac{1}{c} \int \frac{d\mathbf{r}'}{|\mathbf{r} - \mathbf{r}'|} \mathbf{j}(\mathbf{r}', t_r), \qquad (6.147)$$

where $t_r = t - |\mathbf{r} - \mathbf{r}'| / c$ is the retarded time, $\rho(\mathbf{r}, t)$ is the charge density and $\mathbf{j}(\mathbf{r}, t)$ current density distributions. These equations, also known as d'Alembert equations, provide a formal solution of the Maxwell equations given the sources, i.e. the charges and associated with their currents. The electric $\mathbf{E}(\mathbf{r}, t)$ and magnetic $\mathbf{H}(\mathbf{r}, t)$ fields can be obtained from the scalar and vector potentials via Eqs. (6.144) and (6.143). Using the explicit expressions for the scalar and vector potentials as given above in Eqs. (6.146) and (6.147), derive the following equations for the fields:

$$\mathbf{E}(\mathbf{r}, t) = \int \left\{ \frac{\mathbf{r} - \mathbf{r}'}{|\mathbf{r} - \mathbf{r}'|^3} \rho(\mathbf{r}', t_r) + \frac{1}{c} \frac{\mathbf{r} - \mathbf{r}'}{|\mathbf{r} - \mathbf{r}'|} \left(\frac{\partial \rho(\mathbf{r}', t)}{\partial t} \right)_{t=t_r} - \frac{1}{c^2} \frac{1}{|\mathbf{r} - \mathbf{r}'|} \left(\frac{\partial \mathbf{j}(\mathbf{r}', t)}{\partial t} \right)_{t=t_r} \right\} d\mathbf{r}',$$

$$\mathbf{H}(\mathbf{r}, t) = -\frac{1}{c} \int \left\{ \frac{\mathbf{r} - \mathbf{r}'}{|\mathbf{r} - \mathbf{r}'|^3} \times \mathbf{j}(\mathbf{r}', t_r) + \frac{1}{c} \frac{\mathbf{r} - \mathbf{r}'}{|\mathbf{r} - \mathbf{r}'|^2} \times \left(\frac{\partial \mathbf{j}(\mathbf{r}', t)}{\partial t} \right)_{t=t_r} \right\} d\mathbf{r}'.$$

Here the sources (the charge and current densities) and their time derivatives are calculated at the retarded time t_r.

6.9.2 Diffusion and Heat Transport Equations

Consider diffusion of some particles of mass m in a media. This could be the case of solute molecules dissolved in a solvent (a solution), or some interstitial foreign atoms in a crystal lattice. If the concentration of solute molecules or foreign atoms

---

[17] Oleg Jefimenko (1922–2009).

is not homogeneous across the system, there will be a diffusion flux in the direction from higher to lower concentration until equilibrium is established (if there are no external fields, e.g. no electrostatic potential, then equilibrium would correspond to the homogeneous distribution of the particles). The flow of particles per unit time through a surface area $d\mathbf{A} = \mathbf{n}\,dA$ with the normal $\mathbf{n}$ is proportional to the directional derivative of the concentration $C\,(\mathbf{r}, t)$ of the particles, $\partial C/\partial n$, in the direction normal to the surface dA (see Sect. 6.3.3). The total mass passed through dA is

$$dm = -D\frac{\partial c}{\partial n}\,dA\,dt = -D\,(\nabla C \cdot \mathbf{n})\,dA\,dt = -D\,(\nabla C \cdot d\mathbf{A})\,dt\,, \qquad (6.148)$$

where we have made use of the fact that the directional derivative can be related to the gradient, see Eq. (5.71). Above, D is the diffusion constant. This expression is a well-known Fick's first law. The minus sign is required to make the quantity dm positive since the partial derivative $\partial c/\partial n < 0$ (the flow is expected from the regions of higher concentration to the regions of smaller concentration, so the concentration decreases across the surface dA in the direction of $\mathbf{n}$). The total mass passed through a *closed* surface S is then

$$dM = -dt \oiint_S D\nabla C \cdot d\mathbf{A}\,;$$

here the quantity dM is positive.

On the other hand, the flow of the particles out of V results in a decrease of the total mass $M(t) = \int \int \int_V C\,dV$ inside the volume by the following positive amount:

$$dM = M(t) - M(t + dt) = -\frac{\partial M}{\partial t}\,dt = -dt \int \int \int_V \frac{\partial C}{\partial t}\,dV\,.$$

Here, to ensure that $dM > 0$, we subtracted the mass at the later time $t + dt$ (which is smaller) from the mass at time t (that is larger). The two expressions for the dM must be equal to each other due to conservation of mass. Using the divergence theorem for the surface integral, we then obtain the diffusion equation:

$$\mathrm{div}\,(D\,\mathrm{grad}\,C) = \frac{\partial C}{\partial t}\,. \qquad (6.149)$$

This equation describes the time and space dependence of the concentration of the particles from the initial non-equilibrium situation towards equilibrium. If the diffusion coefficient is constant throughout the entire system, then it can be taken out of the divergence, and we obtain in this case

$$\Delta C = \frac{1}{D}\frac{\partial C}{\partial t}\,, \qquad (6.150)$$

as div grad $C = \Delta C$, where Δ is the Laplace operator.

Problem 6.115 The reader may have noticed that our derivation of the diffusion equation was very similar to that of the continuity equation in Sect. 6.7.1. Derive Eq. (6.149) directly from the continuity equation (6.99) or (6.104), extracting the needed flux **j** of mass from the Fick's law, Eq. (6.148).

Problem 6.116 Consider one-dimensional (along x) diffusion of particles, e.g., in a homogeneous solution (D is constant). Check that the function

$$C(x, t) = \frac{1}{\sqrt{4\pi D t}} e^{-x^2/4Dt} \qquad (6.151)$$

is a solution of the corresponding diffusion equation. Sketch this distribution as a function of x at different times and analyse your results. This solution corresponds to the spread of particles distribution with time when initially (at $t = 0$) the particles were all at $x = 0$.

The heat transport in a media can be considered along the same lines. This time, heat Q is transferred from regions of higher temperature $T(\mathbf{r}, t)$ to the regions of smaller temperature. We start by stating an experimental fact (Fourier's law) about the heat conduction: the amount of heat passing through a surface dA per unit time is proportional to the gradient of temperature there:

$$dQ = -\kappa \frac{\partial T}{\partial n} dA dt \ ,$$

where κ is the thermal conductivity, $\partial T / \partial n = \mathbf{n} \cdot \operatorname{grad} T$ is the directional derivative showing the change of temperature in the direction perpendicular to the normal **n** to the surface dA. The minus sign is needed to make the quantity dQ positive since the energy flow goes across dA in the direction along its normal **n** from the region with higher to the region with smaller temperature, hence $\partial T / \partial n < 0$.

Consider now a finite and arbitrary volume V within our system, with the surface S. Then the total amount of heat given away by the volume V to the environment around it over time dt is

$$dQ = -dt \oiint_S \kappa \frac{\partial T}{\partial n} dA = -dt \oiint_S \kappa \operatorname{grad} T \cdot d\mathbf{A}$$
$$= -dt \int\int\int_V \operatorname{div} (\kappa \operatorname{grad} T) dV \ ,$$

the divergence theorem was used in the last passage. Note that κ may depend on the spatial position in general and hence is kept inside the divergence.

On the other hand, the heat dQ is the one lost by the volume over time dt due to decrease in its temperature. If we take a small volume dV inside V, then this volume lost

$$-C_V (dT)\rho dV = -C_V \frac{\partial T}{\partial t}\rho dV dt$$

of heat (again, this quantity is positive as dT and, correspondingly the time derivative of the temperature are negative). Here, C_V is the heat capacity per unit mass and ρ is the mass density. The total loss in the volume is then

$$dQ = -dt \int\int\int_V C_V \frac{\partial T}{\partial t}\rho dV .$$

The two expressions for dQ must be equal, which results in the heat transport equation:

$$C_V \rho \frac{\partial T}{\partial t} = \text{div}\,(\kappa\,\text{grad}\,T) . \qquad (6.152)$$

This heat transport equation, given the appropriate boundary and initial conditions, should provide us with the temperature distribution in the system $T(\mathbf{r}, t)$ over time. For a homogeneous system, the constant κ does not depend on the spatial variables and can be taken out of the divergence. Since the divergence of the gradient is the Laplacian, we arrive at the more familiar form of the heat transport equation:

$$\frac{1}{D}\frac{\partial T}{\partial t} = \Delta T , \qquad (6.153)$$

where $D = \kappa/(C_V \rho)$ is the thermal diffusivity. Comparing the two equations (6.149) and (6.152), we see that these are practically identical. Therefore, from the mathematical point of view, the solution of these equations would be the same.

6.9.3 Hydrodynamic Equations of Ideal Liquid (Gas)

Let us consider an *ideal liquid* (it could also be a gas), i.e. we assume that the forces applied to its any finite volume V with the boundary surface S can be expressed via pressure P due to the external (with respect to the chosen volume) part of the liquid. The latter exerts a force acting inside the volume, i.e. in the direction opposite to the surface normal $\mathbf{n}$ which is assumed to be directed out of the volume. In other words, the total force acting on volume V is

$$-\oiint_S P\mathbf{n}dA = -\int\int\int_V \text{grad}\,P\,dV , \qquad (6.154)$$

where we have used formula (6.93).

We can also have some *external* force density $\mathbf{F}$ acting on the liquid (i.e. there is the force $\mathbf{F}dm = \mathbf{F}\rho dV$ acting on the volume element dV of mass dm), so that the total force acting on volume V will then be

$$\mathbf{f} = \int\int\int_V \mathbf{F}\rho dV - \int\int\int_V \text{grad}\,P\,dV = \int\int\int_V (\rho\mathbf{F} - \text{grad}\,P)\,dV .$$

According to Newton's second law, this force results in the acceleration of the liquid volume given by

$$\int \int \int_V \rho \frac{d\mathbf{v}}{dt} dV \, ,$$

where $\mathbf{v}(\mathbf{r}, t)$ is the velocity of the liquid depending on time t and the spatial point $\mathbf{r}$. Therefore, one can write:

$$\int \int \int_V \rho \frac{d\mathbf{v}}{dt} dV = \int \int \int_V (\rho \mathbf{F} - \mathrm{grad}\, P)\, dV \quad \Longrightarrow \quad \rho \frac{d\mathbf{v}}{dt} = \rho \mathbf{F} - \mathrm{grad}\, P \, .$$
$$(6.155)$$

Note that here the acceleration on the left-hand side is given by the *total derivative* of the velocity. Indeed, $\mathbf{v}$ is the function of both time and the three coordinates, and the liquid particles move in space so that their coordinates change as well. Therefore, similarly to our consideration of the derivative of the distribution function of a gas in Sect. 5.6 (see Eq. (5.26)), we can write:

$$\frac{d\mathbf{v}}{dt} \longmapsto \left(\frac{d\mathbf{v}}{dt} \right)_{tot} = \frac{\partial \mathbf{v}}{\partial t} + \frac{\partial \mathbf{v}}{\partial x} \frac{\partial x}{\partial t} + \frac{\partial \mathbf{v}}{\partial y} \frac{\partial y}{\partial t} + \frac{\partial \mathbf{v}}{\partial z} \frac{\partial z}{\partial t} \equiv \frac{\partial \mathbf{v}}{\partial t} + \frac{\partial \mathbf{v}}{\partial x} v_x + \frac{\partial \mathbf{v}}{\partial y} v_y + \frac{\partial \mathbf{v}}{\partial z} v_z \, .$$

The coordinate derivatives term above is normally written in the following short form as a dot product:

$$\frac{\partial \mathbf{v}}{\partial x} v_x + \frac{\partial \mathbf{v}}{\partial y} v_y + \frac{\partial \mathbf{v}}{\partial z} v_z = \mathbf{v} \cdot \mathrm{grad}\, \mathbf{v} \, ,$$

where the gradient is understood to be taken separately for each component of the velocity field. Finally, we arrive at the following equation:

$$\frac{\partial \mathbf{v}}{\partial t} + \mathbf{v} \cdot \mathrm{grad}\, \mathbf{v} = \mathbf{F} - \frac{1}{\rho} \mathrm{grad}\, P \, . \qquad (6.156)$$

This equation is to be supplemented with the continuity equation (6.99) or (6.102) for the liquid which we derived before. Given the equation of states for the liquid, $P = f(\rho, T)$, i.e. how the pressure depends on the density and the temperature T, one can solve these equations to obtain the velocity field in the liquid under the applied external forces $\mathbf{F}$ and the temperature T.

Infinite Numerical and Functional Series

7

In Sect. 1.12 we considered summation of finite series. In practice it is often necessary to deal with *infinite* series which contain an infinite number of terms a_n. Here a_n is a general (we say "the n-th") term of the series which is constructed via some kind of formula that depends on an integer $n = 1, 2, 3, \ldots$. For instance, the rule $a_n = a_0 q^n$ with $n = 0, 1, 2, 3, \ldots$ corresponds to an infinite geometric progression

$$a_0 + a_0 q + a_0 q^2 + \ldots = a_0 \left(1 + q + q^2 + \ldots \right) .$$

Note that infinite numerical series were already mentioned briefly in Sect. 2.2.3. Here we shall consider this question in more detail. Our interest here is in understanding whether or not one can define a *sum* of such an infinite series,

$$S = \sum_{n=1}^{\infty} a_n , \qquad (7.1)$$

i.e. whether the sum like this one converges to a well-defined number S. It is also possible that each term of the series, $a_n(x)$, is a function of some variable x, and then we should discuss for which values of the x the series converges, and if it does, what are the properties of the function $S(x)$ of the sum. For instance, one may ask if it is possible to integrate or differentiate the series term-by-term, i.e. would the series generated that way converge for the same values of the x and, if it does, whether the result can simply be related to integrating or differentiating, respectively, the function $S(x)$ of the sum itself.

© The Author(s), under exclusive license to Springer Nature Switzerland AG 2022
L. Kantorovich, *Mathematics for Natural Scientists*, Undergraduate Lecture Notes in Physics, https://doi.org/10.1007/978-3-030-91222-2_7

7.1   Infinite Numerical Series

We say that the series (7.1) converges if the limit of its partial sum containing N terms,

$$S_N = \sum_{n=1}^{N} a_n \, , \tag{7.2}$$

has a well-defined limit as $N \to \infty$. The partial sums S_1, S_2, S_3, etc. form a numerical sequence which we considered in Sect. 2.2. Therefore, the numerical series (7.1) converges if the corresponding sequence of its partial sums, S_N, has a well-defined limit at $N \to \infty$.

An infinite geometrical progression, when $a_n = a_0 q^n$, has already been considered in Sect. 2.2.3, and can serve as a simple example. It converges when $|q| < 1$, and diverges otherwise.

As another example, the series of Eq. (1.152) converges to one since

$$\lim_{N \to \infty} S_N = \lim_{N \to \infty} \sum_{n=1}^{N} \frac{1}{n(n+1)} = \lim_{N \to \infty} \left(1 - \frac{1}{N+1} \right) = 1 \, .$$

As an example of a diverging series, let us consider the so-called *harmonic* series

$$1 + \frac{1}{2} + \frac{1}{3} + \frac{1}{4} + \ldots = \sum_{n=1}^{\infty} \frac{1}{n} \, . \tag{7.3}$$

If S_N is its partial sum, then for any N:

$$S_{2N} - S_N = \sum_{n=1}^{2N} \frac{1}{n} - \sum_{n=1}^{N} \frac{1}{n} = \sum_{n=N+1}^{2N} \frac{1}{n} > \sum_{n=N+1}^{2N} \frac{1}{2N} = \frac{N}{2N} = \frac{1}{2} \, ,$$

since for any $n \leq 2N$ we have $1/n \geq 1/2N$ (the equal sign only when $n = 2N$). On the other hand, if we assume that the series converges to S, then $\lim_{N \to \infty} (S_{2N} - S_N) = S - S = 0$, which is impossible as it has been shown above that $S_{2N} - S_N > 1/2$ and hence cannot have the zero limit. This proves that the harmonic series actually diverges.

Problem 7.1 Find the sum of the following infinite series:

$$\sum_{n=1}^{\infty} \frac{1}{n^2 - 1} \, .$$

[Hint: *the partial sum is given by Eq.* (1.150). Answer : 3/4.]

Problem 7.2 Prove that a combination of a geometric and arithmetic progressions, see Eq. (1.151), has a well-defined limit for $|q| < 1$ and any r:

$$\sum_{i=0}^{\infty} (a_0 + ir)q^i = \frac{a_0}{1-q} + \frac{rq}{(1-q)^2} \, .$$

There are several simple theorems which establish essential properties of converging infinite numerical series.

Theorem 7.1 *If the series (7.1) converges to S, one can remove any first m terms in the series, and the rest of the series,*

$$\sum_{n=m+1}^{\infty} a_n \, ,$$

would still converge.

Theorem 7.2 *If the series (7.1) of a_n converges to S, then the series of $b_n = ca_n$ converges to cS, where c is (generally complex) number.*

Theorem 7.3 *If two series $\sum_n a_n$ and $\sum_n b_n$ converge to S_a and S_b, respectively, then the series $\sum_n (\alpha a_n + \beta b_n)$ also converges to $\alpha S_a + \beta S_b$.*

Problem 7.3 *Prove the above theorems.* [Hint: *consider the limits of the appropriate partial sums.*]

Theorem 7.4 (Necessary Condition for Convergence) *If the series $\sum_n a_n$ converges, then $a_n \to 0$ as $n \to \infty$.*

Proof: if S_N is a partial sum of the series, then $a_N = S_N - S_{N-1}$. Taking the limit $N \to \infty$, we obtain the required result. **Q.E.D.**

This is not a sufficient condition, only a necessary one. Indeed, in the case of the harmonic series considered above, $a_n = 1/n$ and it does tend to zero when $n \to \infty$. However, we know that the series diverges. Therefore, to establish convergence of a particular series encountered in practical problems, it is necessary to investigate the question of convergence in more detail; in particular, we need to develop sufficiency criteria as well.

7.1.1 Series with Positive Terms

It is convenient to start a more detailed analysis from a particular case of the series which have all their terms positive, $a_n > 0$.

Theorem 7.5 (Necessary and Sufficient Condition for Convergence) *The series converges if and only if its numerical sequence of the partial sums $\{S_N\}$ is bounded from above.*

Proof: To prove sufficiency, we assume that the sequence of partial sums is bounded from above, i.e. $S_N \le M$, where $M > 0$, and then we need to see if from this follows that the sequence converges. Indeed, because the sequence contains exclusively positive terms, the partial sums form an increasing sequence $0 < S_1 \le S_2 \le S_3 \le \ldots$. It has to reach a well-defined limit since, according to our assumption, it is bounded from above (and hence cannot increase indefinitely). Therefore, once S_N has a limit, the numerical sequence converges to that limit.

Now, to prove the necessity, we first assume that the series converges to S, and then show that from this follows that it is bounded from above. Indeed, since the sequence of partial sums $\{S_N\}$ converges, it has a well-defined limit S. On the other hand, since the numerical sequence of the partial sums is increasing, it must be also bounded from above by S (otherwise, it would never converge, see Sect. 2.2.2). **Q.E.D.**

There are several sufficiency criteria to check whether the series converges or not.

Theorem 7.6 (Sufficient Criterion for Convergence) *Consider two series, both with all terms positive, $\sum_n a_n$ and $\sum_n b_n$, such that $a_n \le b_n$ for any $n = 1, 2, 3, \dots$. Then, if the second series $\sum_n b_n$ converges, then the first one, $\sum_n a_n$, does as well; if the first series diverges, so does the second.*

Proof: Let A_N and B_N be the corresponding partial sums for the two series, respectively; since for all n we have $a_n \le b_n$, then obviously $A_N \le B_N$ for any N. Let us now consider the first part of the theorem based on the assumption that the series $\sum_n b_n$ converges to some value B. Hence, $B_N \le B$. Therefore, $A_N \le B_N \le B$, i.e. the series $\sum_n a_n$ is bounded from above and hence (according to Theorem 7.5) also converges. Now we turn to the second statement where the first series diverges. Let us assume that the second series converges. Then, according to the first part of the theorem, the first series should converge as well, which contradicts our assumption; therefore, if the first series diverges, then so does the second. **Q.E.D.**

As an example, consider the series with $a_n = 1/\sqrt{n}$. Since $1/n < 1/\sqrt{n}$ for any $n > 1$, and the series with $b_n = 1/n$ diverges, then the series $\sum_n n^{-1/2}$ diverges as well. In fact, the series $\sum_n n^{-\alpha}$ diverges for any $0 < \alpha \le 1$.

Problem 7.4 Prove that the series with $a_n = \left(1 + 1/n^2\right)^n$ converges.

Problem 7.5 Prove that the series with $a_n = n^{-n}$ converges. [Hint: *use the fact that* $n^{-n} \le 2^{-n}$ *for any* $n > 1$.]

Theorem 7.7 (The Ratio Test[1]) *If*

$$\lambda = \lim_{n \to \infty} \frac{a_{n+1}}{a_n} < 1 ,$$

the series converges; if $\lambda > 1$, it diverges; if $\lambda = 1$, then this test is inconclusive (the series may either converge or diverge, more investigations are required).

---

[1] This test is due to Jean-Baptiste le Rond d'Alembert and is normally bears his name.

Proof: Let us first consider the case of $\lambda < 1$. Since the limit of a_{n+1}/a_n exists (and equal to λ), then for any $\epsilon > 0$ there exists an integer N such that for any $n > N$ we have $|a_{n+1}/a_n - \lambda| < \epsilon$, i.e. $\lambda - \epsilon < a_{n+1}/a_n < \lambda + \epsilon$. What is essential for us here is the second part: $a_{n+1}/a_n < \lambda + \epsilon$. Indeed, since $\lambda < 1$, one can always find ϵ such that $\lambda + \epsilon$ is still smaller than 1, i.e. for all $n > N$ we would have $a_{n+1}/a_n < \rho < 1$, or $a_{n+1} < \rho a_n$. Therefore, $a_{N+1} < \rho a_N$, $a_{N+2} < \rho a_{N+1} < \rho^2 a_N$, etc. In general, $a_{N+r} < \rho^r a_N$, and hence,

$$\sum_{n=N+1}^{\infty} a_n = \sum_{r=1}^{\infty} a_{N+r} < \sum_{r=1}^{\infty} \rho^r a_N = a_N \sum_{r=1}^{\infty} \rho^r = \frac{a_N \rho}{1 - \rho},$$

where we have made use of the sum of the geometrical progression (recall that $0 < \rho < 1$). Therefore, the series without the first N terms converges, and so does the whole series (Theorem 7.1).

Now consider the case of $\lambda > 1$. We can again write that $\lambda - \epsilon < a_{n+1}/a_n < \lambda + \epsilon$, or $\lambda - \epsilon < a_{n+1}/a_n$. Since $\lambda > 1$, one can always find such $\epsilon > 0$ that $\lambda - \epsilon > 1$. Therefore, $a_{n+1}/a_n > 1$ or $a_{n+1} > a_n$. This means that a_n cannot tend to zero as $n \to \infty$, which is the necessary criterion for convergence (Theorem 7.4), i.e. the original series indeed diverges. **Q.E.D.**

Nothing can be said about the case of $\lambda = 1$, the corresponding series may either diverge or converge, more powerful criteria must be used to establish convergence. For instance, consider the series with $a_n = 1/\sqrt{n}$ which, as we know from the analysis above, diverges. Using the ratio test, we have:

$$\lambda = \lim_{n \to \infty} \frac{a_{n+1}}{a_n} = \lim_{n \to \infty} \frac{1/\sqrt{n+1}}{1/\sqrt{n}} = \lim_{n \to \infty} \sqrt{\frac{n}{n+1}} = 1.$$

Therefore, the ratio test is not powerful enough to establish convergence or divergence of this particular series.

Problem 7.6 Prove that $\lambda = 1$ as well for the series with $a_n = 1/n^2$ which (as we shall learn later on) converges. Therefore, the case of $\lambda = 1$ is also inconclusive in this case.

Problem 7.7 Prove using the ratio test that the series with $a_n = x^n/n!$ converges for all values of x (we shall see later on that this series gives e^x).

Problem 7.8 Prove using the ratio test that the series with $a_n = n^n/n!$ diverges. [Hint: *you will need the result* (2.26).]

Theorem 7.8 (The Root Test[2]) *If a_n, starting from some $n \geq N \geq 1$, satisfies the inequality*

$$\sqrt[n]{a_n} \leq q < 1 , \qquad (7.4)$$

then the series converges; if, however, for all $n \geq M \geq 1$ we have $\sqrt[n]{a_n} > 1$, then the series diverges.

Proof: In the first case, $a_n \leq q^n$ with $q < 1$ for any $n \geq N$. The series with $b_n = q^n$ is a geometric progression with the q^N being its first term, and it converges for $q < 1$. Therefore, the series with a_n and $n \geq N$ converges as well because of Theorem 7.6. As a finite number of terms before the a_N term do not effect the convergence (Theorem 7.1), the whole series converges. In the second case, $a_n > 1$ and hence a_n does not tend to zero as $n \to \infty$ yielding divergence of the series (Theorem 7.4). **Q.E.D.**

The root test is stronger than the ratio test, i.e. in some cases when the latter test is inconclusive or cannot be applied, the root test may work. However, it may not be straightforward to apply the root test, and hence it is used less frequently than the ratio test. Both tests are inconclusive when $\sqrt[n]{a_n}$ and a_{n+1}/a_n tend to one when $n \to \infty$.

Problem 7.9 Prove that the series with $a_n = e^{-\alpha n} \sin^2 (\beta n)$ converges for any β and $\alpha > 0$. Try both the ratio and root tests. You will find that although the root test immediately gives the expected answer, application of the ratio test is inconclusive as the corresponding limit does not exist.

Theorem 7.9 (The Integral Test[3]) *Consider a series with positive non-increasing terms, i.e.*

$$a_1 \geq a_2 \geq a_3 \geq \cdots \geq a_n \geq a_{n+1} \geq \cdots > 0 .$$

Next let us define a function $y = f(x)$ such that at integer values of the x it would be equal exactly to the terms of the series, i.e. $f(n) = a_n$ for any $n = 1, 2, 3 \ldots$. One can always define such a function by replacing n by x in a_n,

---

[2] Due to Augustin-Louis Cauchy.
[3] Also due to Augustin-Louis Cauchy.

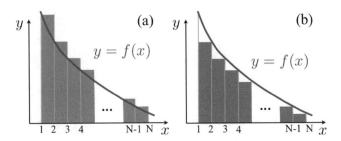

Fig. 7.1 To the proof of the integral convergence test: **a** the integral under the curve between 1 and infinity is smaller than the sum of functions at points 1, 2, 3, etc., while in **b** the integral is larger than the sum of functions at points 2, 3, 4, etc.

i.e. $a_n \to f_n \to f(n) \to f(x)$, and it is clear that this function would be non-increasing. Then the question of whether the series converges can be concluded depending on whether the integral

$$A = \int_1^\infty f(x)dx \qquad (7.5)$$

converges or not.

Proof: Consider the partial sum of our series[4] :

$$S_{N-1} = f(1) + f(2) + f(3) + \cdots + f(N-1) \, .$$

It can be thought of as a sum of $N - 1$ rectangular areas shown in Fig. 7.1a since the corresponding width along the x-axis of each rectangle is exactly equal to one. We also see from the Figure that S_{N-1} will definitely be larger than the area A_N under the curve of the function $f(x)$ between the points 1 and N, i.e.

$$S_{N-1} > A_N = \int_1^N f(x)dx$$

(we can call A_N a partial integral). On the other hand, let us consider the sum

$$S_N - f(1) = f(2) + f(3) + f(4) + \cdots + f(N) \, .$$

This sum can also be interpreted as a total area of the coloured rectangles in Fig. 7.1b, but this time it is obviously smaller than the area under the integral, i.e. $S_N - f(1) =$

---

[4] Note that some ideas of the proof are similar to those we encountered when considering Darboux sums in Sect. 4.2 (specifically, see Fig. 4.2).

$S_N - a_1 < A_N$. Hence, we can write:

$$S_N - a_1 < \int_1^N f(x)dx < S_{N-1} = S_N - a_N , \qquad (7.6)$$

where S_N is the partial sum of N terms of our series. If the integral (7.5) converges, then $A_N < A$ (since the function $f(x) > 0$, the partial integral is an increasing function, i.e. $A_{N+1} > A_N$, and A_N should be bounded from above[5]) and hence $S_N - a_1 < A_N < A$, i.e. $S_N < A + a_1$. This means that the partial sum is bounded from above, and then, according to Theorem 7.5, converges.

Now, let us assume that the integral (7.5) is equal to infinity (diverges). From the second part of the inequality (7.6) it then follows that $S_N > A_N + a_N$. However, the integral diverges and hence its partial integral A_N from some N can be made larger than any positive number. Therefore, S_N as well can be made arbitrarily big, i.e. it does not have a limit. The series diverges. **Q.E.D.**

As an example, let us consider the series with $a_n = 1/n^\alpha = n^{-\alpha}$, where $\alpha > 0$. This series was mentioned above in Problem 7.6 for $\alpha = 2$. Here $f(x) = x^{-\alpha}$ and therefore the convergence of our series depends on whether the integral $\int_1^\infty x^{-\alpha}dx$ converges or diverges. This latter question was answered in Sect. 4.5.3: the integral converges only for $\alpha > 1$ and diverges otherwise. Therefore, the series converges for $\alpha > 1$ and diverges otherwise. In particular, the harmonic series (7.3) corresponding to $\alpha = 1$ diverges, but the series

$$1 + \frac{1}{2^2} + \frac{1}{3^2} + \frac{1}{4^2} + \dots \qquad (7.7)$$

converges ($\alpha = 2$).[6]

Problem 7.10 Show that the root test is inconclusive for the series of inverse squares (7.7). [Hint: *in applying the root test, convert $1/n^2$ into an exponential and then consider the limit of the exponent as $n \to \infty$.*]

Problem 7.11 Using the integral test, prove that the series with $a_n = (n \ln n)^{-1}$ diverges.

Problem 7.12 Using the integral test, prove that the series with $a_n = ne^{-n}$ converges.

---

[5] This follows from the general fact, see Sect. 4.5.3, that improper integrals are understood as limits.
[6] We shall see, e.g. in Chap. 3 of Volume II that this series converges to $\pi^2/6$.

7.1.2 Multiple Series

Theorem 7.9 can also be generalized to multiple series. Suppose, we are given a double series

$$S = \sum_{n=1}^{\infty} \sum_{m=1}^{\infty} a_{m,n}$$

with positive and non-increasing terms $a_{m,n}$, i.e. for any m we have $a_{m,n+1} \leq a_{m,n}$ and for any n we have $a_{m+1,n} \leq a_{m,n}$. Then, the convergence of the series S can be characterized by the convergence of the double integral

$$A = \int_{1}^{\infty} dx \int_{1}^{\infty} dy \, f(x, y),$$

where the function $f(x, y)$ is defined in such a way that $f(m, n) = a_{m,n}$. The proof of this theorem is analogous to the one-dimensional case considered above, but requires working in the 3D space, with the function $f(x, y)$ forming the third dimension. Similarly, one can also consider a triple series

$$S = \sum_{n=1}^{\infty} \sum_{m=1}^{\infty} \sum_{k=1}^{\infty} a_{m,n,k}$$

and characterize its convergence by the appropriate triple integral of a function $f(x, y, z)$ defined on the integer values of its variables by the terms $a_{m,n,k}$ of the series.

As an illustration of this generalization of the integral test, consider the so-called lattice sum that are met in solid state physics:

$$S = \sum_{\mathbf{L} \neq 0} \frac{1}{|\mathbf{L}|^n} = \frac{1}{a^n} \sum_{n_1,n_2,n_3 \, (\neq 0)} \frac{1}{\left(n_1^2 + n_2^2 + n_3^2\right)^{n/2}},$$

where n is a positive integer and $\mathbf{L} = a(n_1\mathbf{i} + n_2\mathbf{j} + n_3\mathbf{k})$ is the lattice vector corresponding to the so-called simple cubic lattice with the lattice constant a. The numbers n_1, n_2 and n_3 run through all possible integers between $-\infty$ and ∞, excluding the zero lattice vector (when all three integers are equal to zero at the same time). Due to symmetry, the lattice sum can be written only via positive values of the integers,

$$S = \frac{2^3}{a^n} \sum_{n_1,n_2,n_3=1}^{\infty} \frac{1}{\left(n_1^2 + n_2^2 + n_3^2\right)^{n/2}}.$$

It is seen that the terms of the series, $a_{n_1,n_2,n_3} = \left(n_1^2 + n_2^2 + n_3^2\right)^{-n/2}$ are all positive and decrease whenever any of their indices is increased. Hence, the series satisfies the conditions of the integral test. Hence, applying the test, we can find the values of n for which this triple series converges. Indeed, consider the triple integral

$$A = \int_1^\infty dx \int_1^\infty dy \int_1^\infty dz \, \frac{1}{\left(x^2 + y^2 + z^2\right)^{n/2}}.$$

We have to verify when this integral converges and when it does not. Here we integrate over the eighth part of the 3D space with positive coordinates from which a cube with sides between 0 and 1 has been cut off. Instead of this integral, it is convenient to consider another one, A', which is taken over the same octant in which, however, the eighth of the sphere of unit radius centred at the origin has been cut off. Obviously, the convergence (or divergence) of both integrals is the same because they differ merely by a finite constant. Then, using the spherical coordinates (r, θ, ϕ), we obtain:

$$A' = \int_1^\infty r^2 dr \int_0^{\pi/2} \sin \theta d\theta \int_0^{\pi/2} d\phi \, \frac{1}{r^n} = \frac{\pi}{2} \int_1^\infty r^{2-n} dr.$$

Obviously, the integral diverges at infinity for $2 - n \geq -1$, i.e. for $n \leq 3$. It converges for $n \geq 4$.

7.1.3 Euler-Mascheroni Constant

We mentioned above that the harmonic series (7.3) diverges. Interestingly, the closely related numerical sequence

$$u_n = 1 + \frac{1}{2} + \frac{1}{3} + \cdots + \frac{1}{n} - \ln n \tag{7.8}$$

has a certain limit which is called Euler-Mascheroni constant:

$$\gamma = \lim_{n \to \infty} \left(1 + \frac{1}{2} + \frac{1}{3} + \cdots + \frac{1}{n} - \ln n \right) = 0.5772\ldots . \tag{7.9}$$

To show that the sequence u_1, u_2, etc. converges, let us consider the function $y = 1/x$ for $1 \leq x \leq n$. At integer values $x_k = k$, where $k = 1, \ldots, n$, the function $y(x_k) = 1/k$. Let us calculate the area under the curve between $x = 1$ and $x = n$ using two methods, similarly to what we did in the previous Section. Indeed, the exact area under the curve is obviously

$$S_n = \int_1^n y(x)dx = \int_1^n \frac{dx}{x} = \ln n .$$

By choosing rectangles which go above the curve, see Fig. 7.2, and summing up their areas (note that the base of each rectangular is equal exactly to one), we shall construct an approximation,

$$S_n^+ = 1 + \frac{1}{2} + \frac{1}{3} + \cdots + \frac{1}{n-1} ,$$

Fig. 7.2 The area under the
curve $y = 1/x$ can be
approximated either by
rectangles which go above or
below the curve (cf. Fig. 7.1)

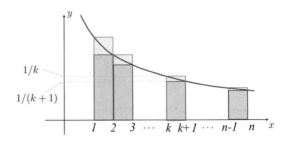

which is larger than the actual area by

$$v_n = S_n^+ - S_n = 1 + \frac{1}{2} + \frac{1}{3} + \cdots + \frac{1}{n-1} - \ln n \equiv u_n - \frac{1}{n}.$$

Obviously, v_n is positive and forms an increasing sequence as positive extra areas
are always added due to new rectangles when n is increased.

On the other hand, the sequence $\{v_n\}$ is limited from above. Indeed, by choosing
the lower rectangles, we can construct an approximation to the area

$$S_n^- = \frac{1}{2} + \frac{1}{3} + \cdots + \frac{1}{n} < S_n = \ln n \implies S_n^- - \ln n < 0.$$

Since

$$v_n = 1 + \underbrace{\frac{1}{2} + \frac{1}{3} + \cdots + \frac{1}{n-1} + \frac{1}{n}}_{S_n^-} - \frac{1}{n} - \ln n = 1 - \frac{1}{n} + \left(S_n^- - \ln n \right),$$

we therefore can write:

$$v_n < 1 - \frac{1}{n} < 1.$$

Therefore, according to Theorem 7.5 the sequence v_n has a limit. Since $u_n = v_n + 1/n$, both sequences have the same limit, which is the required Euler-Mascheroni
constant. Its numerical value is given above.

7.1.4 Alternating Series

So far, we have considered series with all terms positive; this consideration can
of course be extended for the series with all terms negative. Before we touch upon
general series where the sign of terms may change in a general way, it is instructive to
consider another special case of numerical series in which the sign of terms alternates,
i.e.

$$a_1 + a_2 + a_3 + \cdots + a_n + \ldots = |a_1| - |a_2| + |a_3| - \ldots \pm |a_n| \mp \ldots , \quad (7.10)$$

i.e. $a_n = (-1)^{n+1} |a_n|$ (assuming, without loss of generality, that $a_1 > 0$).

For this type of series there exists a very simple sufficient convergence test which
we formulate as the following theorem:

> **Theorem 7.10** [7] *For the series (7.10) to converge, it is sufficient that its terms do not increase in their absolute values, i.e.*
>
> $$|a_1| \geq |a_2| \geq |a_3| \geq \cdots \geq |a_n| \geq \cdots > 0 \,,$$
>
> *and also* $|a_n| \to 0$ *when* $n \to \infty$.

Proof: Assume that $a_1 > 0$ and hence all odd terms in the series are positive; correspondingly, all even terms are negative. Then, consider a partial sum with an even number of terms:

$$S_{2N} = a_1 + a_2 + \cdots + a_{2N} = (a_1 - |a_2|) + (a_3 - |a_4|) + \cdots + \left(a_{2N-1} - |a_{2N}|\right) \,. \tag{7.11}$$

Since terms in the series do not increase in their absolute value, each difference within the round parentheses is positive, and hence the partial sums S_{2N} form an increasing numerical sequence. On the other hand, one can also write S_{2N} as follows:

$$S_{2N} = a_1 - (|a_2| - a_3) - (|a_4| - a_5) - \cdots - (|a_{2N-2}| - a_{2N-1}) - |a_{2N}|$$
$$= a_1 - [\underbrace{(|a_2| - a_3)}_{\geq 0} + \underbrace{(|a_4| - a_5)}_{\geq 0} + \cdots + \underbrace{(|a_{2N-2}| - a_{2N-1})}_{\geq 0} + |a_{2N}|] < a_1 \,,$$

since the expression in the square brackets is definitely positive (since $|a_{2N}| > 0$). Hence, S_{2N} is bounded from above by its first element, and, therefore, the numerical sequence of S_{2N} has a limit.

We still need to consider the partial sum of an odd number of terms:

$$S_{2N+1} = S_{2N} + a_{2N+1} \implies \lim_{N \to \infty} S_{2N+1} = \lim_{N \to \infty} S_{2N} + \lim_{N \to \infty} a_{2N+1} = \lim_{N \to \infty} S_{2N} \,,$$

since the terms in the series tend to zero as their number goes to infinity according to the second condition of the theorem. Since both partial sums have their limits and these coincide, this must be the limit of the whole numerical series. **Q.E.D.**

It also follows from the above proof that the overall sign of the sum of the series $S = \lim_{N \to \infty} S_N$ is entirely determined by the sign of its first term, a_1. Indeed, assuming that $a_1 > 0$, we have from Eq. (7.11) that $S_{2N} > 0$ as the sum of positive terms. Since the sequences S_{2N} and S_{2N+1} converge to the same limit (if that limit exists), this proves the statement that was made. If $a_1 < 0$, then we similarly have the sum $S < 0$.

---

[7] Due to Leibniz.

Theorem 7.11 *Consider again the alternating series from the previous theorem.*
Let us assume that all terms starting from $N + 1$ are dropped. Then, the absolute
value of the sum of all dropped terms in the series does not exceed the first
dropped term.

Proof: Let us consider the first N terms of the series. The sum of all dropped terms
is

$$S_{>N} = a_{N+1} + a_{N+2} + \dots$$
$$= |a_{N+1}| - \underbrace{(|a_{N+2}| - |a_{N+3}|)}_{\geq 0} - \underbrace{(|a_{N+4}| - |a_{N+5}|)}_{\geq 0} - \dots$$
$$< |a_{N+1}| = a_{N+1} ,$$

where we have assumed that the first dropped term, a_{N+1}, is positive. The inequality
holds since all expressions in the round brackets are positive. Since the first term,
a_{N+1}, is positive, then we can also write:

$$S_{>N} = \underbrace{(a_{N+1} - |a_{N+2}|)}_{\geq 0} + \underbrace{(a_{N+3} - |a_{N+4}|)}_{\geq 0} + \dots > 0 ,$$

i.e. we obtain $0 < S_{>N} < a_{N+1}$.

If the first dropped term is negative instead, we similarly write:

$$S_{>N} = a_{N+1} + a_{N+2} + \dots$$
$$= -|a_{N+1}| + \underbrace{(|a_{N+2}| - |a_{N+3}|)}_{\geq 0} + \underbrace{(|a_{N+4}| - |a_{N+5}|)}_{\geq 0} + \dots$$
$$> -|a_{N+1}| = a_{N+1} ,$$

and also

$$S_{>N} = -\underbrace{(|a_{N+1}| - |a_{N+2}|)}_{\geq 0} - \underbrace{(|a_{N+3}| - |a_{N+4}|)}_{\geq 0} + \dots < 0 .$$

Therefore, $a_{N+1} < S_{>N} < 0$.

Combining both cases, we can write that $|S_{>N}| < |a_{N+1}|$, as required. **Q.E.D.**

As an example, we shall consider the series

$$1 - \frac{1}{2} + \frac{1}{3} - \frac{1}{4} + \dots = \sum_{n=1}^{\infty} \frac{(-1)^{n+1}}{n} . \tag{7.12}$$

This series looks like the harmonic series (7.3), however, the signs of the terms in the series above alternate (it is sometimes called the alternating harmonic series). We recall that the harmonic series diverges. However, the series (7.12) actually converges as it fully satisfies the conditions of Theorem 7.10: $1/n > 1/(n+1)$ and $(-1)^{n+1}/n \to 0$ when $n \to \infty$. As will be shown in Sect. 7.3.3, the series (7.12) corresponds to Taylor's expansion of $\ln(1+x)$ at $x = 1$ and hence converges to $\ln 2$ (see also discussion in Sect. 7.3.2).

7.1.5 General Series: Absolute and Conditional Convergence

Here we shall consider a general infinite series in which signs of its terms neither are the same nor they are alternating:

$$a_1 + a_2 + a_3 + \ldots = \sum_{n=1}^{\infty} a_n . \qquad (7.13)$$

Theorem 7.12 (Sufficient Criterion) *If the series*

$$A \equiv |a_1| + |a_2| + |a_3| + \ldots = \sum_{n=1}^{\infty} |a_n| , \qquad (7.14)$$

constructed from absolute values of the terms of the original series, converges, then so does the original series.

Proof: Consider the first N terms of the series (7.13) and (7.14) with the corresponding partial sums being S_N and A_N. Terms of the series (7.13) may have different signs. Let us collect all positive terms into the partial sum $S'_N > 0$ and the *negative* values of all negative terms into the partial sum $S''_N > 0$. Obviously, $S_N = S'_N - S''_N$ and $A_N = S'_N + S''_N$. Positive terms form a subsequence of the whole sequence with the partial sum S'_N. Similarly, negative values of the negative terms also form a subsequence with the partial sum S''_N. Since we assume that the series (7.14) converges, i.e. A_N has a limit, then both partial sums S'_N and S''_N must be bounded from above (both contain positive terms): $S'_N < A_N < A$ and $S''_N < A_N < A$. Therefore, both partial sums have their limits: $S'_N \to S'$ and $S''_N \to S''$ (see also Theorem 2.4 on subsequences). Therefore, there is also a well-defined limit for the original series: $\lim_{N \to \infty} S_N = S' - S''$. **Q.E.D.**

Problem 7.13 Prove that the series

$$1 - \frac{1}{2^\alpha} + \frac{1}{3^\alpha} - \cdots = \sum_{n=1}^{\infty} \frac{(-1)^{n+1}}{n^\alpha}$$

converges (absolutely) for any value of $\alpha > 1$.

Convergence tests considered in Sect. 7.1.1 can now be applied to general series with the only change that absolute values of the terms of the series are to be used. Indeed, according to the ratio test, the series of absolute values (7.14) converges if

$$\lim_{n\to\infty} \frac{|a_{n+1}|}{|a_n|} = \lim_{n\to\infty} \left| \frac{a_{n+1}}{a_n} \right| < 1$$

and diverges if the limit of the ratio of the absolute values is larger than one. Similarly for the root test: the series converges if $\sqrt[n]{|a_n|} < 1$, and diverges if the root is larger than one. However, convergence of the series (7.14) guarantees the convergence of the original series (7.13), and this proves the statement made.

Problem 7.14 Using the ratio test, show that the series $\sum_{n=1}^{\infty} x^n/n$ converges for $|x| < 1$ and diverges for $|x| > 1$.

Problem 7.15 Using the ratio test, show that the series with $a_n = x^n/(n^2 + n)$ converges for $|x| < 1$ and diverges for $|x| > 1$.

Theorem 7.12 provides only a sufficient condition for convergence. If a series converges together with the series of its absolute values, it is said that the series converges *absolutely*. However, the series may converge, but the corresponding series of absolute values of its terms may diverge. In this case it is said that the original series converges *conditionally*. For instance, the series (7.12) converges conditionally. This distinction has a very important meaning based on the following

Theorem 7.13 *If a series converges absolutely, terms in the series may be permuted arbitrarily without affecting the value of the sum.*

Proof: Consider first a series $\sum_n a_n$ of positive terms $a_n > 0$ which partial sum, A_N, converges to A, i.e. $\lim_{N \to \infty} A_N = A$. Let us arbitrarily permute an *infinite* number of terms in the series.[8] The new series created in this way, $\sum_n b_n$, consists of elements $b_n > 0$. We stress again that these are the same as the elements of the first series, but positioned in a different order.

Consider the first N terms in the second series, its partial sum we shall denote B_N. No matter how big N is, all first N terms b_n with $n \leq N$ of the second series will be contained within the first M terms of the first series, where M is some integer. Since amongst the M terms there will most likely be some other terms as well, and all elements of the series are positive, one can write: $B_N \leq A_M$. But the first series converges to A, and hence $A_M \leq A$, i.e. we conclude that $B_N \leq A$, i.e. the partial sum of the second series is bounded from above and hence converges to some value $B \leq A$.

Conversely, let us consider the first N' elements of the first series; the second series would contain these elements within its first M' terms, and hence $A_{N'} \leq B_{M'}$. Since we already established that the second series converges to B, then $B_{M'} \leq B$, and hence $A_{N'} \leq B_{M'} \leq B$. This yields $A \leq B$. Consequently, $A = B$. So, we conclude that if all terms are positive,[9] then one can indeed permute terms in the series, and the new series created in this way would converge to the same sum.

It is also clear that if there are two series with non-negative elements, one can sum up or subtract them from each other term-by-term, i.e. $A \pm B = \sum_n (a_n \pm b_n)$, since this would simply correspond to a permutation of elements of the combined series $\sum_n a_n \pm \sum_n b_n = A \pm B$.

Now we consider an absolutely converging series $\sum_n a_n$ with terms a_n of arbitrary signs. We permute an infinite number of its elements arbitrarily and obtain a new series $\sum_n b_n$. Two auxiliary series, both containing positive terms, can then be constructed out of it:

$$S_1 = \sum_n \frac{1}{2} (b_n + |b_n|) \quad \text{and} \quad S_2 = \sum_n \frac{1}{2} (|b_n| - b_n) \ . \qquad (7.15)$$

Since either of the series is built from only positive (more precisely, non-negative) elements, we can apply to them the statements made above, i.e. we can permute their elements arbitrarily without affecting their sums, and manipulate them. In particular, $S_1 - S_2 = \sum_n b_n$.

Next, we shall permute the elements of the b_n-series precisely in the opposite way to the one we have used initially when deriving the b_n-series from the a_n-series; then the order of terms would correspond to the original a_n series. Two new series of non-negative terms can then be constructed,

$$S_1' = \sum_n \frac{1}{2} (a_n + |a_n|) \quad \text{and} \quad S_2' = \sum_n \frac{1}{2} (|a_n| - a_n) \ ,$$

---

[8] Obviously, if a finite number of terms is permuted, these terms can be removed from the series without affecting its convergence.

[9] In fact, of the same sign.

which contain the same elements as above in Eq. (7.15), but in different order. Since the terms are non-negative, $S'_1 = S_1$ and $S'_2 = S_2$. Subtracting the two series term-by-term yields $S_1 - S_2 = \sum_n a_n$, which is the same as $\sum_n b_n$. **Q.E.D.**

A simple corollary from this theorem is that two absolutely converging series can be algebraically summed or subtracted from each other term-by-term. Indeed, this would simply correspond to permuting terms in the absolutely converging series:

$$A \pm B = \sum_n a_n \pm \sum_n b_n = \sum_n (a_n \pm b_n) \ .$$

Another simple corollary is that one can collect all positive and negative terms of an absolutely converging series into separate sums $A_> = \sum_n^> a_n$ and $A_< = \sum_n^< a_n = -\sum_n^< |a_n|$. Here $A_<$ includes all negative terms, while $A_>$ all positive; both sums converge independently.

One can also show that it is possible to multiply the two series term-by-term yielding a series converging to the product of the two sums, i.e.

$$\left(\sum_n a_n \right) \left(\sum_n b_n \right) \equiv a_1 b_1 + a_2 b_1 + a_1 b_2 + a_1 b_3 + a_2 b_2 + a_3 b_1 + \ldots = AB \ ,$$

$$(7.16)$$

as stated by the following theorem:

Theorem 7.14 *One can multiply terms of two absolutely converging series $A = \sum_n a_n$ and $B = \sum_n b_n$ term-by-term, and the resulting series would converge to the product of the two sums, AB.*

Proof: Let us start by assuming that both series consist of positive elements only. Consider a particular order in which two series are multiplied, e.g. the one in Eq. (7.16), and let us call the elements of the product series c_n, i.e. $c_n = a_k b_l$ with some indices k and l of the first and the second series, respectively. The new series converges to $C = AB$. Indeed, consider first N elements of the product series with the partial sum C_N. We can always find a number M such that all elements a_k and b_l we find in C_N are contained in the partial sums A_M and B_M. However, either of the latter sums may contain extra elements, i.e. $C_N \leq A_M B_M \leq AB$. The last inequality follows from the convergence of the two series, i.e. $A_N \leq A$ and $B_N \leq B$. Since C_N is bounded from above, it converges to $C \leq AB$.

Conversely, let us consider now two partial sums $A_{N'}$ and $B_{N'}$ with the first N' elements of the two original series, and construct the product $A_{N'} B_{N'}$ of all their elements. We can always find a number M' of terms in the product series (probably large) such that the latter contains all terms from the product $A_{N'} B_{N'}$. However, it would also contain other elements, which means that $A_{N'} B_{N'} \leq C_{M'} \leq C$. This yields $AB \leq C$. Combining the two inequalities we get $AB = C$.

Now consider two absolutely converging series $A = \sum_n a_n$ and $B = \sum_n b_n$ with general elements (i.e. of any sign), and let us construct a product series $\sum_n c_n$. According to the above proof, the sum of its absolute values converges to $|A| \cdot |B| = |AB|$, which means that the original product series converges absolutely. Therefore, we can collect all positive and negative terms in the three series into the sums $A_<, A_>, B_<, B_>, C_<$ and $C_>$. Since the product series converges absolutely, according to the previous Theorem 7.13, we are allowed to permute its terms arbitrarily without affecting the sum, i.e. the following manipulation is legitimate:

$$\sum_n c_n = \sum_k a_k \left(\sum_l b_l \right) = \sum_k a_k \left(B_> + B_< \right) = \sum_k a_k B = B \sum_n a_n = BA \; ,$$

as required. **Q.E.D**

Therefore, we have proved that if two series converge absolutely to the values S_1 and S_2, then: (i) a series obtained by summing (or subtracting) the corresponding terms of the two series converges absolutely to $S_1 + S_2$ (or, correspondingly, $S_1 - S_2$); (ii) a series obtained by multiplication of the two series converges absolutely to $S_1 S_2$; (iii) the order of terms in a series can be changed arbitrarily without affecting its sum. In other words, absolutely converging series can be *manipulated algebraically*.

Interestingly, this is not the case for conditionally converging series. Moreover, as was proven by Riemann, by permuting terms in the series in a specific way one can obtain a series converging to any given value; the series may even become diverging. To illustrate the point of the dependence of the sum of a conditionally converging series on the order of terms in it, consider the series (7.12). Let us reorder the terms in it in such a way that after each positive term we shall put two negative ones picked up in the correct order:

$$\begin{aligned}
S &= 1 - \frac{1}{2} + \frac{1}{3} - \frac{1}{4} + \ldots \\
&= \left(1 - \frac{1}{2} - \frac{1}{4} \right) + \left(\frac{1}{3} - \frac{1}{6} - \frac{1}{8} \right) + \left(\frac{1}{5} - \frac{1}{10} - \frac{1}{12} \right) + \left(\frac{1}{7} - \frac{1}{14} - \frac{1}{16} \right) + \ldots \\
&= \left(\frac{1}{2} - \frac{1}{4} \right) + \left(\frac{1}{6} - \frac{1}{8} \right) + \left(\frac{1}{10} - \frac{1}{12} \right) + \left(\frac{1}{14} - \frac{1}{16} \right) + \ldots \\
&= \frac{1}{2} \left(1 - \frac{1}{2} + \frac{1}{3} - \frac{1}{4} + \frac{1}{5} - \frac{1}{6} + \frac{1}{7} - \frac{1}{8} + \ldots \right) = \frac{1}{2} S \; ,
\end{aligned}$$

i.e. S can be made equal to the half of itself. Therefore, conditionally converging series should be considered with great care.

The Coulomb potential of a lattice of point charges (the so-called Hartree potential) is an example of a conditionally converging series. Indeed, consider periodically arranged atoms with charges $Q > 0$ and $-Q$ along a single dimension as shown in Fig. 7.3. The Coulomb potential at some point x away from the charges is

$$V_{Coul}(x) = \sum_{n=-\infty}^{+\infty} \left[\frac{Q}{|x - an - X_+|} - \frac{Q}{|x - an - X_-|} \right] \; ,$$

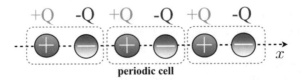

+Q   -Q   +Q   -Q   +Q   -Q

periodic cell

Fig. 7.3 A one-dimensional infinite periodic crystal of alternating point charges. The infinite crystal can be generated by repeating the same unit cell (indicated) indefinitely in both directions. Each such periodic cell contains two oppositely charged atoms with charges $Q > 0$ and $-Q < 0$. The total charge of the unit cell is zero and there is an infinite number of such cells running along the positive and negative directions of the x-axis

where we sum over all unit cells using the unit cell number n, a is the cell length (the distance between nearest charges of the same sign) and X_+ and X_- are positions of the positive and negative atoms, respectively, within the zero cell (with $n = 0$). This series does not converge absolutely. Indeed, the series with all terms positive would correspond to an infinitely long one-dimensional array of charges which are all positive, and hence on physical grounds the potential of such an infinite chain of positive atoms would be infinite. However, the potential of positive and negative charges does converge conditionally, although the result would depend on the order of the terms in the sum above.

Problem 7.16 Consider the Coulomb potential

$$V_{Coul}(x) = \sum_{n=-\infty, n\neq 0}^{+\infty} \left[\frac{Q}{|an + X_+|} - \frac{Q}{|an + X_-|} \right],$$

on the plus atom of the 0-th cell of the 1D lattice shown in Fig. 7.3, i.e. at $x = 0$; we have also excluded the $n = 0$ term from the sum corresponding to removing that atom from producing the potential on itself (which would be an infinity and does not make sense either!). Show that a certain choice of the parameters X_+ and X_- results in the potential to be directly related to the alternating harmonic series of Eq. (7.12).

7.2   Functional Series: General

Now we shall turn our attention to specific infinite series, called functional series, in which the general form of the n-th term a_n is a function $a_n(x)$ of a real variable x defined in some interval $a < x < b$. Correspondingly, as each term of the series is a function of x, i.e. $a_n(x)$, the sum of the series (if exists) will also become a function of x:

$$S(x) = a_0(x) + a_1(x) + a_2(x) + \ldots = \sum_{n=0}^{\infty} a_n(x) \ . \qquad (7.17)$$

For a given value of x the series above is just an infinite numerical series we studied in the previous section. The difference now is that we deal with not just a single value of the x or even its finite set of values, but with x defined continuously in the interval $a < x < b$. Therefore, we shall be interested in the obvious questions related to the series being a function, such as at which values of x the series converges and, if it does, whether the sum of the series is a continuous function. But the range of questions is much wider than that in fact. Can one take a limit inside the sign of the sum, or, in other words, would the limit of the sum be equal to the sum of the limits? Can one integrate or differentiate the series term-by-term, i.e. would the new series generated that way converge to a function obtained from the sum $S(x)$ after integration or differentiation? Finally, we shall consider a special and very important class of functional series called power series and show how functions can be expanded into such series, the so-called Taylor series. This point would conclude our discussion of the Taylor theorem stated in Sect. 3.8. We shall only consider here the series which are functions of a single variable. The treatment of this simple case can be extended to functions of several variables almost without change.

7.2.1 Uniform Convergence

Because functional series have a parameter, x, their convergence may be different for different values of it; for some x the series may converge faster, for some other values slower. It is also possible that convergence within some interval of x is uniform, i.e. it does not depend on x (cf. Sect. 6.1.3). This class of series has a very special significance as for these series all the questions asked above can be answered satisfactorily. Therefore, we shall only be considering this case in what follows.

But before, let us recall the definition of the convergence of a numerical series (see Sect. 2.2.3): it is said that the series $\sum_n a_n$ converges to S if for any positive ϵ one can always find a number $\mathcal{N}$ such that for any $N > \mathcal{N}$ the partial sum S_N would be different from S by no more than ϵ, i.e. $|S_N - S| < \epsilon$. As in our case the terms of the series depend on x, for the given ϵ the number $\mathcal{N}$ may be different for different values of x; in other words, if we would like to approximate our series by its partial sum, different minimum number of terms $\mathcal{N}$ need to be taken for different values of x to achieve the same "precision" ϵ, i.e. $\mathcal{N} \implies \mathcal{N}(x)$. If the convergence is *uniform*, however, then the same $\mathcal{N}$ can be used for *all values* of x, i.e. $\mathcal{N}$ does not depend on x. This is the idea of this essential new notion.

To illustrate this important concept, let us consider an alternating series

$$S(x) = \sum_{n=1}^{\infty} \frac{(-1)^n}{x + n}$$

for $x > 0$. It is known from Theorem 7.10 that this series converges; moreover, if we consider the first N terms, the sum of all dropped terms would not exceed the first dropped term (Theorem 7.11), i.e.

$$|S(x) - S_N(x)| \leq \left| \frac{1}{x + N + 1} \right| \leq \frac{1}{N + 1} < \frac{1}{N} \, ,$$

where we have made use of the fact that $x > 0$. The above means that if we take $\epsilon = 1/N$, then for any $N > \mathcal{N}$ we should have $|S(x) - S_N(x)| \leq 1/(N + 1) < 1/\mathcal{N} = \epsilon$. We see that the same value of $\mathcal{N}$ can be chosen for any $x > 0$. Therefore, this series converges uniformly in the interval $0 < x < \infty$.

As another example, we consider a geometric progression

$$\sum_{n=0}^{\infty} x^n = \frac{1}{1 - x} \, ,$$

which converges for all x between -1 and 1 (excluding the boundary points). Consider the first N terms of it. The residue (the sum of all dropped terms) is

$$S(x) - S_N(x) = \sum_{n=N+1}^{\infty} x^n = x^{N+1} \sum_{n=0}^{\infty} x^n = \frac{x^{N+1}}{1 - x} \, .$$

If we now require that the residue is smaller than ϵ, it should be clear that this would not be possible to achieve for all values of x between -1 and 1 since near $x = 1$ the residue steeply increases to infinity. In this case, therefore, the series converges non-uniformly.

Uniform convergence is sufficient in establishing a number of important properties of the functional series. But before discussing this, we shall provide one simple test for the uniform convergence.

Theorem 7.15 [10] *If all terms $a_n(x)$ of a functional series for all values of x within some interval satisfy the inequality*

$$|a_n(x)| \leq c_n \, ,$$

where numbers $c_n > 0$ ($n = 1, 2, 3, \ldots$) form a converging numerical series, then the functional series converges uniformly for all values of x from this interval.

---

[10] Due to Weierstrass.

Proof: Since the series $\sum_n c_n$ converges, it means that for any $\epsilon > 0$ one can always find $\mathcal{N}$ such that for any $N > \mathcal{N}$ the residue of the series,

$$\Delta C_N = \sum_{n=N+1}^{\infty} c_n \, ,$$

satisfies $|\Delta C_N| < \epsilon$. Consider now the absolute value of the residue of our functional series for an arbitrary x within the given interval:

$$|\Delta S_N(x)| = |S(x) - S_N(x)| = \left| \sum_{n=N+1}^{\infty} a_n(x) \right|$$

$$\leq \sum_{n=N+1}^{\infty} |a_n(x)| < \sum_{n=N+1}^{\infty} c_n \equiv \Delta C_N < \epsilon \, ,$$

where we have used the fact here that the terms of the numerical series are positive and hence so is their residue ΔC_N. Note that here we chose $\mathcal{N}$ by the numerical series $\sum_n c_n$, and hence it is the same for all values of x from the interval. Hence, we established that for the given ϵ the residue of the functional series is less than ϵ for any $N > \mathcal{N}$ with $\mathcal{N}$ not depending on x, and this, according to our definition of the uniform convergence, proves the theorem. **Q.E.D.**

In the so-called Fourier series $a_n(x)$ is given either as $\alpha_n \cos \frac{\pi n x}{l}$ or $\beta_n \sin \frac{\pi n x}{l}$, where numbers α_n and β_n form some numerical sequences, and l is a positive constant. The above theorem then establishes the uniform convergence of the two series based on the convergence of the series composed of the pre-factors α_n and β_n, since

$$\left| \alpha_n \cos \frac{\pi n x}{l} \right| \leq |\alpha_n| \quad \text{and} \quad \left| \beta_n \sin \frac{\pi n x}{l} \right| \leq |\beta_n| \, .$$

It also follows from Theorem 7.15 that the series composed of the absolute values of the terms of the original series, $|a_n(x)|$, would also converge uniformly; moreover, the given series $\sum_n a_n(x)$ would converge absolutely.

7.2.2 Properties: Continuity

Here we shall consider an important question about the continuity of the sum $S(x)$ of a uniformly converging series $\sum_n a_n(x)$.

Theorem 7.16 *Consider a uniformly converging series $\sum_n a_n(x)$ in the interval $a < x < b$. If the terms $a_n(x)$ of the series are continuous functions at a point x_0 belonging to the interval, then the sum of the series $S(x)$ is also a continuous function at this point.*

Proof: Let $\Delta S_N(x) = S(x) - S_N(x)$ be the residue of the series. Then,

$$
\begin{aligned}
|S(x) - S(x_0)| &= |S_N(x) - S_N(x_0) + \Delta S_N(x) - \Delta S_N(x_0)| \\
&\leq |S_N(x) - S_N(x_0)| + |\Delta S_N(x)| + |\Delta S_N(x_0)| .
\end{aligned}
\tag{7.18}
$$

Since the series converges uniformly in the interval of interest (which includes the point x_0), then one can always find such N for the given $\epsilon > 0$ that $|\Delta S_N(x)| < \epsilon/3$ for all x between a and b, including x_0. Each of the functions $a_n(x)$ is assumed to be continuous by the conditions of the theorem, and so is their partial sum. Therefore, for any $\epsilon/3$ one should be able to find such $\delta > 0$ so that from $|x - x_0| < \delta$ follows $|S_N(x) - S_N(x_0)| < \epsilon/3$. Combining all these estimates, we can manipulate the inequality (7.18) into:

$$
\begin{aligned}
|S(x) - S(x_0)| &\leq |S_N(x) - S_N(x_0)| + |\Delta S_N(x)| + |\Delta S_N(x_0)| \\
&< \frac{\epsilon}{3} + \frac{\epsilon}{3} + \frac{\epsilon}{3} = \epsilon ,
\end{aligned}
$$

which proves the continuity of the sum of the series at point x_0. **Q.E.D.**

It is worth noting that the condition of the uniform convergence of the series on the interval $a < x < b$ was essential for the proof of the theorem.

The other question, closely related to the notion of continuity of the functional series, is the question of the limit: can one interchange the limit of the series at $x \to x_0$ with the sign of the sum? In other words, we inquire when the limit of the sum $S(x)$ is equal to the sum of the limits taken term-by-term:

$$
\lim_{x \to x_0} \sum_n a_n(x) = \sum_n \lim_{x \to x_0} a_n(x) .
\tag{7.19}
$$

This question is addressed by the following theorem proof of which is very similar to the one we have just considered.

Theorem 7.17 *Suppose the series $\sum_n a_n(x)$ converges uniformly to a function $S(x)$ on the interval $a < x < b$, and the terms of the series, $a_n(x)$, have well defined limits at $x \to x_0$,*

$$\lim_{x \to x_0} a_n(x) = b_n .$$

Then $\lim_{x \to x_0} S(x) = B$, where B is the sum of the converging series $\sum_n b_n$. Basically, Eq. (7.19) is true.

Proof: We shall first prove one statement of the theorem, namely, that the series $\sum_n b_n$ converges. We shall weaken the proof a bit by assuming that the uniform convergence is guaranteed by an auxiliary series of numbers c_n such that $|a_n(x)| \le c_n$ (Weierstrass test). Taking the limit $x \to x_0$ in this inequality and employing the fact that any function $a_n(x)$ has the limit, we get $|b_n| \le c_n$. Hence, since the series $\sum_n c_n$ converges, the series $\sum_n b_n$ converges as well and converges absolutely.

Let $B = B_N + \Delta B_N$ be the sum of the latter series with the corresponding partial sum B_N. Then,

$$\begin{aligned}
|S(x) - B| &= |S_N(x) - B_N + \Delta S_N(x) - \Delta B_N| \\
&\le |S_N(x) - B_N| + |\Delta S_N(x)| + |\Delta B_N| .
\end{aligned} \tag{7.20}$$

Because of the uniform convergence, we can state that $|\Delta S_N(x)| < \epsilon/3$ for all x; since the $\sum_n b_n$ series converges, then one can always find such N that $|\Delta B_N| < \epsilon/3$. Finally, because the functions $a_n(x)$ all have a well-defined limit at $x \to x_0$, then for any ϵ one can always find such $\delta > 0$ that from $|x - x_0| < \delta$ follows $|a_n(x) - b_n| < \epsilon/3N$. Therefore,

$$\begin{aligned}
|S_N(x) - B_N| &= \left| \sum_{n=1}^{N} a_n(x) - \sum_{n=1}^{N} b_n \right| = \left| \sum_{n=1}^{N} [a_n(x) - b_n] \right| \\
&\le \sum_{n=1}^{N} |a_n(x) - b_n| < \sum_{n=1}^{N} \frac{\epsilon}{3N} = \frac{\epsilon}{3} .
\end{aligned}$$

Collecting all results in Eq. (7.20), we finally obtain that from $|x - x_0| < \delta$ follows:

$$|S(x) - B| \le |S_N(x) - B_N| + |\Delta S_N(x)| + |\Delta B_N| < \frac{\epsilon}{3} + \frac{\epsilon}{3} + \frac{\epsilon}{3} = \epsilon . \tag{7.21}$$

meaning that the series converges to B at $x \to x_0$. **Q.E.D.**

7.2.3   Properties: Integration and Differentiation

Here we shall investigate operations of term-by-term integration and differentiation of the uniformly converging series and will show that these operations are in fact legitimate.

Theorem 7.18 *Consider a uniformly converging series $S(x) = \sum_n a_n(x)$ on the interval $a < x < b$. Then, the integral of the sum is equal to the sum of the integrals of each individual term,*

$$\int_c^d S(x)\,dx = \sum_n \int_c^d a_n(x)dx \; , \tag{7.22}$$

where c and $d > c$ lie within the interval between a and b.

Proof: Since the series converges, we can write:

$$S(x) = \sum_{n=1}^{N} a_n(x) + \Delta S_N(x) \; .$$

This equality contains a finite number of terms and hence can be integrated between c and d term-by-term:

$$\int_c^d S(x)dx = \sum_{n=1}^{N} \int_c^d a_n(x)dx + \int_c^d \Delta S_N(x)dx \; . \tag{7.23}$$

On the other hand, since the series converges uniformly on the interval $a < x < b$ (which includes completely the integration interval), one can always find such N that $|\Delta S_N(x)| < \epsilon$ for all x. Therefore, we can estimate the last integral as follows:

$$\left| \int_c^d \Delta S_N(x)dx \right| \leq \int_c^d |\Delta S(x)|\,dx < \int_c^d \epsilon dx = \epsilon\,(d-c) \equiv \epsilon' \; .$$

This means that the integral of the residue $\Delta S_N(x)$ tends to zero as $N \to \infty$ by the definition of the limit. This proves that at this limit the last term in Eq. (7.23) tends to zero and hence the integral of the sum is equal to the sum of the integrals, i.e. the integration and summation can be interchanged in the case of the uniformly converging series. **Q.E.D.**

Now we shall prove a similar statement for differentiation.

Theorem 7.19 *Consider a uniformly converging series* $S(x) = \sum_n a_n(x)$ *on the interval* $a < x < b$, *and let us assume that functions* $a_n(x)$ *have continuous derivatives* $a'_n(x)$. *If the series containing derivatives of the terms of the original series,* $\sum_n a'_n(x)$, *converges and converges also uniformly, then the sum* $S(x)$ *of the original series has a well-defined derivative*

$$S'(x) = \sum_n a'_n(x) . \tag{7.24}$$

Proof: The series $\sum_n a'_n(x)$ converges uniformly and hence its sum, $S_1(x)$, will be a continuous function (Theorem 7.16). According to Theorem 7.18, we can integrate the series term-by-term between c and some $x > c$, both lying somewhere between a and b:

$$\int_c^x S_1(x')\,dx' = \sum_n \int_c^x a'_n(x')\,dx' \implies \int_c^x S_1(x')\,dx' = \sum_n [a_n(x) - a_n(c)] ,$$

where above we calculated the integral of $a'_n(x')$ explicitly. Using the fact that the original series of $a_n(x)$ converges to $S(x)$, we can write for the right-hand side of the above equation:

$$\sum_n [a_n(x) - a_n(c)] = \sum_n a_n(x) - \sum_n a_n(c) = S(x) - S(c) ,$$

yielding

$$\int_c^x S_1(x')\,dx' = S(x) - S(c) . \tag{7.25}$$

Let us now differentiate both sides of Eq. (7.25) with respect to x (recall Eq. (4.34) or a more general Eq. (4.68)), and we obtain: $S_1(x) = S'(x)$, as required. **Q.E.D.**

So, uniformly converging functional series can be both integrated and differentiated term-by-term (in the latter case the series of derivatives must also converge uniformly).

Problem 7.17 In quantum statistical mechanics a single harmonic oscillator of frequency ω has an infinite set of discrete energy levels $\epsilon_n = \hbar\omega(n + 1/2)$, where $\hbar = h/2\pi$ is the "h-bar" Planck constant, and the integer $n = 0, 1, 2, \ldots$. The probability for an oscillator to be in the n-th state is given by $\rho_n = Z^{-1}\exp(-\epsilon_n/k_B T)$, where k_B is the Boltzmann's constant, T absolute temperature and

$$Z(\xi) = \sum_{n=0}^{\infty} e^{-\epsilon_n/k_B T}$$

is the so-called partition function Z of the oscillator. First, show that $Z(\xi)$ is given by:

$$Z(\xi) = \sum_{n=0}^{\infty} e^{-\xi(n+1/2)} = \frac{e^{\xi/2}}{e^{\xi} - 1} \,, \quad \text{where} \quad \xi = \hbar\omega/k_B T \,.$$

Next, argue that the series for $Z(\xi)$ converges uniformly with respect to ξ. Therefore, prove the following formula:

$$\sum_{n=0}^{\infty} n e^{-\xi n} = -\frac{d}{d\xi}\left(\sum_{n=0}^{\infty} e^{-\xi n}\right) = \frac{e^{\xi}}{\left(e^{\xi} - 1\right)^2} \,,$$

so that the average number of oscillators in the state n is given by

$$\langle n \rangle = \sum_{n=0}^{\infty} n \rho_n = \left(e^{\xi} - 1\right)^{-1} \,,$$

which is called the Bose–Einstein distribution (statistics).

7.2.4   Uniform Convergence of Improper Integrals Depending on a Parameter

The theory of uniformly converging series that we have developed here is instrumental in discussing a related topic of improper integrals (see Sect. 4.5.3), which depend on the parameter t and whose convergence is also called *uniform*. Consider for definiteness the improper integral $F(t) = \int_a^{\infty} f(x, t)dx$ (with $-\infty < a < \infty$) with the infinite upper limit.

As we have seen in Sect. 4.5.3, the integral converges to $F(t)$ if for any $\epsilon > 0$ one can find such $A > a$ that

$$\left| F(t) - \int_a^A f(x, t)dx \right| = \left| \int_A^{\infty} f(x, t)dx \right| < \epsilon \,. \tag{7.26}$$

Since the function $f(x, t)$ depends on the parameter t, the constants ϵ and A would in general depend on it, $\epsilon = \epsilon(t)$ and $A = A(t)$. If, however, they do not, i.e. one can use the same values of ϵ and A for any value of t within a certain interval, then it is said that the integral converges to $F(t)$ uniformly for the values of t in this interval.

To check if the given improper integral depending on a parameter t converges uniformly or not, one can use the tests developed at the end of Sect. 4.5.3. Namely, the Cauchy test tells us that the integral converges if one can show that $|f(x, t)| < M/x^{\lambda}$ with $\lambda > 1$. Then, the integral would converge uniformly, if the constants M and λ do not depend on t.

Uniformly converging improper integrals realize functions $F(t)$ which have a number of important properties similar to the ones established above for the uniformly converging series. This follows from the fact that it is possible to construct an infinite series associated with the given integral. Indeed, let us choose an arbitrary infinite sequence of numbers

$$a < b_1 < b_2 < \cdots < b_n < \ldots,$$

that tend to infinity, i.e. $\lim_{n \to \infty} b_n = \infty$. Then, the integral can be written as a functional series by splitting its integration interval as follows:

$$\int_a^{\infty} f(x, t)dx = \int_a^{b_1} f(x, t)dx + \int_{b_1}^{b_2} f(x, t)dx + \cdots + \int_{b_n}^{b_{n+1}} f(x, t)dx + \ldots$$
$$= u_1(t) + u_2(t) + \cdots + u_{n+1}(t) + \ldots .$$
(7.27)

Hence, if the obtained series converges uniformly, then so is the integral in the left-hand side. Therefore, we can directly transfer our results for the uniformly converging functional series of Sects. 7.2.2 and 7.2.3 to the improper integrals depending on a parameter. Let us formulate these results explicitly assuming that the function $f(x, t)$ is continuous for any $x > a$.

Theorem 7.20 (cf. Theorem 7.16) *If the improper integral $F(t) = \int_a^{\infty} f(x, t)dx$ converges uniformly for all values of the parameter t within a certain interval, then the function $F(t)$ is a continuous function of t within that interval.*

Theorem 7.21 (cf. Theorem 7.18) *If the improper integral $F(t) = \int_a^{\infty} f(x, t)dx$ converges uniformly for all values of the parameter t within a certain interval $A \leq t \leq B$, then when integrating $F(t)$ with respect to t over that interval one can swap the integrals:*

$$\int_A^B dt \left[\int_a^{\infty} dx\, f(x, t) \right] = \int_a^{\infty} dx \left[\int_A^B dt\, f(x, t) \right].$$

Theorem 7.22 (cf. Theorem 7.19) *If the improper integral*

$$\int_a^\infty \frac{\partial f(x,t)}{\partial t} dx$$

converges uniformly for all values of the parameter t within a certain interval, the functions $f(x,t)$ and $\partial f(x,t)/\partial t$ are continuous, then

$$\frac{d}{dt} \int_a^\infty f(x,t)dx = \int_a^\infty \frac{\partial f(x,t)}{\partial t} dx,$$

i.e. one can differentiate the integrand over the parameter when differentiating the improper integral.

A simple test on uniform convergence can be formulated which is analogous to the Weierstrass Theorem 7.15 for the functional series.

Theorem 7.23 *The integral $F(t) = \int_a^\infty f(x,t)dx$ converges uniformly (and absolutely) for the values of t within a certain interval $A \le t \le B$, if one can bound the absolute value of the function $f(x,t)$ by a positive function $\phi(x)$ (that does not depend on t) for all values of t within the same interval, i.e. $|f(x,t)| \le \phi(x)$, and the integral $\int_a^\infty \phi(x)dx$ converges.*

Proof: Indeed, we can write:

$$\left| \int_a^\infty f(x,t)dx \right| \le \int_a^\infty |f(x,t)| \, dx \le \int_a^\infty \phi(x)dx.$$

Because the improper integral of $\phi(x)$ converges, the values of ϵ and A, see Eq. (7.26), used to check its convergence do not depend on t. At the same time, due to the inequality obtained above, the same values of ϵ and A can then be used to check the convergence of the original integral of $f(x,t)$. This proves its uniform convergence. **Q.E.D.**

Now we should be able to justify differentiation over β of the integral

$$F(\beta) = \int_0^\infty e^{-\alpha x} \frac{\sin(\beta x)}{x} dx.$$

in Problem 4.47. Indeed, this integral can be split into two integrals, $F_1(\beta)$ with the limits from 0 to some $A > 0$, and $F_2(\beta)$ taken between A and ∞. The first integral is an ordinary one, it has no singularity at $x = 0$ (since $x^{-1}\sin(\beta x)$ tends to β when x tends to zero) and in it the derivative $d/d\beta$ can be taken inside the integral. Concerning the second integral, it is an improper one depending on the parameter β. However, for this integral we can write an estimate

$$\left| e^{-\alpha x}\frac{\sin(\beta x)}{x} \right| \leq \frac{1}{x}e^{-\alpha x}$$

with the function $\phi(x) = x^{-1}e^{-\alpha x}$ that is integrable at the infinite upper limit, i.e. the integral $\int_A^\infty x^{-1}e^{-\alpha x}dx$ exists for any $A > 0$, this is because

$$\left| \int_A^\infty e^{-\alpha x}\frac{1}{x}dx \right| \leq \frac{1}{A}\int_A^\infty e^{-\alpha x}dx = \frac{1}{A\alpha}e^{-\alpha A}.$$

Hence, $F(\beta)$ converges uniformly (and absolutely) and the differentiation with respect to β under the integral sign must be legitimate.

Problem 7.18 Prove that the Gaussian integral (cf. Eq. (6.16))

$$G(\alpha) = \int_0^\infty e^{-\alpha x^2}dx = \frac{1}{2}\sqrt{\frac{\pi}{\alpha}}$$

converges uniformly with respect to the parameter $\alpha > 0$ and hence can be differentiated with respect to it any number of times. Correspondingly, show that

$$\int_0^\infty x^2 e^{-\alpha x^2}dx = \frac{\sqrt{\pi}}{4}\alpha^{-3/2} \quad \text{and} \quad \int_0^\infty x^4 e^{-\alpha x^2}dx = \frac{3\sqrt{\pi}}{8}\alpha^{-5/2} \quad (7.28)$$

by differentiating $G(\alpha)$ over the parameter α the necessary number of times[11] [Hint: *we only need to consider large values of $x > x_0$ (with an arbitrary finite x_0) to be able to prove the uniform convergence; argue then that for any α one can always find such β that for any $x > \beta/\alpha$ we have $e^{-\alpha x^2} < e^{-\beta x}$, which enables one to apply the Weierstrass theorem.*]

---

[11] In Chap. 4 of Vol. II we shall develop a simpler method of calculating this type of integrals by expressing them via a special function called the Gamma function.

7.2.5 Lattice Sums

The consideration of the specific lattice sum studied in Sect. 7.1.2 can actually be generalized to any type of lattices and even to a more general lattice sum type:

$$S(\mathbf{r}) = \sum_{\mathbf{L}} \frac{1}{|\mathbf{L} - \mathbf{r}|^n}, \tag{7.29}$$

where $\mathbf{r}$ is a vector and $\mathbf{L} = n_1\mathbf{a}_1 + n_2\mathbf{a}_2 + n_3\mathbf{a}_3$ is the so-called lattice translation; non-coplanar vectors $\mathbf{a}_1$, $\mathbf{a}_2$, $\mathbf{a}_3$ (called primitive translations) form a unit cell in the form of a parallelepiped. By running all possible values of the integers n_i ($i = 1, 2, 3$) the whole space will be covered by such unit cells. The conclusion made in Sect. 7.1.2 for the case of the simple cubic lattice (for which the vectors $\mathbf{a}_1$, $\mathbf{a}_2$, $\mathbf{a}_3$ are mutually orthogonal and of the same length a) remains unchanged: the sum converges only for $n \geq 4$. Below we shall briefly sketch main ideas of the proof.

Let us first consider the case of $\mathbf{r} = \mathbf{0}$. In this case we have to exclude the integers n_i being all equal to zero at the same time (when $\mathbf{L} = \mathbf{0}$). Then we shall have a triple sum of $1/\left(\mathbf{L}^2\right)^{n/2}$ taken over all $\mathbf{L} \neq \mathbf{0}$, whose convergence can be assessed by the appropriate triple integral (the integral test)

$$A = \int' \frac{dx_1 dx_2 dx_3}{\left|(x_1\mathbf{a}_1 + x_2\mathbf{a}_2 + x_3\mathbf{a}_3)^2\right|^{n/2}} .$$

When integrating, we can exclude an arbitrary finite region around the centre of the coordinate system (which is indicated by the prime), this would correspond to avoiding the zero lattice vector $\mathbf{L} = \mathbf{0}$. Indeed, the convergence of the infinite sum is related to its behaviour for large absolute values of the summation indices n_i, and this is related to the convergence of the triple integral at infinite distance from the centre $x_1 = x_2 = x_3 = 0$; a region around the centre of the coordinate system has nothing to do with it.

Next, the expressions in the denominator

$$(x_1\mathbf{a}_1 + x_2\mathbf{a}_2 + x_3\mathbf{a}_3)^2 = \sum_{i,j=1}^{3} \left(\mathbf{a}_i \cdot \mathbf{a}_j\right) x_i x_j = \sum_{i,j=1}^{3} \alpha_{ij} x_i x_j$$

represent the so-called quadratic form with respect to the coordinates x_i with the symmetric coefficients $\alpha_{ij} = \alpha_{ji}$ (that form a symmetric 3×3 matrix α). Hence, the convergence of the lattice sum is related to the behaviour at infinity of the integral

$$A = \int' \frac{dx_1 dx_2 dx_3}{\left[\sum_{i,j=1}^{3} \alpha_{ij} x_i x_j\right]^{n/2}} .$$

The quadratic form in the denominator, $\sum_{i,j=1}^{3} \alpha_{ij} x_i x_j$, via a linear transformation $y_i = \sum_{i=1}^{3} \beta_{ij} x_j$ with some coefficients forming a 3×3 matrix $\beta = \{\beta_{ij}\}$, can always be transformed into the so-called "diagonal" form[12]:

$$\sum_{i,j=1}^{3} \alpha_{ij} x_i x_j = \epsilon_1 y_1^2 + \epsilon_2 y_2^2 + \epsilon_3 y_3^2$$

with respect to the new variables y_i; note that coefficients ϵ_i are non-negative to ensure the positiveness of the quadratic form representing $(x_1 \mathbf{a}_1 + x_2 \mathbf{a}_2 + x_3 \mathbf{a}_3)^2 \geq 0$. Therefore, using the appropriate Jacobian,

$$J = \frac{\partial(x_1, x_2, x_3)}{\partial(y_1, y_2, y_3)} = \left| \frac{\partial x_i}{\partial y_j} \right|,$$

which is a 3×3 determinant of the partial derivatives $\partial x_i / \partial y_j$ and is just a number,[13] one can change the integration variables $(x_1, x_2, x_3) \to (y_1, y_2, y_3)$ and arrive at

$$A = J \int' \frac{dy_1 dy_2 dy_3}{\left[\epsilon_1 y_1^2 + \epsilon_2 y_2^2 + \epsilon_3 y_3^2 \right]^{n/2}}. \tag{7.30}$$

Finally, we can elongate/contract the new variables by introducing yet another variable set, $z_i = \sqrt{\epsilon_i} y_i$ ($i = 1, 2, 3$), such that the integral goes into

$$A = \frac{J}{\sqrt{\epsilon_1 \epsilon_2 \epsilon_3}} \int' \frac{dz_1 dz_2 dz_3}{\left[z_1^2 + z_2^2 + z_3^2 \right]^{n/2}}, \tag{7.31}$$

which is exactly of the same form as for the simple cubic lattice considered previously if the region that we cut off around the centre of the coordinate system is chosen as a small sphere. We have proved in Sect. 7.1.2 that this integral converges for $n \geq 4$.

It is left to consider the case of a non-zero vector $\mathbf{r}$. In fact, we should consider the vector $\mathbf{r}$ that does not coincide with any of the lattice vectors $\mathbf{L}$. The vector $\mathbf{r}$ can be expanded via the primitive translations, $\mathbf{r} = \sum_{i=1}^{3} r_i \mathbf{a}_i$, in which case

$$\mathbf{L} - \mathbf{r} = \sum_{i=1}^{3} (n_i - r_i) \mathbf{a}_i \implies |\mathbf{L} - \mathbf{r}|^2 = \sum_{i,j=1}^{3} \alpha_{ij} (n_i - r_i)(n_j - r_j).$$

---

[12] This aspect will be considered in the linear algebra chapter of Vol. II after discussion of eigenvectors and eigenvalues of matrices.

[13] The inverse transformation from y_i to x_i is given by the inverse matrix $\gamma = \beta^{-1}$ of the coefficients β_{ij}, i.e. $x_i = \sum_{j=1}^{3} \gamma_{ij} y_j$, so that $\left| \frac{\partial x_i}{\partial y_j} \right| = |\gamma_{ij}| = \det \gamma$ is the determinant of this inverse matrix γ.

Clearly, similarly to the previous case, this quadratic form is diagonalized via a linear transformation $y_i = \sum_{i=1}^{3} \beta_{ij} \left(x_j - r_j \right)$, and we eventually arrive at the integral (7.30), i.e. the same conclusion applies.

Let us now prove that the lattice sum (7.29) converges uniformly. Firstly, we can always write that

$$|\mathbf{L} - \mathbf{r}|^2 = L^2 + r^2 - 2Lr \cos\theta \geq L^2 + r^2 - 2Lr = |L - r|^2,$$

where $L = |\mathbf{L}|$, $r = |\mathbf{r}|$ and θ is the angle between $\mathbf{L}$ and $\mathbf{r}$. Hence, $1/|\mathbf{L} - \mathbf{r}|^n \leq 1/|L - r|^n$. Next, we can always choose the centre of the coordinate system in the centre of the unit cell and consider vectors $\mathbf{r}$ inside that cell. It is always sufficient to limit ourselves with vectors $\mathbf{r}$ inside one cell since we sum over all lattice vectors; indeed, choosing $\mathbf{r} = \mathbf{r}' + \mathbf{L}'$, where $\mathbf{r}'$ is inside the cell and $\mathbf{L}'$ is some translation, results in

$$\sum_{\mathbf{L}} \frac{1}{|\mathbf{L} - \mathbf{r}|^n} = \sum_{\mathbf{L}} \frac{1}{|\mathbf{L} - \mathbf{L}' - \mathbf{r}'|^n} = \sum_{\mathbf{L}''} \frac{1}{|\mathbf{L}'' - \mathbf{r}'|^n} \Longrightarrow \sum_{\mathbf{L}} \frac{1}{|\mathbf{L} - \mathbf{r}'|^n},$$

since we sum over all lattice vectors anyway and hence the vectors $\mathbf{L}'' = \mathbf{L} - \mathbf{L}'$ run through the same lattice as the vectors $\mathbf{L}$. Hence, the vector $\mathbf{r}$ can be chosen within the central cell and one can always choose a circle of radius R such that $r < R + \epsilon < \min(\mathbf{L})$, where $\epsilon \to +0$. This allows us to write $(r + R)/2 \leq \min(\mathbf{L}) \leq |\mathbf{L}| = L$. It is easy to see that this inequality is equivalent to (see also Problem 1.20)

$$(L - r)^2 \geq (L - R)^2 \Longrightarrow \frac{1}{|L - r|} \leq \frac{1}{|L - R|}.$$

Hence, we can run an estimate:

$$\sum_{\mathbf{L}} \frac{1}{|\mathbf{L} - \mathbf{r}|^n} \leq \sum_{\mathbf{L}} \frac{1}{|L - r|^n} < \sum_{\mathbf{L}} \frac{1}{|L - R|^n}.$$

The lattice sum in the right-hand side converges as can be verified using the integral test similarly to the method sketched above in this Section. After an appropriate change of the variables the integral to consider becomes

$$\int' \frac{dz_1 dz_2 dz_3}{\left[\left(z_1^2 + z_2^2 + z_3^2 \right)^{1/2} - R \right]^n}.$$

To avoid zero in the denominator, the integration is performed over all space apart from a sphere centred at the origin and of the radius $R + \epsilon$; this is indicated by the prime. Cutting this region out should not effect the test as we are considering convergence at infinity anyway. Using spherical coordinates, this integral is transformed into

$$4\pi \int_{R+\epsilon}^{\infty} \frac{r^2 dr}{(r - R)^n} = 4\pi \int_{\epsilon}^{\infty} \frac{(t + R)^2}{t^n} dt.$$

It is easy to see that for $n \geq 4$ this integral has a finite value; otherwise, it diverges at the upper limit (at $r \to \infty$).

Consider now a slightly different, although related, type of the lattice sums, namely,

$$S_\alpha(\mathbf{r}) = \sum_\mathbf{L} \frac{L_\alpha - r_\alpha}{|\mathbf{L} - \mathbf{r}|^n}, \quad S_{\alpha\beta}(\mathbf{r}) = \sum_\mathbf{L} \frac{(L_\alpha - r_\alpha)(L_\beta - r_\beta)}{|\mathbf{L} - \mathbf{r}|^n}, \tag{7.32}$$

and so on. Here Greek indices indicate Cartesian components of vectors. Since

$$\left| \frac{L_\alpha - r_\alpha}{|\mathbf{L} - \mathbf{r}|} \right| \leq 1,$$

then

$$|S_\alpha(\mathbf{r})| \leq \sum_\mathbf{L} \left| \frac{L_\alpha - r_\alpha}{|\mathbf{L} - \mathbf{r}|^n} \right| = \sum_\mathbf{L} \frac{1}{|\mathbf{L} - \mathbf{r}|^{n-1}} \left| \frac{L_\alpha - r_\alpha}{|\mathbf{L} - \mathbf{r}|} \right| \leq \sum_\mathbf{L} \frac{1}{|\mathbf{L} - \mathbf{r}|^{n-1}},$$

$$|S_{\alpha\beta}(\mathbf{r})| \leq \sum_\mathbf{L} \frac{1}{|\mathbf{L} - \mathbf{r}|^{n-2}},$$

and so on, i.e. $S_\alpha(\mathbf{r})$ converges (and converges uniformly) for $n \geq 5$, $S_{\alpha\beta}(\mathbf{r})$ converges (also uniformly) for $n \geq 6$, and so on.

Finally, consider another modification of the lattice sums considered above which contain an exponential term in it, e.g.

$$S(\mathbf{r}) = \sum_\mathbf{L} \frac{1}{|\mathbf{L} - \mathbf{r}|^n} e^{-\gamma|\mathbf{L}-\mathbf{r}|^2}$$

with $\gamma > 0$. Because of the exponential factor $e^{-\gamma|\mathbf{L}-\mathbf{r}|^2}$, this series converges, and converges uniformly, for any n which must be obvious after applying the integral test. All other variants of the lattice sums considered above, when appended with this exponential term, would also converge uniformly. The results obtained here we shall find useful in Sect. 7.4.2, when discussing the Coulomb potential inside a large, but finite, chunk of a periodic crystal.

7.3 Power Series

Now we shall turn our attention to an important special case of functional series, when $a_n(x) = \alpha_n x^n$ forming an infinite numerical sequence. Here α_n are some coefficients. Correspondingly, we shall be considering the so-called *power series* of the form:

$$S(x) = \alpha_0 + \alpha_1 x + \alpha_2 x^2 + \ldots = \sum_{n=0}^{\infty} \alpha_n x^n. \tag{7.33}$$

This power series is said to be run around point $x = 0$. A more general power series can be considered around an arbitrary point $x = a$,

$$S(x) = \alpha_0 + \alpha_1 (x - a) + \alpha_2 (x - a)^2 + \ldots = \sum_{n=0}^{\infty} \alpha_n (x - a)^n . \qquad (7.34)$$

However, this one can be constructed from the first one in Eq. (7.33) by a simple shift and hence this generalization is not really necessary for proving general statements and theorems, and therefore will be ignored in what follows.

7.3.1   Convergence of the Power Series

We shall start by proving two important theorems related to convergence of the power series, both due to Abel.

Theorem 7.24 *If the power series (7.33) converges at some x_0 (which is not equal to zero), then it converges absolutely for any x satisfying $|x| < |x_0|$, i.e. for $-|x_0| < x < |x_0|$.*

Proof: Since we know that the series with $a_n = \alpha_n x_0^n$ converges, then $\alpha_n x_0^n \to 0$ when $n \to \infty$, and therefore, since the limit exists, a general term of the series is bounded by some real positive number M_0, i.e. $\left| \alpha_n x_0^n \right| < M_0$. Consider now the series constructed of absolute values of the terms of the original series taken at the value of x satisfying the inequality $|x| < |x_0|$:

$$\sum_{n=0}^{\infty} \left| \alpha_n x^n \right| = \sum_{n=0}^{\infty} \left| \alpha_n x_0^n \right| \left| \frac{x}{x_0} \right|^n < \sum_{n=0}^{\infty} M_0 \left| \frac{x}{x_0} \right|^n = M_0 \sum_{n=0}^{\infty} q^n = \frac{M_0}{1 - q} ,$$

where $q = |x/x_0| < 1$ and the sum of its powers forms a converging geometrical progression which sum we know well, Eq. (2.7). This means that our series converges absolutely for any x between $-|x_0|$ and $|x_0|$, i.e. $-|x_0| < x < |x_0|$, as required. **Q.E.D.**

Theorem 7.25 *If, however, the series (7.33) diverges at x_0, it diverges for any x satisfying $|x| > |x_0|$, i.e. for $x < -|x_0|$ and $x > |x_0|$.*

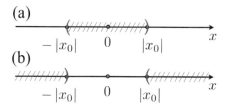

Fig. 7.4 **a** If a series converges at x_0, it also converges (and converges absolutely) within the interval shaded red; **b** if the series diverges at x_0, it also diverges for any x in the two intervals shaded blue

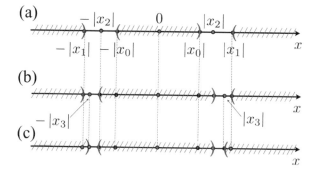

Fig. 7.5 For the proof of Theorem 7.26. **a** The series is known to converge everywhere in the red-shaded interval and to diverge everywhere in the blue-shaded ones. **b** Assuming that it also converges at some x_2 which lies between $|x_0|$ and $|x_1|$, would inevitably widen the convergence interval according to the Abel's theorem

Problem 7.19 Prove the above theorem using the method of contradiction.

Schematically the convergence and divergence intervals are shown in Fig. 7.4.

Theorem 7.26 *If the series converges at some* $x_0 \neq 0$ *and also is known to diverge at some* x_1 *(obviously,* $|x_1|$ *must be larger than* $|x_0|$, *otherwise it would contradict Abel's theorems), then there exists such positive R that the series converges for any* $|x| < R$ *and diverges for any* $|x| > R$ *(nothing can be said about x* $= R$ *though).*

Proof: Since the series converges at some x_0, it must converge at any x within the interval $-|x_0| < x < |x_0|$ according to Theorem 7.24; moreover, since it diverges at some x_1, it will also diverge for any x satisfying $x < -|x_0|$ and $x > |x_0|$ (Theorem 7.25). Assume then that the series converges at some point x_2 which absolute value is between $|x_0|$ and $|x_1|$, see Fig. 7.5a. Then, according to the Abel's theorem, it will also converge for any x satisfying $-|x_2| < x < |x_2|$, i.e. in the interval which completely includes the previous convergence interval $-|x_0| < x < |x_0|$, see Fig. 7.5b. Similarly, one can consider a point x_3 lying between $|x_2|$ and $|x_1|$. Assuming that the series diverges at x_3, we have to conclude from Abel's theorem that it will diverge anywhere when $x > |x_3|$ and $x < -|x_3|$, i.e. the new divergence intervals would completely incorporate the previous ones, and the boundaries of the diverging and converging intervals that face each other would move closer. Continuing this process, we shall come to a point where the boundaries meet at some point $R > 0$ and $-R < 0$, so that for any x between $-R$ and R the series converges, and for any $x < -R$ and $x > R$ it diverges. Since the intervals are exclusive of their boundary points, nothing can be said about the points $x = \pm R$. **Q.E.D.**

The positive number R is called the *radius of convergence*. If the series converges everywhere, it is said that the radius of convergence is infinite, $R = \infty$. If the series converges only at $x = 0$, then $R = 0$.

It is easy to see now that the radius of convergence can be calculated via

$$R = \lim_{n \to \infty} \left| \frac{\alpha_n}{\alpha_{n+1}} \right|. \tag{7.35}$$

If the limit above is infinite, then $R = \infty$, while for the zero limit we have $R = 0$.

Problem 7.20 Prove the above formula by applying the ratio test.

Problem 7.21 Prove by applying the root test that R can also be written as

$$\frac{1}{R} = \mathrm{Sup}_{n \to \infty} \sqrt[n]{|\alpha_n|}, \tag{7.36}$$

where Sup means the maximum value achieved after taking the limit.

Problem 7.22 Show that $R = \infty$ for the series with $\alpha_n = 1/n!$.

Problem 7.23 Show that $R = \infty$ for the series with $a_n = (-1)^n x^{2n+1}/(2n+1)!$.

Problem 7.24 Show that $R = \infty$ for the series with $a_n = (-1)^n x^{2n}/(2n)!$.

Problem 7.25 Show that $R = 1$ for the series with $\alpha_n = 1/n$.

For the latter problem ($\alpha_n = 1/n$) nothing can be said about the points $x = \pm 1$ from the ratio test. However, we studied both series before: at $x = 1$ the power series becomes the harmonic series about which we already know that it diverges; at $x = -1$

the alternating harmonic series appears for which we also know that it converges (conditionally). The situation is opposite for the series with $\alpha_n = (-1)^n /n$.

7.3.2 Uniform Convergence and Term-by-Term Differentiation and Integration of Power Series

Power series present the simplest and very important example of the uniformly converging functional series and hence can be integrated and differentiated term-by-term. Indeed, the following theorem proves the uniform convergence.

Theorem 7.27 *If R is the radius of convergence of the power series (7.33), then the series converges uniformly within the radius of $0 < r < R$, i.e. for any x satisfying $-r \leq x \leq r$.*

Proof: Since $r < R$, then the series $\sum_n \alpha_n r^n$ converges absolutely, i.e. the series $\sum_n |\alpha_n r^n|$ converges. Also, for any $|x| < r$ we have

$$|a_n(x)| = |\alpha_n| |x|^n < |\alpha_n| r^n ,$$

which, by virtue of the uniform convergence test (Theorem 7.15), means that our series indeed converges uniformly. **Q.E.D.**

This in turn means that the power series represents a continuous function for any $-r \leq x \leq r$ with the positive $r < R$ (Theorem 7.16). Note that nothing can be said about the boundary points $x = \pm R$ where a special investigation is required. It can be shown, for instance, that if the series converges at $x = R$, it converges uniformly for all x satisfying $0 \leq x \leq R$. For instance, the series

$$x - \frac{x^2}{2} + \frac{x^3}{3} - \ldots = \sum_{n=1}^{\infty} \frac{(-1)^{n+1}}{n} x^n \tag{7.37}$$

at $x = 1$ becomes the alternating harmonic series (7.12) which, as we know from Sect. 7.1.4, converges (although not absolutely). This means that the series (7.37) converges uniformly for any $0 \leq x \leq 1$, and hence its sum is a continuous function for all these values of x. We shall see explicitly in the following that this series represents a power series of $\ln(1 + x)$ and hence at $x = 1$ converges to $\ln 2$. Note that the statement made above comes from the fact that if $S(x) = \sum_n \alpha_n x^n$ uniformly converges for any x from the semi-open interval $0 \leq x < R$, where the function $S(x)$ is continuous, and since the series converges at the right boundary $x = R$, one can take the limit of $S(x)$ at $x \to R - 0$ (from the left) and then arrive at the converging series at $x = R$. So, the function $S(x)$ will be continuous in the closed interval $0 \leq x \leq R$.

If the power series converges uniformly in the open, semi-open or closed interval between $-R$ and R, then, according to Theorems 7.18 and 7.19 of Sect. 7.2.3 it can be integrated and differentiated term-by-term.

One point still needs clarifying though: it is that, according to Theorem 7.19, the series containing derivatives of the terms of the original series should also converge uniformly. According to Theorem 7.27, for this it is sufficient to demonstrate that the series of derivatives has the same radius of convergence as the original power series. To this end, we can apply the ratio test to the series of derivatives, which has the general term $b_n = n a_n x^{n-1}$:

$$\lim_{n \to \infty} \left| \frac{b_{n+1}}{b_n} \right| = \lim_{n \to \infty} \left| \frac{(n+1) a_{n+1} x^n}{n a_n x^{n-1}} \right| = \lim_{n \to \infty} \left| \left(1 + \frac{1}{n} \right) \frac{a_{n+1} x}{a_n} \right| = \lim_{n \to \infty} \left| \frac{a_{n+1} x}{a_n} \right| .$$

But the original series of $\sum_n a_n x^n$ converges for any x between $-R$ and R and hence, according to the same ratio test, the last limit is smaller than one:

$$\lim_{n \to \infty} \left| \frac{a_{n+1} x^{n+1}}{a_n x^n} \right| = \lim_{n \to \infty} \left| \frac{a_{n+1} x}{a_n} \right| < 1 .$$

Hence, the ratio test establishes convergence of the series of derivatives within the same interval of the original power series.

Since the series of derivatives is also a power series, it can be also differentiated. It is seen from this discussion, that the power series can be differentiated any number of times; each series thus obtained would have the same radius of convergence.

7.3.3   Taylor Series

The properties of the power series considered above allow us to formulate a very powerful method enabling one to expand a function $f(x)$ into a power series (cf. Sect. 3.8).

First of all, we prove that if $f(x)$ has a power series expansion,

$$f(x) = \sum_{n=0}^{\infty} \alpha_n x^n = \alpha_0 x^0 + \alpha_1 x^1 + \alpha_2 x^2 + \alpha_3 x^3 + \dots , \qquad (7.38)$$

then this expansion is unique. Indeed, suppose there exists another expansion of the same function,

$$f(x) = \sum_{n=0}^{\infty} \beta_n x^n = \beta_0 x^0 + \beta_1 x^1 + \beta_2 x^2 + \beta_3 x^3 + \dots . \qquad (7.39)$$

Since both series converge within some interval between $-R$ and R, they converge uniformly, and hence the series can be differentiated term-by-term. In fact, they can be differentiated many times:

$$f^{(1)}(x) = \sum_{n=1}^{\infty} n\alpha_n x^{n-1} = \sum_{n=1}^{\infty} \frac{n!}{(n-1)!}\alpha_n x^{n-1}$$
$$= \alpha_1 x^0 + 2\alpha_2 x^1 + 3\alpha_3 x^2 + \dots ,$$

(7.40)

$$f^{(2)}(x) = \sum_{n=2}^{\infty} n(n-1)\alpha_n x^{n-2} = \sum_{n=2}^{\infty} \frac{n!}{(n-2)!}\alpha_n x^{n-2}$$
$$= 2 \cdot 1\alpha_2 x^0 + 3 \cdot 2\alpha_3 x^1 + 4 \cdot 3\alpha_4 x^2 + \dots ,$$

(7.41)

$$f^{(3)}(x) = \sum_{n=3}^{\infty} n(n-1)(n-2)\alpha_n x^{n-3} = \sum_{n=3}^{\infty} \frac{n!}{(n-3)!}\alpha_n x^{n-3}$$
$$= 3 \cdot 2 \cdot 1\alpha_3 x^0 + 4 \cdot 3 \cdot 2\alpha_4 x^1 + 5 \cdot 4 \cdot 3\alpha_5 x^2 + \dots ,$$

(7.42)

and so on. The k-th term has the form:

$$f^{(k)}(x) = \sum_{n=k}^{\infty} n(n-1) \cdots (n-k+1)\alpha_n x^{n-k} = \sum_{n=k}^{\infty} \frac{n!}{(n-k)!}\alpha_n x^{n-k}$$
$$= k!\alpha_k x^0 + \frac{(k+1)!}{1!}\alpha_{k+1} x^1 + \frac{(k+2)!}{2!}\alpha_{k+2} x^2 + \dots ,$$

(7.43)

and similarly for the other series (7.39). Notice that after the differentiation of the original series the constant term in the series (the term with x^0) disappears and the series starts from the next term; correspondingly, the series for the k-th derivative of $f(x)$ starts from the $n = k$ term. Let us now put $x = 0$ in Eqs. (7.38) and (7.39); we immediately obtain that $\alpha_0 = \beta_0$. If we now do the same in the two expressions for the first derivative of $f(x)$, we obtain $\alpha_1 = \beta_1$; continuing this process, we progressively have $\alpha_2 = \beta_2$, $\alpha_3 = \beta_3$, etc., i.e. $\alpha_k = \beta_k$ for any $k = 0, 1, 2$, etc. This proves the statement made above, as both series coincide that contradicts the assumption made.

Next, we can relate the coefficients α_0, α_1, α_3, etc. to the function $f(x)$ itself and its derivatives. Indeed, putting $x = 0$ again in Eqs. (7.38), (7.40)–(7.42) yields, respectively:

$$f(0) = \alpha_0 , \quad f^{(1)}(0) = \alpha_1 , \quad f^{(2)}(0) = 2!\alpha_2 , \quad f^{(3)}(0) = 3!\alpha_3 , \quad \dots ,$$

and from (7.43) we generally have $f^{(k)}(0) = k!\alpha_k$, i.e. the expansion of $f(x)$, if exists, should necessarily have the form:

$$f(x) = f(0) + f^{(1)}(0)x + \frac{f^{(2)}(0)}{2!}x^2 + \frac{f^{(3)}(0)}{3!}x^3 + \dots = \sum_{n=0}^{\infty} \frac{f^{(n)}(0)}{n!}x^n ,$$

(7.44)

which is called Maclaurin series, and, of course, this is not accidental. After all, we
can write the Maclaurin formula (3.86)

$$f(x) = f(0) + \frac{f'(0)}{1!}x + \frac{f''(0)}{2!}x^2 + \frac{f'''(0)}{3!}x^2 + \cdots + \frac{f^{(n)}(0)}{n!}x^n + R_{n+1}(x) ,$$

$$(7.45)$$

when representing any n-times differentiable function $f(x)$, where

$$R_{n+1}(x) = \frac{f^{(n+1)}(\xi)}{(n+1)!}x^{n+1} = \frac{f^{(n+1)}(\vartheta x)}{(n+1)!}x^{n+1} , \qquad (7.46)$$

is the reminder term with $0 < \vartheta < 1$. Comparing the two results, Eqs. (7.44) and
(7.45), we immediately notice that the first n terms in both of them coincide. One
may say then that the Maclaurin series is obtained from the Maclaurin formula of the
infinite order ($n \to \infty$). This close relationship between the two results allows one
to formulate the necessary and sufficient conditions at which the Maclaurin series of
the function $f(x)$ converges exactly to the function $f(x)$ itself.

Theorem 7.28 *In order for the Maclaurin series (7.44) to converge to $f(x)$
within the interval $-R < x < R$, it is necessary and sufficient that the remainder
term (7.46) of the Maclaurin formula tends to zero as $n \to \infty$.*

Proof: We shall first prove sufficiency, i.e. we start by assuming that $\lim_{n\to\infty} R_{n+1}(x) = 0$. Then $f(x) = S_{n+1}(x) + R_{n+1}(x)$, where $S_{n+1}(x)$ contains the first $(n + 1)$
terms of the Maclaurin formula, and, at the same time, coincides with the partial sum
of the series with $(n + 1)$ first terms. Since $\lim_{n\to\infty} R_{n+1}(x) = 0$ by our assumption,
then

$$\lim_{n\to\infty} R_{n+1}(x) = \lim_{n\to\infty} \left[f(x) - S_{n+1}(x) \right] = f(x) - \lim_{n\to\infty} S_{n+1}(x) = 0$$

$$\implies \lim_{n\to\infty} S_{n+1}(x) = f(x) ,$$

which immediately proves the required statement of the convergence of the infinite
power series to the function $f(x)$ at each point x between $-R$ and R.

Now let us prove the necessary condition. We start from the assumption that the
series (7.44) converges to $f(x)$, and hence $\lim_{n\to\infty} S_{n+1}(x) = f(x)$ for $-R < x < R$. Correspondingly,

$$\lim_{n\to\infty} S_{n+1}(x) = \lim_{n\to\infty} \left[f(x) - R_{n+1}(x) \right] = f(x) - \lim_{n\to\infty} R_{n+1}(x) = f(x)$$

$$\implies \lim_{n\to\infty} R_{n+1}(x) = 0 ,$$

as required. **Q.E.D.**

Therefore, convergence of the series (7.44) requires in each case investigating the limiting behaviour of the reminder term (7.46). In this respect it is expedient to notice that an important n-dependent part of a general term in Taylor's series tends to zero as $n \to \infty$, i.e.

$$\lim_{n \to \infty} \frac{x^n}{n!} = 0 \qquad (7.47)$$

for any real x. This can be seen, e.g. by applying the ratio test to the sequence $a_n(x) = x^n/n!$ (which is the Maclaurin expansion of the e^x as we shall immediately see below). Since this series converges everywhere (the ratio test is a sufficient condition), then necessarily (Theorem 7.4) $a_n = x^n/n! \to 0$ as $n \to \infty$. Of course, condition (7.47) is only necessary and does not mean that for any $f(x)$ the series (7.44) converges for any real x since the behaviour of the pre-factors $f^{(n)}(0)$ may still affect the convergence region (the value of R).

As an example, let us consider the expansion of the function $f(x) = e^x$. The remainder term

$$R_{n+1}(x) = x^{n+1} \left[\frac{(e^x)^{(n+1)}}{(n+1)!} \right]_{x \to \vartheta x} = \frac{e^{\vartheta x} x^{n+1}}{(n+1)!} = e^{\vartheta x} \left[\frac{x^{n+1}}{(n+1)!} \right] \to 0$$

as $n \to \infty$ because of the necessary condition (7.47). Therefore, the expansion of the exponential function has the form:

$$e^x = 1 + x + \frac{x^2}{2!} + \ldots = \sum_{n=0}^{\infty} \frac{x^n}{n!} . \qquad (7.48)$$

The series converges to e^x for any x from $-\infty < x < \infty$ (in other words, the radius of convergence $R = \infty$).

Similarly one can consider expansions of other elementary functions into the Maclaurin series and investigate their convergence. The form of the series themselves can be borrowed from the corresponding Taylor's formulae (3.90)–(3.93):

$$\ln(1+x) = x - \frac{x^2}{2} + \frac{x^3}{3} - \cdots + (-1)^{n+1} \frac{x^n}{n} + \ldots = \sum_{n=1}^{\infty} (-1)^{n+1} \frac{x^n}{n} ;$$
$$(7.49)$$

$$\sin x = x - \frac{x^3}{3!} + \frac{x^5}{5!} - \frac{x^7}{7!} + \cdots + \frac{(-1)^{n+1} x^{2n-1}}{(2n-1)!} + \ldots = \sum_{n=1}^{\infty} \frac{(-1)^{n+1} x^{2n-1}}{(2n-1)!} ;$$
$$(7.50)$$

$$\cos x = 1 - \frac{x^2}{2!} + \frac{x^4}{4!} - \cdots + \frac{(-1)^n x^{2n}}{(2n)!} + \ldots = \sum_{n=0}^{\infty} \frac{(-1)^n x^{2n}}{(2n)!} ; \qquad (7.51)$$

$$(1+x)^\alpha = 1 + \alpha x + \frac{\alpha(\alpha-1)}{2}x^2 + \cdots + \frac{\alpha(\alpha-1)(\alpha-2)\cdots(\alpha-n+1)}{n!}x^n$$

$$+ \ldots = \sum_{n=0}^{\infty} \binom{\alpha}{n} x^n ,$$

$$(7.52)$$

where the generalized binomial coefficients $\binom{\alpha}{n}$ are the same as in Eq. (3.94).

Further, using Eqs. (7.53)–(7.55), we can also write:

$$\arcsin(x) = \sum_{n=0}^{\infty} \frac{(2n)!}{4^n (2n+1) (n!)^2} x^{2n+1} ; \qquad (7.53)$$

$$\arccos(x) = \frac{\pi}{2} - \sum_{n=0}^{\infty} \frac{(2n)!}{4^n (2n+1) (n!)^2} x^{2n+1} ; \qquad (7.54)$$

$$\arctan(x) = \sum_{n=0}^{\infty} \frac{(-1)^n}{2n+1} x^{2n+1} . \qquad (7.55)$$

Note that the series for the inverse sine, cosine and tangent functions also converge at $x = \pm 1$ and hence the formulae (7.53)–(7.55) are valid for these values as well (in the former two cases this can be shown with the ratio test, a more powerful test should be used for the arctangent).

Problem 7.26 Consider an expression $(1+x)^\alpha (1+x)^\beta = (1+x)^{\alpha+\beta}$. By expanding the left- and the right-hand sides using Eq. (7.52) and then comparing the coefficients on both sides to the same power of x, prove the following identity:

$$\sum_{k=0}^{n} \binom{\alpha}{k}\binom{\beta}{n-k} = \binom{\alpha+\beta}{n} . \qquad (7.56)$$

Problem 7.27 Show (using, e.g. the ratio test) that the radius of convergence of the $\ln(1+x)$ expansion is $R = 1$. Explain why the expansion also converges (to $\ln 2$) at $x = 1$, but diverges at $x = -1$.

Problem 7.28 Show that $R = \infty$ for the sine and cosine functions.

Problem 7.29 Show that $R = 1$ for the expansion of $(1+x)^\alpha$.

Problem 7.30 Show that $R = 1$ for the inverse sine and cosine functions.

Problem 7.31 Consider the geometric progression for $(1 + x)^{-1}$. Obtain the Maclaurin expansion of $\ln(1 + x)$ by integrating the progression term-by-term.

Problem 7.32 Consider the geometric progression for $(1 + x^2)^{-1}$. Obtain the Maclaurin expansion of $\arctan x$ by integrating the progression term-by-term.

Problem 7.33 Consider the expansion of $(1 - x^2)^{-1/2}$. Obtain the Maclaurin expansion of $\arcsin x$ by integrating the expansion term-by-term.

Problem 7.34 Obtain the Maclaurin series of $(1 - x)^{-2}$ by differentiating term-by-term the Maclaurin series of $(1 - x)^{-1}$.

Problem 7.35 Prove the following limit

$$\lim_{x \to 0} \frac{1 - \cos x^{\gamma}}{(1 - \cos x)^{\gamma}} = 2^{\gamma - 1},$$

where $\gamma \neq 0$ is a real number, by expanding into the Maclaurin series its denominator and numerator. Note that due to the general nature of the power γ L'Hpital's rule (Sect. 3.10) is useless here!

Problem 7.36 Determine the first five terms of the Maclaurin series for the following functions of t:

$$\frac{1}{\sqrt{1 - 2tx + t^2}} \; ; \quad e^{2xt - t^2} \; ; \quad \frac{1}{1 - t} \exp\left(-\frac{xt}{1 - t}\right) \; ; \quad \frac{1 - tx}{1 - 2xt + t^2} \cdot$$

The corresponding coefficients of the expansion are polynomials in x, and are related, respectively, to the so-called Legendre, Hermite, Laguerre and Chebyshev polynomials of the first kind.

Problem 7.37 Prove for all the cases above in Eqs. (7.49)–(7.55) that the corresponding reminder $R_{n+1} \to 0$ as $n \to \infty$. Explicit expressions for the remainder term in each case are given at the end of Sect. 3.8.

Problem 7.38 Show that

$$\lim_{x \to 0} \left(\frac{\partial}{\partial x} \right)^n e^{\sigma x^2/2} = \frac{n!}{2^{n/2}\,(n/2)!} \sigma^{n/2} \,,$$

if n is even, and that it is equal to zero if n is odd. [Hint: *expand first the exponential function into the Maclaurin series.*]

Maclaurin series is frequently used for approximate calculations. In particular, one can obtain expansions into a power series of some integrals which cannot be calculated analytically. This is illustrated by the following problems.

Problem 7.39 Using the expansion of $\sin x / x$, prove the following formula:

$$\int_0^x \frac{\sin t}{t} dt = x - \frac{x^3}{3! \cdot 3} + \frac{x^5}{5! \cdot 5} + \ldots = \sum_{n=0}^{\infty} \frac{(-1)^n\, x^{2n+1}}{(2n+1)!\,(2n+1)} \,. \quad (7.57)$$

Problem 7.40 Using the expansion of e^{-x^2}, derive the following expansion of the integral of it (called the error function):

$$\mathrm{erf}(x) = \frac{2}{\sqrt{\pi}} \int_0^x e^{-t^2} dt = \frac{2}{\sqrt{\pi}} \left(x - \frac{x^3}{3 \cdot 1!} + \frac{x^5}{5 \cdot 2!} - \ldots \right)$$

$$= \frac{2}{\sqrt{\pi}} \sum_{n=0}^{\infty} \frac{(-1)^n\, x^{2n+1}}{(2n+1)n!} \,. \quad (7.58)$$

Problem 7.41 Using the expansion of the sine function, obtain the following power expansion for the Fresnel integrals:

$$C(x) = \int_0^x \cos\left(t^2\right) dt = \sum_{n=0}^{\infty} \frac{(-1)^n}{(4n+1)\,(2n)!} x^{4n+1} \,, \quad (7.59)$$

$$S(x) = \int_0^x \sin\left(t^2\right) dt = \sum_{n=0}^{\infty} \frac{(-1)^n}{(4n+3)\,(2n+1)!} x^{4n+3} \,. \quad (7.60)$$

Problem 7.42 By representing the integrand as an exponential that is expanded into the Maclaurin series, prove the formula:

$$\int_0^1 x^{-x} dx = \sum_{n=1}^{\infty} \frac{1}{n^n}.$$

[Hint: *you may find the result* (4.51) *helpful.*]

Problem 7.43 Generalize the result of the previous problem:

$$\int_0^1 x^{\alpha x} dx = \sum_{n=1}^{\infty} \frac{(-\alpha)^{n-1}}{n^n}.$$

Problem 7.44 In theory of magnetism of atoms the following so-called Brillouin function is used:

$$B_j(x) = \left(1 + \frac{1}{2j}\right) \coth\left(\left(j + \frac{1}{2}\right)x\right) - \frac{1}{2j} \coth\left(\frac{x}{2}\right).$$

Show that its derivative at $x = 0$ is equal to $(j + 1)/3$. [Hint: *direct calculation of the derivative results in a difference of infinities; therefore, after taking the derivative and before applying the limit expand the functions in the Maclaurin series and then take the limit.*]

In many cases the full Taylor (or Maclaurin) expansion is not needed; instead, only a few first terms of such an expansion would suffice. If the function in question is a composition of several functions, than instead of applying the general formula (7.44) (which could be rather cumbersome) one may use a simpler approach based on expanding each function separately. We shall illustrate this method by obtaining an expansion of the function $f(x) = \ln \cosh(x)$ for small x keeping terms up to the fourth power of x. We shall start from the expansion of the logarithm (7.49) and of the hyperbolic cosine (which one can easily derive from the expansion of the exponential function):

$$\ln(1 + y) = y - \frac{y^2}{2} + \frac{y^3}{3} - \cdots,$$

$$\cosh(x) = \frac{1}{2}\left(e^x + e^{-x}\right) = 1 + \underbrace{\left(\frac{x^2}{2} + \frac{x^4}{24} + \cdots\right)}_{y}.$$

We indicated by y the part of the $\cosh(x)$ that should go into the logarithm. When substituting the expansion of $\cosh(x)$ into the logarithm, we need to make sure that we keep enough terms in it so that all contributions up to x^4 would be retained. It is clear that it is sufficient to keep terms up to x^4 in the expansion of the $\cosh(x)$ and terms up to the order y^2 in the logarithm if we are only interested in the terms up to x^4 overall:

$$\ln\cosh(x) \approx \left(\frac{x^2}{2} + \frac{x^4}{24}\right) - \frac{1}{2}\left(\frac{x^2}{2} + \frac{x^4}{24}\right)^2 .$$

In calculating the square term, we only keep a single term which is of the fourth power in x, i.e.

$$\ln\cosh(x) \approx \left(\frac{x^2}{2} + \frac{x^4}{24}\right) - \frac{1}{2}\left(\frac{x^2}{2}\right)^2 = \frac{x^2}{2} - \frac{x^4}{12} .$$

Note that the correct coefficient to x^4 requires considering two first terms in the expansion of the logarithm. Keeping only the first term would result in the coefficient $1/24$, which is incorrect. The method used is elementary and is to be preferred in many such cases when direct calculation of the derivatives of the given function is not convenient and only first terms of the expansion are sufficient.

Problem 7.45 Confirm the first few terms of the Maclaurin expansion of the following functions:

$$e^{2(\cos x - 1)} = 1 - x^2 + \frac{7x^4}{12} + \dots ;$$

$$\sin\left(\pi + x^2\right) = -x^2 + \frac{x^6}{6} + \dots ;$$

$$\frac{2x + 1}{\left(x^2 - 4x + 1\right)^3} = 1 + 14x + 117x^2 + \dots .$$

Problem 7.46 Expand $\sin^2 x$ into the Maclaurin series by expressing the sine squared via another trigonometric function and expanding the latter instead. Then confirm the first few terms in this expansion,

$$\sin^2 x = x^2 - \frac{x^4}{3} + \dots ,$$

by squaring the expansion of the sine function (7.50) and keeping the appropriate terms.

Problem 7.47 Obtain an additional term (of the order of x^6) in the expansion of $\ln \cosh(x)$:

$$\ln \cosh(x) \approx \frac{x^2}{2} - \frac{x^4}{12} + \frac{x^6}{45}.$$

Check this result by directly differentiating the function $f(x) = \ln \cosh(x)$.

Problem 7.48 Obtain the first terms of the following power expansion:

$$\frac{x}{e^x - 1} = 1 - \frac{1}{2}x + \frac{1}{12}x^2 - \frac{1}{720}x^4 + \frac{1}{30240}x^6 - \frac{1}{1209600}x^8 + \dots. \tag{7.61}$$

[Hint: *calculating multiple derivatives of the function in the left-hand side is cumbersome as this requires careful resolution of $0/0$ type singularities; a simpler method is in expanding first $e^x - 1$, taking out the common pre-factor x that will cancel with the x in the numerator of the function; the obtained expression will be of the form $1/(1 + z(x))$ that can then be expanded in the powers of z; the expression for $z(x)$ is a power series, and when calculating powers of z, you only need to keep terms that contribute to the required power of x in the expansion* (7.61).]

Problem 7.49 The so-called cumulants K_n and moments M_n of a random process are defined by the following connection relation:

$$1 + \sum_{n=1}^{\infty} \frac{u^n}{n!} M_n = \exp\left(\sum_{n=1}^{\infty} \frac{u^n}{n!} K_n\right). \tag{7.62}$$

By expanding the exponential function in the right-hand side and comparing the coefficients at the same powers of the parameter u in both sides show that:

$$M_1 = K_1 ; \quad M_2 = K_2 + K_1^2 ; \quad M_3 = K_3 + 3K_1 K_2 + K_1^3 .$$

Problem 7.50 Take the logarithm of both sides of Eq. (7.62), expand the logarithm and show that the low order cumulants K_n are related to the moments M_n via

$$K_1 = M_1 ; \quad K_2 = M_2 - M_1^2 ; \quad K_3 = M_3 - 3M_1 M_2 + 2M_1^3 .$$

7.3.4   Bernoulli Numbers and Summation of Powers of Integers

In Sect. 1.14 we have introduced a set of numbers (1.169), called Bernoulli numbers, and gave formula (1.168) for summing up the first n integers in power p,

$$K_p(n) = \sum_{m=1}^{n} m^p = 1^p + 2^p + \cdots + n^p \, , \tag{7.63}$$

via these numbers. Here we shall define Bernoulli numbers and derive formula (1.168) using the method of generating functions.

Let us define Bernoulli numbers as coefficients in the Maclaurin expansion of the following function:

$$\frac{x}{e^x - 1} = \sum_{k=0}^{\infty} B_k \frac{x^k}{k!} \, . \tag{7.64}$$

The function on the left-hand side is called *generating function* of the Bernoulli numbers. We have already considered the Maclaurin expansion of this function in Problem 7.48. Using this expansion, a few Bernoulli numbers can easily be worked out as given by Eq. (1.169). It is seen from Eq. (7.61) that the expansion of this function contains all terms with even powers of x apart from the linear term. If this is indeed the case, then all Bernoulli numbers B_n with odd values of n, apart from B_1, are equal to zero. For initial values of n we have indeed seen this in Sect. 1.14, but is this true generally for any B_n with odd $n \geq 3$? The answer is in checking explicitly whether the function

$$f(x) = \frac{x}{e^x - 1} + \frac{x}{2} = \frac{x \, e^x + 1}{2 \, e^x - 1}$$

is even. This is easy to check proving that our hypothesis is indeed correct (do it!). Hence, in its Maclaurin expansion only even powers of x will appear, proving positively that all Bernoulli numbers B_n with odd indices n, starting from B_3, are equal to zero; only B_n with even indices $n \geq 2$ remain.

Our task now is to relate Bernoulli numbers to the sum (7.63) of integers in power p. To this end, let us introduce another generating function, specifically for the sum of powers in integers:

$$G(x, n) = \sum_{p=0}^{\infty} K_p(n) \frac{x^p}{p!} \, . \tag{7.65}$$

It is seen that the coefficients in this series are chosen to be the required finite sums of the first n integers each raised in power p. The beauty of this generating function is that it can actually be calculated analytically. Indeed, using the definition (7.63) of $K_p(n)$, we can write:

$$G(x, n) = \sum_{p=0}^{\infty} \left(\sum_{m=1}^{n} m^p \right) \frac{x^p}{p!} = \sum_{m=1}^{n} \left[\sum_{p=0}^{\infty} \frac{(mx)^p}{p!} \right] = \sum_{m=1}^{n} e^{mx} \, ,$$

where we have made use of the fact that the expression in the square brackets is the Maclaurin expansion of the exponential function e^{mx}. The obtained expression represents a finite geometric progression since $e^{mx} = (e^x)^m$, and hence the sum over m can be calculated using Eq. (1.143):

$$G(x, n) = \sum_{m=1}^{n} (e^x)^m = e^x \frac{e^{xn} - 1}{e^x - 1} = \left(\frac{e^{xn} - 1}{x} \right) \left(\frac{-x}{e^{-x} - 1} \right).$$

In the second factor in the right-hand side one can recognize the expansion (7.64) written for $-x$ that contains Bernoulli numbers. The first factor we can expand in the Maclaurin series:

$$G(x, n) = \left(\frac{1}{x} \sum_{k=1}^{\infty} \frac{(xn)^k}{k!} \right) \left(\sum_{j=0}^{\infty} B_j \frac{(-x)^j}{j!} \right) = \frac{1}{x} \sum_{j=0}^{\infty} \sum_{k=1}^{\infty} \left((-1)^j B_j \frac{n^k}{k! j!} \right) x^{j+k}.$$

$$(7.66)$$

Now we shall introduce a new summation index $m = j + k$ that changes between 1 and ∞. The following identity can be easily checked, by writing out terms on both sides explicitly:

$$\sum_{j=0}^{\infty} \sum_{k=1}^{\infty} a_{k,j} x^{j+k} = \sum_{m=1}^{\infty} \left(\sum_{l=0}^{m-1} a_{m-l,l} \right) x^m.$$

Using this identity, Eq. (7.66) is manipulated into:

$$G(x, n) = \sum_{m=1}^{\infty} \left[\sum_{l=0}^{m-1} (-1)^l B_l \binom{m}{l} n^{m-l} \right] \frac{x^{m-1}}{m!}$$

$$= \sum_{p=0}^{\infty} \left[\frac{1}{p+1} \sum_{l=0}^{p} (-1)^l B_l \binom{p+1}{l} n^{p+1-l} \right] \frac{x^p}{p!},$$

where in the last passage we have changed the summation index $m \to p = m - 1$. The obtained expansion of the generating function can now we compared with its definition in Eq. (7.65), and the required result (1.168) for $K_p(n)$ immediately follows.

7.3.5 Fibonacci Numbers

Here we shall return to this remarkable sequence of numbers F_n (where $n = 0, 2, 3, \ldots$), called the Fibonacci sequence, considered in Sect. 1.7, and find a simple generating function $G(x)$ for them by defining the following power series:

$$G(x) = \sum_{n=0}^{\infty} F_n x^n. \qquad (7.67)$$

To calculate $G(x)$, we must use the definition (1.78) of the Fibonacci numbers: $F_n = F_{n-1} + F_{n-2}$. Then, we can write:

$$G(x) = F_0 + F_1 x + \sum_{n=2}^{\infty} F_n x^n = x + \sum_{n=2}^{\infty} F_n x^n = x + \sum_{n=2}^{\infty} (F_{n-1} + F_{n-2}) x^n$$

$$= x + x \sum_{n=2}^{\infty} F_{n-1} x^{n-1} + x^2 \sum_{n=2}^{\infty} F_{n-2} x^{n-2} .$$

In the first sum in the right-hand side we change the summation index $n \to k = n - 1$, while in the second $n \to k = n - 2$. This yields

$$G(x) = x + x \underbrace{\sum_{k=1}^{\infty} F_k x^k}_{G(x) - F_0 = G(x)} + x^2 \underbrace{\sum_{k=0}^{\infty} F_k x^k}_{G(x)} = x + G(x) \left(x + x^2 \right) ,$$

solving which we finally obtain

$$G(x) = \frac{x}{1 - x - x^2} . \tag{7.68}$$

Problem 7.51 Obtain the first few terms of the Maclaurin series of the function $G(x) = x/\left(1 - x - x^2\right)$ to show that the coefficients of the series indeed form the Fibonacci sequence.

Problem 7.52 Consider a general sequence $F_n = a F_{n-1} + b F_{n-2}$ with $F_0 = 0$, $F_1 = 1$ and a and b being some arbitrary integers (cf. Problem 1.103). Show that for this sequence the generating function is $G(x) = x/\left(1 - ax - bx^2\right)$.

Problem 7.53 Obtain the first 5 terms of the Maclaurin series of the function $G(x) = x/\left(1 - 2x - x^2\right)$ to show that the coefficients of the series form the generalized Fibonacci sequence for $a = 2$ and $b = 1$ (see Problem 1.104).

7.3.6 Complex Exponential

In Sect. 1.11 we have introduced complex numbers and some of the elementary operations with them such as summation, subtraction, multiplication and division. We have also introduced a complex exponential function in Sect. 3.4.4 and derived famous Euler's formulae (3.42) and (3.43). Here we shall give a different derivation of these identities based on Maclaurin series of sine, cosine and exponential functions.

Consider the function $f(x) = \cos x + i \sin x$ of a real variable x. Let us expand both sine and cosine functions in the Maclaurin series using Eqs. (7.50) and (7.51):

$$f(x) = \cos x + i \sin x = \sum_{n=0}^{\infty} \frac{(-1)^n x^{2n}}{(2n)!} + i \sum_{n=1}^{\infty} \frac{(-1)^{n-1} x^{2n-1}}{(2n-1)!}.$$

We notice that the expansion of the sine function contains only even powers of x, while the expansion of the cosine - odd powers. In addition, we notice that $-1 = i^2$ and hence the sign pre-factors in both expansions can be expressed via the complex i as $(-1)^n = i^{2n}$ and $(-1)^{n-1} = i^{2n-2}$, leading to

$$f(x) = \sum_{n=0}^{\infty} \frac{(ix)^{2n}}{(2n)!} + \sum_{n=1}^{\infty} \frac{(ix)^{2n-1}}{(2n-1)!} = \sum_{k=0}^{\infty} \frac{(ix)^k}{k!}.$$

In the last step we have combined both sums into one introducing appropriately the summation index. The obtained series is identical to the one of the exponential, Eq. (7.48), written for the argument ix. Hence, formally, one can write $f(x) = e^{ix}$ as well, which is Euler's formula (3.42).

Problem 7.54 Consider the following final sum containing N terms:

$$S_N(k) = \frac{1}{N} \sum_{n=1}^{N} e^{i2\pi nk/N},$$

where k is an integer. Show that

$$S_N(k) = \sum_{m=-\infty}^{\infty} \delta_{k,mN} = \delta_{k,0} + \delta_{k,N} + \delta_{k,-N} + \delta_{k,2N} + \delta_{k,-2N} \ldots,$$

i.e. $S_N(k)$ is an infinite sum of Kronecker delta symbols for all possible k that are multiples of N, including $k = 0$. This formula tells us that $S_N(k)$ is equal to one only when k is a multiple of N; in all other cases it is equal to zero. [Hint: *consider first the case of $k = mN$, i.e. when k is a multiple of N; then consider any different value of k and perform the summation explicitly.*]

7.3.7   Taylor Series for Functions of Many Variables

So far, we have discussed Taylor series for a function of a single variable. Here we shall generalize this result for a function of many variables.

Let $f(x_1, \ldots, x_n)$ be a function of n variables that we would like to expand around their values $x_1^0, x_2^0, \ldots, x_n^0$ into a power series. Writing $x_i = x_i^0 + \Delta x_i$ for every of the variables $(i = 1, \ldots, n)$, we anticipate that the Taylor series would contain terms with a product of deviations Δx_i in different powers, i.e. $(\Delta x_1)^{k_1} (\Delta x_2)^{k_2} \ldots (\Delta x_n)^{k_n}$, where $k_1, k_2, \ldots, k_n \geq 0$.

Similarly to the function of a single variable considered above, one can easily prove that such an expansion is unique. For this, we shall employ the same auxiliary function

$$F(t) = f\left(x_1^0 + t\Delta x_1, x_2^0 + t\Delta x_2, \ldots, x_n^0 + t\Delta x_n\right)$$

as in Sect. 5.9 where we considered the Taylor formula. This is a function of a single variable t and hence its Maclaurin expansion around $t = 0$ is unique as was proven in Sect. 7.3.3. Moreover, we can start from the Taylor formula (5.80) (but written for n variables),

$$f(x_1, \ldots, x_n) = f(x_1^0, \ldots, x_n^0) + F'(0) + \frac{F''(0)}{2} + \cdots + \frac{F^{(N)}(0)}{N!} + R_{N+1}(x_1, \ldots, x_n)$$

$$= S_N(x_1, \ldots, x_n) + R_{N+1}(x_1, \ldots, x_n),$$

where S_N is the partial sum, R_{N+1} the remainder, while

$$F^{(k)}(t) = \left(\sum_{i=1}^{n} \Delta x_i \frac{\partial}{\partial x_i}\right)^k F(t), \tag{7.69}$$

and consider the infinite series in which $N \to \infty$:

$$f(x_1, \ldots, x_n) = f(x_1^0, \ldots, x_n^0) + \sum_{N=0}^{\infty} \frac{F^{(N)}(0)}{N!}. \tag{7.70}$$

Theorem 7.28 formulated for a function of a single variable goes practically without change here: for the series (7.70) to converge, it is necessary and sufficient that the remainder tends to zero as $N \to \infty$.

We could have stopped here, but it is worthwhile rewriting the Taylor expansion in a more convenient form that resembles more the familiar result for a single variable function. To this end, we apply the multinomial expansion (1.177) to the power operator term of Eq. (7.69):

$$f(x_1, \ldots, x_n) = f(x_1^0, \ldots, x_n^0) + \sum_{N=0}^{\infty} \frac{1}{N!} \sum_{k_1+k_2+\cdots+k_N=N} \binom{N}{k_1, k_2, \ldots, k_N}$$

$$\times \left[\left(\frac{\partial}{\partial x_1}\right)^{k_1} \left(\frac{\partial}{\partial x_n}\right)^{k_2} \cdots \left(\frac{\partial}{\partial x_n}\right)^{k_n} F(t)\right]_{t=0} (\Delta x_1)^{k_1} (\Delta x_2)^{k_2} \ldots (\Delta x_n)^{k_n},$$

where the second sum in the right-hand side corresponds actually to N sums over all possible integers $k_1, k_2, \ldots, k_n \geq 0$ such that their sum is equal to N. Next, we shall use the explicit expression (1.176) for the multinomial coefficients, and replace $F(t)$ with the actual function:

$$
f(x_1, \ldots, x_n) = f(x_1^0, \ldots, x_n^0)
$$
$$
+ \sum_{N=0}^{\infty} \sum_{k_1 + \cdots + k_N = N} \frac{(\Delta x_1)^{k_1} \ldots (\Delta x_n)^{k_n}}{k_1! \cdots k_n!} \left[\frac{\partial^N f(x_1, \ldots, x_n)}{\partial x_1^{k_1} \ldots \partial x_n^{k_n}} \right]_0 ,
$$

where the subscript 0 indicates that the multiple derivative is calculated at $x_1^0, x_2^0, \ldots, x_n^0$. It is now easy to see that by taking all possible values of $N \geq 0$, all possible powers of the deviations are to be encountered. Hence, finally, we can write:

$$
f(x_1, \ldots, x_n) = f(x_1^0, \ldots, x_n^0)
$$
$$
+ \sum_{k_1=0}^{\infty} \cdots \sum_{k_n=0}^{\infty} \left[\frac{\partial^N f(x_1, \ldots, x_n)}{\partial x_1^{k_1} \ldots \partial x_n^{k_n}} \right]_0 \frac{\left(x_1 - x_1^0\right)^{k_1} \ldots \left(x_n - x_n^0\right)^{k_n}}{k_1! \cdots k_n!} .
$$
$$(7.71)$$

As an example, consider expanding the function $f(x, y) = e^x \cos y$ into the power series around $x = y = 0$ up to the second order. Denoting partial derivatives of f by the appropriate variables, we have: $f(0, 0) = 1$, $f_x(0, 0) = (e^x \cos y)_{x=y=0} = 1$, $f_y(0, 0) = 0$, $f_{xx}(0, 0) = 1$, $f_{xy}(0, 0) = 0$, $f_{yy}(0, 0) = -1$, so that we obtain:

$$
f(x, y) = 1 + f_x(0, 0)x + f_y(0, 0)y + \frac{f_{xx}(0, 0)}{2!0!}x^2 + \frac{f_{xy}(0, 0)}{1!1!}xy + \frac{f_{yy}(0, 0)}{0!2!}y^2 + \ldots
$$
$$
= 1 + x + \frac{x^2}{2} - \frac{y^2}{2} + \ldots .
$$

Problem 7.55 Work out an expansion of $f(x, y) = e^x \cos y$ around $x = y = 0$ to terms up to the fourth order:

$$
e^x \cos y = 1 + x + \frac{x^2}{2} - \frac{y^2}{2} + \frac{x^3}{6} - \frac{xy^2}{2} + \frac{x^4}{24} - \frac{x^2 y^2}{4} + \frac{y^4}{24} + \ldots .
$$

Also check that the same result is obtained if the appropriate expansions of the e^x and $\cos y$ are multiplied term-by-term.

7.4   Applications in Physics

7.4.1   Diffusion as a Random Walk

We have already discussed diffusion on a number of occasions (Sects. 6.3.3 and 6.9.2 and Problems 1.196, 5.11, 6.13, and 6.23); Here we shall approach this phenomenon from the viewpoint of a random walk.

Here we shall discuss Brownian motion whereby a large particle diffuses around in a liquid formed by smaller particles (e.g. a pollen in water as observed by the Scottish botanist Robert Brown in 1827). To describe diffusion of this larger particle, we shall adopt a simple random walk model.

Consider a periodic one-dimensional lattice with sites placed at equal distance a from each other, and let us assume that over time interval τ our particle performs jumps from the current site to either of the two nearest ones, on the right or left, with equal probabilities equal to 1/2. It may only jump between nearest sites performing a random 1D walk. The idea is to describe diffusion via such jumps first, and then make a transition towards the continuous limit by tending the lattice constant a and the time step τ both to zero at the same time. This would enable us to consider continuous diffusion of the Brownian particle.

However, before we indulge ourselves into this treatment, we have to discuss in a bit more detail the peculiarity associated with the limit. For this, let us consider ideas of a simple kinetic theory of 1D diffusion, see Fig. 7.6. Let us estimate the flux of particles through the cross section of area ΔS (the dark grey oval) in the perpendicular direction along the x-axis. Only particles between the light and dark grey ovals on both sides can reach the cross section. The number of particles diffusing to the right

$$\Delta N_{\rightarrow} = \frac{1}{2}n(x - \lambda)\,\Delta S \bar{v}\tau = \frac{1}{2}n(x - \lambda)\,\Delta S\lambda,$$

where $\bar{v} = \lambda/\tau$ is the particles average velocity, τ is the average collision time and $\lambda = \bar{v}\tau$ is the average mean free path between two consecutive collisions, while $n(x)$ is the concentration of the particles, and above it is calculated at $x - \lambda$. We assume that $n(x)$ decays along the positive x direction, so that the net diffusion is expected in this direction. Correspondingly, the number of particles diffusing in the opposite

Fig. 7.6 Particles on the left and the right of the cross section ΔS can cross it if they are within the mean free path length λ from it on both sides. The initial distribution $n(x)$ of the particles along the x-axis is also shown

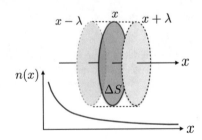

direction,

$$\Delta N_{\leftarrow} = \frac{1}{2} n(x + \lambda) \, \Delta S \lambda,$$

where $n(x)$ is calculated at $x + \lambda$, must be smaller, so that the net flux (the number of particles, crossing ΔS per unit time)

$$J = \frac{\Delta N_{\rightarrow} - \Delta N_{\leftarrow}}{\tau \Delta S} = \frac{\lambda}{2\tau} \left[n(x - \lambda) - n(x + \lambda) \right] \approx -\frac{\lambda}{2\tau} \left(2\frac{dn}{dx} \lambda \right) = -\frac{\lambda^2}{\tau} \frac{dn}{dx}$$

is positive in the direction to the right. Indeed, as the derivative $dn/dx < 0$, the flux in the positive x direction is positive, as expected. The diffusion coefficient D is defined as the proportionality constant in the relation for the flux, Sect. 6.3.3,

$$J = -D\frac{dn}{dx},$$

and hence we obtain $D \propto \lambda^2/\tau$.

Returning back to our random walk consideration, we may associate the mean free path λ with the distance a between nearest sites to which the particle can jump from the given site, and τ with the time of the jump. Hence, by taking the $a \to 0$ and $\tau \to 0$ limits, we have to keep the ratio a^2/τ finite.

Let $P(na, j\tau)$ be the probability to find the particle at the site number n (at the coordinate $x = na$) at the time step $j = 0, 1, 2, \ldots$ (at the time $t = j\tau$). Since the particle may only get onto that site from the nearest sites $n \pm 1$, and with equal probabilities, one can write:

$$P(na, j\tau) = \frac{1}{2} P\left((n - 1)a, (j - 1)\tau\right) + \frac{1}{2} P\left((n + 1)a, (j - 1)\tau\right),$$

where the first term corresponds to the particle jumping from the left site number $n - 1$, while the second term—from the right site $n + 1$. Thinking in advance of taking the continuous limit, we can also write this expression via the coordinate x and time t:

$$P(x, t) = \frac{1}{2} P(x - a, t - \tau) + \frac{1}{2} P(x + a, t - \tau). \tag{7.72}$$

Using just this so-called master equation, it is possible to find a partial differential equation the probability $P(x, t)$ must satisfy. Anticipating the limits $a \to 0$ and $\tau \to 0$, we can expand the right-hand side of the above equation with respect to "small" a and τ into the Taylor series. As $P(x \pm a, t - \tau)$ is the function of two variables, we have to apply the theory developed in Sect. 7.3.7:

$$P(x - a, t - \tau) = \sum_{n=0}^{\infty} \frac{1}{n!} \left(a\frac{\partial}{\partial x} + \tau\frac{\partial}{\partial t} \right)^n (-1)^n P(x, t).$$

Using the binomial expansion for the operator expression in the brackets raised into the n-th power, we obtain:

$$P(x - a, t - \tau) = P(x, t)$$
$$+ \sum_{n=1}^{\infty} \frac{(-1)^n}{n!} \sum_{m=0}^{n} \binom{n}{m} a^m \tau^{n-m} \left[\left(\frac{\partial}{\partial x} \right)^m \left(\frac{\partial}{\partial t} \right)^{n-m} P(x, t) \right],$$

where we have specifically extracted the $n = 0$ term giving simply $P(x, t)$. Similarly,

$$P(x + a, t - \tau) = P(x, t)$$
$$+ \sum_{n=1}^{\infty} \frac{(-1)^n}{n!} \sum_{m=0}^{n} \binom{n}{m} (-a)^m \tau^{n-m} \left[\left(\frac{\partial}{\partial x} \right)^m \left(\frac{\partial}{\partial t} \right)^{n-m} P(x, t) \right],$$

so that in Eq. (7.72) $P(x, t)$ cancels out on both sides leading to the equation

$$\sum_{n=1}^{\infty} \frac{(-1)^n}{n!} \sum_{m=0}^{n} \binom{n}{m} \left[1 + (-1)^m \right] a^m \tau^{n-m-1} \left[\left(\frac{\partial}{\partial x} \right)^m \left(\frac{\partial}{\partial t} \right)^{n-m} P(x, t) \right] = 0.$$

For convenience, both sides of this equation we have divided by τ. It is seen that only even values of m contribute, because the factor $1 + (-1)^m$ is zero for odd values of m. Hence, we can simplify:

$$\sum_{n=1}^{\infty} \frac{(-1)^n}{n!} \sum_{k=0}^{[n/2]} \binom{n}{2k} a^{2k} \tau^{n-2k-1} \left[\left(\frac{\partial}{\partial x} \right)^{2k} \left(\frac{\partial}{\partial t} \right)^{n-2k} P(x, t) \right] = 0, \quad (7.73)$$

where we replaced the summation over $m = 2k$ by the summation over k between 0 and the even part $[n/2]$ of $n/2$.

Now we can try applying the limits $a \to 0$ and $\tau \to 0$ with the condition that a^2/τ must be kept finite. To this end, we need to be mindful of the expression $a^{2k} \tau^{n-2k-1} = \left(a^2/\tau \right)^k \tau^{n-k-1}$. We shall analyse the terms one by one. For $n = 1$ there is only one value of $k = 0$, leading to the contribution

$$\frac{-1}{1!} \binom{1}{0} a^0 \tau^0 \left(\frac{\partial}{\partial x} \right)^0 \left(\frac{\partial}{\partial t} \right)^{1-0} P(x, t) = -\frac{\partial}{\partial t} P(x, t).$$

When $n = 2$, two values of k are possible: $k = 0$ and $k = 1$. At $k = 0$ the corresponding term contains $\tau^{2-0-1} = \tau^1$ and tends to zero in the limit; at $k = 1$ we obtain the contribution

$$\frac{(-1)^2}{2!} \binom{2}{2} \left(\frac{a^2}{\tau} \right)^1 \tau^{2-1-1} \left(\frac{\partial}{\partial x} \right)^2 \left(\frac{\partial}{\partial t} \right)^{2-2} P(x, t) = D \frac{\partial^2}{\partial x^2} P(x, t),$$

where $D = a^2/2\tau$ is (as we shall see shortly) is the diffusion coefficient (constant). At larger values of n ($n \geq 3$) the expression $n - k - 1$ is always a positive integer (since $k \leq [n/2]$) leading to the $\tau^{n-k-1} \to 0$ in the limit. Therefore, in Eq. (7.73) only two terms contribute after taking the limits, and we finally obtain the diffusion equation:

$$\frac{\partial}{\partial t} P(x, t) = D \frac{\partial^2}{\partial x^2} P(x, t), \qquad (7.74)$$

in which D is the diffusion constant. This equation describes the time evolution of the probability $P(x, t)$ to find a particle at position x at time t following their specific initial distribution; the concentration of the particles $n(x, t) = n_0 P(x, t)$ will obviously satisfy the same equation (n_0 is the average concentration of the particles).

Problem 7.56 Derive the diffusion equation (7.74) starting from the master equation (7.72) written for the time $t \to t + \tau$, i.e.

$$P(x, t + \tau) = \frac{1}{2} P(x - a, t) + \frac{1}{2} P(x + a, t). \qquad (7.75)$$

In this calculation you would only need the Taylor expansion for a function of a single variable (with respect to t in the left and with respect to x in the right-hand sides).

Problem 7.57 Consider now the case in which the probabilities to the right and left, p and q, are different, while, of course, $p + q = 1$. This corresponds to the case when, apart from the diffusive (stochastic) motion of the particles there is also a translational motion, either to the right or left (depending on the relationship between the two probabilities) due to, e.g. an external field. You may use either the modified Eq. (7.72) or (7.75) for your calculation. Note that in this case, apart from considering the ratio a^2/τ being finite when taking the limits, the ratio a/τ (entering the velocity $v = (p - q)a/\tau$ of the translational motion) must also be considered finite. Show that in this case the diffusion equation reads

$$\frac{\partial}{\partial t} P(x, t) = -v \frac{\partial}{\partial x} P(x, t) + D \frac{\partial^2}{\partial x^2} P(x, t). \qquad (7.76)$$

Problem 7.58 Generalize your results for the case of N dimensions (e.g. $N = 2, 3$). Show that in this case the diffusion equation has the form

$$\frac{\partial}{\partial t} P(x, t) = D \Delta_N P(x, t),$$

where Δ_N is the N-dimensional Laplacian and $D = a^2 / (k\tau)$, with k being the number of nearest neighbours (4 and 6 in the 2D and 3D cases, respectively).

7.4.2   Coulomb Potential in a Periodic Crystal

Consider a finite but very large chunk of a crystal composed of N identical neutral unit cells ($N \gg 1$). Each unit cell contains a certain number of atoms at positions $\mathbf{X}_s$ with charges q_s; here the so-called basis index $s = 1, \ldots, n$ numbers all atoms in each unit cell. As the cells are neutral, $\sum_s q_s = 0$. Each unit cell is constructed as a parallelepiped based on three non-coplanar vectors $\mathbf{a}_1$, $\mathbf{a}_2$ and $\mathbf{a}_3$ (basic translations), and hence each unit cell can be characterized by three integers $n_i = 0, \pm 1, \pm 2, \ldots$ ($i = 1, 2, 3$) and correspondingly by a vector (the lattice translation) $\mathbf{L} = n_1 \mathbf{a}_1 + n_2 \mathbf{a}_2 + n_3 \mathbf{a}_3$. The three integers n_1, n_2, n_3 take all the necessary values to up to some very large positive and negative values sufficient to reach each and every unit cell in the crystal, see a 2D illustration in Fig. 7.7. We shall assume that the centre of the coordinate system is chosen exactly in the centre of a parallelepiped of the 0-th unit cell (corresponds to the zero lattice vector, $\mathbf{L} = \mathbf{0}$) that is located somewhere close to the centre of the whole crystal. Therefore, an atom of type s in the unit cell (n_1, n_2, n_3) has the position

$$\mathbf{L} + \mathbf{X}_s = n_1 \mathbf{a}_1 + n_2 \mathbf{a}_2 + n_3 \mathbf{a}_3 + \mathbf{X}_s.$$

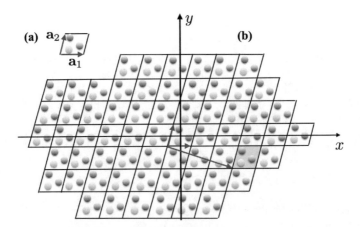

Fig. 7.7 A finite fragment of a hypothetical 2D periodic crystal of a certain shape in **b** composed of identical unit cells shown in **a**, each containing three different atoms (shown by different colours). The basic translations $\mathbf{a}_1$ and $\mathbf{a}_2$ are shown by blue vectors. One of the lattice translations $\mathbf{L} = n_1 \mathbf{a}_1 + n_2 \mathbf{a}_2$ (in red) to a particular unit cell (shaded) is also indicated as an example (for $n_1 = 3$ and $n_2 = -1$). The centre of the coordinate system is chosen in the centre of the 0-th unit cell corresponding to $n_1 = n_2 = 0$

We would like to calculate the Coulomb (electrostatic) potential due to all atoms somewhere well inside our sample crystal, i.e. at a point $\mathbf{r}$ in the central region of it near the centre of the coordinate system,

$$\phi_N(\mathbf{r}) = \sum_{\{L\}} \sum_s \frac{q_s}{|\mathbf{L} + \mathbf{X}_s - \mathbf{r}|}, \tag{7.77}$$

where the second summation runs over all atoms in the unit cell, while the first one is performed over all possible values of the integers $\{n_1, n_2, n_3\}$ required for the given finite crystal to describe each and every unit cell. Therefore, the first summation runs over all possible lattice vectors $\mathbf{L}$ corresponding to the particular size and shape of our finite crystal (indicated by the curly brackets), and hence is actually finite.

As the summation is over a finite number of terms, it definitely has a well-defined value. However, calculating this summation by adding terms one by one is impractical due to a very large number of the unit cells (lattice vectors), i.e. the first summation from the practical point of view is basically infinite. On the other hand, the first summation for the given value of s resembles the harmonic series. In fact, for $\mathbf{r} = 0$ and excluding the $n_1 = 0$ term it becomes exactly the one in the 1D case:

$$q_s \sum_{\{L \neq 0\}} \frac{1}{|\mathbf{L}|} = \frac{q_s}{|\mathbf{a}_1|} \sum_{\{n_1 \neq 0\}} \frac{1}{|n_1|}.$$

This sum, if considered as an infinite series, does not converge (Sect. 7.1), and hence we may anticipate that care is needed in performing the lattice summation in Eq. (7.77) in the general 3D case as well. We shall show in the following that the potential in the central region of the crystal can actually be presented as a sum of two types of contributions: one that is represented by quickly convergent series (the so-called Ewald sum) and another one that is related to the shape of the finite crystal fragment, i.e. corresponds to surface effects.[14]

We shall start by employing the following identity associated with the so-called error function $\mathrm{erf}(x)$ introduced in Eq. (7.58) and its complementary function

$$\mathrm{erfc}(x) = 1 - \mathrm{erf}(x) = \frac{2}{\sqrt{\pi}} \int_x^\infty e^{-t^2} dt. \tag{7.78}$$

With these definitions, one can write:

$$\frac{1}{x} = \frac{\mathrm{erfc}(Gx) + \mathrm{erf}(Gx)}{x}$$

$$= \frac{\mathrm{erfc}(Gx)}{x} + \frac{2}{\sqrt{\pi}} \int_0^\epsilon e^{-t^2 x^2} dt + \frac{2}{\sqrt{\pi}} \int_\epsilon^G e^{-t^2 x^2} dt,$$

---

[14] We shall closely follow here the treatment presented in I. I. Tupizin and I. V. Abarenkov - Phys. Stat. Sol. (b) **82**, p. 99 (1977) and L. N. Kantorovich and I. I. Tupitsyn - J. Phys.: Condens. Matter, **11**, p. 6159 (1999).

where $\epsilon > 0$ is some small constant and G a positive parameter. Above, we have split the integral $\int_0^{Gx}$ of erf(x) into two integrals, $\int_0^{\epsilon x} + \int_{\epsilon x}^{Gx}$, and then have made an appropriate change of the integration variable $(t \to t/x)$ to eliminate the $1/x$ prefactor in each case. At the end of the calculation we intend to take the limit $\epsilon \to +0$. Using this identity, the potential (7.77) can be written as follows:

$$
\phi_N(\mathbf{r}) = \sum_s q_s \sum_{\{\mathbf{L}\}} \frac{\text{erfc}\,(G\,|\mathbf{r} - \mathbf{L} - \mathbf{X}_s|)}{|\mathbf{r} - \mathbf{L} - \mathbf{X}_s|}
$$

$$
+ \sum_s q_s \frac{2}{\sqrt{\pi}} \int_\epsilon^G \left(\sum_{\{\mathbf{L}\}} e^{-t^2|\mathbf{r} - \mathbf{L} - \mathbf{X}_s|^2} \right) dt + \phi_N^{ext}(\mathbf{r}),
$$

(7.79)

where

$$
\phi_N^{ext}(\mathbf{r}) = \frac{2}{\sqrt{\pi}} \sum_{\{\mathbf{L}\}} \sum_s q_s \int_0^\epsilon e^{-t^2|\mathbf{r} - \mathbf{L} - \mathbf{X}_s|^2} dt.
$$

(7.80)

The first two terms in the right-hand side of Eq. (7.79) represent very quickly converging sums over $\mathbf{L}$ due to a very fast decay of the complementary error function (erfc$(x) \sim e^{-x^2}$ at large x) in the first term and of the exponential function in the second (note that t there is positive and strictly non-zero). Both summations can therefore be replaced by infinite series $(N \to \infty)$ over all possible lattice vectors $\mathbf{L}$ (with n_i for $i = 1, 2, 3$ ranging between $\pm\infty$). In fact, the second term in the $N \to \infty$ limit can be conveniently rewritten into a different form using the so-called reciprocal summation; moreover, the t-integration becomes possible analytically after which taking the $\epsilon \to 0$ limit becomes straightforward. Since performing this calculation is based on theory of Fourier series, we shall not work it out here.[15]

Hence,

$$
\phi_N(\mathbf{r}) = \sum_s q_s \phi_{Ew}(\mathbf{X}_s - \mathbf{r}) + \phi_N^{ext}(\mathbf{r}),
$$

where $\phi_{Ew}(\mathbf{r})$ arises from the first two terms in Eq. (7.79); it is the so-called Ewald potential that is a well-defined function of $\mathbf{r}$ (note that at $\mathbf{r} = \mathbf{0}$ one has to remove the self-action of the atom $s = 1$ in the zero unit cell). From now on we shall only be concerned with the correction term $\phi_N^{ext}(\mathbf{r})$ of Eq. (7.80) that requires a rather careful consideration.

Making a substitution $\lambda = t\,|\mathbf{r} - \mathbf{L}|$ in the integral and introducing the shortcuts $\mathbf{z} = \mathbf{L} - \mathbf{r}$ and

$$
\sigma = \frac{1}{z^2} \left(X_s^2 + 2\mathbf{X}_s \cdot \mathbf{z} \right),
$$

(7.81)

---

[15] As the sum over the lattice in the round brackets in Eq. (7.79) can be extended to the whole infinite lattice, the function of $\mathbf{r}$ becomes periodic with respect to any lattice translation $\mathbf{L}'$, i.e. replacing $\mathbf{r} \to \mathbf{r} + \mathbf{L}'$ does not change the sum; hence, it can then be expanded into a Fourier series. The Fourier series will only be considered in Chap. 3 of Vol. II of this course.

we obtain:

$$\phi_N^{ext}(\mathbf{r}) = \frac{2}{\sqrt{\pi}} \sum_{\{L\}} \sum_s q_s \frac{1}{z} \int_0^{\epsilon z} e^{-\lambda^2} e^{-\sigma \lambda^2} d\lambda.$$

The exponential $e^{-\sigma \lambda^2}$ can be expanded into the Maclaurin series that (for the exponential function) converges for any values of its variable:

$$\phi_N^{ext}(\mathbf{r}) = \frac{2}{\sqrt{\pi}} \sum_{\{L\}} \sum_s q_s \frac{1}{z} \int_0^{\epsilon z} e^{-\lambda^2} \left[1 - \sigma \lambda^2 + \frac{\sigma^2}{2} \lambda^4 - \dots \right] d\lambda$$

$$= \frac{2}{\sqrt{\pi}} \sum_{\{L\}} \sum_s q_s \frac{1}{z} \int_0^{\epsilon z} e^{-\lambda^2} \left[-\sigma \lambda^2 + \frac{\sigma^2}{2} \lambda^4 - \dots \right] d\lambda$$

$$= \frac{2}{\sqrt{\pi}} \sum_{\{L\}} \sum_s q_s \int_0^{\epsilon z} e^{-\lambda^2} \left[-\frac{X_s^2 + 2\mathbf{X}_s \cdot \mathbf{z}}{z^3} \lambda^2 + \frac{(X_s^2 + 2\mathbf{X}_s \cdot \mathbf{z})^2}{2z^5} \lambda^4 - \dots \right] d\lambda,$$

$$(7.82)$$

where the very first term inside the square brackets on the first line does not contribute due to electroneutrality of the unit cell. Substituting here the explicit expression for σ, Eq. (7.81), as has been done in the last line, various lattice sums have emerged. For instance, we have to deal with the lattice sum

$$\sum_{\{L\}} \frac{1}{z^n} \int_0^{\epsilon z} e^{-\lambda^2} \lambda^m d\lambda = \sum_{\{L\}} \frac{1}{|\mathbf{L} - \mathbf{r}|^n} \int_0^{\epsilon |\mathbf{L} - \mathbf{r}|} e^{-\lambda^2} \lambda^m d\lambda \qquad (7.83)$$

for various positive integers n and m. Since the integral here is a monotonously increasing function of its upper limit and the lattice sum contains only positive terms, its sum can be estimated by replacing the upper integral limit with infinity,

$$\sum_{\{L\}} \frac{1}{z^n} \int_0^{\epsilon z} e^{-\lambda^2} \lambda^m d\lambda < \sum_{\{L\}} \frac{1}{z^n} \int_0^{\infty} e^{-\lambda^2} \lambda^m d\lambda = \left(\int_0^{\infty} e^{-\lambda^2} \lambda^m d\lambda \right) \sum_{\{L\}} \frac{1}{z^n}.$$

As we learned in Sect. 7.2.5, the lattice sum in the right-hand side converges uniformly for $n \geq 4$. Therefore, when considering different lattice sums (7.83) that appear in Eq. (7.82), it is possible to extend the summation over lattice vectors $\mathbf{L}$ to the whole infinite lattice when $n \geq 4$ (as the lattice sums converge and the crystal is very large); moreover, since these lattice sums converge uniformly, we can apply the limit $\epsilon \to +0$ to each term of the sum (Theorem 7.17). However, the integral in Eq. (7.83) tends to zero in this limit, which means that for $n \geq 4$ the lattice sums are zero in the limit. A similar consideration is applied to other lattice sums that appear in Eq. (7.82), such as, for instance,

$$\sum_{\{L\}} \frac{z_\alpha}{z^n} \int_0^{\epsilon z} e^{-\lambda^2} \lambda^m d\lambda, \quad \sum_{\{L\}} \frac{z_\alpha z_\beta}{z^n} \int_0^{\epsilon z} e^{-\lambda^2} \lambda^m d\lambda,$$

$$\sum_{\{L\}} \frac{z_\alpha z_\beta z_\gamma}{z^n} \int_0^{\epsilon z} e^{-\lambda^2} \lambda^m d\lambda,$$

$$(7.84)$$

and so on (Greek indices correspond to Cartesian components of vectors). These can also be estimated in a similar way by uniformly converging lattice sums (Sect. 7.2.5). Hence, the first lattice sum in Eq. (7.84) converges uniformly for $n \geq 5$ and can be replaced by zero in the $\epsilon \to +0$ limit, the second sum can similarly be replaced by zero whenever $n \geq 6$, the third - when $n \geq 7$, and so on.

Guided by this discussion, one can easily establish that only the linear in σ terms in Eq. (7.82) and just one appearing in σ^2 will contribute; all other (higher order) terms will not.

Problem 7.59 Consider in detail all three terms arising due to σ^2 and all four terms due to σ^3 and make sure that there is only one term contributing in the former and none in the latter. Then consider the general term coming from σ^n with $n \geq 4$ and prove that none of the terms contribute in the $\epsilon \to +0$ limit.

Introducing auxiliary functions

$$\xi(z) = \frac{1}{z^3} \int_0^{\epsilon z} e^{-\lambda^2} \lambda^2 d\lambda, \quad \eta_\alpha(\mathbf{z}) = z_\alpha \xi(z)$$

and

$$\chi_{\alpha\beta}(\mathbf{z}) = \frac{z_\alpha z_\beta}{z^5} \int_0^{\epsilon z} e^{-\lambda^2} \lambda^4 d\lambda = \frac{z_\alpha z_\beta}{2z^2} \left(3\xi(z) - \epsilon^3 e^{-\epsilon^2 z^2} \right) \tag{7.85}$$

(we have used integration by parts in the last line), we obtain:

$$\phi_N^{ext}(\mathbf{r}) = \frac{2}{\sqrt{\pi}} \sum_{\{L\}} \sum_s q_s \left[-X_s^2 \xi(z) - 2\sum_\alpha X_{s\alpha} \eta_\alpha(\mathbf{z}) + 2\sum_{\alpha\beta} X_{s\alpha} X_{s\beta} \chi_{\alpha\beta}(\mathbf{z}) \right]$$

$$= \frac{2}{\sqrt{\pi}} \sum_{\{L\}} \left[-Q_0 \xi(z) - 2\sum_\alpha P_\alpha \eta_\alpha(\mathbf{z}) + 2\sum_{\alpha\beta} \tilde{Q}_{\alpha\beta} \chi_{\alpha\beta}(\mathbf{z}) \right],$$
$$\tag{7.86}$$

where $Q_0 = \sum_s q_s X_s^2$ is the unit cell spheropol, $\mathbf{P} = \sum_s q_s \mathbf{X}_s$ its dipole moment, and $\tilde{Q}_{\alpha\beta} = \sum_s q_s X_{s\alpha} X_{s\beta}$ is related to the quadrupole moment

$$Q_{\alpha\beta} = \sum_s q_s \left(3X_{s\alpha} X_{s\beta} - \delta_{\alpha\beta} X_s^2 \right) = 3\tilde{Q}_{\alpha\beta} - \delta_{\alpha\beta} Q_0 \tag{7.87}$$

of the unit cell.

In the following steps we shall manipulate the lattice sums into volume integrals, and will have to do it separately for each term appearing in Eq. (7.86). Let us introduce a vector $\mathbf{x}$ within the 0-th unit cell, and then write: $\mathbf{z} = \mathbf{L} - \mathbf{r} = (\mathbf{L} + \mathbf{x}) - (\mathbf{r} + \mathbf{x})$.

In the first term of Eq. (7.86) we shall expand $\xi(z)$ in the power series around $\mathbf{z}_0 = \mathbf{L} + \mathbf{x}$:

$$\xi(z) = \xi(z_0) - \sum_\alpha (r_\alpha + x_\alpha)\xi_\alpha(\mathbf{z}_0) + \frac{1}{2}\sum_{\alpha\beta}(r_\alpha + x_\alpha)(r_\beta + x_\beta)\xi_{\alpha\beta}(\mathbf{z}_0) - \ldots,$$

where

$$\xi_\alpha(\mathbf{x}) = \frac{\partial\xi(x)}{\partial x_\alpha} = \frac{d\xi(x)}{dx}\frac{\partial x}{\partial x_\alpha} = \frac{d\xi(x)}{dx}\frac{x_\alpha}{x}$$

$$= -\frac{3x_\alpha}{x^5}\int_0^{\epsilon x} e^{-\lambda^2}\lambda^2 d\lambda + \epsilon^3 \frac{x_\alpha}{x^2}e^{-\epsilon^2 x^2}$$

and similarly for $\xi_{\alpha\beta}(\mathbf{x}) = \frac{\partial^2}{\partial x_\alpha \partial x_\beta}\xi(x)$ and any higher order terms. All these contain either a λ-integral with the corresponding $1/z_0^n = 1/|\mathbf{L} + \mathbf{x}|^n$ term of a sufficiently large value of n or the exponential term $\exp(-\epsilon^2|\mathbf{L} + \mathbf{x}|^2)$. It is easy to see that all these terms result in uniformly converging sums and hence tend to zero in the $\epsilon \to +0$ limit. We also note that the function $\xi(z)$ behaves well at $z = 0$, which can be easily checked by making the substitution $t = \lambda/z$ in the integral in its definition above. Therefore, in the lattice sum in the first term in Eq. (7.86) no need to consider the case of $\mathbf{L} = 0$ separately.

Hence, only $\xi(z_0)$ in this term contributes to the lattice sum in the limit, i.e. we have to consider

$$\sum_{\{\mathbf{L}\}} \xi(z) = \sum_{\{\mathbf{L}\}} \xi(|\mathbf{L} + \mathbf{x}|).$$

Integrating both sides of this expression with respect to $\mathbf{x}$ over the volume of the unit cell (the left-hand side does not depend on $\mathbf{x}$), we obtain:

$$\sum_{\{\mathbf{L}\}} \xi(z) = \frac{1}{v_c}\sum_{\{\mathbf{L}\}}\int_{cell}\xi(|\mathbf{L} + \mathbf{x}|)d\mathbf{x} = \frac{1}{v_c}\int_{V_N}\xi(\mathbf{x})d\mathbf{x},$$

where $v_c = [\mathbf{a}_1, \mathbf{a}_2, \mathbf{a}_3] = |\mathbf{a}_1 \cdot [\mathbf{a}_2 \times \mathbf{a}_3]|$ is the volume of the unit cell, Eq. (1.111). In the last passage we noticed that the integrals taken over a unit cell are summed over the whole finite crystal, and hence this results in the integral taken over the whole volume V_N of the crystal. So, the first term in Eq. (7.86) contributes

$$\phi_N^{ext}(\mathbf{r})_1 = \frac{-2Q_0}{\sqrt{\pi}v_c}\int_{V_N}\xi(\mathbf{x})d\mathbf{x}. \tag{7.88}$$

Similarly other terms entering Eq. (7.86) are considered. When expanding

$$\chi_{\alpha\beta}(\mathbf{z}) = \chi_{\alpha\beta}(\mathbf{z}_0) - \sum_\gamma (r_\gamma + x_\gamma)\chi_{\alpha\beta\gamma}(\mathbf{z}_0) + \ldots,$$

where $\chi_{\alpha\beta\gamma}(\mathbf{z}) = \frac{\partial}{\partial z_\gamma}\chi_{\alpha\beta}(\mathbf{z})$, already the linear order term results in the converging series, so only the zero-order term survives in the $\epsilon \to +0$ limit, as in the previous case. Hence, we can similarly write

$$\sum_{\{\mathbf{L}\}} \chi_{\alpha\beta}(\mathbf{z}) = \frac{1}{v_c} \int_{V_N} \chi_{\alpha\beta}(\mathbf{x})d\mathbf{x},$$

so that the contribution of the third term in Eq. (7.86) is:

$$\phi_N^{ext}(\mathbf{r})_3 = \frac{2}{\sqrt{\pi}v_c} \sum_{\alpha\beta} 2\tilde{Q}_{\alpha\beta} \int_{V_N} \chi_{\alpha\beta}(\mathbf{x})d\mathbf{x}. \tag{7.89}$$

Here as well we check explicitly that $\chi_{\alpha\beta}(\mathbf{z})$ behaves well when $\mathbf{z} \to \mathbf{0}$ and hence the $\mathbf{L} = \mathbf{0}$ contribution did not require any special treatment.

Finally, consider the second contribution into Eq. (7.86). Here,

$$\eta_\alpha(\mathbf{z}) = \eta_\alpha(\mathbf{z}_0) - \sum_\beta (r_\beta + x_\beta)\eta_{\alpha\beta}(\mathbf{z}_0) + \frac{1}{2} \sum_{\beta\gamma}(r_\beta + x_\beta)(r_\gamma + x_\gamma)\eta_{\alpha\beta\gamma}(\mathbf{z}_0) - \dots,$$

where, e.g.

$$\eta_{\alpha\beta}(\mathbf{x}) = \frac{\partial}{\partial x_\beta}\eta_\alpha(\mathbf{x}) = x^3 T_{\alpha\beta}(\mathbf{x})\xi(x) + \epsilon^3 \frac{x_\alpha x_\beta}{x^2}e^{-\epsilon^2 x^2} \tag{7.90}$$

with $T_{\alpha\beta}(\mathbf{x}) = \delta_{\alpha\beta}x^{-3} - 3x_\alpha x_\beta x^{-5}$. In this case we have to keep the zero- and first-order terms, all other terms vanish in the $\epsilon \to +0$ limit. Hence, the corresponding lattice sum to consider is

$$\sum_{\{\mathbf{L}\}} \eta_\alpha(\mathbf{z}) = \sum_{\{\mathbf{L}\}} \eta_\alpha(\mathbf{L} + \mathbf{x}) - \sum_\beta (r_\beta + x_\beta) \sum_{\{\mathbf{L}\}} \eta_{\alpha\beta}(\mathbf{L} + \mathbf{x}),$$

which after integrating with respect to $\mathbf{x}$ over the unit cell volume gives

$$\sum_{\{\mathbf{L}\}} \eta_\alpha(\mathbf{z}) = \frac{1}{v_c} \int_{V_N} \eta_\alpha(\mathbf{x})d\mathbf{x} - \frac{1}{v_c} \sum_{\alpha\beta} \left[r_\beta \int_{V_N} \eta_{\alpha\beta}(\mathbf{x})d\mathbf{x} + \sum_{\{\mathbf{L}\}} \int_{cell} x_\beta \eta_{\alpha\beta}(\mathbf{L} + \mathbf{x})d\mathbf{x} \right].$$

The last term is zero due to symmetry if we assume that for each lattice vector $\mathbf{L}$ in our big finite crystal there is also a cell with the vector $-\mathbf{L}$. Indeed,

$$\sum_{\{\mathbf{L}\}} \int_{cell} x_\beta \eta_{\alpha\beta}(\mathbf{L} + \mathbf{x})d\mathbf{x} = \frac{1}{2} \sum_{\{\mathbf{L}\}} \int_{cell} x_\beta \left[\eta_{\alpha\beta}(\mathbf{L} + \mathbf{x}) + \eta_{\alpha\beta}(-\mathbf{L} + \mathbf{x}) \right] d\mathbf{x}$$

$$= \frac{1}{2} \sum_{\{\mathbf{L}\}} \int_{cell} x_\beta \left[\eta_{\alpha\beta}(\mathbf{L} + \mathbf{x}) + \eta_{\alpha\beta}(\mathbf{L} - \mathbf{x}) \right] d\mathbf{x},$$

because $\eta_{\alpha\beta}(\mathbf{x})$ is symmetric with respect to the substitution $\mathbf{x} \to -\mathbf{x}$, i.e. $\eta_{\alpha\beta}(-\mathbf{x}) = \eta_{\alpha\beta}(\mathbf{x})$, see Eq. (7.90). Changing the integration variable in the second term from $\mathbf{x}$ to $-\mathbf{x}$, we obtain[16]

$$\frac{1}{2} \sum_{\{\mathbf{L}\}} \int_{cell} x_\beta \left[\eta_{\alpha\beta}(\mathbf{L} + \mathbf{x}) - \eta_{\alpha\beta}(\mathbf{L} + \mathbf{x}) \right] d\mathbf{x},$$

which is zero. Therefore,

$$\sum_{\{\mathbf{L}\}} \eta_\alpha(\mathbf{z}) = \frac{1}{v_c} \int_{V_N} \eta_\alpha(\mathbf{x}) d\mathbf{x} - \frac{1}{v_c} \sum_\beta r_\beta \int_{V_N} \eta_{\alpha\beta}(\mathbf{x}) d\mathbf{x}$$

and the second contribution to the potential becomes

$$\phi_N^{ext}(\mathbf{r})_2 = \frac{2}{\sqrt{\pi} v_c} \sum_\alpha \int_{V_N} P_\alpha \left[-2\eta_\alpha(\mathbf{x}) + 2 \sum_\beta r_\beta \eta_{\alpha\beta}(\mathbf{x}) \right] d\mathbf{x}. \qquad (7.91)$$

Collecting all three non-zero contributions from Eqs. (7.88), (7.91) and (7.89), we obtain:

$$\phi_N^{ext}(\mathbf{r}) = \frac{2}{\sqrt{\pi} v_c} \int_{V_N} \left\{ -Q_0 \xi(x) - 2 \sum_\alpha P_\alpha \eta_\alpha(\mathbf{x}) + 2 \sum_{\alpha\beta} P_\alpha r_\beta \eta_{\alpha\beta}(\mathbf{x}) \right.$$
$$\left. + 2 \sum_{\alpha\beta} \tilde{Q}_{\alpha\beta} \chi_{\alpha\beta}(\mathbf{x}) \right\} d\mathbf{x}. \qquad (7.92)$$

The first and the last terms within the curly brackets can be combined after replacing $\chi_{\alpha\beta}(\mathbf{x})$ with its detailed expression (7.85) and using the definition (7.87) of the quadrupole moment. It is easy to check then, noting that $\sum_{\alpha\beta} \delta_{\alpha\beta} Q_{\alpha\beta} = \sum_\alpha Q_{\alpha\alpha} = 0$, that the potential becomes

$$\phi_N^{ext}(\mathbf{r}) = \frac{2}{\sqrt{\pi} v_c} \int_{V_N} \left\{ -\frac{1}{3} \sum_{\alpha\beta} Q_{\alpha\beta} x^3 \xi(x) T_{\alpha\beta}(\mathbf{x}) - \sum_{\alpha\beta} \tilde{Q}_{\alpha\beta} \frac{x_\alpha x_\beta}{x^2} \epsilon^3 e^{-\epsilon^2 x^2} \right.$$
$$\left. -2 \sum_\alpha P_\alpha \eta_\alpha(\mathbf{x}) + 2 \sum_{\alpha\beta} P_\alpha r_\beta \eta_{\alpha\beta}(\mathbf{x}) \right\} d\mathbf{x}.$$

$$(7.93)$$

Now, each term here will be dealt with in order of their appearance, one by one. In the first term we write

$$x^3 \xi(x) = \int_0^{\epsilon x} e^{-\lambda^2} \lambda^2 d\lambda = f(0) - f(\epsilon x) = \frac{\sqrt{\pi}}{4} - f(\epsilon x),$$

---

[16] Recall that the centre of the coordinate system was chosen in the centre of the 0-th unit cell parallelepiped.

where $f(x) = \int_x^\infty e^{-\lambda^2} \lambda^2 d\lambda$ is a quickly decaying function of its argument (it is proportional to $e^{-\epsilon^2 x^2}$) and $f(0) = \sqrt{\pi}/4$ according to Eq. (7.28). This yields the contribution

$$-\frac{1}{6v_c} \sum_{\alpha\beta} Q_{\alpha\beta} \overline{T}_{\alpha\beta} + \frac{2}{3v_c\sqrt{\pi}} \sum_{\alpha\beta} Q_{\alpha\beta} \int_{V_N} f(\epsilon x) T_{\alpha\beta}(\mathbf{x}) d\mathbf{x} = -\frac{1}{6v_c} \sum_{\alpha\beta} Q_{\alpha\beta} \overline{T}_{\alpha\beta},$$

$$(7.94)$$

where

$$\overline{T}_{\alpha\beta} = \int T_{\alpha\beta}(\mathbf{x}) d\mathbf{x},$$

since in the integral, containing $f(\epsilon x)$, because of the fast decay of that function, the integration can be extended to the whole space and then the integral is equal to zero for any Cartesian indices due to symmetry.

Problem 7.60 Show that the all space volume integrals $\int \varphi(r) r_\alpha d\mathbf{r}$ and $\int \varphi(r) T_{\alpha\beta}(\mathbf{r}) d\mathbf{r}$, where the function $\varphi(r)$ depends only on the length r of the vector $\mathbf{r}$ and $T_{\alpha\beta}(\mathbf{r}) = \delta_{\alpha\beta} r^{-3} - 3 r_\alpha r_\beta r^{-5}$, are equal to zero for any α, β. [Hint: *use spherical coordinates.*]

The second term in Eq. (7.93) contains the exponential term and hence can be extended to the whole space; then it is manipulated as follows:

$$-\frac{2\epsilon^3}{\sqrt{\pi}v_c} \sum_{\alpha\beta} \tilde{Q}_{\alpha\beta} \int \frac{x_\alpha x_\beta}{x^2} e^{-\epsilon^2 x^2} d\mathbf{x} = \left| \begin{matrix} \mathbf{y} = \epsilon \mathbf{x} \\ d\mathbf{y} = \epsilon^3 d\mathbf{x} \end{matrix} \right| = -\frac{2}{\sqrt{\pi}v_c} \sum_{\alpha\beta} \tilde{Q}_{\alpha\beta} \int \frac{y_\alpha y_\beta}{y^2} e^{-y^2} d\mathbf{y}.$$

Again, as it easily shown by direct calculation in spherical coordinates, the integral is zero for $\alpha \neq \beta$ and is the same for any $\alpha = \beta$. Therefore, one can replace y_α^2 with $\frac{1}{3} y^2$, and we obtain the contribution:

$$-\frac{2}{\sqrt{\pi}v_c} \sum_{\alpha\beta} \tilde{Q}_{\alpha\beta} \delta_{\alpha\beta} \frac{1}{3} \int \frac{y^2}{y^2} e^{-y^2} d\mathbf{y} = -\frac{2}{3\sqrt{\pi}v_c} \sum_\alpha \tilde{Q}_{\alpha\alpha} \int e^{-y^2} d\mathbf{y}.$$

The volume integral is calculated in the spherical coordinates giving $(4\pi) \left(\sqrt{\pi}/4 \right) = \pi\sqrt{\pi}$, so that the contribution here becomes $- (2\pi/3v_c) Q_0$.

In the third term in Eq. (7.93) we write

$$\eta_\alpha(\mathbf{x}) = x_\alpha \xi(x) = \frac{x_\alpha}{x^3} \int_0^{\epsilon x} e^{-\lambda^2} \lambda^2 d\lambda = \frac{x_\alpha}{x^3} \left[\frac{\sqrt{\pi}}{4} - f(\epsilon x) \right].$$

The integration with $f(\epsilon x)$ can be extended to the whole space and then the integral is equal to zero due to symmetry (note that we integrate x_α times some function of

x), see Problem 7.60. Hence, we are left with the contribution

$$-\frac{4}{\sqrt{\pi}v_c}\frac{\sqrt{\pi}}{4}\sum_\alpha P_\alpha \int_{V_N}\frac{x_\alpha}{x^3}d\mathbf{x} = -\frac{1}{v_c}\sum_\alpha P_\alpha \int_{V_N}\frac{x_\alpha}{x^3}d\mathbf{x} = -\frac{1}{v_c}\mathbf{P}\cdot\overline{\mathbf{e}}, \quad (7.95)$$

where

$$\overline{e}_\alpha = \int_{V_N}\frac{x_\alpha}{x^3}d\mathbf{x}.$$

This integral is zero if the crystal sample possesses inversion symmetry $\mathbf{x} \to -\mathbf{x}$, but in any event this term is a constant and in the Coulomb potential can be ignored as not contributing to the electric field (which is the minus gradient of the electrostatic potential).

Finally, let us consider the fourth term in Eq. (7.93). It contains the integral

$$\int_{V_N}\eta_{\alpha\beta}(\mathbf{x})d\mathbf{x} = \int_{V_N}T_{\alpha\beta}(\mathbf{x})\left(\int_0^{\epsilon x}e^{-\lambda^2}\lambda^2 d\lambda\right)d\mathbf{x} + \epsilon^3\int_{V_N}\frac{x_\alpha x_\beta}{x^2}e^{-\epsilon^2 x^2}d\mathbf{x},$$

$$(7.96)$$

where we have made use of Eq. (7.90). In the first term we replace the λ-integral with $\sqrt{\pi}/4 - f(\epsilon x)$, as before. Only the $\sqrt{\pi}/4$ term would contribute leading to $\overline{T}_{\alpha\beta}\sqrt{\pi}/4$ since the contribution due to $f(\epsilon x)T_{\alpha\beta}(\mathbf{x})$ is zero because of symmetry. The second term in Eq. (7.96) is calculated in the same way as analogous terms above: because of the exponential function, the integration can be extended to the whole space and then symmetry can be used. The integral is proportional to $\delta_{\alpha\beta}$ and is the same for $\alpha = \beta$. Hence, replacing x_α^2 with $\frac{1}{3}x^2$ and integrating in spherical coordinates, we obtain

$$\epsilon^3\int\frac{x_\alpha x_\beta}{x^2}e^{-\epsilon^2 x^2}d\mathbf{x} = \delta_{\alpha\beta}\frac{1}{3}\int e^{-y^2}dy = \delta_{\alpha\beta}\frac{1}{3}\pi\sqrt{\pi},$$

leading to the overall contribution of the last term in Eq. (7.93) to be

$$\frac{4}{\sqrt{\pi}v_c}\sum_{\alpha\beta}P_\alpha r_\beta\left(\frac{\sqrt{\pi}}{4}\overline{T}_{\alpha\beta} + \delta_{\alpha\beta}\frac{1}{3}\pi\sqrt{\pi}\right) = \frac{1}{v_c}\sum_{\alpha\beta}P_\alpha\overline{T}_{\alpha\beta}r_\beta + \frac{4\pi}{3v_c}\mathbf{P}\cdot\mathbf{r}. \quad (7.97)$$

Collecting all contributions, we obtain the final result:

$$\phi_N^{ext}(\mathbf{r}) = -\frac{1}{6v_c}\sum_{\alpha\beta}Q_{\alpha\beta}\overline{T}_{\alpha\beta} - \frac{2\pi}{3v_c}Q_0 - \frac{1}{v_c}\mathbf{P}\cdot\overline{\mathbf{e}} + \frac{1}{v_c}\sum_{\alpha\beta}P_\alpha\overline{T}_{\alpha\beta}r_\beta + \frac{4\pi}{3v_c}\mathbf{P}\cdot\mathbf{r}.$$

$$(7.98)$$

The first three terms are constants, two of which depend on the shape of the crystal (via integrals $\overline{T}_{\alpha\beta}$ and $\overline{e}$). Note that the integral $\overline{T}_{\alpha\beta}$ does not depend on the size of the crystal (a linear change of the integration variables does not change the value of

Fig. 7.8 A uniformly
polarized finite crystal
resembles a capacitor with
positive charge accumulated
on its one side and negative -
on another

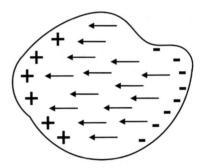

the integral). The last two terms give rise to a macroscopic electric field within the
sample

$$E_\alpha(\mathbf{r}) = -\frac{\partial}{\partial r_\alpha} \phi_N^{ext}(\mathbf{r}) = -\sum_\beta p_\beta \overline{T}_{\beta\alpha} - \frac{4\pi}{3} p_\alpha.$$

The second term here is the well-known Lorentz contribution ($\mathbf{p} = \mathbf{P}/v_c$ is the dipole
moment per unit volume) that is the same for any crystal, while another term is crystal
shape dependent. Both terms appear due to the net dipole moment in the unit cell.

Concluding, in the general case of *arbitrary* crystal shape the electrostatic potential well inside the crystal is composed of the Ewald potential corresponding to
the crystal of infinite size and a correction term. The latter can be shown[17] to have
a pure macroscopic origin and is related to the potential inside a large finite crystal
with a uniform distribution of the dipole moment density $\mathbf{p}$. This result can be easily
understood: a uniformly polarized crystal manifests itself as a capacitor of a certain
geometry, see Fig. 7.8.

---

[17] See L. N. Kantorovich and I. I. Tupitsyn - J. Phys.: Condens. Matter, **11**, p. 6159 (1999).

Ordinary Differential Equations

8

Ordinary differential equations (ODE's in short), or simply differential equations (DE), are the equations of the type

$$F\left(x, y, y', y'', \ldots, y^{(n)}\right) = 0 \,,$$

relating the variable x, a function $y(x)$ of x, and its derivatives $\frac{dy}{dx} \equiv y'$, $\frac{d^2 y}{dx^2} \equiv y''$, etc. with each other. The fundamental task is to find a function $y(x)$n that satisfies this equation. It may be that not one, but many functions form different solutions.

ODE's can be characterized by:

- *order*—order n of the highest derivative $y^{(n)}$ of the unknown function;
- *degree*—the power of its highest order derivative when the equation is written in a polynomial form (e.g. without roots).

A specific class of ODE's are *linear* ODE's if the unknown function and its derivatives are there in the first power without any cross terms. We shall see in this chapter that a significant progress can be made for this particular class of ODE's. Linear ODE's are also very often met in practical applications in physics, biology, chemistry and engineering.

Similar to integration, a general method for solving an arbitrary ODE does not exist, and whether the particular equation can be solved depends in many cases on fortune to find the right trick. There are certain cases, however, for which the necessary tricks have been developed and we shall consider some of them in this chapter.

An ODE may have a *general solution* that contains one or more *arbitrary constants* (that we shall write as C, C_1, C_2, etc. in what follows). The value(s) of the constant(s) can be fixed by specifying some information about the unknown function, e.g. its value $y(x_0)$ at some point x_0, and/or the value of its derivative(s). If the same point x_0 is used, these additional conditions are called *initial conditions* and the problem *initial value problem*. If different points x_1, x_2, etc. are used, such conditions are usually called *boundary conditions* since these points normally correspond to the boundary of the system in question. In this chapter we shall be only concerned with the former case; examples of the latter are met, e.g. when solving partial differential equations of mathematical physics which we do not consider here (see Vol. II, Chap. 8). The solution in which no arbitrary constant(s) is (are) present (and hence some specific information about the unknown function was already applied) is called a *particular solution*.

8.1 First-Order First Degree Differential Equations

Consider an equation $F\left(x, y, y'\right) = 0$, in which only the first derivative of an un-known function $y(x)$ is present. Suppose one can solve for the derivative y' in this (algebraic) equation. Then we have to study the ODE of the following simplest general form:

$$\frac{dy}{dx} = f(x, y) \,. \tag{8.1}$$

In this section several methods will be considered which enable one to find the function(s) $y(x)$ satisfying this equation. In some cases the solution is obtained in an explicit form as a function $y(x)$ (it would also depend on an arbitrary constant C); in others the solution is obtained in an implicit form via a transcendental equation $G(x, y) = 0$. Even though we might not be able to solve this algebraic equation with respect to y and hence write it down as an explicit function of x, this is considered a solution as the function G does not contain the derivative $y'(x)$ anymore.

8.1.1 Separable Differential Equations

If $f(x, y) = g(x)/h(y)$, then (8.1) is called *separable ODE* because x and y can be separated, i.e. all terms containing x can be placed on one sides of the equal sign, while all the terms containing y—on the other. Indeed, rearranging both sides, we obtain:

$$\frac{dy}{dx} = \frac{g(x)}{h(y)} \quad \Longrightarrow \quad h(y)dy = g(x)dx \,. \tag{8.2}$$

The next step needs some explanations. Recall that differentials can not only be written for a single variable, like dx, but also for a function $f(x)$, like $df = \frac{df}{dx}dx = f'(x)dx$. Here $df = f(x + dx) - f(x)$ corresponds to the change of $f(x)$ associated with the variable x changing between x and $x + dx$. This means that an expression like $g(x)dx$ in the second step in (8.2) can be written as the differential $dG(x) = g(x)dx$ of some function $G(x) = \int g(x)dx$ of x, while the expression $h(y)dy$ there can also be written as a differential $dH(y) = h(y)dy$ of some other function $H(y) = \int h(y)dy$, with respect to the y. Therefore, the line $h(y)dy = g(x)dx$ is equivalent to saying that these two differentials are equal, $dH(y) = dG(x)$, which means that the functions themselves differ only by an arbitrary constant C, i.e. $H(y) = G(x) + C$. In other words, we obtain the final result which formally solves the ODE in question:

$$\int h(y)dy = \int g(x)dx + C ,$$

where the two integrals are understood as indefinite ones. It is seen that the above solution can formally be obtained by integrating both sides of the equation for differentials:

$$h(y)dy = g(x)dx \implies \int h(y)dy = \int g(x)dx ,$$

and introducing a constant C after the integration. This gives the general solution of the ODE. If the initial condition is known, $y(x_0) = y_0$, then integration can be performed from y_0 to y in the left, and from x_0 to x in the right integrals, respectively, in which case a particular integral of the ODE is obtained.

It is instructive to discuss another, probably simpler, but closely related method that is based on the following observation. Let $G'(x) = g(x)$ and $H'(y) = h(y)$, i.e. they are indefinite integrals of $g(x)$ and $h(y)$, respectively. Then we can rewrite (8.2) as

$$h(y)\frac{dy}{dx} - g(x) = 0 \implies H'(y)\frac{dy}{dx} - G'(x) = 0$$

$$\implies \frac{d}{dx}[H(y(x)) - G(x)] = 0 ,$$

from which it follows that $H(y) - G(x)$ is a constant, the same result as before.

As an example, consider equation

$$xy' - xy = y ,$$

subject to the initial conditions $y(1) = 1$. Regrouping the terms, we can write:

$$x\frac{dy}{dx} = y + xy \implies x\frac{dy}{dx} = y(1+x) \implies \frac{dy}{y} = \frac{1+x}{x}dx$$

$$\implies \int \frac{dy}{y} = \int \left(\frac{1}{x} + 1\right) dx ,$$

which gives the general solution as

$$\ln y = \ln x + x + C_1 \implies \frac{y}{x} = \exp(x + C_1) = e^{C_1}e^x$$

$$\implies y = Cxe^x .$$

Here we have come across a typical situation related to the arbitrary constant: after integration we introduced a constant C_1 which upon rearrangement turned into an expression e^{C_1} in the formula for the solution $y(x)$. Since C_1 is a constant, so is e^{C_1}. There is no need to keep this complicated form for an unknown constant, and we simply replaced the whole expression e^{C_1} with C in the final form.

Applying an initial condition would fix the value of such a constant (the important points of existence and uniqueness of the solutions of the initial value problems will be discussed later on). For instance, if the initial condition is $y(1) = 1$, we write: $1 = Ce$ yielding $C = e^{-1}$. Hence, the particular integral (solution) reads $y = xe^{x-1}$. This function satisfies both the differential equation and the initial conditions, i.e. it is the required solution of the initial value problem

$$xy' - xy = y, \quad y(1) = 1.$$

Problem 8.1 Obtain solutions of the following separable ODE:

$$x\sqrt{1 - y^2}dx + y\sqrt{1 - x^2}dy = 0,$$

subject to the initial condition $y(1) = 0$. [Answer: $\sqrt{1 - x^2} + \sqrt{1 - y^2} = 1$]

Problem 8.2 Obtain solutions of

$$x^2\frac{dy}{dx} = y^2\frac{dx}{dy}.$$

What its particular solutions would be if $y(1) = 1$? [Answer: $y = x$ and $y = 1/x$.]

Problem 8.3 Consider an evolution of the population $N(t)$ of some species in time. The rate of change, dN/dt, of the population must be proportional to N,

$$\frac{dN}{dt} = rN,$$

where r is the rate of growth, and we assume the initial conditions $N(0) = N_0 > 0$. The rate will depend on available resources required for the population to grow. If N is smaller than a certain critical value N_*, the rate will be positive as the available resources are still sufficient to maintain growth, while if $N > N_*$, the resources are insufficient and, as the consequence, the rate becomes negative leading to decrease of the population (species die) until the situation becomes sustainable again. In the simplest model, r is chosen to be proportional to $N_* - N$, i.e. this model leads to a non-linear differential equation for the evolution of the population,

$$\frac{dN}{dt} = \beta (N_* - N) N,$$

where $\beta > 0$ is a parameter. By virtue of rescaling the population, $y = N/N_*$, the differential equation can be rewritten in a form of the so-called *logistic ODE*

$$\frac{dy}{dt} = \alpha(1 - y)y, \quad y(0) = y_0 = \frac{N_0}{N_*}.$$

Show that its solution is

$$y(t) = \frac{y_0}{y_0 + (1 - y_0)e^{-\alpha t}}.$$

As you can see, at long times $y(t) \to 1$, i.e. as $t \to \infty$ the population N approaches the sustainable value N_*.

Problem 8.4 Let us work out how the air density and correspondingly the air pressure decrease with the height z from the surface of our planet ($z = 0$ corresponds to the Earth's surface and we consider the heights that are much smaller than the Earth's radius). Consider a horizontal layer of air of width dz at the height z. The horizontal size of the layer is assumed being not too wide so that we could neglect the Earth's curvature. The pressure of the air P would decay with z according to the Archimedes principle as $dP = -\rho(z)gdz$, where $\rho = mdN/dV$ is the air density due to dN identical molecules of mass m contained in the volume dV of the layer, and g the gravitational constant on the Earth surface. Use the ideal gas equation $PdV = k_B TdN$, where k_B is the Boltzmann's constant and T absolute temperature, assumed constant, to show that the density is proportional to the pressure, $\rho = mP/k_B T$. Hence, find the following differential equation for the pressure:

$$\frac{dP}{dz} = -\frac{mg}{k_B T} P.$$

Show that the solution of this equation, subject to the initial condition $P(z = 0) = P_0$, is

$$P(z) = P_0 \exp(-mgz/k_B T).$$

This result shows that the pressure (and hence the air density) decay exponentially from the Earth's surface.

Problem 8.5 In the previous problem we have assumed that the temperature of the air remains constant with z. Here we shall relax this condition assuming that the air temperature, for not too large heights above the Earth's surface, decreases linearly with z as $T(z) = T_0 - \alpha z$. Show that in this case

$$P(z) = P_0 \left(\frac{T(z)}{T_0} \right)^{mg/k_B \alpha} .$$

By taking the limit of $\alpha \to 0$, demonstrate explicitly using (2.85) that this formula goes over to the one from the previous problem when it was assumed that the temperature $T(z) = T_0$ is constant.

8.1.2 "Exact" Differential Equations

If the ODE can be written in the form

$$A(x, y)dx + B(x, y)dy = 0 , \tag{8.3}$$

in which the left-hand side is the exact differential (see Sect. 5.5) of a function $F(x, y)$ of two variables x and y, i.e.

$$dF = \frac{\partial F}{\partial x}dx + \frac{\partial F}{\partial y}dy \equiv A(x, y)dx + B(x, y)dy , \tag{8.4}$$

then this equation can be integrated and presented in an algebraic form $F(x, y) = C$ that contains no derivatives. As such, this form has to be considered as a *solution* of the ODE even though this equation may be a transcendental one with respect to y. Of course, if the equation $F(x, y) = C$ can be solved for $y(x)$, then the solution is obtained explicitly. However, in many cases this may not be possible or such a representation is cumbersome or inconvenient, as illustrated by the examples below.

In order for the left-hand side of (8.3) to be the exact differential, we should obviously have:

$$\frac{\partial F}{\partial x} = A(x, y) \quad \text{and} \quad \frac{\partial F}{\partial y} = B(x, y) , \tag{8.5}$$

which inevitably result in the necessary condition (Sect. 5.5)

$$\frac{\partial B}{\partial x} = \frac{\partial A}{\partial y} \tag{8.6}$$

for the two functions $A(x, y)$ and $B(x, y)$ to satisfy. If this equation is satisfied, the ODE is called an exact DE and the solution is obtained by integrating either of the (8.5) using, e.g. the method explained in detail in Sect. 5.5. Briefly, consider equation

$\partial F / \partial x = A(x, y)$ in which the variable y is fixed. It can be written as $dF = Adx$ (keeping in mind that y is fixed, we can temporarily use d instead of ∂ here), which is separable and hence can be integrated:

$$F(x, y) = \int A(x, y)dx + C_1(y) , \qquad (8.7)$$

where the "arbitrary constant" $C_1(y)$ may still (and most probably will) be a function of the other variable y. This unknown function $C_1(y)$ is then obtained from the other condition $\partial F / \partial y = B(x, y)$. Once the function $F(x, y)$ is obtained, one has to set it to a constant since $dF(x, y) = 0$ according to the DE (8.3).

As an example of using this method, consider the equation:

$$\frac{dy}{dx} = -\frac{2xy}{x^2 + y^2} .$$

First of all, we rewrite this ODE in the form of (8.3):

$$2xydx + \left(x^2 + y^2\right) dy = 0 .$$

Next, we should check if the condition (8.6) is satisfied:

$$A(x, y) = 2xy, \quad B(x, y) = x^2 + y^2 \implies \frac{\partial A}{\partial y} = 2x \text{ and } \frac{\partial B}{\partial x} = 2x ,$$

i.e. the condition is indeed satisfied, and thus we can integrate the equation $\partial F / \partial x = A(x, y) = 2xy$ with respect to x to get

$$F = \int 2xydx = 2y\frac{x^2}{2} + C_1(y) = x^2y + C_1(y)$$

(note that y is a constant in this integral and hence must be considered as a parameter!). The unknown function $C_1(y)$ is obtained from $\partial F / \partial y = B(x, y) = x^2 + y^2$ as follows:

$$\frac{\partial F}{\partial y} = \frac{\partial}{\partial y} \left(x^2y + C_1(y)\right) = x^2 + \frac{dC_1}{dy} \implies x^2 + \frac{dC_1}{dy} = x^2 + y^2$$

$$\implies \frac{dC_1}{dy} = y^2 ,$$

that is easily integrated (since it is a separable equation) to give:

$$dC_1 = y^2dy \implies C_1(y) = \int y^2dy = \frac{y^3}{3} + C ,$$

with C being an arbitrary constant. Thus, the function $F(x, y)$ is now completely defined (up to the constant C, of course!) as

$$F(x, y) = x^2 y + \frac{y^3}{3} + C \, .$$

Since the total differential of $F(x, y)$, according to (8.3), is zero, the function F should be equal to a constant, say C_1. Therefore, we obtain:

$$F(x, y) = x^2 y + \frac{y^3}{3} + C = C_1 \, ,$$

i.e. the solution $y(x)$ is obtained by solving the following algebraic equation in y:

$$x^2 y + \frac{y^3}{3} = C_2 \, ,$$

where $C_2 = C_1 - C$ is an arbitrary constant to be determined from initial conditions or some other additional information available for $y(x)$. Note that the solution obtained can be checked by differentiating both sides of the equation above with respect to x (remember to treat y as a function of x):

$$\frac{d}{dx} \left(x^2 y + \frac{y^3}{3} \right) = 2xy + x^2 \frac{dy}{dx} + \frac{3y^2}{3} \frac{dy}{dx} = 2xy + \left(x^2 + y^2 \right) \frac{dy}{dx} \equiv 0 \, ,$$

which is the original equation.

The other, closely related, methods that can also be used for finding solutions of the exact differential equations follow from the results of Sect. 6.4.4: given the exact differential equation (8.3), the function $F(x, y)$ giving rise to the exact differential $dF = A dx + B dy$ is obtained via the line integral:

$$F(x, y) = \int_{(x_0, y_0)}^{(x, y)} A(x, y) dx + B(x, y) dy \, ,$$

that does not depend on the path taken to connect the two points (x_0, y_0) and (x, y) in the $x - y$ plane (the initial point (x_0, y_0) can be chosen arbitrarily). By choosing conveniently the integration path, calculating this function and then setting it to a constant, $F(x, y) = C$, one obtains the required solution $y = y(x)$ as a function of that constant. It is easy to see that by choosing the path along the coordinate axes, $(x_0, y_0) \rightarrow (x, y_0) \rightarrow (x, y)$, we essentially arrive at the method described previously.

Problem 8.6 Show that the following ODE can be solved using the method of exact differentials:

$$\frac{dy}{dx} = -\frac{y + \sin(x)}{x - 2\cos(y)}$$

and then solve it using both methods explained above. [**Answer:** $yx - \cos x - 2\sin y = C$]

8.1.3 Method of an Integrating Factor

Sometimes, if an ODE (8.3) does not satisfy the condition (8.6), it can be forced to do so. This can be done by multiplying the equation with a function $I(x, y)$, called an *integrating factor*, so that the functions $A(x, y)$ and $B(x, y)$ are replaced by $A(x, y) \rightarrow I(x, y)A(x, y)$ and $B(x, y) \rightarrow I(x, y)B(x, y)$. Correspondingly, the new necessary condition

$$\frac{\partial}{\partial x}(I(x, y)B(x, y)) = \frac{\partial}{\partial y}(I(x, y)A(x, y)) \tag{8.8}$$

is to be satisfied and the original ODE is turned into an equation of the exact differential form. It can then be solved using the method developed above. Note that the original ODE (8.3) can always be multiplied by any function in the left-hand side as we have zero in the right-hand side.[1]

As an example of using this method, let us try to solve the following equation:

$$\frac{dy}{dx} = \frac{y}{x}.$$

Of course, this is a separable equation for which the solution is easily obtained:

$$\frac{dy}{y} = \frac{dx}{x} \implies \int \frac{dy}{y} = \int \frac{dx}{x} \implies \ln|y| = \ln|x| + \ln|C|$$

$$\implies \ln\left|\frac{y}{Cx}\right| = 0 \implies y = Cx.$$

(We had a liberty here to write an arbitrary constant as $\ln|C|$ for convenience.)

However, it is instructive to solve this equation using the integrating factor method as well. We write the equation in the unfolded form first as

$$ydx - xdy = 0,$$

---

[1] This operation may lead to spurious solutions of the ODE that are roots of the equation $I(x, y) = 0$. Hence, the obtained solutions must be substituted back into the original ODE to check if they satisfy it.

with $A = y$ and $B = -x$, that obviously do not satisfy the necessary condition
(8.6): $\partial A / \partial y = 1$, but $\partial B / \partial x = -1$. Let us try to find an integrating factor as
$I(x, y) = x^{\alpha} y^{\beta}$. The new functions A and B become: $A(x, y) = x^{\alpha} y^{\beta+1}$ and
$B(x, y) = -x^{\alpha+1} y^{\beta}$. The idea is to find such α and β that the two newly defined
functions A and B would satisfy the necessary condition:

$$\frac{\partial A}{\partial y} = (\beta + 1) x^{\alpha} y^{\beta} \quad \text{should be equal to} \quad \frac{\partial B}{\partial x} = -(\alpha + 1) x^{\alpha} y^{\beta} .$$

Comparing the two expressions, we get $\beta + 1 = -\alpha - 1$ or $\beta = -\alpha - 2$, a single
condition for two constants to be determined, α and β; for instance, one can take
$\alpha = -2$ and then get $\beta = 0$, so that the required integrating factor becomes $I(x, y) = x^{-2}$. Now the equation becomes the one containing the exact differential:

$$x^{-2} y dx - x^{-1} dy = 0, \quad \text{where} \quad \frac{\partial}{\partial y} \left(x^{-2} y \right) = x^{-2} = \frac{\partial}{\partial x} \left(-x^{-1} \right) ,$$

and hence can be integrated using the method of the previous subsection:

$$F(x, y) = \int x^{-2} y dx = y \frac{x^{-2+1}}{-2 + 1} + C_1(y) = -yx^{-1} + C_1(y)$$

and

$$\frac{\partial F}{\partial y} = \frac{\partial}{\partial y} \left(-yx^{-1} + C_1(y) \right) = -x^{-1} + \frac{dC_1}{dy} \equiv -x^{-1} ,$$

so that $dC_1 / dy = 0$ yielding $C_1(y) = C$, which is a constant. Therefore, $F(x, y) = -yx^{-1} + C$, and, since it must be a constant itself, this yields $-yx^{-1} = C_1$ or
$y = -C_1 x$ with C_1 being another arbitrary constant. We have obtained the same
general solution as above using this method.

The choice of an integrating factor in every case is a matter of some luck, but
special methods have been developed for certain types of differential equations. For
instance, the factor $I(x, y) = x^{\alpha} y^{\beta}$ used above may work if both original functions
$A(x, y)$ and $B(x, y)$ have a polynomial form in x and y.

Problem 8.7 Find the integrating factor for the DE

$$-x \frac{dy}{dx} + y = 2x .$$

[Hint: assume that the integrating factor is a function of x only and obtain
a separable ODE for it. Answer: the integrating factor may be chosen as
$\mu(x) = x^{-2}$.]

Problem 8.8 Find the integrating factor $I(x, y)$ to turn the ODE

$$(1 - xy)\frac{dy}{dx} + y^2 + 3xy^3 = 0$$

into the exact DE. [Hint: *Try $I(x, y) = x^\alpha y^\beta$ and determine α and β. There is only one possibility here.* Answer: $\alpha = 0$, $\beta = -3$.]

Problem 8.9 Solve the DE of the previous problem using the integrating factor $I(x, y) = y^{-3}$. [Answer: $2xy + 3x^2y^2 - 1 = Cy^2$.]

Problem 8.10 Solve the ODE

$$\frac{dy}{dx} = \frac{xy}{x^2 - y^2}$$

using the integrating factor $I(y) = y^\alpha$. [Answer: $\alpha = -3$ *and then* $y(x)$ *is determined from the equation* $-x^2/2y^2 = \ln|y| + C$.]

There are also special cases in which the integrating factor can be deduced explicitly. Rewriting (8.8) by performing the differentiation and dividing by $I(x, y)$, we obtain:

$$\frac{\partial A}{\partial y} - \frac{\partial B}{\partial x} = \frac{1}{I}\frac{\partial I}{\partial x}B - \frac{1}{I}\frac{\partial I}{\partial y}A. \tag{8.9}$$

Let us try to work out a necessary condition for the integrating factor to only depend on x. Divide both sides of the above equation by $B(x, y)$:

$$\left(\frac{\partial A}{\partial y} - \frac{\partial B}{\partial x}\right)\frac{1}{B} = \frac{1}{I}\frac{\partial I}{\partial x} - \frac{1}{I}\frac{\partial I}{\partial y}\frac{A}{B}.$$

If $I \equiv I(x)$ only depends on x, then the second term in the right-hand side is identically zero, while the first term can be rewritten as $\frac{d}{dx}\ln I$. Since this expression is a function of x only, then the left-hand side must also be some function

$$f(x) = \left(\frac{\partial A}{\partial y} - \frac{\partial B}{\partial x}\right)\frac{1}{B}$$

that only depends on x. This enables one to explicitly calculate the integrating factor:

$$f(x) = \frac{1}{I}\frac{dI}{dx} \implies I(x) = \exp\left(\int f(x)dx\right).$$

Problem 8.11 Show that if instead an expression $\left(\frac{\partial A}{\partial y} - \frac{\partial B}{\partial x}\right)\frac{1}{A}$ is a function $g(y)$ of y only, then the integrating factor

$$I(y) = \exp\left(\int g(y)dy\right)$$

can be chosen as a function of y only.

We shall apply this method when solving a general linear first-order ODE in Sect. 8.1.5.

Next, we shall derive a necessary condition for the integrating factor to only depend on the product $t = xy$. In this case $\partial I/\partial x = yI'(t)$ and $\partial I/\partial y = xI'(t)$, so that condition (8.9) is transformed into

$$\frac{\partial A/\partial y - \partial B/\partial x}{yB - xA} = \frac{I'(t)}{I(t)}.$$

Hence, for the integrating factor to depend only on the product $t = xy$, the left-hand side of the above equation must also depend on $t = xy$. And again, if we denote the left-hand side as $h(t)$, then $I(t) = \exp\left(\int h(t)dt\right)$, i.e. can be explicitly calculated.

Problem 8.12 Using this method, show that the integrating factor for the ODE

$$\left(2y - xy^2\right)dx + \left(2x + x^2y\right)dy = 0$$

is $I(x, y) = 1/(xy)^2$.

Problem 8.13 In the same vein, show that if

$$\frac{\partial A/\partial y - \partial B/\partial x}{B - A} = f(x + y)$$

depends only on $t = x + y$, then $I(t) = \exp\left(\int f(t)dt\right)$ can be chosen as depending on $x + y$. Using this method, show that the integrating factor for the ODE

$$(1 - 2y - 2xy)dx + (1 - 2x - 2xy)dy = 0$$

is $I(x, y) = e^{x+y}$.

Problem 8.14 Show that if

$$\frac{\partial A/\partial y - \partial B/\partial x}{y^2 B - 2xyA} = f(xy^2)$$

depends only on $t = xy^2$, then the integrating factor $I(t) = \exp\left(\int f(t)dt\right)$, i.e. can also be chosen as depending on xy^2.

The reader can appreciate that similarly one can develop conditions for the integrating factor to depend on other combinations of the variables x and y.

8.1.4   Homogeneous Differential Equations

Some ODE's can be solved by a simple substitution. Specifically, consider an ODE of the form of (8.3) (not necessarily the exact differential equation!) in which functions $A(x, y)$ and $B(x, y)$ are *homogeneous functions* of the same degree. This means that if we scale x and y in both A and B by some λ, we will get the same factor of λ^n to come out:

$$A(\lambda x, \lambda y) = \lambda^n A(x, y) \ \text{ and } \ B(\lambda x, \lambda y) = \lambda^n B(x, y) \ . \tag{8.10}$$

If this is the case for the given equation, it can be solved by the substitution $y(x) \implies z(x) = y(x)/x$. Indeed, consider an ODE:

$$A(x, y)dx + B(x, y)dy = 0 \ ,$$

and let us make the substitution $y(x) = z(x)x$ in it:

$$dy = zdx + xdz \ , \ A(x, y) = A(x, zx) = x^n A(1, z)$$
$$\text{and } \ B(x, y) = B(x, zx) = x^n B(1, z) \ ,$$

where the role of λ was played by x in both $A(x, zx)$ and $B(x, zx)$. Using these results, we are now able to perform simple manipulations in the original ODE to obtain an equation which is separable:

$$x^n A(1, z)dx + x^n B(1, z)(zdx + xdz) = 0$$
$$\implies \ A(1, z)dx + B(1, z)(zdx + xdz) = 0$$
$$\implies \ (B(1, z)z + A(1, z))dx + xB(1, z)dz = 0$$
$$\implies \ \frac{dx}{x} = -\frac{B(1, z)}{zB(1, z) + A(1, z)}dz \ ,$$

i.e. it is indeed separable now and hence can be integrated, at least, in principle.

Problem 8.15 Show that homogeneous differential equations belong to the class of equations $y' = f(y/x)$.

As an example, consider equation

$$\frac{dy}{dx} = \frac{x^2 y}{x^3 + y^3} .$$

In the unfolded form the equation becomes

$$-x^2 y \, dx + \left(x^3 + y^3\right) dy = 0 .$$

It is seen that both functions $A(x, y) = -x^2 y$ and $B(x, y) = x^3 + y^3$ are homogeneous functions of the order 3:

$$A(\lambda x, \lambda y) = - (\lambda x)^2 (\lambda y) = \lambda^3 \left(-x^2 y\right) = \lambda^3 A(x, y) ,$$
$$B(\lambda x, \lambda y) = (\lambda x)^3 + (\lambda y)^3 = \lambda^3 \left(x^3 + y^3\right) = \lambda^3 B(x, y) ,$$

and thus the substitution $z = y/x$ should be appropriate: $y = zx$ and $dy/dx = (dz/dx) x + z$, so that

$$\frac{dy}{dx} = \frac{x^2 y}{x^3 + y^3} \implies \frac{dz}{dx} x + z = \frac{x^3 z}{x^3 + z^3 x^3} \implies \frac{dz}{dx} x + z = \frac{z}{1 + z^3}$$

$$\implies \frac{dz}{dx} x = \frac{z}{1 + z^3} - z \implies \frac{1 + z^3}{z^4} dz = -\frac{dx}{x} ,$$

which is integrated to give

$$\int \frac{1 + z^3}{z^4} dz = - \int \frac{dx}{x} \implies -\frac{1}{3z^3} + \ln |z| = -\ln |x| - \ln |C| = \ln \left| \frac{1}{Cx} \right|$$

$$\implies -\frac{1}{3z^3} = \ln \left| \frac{1}{Czx} \right| .$$

Substituting back $z = y/x$, we obtain:

$$-\frac{x^3}{3y^3} = \ln \left| \frac{1}{Cy} \right| \quad \text{or} \quad -\frac{x^3}{3y^3} = -\ln |y| + C_1 ,$$

where C_1 is an arbitrary constant. The obtained solution is a transcendental algebraic equation with respect to y. In spite of the fact that this equation cannot be solved analytically with respect to y (it contains both the logarithm of y and its cube), it serves as the solution of the ODE since it does not contain anymore derivatives of y. Note that after applying the corresponding initial conditions to determine the constant C_1, this equation can easily be solved numerically.

Problem 8.16 Solve the following homogeneous DE:

$$x^2 dy + (y^2 - xy)dx = 0 .$$

[Answer: $y = x/(C + \ln|x|)$]

Problem 8.17 Show that the ODE

$$\frac{dy}{dx} = \frac{xy}{x^2 - y^2}$$

is of the homogeneous type and then solve it (compare with Problem 8.10).

8.1.5 Linear First-Order Differential Equations

Consider the following first-order linear (first degree) differential equation

$$\frac{dy}{dx} + p(x)y = q(x) , \quad a < x < b , \tag{8.11}$$

where $p(x)$ and $q(x)$ are some functions of x. We shall show that this equation can be solved for any functions p and q using two methods: the method of an integrating factor of Sect. 8.1.3 and another method called variation of parameters.

8.1.5.1 Using the Method of an Integrating Factor
Our ODE, if rewritten in the required (unfolded) form (8.3) as

$$[-q(x) + p(x)y] dx + dy = 0$$

with $A(x, y) = -q(x) + yp(x)$ and $B(x, y) = 1$, obviously does not satisfy condition (8.6) for arbitrary $p(x)$ and $q(x)$. Let us check if the method of the integrating factor may help in solving this equation by considering the function

$$\left(\frac{\partial A}{\partial y} - \frac{\partial B}{\partial x}\right) \frac{1}{B} = \frac{\partial A}{\partial y} = p(x) .$$

It is a function of x only and hence, according to the method discussed as the end of Sect. 8.1.3, we can choose the integrating factor to be only x dependent. Moreover, we can immediately write down its explicit expression as

$$I(x) = \exp\left[\int p(x)dx\right] . \tag{8.12}$$

The integral here is an *indefinite* integral which is considered as a function of x.

Once the integrating factor has been determined, we can solve the original equation as explained in Sect. 8.1.2. We have for the new functions A and B:

$$A(x, y) = I(x)[-q(x) + p(x)y] \quad \text{and} \quad B(x, y) = I(x) ,$$

so that

$$F(x, y) = \int B(x, y)dy = I(x)y + C_1(x) ,$$

$$\frac{\partial F(x, y)}{\partial x} = \frac{dI}{dx}y + \frac{dC_1}{dx} \equiv A(x, y) = I(x)[-q(x) + p(x)y] .$$

However, the function $I(x)$ satisfies $dI/dx = p(x)I(x)$, so that, after simplification, we get:

$$\frac{dC_1}{dx} = -I(x)q(x) ,$$

which is in a separable form and hence accepts the solution (again, via an indefinite integral):

$$C_1(x) = -\int I(x)q(x)dx + C_2 .$$

Thus, we have found the function completely:

$$F(x, y) = I(x)y + C_1(x) = I(x)y - \int I(x)q(x)dx + C_2 .$$

Since it must be a constant, we can write:

$$I(x)y - \int I(x)q(x)dx = C ,$$

which can be solved for y to get (C is the final arbitrary constant):

$$y = I(x)^{-1}\left[C + \int I(x)q(x)dx\right] . \tag{8.13}$$

Problem 8.18 Check that if each integral in the above formula is understood as a definite integral with the upper limit being the corresponding variable, the bottom limits do not matter.

As an example, let us solve the following ODE

$$\frac{dy}{dx} + y = e^x \tag{8.14}$$

subject to the condition $y(0) = 1$. Here $p(x) = 1$ and $q(x) = e^x$. Therefore, the function $I(x)$ from (8.12) is simply

$$I(x) = e^{\int dx} = e^x ,$$

and thus the solution is obtained as

$$y(x) = e^{-x} \left[C + \int e^x e^x dx \right] = Ce^{-x} + e^{-x} \int e^{2x} dx$$

$$= Ce^{-x} + e^{-x} \frac{1}{2} e^{2x} = Ce^{-x} + \frac{1}{2} e^x ,$$

which is the required solution (can be checked by a direct substitution).

Problem 8.19 Consider two infinite power series

$$S_e(x) = \sum_{n=0 \,(\text{even})}^{\infty} \frac{x^n}{n!} = \sum_{m=0}^{\infty} \frac{x^{2m}}{(2m)!} ,$$

$$S_o(x) = \sum_{n=1 \,(\text{odd})}^{\infty} \frac{x^n}{n!} = \sum_{m=1}^{\infty} \frac{x^{2m-1}}{(2m-1)!} .$$

Using the obvious fact that $S_e(x) + S_o(x) = e^x$, see (7.48), show that both functions satisfy the same DE (8.14). Using the initial conditions $S_e(0) = 1$ and $S_o(0) = 0$, solve the DE to show that $S_e(x) = \cosh x$ and $S_o(x) = \sinh x$, i.e. these functions represent the corresponding Maclaurin series of the hyperbolic exponential functions (the result to be expected since when summing up the Maclaurin series of e^x and e^{-x} all odd power terms will cancel out, while, when subtracting them, all even terms cancel out).

Problem 8.20 Obtain the general solutions of the following linear ODEs:

$$(a) \; y' + y \cos(x) = \frac{1}{2} \sin(2x) \; ; \quad (b) \; \frac{dy}{dx} + \frac{2}{x} y = 2 + x^2 .$$

[Answer: *(a)* $y = C \exp(-\sin x) + \sin x - 1$; *(b)* $y(x) = 2x/3 + x^3/5 + C/x^2$.]

Problem 8.21 Show that the general solution of ODE

$$\frac{dy}{dx} + \frac{a}{x}y = x^b$$

is

$$y(x) = Cx^{-a} + \frac{x^{b+1}}{a+b+1} .$$

Given particular initial conditions, the solution (8.13) is unique; in other words, a linear first-order initial value problem has a unique solution. Indeed, suppose there exist two different solutions y_1 and y_2, satisfying the same ODE (8.11): $y_1' + py_1 = q$ and $y_2' + py_2 = q$. Subtracting one equation from the other and introducing the function $v(x) = y_1(x) - y_2(x)$ of their difference, we obtain the ODE $v' + pv = 0$ whose solution is

$$v(x) = C \exp\left(-\int_a^x p(x')dx'\right) ,$$

where we conveniently have chosen the bottom limit of the integral to be a, the point where the initial condition $y(a) = y_0$ is defined. Since both solutions satisfy the same initial condition at $x = a$, obviously, $v(a) = 0$ and hence the arbitrary constant $C = 0$ as well. Hence, $v(x) = 0$ leading to the contradiction $y_1 = y_2$. Hence, the solution given by (8.13) is indeed unique.

8.1.5.2 Using the Method of Variation of Parameters
The solution of the general linear first-order ODE (8.11) can also be obtained using the *method of variation of parameters* which we shall find especially useful in a more general case of linear higher order ODE (the case of second-order ODEs will be considered in Sect. 8.2.3.2). The idea of this method is to first solve the *homogeneous* equation (without the right-hand side):

$$\frac{dy}{dx} + p(x)y = 0 \quad \Longrightarrow \quad \frac{dy}{y} = -pdx$$

$$\Longrightarrow \quad y(x) = C \exp\left(-\int p(x)dx\right) = C\, I(x)^{-1} , \qquad (8.15)$$

where C is an arbitrary constant. To solve the *nonhomogeneous* equation, i.e. with the right hand side, our original equation (8.11), we seek the solution in the form (8.15), but with the constant C replaced by a function $C(x)$:

$$y(x) = C(x) \exp\left(-\int p(x)dx\right) = C(x)I(x)^{-1} .$$

We need to substitute this trial solution into the ODE (8.11) to find an equation for the unknown function $C(x)$. This gives:

$$\left[C'(x)I(x)^{-1} - C(x)I(x)^{-1}p(x) \right] + p(x)\left[C(x)I(x)^{-1} \right] = q(x) \,,$$

where we made use of the fact that $I(x)^{-1}$ satisfies the equation

$$\frac{d}{dx}I(x)^{-1} = -p(x)I(x)^{-1} \,,$$

which follows from (8.12). Note that when differentiating the exponential term, we treated the integral as $\int^x p\left(x'\right) dx'$, i.e. with the x appearing in the upper limit, and hence its derivative is $p(x)$. After simplification, we obtain:

$$C'(x)I(x)^{-1} = q(x) \quad \Longrightarrow \quad C'(x) = q(x)I(x)$$

$$\Longrightarrow \quad C(x) = \int q(x)I(x)dx + C_1 \,,$$

where C_1 is an arbitrary constant. It is easy to see now that $y(x) = C(x)/I(x)$ is exactly the same as in (8.13).

As we shall see later on when considering a general theory of linear nonhomogeneous ODEs in Sect. 8.2, another viewpoint on this method is actually more helpful: when solving for the function $C(x)$, the arbitrary constant C_1 is disregarded. Instead, it is stated that the general solutions of the ODE is obtained by adding the solution of the nonhomogeneous equation,

$$y_p(x) = C(x)I(x)^{-1} = I(x)^{-1}\int q(x)I(x)dx$$

(the particular integral), to the complementary solution $y_c(x) = C_2 I(x)^{-1}$ of the homogeneous equation with C_2 being a constant. It is easy to see that $y = y_c + y_p$ coincides with the full solution we have obtained above.

8.1.6 Examples of Non-linear ODEs

As our first example, we shall consider the so-called Bernoulli equation:

$$\frac{dy}{dx} + p(x)y + g(x)y^\alpha = 0 \,. \tag{8.16}$$

This ODE is non-linear for any value of $\alpha \neq 1$. Nevertheless, it can easily be solved by changing the function $y(x)$. Indeed, let us make the substitution $y(x) = u(x)^\beta$ and try to find such value of the parameter β that would enable us to find the solution. Substituting $y = u^\beta$ into the ODE, we obtain:

$$\beta u^{\beta-1}\frac{du}{dx} + pu^\beta + gu^{\beta\alpha} = 0 \,,$$

$$\beta \frac{du}{dx} + pu + gu^{\beta\alpha-\beta+1} = 0.$$

The obtained equation becomes of the linear nonhomogeneous type (8.11) if $\beta\alpha - \beta + 1 = 0$ leading to $\beta = 1/(1-\alpha)$. Hence, the substitution $y = u^{1/(1-\alpha)}$ brings Bernoulli equation to the solvable form.

Our second example is the (generally) non-linear equation

$$\frac{dy}{dx} = f(ax + by + c), \quad b \neq 0. \tag{8.17}$$

This ODE can be solved by introducing another unknown function $u(x) = ax + by(x) + c$. Since $u' = a + by'$, we can rewrite this ODE as

$$\frac{1}{b}u' - \frac{a}{b} = f(u) \implies u' = bf(u) + a,$$

which is of a separable form.

Next, consider

$$\frac{dy}{dx} = f\left(\frac{ax + by + c}{px + qy + r}\right). \tag{8.18}$$

Several cases need to be considered.

(a) If $\Delta = qa - pb \neq 0$, one can make a substitution $x \to u$ and $y(x) \to v(u)$ using the following connection relations:

$$\begin{cases} au + bv = ax + by + c \\ pu + qv = px + qy + r \end{cases}.$$

The idea is to eliminate the free constants c and r in the argument of the function f in the right-hand side. These linear algebraic equations can be easily solved (e.g. by substitution) with respect to u and v yielding

$$u = x + \frac{qc - br}{\Delta} \quad \text{and} \quad v = y + \frac{ar - pc}{\Delta}.$$

Therefore,

$$\frac{dv}{du} = \frac{dv}{dy}\frac{dy}{du} = \frac{dy}{du} = \frac{dy}{dx}\frac{dx}{du} = \frac{dy}{dx},$$

since $dv/dy = du/dx = 1$. Hence, (8.18) takes the form:

$$\frac{dv}{du} = f\left(\frac{au + bv}{pu + bv}\right).$$

As the argument of the function f does not have anymore free constants, this ODE becomes of the homogeneous form considered in Sect. 8.1.4. Indeed, in the right-hand side we have a function that depends on v/u, see Problem 8.15. Hence this equation can be solved by the substitution $v = uz(u)$, where $z(u)$ is a new unknown function.

(b) $\Delta = 0$, but $b \neq 0$. In this case, we only replace the function $y(x) \to v(x) = ax + by + c$ leading to $y = (v - ax - c)/b$ and hence

$$\frac{dy}{dx} = \frac{1}{b}\left(\frac{dv}{dx} - a\right).$$

Also,

$$px + qy + r = px + \frac{q}{b}(v - ax - c) + r = \frac{1}{b}(qv - qc + rb)$$

since the condition $\Delta = 0$ eliminates the x terms. Consequently we obtain the equation

$$\frac{dv}{dx} = a + bf\left(\frac{bv}{qv - qc + rb}\right),$$

which is of a separable form $dv/dx = g(v)$.

(c) $\Delta = 0$, but $p \neq 0$. Similarly to the previous case, we replace the function $y(x) \to v(x) = px + qy + r$.

Problem 8.22 Show that in this case we obtain a separable ODE

$$\frac{dv}{dx} = p + qf\left(\frac{bv - br + cq}{qv}\right).$$

Our final example we shall formulate as a problem.

Problem 8.23 Show that the ODE

$$\frac{dy}{dx} = \frac{y}{x} + g(x)f\left(\frac{y}{x}\right)$$

is taken into a separable form $xz'(x) = g(x)f(z)$ by the substitution $y(x) = xz(x)$.

8.1.7   Non-linear ODEs: Existence and Uniqueness of Solutions

We have proven in Sect. 8.1.5.1 that a linear first-order ODE has a unique solution. An important point about non-linear ODEs is that their solutions might not be unique. Examples could be provided, e.g. by Problems 8.1 and 8.2. Here we shall briefly

discuss the existence and uniqueness of solutions of a non-linear ODE of the form
(8.1) that is resolved with respect to the derivative dy/dx.

It can be shown that this ODE has at least one solution $y(x)$ if the function $f(x, y)$
is continuous with respect to both of its variables within a region in the $x - y$ plane
(e.g. in a rectangular $a \leq x \leq b$ and $c \leq y \leq d$). This condition is not sufficient,
however, to guarantee the uniqueness of the solution. For that, it is necessary that
$f(x, y)$ be continuously differentiable, i.e. its partial derivatives $\partial f/\partial x$ and $\partial f/\partial y$
be continuous in the same region of the $x - y$ plane. We shall not prove these
statements, but will just illustrate the latter point with a simple (albeit not very
rigorous) consideration.

Consider the initial value problem

$$\frac{dy}{dx} = f(x, y), \quad y(x_0) = y_0.$$

If $f(x, y)$ is continuous and its first derivative with respect to y is also continuous,
then in the vicinity of the point y_0 one can write

$$f(x, y) \approx f(x, y_0) + \left(\frac{\partial f(x, y)}{\partial y}\right)_{y=y_0} (y - y_0),$$

leading to the linear ODE

$$\frac{dy}{dx} + \underbrace{\left[-\left(\frac{\partial f(x, y)}{\partial y}\right)_{y=y_0}\right]}_{p(x)} y = \underbrace{\left[f(x, y_0) - \left(\frac{\partial f(x, y)}{\partial y}\right)_{y=y_0} y_0\right]}_{q(x)}$$

of the general form (8.11) that we know has a unique solution. If $f(x, y)$ is continuously differentiable in a certain region of the $x - y$ plane, the same expansion can
be made about any point in that region leading to uniqueness in the whole region.

Let us consider as an example the following non-linear initial value problem:

$$\frac{dy}{dx} = \frac{7}{3}y^{4/7}, \quad y(0) = 0.$$

It is easy to see that $y = 0$ is a solution of the ODE and of the initial conditions.
On the other hand, by simple integration of this separable ODE one also obtains the
general solution

$$\int y^{-4/7} dy = \frac{7}{3}\int dx \implies \frac{y^{3/7}}{3/7} = \frac{7}{3}x + C_1 \implies y(x) = (x + C)^{7/3},$$

which after using the initial conditions yields the second solution $y = x^{7/3}$. Hence,
this initial value problem has at least two solutions. In fact, it appears that it has an

infinite number of solutions. Indeed, consider the following concatenated function:

$$y(x) = \begin{cases} (x-a)^{7/3}, & x \le a < 0 \\ 0, & a \le x \le b \\ (x-b)^{7/3}, & x \ge b > 0 \end{cases}$$

with $a < 0$ and $b > 0$ that are otherwise arbitrary. Note that this function is continuous. It is easy to see that since $x = 0$ is contained in the interval $a \le x \le b$, the
initial conditions are satisfied. Also, the ODE is satisfied by any choice of a and b.
Hence, all these functions form solutions to the initial value problem.

It is easy to see why this is so. Indeed, in this case $f(x, y) = \frac{7}{3} y^{4/7}$ and its partial
derivative $\partial f / \partial y = \frac{4}{3} y^{-3/7}$ at $y = 0$ is not continuous (it is equal to $\pm\infty$ on both
sides of $y = 0$).

> **Problem 8.24** Explain why the same ODE but with the initial condition
> $y(x_0) = 0$ in which $x_0 \ne 0$ has a unique solution.

Consider now another example for which the conditions of the uniqueness are
satisfied:

$$\frac{dy}{dx} = y^2, \quad y(0) = 0.$$

This ODE is also non-linear and has a trivial constant solution $y = 0$. On the other
hand, integrating this separable ODE, we easily get the general solution

$$y(x) = (C - x)^{-1}.$$

It may seem that in this case as well we get at least two solutions. However, this is
proved to be wrong once we use the initial conditions: $C = \infty$. Hence, for all x we
obtain $y(x) = 0$ again! The solution of the initial value problem is unique in this
case since $f(x, y) = y^2$ is continuously differentiable as $\partial f / \partial y = 2y$.

There is a number of other interesting and important for practical applications
points one might consider. We shall first briefly mention stability. The question we
might ask is whether a small change in the initial conditions may result in a large
deviation of the solution(s). Consider for instance the so-called autonomous equation
$\dot{y} = f(y)$ for the function $y(t)$ of time. One may envisage three situations: (i) the
solution is unstable, i.e. even a tiny change in the initial condition results in ever
growing deviation of the solution in time; (ii) the solution is stable, this happens
when a small enough change in the initial conditions results in the solution that
still remains close to the former one for all times, and (iii) asymptotically stable,
when, given a certain deviation in the initial conditions, at long times the solution
asymptotically approaches the former one.

Another important point to have in mind when solving non-linear ODEs is that
some algebraic manipulations (e.g. the squaring) might lead to spurious solutions the

original ODE does not have. Therefore, when solving a non-linear ODE, one has to substitute each solution into the original differential equation to check if it actually satisfies it. To illustrate this point, consider an ODE

$$\sqrt{y' + 1} = -y.$$

Squaring both sides and rearranging, we get:

$$y' + 1 = y^2 \quad \Longrightarrow \quad y' = y^2 - 1.$$

Problem 8.25 Show that the solution of this ODE is

$$y(x) = \frac{1 - Ce^{2x}}{1 + Ce^{2x}}$$

with C being an arbitrary constant.

However, it is easy to see that this function satisfies a different ODE, namely, $\sqrt{y' + 1} = y$. Hence, we have obtained a spurious solution caused by the loss of the minus sign in the right-hand side when squaring.

Finally, one has to be careful to not lose solutions when performing certain other algebraic manipulations. For instance, when solving a separable ODE

$$\frac{dy}{dx} = \frac{2y + 1}{2x + 1}$$

one would divide both sides by $2y + 1$ to transform this equation to a form which is integrable,

$$\frac{dy}{2y + 1} = \frac{dx}{2x + 1} \quad \Longrightarrow \quad \int \frac{dy}{2y + 1} = \int \frac{dx}{2x + 1}$$

$$\Longrightarrow \quad \ln \left| y + \frac{1}{2} \right| = \ln \left| x + \frac{1}{2} \right| + C_1$$

$$\Longrightarrow \quad \frac{y + 1/2}{x + 1/2} = e^{C_1} \quad \Longrightarrow \quad y = C \left(x + \frac{1}{2} \right) - \frac{1}{2},$$

where $C = e^{C_1}$. At the same time, $y = -1/2$ is also a solution of this ODE. Even if formally this solution is contained in the above one when setting $C = 0$, strictly speaking, this is not legitimate as $C = e^{C_1}$ cannot be zero. Hence, we lost the constant solution by dividing both sides of the equation by $2y + 1$. This simple example shows that one has to carefully monitor all the algebraic manipulations when transforming the ODE to a solvable form and check the made assumptions in order not to lose possible solutions.

Problem 8.26 Show that the initial value problem

$$y' = x^2 y^{1/3}, \quad y(a) = 0, \quad a > 0,$$

must have multiple solutions. Then check that $y_1(x) = 0$ and $y_2(x) = \frac{2\sqrt{2}}{27}(x^3 - a^3)^{3/2}$ are its two solutions.

Problem 8.27 Explain why the ODEs

$$(y')^4 + x^2 + 1 = 0 \quad \text{and} \quad (y')^2 + 2y^4 + \frac{3}{2} = 0$$

have no real valued solutions.

8.1.8   Picard's Method

This is a technique that enables one to obtain an approximate solution of a general initial value problem

$$\frac{dy}{dx} = f(x, y), \quad y(x_0) = y_0$$

by successive steps called Picard's iterations.[2]
   We first integrate both sides of the equation between x_0 and x to obtain:

$$y(x) = y_0 + \int_{x_0}^{x} f(t, y(t)) \, dt.$$

What we have obtained is called an integral equation. To obtain a solution, iterations can be built by choosing $y_0(x) = y_0$ and applying the recurrence formula

$$y_{n+1}(x) = y_0 + \int_{x_0}^{x} f(t, y_n(t)) \, dt, \quad n = 0, 1, 2, \dots.$$

This way we shall obtain:

$$y_1(x) = y_0 + \int_{x_0}^{x} f(t, y_0) \, dt,$$

$$y_2(x) = y_0 + \int_{x_0}^{x} f(t, y_1(t)) \, dt,$$

---

[2] Charles Émile Picard (1856–1941) was a French mathematician.

and so on. It can be shown that this sequence $\{y_1, y_2, \ldots\}$ of the functions converges to the exact solution, $y_n(x) \to y(x)$ as $n \to \infty$.

Consider a simple example:

$$y' = x + y, \quad y(0) = 0,$$

whose exact solution is easily obtained to be (this is a linear first order DE, Sect. 8.1.5)

$$y(x) = e^x - x - 1.$$

On the other hand, using Picard's method, we have $x_0 = y_0 = 0$ and $f(x, y) = x + y$, leading to the recurrence relation

$$y_{n+1}(x) = \int_0^x (t + y_n(t))dt, \quad n = 0, 1, 2, 3, \ldots$$

We have:

$$y_1 = \int_0^x (t + y_0)\, dt = \int_0^x t\, dt = \frac{x^2}{2},$$

$$y_2 = \int_0^x (t + y_1)\, dt = \int_0^x \left(t + \frac{t^2}{2}\right) dt = \frac{x^2}{2} + \frac{x^3}{6},$$

$$y_3 = \int_0^x \left(t + \frac{t^2}{2} + \frac{t^3}{6}\right) dt = \frac{x^2}{2} + \frac{x^3}{6} + \frac{x^4}{24},$$

$$y_4 = \int_0^x \left(t + \frac{t^2}{2} + \frac{t^3}{6} + \frac{t^4}{24}\right) dt = \frac{x^2}{2} + \frac{x^3}{6} + \frac{x^4}{24} + \frac{x^5}{120},$$

and so on. It is easy to see that the obtained at the last step expression corresponds to the first four terms in the Maclaurin series of the exact solution.

Problem 8.28 Using Picard's method, obtain the first three successive approximations for the initial value problem

$$y' = x + y^2, \quad y(0) = 0.$$

[Answer: $y_3(x) = x^2/2 + x^5/20 + x^8/160 + x^{11}/4400$.] Note that the term with x^{11} will modify its pre-factor as you go to the next fourth order.

8.1.9 Orthogonal Trajectories

Let us consider one nice geometrical problem that has a practical utility, e.g. in engineering.[3] Suppose, we are given a set of curves $F(x, y) = k$, each curve corresponds to a particular value of the parameter k. The problem is to find another family of curves that at each point would be perpendicular to the given family. These are called their orthogonal trajectories.

Differentiating the equation $F(x, y) = k$ with respect to x, we can calculate the derivative $y'(x)$ at x for the given family of curves as

$$\frac{\partial F}{\partial x} + \frac{\partial F}{\partial y}\frac{dy}{dx} = 0 \implies \frac{dy}{dx} = -\frac{\partial F/\partial x}{\partial F/\partial y}.$$

This gives a slope of the given curve (for the particular value of k). The curve that is perpendicular to this one and passing through the point (x, y) on the $x - y$ plane has the slope given by $-1/y'(x)$ (see Sect. 3.11 and also Problem 1.217). Hence, the required equation for the curves orthogonal to the given curves is

$$\frac{dy}{dx} = \frac{\partial F/\partial y}{\partial F/\partial x}. \tag{8.19}$$

Let us consider a simple example of the given curves being circles $x^2 + y^2 = R^2$ of radius R and centred at the origin. In this case $F(x, y) = x^2 + y^2$ and $k = R^2$. The equation (8.19) is easily solved,

$$\frac{dy}{dx} = \frac{2y}{2x} \implies \int \frac{dy}{y} = \int \frac{dx}{x} \implies y = Cx,$$

to give straight lines passing through the origin. As expected, these will at any point be perpendicular to the circles. The parameter C depends on the particular point x_0 at which we would like to have the intersection: if the intersection is expected at the point (x_0, y_0) of the circle of the particular radius R, then replacing $y_0 = Cx_0$ in the equation for the circle we obtain $C = \sqrt{(R/x_0)^2 - 1}$.

Problem 8.29 Show that orthogonal trajectories to hyperbolas $x^2 - y^2 = k$ are provided by the hyperbolas $y = C/x$, where $C = \pm x_0\sqrt{x_0^2 - k}$.

---

[3] Adapted from the book R. K. Nagle, E. B. Saff and A. D. Snider "Fundamentals of Differential Equations", Addison-Wesley, 8-th edition (2012).

Problem 8.30 Show that the orthogonal trajectories to the exponential functions $y = e^{kx}$ are provided by the curves $y(x)$ satisfying the implicit equation

$$y^2 (1 - 2 \ln y) - 2x^2 = C \,.$$

8.2   Linear Second-Order Differential Equations

8.2.1   General Consideration

Consider a general n-th order linear differential equation:

$$a_n(x) \frac{d^n y}{dx^n} + a_{n-1}(x) \frac{d^{n-1} y}{dx^{n-1}} + \cdots + a_1(x) \frac{dy}{dx} + a_0(x)y = f(x) \,. \qquad (8.20)$$

This equation contains a function $f(x)$ in the right-hand side depending only on x, and is called *nonhomogeneous*. *Homogeneous* equations have zero instead.

A very general property of this equation is that it is *linear*. This has a very profound effect on the solutions. Indeed, if we know two solutions $y_1(x)$ and $y_2(x)$ of this equation, then any of their linear combinations with *arbitrary* constants C_1 and C_2, i.e.

$$y(x) = C_1 y_1(x) + C_2 y_2(x) \,,$$

will also be a solution. Generally, any n-th order ODE like (8.20) has n linearly independent solutions $y_1(x), \ldots, y_n(x)$, and a *general solution*

$$y(x) = C_1 y_1(x) + C_2 y_2(x) + \cdots + C_n y_n(x) = \sum_{i=1}^{n} C_i y_i(x) \qquad (8.21)$$

is constructed as their linear combination with arbitrary coefficients C_1, C_2, etc. These are determined using additional information about the solution, e.g. *initial conditions*, i.e. known values of $y(x_0)$, $y^{(1)}(x_0)$, ..., $y^{(n-1)}(x_0)$ at some value x_0 of x, where $y^{(i)}(x) = d^i y/dx^i$.

Problem 8.31 Prove that the linear combination (8.21) of the solutions of the n-th order ODE (8.20) is also its solution.

Problem 8.32 Consider a general linear second-order ODE

$$a_2(x)y'' + a_1(x)y' + a_0(x)y = 0 .$$ (8.22)

Show that the function $f(x)$ which brings this ODE into the form $(a_2 f y')' + a_0 f y = 0$ satisfies the ODE $f'/f = a_1/a_2 - a_2'/a_2$, and obtain its solution. [Answer: $f(x) = \exp\left(\int (a_1/a_2)\,dx\right)/a_2$.].

From now on in this section we shall only be concerned with linear *second-order* ODEs. There must be *two linearly independent solutions* of the homogenous linear second-order DE. This fundamental result will be (partially) proven later on in this section. We shall start from one important general result that states that the second solution of any general linear second-order ODE can always be found if the first solution is known.

Theorem 8.1 *The second solution $y_2(x)$ of a linear second-order ODE (8.22) can always be found if the first solution $y_1(x)$ is known.*

Proof: Since both y_1 and y_2 satisfy the ODE (8.22), we have

$$a_2 y_1'' + a_1 y_1' + a_0 y_1 = 0 \text{ and } a_2 y_2'' + a_1 y_2' + a_0 y_2 = 0 .$$

Next, we multiply the first equation by y_2 and the second by y_1, and subtract the resulting equations from each other:

$$y_1 \left(a_2 y_2'' + a_1 y_2' + a_0 y_2\right) - y_2 \left(a_2 y_1'' + a_1 y_1' + a_0 y_1\right) = 0 ,$$

which after an obvious simplification yields

$$a_2 \left(y_1 y_2'' - y_2 y_1''\right) + a_1 \left(y_1 y_2' - y_2 y_1'\right) = 0 .$$ (8.23)

Consider now the determinant (called *Wronskian*)

$$W = \begin{vmatrix} y_1 & y_2 \\ y_1' & y_2' \end{vmatrix} = y_1 y_2' - y_2 y_1' .$$ (8.24)

Its derivative

$$W' = \left(y_1' y_2' + y_1 y_2''\right) - \left(y_2' y_1' + y_2 y_1''\right) = y_1 y_2'' - y_2 y_1'' .$$

Therefore, (8.23) can now be rewritten as the first-order ODE with respect to the function $W(x)$:

$$a_2(x)\frac{dW(x)}{dx} + a_1(x)W(x) = 0 ,$$

which is separable and thus can be easily integrated:

$$\frac{dW}{W} = -\frac{a_1(x)}{a_2(x)}dx \implies \int \frac{dW}{W} = -\int \frac{a_1(x)}{a_2(x)}dx$$

$$\implies \ln|W(x)| = -\int \frac{a_1(x)}{a_2(x)}dx ,$$

where an indefinite integral has been used. Hence, the Wronskian becomes

$$W(x) = \exp\left(-\int \frac{a_1(x)}{a_2(x)}dx\right) . \tag{8.25}$$

This result is called *Abel's formula*.

Once the function $W(x)$ has been calculated, it is now possible to consider (8.24) as a first-order ODE for $y_2(x)$:

$$y_2'(x) - \frac{y_1'(x)}{y_1(x)}y_2(x) = \frac{W(x)}{y_1(x)} .$$

This ODE is exactly of the type (8.11) we considered before in Sect. 8.1.5, and its solution is given by (8.13) with $p(x) = -y_1'(x)/y_1(x)$ and $q(x) = W(x)/y_1(x)$.

This way of finding y_2, however, appears a bit tricky as one has to deal carefully with the modulus of $y_1(x)$. A much simpler and straightforward way of obtaining $y_2(x)$ is to try a substitution $y_2(x) = u(x)y_1(x)$ directly in the above equation:

$$\left(u'y_1 + uy_1'\right) - \frac{y_1'}{y_1}(uy_1) = \frac{W}{y_1} \implies u'y_1 = \frac{W}{y_1} \implies u'(x) = \frac{W(x)}{y_1^2(x)} ,$$

which, after integration, gives immediately

$$u(x) = \int \frac{W(x)}{y_1^2(x)}dx . \tag{8.26}$$

This should be understood as an indefinite integral without an arbitrary constant: this is because $y_2 = uy_1$ and a constant C in $u(x)$ would simply give an addition of Cy_1 in y_2 that will be absorbed in the general solution $y = C_1y_1 + C_2y_2$. Hence,

$$y_2(x) = u(x)y_1(x) = y_1(x)\int \frac{W(x)}{y_1^2(x)}dx$$

$$= y_1(x)\int \frac{1}{y_1^2(x)}\exp\left(-\int \frac{a_1(x)}{a_2(x)}dx\right)dx . \tag{8.27}$$

Thus, if one solution of the second-order linear ODE is known, the other one can always be obtained using the above formula. **Q.E.D.**

As an example, let us find the second solution of the equation

$$y'' + y = 0 ,$$

if it is known that the first one is $y_1(x) = \sin x$. Here $a_2 = 1$, $a_1 = 0$ and $a_0 = 1$, so that, according to (8.25), $W(x) = 1$ and, using (8.27), we get:

$$y_2 = \sin x \int \frac{dx}{\sin^2 x} = \sin x \, (-\cot x) = -\cos x ,$$

which, as can be easily checked, is the correct solution.[4]

The theorem considered above assumes that the Wronskian is not equal to zero. It is easy to see, however, that it is equal to zero for *all* values of x if and only if the two functions $y_1(x)$ and $y_2(x)$ are linearly dependent. Hence, if we have two *linearly independent solutions* of the ODE (8.22), i.e. when $y_2(x) \neq Cy_1(x)$ with C being a constant, then we are guaranteed that $W(x) \neq 0$.

Theorem 8.2 (Linear Dependence of Two Functions) *The Wronskian (8.24) is equal to zero if and only if the two functions $y_1(x)$ and $y_2(x)$ it is constructed from are linearly dependent, i.e. if $y_2(x) = Cy_1(x)$ within some interval of x and with some constant $C \neq 0$.*

Proof: Indeed, if two functions are linearly dependent, $y_2(x) = Cy_1(x)$, then a simple calculation shows that $W = 0$ for all values of x. Assume now that

$$W(x) = y_1 y_2' - y_2 y_1' = 0 \implies y_1 y_2' = y_2 y_1' .$$

Disregarding the trivial case of either of the two solutions to be identically equal to zero, we divide both sides of the above equation by $y_1 y_2$ and rearrange:

$$\frac{y_2'}{y_2} = \frac{y_1'}{y_1} \implies \frac{d}{dx} (\ln y_2) = \frac{d}{dx} (\ln y_1) ,$$

which after integration gives $y_2(x) = C \, y_1(x)$, i.e. the linear dependence. **Q.E.D.**

---

[4] The minus sign does not matter as in the general solution y_2 will be multiplied by an arbitrary constant anyway.

Theorem 8.3 (Existence of Two Linearly Independent Solutions) *A general linear second-order ODE (8.22) always has two linearly independent solutions.*

Proof: Indeed, consider two initial value problems:

$$a_2(x)y'' + a_1(x)y' + a_0(x)y = 0 , \quad y(x_0) = 1 \text{ and } y'(x_0) = 0 \qquad (8.28)$$

and

$$a_2(x)y'' + a_1(x)y' + a_0(x)y = 0 , \quad y(x_0) = 0 \text{ and } y'(x_0) = 1 . \qquad (8.29)$$

Due to existence (which we do not prove here[5]), we are guaranteed that both initial value problems have unique solutions. Let us call them $y_1(x)$ and $y_2(x)$, respectively. The Wronskian $W(x_0)$ at x_0 is not equal to zero,

$$W(x_0) = y_1(x_0)y_2'(x_0) - y_2(x_0)y_1'(x_0) = 1 \cdot 1 - 0 \cdot 0 = 1 .$$

Moreover, the Wronskian at any x (in the interval where the functions $a_0(x)$, $a_1(x)$ and $a_2(x)$ are continuous) is non-zero. Indeed, for it to be zero, according to Theorem 8.2, the two functions y_2 and y_1 must be linearly dependent, i.e. $y_2(x) = Cy_1(x)$ with $C \neq 0$. This however cannot be true since in this case they would not satisfy the specific initial conditions (8.28) and (8.29). Therefore, $W(x) \neq 0$ for any x (in fact, $W(x)$ is given by (8.25)) and hence the two solutions are linearly independent. **Q.E.D.**

These simple results also ensure that the initial value problem

$$a_2(x)y'' + a_1(x)y' + a_0(x)y = 0 , \quad y(x_0) = y_0 \text{ and } y'(x_0) = y_1 \qquad (8.30)$$

has always a unique solution.

Theorem 8.4 (Uniqueness of the Solution) *The initial value problem (8.30) always has a unique solution.*

---

[5] All essential theorems related to uniqueness and existence of solutions of ODEs and their systems are given, e.g. in Chap. 6 of "*An introduction to ordinary differential equations*" by E. A. Coddington, Dover, N.Y. 1961.

Proof: Indeed, since the general solution of the ODE is $y(x) = C_1 y_1(x) + C_2 y_2(x)$, hence, to determine the two arbitrary constants we obtain a system of two linear algebraic equations:

$$\begin{cases} C_1 y_1(x_0) + C_2 y_2(x_0) = y_0 \\ C_1 y_1'(x_0) + C_2 y_2'(x_0) = y_1 \end{cases}. \tag{8.31}$$

Let us choose as two linearly independent solutions $y_1(x)$ and $y_2(x)$ the solutions of the initial value problems (8.28) and (8.29). Then, it is easy to see that the above system of equations

$$\begin{cases} C_1 y_1(x_0) + C_2 y_2(x_0) = C_1 \cdot 1 + C_2 \cdot 0 = y_0 \\ C_1 y_1'(x_0) + C_2 y_2'(x_0) = C_2 \cdot 0 + C_2 \cdot 1 = y_1 \end{cases}$$

has the unique solution of $C_1 = y_0$ and $C_2 = y_1$. As was noted above, the Wronskian constructed out of these two specific solutions is not zero at x_0.

One can also choose, if desired, as the two solutions $y_1(x)$ and $y_2(x)$ in the linear combination *any* two functions that satisfy the ODE and are linearly independent. In this case, the unique solution follows from the fact that the system of two algebraic equations (8.31) has a unique solution with respect to the constants C_1 and C_2 since the determinant made up of the coefficients of the linear system,

$$\begin{vmatrix} y_1(x_0) & y_2(x_0) \\ y_1'(x_0) & y_2'(x_0) \end{vmatrix}$$

is not equal to zero, see Sect. 1.3.6. Indeed, this determinant is the Wronskian $W(x_0)$ calculated at the initial point, and it cannot be equal to zero since the two functions y_1 and y_2 are linearly independent for any x including x_0. This proves the above made statement. **Q.E.D.**

Problem 8.33 Consider the following linear second-order ODE

$$y'' + 4y' + 4y = 0, \tag{8.32}$$

which first solution is $y_1 = e^{-2x}$. Show that the second solution is $y_2 = x y_1 = x e^{-2x}$.

Problem 8.34 Consider a one-particle one-dimensional Schrödinger equation

$$-\frac{\hbar^2}{2m}\psi''(x) + V(x)\psi(x) = E\psi(x),$$

where $\psi(x)$ is the particle wavefunction, $V(x)$ external potential, m particle mass, and $\hbar$ Planck's constant. Consider a discrete energy spectrum E for which

$\psi(\pm\infty) = 0$, meaning that the particle is localized in some region of space (as opposite to the continuum spectrum when there is a non-zero probability to find the particle anywhere in space). Show that the energies E in the discrete spectrum cannot be degenerate, i.e. there can only be a single state $\psi(x)$ for each energy E. [Hint: *prove by contradiction, assuming there are two such solutions, ψ_1 and ψ_2; then, argue that ψ''/ψ for both should be the same; then use integration by parts.*]

Problem 8.35 Find the second solution of the ODE

$$xy'' - (x + 1)y' + y = 0,$$

if the first solution is known to be $y_1 = e^x$. [Answer: $y_2 = -x - 1$.]

Problem 8.36 Consider the Bessel equation of order one-half:

$$x^2 y'' + xy' + \left(x^2 - \frac{1}{4}\right) y = 0.$$

Its first solution is $y_1(x) = x^{-1/2}\cos x$. Show that its second solution is $y_2(x) = x^{-1/2}\sin x$.

Problem 8.37 Consider the Legendre equation:

$$\left(1 - x^2\right) y'' - 2xy' + 2y = 0.$$

Its first solution is $y_1(x) = x$. Show that its second solution is

$$y_2(x) = -1 + \frac{x}{2}\ln\left|\frac{1 + x}{1 - x}\right|$$

(where $-1 < x < 1$).

Problem 8.38 Consider a Cauchy–Euler equation

$$x^2 y'' + xy' - y = 0.$$

Given its first solution, $y_1 = x$, show that the second solution can be chosen as $y_2 = x^{-1}$.

Note that both solutions of the homogeneous Cauchy–Euler equation

$$ax^2 y'' + bxy' + cy = 0$$

are obtained by trying $y(x) = x^p$ since after substitution into the equation we obtain a quadratic equation for the parameter p:

$$ap(p-1) + bp + c = 0 \implies ap^2 + (b-a)p + c = 0.$$

For instance, in the case of the previous problem the quadratic equation $p^2 - 1 = 0$ is obtained giving immediately $p = \pm 1$, so that its two linearly independent solutions are $y_1 = x$ and $y_2 = x^{-1}$.

Of course, this method would only result in two linearly independent solutions if this quadratic equation gives two distinct roots. If this is not the case, the above method based on the Wronskian can be used to find the second solution as illustrated by the next Problem.

Problem 8.39 Consider the following homogeneous Cauchy–Euler equation

$$x^2 y'' - 3xy' + 4y = 0.$$

Using the trial solution $y = x^p$, show that the first solution of this ODE is $y_1 = x^2$. Then, show that the second solution $y_2 = x^2 \ln x$.

8.2.2   Homogeneous Linear Differential Equations with Constant Coefficients

We have learned in the previous section that the first-order linear ODE's always allow for a general solution. However, it is not possible to obtain such a solution in the cases of higher order ODEs containing variable coefficients. One specific case that allows for a general solution is that of a higher order ODE with *constant coefficients*. We shall again limit ourselves with the ODEs of the second order; higher order ODE's can be considered along the same lines.

We shall consider equations of the form:

$$a_2 \frac{d^2 y}{dx^2} + a_1 \frac{dy}{dx} + a_0 y = 0, \tag{8.33}$$

where a_0, a_1 and $a_2 \neq 0$ are some constant coefficients. This equation is generally solved using an exponential trial solution e^{px}. Indeed, substituting this trial function into the ODE above, we have:

$$\left(a_2 p^2 + a_1 p + a_0\right) e^{px} = 0.$$

Since we want this equation to be satisfied for any x, we must have p to satisfy the following *characteristic equation*:

$$a_2 p^2 + a_1 p + a_0 = 0, \tag{8.34}$$

that yields either one or two solutions for p via

$$p_{1,2} = \frac{1}{2a_2}\left[-a_1 \pm \sqrt{a_1^2 - 4a_0a_2}\right].$$

If there are *two (different)* solutions p_1 and p_2, then two functions e^{p_1x} and e^{p_2x} satisfy the same equation; note that the two functions are linearly independent in this case. Since the equation is linear, their linear combination would also satisfy it:

$$y = C_1e^{p_1x} + C_2e^{p_2x}. \tag{8.35}$$

Therefore, this must be the required general solution of the homogeneous equation (8.33).

The above consideration can be put on a firmer footing with the following discussion. Let us rewrite our ODE using the operator notations of Sect. 3.7.4:

$$\left(a_2D^2 + a_1D + a_0\right)y(x) = 0.$$

The operator acting on the unknown function $y(x)$ is a square polynomial in D of the same form as the left-hand side of the quadratic equation (8.34). Therefore, if p_1 and p_2 are its different roots, the operator can be equivalently rewritten via its roots, i.e. we instead have the equation:

$$a_2\left(D - p_2\right)\left(D - p_1\right)y(x) = 0.$$

Obviously, this equation is solved by $y(x)$ that satisfies

$$\left(D - p_1\right)y = 0 \implies y' = p_1y,$$

whose solution is an exponential, e^{p_1x}. At the same time, the operators $D - p_1$ and $D - p_2$ commute, i.e.

$$\left(D - p_2\right)\left(D - p_1\right) = D^2 - \left(p_1 + p_2\right)D + p_1p_2 = \left(D - p_1\right)\left(D - p_2\right),$$

so that the ODE can equivalently be rewritten as

$$a_2\left(D - p_1\right)\left(D - p_2\right)y(x) = 0.$$

In the same vein, its solution is also given by functions $y(x)$ that satisfy the equation $\left(D - p_2\right)y(x) = 0$, and the exponential e^{p_2x} is its solution. Since the ODE is linear, an arbitrary linear combination of the two solutions is its general solution, and we arrive at (8.35).

Let us now consider some simple examples. To solve the equation

$$\frac{d^2y}{dx^2} - 8\frac{dy}{dx} + 12y = 0,$$

the trial solution e^{px} is substituted into the equation to give $p^2 - 8p + 12 = 0$. Hence, two solutions, $p_1 = 2$ and $p_2 = 6$, are obtained for p. Hence the general solution is

$$y(x) = C_1 e^{2x} + C_2 e^{6x} .$$

To solve the equation

$$\frac{d^2 y}{dx^2} + 4y = 0 ,$$

we substitute the trial solution e^{px} into the equation to get $p^2 + 4 = 0$, which has two complex solutions: $p_1 = 2i$ and $p_2 = -2i$. Therefore, a general solution is

$$y(x) = C_1 e^{i2x} + C_2 e^{-i2x} .$$

This can actually be rewritten via sine and cosine functions using Euler's identities (3.43):

$$y(x) = C_1 [\cos(2x) + i \sin(2x)] + C_2 [\cos(2x) - i \sin(2x)]$$
$$= A_1 \cos(2x) + A_2 \sin(2x) ,$$

where $A_1 = C_1 + C_2$ and $A_2 = i (C_1 - C_2)$ are two new arbitrary (and in general complex) constants.

Let us now solve the equation

$$\frac{d^2 y}{dx^2} - 4\frac{dy}{dx} + 8y = 0 .$$

The corresponding equation for p is $p^2 - 4p + 8 = 0$, that has two complex solutions: $p_1 = 2 + 2i$ and $p_2 = 2 - 2i$. Therefore, a general solution is

$$y(x) = C_1 e^{(2+2i)x} + C_2 e^{(2-2i)x} = e^{2x} \left(C_1 e^{i2x} + C_2 e^{-i2x} \right)$$
$$= e^{2x} [A_1 \cos(2x) + A_2 \sin(2x)] .$$

If, however, there is only *one* root in (8.34), i.e. $p_1 = p_2$ (the case of repeated roots), then only one solution of the ODE can be immediately written as an exponential function, $y_1 = \exp(p_1 x)$. To find the second solution, we recall that, as was found in Theorem 8.1, the second solution can always be found from the first via (8.27) in the form of $y_2(x) = u(x)y_1(x)$, where the function $u(x)$ is to be determined from (8.26). The Wronskian (8.25) in this case is

$$W(x) = \exp\left(-\frac{a_1}{a_2} \int dx \right) = \exp\left(-\frac{a_1}{a_2} x \right) .$$

Since the roots of the characteristic equation (8.34) are the same, we must have $a_1^2 = 4a_0a_2$ and $p_1 = -a_1/2a_2$, i.e. $W(x) = \exp(2p_1x)$. Using this now in (8.26), we easily obtain:

$$u(x) = \int \frac{W}{y_1^2}dx = \int \frac{e^{2p_1x}}{e^{2p_1x}}dx = x \,,$$

and hence the second solution is to be taken in the form: $y_2 = xy_1 = x\exp(p_1x)$.

Problem 8.40 Consider equation

$$y'' - 4y' + 4y = 0 \,,$$

which has repeated roots $p = 2$ in the characteristic equation $p^2 - 4p + 4 = 0$, i.e. $y_1(x) = e^{2x}$. To obtain the second solution, try directly the substitution $y = u(x)e^{2x}$ in the original DE. Show that $u'' = 0$, and hence $u(x) = A_1x + A_2$. Finally, argue that it is sufficient to take $y_2(x) = xe^{2x}$ as the second independent solution.

Problem 8.41 Consider the case of two distinct but very close roots of the characteristic equation: p_1 and $p_2 = p_1 + \delta$. Prove that the same result for the second solution y_2 is obtained in the limit of $\delta \to 0$.

Problem 8.42 Consider the case of two distinct roots $p_1 \neq p_2$ of the characteristic equation. Show explicitly using Theorem 8.1 that from $y_1 = e^{p_1x}$ follows $y_2 = e^{p_2x}$.

Problem 8.43 Obtain the general solutions of:

$$\text{(a)} \quad y'' - 4y = 0; \quad \text{(b)} \quad y'' - 6y' + 9y = 0 \,.$$

[Answer: (a) $y = C_1e^{2x} + C_2e^{-2x}$; (b) $y = (C_1x + C_2)e^{3x}$.].

8.2.3   Nonhomogeneous Linear Differential Equations

Here we shall consider nonhomogeneous equations (8.20), i.e. with non-zero right-hand side function $f(x) \neq 0$. Whatever the order of the ODE and the form of the functions-coefficients $a_n(x), \ldots, a_0(x)$, the construction of the general solution of the equation is based on the following theorem:

> **Theorem 8.5** *The general solution of a linear nonhomogeneous equation is given by*
>
> $$y(x) = y_c(x) + y_p(x) \,, \tag{8.36}$$
>
> *where*
>
> $$y_c(x) = C_1 y_1(x) + \cdots + C_n y_n(x)$$
>
> *is the general solution of the corresponding homogeneous equation, called the* complementary solution, *and* $y_p(x)$ *is the so-called* particular solution *(integral) of the nonhomogeneous equation.*

Proof: Let us start by introducing a shorthand notation for the left-hand side of (8.20) by defining an *operator*

$$\mathcal{L} = a_n(x)\frac{d^n}{dx^n} + a_{n-1}(x)\frac{d^{n-1}}{dx^{n-1}} + \cdots + a_1(x)\frac{d}{dx} + a_0(x) \,, \tag{8.37}$$

which, when acting on a function $g(x)$ standing on the *right* of it, produces the following action:

$$\mathcal{L}g(x) \equiv a_n(x)\frac{d^n g}{dx^n} + a_{n-1}(x)\frac{d^{n-1}g}{dx^{n-1}} + \cdots + a_1(x)\frac{dg}{dx} + a_0(x)g \,.$$

The operator $\mathcal{L}$ is linear, i.e.

$$\mathcal{L}(c_1 g_1 + c_2 g_2) = c_1 \mathcal{L}g_1 + c_2 \mathcal{L}g_2 \,, \tag{8.38}$$

where $g_1(x)$ and $g_2(x)$ are two arbitrary functions. Then, (8.20) can be rewritten in the following short form:

$$\mathcal{L}y = f(x) \,. \tag{8.39}$$

After these notations, we see that the complementary and particular solutions are to satisfy two different equations:

$$\mathcal{L}y_c = 0 \ \text{ and } \ \mathcal{L}y_p = f(x) \,. \tag{8.40}$$

Let y be a solution of the nonhomogeneous ODE. Acting wit the operator $\mathcal{L}$ on the difference $y - y_p$, we get:

$$\mathcal{L}\left(y - y_p\right) = \mathcal{L}y - \mathcal{L}y_p = f(x) - f(x) = 0 \,, \tag{8.41}$$

i.e. $y - y_p$ must be the general solution y_c of the homogeneous equation $\mathcal{L}y_c = 0$, i.e. any solution has the form $y = y_c + y_p$. Note that in the first passage in (8.41) we have made use of the fact that the operator $\mathcal{L}$ is linear. **Q.E.D.**

Let us find a general solution of the equation

$$y'' + y = e^x .$$

According to the theorem proven above, we need first to obtain the complementary solution $y_c(x)$ which is the general solution of the homogeneous equation $y'' + y = 0$. This task must be trivial and yields $y_c = C_1 \cos x + C_2 \sin x$. Trying a particular solution in the form $y_p = Ae^x$ with some unknown constant A, we get after substitution in the equation:

$$Ae^x + Ae^x = e^x \implies 2A = 1 \implies A = \frac{1}{2} \implies y_p = \frac{1}{2}e^x ,$$

so that the general solution is

$$y = y_c + y_p = C_1 \cos x + C_2 \sin x + \frac{1}{2}e^x .$$

Note that if we take two specific solutions, say, $2 \sin x + \frac{1}{2}e^x$ and $-3 \cos x + \frac{1}{2}e^x$, obtained by choosing particular values for the arbitrary constants C_1 and C_2, then the difference $2 \sin x + 3 \cos x$ of the two solutions is in fact a complementary solution of the homogeneous equation with the particular choice of the constants $C_1 = 3$ and $C_2 = 2$, as expected.

Can we ensure that the initial value problem

$$a_2(x)y'' + a_1(x)y' + a_0(x)y = f(x) , \quad y(x_0) = y_0 \text{ and } y'(x_0) = y_1$$

has a unique solution? Similarly to the case of the homogeneous equation considered in Sect. 8.2.1, let us assume that two solutions $y_1(x)$ and $y_2(x)$ are linearly independent and that $y_p(x)$ is a particular integral satisfying the nonhomogeneous ODE above. Then, the general solution is

$$y(x) = y_p(x) + C_1 y_1(x) + C_2 y_2(x)$$

and therefore the arbitrary constants must satisfy algebraic equations

$$\begin{cases} C_1 y_1(x_0) + C_2 y_2(x_0) = y_0 - y_p(x_0) \\ C_1 y_1'(x_0) + C_2 y_2'(x_0) = y_1 - y_p'(x_0) \end{cases} .$$

We know that this system of two linear algebraic equation has a unique solution if the Wronskian $W(x_0) \neq 0$. However, as was shown above in Sect. 8.2.1, this condition is necessary for the two solutions of the homogeneous equation be linearly independent. Hence, the solution of the nonhomogeneous initial value problem is unique.

How can one choose the particular solution? We can see from the above example that a wise choice might be to try the form similar to that of the function $f(x)$ itself. In the case of a linear equation with constant coefficients this so-called *method of undetermined coefficients* may be successful in some cases and will be considered next.

8.2.3.1 Method of Undetermined Coefficients

Consider a second-order linear nonhomogeneous ODE with *constant* coefficients:

$$a_2 y'' + a_1 y' + a_0 y = f(x) . \tag{8.42}$$

We stress that this method does not work in the case of the variable coefficients, i.e. when $a_0(x)$, $a_1(x)$ and/or $a_2(x)$ are functions of x. Hence, this method is rather limited.

Four cases of $f(x)$ can be considered:

1. If $f(x)$ is a polynomial $P^{(n)}(x) = b_n x^n + \cdots + b_1 x + b_0$ of the order n, then the particular solution should be sought in the form of a polynomial $Q^{(n)}(x)$ of the same order with unknown coefficients that must be determined by substituting it into the equation and comparing terms with the like powers of x on both sides. This prescription is easy to understand. Indeed, we are seeking a particular solution y_p of the (8.42). Let us rewrite this equation in a shorthand form as $\mathcal{L} y_p = f(x)$, where

$$\mathcal{L} = a_2 \frac{d^2}{dx^2} + a_1 \frac{d}{dx} + a_0$$

is an operator that appears in the left-hand side of our differential equation. Then, an action of this operator on the polynomial $Q^{(n)}(x)$, i.e. the expression

$$\mathcal{L} Q^{(n)}(x) = a_2 \frac{d^2}{dx^2} Q^{(n)}(x) + a_1 \frac{d}{dx} Q^{(n)}(x) + a_0 Q^{(n)}(x) ,$$

is also a polynomial of the same order that should be equal to the polynomial $P^{(n)}(x)$, again of the same order, in the right-hand side of the DE. Clearly, this is easy to satisfy by equating coefficients to the same powers of x of the polynomials on both sides of the DE: this procedure will lead to a system of $n + 1$ linear algebraic equations (as there are $n + 1$ different powers x^m on both sides of the equation, where $m = 0, 1, 2, \ldots, n$) for the $n + 1$ coefficients of the polynomial $Q^{(n)}(x)$; this must be enough to determine them.

2. If $f(x) = A \cos(\omega x) + B \sin(\omega x)$ is a linear combination of trigonometric functions, try

$$y_p(x) = A_1 \cos(\omega x) + B_1 \sin(\omega x)$$

in the same form and with identical ω, but with some unknown coefficients A_1 and B_1. These unknown coefficients are obtained by substitution into the ODE and comparing coefficients to $\sin(\omega x)$ and $\cos(\omega x)$ on both sides. Note that even if $A = 0$ or $B = 0$ in $f(x)$, still *both* sine and cosine terms must be present in $y_p(x)$. Indeed, it is easy to see that both $\mathcal{L} \sin(\omega x)$ and $\mathcal{L} \cos(\omega x)$ are linear combinations of the sine and cosine functions, $\alpha \sin(\omega x) + \beta \cos(\omega x)$, so that when we substitute y_p into the equation, $\mathcal{L} y_p(x) = f(x)$, we shall have two equation for the two coefficients A_1 and B_1 when comparing coefficients to $\sin(\omega x)$ and $\cos(\omega x)$ on both sides of the DE.

3. If $f(x)$ is an exponential $Be^{\lambda x}$, then one has to use $y_p(x) = Ae^{\lambda x}$ as the particular solution with some constant A to be determined by substitution into the equation. In this case $\mathcal{L}e^{\lambda x}$ gives $e^{\lambda x}$ with some purely numerical pre-factor proportional to A, so that when equating this to the right-hand side and cancelling on the exponential function, we immediately obtain an equation for the unknown constant A.

4. If, more generally, $f(x)$ is given as a combination of all these functions, i.e.

$$f(x) = P^{(n)}(x)e^{\lambda x}\left[A\cos(\omega x) + B\sin(\omega x)\right] ,$$

then the particular solution is sought in a similar form keeping the same λ and ω, but with different coefficients in place of A, B and those in the polynomial; the coefficients are then obtained by substituting into the differential equation and comparing coefficients to the like powers of x and sine and cosine (the exponential function will cancel out). This case is a combination of the three previous cases.

There are exceptions to the prescriptions introduced above, these will be considered later on. Also note that if $f(x)$ is a sum of functions, particular integrals could be found individually for each of them; the final particular integral will be given as a sum of all these individual ones.

Examples below should illustrate the idea of the method.

Example 8.1 ▶ Solve the equation

$$y'' - 5y' + 6y = 1 + x .$$

Solution. The complementary solution reads $y_c = C_1 e^{2x} + C_2 e^{3x}$. Since the right-hand side is the first-order polynomial, the particular integral is sought as a first-order polynomial as well, $y_p = Ax + B$. Substituting it into the equation and comparing coefficients to the same powers of x, gives:

$$0 - 5A + 6(Ax + B) = 1 + x \implies \begin{cases} 6A = 1 \\ -5A + 6B = 1 \end{cases} \implies \begin{cases} A = 1/6 \\ B = 11/36 \end{cases},$$

so that the general solution

$$y = C_1 e^{2x} + C_2 e^{3x} + \frac{1}{6}\left(x + \frac{11}{6}\right). \blacktriangleleft$$

Example 8.2 ▶ Find the particular solution of the equation

$$y'' - 5y' + 6y = \sin(4x) + 3\cos x .$$

Solution. We must find the particular solutions individually for each of the terms in the right-hand side since they have different "frequencies". For $\sin(4x)$ we try the function $A\sin(4x) + B\cos(4x)$, that gives upon substitution:

$$[-16A\sin(4x) - 16B\cos(4x)] - 5[4A\cos(4x) - 4B\sin(4x)]$$
$$+ 6[A\sin(4x) + B\cos(4x)] = \sin(4x),$$

that, after comparing pre-factors to the sine and cosine functions, yields

$$\begin{cases} -16A + 20B + 6A = 1 \\ -16B - 20A + 6B = 0 \end{cases} \implies \begin{cases} A = -1/50 \\ B = 2/50 \end{cases},$$

so that the first part of the particular solution is

$$y_{p1} = -\frac{1}{50}\sin(4x) + \frac{2}{50}\cos(4x).$$

The second part is obtained by trying $C\sin x + D\cos x$ that similarly gives:

$$[-C\sin x - D\cos x] - 5[C\cos x - D\sin x] + 6[C\sin x + D\cos x] = 3\cos x,$$

$$\begin{cases} -C + 5D + 6C = 0 \\ -D - 5C + 6D = 3 \end{cases} \implies \begin{cases} C = -3/10 \\ D = 3/10 \end{cases} \implies y_{2p} = -\frac{3}{10}\sin x + \frac{3}{10}\cos x.$$

Therefore, the particular solution of the whole equation is

$$y_p = -\frac{1}{50}\sin(4x) + \frac{2}{50}\cos(4x) - \frac{3}{10}\sin x + \frac{3}{10}\cos x,$$

which can be checked by direct substitution. ◄

Example 8.3 ► Find the particular solution of the equation:

$$y'' - 5y' + 6y = 5xe^x.$$

Solution. Since the right-hand side is a product of the first-order polynomial and the exponential function, we try $y_p = (Ax + B)e^x$. Substituting into the equation, this gives:

$$(Ax + 2A + B)e^x - 5(Ax + A + B)e^x + 6(Ax + B)e^x = 5xe^x$$
$$\implies (Ax + 2A + B) - 5(Ax + A + B) + 6(Ax + B) = 5x$$
$$\implies \begin{cases} A - 5A + 6A = 5 \\ 2A + B - 5(A + B) + 6B = 0 \end{cases} \implies \begin{cases} A = 5/2 \\ B = 15/4 \end{cases}$$
$$\implies y_p = \left(\frac{5}{2}x + \frac{15}{4}\right)e^x \quad ◄$$

As was mentioned earlier, there are exceptions to the prescriptions 2, 3 and 4 given above (when the right-hand side contains exponential and/or trigonometric functions), i.e. in some special cases the method requires modification. The case of the sine and cosine functions can be considered on the same footing with the case of the exponential function because of the fact that these trigonometric functions, by virtue of Euler's equations Sect. 3.4.4, can be related to the exponential function with a complex exponent. Hence, it is sufficient to consider only the case of the exponential function in the right-hand side of the DE.

The special cases appear when $f(x)$ happens to be *related* to one (or even both) solutions of the corresponding homogeneous equation. Two special cases are possible:

- A single "resonance", when the exponent λ (a complex number in general) of $f(x) = Be^{\lambda x}$ coincides with one of the two distinct solutions p_1 or p_2 of the characteristic equation (8.34) of the homogeneous DE ($p_1 \neq p_2$). In this case y_p should be sought as $y_p = Axe^{\lambda x}$, i.e. the previous recipe is modified by introducing an extra x.
- A double "resonance", when $p_1 = p_2 = \lambda$, in which case $y_p = Ax^2 e^{\lambda x}$, i.e. the usual recipe is modified by introducing an extra x^2.

A justification for these modifications will be given in the next subsection. Here we shall only illustrate these special cases by considering a few examples.

Example 8.4 ▶ Find the particular solution of the equation:

$$y'' - 5y' + 6y = e^{2x} .$$

Solution. If we try $y_p = Ae^{2x}$ following the function type in the right-hand side, then we get zero in the left-hand side since e^{2x}, as can easily be checked, is a solution of the homogeneous equation. Obviously, this trial solution does not work. However, the required modification is actually very simple: if we multiply y_p tried above by x, it will work. Indeed, try $y_p = Axe^{2x}$. Then, after substitution into the equation, we obtain:

$$(4A + 4Ax) e^{2x} - 5 (A + 2Ax) e^{2x} + 6Axe^{2x} = e^{2x} ,$$

$$4A + 4Ax - 5A - 10Ax + 6Ax = 1 \quad \text{or} \quad A = -1 ,$$

so that $y_p = -xe^{2x}$ in this case. ◀

The considered case corresponds to the single "resonance" as the exponent in the right-hand side coincides with that of only *one* of the solutions of the homogeneous equation. If however *both* complementary exponents are the same, $p_1 = p_2$, and the same exponent happens to be present in the right-hand side, we have the case of the double "resonance". In that case the y_p constructed according to the rules stated above should be additionally multiplied by x^2 instead.

Example 8.5 ▶ Find the particular solution of the equation:

$$y'' - 4y' + 4y = xe^{2x} .$$

Solution. The complementary solution $y_c = C_1 e^{2x} + C_2 x e^{2x}$, and we see that the same exponent is used in the right-hand side. Substituting $y_p = (Ax + B) e^{2x}$ constructed according to the usual rules gives zero in the left-hand side as it is basically the same as y_c; not good. Let us now try $y_p = x (Ax + B) e^{2x}$ instead, which gives:

$$\left[2A + 4(2Ax + B) + 4\left(Ax^2 + Bx\right)\right] e^{2x} - 4\left[(2Ax + B) + 2\left(Ax^2 + Bx\right)\right] e^{2x}$$
$$+ 4\left(Ax^2 + Bx\right) e^{2x} = xe^{2x}$$

and, after canceling on the exponential function,

$$\left[2A + 4(2Ax + B) + 4\left(Ax^2 + Bx\right)\right] - 4\left[(2Ax + B) + 2\left(Ax^2 + Bx\right)\right]$$
$$+ 4\left(Ax^2 + Bx\right) = x .$$

Comparing coefficients to x^0, we get $A = 0$, good; however, coefficients to x^1 all cancel out in the left-hand side leading to an impossible $0 = 1$, i.e. we obtained a contradiction.

Now let us try $y_p = x^2 (Ax + B) e^{2x}$, which results (after canceling on e^{2x}) in:

$$\left[4Ax^3 + 4(3A + B) x^2 + 2(3A + 4B) x + 2B\right]$$
$$-4\left[2Ax^3 + (3A + 2B) x^2 + 2Bx\right] + 4\left(Ax^3 + Bx^2\right) = x .$$

The terms containing x^3 and x^2 in the left-hand side cancel out; the terms with x yield $A = 1/6$, while the terms with x^0 result in $B = 0$, i.e. $y_p = \left(x^3/6\right) e^{2x}$, which as easily checked by direct substitution is the correct particular integral of this ODE. ◀

Problem 8.44 Find the particular integral solutions of the ODE $y'' + 3y' + 2y = f(x)$ with the following right hand side function $f(x)$:

(a) $6 + x^2$; (b) $2e^{3x}$; (c) $6 \sin x$; (d) $6e^{-3x} \sin x$.

[Answer: (a) $y_p = 19/4 - 3x/2 + x^2/2$; (b) $y_p = e^{3x}/10$; (c) $y_p = (3/5) \sin x - (9/5) \cos x$; (d) $y_p = (3/5) e^{-3x} (\sin x + 3 \cos x)$.]

Problem 8.45 Obtain the general solution of the previous problem with $f(x) = xe^{-x}$. [Answer: $y = C_1 e^{-2x} + C_2 e^{-x} + x(x/2 - 1)e^{-x}$.]

Problem 8.46 Find the general solutions of the ODE $y'' - 4y' + 3y = f(x)$ for

$$\text{(a)} \quad f(x) = 3e^{2x} ; \quad \text{(b)} \quad f(x) = 5e^{x} .$$

[Answer: $(a)\, y = C_1 e^{3x} + C_2 e^{x} - 3e^{2x}; (b)\, y = C_1 e^{3x} + C_2 e^{x} - (5/2)xe^{x}$.]

Problem 8.47 Find the general solutions of the equations:

$$\text{(a)} \ y'' - 6y' + 9y = \left(3x^2 + 1\right)e^{3x} ; \quad \text{(b)} \ y'' + 6y' + 9y = x\left(e^{-3x} - 1\right).$$

[Answer: $(a)\, y = \left(C_1 + C_2 x + x^2/2 + x^4/4\right)e^{3x}; (b)\, y = \left(C_1 + C_2 x + x^3/6\right)$ $e^{-3x} + 2/27 - x/9$.]

Problem 8.48 Solve the following ODEs:

$$\text{(a)} \quad y'' + 9y = A\sin(3x) ; \quad \text{(b)} \quad y'' + 9y = 3\cos x + xe^{x} .$$

[Answer: $(a)\, y = C_1 \cos(3x) + C_2 \sin(3x) - (Ax/6)\cos(3x) + (A/6)\sin(3x);$ $(b)\, y = C_1 \cos(3x) + C_2 \sin(3x) + (3/8)\cos x + e^{x}(5x - 1)/50.$]

Problem 8.49 Show that the solution of

$$y'' - 4y' + 4y = e^{2x}\left(2x^2 + 3x + 1\right)$$

is $y = e^{2x}\left(C_1 + C_2 x + x^2/2 + x^3/2 + x^4/6\right)$.

8.2.3.2 Method of Variation of Parameters

Above we considered a rather simple case when the coefficients in the second-order ODE are constants. Here we shall discuss a general case of the ODE with variable coefficients,

$$a_2(x)y'' + a_1(x)y' + a_0(x)y = f(x) , \tag{8.43}$$

and will show that there is a technique for obtaining the particular integral of this nonhomogeneous ODE for any right-hand side $f(x)$ provided the two solutions $y_1(x)$ and $y_2(x)$ of the corresponding homogeneous equation are known. In other words, the method works if we can solve the homogeneous equation (with $f(x) = 0$) and hence obtain the corresponding complementary solution:

$$y_c(x) = C_1 y_1(x) + C_2 y_2(x) \,, \tag{8.44}$$

where C_1 and C_2 are arbitrary constants. The idea of the method is to search for the particular solution of the original equation in the form (cf. the end of Sect. 8.1.5)

$$y_p(x) = u_1(x) y_1(x) + u_2(x) y_2(x) \,, \tag{8.45}$$

which is basically obtained from (8.44) by replacing the constants C_1 and C_2 with unknown functions $u_1(x)$ and $u_2(x)$. Substituting the trial solution (8.45) in (8.43), we obtain:

$$a_2 \left(u_1'' y_1 + 2 u_1' y_1' + u_1 y_1'' + u_2'' y_2 + 2 u_2' y_2' + u_2 y_2'' \right)$$
$$+ a_1 \left(u_1' y_1 + u_1 y_1' + u_2' y_2 + u_2 y_2' \right) + a_0 \left(u_1 y_1 + u_2 y_2 \right) = f \,,$$

or after some trivial manipulation:

$$u_1 \left(a_2 y_1'' + a_1 y_1' + a_0 y_1 \right) + u_2 \left(a_2 y_2'' + a_1 y_2' + a_0 y_2 \right)$$
$$+ a_2 \left(u_1'' y_1 + u_1' y_1' + u_2'' y_2 + u_2' y_2' \right) + a_2 \left(u_1' y_1' + u_2' y_2' \right)$$
$$+ a_1 \left(u_1' y_1 + u_2' y_2 \right) = f \,.$$

The first two terms vanish since y_1 and y_2 satisfy the homogeneous equation by construction. The expression in the parentheses in the 3rd term can be written as

$$u_1'' y_1 + u_1' y_1' + u_2'' y_2 + u_2' y_2' = \frac{d}{dx} \left(u_1' y_1 + u_2' y_2 \right) \,.$$

Therefore, we obtain:

$$a_2 \frac{d}{dx} \left(u_1' y_1 + u_2' y_2 \right) + a_2 \left(u_1' y_1' + u_2' y_2' \right) + a_1 \left(u_1' y_1 + u_2' y_2 \right) = f \,. \tag{8.46}$$

Both functions $u_1(x)$ and $u_2(x)$ are to be determined. However, we have to find just one particular solution. This means that we are free to constrain the two functions $u_1(x)$ and $u_2(x)$ in some way, and this can be done to simplify (8.46), so that it can be solved. The simplest choice seems to be

$$u_1' y_1 + u_2' y_2 = 0 \,, \tag{8.47}$$

because it would remove two terms from (8.46) yielding simply

$$a_2 \left(u_1' y_1' + u_2' y_2' \right) = f \,. \tag{8.48}$$

Equations (8.47) and (8.48) are linear algebraic equations with respect to two unknown functions u_1' and u_2', and can easily be solved to give:

$$u_1' = -\frac{y_2 f}{a_2 W} \quad \text{and} \quad u_2' = \frac{y_1 f}{a_2 W} \,, \tag{8.49}$$

where $W = y_1 y_2' - y_1' y_2$ is the Wronskian of y_1 and y_2, see (8.24). We know that $W \neq 0$ if y_1 and y_2 are linearly independent solutions of the homogeneous equation (see Sect. 8.2.1). Thus, integrating the above equations, the functions u_1 and u_2 are found and the particular solution is obtained. Note that when integrating these equations, arbitrary constants can be dropped due to the specific form of (8.45) in which the particular solution is sought: the terms to be originated from those constants will simply be absorbed in the complementary solution (8.44).

As an example, let us find the general solution of the equation:

$$y'' - 2y' + y = x^{-2} e^x .$$

Note that the right-hand side here has a form which is different from any type we could handle using the familiar method of undetermined coefficients. The complementary solution is:

$$y_c = C_1 e^x + C_2 x e^x ,$$

since the characteristic equation gives repeated roots. The Wronskian of $y_1 = e^x$ and $y_2 = x e^x$ is trivially found to be

$$W = y_1 y_2' - y_1' y_2 = e^x \left(e^x + x e^x \right) - e^x x e^x = e^{2x} .$$

Then, Eq. (8.49) can be written explicitly as

$$u_1' = - \frac{(x e^x)\left(x^{-2} e^x \right)}{e^{2x}} = -x^{-1} \implies u_1(x) = - \int \frac{dx}{x} = - \ln|x| ,$$

$$u_2' = \frac{e^x \left(x^{-2} e^x \right)}{e^{2x}} = x^{-2} \implies u_2(x) = -x^{-1} .$$

Therefore, the particular solution

$$y_p = - \left(\ln|x| \right) e^x + \left(-x^{-1} \right) x e^x = -e^x \ln|x| - e^x .$$

In fact, the second term, e^x, will be absorbed by the complementary solution, so that only the first term may be kept. Thus, the general solution is

$$y = y_c + y_p = C_1 e^x + C_2 x e^x - e^x \ln|x| ,$$

as can be checked by direct substitution.

Problem 8.50 Show that the particular integral of the ODE

$$y'' - 2y' + y = x^{-1} e^x$$

is $y_p(x) = x e^x \ln|x|$.

Problem 8.51 Show that the particular integral of the ODE

$$y'' + y = \cot x$$

is

$$y_p(x) = \ln \left| \tan \frac{x}{2} \right| \sin x .$$

Problem 8.52 Prove prescription 1 of the method of undetermined coefficients from Sect. 8.2.3.1 (when the right-hand side $f(x)$ is a polynomial $P^{(n)}(x)$) using the method of this section. You will have to consider two cases, when the roots of the characteristic equation (8.34) of the homogeneous equation are distinct, $p_1 \neq p_2$, and hence $y_1 = e^{p_1 x}$ and $y_2 = e^{p_2 x}$, and when they are the same, in which case $y_2 = x y_1$. Note that an integral $\int Q^{(n)}(x) e^{\lambda x} dx$, where $Q^{(n)}(x)$ is a polynomial of order n, is equal to the same exponential function times another polynomial of the same order. This can easily be established by a repeated integration by parts or by induction (Problem 4.24). [Hint: *when considering the second case ($p_1 = p_2$), it will be necessary to separate out the highest order term in the polynomial and perform an integration by parts.*]

Problem 8.53 Consider now the right-hand side $f(x) = Be^{\lambda x}$ with a complex λ, and prove prescriptions 2 and 3 of the method of undetermined coefficients from Sect. 8.2.3.1. Consider four cases:

- the roots of the characteristic equation (8.34) of the homogeneous equation are distinct, $p_1 \neq p_2$, and λ does not coincide with any of them, then $y_p = Ae^{\lambda x}$;
- $p_1 = p_2$ and $\lambda \neq p_1$, then $y_p = Ae^{\lambda x}$;
- $p_1 \neq p_2$, but $\lambda = p_1$, then $y_p = Axe^{\lambda x}$;
- $p_1 = p_2 = \lambda$, then $y_p = Ax^2 e^{\lambda x}$.

Consider now a second-order DE with constant coefficients and a general Right-hand side $f(x) = P^{(n)}(x) e^{\lambda x}$ assuming a complex λ. This is a generalization of the previous problem. Our objective is to prove prescription 4 of the method of undetermined coefficients from Sect. 8.2.3.1 including two special cases of a single and double resonance.

Consider first the case of different roots $p_1 \neq p_2$ of the corresponding characteristic equation. In this case $y_1 = e^{p_1 x}$, $y_2 = e^{p_2 x}$ and $W = (p_2 - p_1) e^{(p_1 + p_2)x}$. If λ does not coincide with any of the two roots, then we have from the first (8.49):

$$u_1' = -\frac{1}{a_2 (p_2 - p_1)} P^{(n)}(x) e^{(\lambda - p_1)x} .$$

Integrating, we obtain

$$u_1(x) = -\frac{1}{a_2(p_2 - p_1)} \int P^{(n)}(x)e^{(\lambda - p_1)x}dx = A^{(n)}(x)e^{(\lambda - p_1)x},$$

where $A^{(n)}(x)$ is some n-th order polynomial. Similarly,

$$u_2' = \frac{1}{a_2(p_2 - p_1)}P^{(n)}(x)e^{(\lambda - p_2)x},$$

leading after integration to

$$u_2(x) = B^{(n)}(x)e^{(\lambda - p_2)x},$$

with $B^{(n)}(x)$ being a polynomial of order n. The particular integral (8.45) then becomes

$$y_p(x) = u_1(x)e^{p_1 x} + u_2(x)e^{p_2 x}. \tag{8.50}$$

After substituting the obtained auxiliary functions u_1 and u_2, we get for $y_p(x)$ an n-th order polynomial times $e^{\lambda x}$, as required by the prescription.

If $\lambda = p_1$ and $p_1 \neq p_2$, then we have the same y_1, y_2 and W, and

$$u_1' = \frac{1}{a_2(p_2 - p_1)}P^{(n)}(x) \implies u_1(x) = A^{(n+1)}(x),$$

i.e. $u_1(x)$ is an $(n+1)$-th order polynomial, and for $u_2(x)$ we obtain the same expression as before. Combining both auxiliary functions in (8.50), we shall obtain that $y_p(x)$ is an $(n+1)$-th order polynomial times $e^{\lambda x}$, as required. Hence, in the case of a single resonance one has to increase the order of the polynomial by one.

Consider now the case of $p_1 = p_2$. Here $y_1 = e^{p_1 x}$, $y_2 = xy_1$, and $W = e^{2p_1 x}$. Assume first that $\lambda \neq p_1$. Then,

$$u_1' = -\frac{1}{a_2}x P^{(n)}(x)e^{(\lambda - p_1)x}.$$

To obtain $u_1(x)$, we shall integrate by parts once:

$$u_1(x) = -\frac{1}{a_2}\int \underbrace{x P^{(n)}(x)}_{u} \underbrace{e^{(\lambda - p_1)x}dx}_{dv}$$

$$= -\frac{1}{a_2(\lambda - p_1)}\left\{ x P^{(n)}(x)e^{(\lambda - p_1)x} - \int \left[P^{(n)}(x) + x\frac{dP^{(n)}(x)}{dx}\right]e^{(\lambda - p_1)x}dx\right\}$$

$$= -\frac{1}{a_2(\lambda - p_1)}x P^{(n)}(x)e^{(\lambda - p_1)x} + A^{(n)}(x)e^{(\lambda - p_1)x}.$$

The calculation of $u_2(x)$ is the same as before leading to

$$u_2(x) = \frac{1}{a_2}\int P^{(n)}(x)e^{(\lambda - p_1)x}dx.$$

Combining both auxiliary functions in the final expression for y_p according to (8.45), we obtain:

$$y_p(x) = A^{(n)}(x)e^{\lambda x} - \frac{x}{a_2}\left[\frac{e^{\lambda x}}{\lambda - p_1}P^{(n)}(x) - e^{p_1 x}\int P^{(n)}(x)e^{(\lambda - p_1)x}dx\right].$$

$$(8.51)$$

At the first sight it seems that the solution is an $(n+1)$-th order polynomial times the exponential function $e^{\lambda x}$. This, however, is not the case as the terms with x^{n+1} actually cancel out. To see this, let us consider the expression inside the square brackets, where we shall separate out the highest order term in $P^{(n)}(x)$ by writing $P^{(n)}(x) = ax^n + P^{(n-1)}(x)$. An integration by parts of the integral containing ax^n leads to

$$\int ax^n e^{(\lambda - p_1)x}dx = \frac{a}{\lambda - p_1}\left[x^n e^{(\lambda - p_1)x} - n\int x^{n-1}e^{(\lambda - p_1)x}dx\right]$$

and it is easily seen then that the terms containing x^n inside the square brackets in (8.51) cancel out and hence overall the terms containing a product of the exponential $e^{\lambda x}$ and x^{n+1} do not appear. Hence, in this case we again obtain the same result that $y_p(x)$ is an exponential $e^{\lambda x}$ times a polynomial of order n.

Finally, consider the case of $p_1 = p_2 = \lambda$ of the double resonance. Then, $y_1 = e^{\lambda x}$, $y_2 = xy_1$ and $W = e^{2\lambda x}$. We have:

$$u_1' = -\frac{1}{a_2}xP^{(n)}(x) \implies u_1(x) = A^{(n+2)}(x),$$

i.e. it is the $(n+2)$-th order polynomial, and

$$u_2' = \frac{1}{a_2}P^{(n)}(x) \implies u_2(x) = B^{(n+1)}(x),$$

i.e. $u_2(x)$ is the $(n+1)$-th order polynomial. Combining both auxiliary functions in $y_p(x) = u_1 y_1 + u_2 y_2$, we arrive at the result expected that $y_p(x)$ is the exponential $e^{\lambda x}$ multiplied by a polynomial of the order $(n+2)$, as expected. Hence, in the case of the double resonance, the order of the polynomial in y_p is to increased by two.

8.3 Non-linear Second-Order Differential Equations

8.3.1 A Few Methods

Here we shall discuss a few simple tricks which in some cases are useful in solving non-linear second-order ODEs. In all these cases it is possible to lower the order of the ODE and hence obtain a simpler equation which may eventually be solved.

Consider first an ODE of the form $y'' = f(x)$. It does not contain y and y' and is solved by simply noticing that $y'' = (y')'$. Therefore, by introducing a new function

$z(x) = y'(x)$, the first-order ODE $z' = f(x)$ with respect to the function $z(x)$ is obtained which is integrated immediately to give

$$z(x) = \int f(x)dx + C_1 .$$

To obtain $y(x)$, we have to integrate $y' = z(x)$ which is another first-order ODE, solved similarly, yielding finally the solution

$$y(x) = \int z(x)dx + C_2 = \int^x dx_1 \int^{x_1} dx_2 \, f(x_2) + C_1 x + C_2 .$$

Two arbitrary constants appear naturally.

As an example, consider Newton's equation of motion for a particle moving under an external force changing sinusoidally with time:

$$m\ddot{x} = A \sin(\omega t) .$$

Let the particle be initially at rest at the centre of the coordinate system, i.e. $x(0) = \dot{x}(0) = 0$. To solve the equation above, we introduce the velocity $v = \dot{x}$, and then the equation of motion would read simply $\dot{v} = (A/m) \sin(\omega t)$. It is integrated to give

$$v(t) = \frac{A}{m} \int_0^t \sin(\omega \tau) \, d\tau = \frac{A}{m\omega} [1 - \cos(\omega t)] .$$

Integrating the velocity again, we obtain the particle position as a function of time:

$$x(t) = \int_0^t v(\tau) \, d\tau = \frac{A}{m\omega} \left[t - \frac{1}{\omega} \sin(\omega t) \right] .$$

Another class of equations which can be simplified is of the form $y'' = F(y, y')$, i.e. the ODE lacks direct dependence on x. In this case y can be considered instead of x as a variable. Indeed, we denote $z = y'$ and then notice that

$$y'' = \frac{d}{dx}(y') = \frac{dz}{dx} = \frac{dz}{dy}\frac{dy}{dx} = \frac{dz}{dy} z ,$$

which yields

$$z\frac{dz}{dy} = F(y, z) ,$$

which is now a first-order ODE with respect to the function $z(y)$. If it can be solved, we then write $y' = z(y)$ which may be integrated without difficulty via

$$y' = z(y) \quad \Longrightarrow \quad \int \frac{dy}{z(y)} = \int dx \quad \Longrightarrow \quad \int \frac{dy}{z(y)} = x + C_2 .$$

Note that the first arbitrary constant C_1 is contained in the function $z(y)$ after the first integration.

To illustrate this case, let us derive a trajectory $x(t)$ of a particle subject to an external potential $V(x)$. Let x_0 and v_0 be the initial ($t = 0$) position and velocity of the particle, respectively. Newton's equation of motion in this case reads

$$m\ddot{x} = -\frac{dV}{dx} \equiv f(x) \,,$$

where $f(x) = -V'(x)$ is the external force due to the potential $V(x)$. Using the method described above, we introduce the velocity $v = \dot{x}$ and then write the acceleration as $a = \ddot{x} = \dot{v} = v\,(dv/dx)$, which gives

$$mv\frac{dv}{dx} = f(x) \quad \Longrightarrow \quad m\int_0^t v\,dv = \int_0^x f\left(x'\right)dx'$$

$$\Longrightarrow \quad \frac{mv^2}{2} - \frac{mv_0^2}{2} = \int_0^x f\left(x'\right)dx' \,,$$

which is nothing but the expression of the conservation of energy: in the left-hand side we have the change of the particle kinetic energy due to the work by the force given in the right-hand side. Alternatively, in the right-hand side one may recognize the corresponding difference

$$\int_0^x f\left(x'\right)dx' = -\int_0^x \frac{dV}{dx'}dx' = V\,(0) - V\,(x)$$

of the potential energies at two times leading to the energy conservation. Solving this equation with respect to the velocity, we find

$$v(x) = \sqrt{v_0^2 + \frac{2}{m}\int_0^x f\left(x'\right)dx'} \,,$$

which is also dx/dt. Therefore, integrating again, we obtain an equation for the trajectory in a rather general form:

$$\int_{x_0}^x \left(v_0^2 + \frac{2}{m}\int_0^{x''} f\left(x'\right)dx'\right)^{-1/2} dx'' = t \,.$$

Finally, if the ODE $f\left(y'', y', y, x\right) = 0$ is a homogeneous function with respect to its variables y, y' and y'' (treated together, see an example below), then one can introduce instead of $y(x)$ a new function $z(x)$ via

$$y(x) = \exp\left(\int z(x)dx\right) \tag{8.52}$$

to reduce the order of the ODE. Indeed,

$$y' = \exp\left(\int z(x)dx\right) z(x) = yz \quad \text{and} \quad y'' = (yz)' = y'z + yz' = y\left(z^2 + z'\right),$$

so that each derivative of y factorizes into y and some linear combination of z and its first derivative; this allows cancelling out $y(x)$ in the equation as demonstrated by the following example. Let us consider

$$yy'' - y'\left(y + y'\right) = 0.$$

Each of its terms is homogeneous of degree two. Using the substitution (8.52), we obtain:

$$y^2\left(z^2 + z'\right) - yz\left(y + zy\right) = 0 \quad \Longrightarrow \quad z^2 + z' - z\left(1 + z\right) = 0 \quad \Longrightarrow \quad z' = z,$$

which is easily integrated giving $z(x) = C_1 e^x$. Therefore, the function $y(x)$ sought for is found as

$$y(x) = \exp\left(\int C_1 e^x dx\right) = \exp\left(C_1 e^x + C_3\right) = C_2 \exp\left(C_1 e^x\right),$$

where C_1 and C_2 (or C_3, depending of the form of the solution taken) are two arbitrary constants.

Problem 8.54 Solve the following ODE using one of the methods considered above:

$$\text{(a) } y'' = e^x ; \quad \text{(b) } y\left(y'' + y'\right) = \left(y'\right)^2.$$

[Answer: (a) $y = C_1 x + C_2 + e^x$; (b) $y = C_1 \exp\left(C_2 e^{-x}\right)$.]

8.3.2   Curve of Pursuit

Consider[6] the following problem[7] : a big boat (or a rabbit) is moving in the vertical direction from position $(a, 0)$ on the $x - y$ plane, Fig. 8.1a, with a constant speed v_B. A sailboat (or a fox) starts from the centre of the coordinate system together

---

[6] Adapted from the book R. K. Nagle, E. B. Saff and A. D. Snider "Fundamentals of Differential Equations", Addison-Wesley, 8-th edition (2012).
[7] Pursuit curves were considered by the French scientist Pierre Bouguer in 1732. However, the term "pursuit curve" was first coined by George Boole, when he considered this problem in Chap. XI of his "Treatise on differential equations" published in 1859.

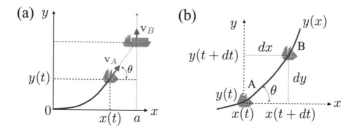

Fig. 8.1 a A sailboat starting at the position $(0, 0)$ sails with a constant velocity v_A in pursuit of a big boat that starts at the position $(a, 0)$ and moves with a constant speed v_B in the vertical direction. **b** Two successive positions of the sailboat at times t (point A) and $t + dt$ (point B)

with the other boat in pursuit of it, also moving with a constant speed $v_A \geq v_B$. It is understood that in doing so at each time t the sailboat's velocity is directed towards the current position of the pursued object. Let us derive the differential equations governing the coordinates of the pursuer as a function of time, and also establish its trajectory. We shall also ask a question of whether the pursuer would ever reach the other object.

At time t the sailboat's position is $(x(t), y(t))$, while the position of the bigger boat is $(a, v_B t)$. Since the velocity of the sailboat is always directed towards the bigger boat, this means that the derivative $y'_x = dy/dx$ corresponds to the tangent of the angle θ the trajectory at the point (x, y) makes with the x-axis, Fig. 8.1a, i.e. we can immediately write:

$$y'_x = \tan \theta = \frac{v_B t - y}{a - x}. \tag{8.53}$$

On the other hand, let us consider the sailboat progression along its trajectory over time dt from time t (position A) to $t + dt$ (position B), see Fig. 8.1b. $dl = AB$ can be considered as a straight line for small dt, hence $dl = v_A dt$. We then have:

$$\frac{dy}{dt} = \frac{y(t + dt) - y(t)}{dt} = \frac{dx \tan \theta}{dt} = \frac{dx}{dt} \tan \theta$$

$$= \sqrt{\left(\frac{dl}{dt}\right)^2 - \left(\frac{dy}{dt}\right)^2} \, y'_x = y'_x \sqrt{v_A^2 - \left(y'_t\right)^2},$$

since $dx = \sqrt{dl^2 - dy^2}$ and $\tan \theta = y'_x$. Replacing the derivative y'_x with its expression (8.53), we shall obtain a differential equation for the function $y(t)$:

$$y'_t = \sqrt{v_A^2 - \left(y'_t\right)^2} \, \frac{v_B t - y}{a - x}. \tag{8.54}$$

Both sides of this equation can be squared to get rid of the root and solve for y'_t; obviously, the right-hand side is positive so that $dy/dt \geq 0$, and hence we should seek the solution with the positive derivative:

$$y'_t = v_A \frac{v_B t - y}{\sqrt{(a-x)^2 + (v_B t - y)^2}} . \tag{8.55}$$

A second equation is necessary to describe fully the trajectory of the sailboat since x depends on time as well. This can be derived via $y'_x = \frac{dy}{dt}\frac{dt}{dx} = y'_t/x'_t$ and using (8.53) and (8.55):

$$x'_t = y'_t \frac{a-x}{v_B t - y} = v_A \frac{a-x}{\sqrt{(a-x)^2 + (v_B t - y)^2}} . \tag{8.56}$$

Solving both (8.55) and (8.56) simultaneously, we should be able to obtain the position of the sailboat in pursuit as a function of time.

This task may seem to be rather challenging for the analytical calculation though; however, the obtained equations are quite appropriate for a numerical analysis since both are solved with respect to the first derivatives x'_t and y'_t. It appears however that it is possible to obtain in a closed form the curve of the trajectory, i.e. the dependence $y(x)$. To this end, we again consider the triangle in Fig. 8.1b and write $dx^2 + dy^2 = dl^2$, where $dl = v_A dt$ and $dy = y'_x dx$, so that

$$dx^2 \left[1 + \left(y'_x\right)^2\right] = v_A^2 dt^2 \implies dt = \frac{1}{v_A}\sqrt{1 + \left(y'_x\right)^2} dx .$$

Integrating both sides, we obtain

$$t = \frac{1}{v_A} \int_0^x \sqrt{1 + \left(y'_x\right)^2} dx .$$

Replacing time in (8.53) with this result leads to a closed equation for the trajectory $y(x)$:

$$y - (x-a) y'_x = \gamma \int_0^x \sqrt{1 + \left(y'_x\right)^2} dx ,$$

where $\gamma = v_B/v_A$. Finally, differentiating it with respect to x yields

$$(x-a)y'' = -\gamma\sqrt{1 + (y')^2} \tag{8.57}$$

with both derivatives assumed to be over x.

The obtained equation is of the type $F(x, y', y'') = 0$ as it does not contain the function y itself. Hence, it can be solved by simply introducing a new function $u(x) = y'$ so that $y'' = u'$. This results in a non-linear separable ODE for $u(x)$.

Problem 8.55 Solve this ODE with respect to $u(x)$ subject to the initial conditions that at $x = 0$ we have the big ship at the $y = 0$ position and hence initially the velocity of the sailboat was directed horizontally, i.e. $y'(0) = u(0) = 0$. Show that

$$u(x) = y'_x = \frac{1}{2}\left[\left(1 - \frac{x}{a}\right)^{-\gamma} - \left(1 - \frac{x}{a}\right)^{\gamma}\right].$$

Problem 8.56 Obtain the trajectory by integrating the ODE $y' = u$ that is also separable. Show that its solution, subject to the initial condition $y(x = 0) = 0$, is

$$y(x) = \frac{a}{2}\left[\frac{1}{1+\gamma}\left(1 - \frac{x}{a}\right)^{1+\gamma} - \frac{1}{1-\gamma}\left(1 - \frac{x}{a}\right)^{1-\gamma}\right] + \frac{a\gamma}{1-\gamma^2},$$

if $\gamma < 1$, and

$$y(x) = \frac{x^2}{4a} - \frac{x}{2} - \frac{a}{2}\ln\left(1 - \frac{x}{a}\right),$$

if $\gamma = 1$.

In the former case of $\gamma < 1$, when the speed of the sailboat is larger than that of the bigger ship, the sailboat will be able to reach the other ship since $y(a) = a\gamma/(1 - \gamma^2)$ has a finite positive value. however, $y(a)$ in the latter case of $\gamma = 1$ (i.e. both vessels have equal speeds) is infinite because of the logarithm, i.e. the two objects will never meet. Obviously, they will never meet if $\gamma > 1$.

8.3.3 Catenary Curve

Catenary is a special curve type,[8] which corresponds to the shape taken by a suspended cable or chain hanging by gravity between two points at both ends where it is fixed, Fig. 8.2a.

Let us derive a differential equation that describes the equilibrium shape of the cable. Let us choose the coordinate system as in Fig. 8.2a so that $x = 0$ corresponds to the lowest point of the cable.

As the cable is at equilibrium, the forces must balance at each point. Consider then a small length of the cable between coordinates x and $x + dx$ (shown blue in Fig. 8.2b). There are three forces: (i) the gravity $\rho g \, dl$ acting downwards, where ρ is

---

[8] Gottfried Leibniz, Christiaan Huygens, and Johann Bernoulli derived the equation for the catenary which was published in *the Acta Eruditorum* in June 1691.

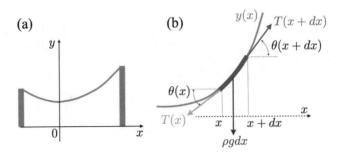

Fig. 8.2 a A cable suspended between two supports takes the form of a catenary. The $x = 0$ is chosen at the lowest point of the cable. **b** Consideration of the balance of the forces acting on a small element (blue) of the cable with endpoints at coordinates x and $x + dx$

the linear density of the cable, g the Earth's gravitational acceleration, and

$$dl = \sqrt{dx^2 + dy^2} = \sqrt{1 + (y')^2}dx \tag{8.58}$$

the arch length of the blue part; (ii) there is a tension force $T(x + dx)$ acting on the right end of the length and (iii) another tension $T(x)$ that acts on its left end as shown in Fig. 8.2b. If $\theta(x + dx)$ and $\theta(x)$ are the angles the direction of the length makes with the x-axis at both ends, the tension forces are directed along these tangential directions. Hence, when projecting the forces on the horizontal direction and equilibrating, we get:

$$T(x + dx) \cos \theta(x + dx) - T(x) \cos \theta(x) = 0, \tag{8.59}$$

while projecting the three forces on the vertical direction, we get instead

$$T(x + dx) \sin \theta(x + dx) - T(x) \sin \theta(x) - \rho g dl = 0. \tag{8.60}$$

Consider first the horizontal equation. Expanding the tension and the angles,

$$T(x + dx) = T(x) + T'(x)dx,$$

$$\cos \theta(x + dx) = \cos \left(\theta + \theta' dx \right) = \cos \theta - \theta' dx \sin \theta,$$

we get, to the first order:

$$T' \cos \theta - T\theta' \sin \theta = 0 \implies \frac{d}{dx} (T \cos \theta) = 0,$$

leading to a simple solution $T(x) \cos \theta(x) = C$. The constant C can be chosen by the tension T_0 at the lowest point of the cable ($x = 0$). There $\theta = 0$ and we obtain $C = T_0$. Hence, the tension and the angle θ are related via

$$T(x) = \frac{T_0}{\cos \theta(x)}.$$

Consider similarly the vertical equation (8.60). Employing this time

$$\sin \theta (x + dx) = \sin \theta + \theta' dx \, \cos \theta ,$$

it results in

$$T' \sin \theta + T \theta' \cos \theta = \rho g dl .$$

The left-hand side can be rewritten as

$$\frac{d}{dx} (T \sin \theta) = \frac{d}{dx} (T_0 \tan \theta) = T_0 \frac{d}{dx} \tan \theta = T_0 y'' ,$$

since $\tan \theta = y'$ gives the slope of the curve of the cable at point A. Finally, recalling expression (8.58) for dl, we obtain the required differential equation for the shape of the cable:

$$y'' = \frac{1}{a} \sqrt{1 + (y')^2} , \quad a = \frac{T_0}{\rho g} . \tag{8.61}$$

This equation is of the type $F(y', y'') = 0$ and can be solved by the same method as in the previous subsection.

Problem 8.57 Integrate (8.61) with the initial conditions $y(0) = a$ and $y'(0) = 0$ to show that

$$y(x) = a \, \cosh \frac{x}{a} . \tag{8.62}$$

The obtained solution corresponds to the catenary. It is easy to see by expanding the hyperbolic cosine for small $x/a \ll 1$, that in this limit the catenary becomes very close to the parabola. This corresponds to stiff cables when the curvature is small.

8.4   Series Solution of Linear ODEs

In many applications, e.g. when solving partial differential equations of mathematical physics, one obtains ordinary second-order differential equations of the type

$$y''(x) + p(x) y'(x) + q(x) y(x) = 0 , \tag{8.63}$$

where $p(x)$ and $q(x)$ are known variable functions of x. This is the so-called *canonical* form of the ODE. Since one can always ensure that the coefficient to the second derivative of y is equal to one by dividing on it both sides of the equation, it is sufficient to study any ODE in its canonical form only.

We know that when the coefficients $p(x)$ and $q(x)$ are both *constants,* it is possible to solve (8.63) and express the solution in terms of elementary functions (exponentials, sine and cosine). However, if the coefficients $p(x)$ and $q(x)$ are *variable* functions then the solution of (8.63) in terms of elementary functions is possible only in special cases; in most cases the solution cannot be expressed in terms of a finite number of elementary functions. It can be shown that it is generally possible to devise a method whereby the solution is expressed as an infinite power series.[9] This can be used either as an analytical tool or as a means of calculating the solution numerically when the series is terminated at a finite number of terms. This section is devoted to developing a method that would enable one to construct a series solution to a linear ODE.

It is absolutely essential first to choose a point $x = a$ around which the series is to be developed. The point $x = a$ is said to be an *ordinary* point of the differential equation (8.63) if *both* $p(x)$ and $q(x)$ are "*well-behaved*" functions around this point, i.e. the functions are single-valued, finite and continuous there possessing derivatives $f^{(k)}(a)$ of *all* orders. If the point $x = a$ is *not* an *ordinary* point then we shall call it a *singular* point.

For instance, the differential equation

$$x^2(x + 3)y''(x) - (x + 3)y'(x) + 3xy(x) = 0 , \tag{8.64}$$

if written in the canonical form, yields the two functions

$$p(x) = -\frac{1}{x^2} \quad \text{and} \quad q(x) = \frac{3}{x(x + 3)} , \tag{8.65}$$

which are singular at points $x = 0$ and -3. All other points in the interval $-\infty < x < \infty$ are therefore *ordinary* points.

Before discussing the most general case, we shall first introduce main ideas of the method for the case when the series is constructed around an ordinary point.

8.4.1 Series Solutions About an Ordinary Point

Suppose that the initial conditions $y(a)$ and $y'(a)$ are given at the point $x = a$ which we assume is an *ordinary* point of (8.63). Then we can, *at least in principle*, calculate all the derivatives $y^{(k)}(a)$ $(k = 1, 2, 3\ldots)$ in terms of $y(a)$ and $y'(a)$, by repeated differentiation of the (8.63) with respect to x. Indeed, $y''(a)$ follows immediately from the equation itself. Differentiating the equation once and setting $x = a$, we can express the third derivative of $y(x)$ via $y(a)$, $y'(a)$ and $y''(a)$. Repeating this process, one can calculate any derivative $y^{(r)}(a)$ of $y(x)$ at $x = a$, where $r = 1, 2, \ldots$. Hence, we can *formally* construct a Taylor *series* about $x = a$ for our function $y(x)$,

$$y(x) = \sum_{r=0}^{\infty} c_r(x - a)^r \quad \text{with} \quad c_r = \frac{y^{(r)}(a)}{r!} , \tag{8.66}$$

---

[9] In some cases the series may be finite.

which forms a solution of the equation; note that the initial conditions are satisfied if $c_0 = y(a)$ and $c_1 = y'(a)$. This series gives us a *general* solution of (8.63) in the neighbourhood of $x = a$ if $c_0 = y(a) \equiv C_1$ and $c_1 = y'(a) \equiv C_2$ are considered as arbitrary constants.

This solution would converge within an interval $|x - a| < R$, where R is the *radius of convergence* of the power series, (8.66). Usually, the radius of convergence R is equal to the distance from the point $x = a$ to the nearest singular point of (8.63). This point follows from a more general discussion based on the theory of functions of complex variables (see Chap. 2, Vol. II), while for functions of a real variable see Theorem 7.26.

The method considered above is a proof that the solution can always be found and is *unique*. However, this method is not convenient for actual applications. In *practice* the coefficients c_r are most easily found by substituting the series (8.66) into the differential equation (8.63), collecting terms with the like powers of x and setting their coefficients to zero. Indeed, $p(x)$ and $q(x)$ can be expanded into the Taylor series around $x = a$ (recall that these functions have no singularities at $x = a$):

$$p(x) = p_0 + p_1 (x - a) + p_2 (x - a)^2 + \dots \, ,$$

$$q(x) = q_0 + q_1 (x - a) + q_2 (x - a)^2 + \dots \, ,$$

which, when substituted into the differential equation (8.63), results in the following infinite power series being set to zero:

$$(2c_2 + p_0c_1 + q_0c_0) u^0 + (3 \cdot 2c_3 + 2p_0c_2 + p_1c_1 + q_0c_1 + q_1c_0) u^1$$
$$+ (4 \cdot 3c_4 + 3p_0c_3 + 2p_1c_2 + p_2c_1 + q_0c_2 + q_1c_1 + q_2c_0) u^2 + \dots = 0 \, ,$$

where $u = x - a$. Since this equation is expected to be satisfied for a continuous set of x values in some vicinity of the point $x = a$ (or for u around zero), and because the functions u^0, u^1, u^2, etc. form a linearly independent set, this can only be possible if all coefficients of the powers of u are equal to zero at the same time, which yields an infinite set of algebraic equations for the unknown coefficients c_2, c_3, c_4, etc.:

$$2c_2 + p_0c_1 + q_0c_0 = 0 \, ,$$

$$3 \cdot 2c_3 + 2p_0c_2 + p_1c_1 + q_0c_1 + q_1c_0 = 0 \, ,$$

$$4 \cdot 3c_4 + 3p_0c_3 + 2p_1c_2 + p_2c_1 + q_0c_2 + q_1c_1 + q_2c_0 = 0 \, ,$$

and so on. The first equation gives c_2, the second equation gives c_3 via c_2, the third c_4 via c_2 and c_3, etc. This way one can obtain recurrence relations for the coefficients which express c_r ($r \geq 3$) via the previous ones: $c_2, c_3, \dots, c_{r-1}$. So, given the values of the initial coefficients c_0 and c_1, it is possible to generate all the coefficients c_r up to any desired maximum value of the index r.

Although at first sight the procedure looks cumbersome, it is actually very convenient in actual applications, especially when the functions $p(x)$ and $q(x)$ are polynomials.

Example 8.6 ▶ Determine the series solution of the differential equation

$$y'' - xy = 0 , \tag{8.67}$$

satisfying the boundary conditions $y(0) = 1$ and $y'(0) = 0$.

Solution. Expanding around an ordinary point $x = 0$ (i.e. for $a = 0$), we can write:

$$y(x) = \sum_{r=0}^{\infty} c_r x^r . \tag{8.68}$$

Differentiating $y(x)$, we find:

$$y'(x) = \sum_{r=1}^{\infty} r c_r x^{r-1} \quad \text{and} \quad y''(x) = \sum_{r=2}^{\infty} r(r-1) c_r x^{r-2} . \tag{8.69}$$

Note that in $y'(x)$ the first term ($r = 0$) in the sum disappears, while the first two terms ($r = 0, 1$) disappear in the sum in the expression for $y''(x)$. That is why the sum for $y'(x)$ starts from $r = 1$, while the sum for $y''(x)$ from $r = 2$. Substituting the series (8.68) and (8.69) into the differential equation (8.67) gives

$$\sum_{r=2}^{\infty} r(r-1) c_r x^{r-2} - \sum_{r=0}^{\infty} c_r x^{r+1} = 0 .$$

Here the first sum contains all powers of x starting from x^0, while the second sum has all powers of x starting from x^1. To collect terms with like powers of x, we need to combine the two sums into one. To this end, it is convenient to separate out the x^0 term in the first sum (corresponding to $r = 2$), so that the two sums would contain identical powers of x:

$$2c_2 + \sum_{r=3}^{\infty} r(r-1) c_r x^{r-2} - \sum_{r=0}^{\infty} c_r x^{r+1} = 0 . \tag{8.70}$$

Indeed, now the first sum contains terms x^1, x^2, etc. and one can easily see that the same powers of x appear in the second sum. Once we have both sums starting from the same powers of x, we can now make them look similar which would eventually enable us to combine them together into a single sum. For instance, this can be done by replacing the summation index r in the first sum with a different index m selected in such a way that the power of x in the first sum would look the same as in the second. Making the replacement $m + 1 = r - 2$ for the summation index in the first sum of (8.70) does the trick, and we obtain:

$$2c_2 + \sum_{m=0}^{\infty} (m+3)(m+2) c_{m+3} x^{m+1} - \sum_{r=0}^{\infty} c_r x^{r+1} = 0 .$$

Note that $m = r - 3$ here, and hence it starts from the zero value corresponding to $r = 3$ of the original first sum. Now the two sums look indeed the same: they have the same powers of x and the summation indices start from the same (zero) value. To stress the similarity even further, we can use in the first sum the same letter r for the summation index instead of m (the summation index is a "dummy index" and any symbol can be used for it!), in which case the two sums are readily combined:

$$2c_2 + \sum_{r=0}^{\infty} (r + 3)(r + 2)c_{r+3}x^{r+1} - \sum_{r=0}^{\infty} c_r x^{r+1} = 0$$

$$\implies 2c_2 + \sum_{r=0}^{\infty} \left[(r + 3)(r + 2)c_{r+3} - c_r\right] x^{r+1} = 0 .$$

(8.71)

The differential equation is satisfied for all values of x provided that

$$c_2 = 0 \tag{8.72}$$

and, at the same time,

$$(r + 3)(r + 2)c_{r+3} = c_r \implies c_{r+3} = \frac{c_r}{(r + 2)(r + 3)}, \quad r = 0, 1, 2, \ldots .$$

(8.73)

From this recurrence relation, we can generate the coefficients c_r:

$$r = 0 \implies c_3 = \frac{c_0}{2 \cdot 3},$$

$$r = 1 \implies c_4 = \frac{c_1}{3 \cdot 4},$$

$$r = 2 \implies c_5 = \frac{c_2}{4 \cdot 5} = 0,$$

$$r = 3 \implies c_6 = \frac{c_3}{5 \cdot 6} = \frac{c_0}{2 \cdot 3 \cdot 5 \cdot 6},$$

$$r = 4 \implies c_7 = \frac{c_4}{6 \cdot 7} = \frac{c_1}{3 \cdot 4 \cdot 6 \cdot 7},$$

$$r = 5 \implies c_8 = \frac{c_5}{7 \cdot 8} = 0,$$

and so on. One can see that three families of coefficients are generated. If we start from c_0, we generate c_3, c_6, etc.; if we start from c_1, we generate c_4, c_7, etc.; finally, starting from c_2 (which is zero), we generate c_5, c_8, etc. which are all equal to zero. Since nothing can be said about the coefficients c_0 and c_1, we have to accept that these can be arbitrary. Therefore, if we recall the general form of the solution,

$$y(x) = c_0 x^0 + c_1 x^1 + c_2 x^2 + \ldots ,$$

then two solutions are obtained, one starting from c_0, and another one from c_1:

$$y_1(x) = c_0 + c_3 x^3 + c_6 x^6 + \ldots$$

$$= c_0 \left[1 + \frac{x^3}{2 \cdot 3} + \frac{x^6}{(2 \cdot 5)(3 \cdot 6)} + \frac{x^9}{(2 \cdot 5 \cdot 8)(3 \cdot 6 \cdot 9)} + \ldots \right],$$

$$y_2(x) = c_1 x + c_4 x^4 + c_7 x^7 + \ldots$$

$$= c_1 \left[x + \frac{x^4}{3 \cdot 4} + \frac{x^7}{(3 \cdot 6)(4 \cdot 7)} + \frac{x^{10}}{(3 \cdot 6 \cdot 9)(4 \cdot 7 \cdot 10)} + \ldots \right].$$

As expected, the coefficients c_0 and c_1 serve as arbitrary constants, so that expressions in the square brackets can in fact be considered as two linearly independent solutions

$$y_1(x) = 1 + \frac{x^3}{2 \cdot 3} + \frac{x^6}{(2 \cdot 5)(3 \cdot 6)} + \frac{x^9}{(2 \cdot 5 \cdot 8)(3 \cdot 6 \cdot 9)} + \ldots,$$

$$y_2(x) = x + \frac{x^4}{3 \cdot 4} + \frac{x^7}{(3 \cdot 6)(4 \cdot 7)} + \frac{x^{10}}{(3 \cdot 6 \cdot 9)(4 \cdot 7 \cdot 10)} + \ldots.$$

while the general solution is

$$y(x) = C_1 y_1(x) + C_2 y_2(x),$$

with arbitrary constants C_1 and C_2. Note that $y_1(0) = 1$ and $y_2(0) = 0$, while $y_1'(0) = 0$ and $y_2'(0) = 1$.

The application of the initial conditions $y(0) = C_1 \equiv 1$, and $y'(0) = C_2 \equiv 0$ gives the constants and hence the required *particular* solution (integral) is:

$$y(x) \equiv y_1(x) = 1 + \frac{x^3}{2 \cdot 3} + \frac{x^6}{(2 \cdot 5)(3 \cdot 6)} + \frac{x^9}{(2 \cdot 5 \cdot 8)(3 \cdot 6 \cdot 9)} + \ldots. \qquad (8.74)$$

The convergence of the series solution (8.74) can be investigated directly using the ratio test with the help of the recurrence relation (8.73). Recall, that an infinite series $\sum_{k=0}^{\infty} a_k$ is *convergent* if $\lim_{k \to \infty} |a_{k+1}/a_k| < 1$, see Sect. 7.1.5. In our case of the series (8.74) the two adjacent terms in the series are $c_{3k} x^{3k}$ and $c_{3k+3} x^{3k+3}$, which gives

$$\lim_{k \to \infty} \left| \frac{c_{3k+3} x^{3k+3}}{c_{3k} x^{3k}} \right| = \lim_{k \to \infty} \frac{c_{3k+3}}{c_{3k}} |x|^3 = |x|^3 \lim_{k \to \infty} \frac{1}{(3k + 2)(3k + 3)} = 0$$

$$(8.75)$$

for any value of x. We see from this result that the series solution (8.74) converges for *all* finite values of x. ◀

8.4.2 Series Solutions About a Regular Singular Point

Let us now suppose that the differential equation (8.63) has a *singular point* at $x = a$. It can be shown (Chap. 2, Vol. II) that one has to distinguish two cases: (i) if the functions $(x - a)p(x)$ and $(x - a)^2 q(x)$ are *both "well-behaved"* at $x = a$, then we say that $x = a$ is a *regular singular point* (RSP) of (8.63). Otherwise, a *singular point* is called an *irregular singular point* (ISP) of the differential equation (8.63). This *classification scheme* for singular points is *important* because it is always possible to obtain two independent solutions of (8.63) in the neighbourhood of a regular singular point $x = a$ by using a *generalized* series method to be introduced below; no such expansion is possible around an irregular singular point.

The introduced classification of singular points we shall illustrate for the differential equation (8.64) for which the functions $p(x)$ and $q(x)$ are given by (8.65). It has two singular points: at $x = 0$ and $x = -3$. For the singular point $a = -3$ we have

$$(x - a)\, p(x) \quad \Longrightarrow \quad (x - (-3))p(x) = -\frac{(x + 3)}{x^2}$$

and

$$(x - a)^2\, q(x) \quad \Longrightarrow \quad (x - (-3))^2 q(x) = 3\frac{(x + 3)}{x} \ .$$

Both these functions are *"well-behaved"* at $x = -3$ which is then a *regular singular point*. For the singular point, $a = 0$, we similarly find that

$$(x - 0)p(x) = -\frac{1}{x} \quad \text{and} \quad (x - 0)^2 q(x) = \frac{3x}{(x + 3)} \ .$$

It is clear that the function $(x - 0)p(x) = -1/x$ is *not* "well-behaved" at $x = 0$, and hence the point $x = 0$ is an *irregular singular point*.

If $x = a$ is a RSP of (8.63) then the Taylor series solution (8.66) will *not be possible* because the functions $p(a)$ and/or $q(a)$ do not exist and hence the derivatives $y^{(r)}(a)$ for $r = 2, 3, 4, \ldots$ cannot be determined directly as we have discussed above in the case of an ordinary point. However, it can be shown that (8.63) always has at least *one* solution of the *Frobenius*[10] type (Chap. 2, Vol. II):

$$y(x) = (x - a)^s \sum_{r=0}^{\infty} c_r (x - a)^r = \sum_{r=0}^{\infty} c_r (x - a)^{r+s} \ , \tag{8.76}$$

where the exponent s can be *negative* and/or even *non-integer*. The radius of convergence R of the series (8.76) is at least as large as the distance to the nearest singular point of the differential equation (8.63).

In *practice* the exponent s and the coefficients c_r are found by exactly the same method as for an ordinary point discussed above: by substituting the series (8.76) into

---

[10] Ferdinand Georg Frobenius (1849–1917) was a German mathematician.

the differential equation, collecting terms with the like powers of x and equating the corresponding coefficients to zero. This procedure leads to an *indicial equation* for s and a *recurrence relation* for the coefficients c_r. It is more transparent to illustrate this method considering an example.

Example 8.7 ► Determine the general series solution of the differential equation

$$2xy'' + y' + y = 0 \tag{8.77}$$

about $x = 0$.

Solution. In this differential equation $p(x) = q(x) = 1/2x$, and hence $x = 0$ is a singular point. We should first check if it is regular or irregular. Since the functions $xp(x) = 1/2$ and $x^2q(x) = x/2$ are both *"well-behaved"* at $x = 0$, it follows, therefore, that $x = 0$ is a RSP. We can therefore assume a series solution of the Frobenius type around $x = 0$:

$$y(x) = \sum_{r=0}^{\infty} c_r x^{r+s} \tag{8.78}$$

with some yet to be determined number s. From (8.78) we find that

$$y'(x) = \sum_{r=0}^{\infty} c_r (r+s) x^{r+s-1} \quad \text{and} \quad y''(x) = \sum_{r=0}^{\infty} c_r (r+s)(r+s-1) x^{r+s-2} .$$
$$\tag{8.79}$$

Note that, contrary to Example 8.6 considered above for the case of an ordinary point, in this case the first terms in the sums do *not* disappear after differentiation as s may be non-integer in general, and hence in both cases we have to keep all values of the summation index r starting from zero. The substitution of these results in (8.77) and combining the sums originating from y'' and y', gives:

$$\sum_{r=0}^{\infty} c_r (r+s)(2r+2s-1) x^{r+s-1} + \sum_{r=0}^{\infty} c_r x^{r+s} = 0 .$$

Let us separate out the first ($r = 0$) term in the first sum as the second sum does not have a term with x^{s-1}:

$$c_0 s(2s-1) x^{s-1} + \sum_{r=1}^{\infty} c_r (r+s)(2r+2s-1) x^{r+s-1} + \sum_{r=0}^{\infty} c_r x^{r+s} = 0 , \tag{8.80}$$

Next, we make the index shift $r \mapsto r + 1$ in the first summation in (8.80), i.e. we introduce a new index $r' = r - 1$ in the sum and then write r instead of r' for convenience, so that the first sum undergoes the transformation:

$$\sum_{r=1}^{\infty} c_r (r+s)(2r+2s-1)x^{r+s-1} = \sum_{r=0}^{\infty} c_{r+1}(r+s+1)(2r+2s+1)x^{r+s}.$$

$$= \sum_{r=0}^{\infty} c_{r+1}(r+s+1)(2r+2s+1)x^{r+s}.$$

At the last step we used r instead of r' for the dump index for convenience. The two sums in (8.80) can now be combined into one, which leads us to the equation:

$$c_0 s (2s-1) x^{s-1} + \sum_{r=0}^{\infty} \left[(r+s+1)(2r+2s+1)c_{r+1} + c_r \right] x^{r+s} = 0 . \quad (8.81)$$

This is an infinite sum of different powers of x which is to be equal to zero for all values of the x around $x = 0$. This means that in each and every term in the expansion above the coefficients to the powers of x must vanish, i.e. we should have $c_0 s (2s-1) = 0$ and, at the same time,

$$(r+s+1)(2r+2s+1)c_{r+1} + c_r = 0 \implies c_{r+1} = -\frac{c_r}{(r+s+1)(2r+2s+1)} \quad (8.82)$$

for every $r = 0, 1, 2, \ldots$.

Consider first the recurrence relation (8.82). One can see that c_{r+1} is directly proportional to c_r. More explicitly: c_1 is proportional to c_0, c_2 is proportional to c_1 and hence eventually also to c_0. Discussing along similar lines it becomes apparent that any coefficient c_r for $r \geq 1$ will be proportional to c_0 in the end. Next, let us consider the first equation, $c_0 s (2s - 1) = 0$. It accepts as a solution $c_0 = 0$. However, it is clear from what was said above that in this case all other coefficients will also be equal to zero and hence this solution leads to the trivial solution $y(x) = 0$ of the differential equation and is thus of no interest. Hence, we have to assume that $c_0 \neq 0$.

Hence, as $a_0 \neq 0$, we arrive at an algebraic equation $s(2s - 1) = 0$, which is a quadratic equation for s. It is called *indicial* equation and gives two solutions for s, namely: $s = 1/2$ and $s = 0$. Using these two particular values of s, we can now generate the values of the coefficients c_r via the recurrence relation (8.82) and hence build completely two linearly independent solutions.

Consider first the case of $s = 1/2$. From (8.82) the coefficients c_r satisfy the recurrence relation

$$c_{r+1} = -\frac{c_r}{(r+3/2)(2r+2)} = -\frac{c_r}{(r+1)(2r+3)} \quad \text{for} \quad r = 0, 1, 2, 3, \ldots.$$

It is clear that all coefficients will be proportional to c_0, the latter will therefore eventually become an arbitrary constant. Therefore, for constructing a linearly independent elementary solution, it is sufficient (and convenient) to set $c_0 \equiv 1$. It follows then that the first solution corresponding to $s = 1/2$ becomes:

$$y_1(x) = \sum_{r=0}^{\infty} c_r x^{r+1/2} = \sqrt{x}\left(1 - \frac{x}{1\cdot 3} + \frac{x^2}{1\cdot 2\cdot 3\cdot 5} - \frac{x^3}{1\cdot 2\cdot 3^2\cdot 5\cdot 7} + \cdots\right).$$

$$(8.83)$$

The value of $s = 0$ leads to the recurrence relation

$$c_{r+1} = -\frac{c_r}{(r+1)(2r+1)} \, ,$$

and correspondingly to the second elementary solution (we set $c_0 = 1$ again):

$$y_2(x) = 1 - x + \frac{x^2}{2\cdot 3} - \frac{x^3}{2\cdot 3^2\cdot 5} + \cdots . \qquad (8.84)$$

The general solution is given as a linear combination of the two constructed elementary solutions. ◀

Problem 8.58 By using the recurrence relations for both values of $s = 1/2$ and $s = 0$, work out explicitly the first four terms in the expansions of $y_1(x)$ and $y_2(x)$ and hence confirm expressions (8.83) and (8.84) given above.

The next example shows that the Frobenius method can also be used even in the case of expanding around an ordinary point.

Example 8.8 ▶ Solve the harmonic oscillator equation

$$y''(x) + \omega^2 y(x) = 0 \, ,$$

using the generalized series expansion method.

Solution. This equation does not have any singular points and hence we can expand around $x = 0$ which is an ordinary point using an ordinary Taylor's series. However, to illustrate the power of the Frobenius method, we shall solve the equation using the generalized series expansion assuming some s:

$$y = \sum_{r=0}^{\infty} c_r x^{r+s} \, .$$

After substituting into the DE, we get:

$$\sum_{r=0}^{\infty} c_r (r+s)(r+s-1)x^{r+s-2} + \omega^2 \sum_{r=0}^{\infty} c_r x^{r+s} = 0 .$$

The following powers of x are contained in the first sum: $x^{s-2}, x^{s-1}, x^s, x^{s+1}$, etc., corresponding to the summation index $r = 0, 1, 2, 3$, etc., respectively. At the same time, the second sum contains the terms with x^s, x^{s+1}, etc. Anticipating that we will have to combine the two sums together later on, we separate out the first two "foreign" terms from the first sum:

$$c_0 s(s-1)x^{s-2} + c_1(s+1)sx^{s-1} + \sum_{r=2}^{\infty} c_r(r+s)(r+s-1)x^{r+s-2} + \omega^2 \sum_{r=0}^{\infty} c_r x^{r+s} = 0 .$$

In the first sum we shift the summation index $r - 2 \to r$, so that the two sums can be combined into one:

$$c_0 s(s-1)x^{s-2} + c_1 s(s+1)x^{s-1} + \sum_{r=0}^{\infty} \left[(r+s+2)(r+s+1)c_{r+2} + \omega^2 c_r \right] x^{r+s} = 0 .$$

$$(8.85)$$

It is convenient to start by assuming that $c_0 \neq 0$. Then, from the first term we conclude that $s(s-1) = 0$ (the indicial equation), which gives us two possible values of s, namely, 0 and 1. Therefore, we need to consider two cases.

In the case of $s = 0$ we see that the second term containing x^{s-1} is zero automatically, so that c_1 is arbitrary at this stage. From the last term in (8.85) we obtain the recurrence relation

$$c_{r+2} = -\frac{\omega^2}{(r+2)(r+1)} c_r , \quad \text{where } r = 0, 1, 2, \ldots ,$$

which (assuming $c_0 = 1$) generates the coefficients with even indices: $c_2 = -\omega^2/2$, $c_4 = \omega^4/4!$, etc. It is easy to see that generally

$$c_{2n} = (-1)^n \frac{\omega^{2n}}{(2n)!} , \quad n = 0, 1, 2, \ldots ,$$

which can be proven, e.g. by the method of mathematical induction. Indeed, this is true for $n = 1$. We then assume that it is also true for some value of n. Using the recurrence relation, we get for the next coefficient:

$$c_{2(n+1)} = c_{2n+2} = -\frac{\omega^2}{(2n+2)(2n+1)} c_{2n} = -\frac{\omega^2}{(2n+2)(2n+1)} \cdot (-1)^n \frac{\omega^{2n}}{(2n)!}$$

$$= (-1)^{n+1} \frac{\omega^{2n+2}}{(2n+2)!} ,$$

as required. Combining all terms with even coefficients (and thus even powers of x), we obtain the first elementary solution of the equation:

$$y_1(x) = \sum_{n=0}^{\infty} (-1)^n \frac{\omega^{2n}}{(2n)!} x^{2n} ,$$

which as can easily be seen is Taylor's expansion of the cosine function $\cos(\omega x)$.

Let us now consider the case of $s = 1$. In this case c_0 is arbitrary as $s(s - 1) = 0$ in the first term in (8.85); however, the second term there gives $c_1 = 0$ as $s(s + 1) = 2 \neq 0$. Then the corresponding recurrence relation

$$c_{r+2} = -\frac{\omega^2}{(r + 3)(r + 2)} c_r , \quad r = 0, 1, 2, \ldots$$

leads to

$$y_2(x) = \sum_{r=0}^{\infty} c_r x^{r+1} = \sum_{n=0}^{\infty} (-1)^n \frac{\omega^{2n}}{(2n + 1)!} x^{2n+1}$$

$$= \frac{1}{\omega} \sum_{n=0}^{\infty} (-1)^n \frac{(\omega x)^{2n+1}}{(2n + 1)!} = \frac{1}{\omega} \sin(\omega x) ,$$

which is Taylor's expansion of the corresponding sine function. Note that since $c_1 = 0$ there will be no odd terms at all: $c_3 = c_5 = \cdots = 0$. Thus, by using the second value of s, the expected second solution is generated. Concluding, the final general solution of the harmonic oscillator equation is

$$y(x) = C_1 \sin(\omega x) + C_2 \cos(\omega x) ,$$

as expected, where C_1 and C_2 are arbitrary constants. ◄

Problem 8.59 Above, both solutions were obtained by assuming that $c_0 \neq 0$. Consider now another possible proposition which is assuming that $c_1 \neq 0$. Show that in this case s is equal to either 0 or -1 which generate the same solutions.

Example 8.9 ► Solve the Legendre equation

$$(1 - x^2) y''(x) - 2xy'(x) + l(l + 1)y(x) = 0 , \tag{8.86}$$

using the generalized series method. Here $l > 0$.

Solution. First of all, we consider $p(x)$ and $q(x)$ to find all singular points of the DE and characterize them:

$$p(x) = -\frac{2x}{1-x^2} = -\frac{2x}{(1+x)(1-x)} \quad \text{and} \quad q(x) = \frac{l(l+1)}{(1+x)(1-x)}.$$

Points $x = \pm 1$ are regular singular points since:

$$(1+x)p(x) = -\frac{2x}{1-x} \quad \text{and} \quad (1+x)^2 q(x) = (1+x)\frac{l(l+1)}{1-x}$$

are both regular at $x = -1$; similarly for the $x = 1$:

$$(1-x)p(x) = -\frac{2x}{1+x} \quad \text{and} \quad (1-x)^2 q(x) = (1-x)\frac{l(l+1)}{1+x}$$

are both regular at $x = 1$ and hence this point is also a RSP.

Correspondingly, if we seek the series expansion around $x = 0$ exactly in the middle of the interval $-1 < x < 1$, in this case the series is supposed to converge for any x within this interval, $-1 < x < 1$. This follows from the fact, mentioned above, that the interval of convergence of the generalized series expansion is determined by the distance from the point a (which is 0 in our case) to the nearest singular point (which is either $+1$ or -1).

Substituting the expansion

$$y = \sum_{r=0}^{\infty} c_r x^{r+s}$$

into the differential equation, we get:

$$\sum_{r=0}^{\infty} c_r (r+s)(r+s-1)\left(1-x^2\right) x^{r+s-2} - 2\sum_{r=0}^{\infty} c_r (r+s) x^{r+s}$$

$$+ l(l+1) \sum_{r=0}^{\infty} c_r x^{r+s} = 0. \tag{8.87}$$

The first term leads to two sums due to $\left(1 - x^2\right)$. The first one,

$$1\text{st} = \sum_{r=0}^{\infty} c_r (r+s)(r+s-1) x^{r+s-2}$$

$$= c_0 s(s-1) x^{s-2} + c_1 s(s+1) x^{s-1} + \sum_{r=2}^{\infty} c_r (r+s)(r+s-1) x^{r+s-2},$$

after changing the summation index $r - 2 \to r' \to r$, is worked out into

$$\text{1st} = c_0 s(s - 1)x^{s-2} + c_1 s(s + 1)x^{s-1} + \sum_{r=0}^{\infty} c_{r+2}(r + s + 2)(r + s + 1)x^{r+s} \,,$$

while the second sum is

$$\text{2nd} = -\sum_{r=0}^{\infty} c_r (r + s)(r + s - 1)x^{r+s} \,.$$

We separated out the first two terms in the first sum above in order to be able to combine the two sums together into a single sum. Moreover, all other sums in (8.87) have now the same structure and can be combined all together:

$$c_0 s(s - 1)x^{s-2} + c_1 s(s + 1)x^{s-1} + \sum_{r=0}^{\infty} \left[(r + s + 2)(r + s + 1)c_{r+2} \right.$$
$$\left. -(r + s)(r + s - 1)c_r - 2(r + s)c_r + l(l + 1)c_r \right] x^{r+s} = 0 \,,$$

which after rearranging yields:

$$c_0 s(s - 1)x^{s-2} + c_1 s(s + 1)x^{s-1} + \sum_{r=0}^{\infty} \left[(r + s + 2)(r + s + 1)c_{r+2} \right.$$
$$\left. - \left[(r + s)(r + s + 1) - l(l + 1) \right] c_r \right] x^{r+s} = 0 \,, \tag{8.88}$$

The coefficients to x in the first two terms should be zero, so that, assuming for definiteness that c_0 is arbitrary, we have the indicial equation $s(s - 1) = 0$, resulting in two values of s. We must consider both cases. If $s = 0$, then c_1 must be arbitrary as well because of the second term in (8.88) which is zero for any c_1. Next, the recurrence relation (from the last term with the sum sign) reads:

$$c_{r+2} = \frac{r(r + 1) - l(l + 1)}{(r + 2)(r + 1)} c_r \,. \tag{8.89}$$

We can start either from c_0 or c_1. If we first start from $c_0 = 1$, then we can generate c_{2n} coefficients with even indices (below we use $k = l(l + 1)$):

$$c_2 = \frac{-l(l + 1)}{2 \cdot 1} \equiv -\frac{k}{2!} \,, \quad c_4 = \frac{2 \cdot 3 - k}{4 \cdot 3} c_2 = -\frac{(6 - k)k}{4!} \,,$$

and so on, leading to a series solution

$$y_1(x) = 1 - \frac{k}{2!}x^2 - \frac{(6 - k)k}{4!}x^4 - \cdots \,, \tag{8.90}$$

containing even powers of x.

Starting from c_1, we similarly obtain a series expansion containing only odd powers of x:

$$y_2(x) = x + \frac{2-k}{3!}x^3 + \frac{(12-k)(2-k)}{5!}x^5 + \dots . \tag{8.91}$$

We now consider the other case, $s = 1$. The recurrence relation in this case is:

$$c_{r+2} = \frac{(r+1)(r+2) - k}{(r+3)(r+2)}c_r .$$

Since the second term in (8.88) should be zero, we have to set $c_1 = 0$ since $s(s+1) \neq 0$ when $s = 1$. Then, starting from c_0, we obtain coefficients c_{2n} with even indices and it can easily be checked that this way we arrive at $y_2(x)$ again (note that in the case of $s = 1$ all terms in the expansion $\sum_r c_{2r}x^{2r+1}$ have odd powers, as required). Thus, the second value of s does not lead to any new solutions.

Let us now investigate the convergence of the first series ($s = 0$) using the ratio test and employing the recurrence relation (8.89):

$$\left| \frac{c_{r+2}x^{r+2}}{c_r x^r} \right| = \frac{r(r+1) - k}{(r+2)(r+1)}x^2 = \left[\frac{r}{r+2} - \frac{k}{(r+1)(r+2)} \right]x^2 \to x^2$$

as $r \to \infty$. Therefore, the series $y_1(x)$ converges for $-1 < x < 1$, as expected. Similar analysis is performed for $y_2(x)$:

$$\left| \frac{c_{r+2}x^{r+2}}{c_r x^r} \right| = \frac{r(r+1) - k}{(r+1)(r+2)}x^2 \to x^2$$

as $r \to \infty$, leading therefore to exactly the same interval.

In general, either of the solutions diverges at $x = \pm 1$. However, when l is a positive integer, then one of the solutions contains a finite number of terms becoming a polynomial, and therefore it obviously converges at the boundary points $x = \pm 1$ (as well as at any other x between $\pm\infty$). Indeed, consider first l being even. In the expansion of $y_1(x)$ only coefficients c_r with even indices r are present. Consider then the recurrence relation for the coefficients:

$$c_{r+2} = \frac{r(r+1) - k}{(r+1)(r+2)}c_r = \frac{r(r+1) - l(l+1)}{(r+1)(r+2)}c_r .$$

It is readily seen that when $r = l$ (which is possible as both r and l are even integers) we have $c_{l+2} = 0$ because of the numerator in the equation above. As a result, all other coefficients c_{l+4}, c_{l+6}, etc. will also be equal to zero, i.e. the solution $y_1(x)$ contains only a finite number of terms and is a polynomial of degree l. For, example, if $l = 2$, then $k = 2 \cdot 3 = 6$ and

$$c_2 = \frac{-6}{1 \cdot 2} = -3 , \quad c_4 = -\frac{2 \cdot 3 - 6}{3 \cdot 4}c_2 = 0 , \quad c_6 = c_8 = \dots = 0 ,$$

and thus $y_1(x)$ becomes simply $[y_1(x)]_{l=2} = 1 - 3x^2$. At the same time, it can be seen that the other solution, $y_2(x)$, will not terminate for any even l since in the recurrence relation the numerator reads $r(r+1) - l(l+1)$ and is not equal to zero for any odd r.

If l is odd, however, then $y_2(x)$ is of a polynomial form, while $y_1(x)$ will be an infinite series. The two polynomial solutions are directly proportional to the so-called *Legendre polynomials.* ◄

Problem 8.60 Derive a few more solutions proportional to Legendre polynomials:

$$[y_1(x)]_{l=4} = 1 - 10x^2 + \frac{35}{3}x^4 ; \quad [y_2(x)]_{l=3} = x - \frac{5}{3}x^3 .$$

Problem 8.61 Show that the two independent solutions of the DE

$$2xy'' - y' + 2y = 0$$

are:

$$y_1 = 1 + 2x - \sum_{n=2}^{\infty} \frac{(-2)^n}{n!\,(2n-3)!!}x^n , \quad y_2 = x^{3/2}\sum_{n=0}^{\infty} \frac{3\,(-2)^n}{n!\,(2n+3)!!}x^n ,$$

where $k!! = 1 \cdot 3 \cdot 5 \cdot \ldots \cdot k$ (k is odd) is a product of all odd integers from 1 to k. Using the ratio test, prove that either series converges for any x.

Problem 8.62 Show that the series solutions of the DE

$$8x^2y'' + 10xy' - (1+x)\,y = 0$$

are

$$y_1 = x^{1/4}\left(1 + \frac{x}{14} + \frac{x^2}{616} + \dots\right) , \quad y_2 = x^{-1/2}\left(1 + \frac{x}{2} + \frac{x^2}{40} + \dots\right) .$$

Problem 8.63 Show that the series solutions of the DE

$$x^2y'' + 2x^2y' - 2y = 0$$

are

$$y_1 = -1 + x^{-1}, \quad y_2 = x^2 \left(1 - x + \frac{3x^2}{5} - \frac{4x^3}{15} + \cdots \right).$$

Problem 8.64 Show that the series solutions of the DE

$$3xy'' + (3x + 1)\, y' + y = 0$$

are

$$y_1 = \sum_{n=0}^{\infty} \frac{(-x)^n}{n!} = e^{-x}, \quad y_2 = x^{2/3} \left(1 - \frac{3x}{5} + \frac{(-3x)^2}{5 \cdot 8} + \frac{(-3x)^3}{5 \cdot 8 \cdot 11} + \cdots \right).$$

Problem 8.65 Use the generalized series method to obtain general solutions of the following DEs:

(a) $x^2 y'' + xy' - 9y = 0$;   (b) $x^2 y'' - 6y = 0$;   (c) $x^2 y'' + xy' - 16y = 0$.

[Answer: *(a)* $y = C_1 x^3 + C_2 x^{-3}$; *(b)* $y = C_1 x^{-2} + C_2 x^3$; *(c)* $y = C_1 x^4 + C_2 x^{-4}$.]

Problem 8.66 Show that the two independent power series solutions of the ODE

$$36x^2 y'' + (5 - 9x^2)y = 0$$

are:

$$y_1(x) = x^{5/6} \left[1 + \frac{3}{64} x^2 + \frac{9}{14336} x^4 + \cdots \right],$$

$$y_2(x) = x^{1/6} \left[1 + \frac{3}{32} x^2 + \frac{9}{5120} x^4 + \cdots \right].$$

Problem 8.67 Show that the two independent solutions of the ODE

$$4x^2 y'' - 2x(x + 2)y' + (x + 3)y = 0$$

are:

$$y_1(x) = x^{1/2} \sum_{n=0}^{\infty} \frac{x^n}{2^n n!} \quad \text{and} \quad y_1(x) = x^{3/2} \sum_{n=0}^{\infty} \frac{x^n}{2^n (n + 1)!}.$$

Relate $y_1(x)$ to $y_2(x)$ and hence justify that $y_1(x) = \sqrt{x}$ can actually be chosen instead of its full power series.

Problem 8.68 Show that the two independent power series solutions of the ODE

$$x^2 y'' - (x^2 + 2)y = 0$$

are:

$$y_1(x) = x^2 \left[1 + \frac{x^2}{10} + \frac{x^4}{280} + \ldots \right],$$

$$y_2(x) = x^{-1} \left[1 - \frac{x^2}{2} - \frac{x^4}{8} + \ldots \right].$$

8.4.3 Special Cases

The procedure described above always gives the general series solution of the differential equation (8.63) about a RSP when the roots of the indicial equation differ by more than an integer. If $s_2 = s_1 + m$, where m is a positive integer, including zero, only one series solution $y_1(x)$ corresponding to s_1 can be found; to find the other solution the method must be modified. Here we shall consider this case in some detail.

Adopting the most general expansion of the functions $p(x)$ and $q(x)$ around the point x_0, we can write:

$$p(x) = \frac{p_{-1}}{x - x_0} + p_0 + p_1 (x - x_0) + p_2 (x - x_0)^2 + \ldots , \tag{8.92}$$

$$q(x) = \frac{q_{-2}}{(x - x_0)^2} + \frac{q_{-1}}{x - x_0} + q_0 + q_1 (x - x_0) + q_2 (x - x_0)^2 + \ldots . \tag{8.93}$$

These expressions take account of the fact that the singular point is regular and that the limits

$$\lim_{x \to x_0} (x - x_0) p(x) \quad \text{and} \quad \lim_{x \to x_0} (x - x_0)^2 q(x)$$

both exist, as otherwise the criteria at the beginning of Sect. 8.4.2 would not be satisfied. Correspondingly we shall consider the differential equation

$$(x - x_0)^2 y'' + \left[p_{-1} (x - x_0) + p_0 (x - x_0)^2 + \ldots \right] y'$$
$$+ \left[q_{-2} + q_{-1} (x - x_0) + q_0 (x - x_0)^2 + \ldots \right] y = 0 , \tag{8.94}$$

which we obtained from (8.63) upon multiplication on $(x - x_0)^2$ after employing explicit expansions of $p(x)$ and $q(x)$ given above.

Problem 8.69 Substitute

$$y(x) = (x - x_0)^s \left[1 + c_1 (x - x_0) + c_2 (x - x_0)^2 + \ldots \right]$$

into (8.94). Then, collecting terms with like powers of x and setting the coefficients to zero, derive the following equation for the number s:

$$f(s) \equiv s(s - 1) + p_{-1}s + q_{-2} = s(s - 1 + p_{-1}) + q_{-2} = 0. \quad (8.95)$$

Note that above we have defined a function $f(s)$. Correspondingly, show that the two roots of this quadratic equation, s_1 and s_2, are related via:

$$s_1 + s_2 = 1 - p_{-1}. \quad (8.96)$$

Next, derive a successive infinite set of equations for the coefficients c_1, c_2:

$$c_1 f(s + 1) + p_0 s + q_{-1} = 0,$$

$$c_2 f(s + 2) + c_1 \left[(s + 1) p_0 + q_{-1} \right] + p_1 s + q_0 = 0,$$

and so on; show that an equation for c_n ($n = 1, 2, 3, \ldots$) has the following general form:

$$c_n f(s + n) + F_n(s) = 0, \quad n = 1, 2, 3, \ldots, \quad (8.97)$$

where $F_n(s) = \alpha_n s + \beta_n$ is a linear polynomial in s with the coefficients α_n and β_n which depend on the expansion coefficients of $p(x)$ and $q(x)$, as well as on all the *previous* coefficients $c_1, \ldots, c_{n-1}$.

It is readily seen from (8.97), that it is only possible to determine the coefficients $\{c_n\}$ if the function $f(s + n) \neq 0$ for *any* integer $n = 1, 2, 3, \ldots$ assuming either $s = s_1$ or $s = s_2$. We shall now see that this condition means that the two roots s_1 and s_2 cannot differ by an integer. Indeed, assume that the smallest root is s_2 and the largest $s_1 = s_2 + m$, where m is a positive integer. Since both roots satisfy (8.95), we have $f(s_1) = 0$ and $f(s_2) = f(s_1 - m) = 0$. When considering (8.97) for the coefficients c_n corresponding to the solution associated with s_1, we need $f(s_1 + n) \neq 0$ to be satisfied for any $n = 1, 2, 3, \ldots$. This condition is valid since there could only be two roots of the quadratic equation (8.95) and $s_1 + n > s_2$ for any $n = 1, 2, 3, \ldots$. Consider now if it is possible to construct in the same way a solution corresponding to $s_2 = s_1 - m$. In this case we will have to consider equations (8.97) with the coefficient to c_n being $f(s_2 + n) = f(s_1 - m + n)$. It is seen that for $n = m$ the function $f(s_2 + m) = f(s_1) = 0$ and hence the corresponding coefficient c_n cannot be determined, i.e. the method fails.

Thus, the method needs to be modified when $s_1 = s_2 + m$, where m is a positive integer. We can also include the case of $m = 0$ here as only one value of s is available (as $s_1 = s_2$) and hence only one generalized series solution can be constructed.

The idea of the modified method is based on Theorem 8.1 which states that the second linearly independent solution can always be found using the formula

$$y_2(x) = y_1(x) \int \frac{1}{y_1^2(x)} \exp\left(-\int p(x)dx\right) dx .$$ (8.98)

We shall use this result to find a general form of $y_2(x)$. What we need to do is to use the form (8.92) for $p(x)$ and a general form

$$y_1(x) = (x - x_0)^{s_1} P_1 (x - x_0) \equiv u^{s_1} P_1(u)$$ (8.99)

for the first solution associated with s_1, to investigate the integral in (8.98) and hence find the general form of the second solution. In (8.99) $u = x - x_0$ and $P_1(u)$ is a well-behaved Taylor expansion of some function of u. From this point on we shall be using u instead of x in both integrals in (8.98). We can write:

$$\int p(u)du = p_{-1} \ln u + p_0 u + \frac{1}{2} p_1 u^2 + \cdots = p_{-1} \ln u + P_2(u) ,$$

where $P_2(u)$ is another well-behaved function that also accepts a Taylor's series expansion. Several other such functions will be introduced below, we shall distinguish them with a different subscript. Therefore, the exponential function in (8.98) now reads:

$$\exp\left(-\int p(u)du\right) = u^{-p_{-1}} \exp\left(-P_2(u)\right) = u^{-p_{-1}} P_3(u) ,$$

so that the integrand in (8.98) becomes:

$$\frac{1}{y_1^2} \exp\left(-\int p(u)du\right) = \frac{u^{-p_{-1}} P_3(u)}{u^{2s_1} P_1^2(u)} = u^{-(2s_1+p_{-1})} P_4(u)$$

$$= u^{-(2s_1+p_{-1})} \left(\gamma_0 + \gamma_1 u + \gamma_2 u^2 + \ldots\right) .$$

But, because of (8.96), we have

$$2s_1 + p_{-1} = 2s_1 + (1 - s_1 - s_2) = 1 + s_1 - s_2 = 1 + m ,$$

so that the integrand in (8.98) can be finally rewritten as:

$$\frac{1}{y_1^2} \exp\left(-\int p(u)du\right) = \frac{P_4(u)}{u^{m+1}}$$

$$= \frac{\gamma_0}{u^{m+1}} + \frac{\gamma_1}{u^m} + \cdots + \frac{\gamma_m}{u} + \gamma_{m+1} + \gamma_{m+2}u + \cdots ,$$

which, when integrated, gives

$$\int \frac{1}{y_1^2} \exp\left(-\int p(u)du\right) du = -\frac{\gamma_0/m}{u^m} - \frac{\gamma_1/(m-1)}{u^{m-1}} + \cdots$$

$$-\frac{\gamma_{m-1}}{u} + \gamma_m \ln u + \gamma_{m+1}u + \frac{1}{2}\gamma_{m+2}u^2 + \cdots$$

$$= \gamma_m \ln u + u^{-m}\left[-\frac{\gamma_0}{m} - \frac{\gamma_1}{m-1}u + \cdots\right]$$

$$= \gamma_m \ln u + u^{-m} P_5(u) .$$

Therefore, the second solution should have the following general form:

$$y_2(x) = u^{s_1} P_1(u)\left[\gamma_m \ln u + u^{-m} P_5(u)\right] = \gamma_m u^{s_1} P_1(u) \ln u + u^{s_1-m} P_6(u) ,$$

and hence we can finally write returning back to x:

$$y_2(x) = \gamma_m y_1(x) \ln(x - x_0) + (x - x_0)^{s_2} P_6(x - x_0) . \tag{8.100}$$

We see that the second solution should have, apart from the usual second term with s_2, also an additional term containing the product of the first solution and the logarithm function. In fact, the pre-factor γ_m can always be chosen as equal to one as this would simply rescale the function $y_2(x)$ (the scaling factor is of no importance since when constructing the general solution of the homogeneous equation, y_1 and y_2 are multiplied by arbitrary constants anyway).

Problem 8.70 Obtain both solutions of the zero-order Bessel[11] differential equation

$$xy'' + y' + xy = 0$$

by employing the generalized series method with $x_0 = 0$. (i) First of all, show that only one value of $s = 0$ exists. (ii) Then demonstrate that the corresponding solution is

$$J_0(x) = 1 - \frac{x^2}{2^2} + \frac{x^4}{2^2 4^2} - \frac{x^6}{2^2 4^2 6^2} + \cdots = \sum_{r=0}^{\infty} \frac{(-1)^r x^{2r}}{2^{2r} (r!)^2} .$$

(iii) Write the second solution in the form $K_0(x) = J_0(x) \ln x + P(x)$, substitute into the DE, use the fact that $J_0(x)$ already satisfies it, and hence derive

---

[11] Friedrich Wilhelm Bessel (1784–1846) was a German astronomer, mathematician, physicist and geodesist.

the following nonhomogeneous equation for the function $P(x)$:

$$x P'' + P' + x P = -2J_0'(x) .$$

(iv) Finally, obtain a Taylor series expansion for $P(x)$:

$$P(x) = \sum_{r=1}^{\infty} p_{2r} x^{2r} = \frac{x^2}{4} - \frac{3x^4}{128} + \frac{11x^6}{13824} - \ldots \quad \text{with} \quad p_{2r} = -\frac{p_{2r-2}}{4r^2} + \frac{(-1)^{r+1}}{2^{2r}(r!)^2 r} .$$

Explain, why p_0 must be set to zero. Note that $P(x)$ contains only even powers of x.

The function $J_0(x)$ is called the Bessel function of the first kind of order zero, while $K_0(x)$ the Bessel function of the second kind of the same order.

8.5 Linear Systems of Two Differential Equations

In applications it is often required to solve several differential equations simultaneously. For instance, in the problem of pursuit, Sect. 8.3.2, two ODEs for the time dependent trajectory, (8.55) and (8.56), were obtained. Systems of differential equations are encountered in various applications in physics, chemistry (e.g. chemical kinetics), finances, etc.

In this section we shall consider linear systems of two ODEs with constant coefficients with respect to two functions $x(t)$ and $y(t)$. Linear systems of more than two equations will be discussed in the second volume of this course. We shall not consider systems of non-linear ODEs as this requires developing special general tools that go well beyond this course; the reader is advised to consult special literature.

A linear system of two ODEs has the following general form:

$$\begin{cases} a_1 y'(t) + a_2 y(t) + a_3 x'(t) + a_4 x(t) = f(t) \\ b_1 y'(t) + b_2 y(t) + b_3 x'(t) + b_4 x(t) = g(t) \end{cases} , \qquad (8.101)$$

where a_i and b_i are constant coefficients ($i = 1, \ldots, 4$) and $f(t)$ and $g(t)$ are arbitrary functions.

To solve this equations, we shall use a method very similar to the elimination method we used to solve a system of linear algebraic equations (see Sect. 1.3.6). To simplify the algebraic manipulations, it is convenient to use appropriate notations. Defining the differential operator $D = d/dx$ (Sect. 3.7.4), the system of equations above can be rewritten as follows:

$$\begin{cases} (a_1 D + a_2) y + (a_3 D + a_4) x = f \\ (b_1 D + b_2) y + (b_3 D + b_4) x = g \end{cases} .$$

Next, we shall rewrite these equations in an even simpler form by introducing operators $L_1 = a_1 D + a_2$, $L_2 = a_3 D + a_4$, $M_1 = b_1 D + b_2$ and $M_2 = b_3 D + b_4$. This results in

$$\begin{cases} L_1 y + L_2 x = f \\ M_1 y + M_2 x = g \end{cases}.$$

Multiply both sides of the first equation *from the left* by the operator M_1 and both sides of the second equation, also *from the left*, by L_1:

$$\begin{cases} M_1 L_1 y + M_1 L_2 x = M_1 f \\ L_1 M_1 y + L_1 M_2 x = L_1 g \end{cases}. \tag{8.102}$$

Since the coefficients are constants, it is easy to see that the two operators $M_1 = b_1 D + b_2$ and $L_1 = a_1 D + a_2$ commute, i.e. their action does not depend on the order in which they are applied,[12] i.e. $L_1 M_1 = M_1 L_1$. Indeed, consider a function $h(x)$. Then,

$$\begin{aligned} M_1 L_1 h &= (b_1 D + b_2)(a_1 D + a_2) h \\ &= (b_1 D + b_2)\left(a_1 h' + a_2 h\right) \\ &= b_1 \left(a_1 h'' + a_2 h'\right) + b_2 \left(a_1 h' + a_2 h\right) \\ &= b_1 a_1 h'' + (b_1 a_2 + b_2 a_1) h' + b_2 a_2 h \, ; \end{aligned}$$

at the same time, if we consider

$$L_1 M_1 h = (a_1 D + a_2)(b_1 D + b_2) h \, ,$$

then the same result is obtained.[13] It is worth noting that we can manipulate these expressions containing operators as in algebra, e.g.

$$(b_1 D + b_2)(a_1 D + a_2) h = \left[b_1 a_1 D^2 + (b_1 a_2 + b_2 a_1) D + b_2 a_2 \right] h$$

with $D^2 = DD = d^2/dx^2$ being the second-order derivative (indeed, $D^2 h = D(Dh) = Dh' = h''$).

Subtracting the second equation from the first in (8.102), we shall obtain a second-order ODE just for one function, $x(t)$:

$$(M_1 L_2 - L_1 M_2) x = M_1 f - L_1 g \, . \tag{8.103}$$

Writing it explicitly, we have:

$$[(b_1 D + b_2)(a_3 D + a_4) - (a_1 D + a_2)(b_3 D + b_4)] x = (b_1 D + b_2) f - (a_1 D + a_2) g \, ,$$

---

[12] Generally, two operators A and B commute if $AB = BA$.
[13] If the coefficients $a_i(t)$ and $b_i(t)$ for at least some values of $i = 1, \ldots, 4$ were not constants, i.e. were variable functions of t, the operators $b_1 D + b_2$ and $aD + a_2$ would not commute and the method fails.

or

$$\left[(b_1 a_3 - a_1 b_3)\, D^2 + (b_2 a_3 + b_1 a_4 - a_2 b_3 - a_1 b_4)\, D \right. \\ \left. + (b_2 a_4 - a_2 b_4)\, x\right] = (b_1 D + b_2)\, f - (a_1 D + a_2)\, g \tag{8.104}$$

that is equivalent to

$$(b_1 a_3 - a_1 b_3)\, x'' + (b_2 a_3 + b_1 a_4 - a_2 b_3 - a_1 b_4)\, x' \\ + (b_2 a_4 - a_2 b_4)\, x = b_1 f' + b_2 f - a_1 g' - a_2 g\,.$$

This is a second-order linear nonhomogeneous differential equation with constant coefficients. Solving this ODE using the methods considered in detail in Sect. 8.2, we shall obtain the function $x(t)$ containing two arbitrary constants C_1 and C_2. Substituting the obtained solution for $x(t)$ into any of the original equations, we shall obtain the first-order equation for $y(t)$ that will also be linear and with constant coefficients. For instance, substituting $x(t)$ into the first equation (8.101), we get

$$a_1 y' + a_2 y = f_1$$

with $f_1 = f - (a_3 x' + a_4 x)$ being a known function of t; it contains two constants. This equation is solved using the methods of Sect. 8.1.5. One more constant appears when solving this ODE. Substituting the obtained solution into the second equation of the original set enables one to express the third constant via the previous two. In the end, only two constants are left, as expected, as we have two differential equations of the first order each.

If the operator

$$M_1 L_2 - L_1 M_2 = (b_1 D + b_2)\,(a_3 D + a_4) - (a_1 D + a_2)\,(b_3 D + b_4)$$

in the left-hand side of (8.103) is equal to zero, the system of equation (8.101) is said to be *degenerate*. In this case, similarly to the case of a system of two linear algebraic equations considered in Sect. 1.3.6, we have two cases to consider.

Indeed, if $M_1 L_2 = L_1 M_2$, then in (8.103) we conclude that its right-hand side,

$$M_1 f - L_1 g = (b_1 D + b_2) f - (a_1 D + a_2)g = b_1 f' + b_2 f - a_1 g' - a_2 g\,,$$

must be equal to zero as well. If it is not, then the system of equations (8.101) is incompatible, there are no solutions. If, however, $M_1 f = L_1 g$, then the first equation in (8.102) can be written as

$$M_1 L_1 y + \underbrace{M_1 L_2}_{L_1 M_2} x = \underbrace{M_1 f}_{L_1 g} \quad \Longrightarrow \quad L_1\,(M_1 y + M_2 x - g) = 0\,.$$

We see that the expression in the round brackets coincides with the second equation of the original system (8.101) and therefore the equality to zero is provided automatically. In other words, only one of the equations contains an independent information, the other equation can be disregarded. Hence, the functions $y(t)$ and $x(t)$ are related to each other by either of the two equations. We could take the functions containing say $x(t)$ to the right-hand side,

$$L_1 y(t) = f - L_2 x(t),$$

and solve the obtained linear first-order ODE with respect to $y(t)$, considering $f - L_2 x(t)$ as the right-hand side (nonhomogeneity). Hence, $y(t)$ will be linearly related to $x(t)$ and (possibly) its first derivative (if $a_3 \neq 0$).

We shall now consider a few examples. First, let us solve the system of equations

$$\begin{cases} y'(t) + 2x'(t) + x(t) = t \\ y'(t) + y(t) + x'(t) + x(t) = 2t \end{cases}. \tag{8.105}$$

Rewriting it via operators, we have:

$$\begin{cases} Dy + (2D + 1)x = t \\ (D + 1)y + (D + 1)x = 2t \end{cases}.$$

Multiplying the first equation by $D + 1$ and the second by D and subtracting from each other, we obtain:

$$\left(D^2 + 2D + 1\right)x = t - 1 \quad \text{or} \quad x'' + 2x' + x = t - 1.$$

Problem 8.71 Show that the general solution of this ODE is

$$x(t) = (C_1 + C_2 t)\, e^{-t} + t - 3.$$

At the next step it is easier to substitute this solution into the first equation of (8.105) as this immediately gives a first-order ODE

$$y' = (C_1 - 2C_2 + C_2 t)\, e^{-t} + 1,$$

integration of which gives

$$\begin{aligned} y(t) &= \int \left[(C_1 - 2C_2 + C_2 t)\, e^{-t} + 1\right] dt \\ &= (-C_1 + C_2 - C_2 t)\, e^{-t} + t + C_3 \end{aligned}$$

where C_3 is yet another constant.

Problem 8.72 Substitute the obtained solutions for $x(t)$ and $y(t)$ into the second equation of the system (8.105) to show that $C_3 = 1$.

Hence, both functions $x(t)$ and $y(t)$ have been determined and each contains the same constants C_1 and C_2.

Problem 8.73 Show that the solution of

$$\begin{cases} 2y' - y + 2x' + x = e^{-t} \\ y' + y + x' + x = t \end{cases}$$

is

$$\begin{cases} x(t) = C_1 e^{-t} + \dfrac{1}{2}(t - 3) \\ y(t) = -\dfrac{1}{3}(1 + C_1)e^{-t} + \dfrac{1}{2}(t + 1) \end{cases}.$$

Interestingly, in this case there is only one arbitrary constant since the ODE for $x(t)$ turns out to be of the first, not of the second, order. While solving the ODE for $y(t)$ from the first original equation, the second constant comes in the form of $C_2 e^{t/2}$; however, when substituting this solution into the second equation, this term has to be eliminated ($C_2 = 0$).

Problem 8.74 Show that the solution of

$$\begin{cases} y' + x' = \sin t \\ y + 4x' + 3x = \cos t \end{cases}$$

is

$$\begin{cases} x(t) = C_1 \left(-1 + 3e^{-t/2}\right) - C_2 \left(1 - e^{-t/2}\right) + \dfrac{1}{5}(\cos t + 2\sin t) \\ y(t) = 3C_1 \left(1 - e^{-t/2}\right) + C_2 \left(3 - e^{-t/2}\right) - \dfrac{2}{5}(3\cos t + \sin t) \end{cases}.$$

Fig. 8.3 Electrical circuit
with a resistance R, an
inductance L and a
capacitance C connected to a
voltage source $V(t)$

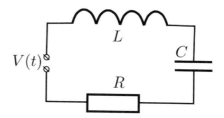

8.6   Examples in Physics

8.6.1   Harmonic Oscillator

In many physical problems it is necessary to find a general solution of the following
second-order linear ODE

$$y'' + \gamma y' + \omega_0^2 y = F_d \sin(\omega t) \,, \tag{8.106}$$

with respect to the function of time $y(t)$.

Depending on the physical problem, the actual meaning of $y(t)$ may be different. A
natural example of a physical problem where this kind of equation plays the principal
role is, for instance, a harmonic oscillator with friction subjected to a sinusoidal
external force. Hence, we shall be calling this a forced harmonic oscillator equation.
A more complex variant of this is encountered, e.g. in modelling oscillations of
an atomic force microscopy (AFM) tip in non-contact AFM experiments (Vol. II,
Chap. 4). Another example is related to electric circuits, for instance, the one shown in
Fig. 8.3. It contains a resistance R, an inductance L and a capacitance C, and a voltage
applied to the circuit is some sinusoidal function of time, e.g. $V(t) = V_0 \cos(\omega t)$.
Using the second Kirchoff's law, we obtain for the voltage drop along the circuit:

$$L\frac{di}{dt} + Ri + \frac{1}{C}\int_{t_0}^{t} i(t_1)dt_1 = V(t) = V_0 \cos(\omega t) \,,$$

where $i(t)$ is the current in the circuit. Differentiating both sides of the equation
above, we get:

$$L\frac{d^2 i}{dt^2} + R\frac{di}{dt} + \frac{1}{C}i = -\omega V_0 \sin(\omega t) \,,$$

which is the forced harmonic oscillator equation (8.106) with $\omega_0 = 1/\sqrt{LC}$, $\gamma = R/L$ and $F_d = -\omega V_0/L$. The term with resistance (the one containing the first
derivative of the current) corresponds to losses in the circuit (dissipation), while the
inductance and capacitance terms work towards current oscillations. The applied
voltage induces oscillations with a different frequency ω. It is sometimes called an
excitation signal.

8.6.1.1 Harmonic Motion

Consider first the equation without resistance and excitation:

$$y'' + \omega_0^2 y = 0 . \qquad (8.107)$$

Its solution,

$$y(t) = A \sin(\omega_0 t) + B \cos(\omega_0 t) ,$$

corresponds to a sinusoidal oscillations with frequency ω_0. Indeed, after some rearrangements,

$$
\begin{aligned}
y(t) &= \sqrt{A^2 + B^2} \left[\frac{A}{\sqrt{A^2 + B^2}} \sin(\omega_0 t) + \frac{B}{\sqrt{A^2 + B^2}} \cos(\omega_0 t) \right] \\
&= \sqrt{A^2 + B^2} \left[\cos(\phi) \sin(\omega_0 t) + \sin(\phi) \cos(\omega_0 t) \right] \\
&= \sqrt{A^2 + B^2} \sin(\omega_0 t + \phi) ,
\end{aligned}
\qquad (8.108)
$$

where $\tan \phi = B/A$. Therefore, the solution of the homogeneous equation (8.107) can always be written as

$$y(t) = D \sin(\omega_0 t + \phi) \qquad (8.109)$$

with two arbitrary constants: D is called the amplitude and ϕ phase. This solution corresponds to, e.g. a *self-oscillation* in a circuit.

8.6.1.2 Driven Oscillator with No Friction

An interesting effect happens when we add an excitation signal with $\omega \neq \omega_0$:

$$y'' + \omega_0^2 y = F_d \sin(\omega t) . \qquad (8.110)$$

This is a harmonically driven oscillator without friction (or a circuit without resistance). In this case the harmonic oscillation (8.109) represents only a complementary solution as there must be a particular solution as well.

> **Problem 8.75** Using the trial function $y = B_1 \sin(\omega t) + B_2 \cos(\omega t)$ for (8.110), show that $B_1 = -F_d / (\omega^2 - \omega_0^2)$ and $B_2 = 0$, so that the general solution becomes
>
> $$y(t) = D \sin(\omega_0 t + \phi) - \frac{F_d}{\omega^2 - \omega_0^2} \sin(\omega t) . \qquad (8.111)$$

We see from the above problem that there are two harmonic motions: one due to self-oscillation (angular frequency ω_0) and another due to applied excitation that happens with the same frequency ω as the excitation signal itself; its amplitude B_1

depends crucially on the difference between the squares of the two frequencies. This dependence is such that when the two frequencies are close, the amplitude of the particular integral solution becomes increasingly large, i.e. the excited oscillation dominates for excitation frequencies ω which are close to ω_0.

If the two frequencies are very close, but are not exactly the same, say $\omega = \omega_0 + \epsilon$ with small ϵ, then the motion overall represents *beats*. Indeed, assume that initially there was no motion in our oscillator, i.e. $y(0) = y'(0) = 0$. Then, from (8.111) it follows that

$$D\sin(\phi) = 0 \quad \text{and} \quad D\omega_0 \cos(\phi) - \frac{F_d\omega}{\omega^2 - \omega_0^2} = 0 .$$

Note that D cannot be equal zero as in this case the second equation above would not be satisfied. Thus, from the first equation we get $\phi = 0$, and hence from the second equation we obtain

$$D = \frac{F_d\omega}{\omega_0\left(\omega^2 - \omega_0^2\right)} ,$$

and thus

$$
\begin{aligned}
y(t) &= \frac{F_d}{\omega_0\left(\omega^2 - \omega_0^2\right)} \left[\omega \sin(\omega_0 t) - \omega_0 \sin(\omega t)\right] \\
&= \frac{F_d}{\omega_0\left((\omega_0 + \epsilon)^2 - \omega_0^2\right)} \left[(\omega_0 + \epsilon) \sin(\omega_0 t) - \omega_0 \sin(\omega_0 t + \epsilon t)\right] \\
&\simeq \frac{F_d}{2\omega_0^2\epsilon} \left[\omega_0 \sin(\omega_0 t) - \omega_0 \sin(\omega_0 t + \epsilon t)\right] = \frac{F_d}{2\omega_0\epsilon} \left[\sin(\omega_0 t) - \sin(\omega_0 t + \epsilon t)\right] \\
&= -\frac{F_d}{\omega_0\epsilon} \sin\left(\frac{\epsilon t}{2}\right) \cos\left(\omega_0 t + \frac{\epsilon t}{2}\right) \simeq -\frac{F_d}{\omega_0\epsilon} \sin\left(\frac{\epsilon t}{2}\right) \cos(\omega_0 t) .
\end{aligned}
$$

(8.112)

To obtain the first expression in the last line we have used the trigonometric identity (2.49). The signal $y(t)$ in (8.112) can be considered as a harmonic signal $B(t) \cos(\omega_0 t)$ of frequency ω_0 and the amplitude $B(t) \sim \sin(\epsilon t/2)$ that also changes in time harmonically with a much smaller frequency (larger period). The effect of this is demonstrated in Fig. 8.4 and is called *amplitude modulation*.

The above consideration is only valid if $\omega \neq \omega_0$. When the excitation frequency is the same as the self-oscillation one, the trial function for the particular solution has to be modified (a "resonance", which we talked about in Sect. 8.2.3.1).

Problem 8.76 Using the trial function

$$y_p(t) = t\left[B_1 \sin(\omega_0 t) + B_2 \cos(\omega_0 t)\right]$$

Fig. 8.4 Amplitude modulation: (8.112) plotted with $-F_d/\omega_0\epsilon = 1$, $\epsilon = 0.004$ and $\omega_0 = 0.2$

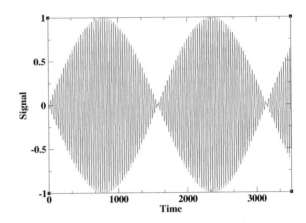

in (8.110) with $\omega = \omega_0$, show that in this case $B_1 = 0$ and $B_2 = -F_d/2\omega_0$, so that the whole solution reads

$$y(t) = D\sin(\omega_0 t + \phi) - \frac{F_d t}{2\omega_0}\cos(\omega_0 t) . \qquad (8.113)$$

We see from (8.113) that the amplitude of the contribution due to the excitation signal increases indefinitely with time, independently of the initial conditions (that are responsible for the values of the arbitrary constants D and ϕ); the system is said to be at *resonance*. The frequency ω_0 is sometimes called the *resonance angular frequency*.

Actually, the infinite increase of the amplitude at resonance frequency can be anticipated from (8.112) as well since

$$\frac{1}{\epsilon}\sin\left(\frac{\epsilon t}{2}\right) = \frac{t}{2}\frac{\sin(\epsilon t/2)}{\epsilon t/2} \to \frac{t}{2} \text{ when } \epsilon \to 0 .$$

Note also that at resonance the second term in the solution (8.113) is shifted by $-\pi/2$ with respect to the driving force itself, compare $F_d\sin(\omega t)$ in (8.110) and $-\cos(\omega_0 t)$ in (8.113). Therefore, at resonance there is a shift of $-\pi/2$ in phase between the driving force and the oscillating signal following it.

8.6.1.3 Harmonic Oscillator with Friction
Now let us consider a self-oscillating system with friction (e.g. the resistance R is present in the circuit in Fig. 8.3):

$$y'' + \gamma y' + \omega_0^2 y = 0 . \qquad (8.114)$$

The characteristic equation $p^2 + \gamma p + \omega_0^2 = 0$ in this case has two solutions:

$$p_{1,2} = \frac{1}{2}\left(-\gamma \pm \sqrt{\gamma^2 - 4\omega_0^2}\right) , \qquad (8.115)$$

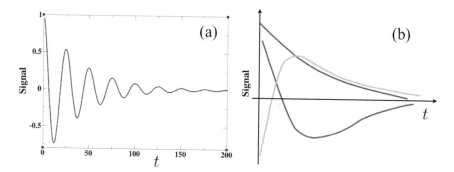

Fig. 8.5 a Under-dumped motion with $A = 1$, $\phi = 0$, $\gamma = 0.1$ s^{-1} and $w = 0.5$ s^{-1}. **b** Over-dumping for various initial conditions and values of the parameters

and hence there are three cases to consider.

Under-damping: If $\gamma^2 - 4\omega_0^2 < 0$, i.e. if the friction is small, $\gamma < 2\omega_0$, then $p_{1,2} = -\gamma/2 \pm iw$ are complex conjugates, where $w = \sqrt{\omega_0^2 - (\gamma/2)^2}$, and the general solution is then

$$y(t) = C_1 e^{(-\gamma/2 + iw)t} + C_2 e^{(-\gamma/2 - iw)t} = e^{-\gamma t/2} \left[C_1 e^{iwt} + C_2 e^{-iwt} \right]$$
$$= A e^{-\gamma t/2} \sin(\omega t + \phi) ,$$

(8.116)

where A and ϕ is a set of arbitrary constants replacing the set C_1 and C_2. In this case the motion is easily recognizable as an oscillation with frequency w (which may be very close to ω_0 if $\gamma \ll \omega_0$); however, its amplitude is exponentially decreasing as shown in Fig. 8.5a providing an exponentially decaying envelope for the sinusoidal motion.

Over-dumping: If $\gamma^2 - 4\omega_0^2 > 0$, i.e. if the friction is large, $\gamma > 2\omega_0$, then both $p_{1,2}$ are real and negative:

$$y(t) = C_1 e^{p_1 t} + C_2 e^{p_2 t} .$$

The oscillations are *not* observed, the motion comes to a halt very quickly. Typical behaviours of the solution in this case are shown in Fig. 8.5b by different curves.

Critical Dumping: When $\gamma^2 - 4\omega_0^2 = 0$, i.e. if the friction value is transitional between the two regimes, the two roots $p_{1,2}$ coincide and the solution has the form

$$y(t) = (C_1 + C_2 t) e^{-\gamma t/2} ,$$

and is similar in shape to the over-dumped case.

8.6.1.4 Driven Harmonic Oscillator with Friction

Finally, we shall consider the most common case of the sinusoidally driven dumped harmonic oscillator described by (8.106). Its complementary solution $y_c(t)$ was considered in the previous subsection and corresponds to either under-dumped, over-dumped or critically dumped case. However, in either case the *transient* motion described by $y_c(t)$ dies away either slowly or quickly (depending on the value of the friction γ). Therefore, if one is interested in a solution corresponding to long times, the complementary solution could be disregarded and one has to consider the particular integral only. The particular solution will survive until, of course, the driving force ceases to be applied.

Problem 8.77 Using the trial function

$$y_p = A \sin(\omega t) + B \cos(\omega t)$$

in (8.106), show that

$$A = \frac{F_d \Delta}{\Delta^2 + (\gamma\omega)^2} \quad \text{and} \quad B = -\frac{F_d \gamma\omega}{\Delta^2 + (\gamma\omega)^2}, \quad \text{where} \ \Delta = \omega_0^2 - \omega^2 .$$

We know from Sect. 8.6.1.1 (see (8.108)) that the sum of sine and cosine functions in y_p can be rewritten using a single sine function as:

$$y_p(t) = D \sin(\omega t + \phi) , \tag{8.117}$$

where the amplitude is

$$D = \sqrt{A^2 + B^2} = \frac{F_d}{\sqrt{\Delta^2 + (\gamma\omega)^2}} = \frac{F_d}{\sqrt{\left(\omega^2 - \omega_0^2\right)^2 + (\gamma\omega)^2}} \tag{8.118}$$

and the phase ϕ is defined via

$$\tan\phi = \frac{B}{A} = -\frac{\gamma\omega}{\Delta} = \frac{\gamma\omega}{\omega^2 - \omega_0^2} . \tag{8.119}$$

Thus, the signal remains sinusoidal with the frequency of the driving force; it is called the *steady-state solution* since it continues to hold without change as long as the driving force is acting.

The situation is similar to the driven harmonic oscillator without friction considered in Sect. 8.6.1.2. Therefore, we may anticipate that the amplitude of the steady-state motion will depend critically on the interplay between the two frequencies ω and ω_0. And, indeed, as shown in Fig. 8.6a, the amplitude demonstrates a maximum (resonance) at some ω that is close to ω_0 at small γ.

Problem 8.78 More precisely, show that D is maximum when ω is equal to the *resonance frequency*

$$\omega_{res} = \sqrt{\omega_0^2 - \frac{\gamma^2}{2}}. \tag{8.120}$$

We see that the resonance frequency is very close to the fundamental frequency ω_0 of the harmonic oscillator when the friction is small ($\gamma \ll \omega_0$).

The phase as a function of ω is shown in Fig. 8.6 (b). It approaches the value of $-\pi/2$ at resonance (note that $\omega_{res} < \omega_0$ so that $\tan\phi$ is negative and in fact very close to $-\infty$ for small γ).

8.6.2 Falling Water Drop

Consider a drop of water of mass m falling down from a cloud under gravity. The equation of motion for the drop is

$$m\frac{dv}{dt} = mg - \gamma v,$$

where γ is a friction coefficient (due to air resistance), g the gravity constant (acceleration in the gravity field of the Earth), v the particle vertical velocity (positive if directed downwards). We shall obtain its general solution and also its particular solution assuming that the initial velocity of the particle was $v_0 = 0$.

This ODE can be solved by rearranging terms and integrating both sides:

$$\frac{dv}{g - \frac{\gamma}{m}v} = dt \implies \int \frac{dv}{g - \frac{\gamma}{m}v} = \int dt \implies -\frac{m}{\gamma}\int \frac{dv}{v - \frac{gm}{\gamma}} = t - C_1$$

$$\implies \frac{m}{\gamma}\ln\left|v - \frac{gm}{\gamma}\right| = -t + C_1 \implies v - \frac{gm}{\gamma} = Ce^{-\gamma t/m},$$

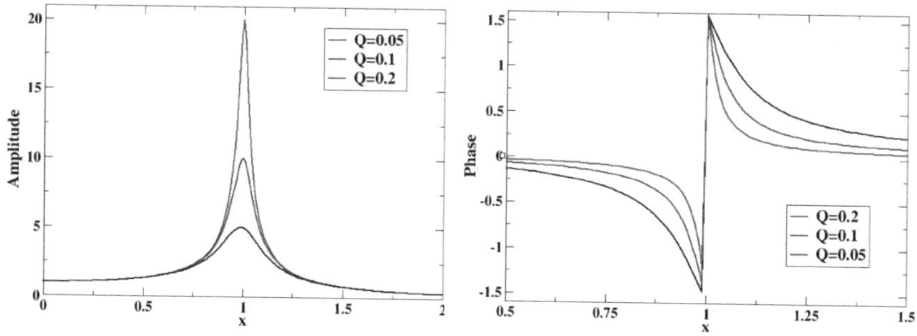

Fig. 8.6 Amplitude D (**a**) and the phase ϕ (**b**) of the steady-state solution (8.117) of the damped driven harmonic oscillator as functions of the relative driving frequency $x = \omega/\omega_0$. Here $Q = \gamma\omega_0$

where $C = e^{\gamma C_1/m}$ is another constant.[14] Thus, a general solution for the velocity of the drop is

$$v(t) = Ce^{-\gamma t/m} + \frac{gm}{\gamma} \ .$$

The particular solution is now obtained if we find the constant C from the additional information given about the velocity, i.e. that we know its initial value: $v(0) = v_0$. This gives:

$$v(0) = v_0 = C + \frac{gm}{\gamma} \implies C = v_0 - \frac{gm}{\gamma} \ .$$

If $v_0 = 0$, then $C = -gm/\gamma$, so that we obtain in this case:

$$v(t) = \frac{gm}{\gamma} \left(1 - e^{-\gamma t/m}\right) \ .$$

Thus, initially the drop has the velocity $v_0 = 0$. Then the velocity is increased; however, after some time the acceleration slows down (due to air resistance) and the velocity approaches a constant value of $v_\infty = gm/\gamma$, called terminal velocity. In fact, the exponential term becomes very small rather quickly, so that almost all the way down the drop will fall approximately with the constant velocity v_∞. Note that it is positive since it is directed downward.

8.6.3   Celestial Mechanics

Consider a mass m (e.g. Earth) orbiting a very heavy mass $M \gg m$ (the Sun) due to the radial gravitational force

$$\mathbf{F(r)} = f(r)\mathbf{e}_r = -G\frac{mM}{r^2}\mathbf{e}_r \ , \tag{8.121}$$

where G is the gravitational constant, r distance between m and the centre of the coordinate system, where the heavy mass is (it is a rather good approximation that the heavy mass is fixed in the centre of the coordinate system), and $\mathbf{e}_r = \mathbf{r}/r$ is the unit vector along $\mathbf{r}$. We consider the motion of the smaller mass within the $x - y$ plane. In this case it will be convenient to consistently use the polar coordinates $x = r\cos\phi$ and $y = r\sin\phi$.

---

[14] As usual, there is no need to keep the old arbitrary constant if it comes inside some function. This will be just another arbitrary constant!

Problem 8.79 Considering $r(t)$ and $\phi(t)$ being functions of time, show that

$$\dot{x} = \dot{r}\cos\phi - r\dot{\phi}\sin\phi, \quad \dot{y} = \dot{r}\sin\phi + r\dot{\phi}\cos\phi, \tag{8.122}$$

$$\ddot{x} = \left(\ddot{r} - r\dot{\phi}^2\right)\cos\phi - \left(r\ddot{\phi} + 2\dot{r}\dot{\phi}\right)\sin\phi \tag{8.123}$$

and

$$\ddot{y} = \left(\ddot{r} - r\dot{\phi}^2\right)\sin\phi + \left(r\ddot{\phi} + 2\dot{r}\dot{\phi}\right)\cos\phi. \tag{8.124}$$

Using obtained equations and the fact that the x and y components of the force acting on the smaller mass are $F_x = f_r\cos\phi$ and $F_y = f_r\sin\phi$, the Newtonian equations of motion $m\ddot{x} = F_x$ and $m\ddot{y} - F_y$ for the mass m become

$$\left(\ddot{r} - r\dot{\phi}^2\right)\cos\phi - \left(r\ddot{\phi} + 2\dot{r}\dot{\phi}\right)\sin\phi = \frac{f_r}{m}\cos\phi,$$
$$\left(\ddot{r} - r\dot{\phi}^2\right)\sin\phi + \left(r\ddot{\phi} + 2\dot{r}\dot{\phi}\right)\cos\phi = \frac{f_r}{m}\sin\phi. \tag{8.125}$$

These equations can be greatly simplified by employing the conservation of the angular momenta of the mass m. The vector of the angular momentum

$$\mathbf{L} = m\,[\mathbf{r} \times \dot{\mathbf{r}}] = m\begin{vmatrix} \mathbf{i} & \mathbf{j} & \mathbf{k} \\ x & y & 0 \\ \dot{x} & \dot{y} & 0 \end{vmatrix} = m\,(x\dot{y} - y\dot{x})\,\mathbf{k} = L_z\mathbf{k}$$

is directed along the z-axis. Using (8.122), the z component of the angular momentum becomes

$$L_z = m\left[r\cos\phi\left(\dot{r}\sin\phi + r\dot{\phi}\cos\phi\right) - r\sin\phi\left(\dot{r}\cos\phi - r\dot{\phi}\sin\phi\right)\right]$$
$$= mr^2\dot{\phi}. \tag{8.126}$$

Since this quantity is conserved,

$$\frac{dL_z}{dt} = mr\left(r\ddot{\phi} + 2\dot{r}\dot{\phi}\right) = 0 \implies r\ddot{\phi} + 2\dot{r}\dot{\phi} = 0.$$

Using this result in the equations of motion (8.125), we can eliminate the second terms in their left-hand sides and arrive at a single equation

$$f_r = m\left(\ddot{r} - r\dot{\phi}^2\right). \tag{8.127}$$

This equation, together with the condition (8.126), fully describes the orbiting of m around M.

To derive the trajectory of m, it is convenient instead of r to consider its inverse $u = 1/r$ as the coordinate that will be a function of the angle ϕ. We have:

$$\frac{du}{d\phi} = \frac{dt}{d\phi}\frac{du}{dt} = \frac{1}{d\phi/dt}\frac{d}{dt}\left(\frac{1}{r}\right) = -\frac{1}{\dot{\phi}}\frac{\dot{r}}{r^2} = -\frac{m}{L_z}\dot{r}$$

and

$$\frac{d^2u}{d\phi^2} = -\frac{m}{L_z}\frac{d\dot{r}}{d\phi} = -\frac{m}{L_z}\frac{d\dot{r}}{dt}\frac{dt}{d\phi} = -\frac{m}{L_z}\frac{\ddot{r}}{\dot{\phi}} = -\frac{\ddot{r}}{(L_z/m)^2\,u^2}\,. \tag{8.128}$$

Using (8.128) one can express $\ddot{r}$ via the second derivative of u with respect to ϕ, while $\dot{\phi}$ can be expressed from (8.126). Hence, the force (8.127) becomes:

$$f_r = -\frac{L_z^2}{m}u^2\left(\frac{d^2u}{d\phi^2} + u\right)\,. \tag{8.129}$$

This equation bears the name of Binet.

Let us now obtain the trajectory for the motion of the small mass. In this case $f_r = -GMmu^2$. Substituting this in (8.129) we obtain a DE for the trajectory:

$$\frac{d^2u}{d\phi^2} + u = \frac{GMm^2}{L_z^2}\,.$$

This is a linear second-order nonhomogeneous DE with constant coefficients and a constant nonhomogeneity, whose solution can be easily written as

$$u(\phi) = A\left[1 + e\cos\left(\phi - \phi_0\right)\right]\,, \tag{8.130}$$

where e and ϕ_0 are the arbitrary constants (to be determined from the initial conditions) and $A = GMm^2/L_z^2$.

The angle ϕ_0 can be set to zero if we measure ϕ from its initial value, leading to the final result,

$$u(\phi) = A\left[1 + e\cos\phi\right]\,, \tag{8.131}$$

that we can analyse. Indeed, we have already came across this kind of equation in Sect. 1.19.3. Four different trajectory types are possible depending on the value of the eccentricity e: (i) a circle if $e = 0$; (ii) an ellipsoid if $0 < e < 1$; (iii) a parabola if $e = 1$, and (iv) a hyperbola if $e > 1$.

Problem 8.80 The total energy of the mass m is given as a sum of its potential and kinetic energies:

$$E = -G\frac{mM}{r} + \frac{m}{2}\left(\dot{x}^2 + \dot{y}^2\right).$$

Show first that the energy E is conserved by directly demonstrating that $dE/dt = 0$. Next, show that

$$E = \frac{m}{2}\left(\frac{GmM}{L_z}\right)^2 (e^2 - 1).$$

It then follows that when $0 \le e < 1$ the total energy is negative, which corresponds to a local motion of the mass m orbiting the heavy mass M. In the other two cases, $e \ge 1$, the mass m would scatter M escaping to infinity at long times.

Problem 8.81 Finally, let us consider an inverse Kepler problem. Let us ask: what must be the radial force f_r such that trajectories are ellipsoidal? Substitute (8.130) into the right-hand side of the Binet equation (8.129) to find that $f_r \sim u^2 \sim 1/r^2$, an expected result.

8.6.4 Finite Amplitude Pendulum

Consider oscillations of a pendulum consisting of a mass m suspended on a rod of length l, see Fig. 8.7. There are two forces acting on the mass: a tension force **T** directed along the rod and the gravity $F_z = mg$ directed along the z-axis. Newtonian equations of motion for the mass read:

$$\begin{cases} m\ddot{x} = -T \sin\theta \\ m\ddot{z} = mg - T\cos\theta \end{cases}.$$

These two equations can be rewritten as a single differential equation with respect to the angle $\theta(t)$. Since $x = l\sin\theta$ and $z = l\cos\theta$, then

$$\ddot{x} = l\left(\ddot{\theta}\cos\theta - \dot{\theta}^2 \sin\theta\right),$$

$$\ddot{z} = -l\left(\ddot{\theta}\sin\theta + \dot{\theta}^2 \cos\theta\right),$$

Fig. 8.7 A simple pendulum: a mass m is connected to a rod of length l

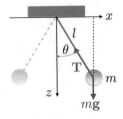

and we obtain the equations of motion:

$$\begin{cases} \ddot{\theta}\cos\theta - \dot{\theta}^2\sin\theta = -\dfrac{T}{ml}\sin\theta \\[2mm] \ddot{\theta}\sin\theta + \dot{\theta}^2\cos\theta = -\dfrac{g}{l} + \dfrac{T}{ml}\cos\theta \end{cases}$$

Multiplying the first equation by $\cos\theta$, the second by $\sin\theta$ and adding them both together, we shall obtain the equation of motion for the pendulum[15]:

$$\ddot{\theta} + \frac{g}{l}\sin\theta = 0. \tag{8.132}$$

It is known that for small oscillations (small deviation angles θ), $\sin\theta \approx \theta$ and we obtain a well-known harmonic differential equation $\ddot{\theta} + \omega^2\theta = 0$ that corresponds to harmonic oscillations with the frequency $\omega = \sqrt{g/l}$. However, if the pendulum deviates significantly from the vertical, this approximation is not valid anymore and one has to solve the original ODE (8.132).

The obtained ODE is of the type considered in Sect. 8.3: the equation does not contain the time t present explicitly, and hence the order of the differential equation can be reduced by introducing a new function $u(\theta) = \dot{\theta}$. Since

$$\ddot{\theta} = \frac{d}{dt}\left(\dot{\theta}\right) = \frac{d\dot{\theta}}{d\theta}\frac{d\theta}{dt} = \frac{d\dot{\theta}}{d\theta}\dot{\theta} = \frac{du}{d\theta}u,$$

we shall obtain a separable first-order ODE

$$u\,du = -\frac{g}{l}\sin\theta\,d\theta,$$

which is trivially integrated to give

$$\frac{u^2}{2} + C = \frac{g}{l}\cos\theta.$$

---

[15] A somewhat simpler method of deriving (8.132) is based on considering an equation of motion $d\mathbf{L}/dt = \mathbf{r}\times\mathbf{F}$ for the angular momentum $\mathbf{L} = m(\mathbf{r}\times\dot{\mathbf{r}})$. Note that only the gravity contributes to the torque $\mathbf{r}\times\mathbf{F}$ as the tension $\mathbf{T}$ is parallel to the vector $\mathbf{r}$.

To proceed, we shall assume the following initial conditions: the pendulum was initially set at some angle θ_0 and then released, i.e. $\theta(0) = \theta_0$ and $\dot{\theta}(0) = u(0) = 0$. Applying both conditions yields $C = (g/l) \cos \theta_0$, and we obtain for $u = \dot{\theta}$:

$$\dot{\theta}^2 = \frac{2g}{l} (\cos \theta - \cos \theta_0) .$$

By taking the square root, we are faced with two possibilities:

$$\frac{d\theta}{dt} = \pm \sqrt{\frac{2g}{l}} \sqrt{\cos \theta - \cos \theta_0} ,$$

which is again a separable ODE:

$$\pm \sqrt{\frac{l}{2g}} \int_{\theta_0}^{\theta} \frac{d\theta'}{\sqrt{\cos \theta' - \cos \theta_0}} = t .$$

Since during its oscillations the angle θ will deviate between $-\theta_0$ and θ_0, i.e. $|\theta| \leq \theta_0$, to ensure a positive time, one has to choose the minus sign in the left-hand side as the integral is negative; this corresponds to the only physically acceptable solution. Therefore, we finally obtain the required solution that gives an implicit dependence of the deviation angle θ on time t:

$$t = \sqrt{\frac{l}{2g}} \int_{\theta(t)}^{\theta_0} \frac{d\theta'}{\sqrt{\cos \theta' - \cos \theta_0}} .$$

This formula can be slightly rearranged. Firstly, we shall notice that $\cos \theta' = 1 - 2 \sin^2 \frac{\theta'}{2}$, so that

$$\cos \theta' - \cos \theta_0 = 2 \left(k^2 - \sin^2 \frac{\theta'}{2} \right) ,$$

where $k = \sin \frac{\theta_0}{2}$. Secondly, we can introduce a more convenient variable ψ such that

$$\sin \frac{\theta'}{2} = k \sin \psi , \quad \theta' = 2 \arcsin (k \sin \psi)$$

and

$$d\theta' = 2 \frac{1}{\sqrt{1 - (k \sin \psi)^2}} k \cos \psi \, d\psi .$$

The initial value of the new variable ψ_0 corresponds to

$$\sin \psi_0 = \frac{1}{k} \sin \frac{\theta_0}{2} = \frac{k}{k} = 1$$

and hence $\psi_0 = \pi/2$. Therefore, we finally obtain:

$$t = \sqrt{\frac{l}{g}} \int_{\psi}^{\pi/2} \frac{d\psi}{\sqrt{1 - k^2 \sin^2 \psi}} \, .$$

To calculate the period T of the oscillations of the pendulum, we have to find first the time of the half of its complete oscillation cycle, when the angle ψ changes between $\pi/2$ and $-\pi/2$, and then multiply the result by two:

$$T = 2\sqrt{\frac{l}{g}} \int_{-\pi/2}^{\pi/2} \frac{d\psi}{\sqrt{1 - k^2 \sin^2 \psi}} = 4\sqrt{\frac{l}{g}} \int_{0}^{\pi/2} \frac{d\psi}{\sqrt{1 - k^2 \sin^2 \psi}} \, ,$$

where in the last equality we made use of the fact that the integrand is an even function of its variable ψ. The obtained integral cannot in general be calculated in quadratures; it is called an elliptic integral of the first kind. Nevertheless, the integration can be easily performed numerically, opening a way to study the oscillations of the pendulum in the general case of the deviation angle being not necessarily small.

8.6.5 Tsiolkovsky's Formula

Consider the ideal rocket equation first derived by Tsiolkovsky. We consider an idealized rocket of initial mass $M_0 = M_R + m_0$, where M_R is the mass of the rocket without fuel and m_0 is the initial mass of all fuel (in fact, normally the fuel determines the initial mass at the launch, i.e. $m_0 \gg M_R$). We assume that the fuel escapes with a constant velocity u with respect to the rocket (called the *"exhaust velocity"*). If $v(t)$ and $M(t) = M_R + m(t)$ are the velocity of the rocket and its total mass at time t ($m(t)$ is the mass of the fuel left at this time), then the equation of motion of the rocket will be given by the second Newton's law $\Delta P/\Delta t = F_{ext}$, where ΔP is the change of the whole system momentum over time Δt and F_{ext} is an external force. Considering system's momentum at two times, t and $t + \Delta t$, we can write for ΔP (see Fig. 8.8):

$$\Delta P = P(t + \Delta t) - P(t) = [(M - \Delta m)(v + \Delta v) - (u - v)\Delta m] - Mv \, ,$$

where $u - v$ stands for the actual velocity of the fuel in a fixed coordinate frame of the Earth (note that u is the *relative* velocity of the escaping fuel with respect to the rocket). Neglecting the second-order term $\Delta m \Delta v$ and keeping only the first order terms, we get

$$\Delta P = M \Delta v - u \Delta m = M \Delta v + u \Delta M \, ,$$

where ΔM (negative) is the change of mass of the whole rocket over time Δt. Balancing this change of the momentum with the external force of Earth gravity, $-Mg$, acting in the opposite direction to the (positive) direction of the rocket movement, we obtain:

$$M\frac{\Delta P}{\Delta t} = -Mg \implies M\frac{\Delta v}{\Delta t} + u\frac{\Delta M}{\Delta t} = -Mg \implies M\frac{dv}{dt} = -u\frac{dM}{dt} - Mg ,$$

(8.133)

which is the equation of motion sought for.

This equation can be solved if we know how the fuel is burned off. Assuming for instance, that it is burned with the constant rate proportional to the initial mass of the rocket, $dM/dt = -\gamma M_0$ with some positive γ, we write

$$\frac{dv}{dt} = \frac{dv}{dM}\frac{dM}{dt} = -\gamma M_0 \frac{dv}{dM} ,$$

which allows us to obtain the equation we can solve:

$$-\gamma M_0 M\frac{dv}{dM} = u\gamma M_0 - Mg .$$

Upon integration between $t = 0$ (when $v(0) = 0$ and $M(0) = M_0$) and t, we obtain:

$$v = \int_{M_0}^{M} \left(-\frac{u}{M} + \frac{g}{M_0\gamma}\right) dM = -u \ln \frac{M}{M_0} + \frac{g}{\gamma M_0} (M - M_0)$$

$$= u \ln \frac{M_0}{M} - \frac{g}{\gamma M_0} (M_0 - M) .$$

This formula is valid until the whole fuel is burned off. The maximum velocity is achieved when the whole fuel is exhausted, in which case $M = M_R$, giving

$$v_{max} = u \ln \frac{M_R + m_0}{M_R} - \frac{g}{\gamma}\frac{m_0}{M_R + m_0} \simeq u \ln \frac{m_0}{M_R} - \frac{g}{\gamma} ,$$

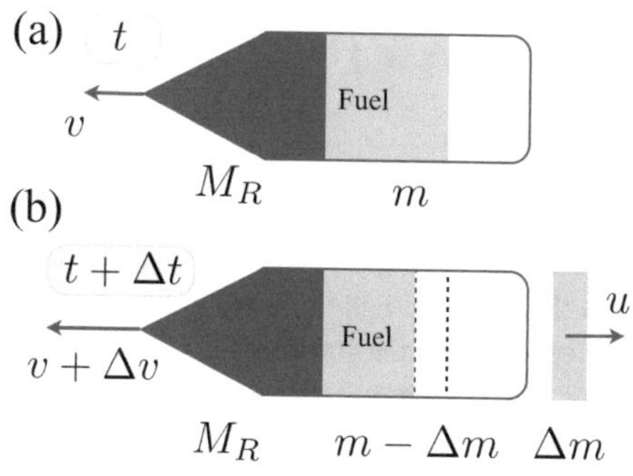

Fig. 8.8 To the derivation of the Tsiolkovsky equation. **a** At time t the rocket has $m(t)$ of fuel, while **b** after time Δt when the mass Δm of the fuel is burned off, the mass of the fuel left will be $m - \Delta m$. The rocket gets a bust Δv in its velocity, while Δm mass of the fuel escapes from the rocket in the opposite direction with the relative velocity u

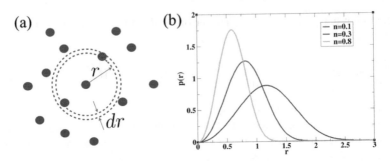

Fig. 8.9 a To the derivation of the probability to find a closest neighbour around a chosen particle in a volume: we draw a spherical shell of radius r and width dr around the particle. **b** Probabilities $p(r)$ for three values of the particles concentration $n = 0.1, 0.3, 0.8$

where in the last passage we used the fact that most of the mass of the rocket is due to the fuel. The last equation also allows one to obtain the amount of fuel,

$$m_0 = M_R \exp\left(\frac{v + g/\gamma}{u}\right) ,$$

required for the rocket to reach the required velocity v on the orbit (e.g. the escape velocity v_{esc} needed to overcome the Earth gravity).

8.6.6   Distribution of Particles

Let us consider a random distribution of particles in a volume as shown in Fig. 8.9a. We would like to calculate the probability $dp = p(r)dr$ of finding a closest particle to a given particle. Let us draw a sphere of radius r around our particle. Then, the probability to find a particle anywhere within that sphere is given by summing up all such probabilities for all radii from zero to r:

$$P(r) = \int_0^r p\left(r'\right) dr' .$$

Now, the probability to find the closest particle within the spherical shell of radius r and width dr will be given by a product of two probabilities: the probability that no particles are found inside the sphere, $Q(r) = 1 - P(r)$, and the probability ndV that there is a particle within the spherical shell. Here $dV = 4\pi r^2 dr$ is the shell volume and n is the particle concentration. On the other hand, this is exactly the definition of the probability $p(r)dr$. Therefore, we can write:

$$p(r)dr = [1 - P\left(r\right)]4\pi r^2 ndr \quad \Longrightarrow \quad \frac{p(r)}{r^2} = \left[1 - \int_0^r p\left(r'\right) dr'\right]4\pi n .$$

$$(8.134)$$

To obtain an equation for the probability $p(r)$ we now differentiate (8.134) with respect to r which yields:

$$\frac{d}{dr}\left(\frac{p}{r^2}\right) = 4\pi n \frac{d}{dr}\left[1 - \int_0^r p\left(r'\right) dr'\right] \implies \frac{1}{r^2}\frac{dp}{dr} - \frac{2p}{r^3} = -4\pi np ,$$

which can be rewritten as

$$\frac{dp}{dr} = \left(\frac{2}{r} - 4\pi r^2 n\right) p . \tag{8.135}$$

This ODE is separable and can be easily integrated:

$$\int \frac{dp}{p} = \int \left(\frac{2}{r} - 4\pi r^2 n\right) dr \implies \ln p = 2\ln r - \frac{4\pi}{3}r^3 n + C$$

$$\implies p = Cr^2 \exp\left(-\frac{4\pi}{3}r^3 n\right) .$$

The arbitrary constant C is determined by normalizing the distribution as there will definitely be a particle found somewhere around the given particle:

$$\int_0^\infty p(r)dr = 1 \implies C = 4\pi n$$

$$\implies p(r) = 4\pi nr^2 \exp\left(-\frac{4\pi}{3}r^3 n\right) . \tag{8.136}$$

This is the final result which is shown in Fig. 8.9b for three values of the particles concentration n. It shows that initially $p(r)$ peaks up, but then decays to zero at large r, i.e. there is always the most probable distance between particles (the maximum of $p(r)$) which is reduced with the increase of the concentration. This distribution is used in various fields of physics including nucleation theory, cosmology (relative distribution of starts in a galaxy), etc.

Problem 8.82 Show that the two-dimensional analogue of formula (8.136) is:

$$p(r) = 2\pi nr \exp\left(-\pi nr^2\right) .$$

8.6.7 Residence Probability

Here we shall derive the *residence kinetics*, i.e. we shall consider a system, which can change its current state (i.e. make a transition to another state) with a certain rate R. For instance, this could be a molecule on a crystal surface that can be in either of two states (conformations), *cis* and *trans* (differing by the orientation of a certain group of atoms). The states are separated by an energy barrier, and hence both are stable; the molecule may change its state, however, by jumping from one state to another, but these events are not deterministic, i.e. they happen with a certain probability depending on the transition rates between the two states and the temperature. Another example is an atomic diffusion on a surface of a crystal. Suppose the lowest energy state of the adatom is above a surface atom; since there are many surface atoms (and hence many available sites), the adatom may occupy many adsorption positions. Each of them is stable, and there must be an energy barrier for the adatom to overcome in order to jump from one position to the other. The rate of such jumps also depends on temperature.

Our goal is to derive a probability $P(t)$ for a system to remain in the current state by the time t. The formula we are about to derive lies in the foundation of a very powerful and popular simulation method called *kinetic Monte Carlo* (KMC).

If R is the rate (probability per unit time) for the system to move away (escape) from the given state it is currently in, then the probability for this to happen over time dt is Rdt. The probability for this *not to happen* over the same time is $1 - Rdt$. Therefore, the probability $P(t + dt)$ for the system to remain in the current state up to the time $t + dt$ is a product of two probabilities corresponding to two events: (i) the system is at this state by the time t with the probability $P(t)$, and (ii) the system remains in the same state over the consecutive time dt, the probability of this is $1 - Rdt$. Therefore,

$$P(t + dt) = P(t)(1 - Rdt) \implies P(t + dt) - P(t) = -RP(t)dt$$

$$\implies \frac{dP}{dt} = -RP .$$

This ODE is easily solved giving $P(t) = C\exp(-Rt)$. Since at $t = 0$ our system is in the current state with probability one, then $C = 1$. Therefore, the required result is:

$$P(t) = \exp(-Rt) .$$

This formula has a very simple meaning: the probability to remain in the current state decreases exponentially with time since the longer we wait the higher the probability for the system to escape from the current state. Eventually the probability to remain in the current state approaches zero, i.e. the system will definitely escape at some point in time.

Problem 8.83 Generalize the previous treatment to the case when the escape rate depends on time, $R = R(t)$. Show that in this case

$$P(t) = \exp\left(-\int_0^t R(\tau)d\tau\right) .$$

8.6.8 Defects in a Crystal

Consider defects in a crystal that can be found in two states A and B; this could be, e.g. a metastable atom. The state A lies higher in energy than B (i.e. the state A is less energetically favorable than B, so that there should be on average more of defects B than A), but it is relatively stable. The energy barriers for a defect to overcome jumping between states, $A \to B$ and $B \to A$, are Δ_{AB} and Δ_{BA}, respectively, so that transition probabilities per unit time (the so-called *transition rates* w_{AB} or w_{BA}) for the two transitions can be assumed to be of the Arrhenius type:

$$w_{AB} = \nu e^{-\Delta_{AB}/k_B T} \text{ and } w_{BA} = \nu e^{-\Delta_{BA}/k_B T} ,$$

where w_{AB} is the transition rate to jump from state A to B, while w_{BA} corresponds to the reverse transition, T is temperature and ν the so-called attempt frequency (a pre-factor). The exponential term gives a probability to overcome the barrier during a single jump (attempt) that is increased with an increase of temperature T or a decrease of the energy barrier (Δ_{AB} or Δ_{BA}), while ν gives the number of attempts per unit time to jump.

The corresponding double well potential energy surface is shown in Fig. 8.10. There will be certain concentrations n_A and n_B of defects in the two states at any given time t; over some time the system would approximately reach the equilibrium state with certain concentrations $n_A(\infty)$ and $n_B(\infty)$ established. Assuming that initially all defects were created in a metastable state A, we can determine their concentration at time t in both states. Indeed, the time dependence of the defects concentrations in the two states should satisfy the following "equations of motion":

$$\frac{dn_A}{dt} = -w_{AB} n_A + w_{BA} n_B \text{ and } \frac{dn_B}{dt} = w_{AB} n_A - w_{BA} n_B . \quad (8.137)$$

The first equation states that the change of the defects concentration in state A is due to two competing processes. Firstly, there is an incoming flux of defects $w_{BA} n_B$ jumping from the state B which is proportional to the existing concentration n_B of defects in state B; this term works to increase n_A (the "gain" term). Secondly, there is also an outgoing flux $-w_{AB} n_A$ from the state A, which is proportional to n_A, that works to reduce their concentration in this state (the "loss" term). The meaning of the second equation is similar, but this time the balance is written for the defects in the other state.

Actually, since the total concentration of the defects $n = n_A + n_B$ is constant (and indeed, summing up the two equations we get that $dn/dt = 0$), the two equations are completely equivalent, so that only one need to be used, e.g. the first:

$$\frac{dn_A}{dt} = -w_{AB} n_A + w_{BA} (n - n_A) . \quad (8.138)$$

It is easily seen that this equation is exactly of the form (8.11):

$$\frac{dn_A}{dt} + w n_A = w_{BA} n \text{ with } w = w_{AB} + w_{BA} .$$

Fig. 8.10 A double well potential energy surface with two minima: A and B. The barriers Δ_{AB} and Δ_{BA} separating these states are also indicated

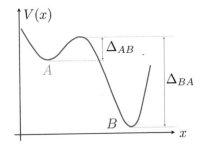

Therefore, using the general method of Sect. 8.1.5, we obtain

$$I(t) = \exp\left(\int w \, dt\right) = e^{wt}$$

and

$$n_A(t) = e^{-wt}\left(C + \int w_{BA} n e^{wt} \, dt\right) = C e^{-wt} + e^{-wt} \frac{w_{BA}}{w} n e^{wt}$$

$$= \frac{w_{BA}}{w} n + C e^{-wt} \; .$$

Since initially $n_A(0) = n$ (all defects were in the A state), then, applying this initial condition, we obtain:

$$n = \frac{w_{BA}}{w} n + C \implies C = n\left(1 - \frac{w_{BA}}{w}\right) \; ,$$

and thus

$$n_A(t) = n\left[\frac{w_{BA}}{w} + \left(1 - \frac{w_{BA}}{w}\right) e^{-wt}\right] \; .$$

This is the final result. It shows that the concentration of defects in state A is reduced over time approaching the equilibrium value (at $t = \infty$) of $n_A(\infty) = n w_{BA}/w$. Consequently, the concentration of defects in state B

$$n_B(t) = n - n_A(t) = [n - n_A(\infty)]\left(1 - e^{-wt}\right)$$

is initially zero, but then it is increased reaching at equilibrium the value of $n - n_A(\infty)$. Note that equilibrium concentrations can also be obtained immediately from (8.137) by setting the time derivatives to zero.

Index

© The Editor(s) (if applicable) and The Author(s), under exclusive license to Springer
Nature Switzerland AG 2022
L. Kantorovich, *Mathematics for Natural Scientists*, Undergraduate Lecture Notes
in Physics, https://doi.org/10.1007/978-3-030-91222-2

Printed in the United States
by Baker & Taylor Publisher Services